# Fracture of Concrete and Rock

Surendra P. Shah  Stuart E. Swartz
Editors

# Fracture of Concrete and Rock

SEM-RILEM International Conference
June 17–19, 1987
Houston, Texas, USA

With 490 Illustrations

Springer-Verlag
New York Berlin Heidelberg
London Paris Tokyo

Surendra P. Shah
NSF Science and Technology Center for
Advanced Cement-Based Materials
Northwestern University
Evanston, IL 60208, USA

Stuart E. Swartz
Department of Civil Engineering
Kansas State University/Seton Hall
Manhattan, KS 66506, USA

Library of Congress Cataloguing-in-Publication Data
SEM-RILEM International Conference (1987: Houston, Texas)
Fracture of concrete and rock: SEM-RILEM International
Conference, June 17-19, 1987, Houston, TX, USA / editors,
Surendra P. Shah, Stuart E. Swartz.
p. cm.
"The International Conference on Fracture of Concrete and
Rock was organized by the Society for Experimental Mechanics
(SEM) subdivision on Fracture of Concrete and Rock and RILEM
Committee 89-FMT Fracture Mechanics of Concrete: Test Methods"--Foreword.
Includes indexes.

ISBN-13: 978-1-4612-8163-4 e-ISBN-13: 978-1-4612-3578-1
DOI: 10.1007/978-1-4612-3578-1

1. Concrete--Fracture--Congresses. 2. Rocks--Fracture-
-Congresses. 3. Fracture mechanics--Congresses. I. Shah, S.P. (Surendra P.)
II. Swartz, Stuart E. III. Society for Experimental Mechanics (U.S.) IV. RILEM
Committee 89-FMT Fracture Mechanics of Concrete Test Methods. V. Title.
TA440.S4576 1987
620.1'366--dc19 89-5917

Printed on acid-free paper



Softcover reprint of the hardcover 1st edition 1989



Camera-ready copy provided by the editors.

9 8 7 6 5 4 3 2 1

## FOREWORD

The International Conference on Fracture of Concrete and Rock was organized by the Society for Experimental Mechanics (SEM) subdivision on Fracture of Concrete and Rock and RILEM Committee 89-FMT Fracture Mechanics of Concrete; Test Methods. The venue was Houston, Texas on June 17-19, 1987 and cooperation was provided by ACI 446, Fracture Mechanics and RILEM 90-FMA Fracture Mechanics of Concrete; Applications. The conference co-chairmen were Professor S. P. Shah, Northwestern University and Professor S. E. Swartz, Kansas State University with the able assistance of Professor K. P. Chong, University of Wyoming.

The conference theme was Fracture Mechanics Applications to Cracking and Fracture of Concrete (plain or reinforced) and Rock Subjected to Uniaxial or Complex Stress States with Static- or Dynamic-Loading Rates. This theme was chosen in recognition of parallel efforts between the rock mechanics community and researchers working in the application of fracture mechanics methods to the problem of cracking and fracture of concrete.

Portland cement concrete is a relatively brittle material. As a result, mechanical behavior of concrete, conventionally reinforced concrete, prestressed concrete, and fiber-reinforced concrete is critically influenced by crack propagation. It is, thus, not surprising that attempts are being made to apply the concepts of fracture mechanics to quantify the resistance to cracking in cementitious composites. A wide range of concerns in design involve fracture in rock masses and rock structures. For example, prediction of the extension or initiation of fracture is important in the design of caverns (underground nuclear-waste isolation) subjected to earthquake shaking and explosion, the production of geothermal and petroleum energy, and predicting and monitoring earthquakes.

The field of fracture mechanics originated in the 1920s with A. A. Griffith's work on fracture of brittle materials such as glass. Its most significant applications, however, have been in controlling brittle fracture and fatigue failure of metallic structures such as pressure vessels, airplanes, ships and pipelines. Considerable development has occurred in the last twenty years in modifying Griffith's ideas or in proposing new concepts to account for the ductility typical of metals. As a result of these efforts, standard testing techniques have been available to obtain fracture parameters for metals, and designs based on these parameters are included in relevant specifications. Many attempts have been made, in the last two decades or so, to apply the fracture-mechanics concepts to rock and concrete. So far, these attempts have not led to a unique set of material parameters which can quantify the resistance of these materials to fracture.

One of the primary reasons for this lack of success is that most of the past work is based on the concept of linear elastic-fracture mechanics. However, it is increasingly being realized that because of the large-scale heterogeneity inherent in the microstructure of concrete and some rock, strain softening, microcracking and large-scale process zone, the classical linear elastic (or the classical elastic-plastic) concepts must be significantly modified to predict crack propagation in concrete and coarse-grained rock.

The recently increased understanding and awareness of unusual aspects of crack growth has resulted, for example, from optically observing crack growth in double-torsion and double-cantilever types of specimens; electron microscopy observations of crack growth in compact-tension specimens; use of infrared spectroscopy, acoustic-emission signal analysis, laser-speckle photography, and optical-interference microscopy to study the process zone; development of finite-element programs to include the nonlinear process zone in structural modeling; theoretical nonlinear fracture mechanics which include more than a single parameter; theoretical analysis which includes the tensile-strain softening in the process zone in front of the crack-tip; application and extension of continuum-damage theory.

These ideas were discussed at the conference which included the themes: Dynamic Fracture Measurements; Fatigue Effects in Concrete Fracture; Mode I Fracture Properties; Analytical Models for Rock Fracture; Fracture Process Zone; Applications to Concrete Design; Size Effects; Testing for Mixed-Mode Fracture Parameters; Dam Analysis; Crack Propagation; and Testing Methods for Mode I. Selected papers from this conference proceedings are contained in this book; keynote papers, invited papers and those papers which were revised according to the reviewer's comments are included. The editors acknowledge the assistance of Dr. Mehmet A. Tasdemir and Auria E. Rosenberg.

S. P. Shah
S. E. Swartz

## TABLE OF CONTENTS

TABLE OF CONTENTS
(continued)

TABLE OF CONTENTS
(continued)

SEM/RILEM International Conference
on Fracture of Concrete and Rock
Houston, Texas, June 1987
Editors: S.P. Shah, S.E. Swartz



# FRACTURE TOUGHNESS OF CEMENT-BASED MATERIALS

**Surendra P. Shah**

Professor of Civil Engineering
Director, Center for Concrete and Geomaterials
Northwestern University
Evanston, IL 60201, USA

## INTRODUCTION

Fracture toughness is a material property which can be defined in terms of critical stress intensity factor. For structural metals, for mode I deformation and small crack-tip plastic deformation the critical stress intensity factor when unstable fracture occurs is designated as $K_{Ic}$. Fracture toughness of metals is used in design to prevent brittle fracture as well as to predict fatigue crack growth. Fracture toughness of metals is a material property distinctly different from the corresponding yield strength. For example, fracture toughness decreases with increasing strain rate, whereas, yield strength shows the opposite effect (Fig. 1). Test methods have been established to determine plane strain fracture toughness ($K_{Ic}$) values for metals (see for example Ref. 1).

Concrete is a relatively brittle material but its fracture toughness is currently not used as a design tool. Concrete structures are usually reinforced so that a designer often neglects the tensile strength of plain concrete, and design is usually based on approximate, empirically based formulae. Even the value of tensile strength of concrete ($f'_t$), is expressed in terms of its compressive strength ($f'_c$) by an equation of the type $f'_t = \beta \sqrt{f'_c}$. The value of $\beta$ is different depending on the type of test (bending vs. split-cylinder). With an increase in efficiency and accuracy possible with the numerical methods, a need to rationally and accurately trace cracking in concrete has become evident. This has led researchers to define and evaluate fracture toughness of concrete.

It is generally accepted that high strength concretes are more brittle than the normal strength concrete. However, we do not yet have a quantitative method of expressing brittleness of concrete. If fracture toughness is expressed using the definition and test methods developed for metal then it is observed that $K_{Ic}$ increases with compressive strength (Fig. 2). The trend shown in Fig. 2 is contrary to the generally accepted observation that high strength concretes are more brittle, that is, less tough. Similar observations have been made for rock. Thus, there is a need to define fracture toughness which is more appropriate for concrete as well as rock. This key-note paper addresses that need.

## PRECRITICAL CRACK GROWTH

To calculate $K_{Ic}$ for metals, fracture mechanics type of specimens such as single edge notched beams are loaded under monotonically increasing load. The load is applied such that a constant rate of increase of crack mouth opening displacement (CMOD) is maintained. If the load-CMOD curve is linear than one can use linear elastic fracture mechanics (LEFM) to calculate $K_{Ic}$ from the measured maximum load and the length of the notch ($a_o$). When $K_{Ic}$ is calculated as described above for concrete (and rock) a significant effect of size and geometry of specimen is observed (Fig. 3). This inapplicability of obtaining a valid fracture toughness parameter from LEFM and laboratory-sized specimens was pointed out by Shah and McGarry [2] in 1970. Since then several researchers have addressed the problem and aspects of this problem have been extensively discussed in recent symposia and technical committees [3-6].

It has now been widely accepted that as a result of heterogeneity inherent in cement paste, mortar, concrete, and rock, there is a substantial crack growth which preceeds the critical (maximum) load. This precritical crack growth is often termed fracture process zone or slow crack growth and is demonstrated from a notched beam test in Fig. 4. The crack growth ahead of the notch was continuously monitored using a specially developed brittle crack gage [7].

There are several aspects of this precritical crack growth which are worth noting. The length of the precritical crack length is different along the width of the specimen. This has been recorded using dye-penetrant technique by Swartz and Go [8] and by Bascoul et al. [9] (Fig. 5). This means that the surface crack measurements are not adequate to describe the fracture process zone.

Fracture process zone in concrete is often compared with plastic zone in metals. For metals, the plastic zone is defined as a zone where material has yielded ahead of a sharp notch. For concrete and rock, it is assumed in such a comparison that microcracking ahead of a notch is analogous to yielding. Such a comparison misses an important aspect of crack growth in concrete and rock: a substantial traction exists behind the crack tip. The traction behind the crack-tip can be attributed to frictional (geometrical or aggregate) interlock and to unbroken ligaments. The existence of such traction has been confirmed by locating sources of acoustic emissions in specimens of concrete and rock [10, 11] (Fig. 6).

The traction forces (also termed closing pressure) behind the crack tip are not constant but their value decreases with an increasing crack opening displacement. This is shown in Fig. 7 where the results of a uniaxial tensile test conducted by Gopalaratnam and Shah [12] are shown. Average (surface) crack opening displacements were measured microscopically.

To accurately calculate fracture toughness, one must include this precritical crack growth, especially when laboratory size specimens are used.

## FICTITIOUS CRACK AND CRACK BAND MODELS

To include the effect of traction forces behind the crack-tip, Hillerborg and his colleagues [13] have developed a fictitious crack model for numerical analysis (Fig. 8). It is assumed that the relationship between crack closing pressure and crack opening displacement (such as shown in Fig. 7) is a material characteristic. A preexisting notch is extended in the direction of principal tensile stresses. Crack closing pressure is calculated by numerical iteration from Fig. 7 and from geometry of the structure. Note that the peak of the curve in Fig. 7 at zero opening is assumed to be equal to the tensile strength of the material and the area under the curve is termed $G_f$ and is taken as the fracture energy of the material.

Assuming that $G_f$, tensile strength and width of the fracture process zone are the material properties, Bazant and his colleagues [14] have developed a crack-band model for finite element analysis (Fig. 8).

Fictious crack model and crack band model have been successfully applied for finite element analysis. However, it should be noted that the model predicted results are sensitive to the assumed value of tensile strength, $G_f$, and the shape of the closing pressure vs. crack opening displacement curve [15, 16].

An excellent summary of various models using the fictitious crack approach has been presented by Hillerborg [17]. He points out that some of the numerical discrepancies in the fictitious crack model can be avoided if a combination of closing pressure and a crack growth criteria based on stress-intensity factor is used. This is the model proposed by Jenq and Shah (Fig. 9) [18]. They showed that when the calculated critical stress intensity at the tip of the fictitious crack is an algebraic sum of far-field stresses and closing pressure, a size-independent fracture toughness parameter value can be obtained. According to their model, the fracture energy is absorbed in part by closing forces and in part at the crack tip. The resultant stress intensity factor in this model can be calculated using a Green's function approach and solving a nonlinear integral equation [19] or by a boundary element method. Jenq and Shah [20] have also proposed a simpler, two parameter fracture model described in the next section: Effective Crack Models.

Note that it has been suggested that the fracture energy, $G_f$ can be used to define fracture toughness. However, the value of $G_f$ depends significantly on the size, geometry and types of specimen [3] and its value increases with an increase in compressive strength (Fig. 10), which makes $G_f$ not very useful as a simple fracture parameter.

## EFFECTIVE CRACK MODELS

The length of the plastic zone for an elastic perfectly brittle material can be estimated by replacing an actual crack $a_o$ with an effective critical crack $a_e$ where $2(a_e-a_o)$ is the plastic zone size. The length $a_e$ is calculated such that the sum of the stress-intensity factors at the tip of the effective crack due to far-field stresses and closing pressure (equal to yield stress) is zero (Fig. 11). This Dugdale-Barenblat type of effective crack is often replaced with Bilby-Cottrell-Swinden-Dugdale crack for material with a limited plasticity (Fig. 11). In this model, the crack surfaces are assumed to be traction free when crack opening displacement exceeds a certain value.

For brittle materials it is perhaps more appropriate to have effective crack models where the local crack-tip stress intensity factor is not zero. An effective Griffith crack can be accurately evaluated from the measurement of the compliance recorded at the point of the critical crack extention, Fig. 12. Such effective crack models have been proposed by Nallathambi and Karihaloo [21] and Refai and Swartz [22]. The effective crack (or the precritical crack growth) depends on the microstructure of the material (maximum grain size of aggregate), and also on the geometry of the specimen (Fig. 13). When $K^s_{Ic}$ is calculated using the effective crack, a size independent value is obtained. Thus the effective crack $a_e$ and $K_{Ic}$ at the tip of the effective crack become two fracture parameters. Empirically based equations relating effective crack length (for notched-beam specimens) with specimen geometry and material properties can be developed from the experimental data.

If such an approach is used one needs empirical equations for effective crack length for each type of specimen (notched-beam, compact tension, center wedge loaded, etc). Also since the effective crack length at the critical extension is dependent on the geometry and size of structure, one cannot use it as an a priori material property for structural analysis.

## TWO PARAMETER FRACTURE MODEL (TPFM)

To generalize the concept of the effective Griffith crack (this term always refers to the critical point) for different types of specimens and for structural analysis, Jenq and Shah [18, 20] have proposed a two parameter fracture model (Fig. 12,14). The two fracture parameters are $K^s_{Ic}$ and $CTOD_c$. The critical stress intensity factor is calculated at the tip of the effective crack whereas the crack tip opening displacement is calculated at the original (that is, preexisiting) crack ($a_0$). Jenq and Shah [18, 20] observed that when $a_e$ was calculated from the compliance measurement for the notched beam specimens, the calculated values (using LEFM) of $CTOD_c$ were more or less independent of size and geometry of beams. Based on these and other observations they have proposed these parameters ($K^s_{Ic}$ and $CTOD_c$) as valid material fracture parameters. Note that effective crack can be calculated for any type of specimen using only these two parameters. These two parameters depend on material composition and strain rate.

Using these parameters (obtained from beam tests) one can calculate uniaxial tensile strength (Fig. 15), split cylinder strength (Fig. 16), size effect of conventional $K_{Ic}$ (Fig. 3a), and size effect on modulus of rupture (Fig. 17,) (also see Refs. 20 and 23).

The effective length calculated using these two fracture parameters depends on size of beam specimens (Fig. 13 and 18), compressive strength of concrete (Fig. 19) and on strain rate (Fig. 20). It is important to note that $a_e$ reduces with increasing compressive strength and increasing strain rate. Note that for elastic perfectly brittle material $CTOD_c$ approaches 0 and $a_e$ becomes $o_e$. Thus the two parameter fracture model correctly predicts that higher strength concrete is more brittle and that concrete is more brittle at higher strain rate.

The application and validity of the two parameter fracture model for predicting mixed mode fracture is detailed elsewhere in these proceedings [24]. The details regarding the high strength and strain rate effects are also given in another paper included in these proceedings [23].

## LASER HOLOGRAPHY AND ACOUSTIC EMISSION MEASUREMENTS

The validity of any theoretical model should be finally based on physical measurements. To study fracture processes at the microscopic level, we are currently studying crack profiles for center-wedge loaded tensile specimens. Crack opening displacement as the crack travels are being measured with an accuracy of 0.1 micron using sandwich holography [25]. Acoustic emission source location and deconvolution techniques are being developed to identify fracture processes and fracture process zone [26]. Comparing these observations with compliance measurements should provide a sound basis for a theoretical model to evaluate fracture toughness of concrete.

The optical arrangement for obtaining split-beam transmission (Lieth-Upatniks) hologram is shown in Fig. 21 [25]. A hologram is produced as a result of the interference of the object beam and the reference beam.. When two holograms of the object at two levels of loading are superimposed, (sandwich holograms) fringes are observed. The number of fringes give the displacement along the sensitivity vector. With this arrangement we can observe crack profile as the crack propagates with an increase in load with an accuracy of about 1/2 micron.

In a separate series of tests we are also testing similar specimens in a closed-loop testing machine using notch-tip opening displacement as a feed back control [27]. During testing, strains and acoustic emissions are being recorded. A comparison of notch-tip opening displacement (NTOD) measured with laser holography and clip gages in similar specimens was found to be satisfactory. The next two figures (Figs. 22 and 23) show the crack profile observed at two different loads using holographic interferometry. Crack profile obtained from linear elastic fracture mechanics using finite element analysis with singular elements [27] is also shown in these figures. For small loads and small crack lengths both profiles are comparable. However, at higher loads, the actual crack front is longer than that predicted by LEFM (both were assumed to have the same NTOD). The observed crack profile at about the peak load is shown in Fig. 24. In addition to the holographic crack profile, 3 other crack profiles are shown: (1) based on compliance measurement: that is, two parameter fracture model; (2) based on conventional compliance method: that is, assuming that the observed crack length is traction free; and (3) based on our acoustic emission source location technique. Holographic and the TPFM crack profile clearly show that closing pressures exist behind the crack tip. Note that acoustic emission measurements and holographic observations agree quite well for these relatively thin specimens.

Changes in crack profile for two different crack lengths are shown in Fig. 25. If the largest observed crack opening displacement for a particular crack length is taken as a critical crack tip opening displacement, then this value is probably a lower bound value of the actual critical CTOD. These values are shown in Fig. 26. Also shown are values obtained using TPFM on beam specimens and using center wedge loaded specimens. The assumption of $CTOD_c$ being a material property seems to be confirmed from these comparisons.

**ACKNOWLEDGEMENT**
This research was supported by a grant from the Air Force Office of Scientific Research (AFOSR-85-0261) under a program managed by Dr. Spencer T. Wu.

**REFERENCES**

1. Barsom, J. M., and Rolfe, S. T., Fracture and Fatigue Control in Structures, Prentice Hall Inc., New Jersey, USA, 1987.

2. Shah, S. P., and McGarry, F. J., "Griffith Fracture Criterion and Concrete," Journal of Engineering Division, ASCE, Dec. 1971, pp. 1663-1675.

3. Wittmann, F. H. (Editor), Fracture Toughness and Fracture Energy of Concrete, Elsevier Science Publishers, The Netherlands, 1986.

4. Bazant, Z. P. (Editor), Mechanics of Geomaterials, John Wiley and Sons, UK, 1985, 611 pp.

5. Shah, S. P. (Editor), Application of Fracture Mechanics to Cementitious Composites, Martinus Nijhoff Publishers, The Netherlands, 1985.

6. Shah, S. P. and Swartz, S. E. (Editors), Proceedings, SEM-RILEM International Conference on Fracture of Concrete and Rock, Houston, USA, 1987.

7. John, R. and Shah, S. P., "Fracture of Concrete Subjected to Impact Loading," Journal of Cement, Concrete and Aggregates, ASTM, Vol. 8, No. 1, pp. 24-32, Summer 1986.

8. Swartz, S. E., and Go, C.-G., "Validity of Compliance Calibration to Cracked Concrete Beams in Bending," Experimental Mechanics, SEM, Vol. 24, No. 2, pp. 129-134, June 1984.

9. Bascoul, A., Kharchi, F., and Maso, J. C., "Concerning the Measurement of the Fracture Energy of a Micro-Concrete According to the Crack Growth in a Three Points Bending Test on Notched Beams," Proceedings, SEM-RILEM International Conference on Fracture of Concrete and Rock, Eds. S. P. Shah and W. E. Swartz, Houston, USA, 1987.

10. Maji, A. K. and Shah, S. P., "Process Zone and Acoustic Emission Measurements in Concrete," Experimental Mechanics, SEM, Paper No. 3609, Accepted for Publication (in press), 1987.

11. Labuz, J., Shah, S. P., and Dowding, C. H., "Using Seismic Techniques to Characterize Fracture in Rock," Experimental Techniques, SEM, pp. 30-32, March 1987.

12. Gopalaratnam, V. S., and Shah, S. P., "Softening Response of Plain Concrete in Direct Tension," ACI Journal, Proceedings, Vol. 82, No. 3, pp. 310-323, May-June 1985.

13. Hillerborg, A., Modeer, M., and Petersson, P. E., "Analysis of Crack Formation and Crack Growth in Concrete by Means of Fracture Mechanics and Finite Elements," Journal of Cement and Concrete Research, Vol. 6, pp. 773-782, 1976.

14. Bazant, Z. P., and Oh, B. H., "Crack Band Theory for Fracture of Concrete," RILEM, Materials and Structures, Vol. 16, No. 93, pp. 155-177, 1983.

15. Roelfstra, P. E. and Wittmann, F. H., "Numerical Method to Link Strain Softening with Failure of Concrete," Fracture Toughness and Fracture Energy, (Ed. F. H. Wittmann), Elsevier Science Publishers, The Netherlands, pp. 163-175, 1986.

16. Rots, J. G., "Strain-Softening Analysis of Concrete Fracture Specimens," Fracture Toughness and Fracture Energy (Ed. F. H. Wittmann), Elsevier Science Publishers, The Netherlands, pp. 137-148, 1986.

17. Hillerborg, A., "Discrete Crack Approach," Fracture Mechanics of Concrete Applications - Part A: RILEM Technical Committee 90-FMA Report, Second Draft, May 1987.

18. Jenq, Y. S. and Shah, S. P., "A Fracture Toughness Criterion for Concrete," Engineering Fracture Mechanics, Vol. 21, No. 5, pp. 1055-1069, 1985.

19. Ballarini, R., Shah, S. P., and Keer, L. M., "Crack Growth in Cement Based Composites," Engineering Fracture Mechanics, Vol. 20, No. 3, pp. 433-445, 1984.

20. Jenq, Y. S., and Shah, S. P., "Two Parameter Fracture Model for Concrete," Journal of Engineering Mechanics, ASCE, Vol. 111, No. 4, pp. 1227-1241, October 1985.

21. Nallathambi, P., and Karihaloo, B. L., "Influence of Slow Crack Growth on the Fracture Toughness of Plain Concrete," Fracture Toughness and Fracture Energy, (Ed. F. H. Wittmann), Elsevier Science Publishers, The Netherlands, pp. 271-180, 1986.

22. Refai, T. M. E., and Swartz, S. E., "Influence of Size Effects on Opening Mode Fracture Parameters for Precracked Concrete Beams in Bending," Proceedings, SEM-RILEM Internal Conference on Fracture of Concrete and Rock, (Eds. S. P. Shah and S. E. Swartz), Houston, USA, 1987.

23. John, R. and Shah, S. P., "Effect of High Strength and Rate of Loading on Fracture Parameters of Concrete," Proceedings, SEM-RILEM International Conference on Fracture of Concrete and Rock, (Eds. S. P. Shah and S. E. Swartz), Houston, USA, 1987.

24. Jenq, Y. S., and Shah, S. ., "Mixed Mode Fracture Parameters of Concrete," Proceedings, SEM-RILEM International Conference on Fracture of Concrete and Rock, (Eds. S. P. Shah and S. E. Swartz), Houston, USA, 1987.

25. Miller, R. A., Shah, S. P., and Bjilkhagen, H. I., "Crack Profiles in Mortar Measured by Holographic Interferometry," Accepted for publication, Experimental Mechanics, SEM, 1988.

26. Maji, A., and Shah, S. P., "Process Zone and Acoustic Measurements in Concrete," Experimental Mechanics, SEM Paper No. 3609, Accepted for Publication, 1988.

27. Alvarado, M., Shah, S. P., and John, R., "Mode I Fracture in Concrete Using Center Cracked Plate Specimens," Accepted for publication, Journal of Engineering Mechanics, ASCE, March 1988.

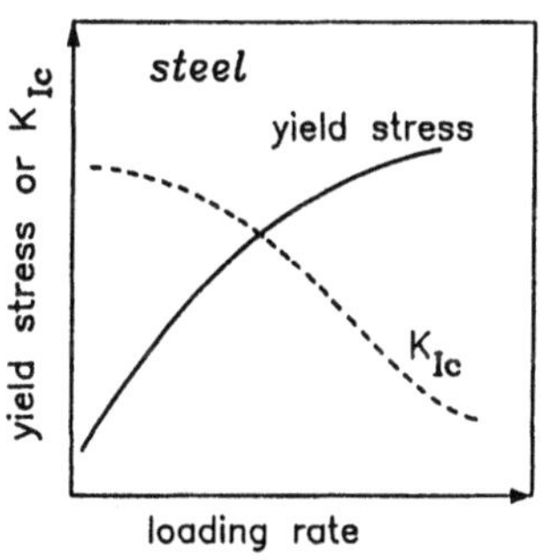


Fig. 1. Effect of loading rate on $K_{IC}$ and yield stress of metals at constant temperature [Ref. 1].

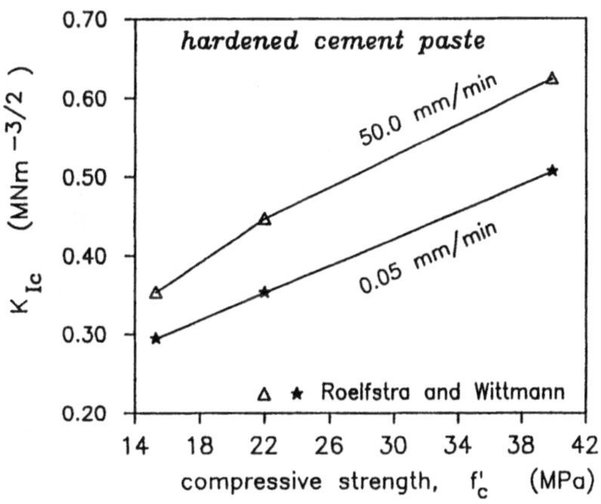


Fig. 2. Effect of compressive strength and rate of

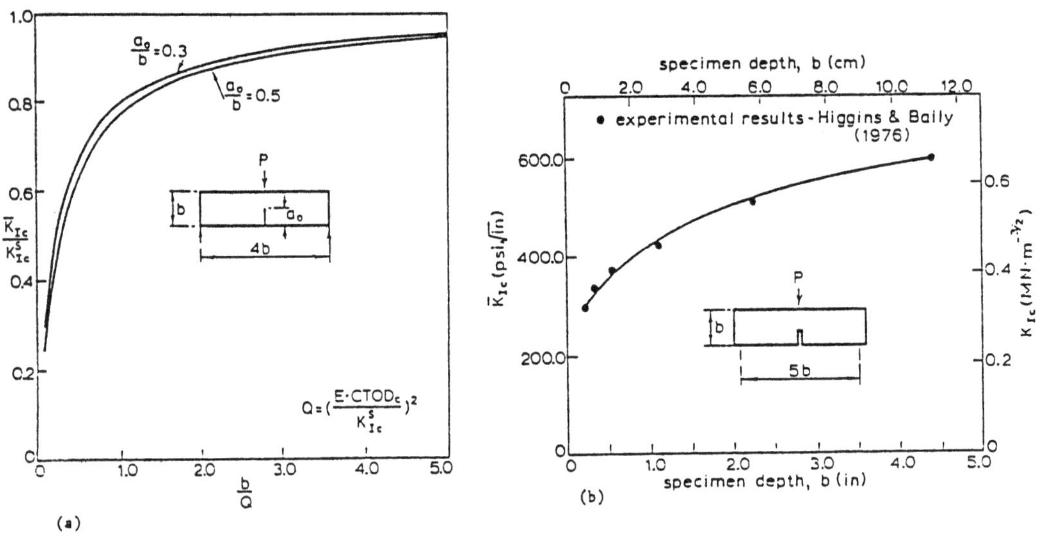


Fig. 3. Size effect on conventional stress intensity factor, $K_{IC}$ : (a) Experimental results and (b) Theoretical [Jeng and Shah].

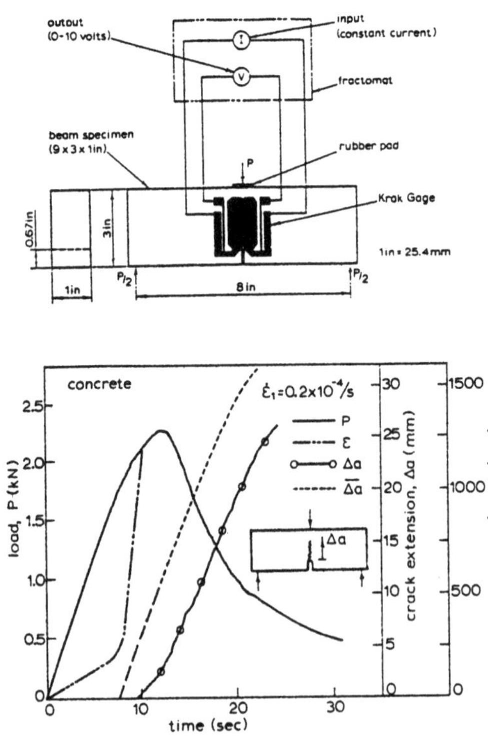


Fig. 4. Typical crack growth test data. Crack growth was measured using 'Krak Gages' [John and Shah].

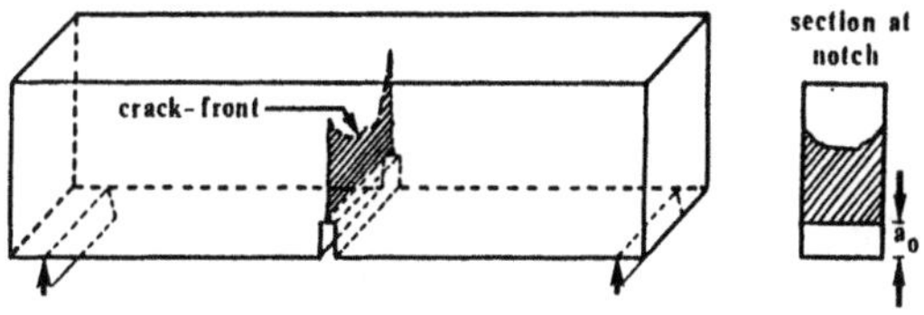


Fig. 5. Variation of crack length along the width of specimen as observed using dye-penetration techniques [Swartz and Gio, and Bascoul et al.].

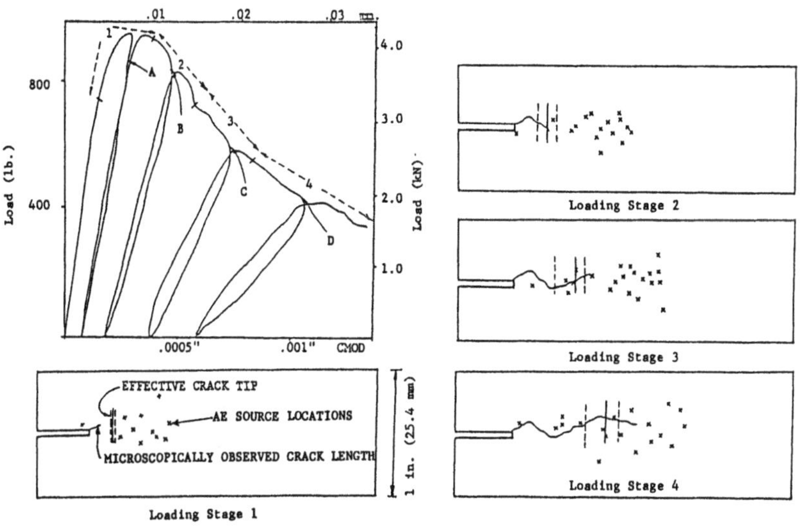


Fig. 6. Crack propagation study using acoustic emission source location technique [Maji and Shah].

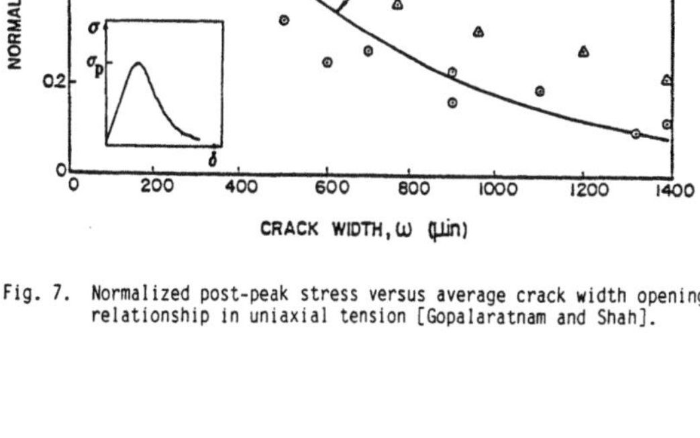


Fig. 7. Normalized post-peak stress versus average crack width opening relationship in uniaxial tension [Gopalaratnam and Shah].

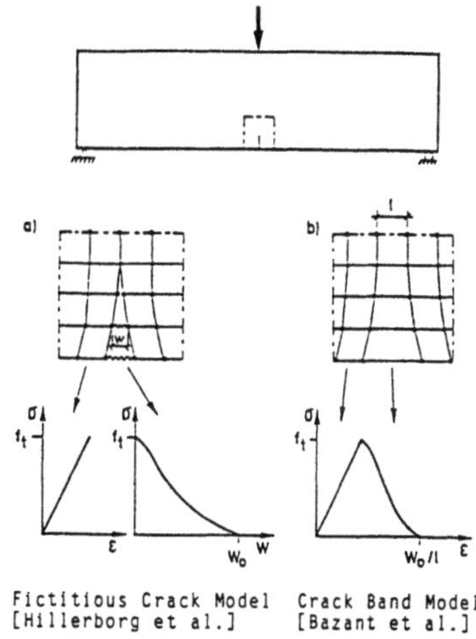


Fig. 8. Fictitious Crack Model and Crack Band Model-Figure reproduced from RILEM committee report by Hillerborg.

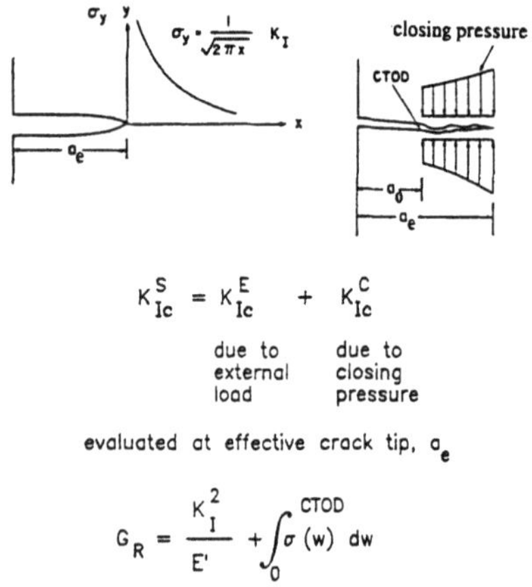


Fig. 9. $K_I$ evaluation and energy considerations assuming crack tip singularity and closing pressure acting behind effective crack tip [Jenq and Shah].

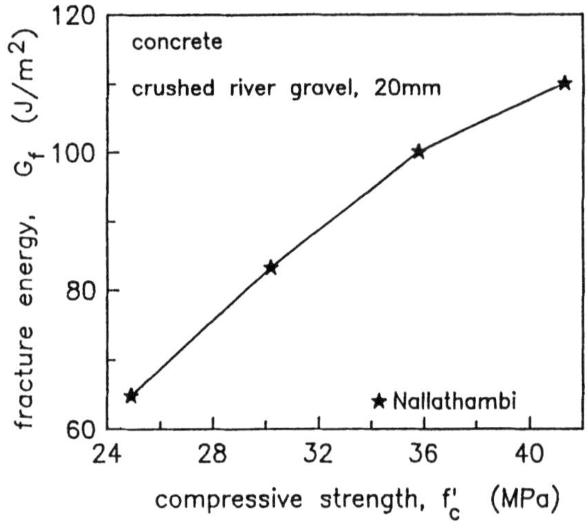


Fig. 10. Effect of compressive strength on fracture energy, $G_f$. $G_f$ was evaluated using RILEM method [Nallathambi].

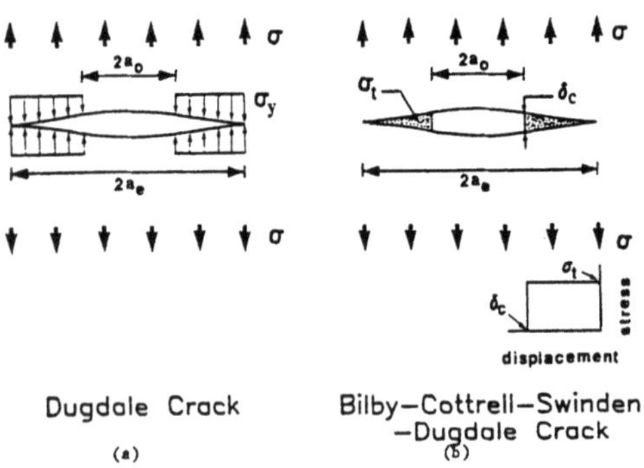


Fig. 11. Schematic of Dugdale crack and Bilby-Cottrell-Swinden-Dugdale crack.

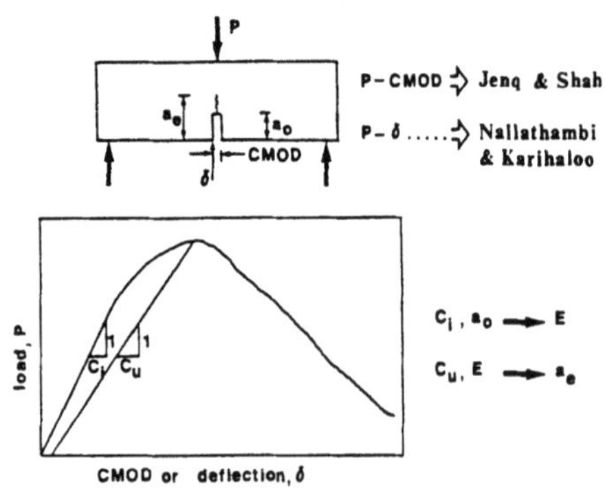


Fig. 12. Effective Crack Models.
$a_e$ = effective crack at peak load.

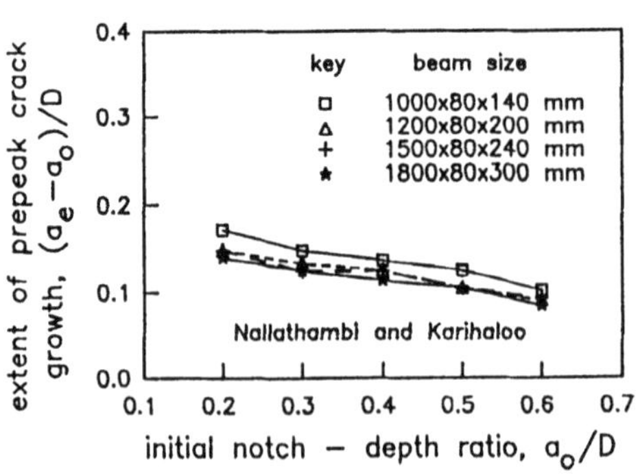


Fig. 13. Extent of prepeak crack growth versus initial notch depth ratio [Nallathambi and Karihaloo].

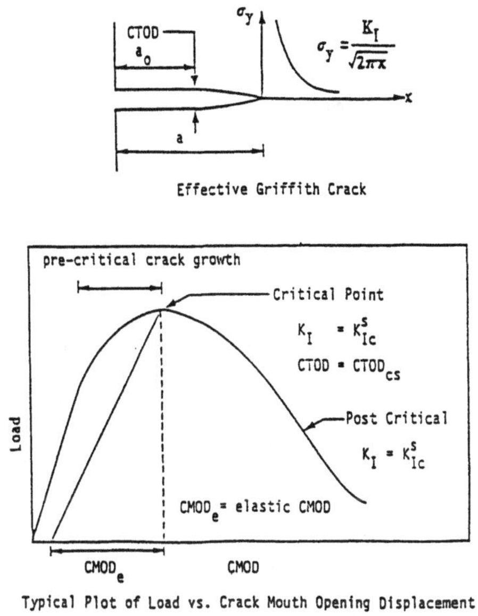


Fig. 14. Two Parameter Fracture Model for Concrete [Jenq and Shah].

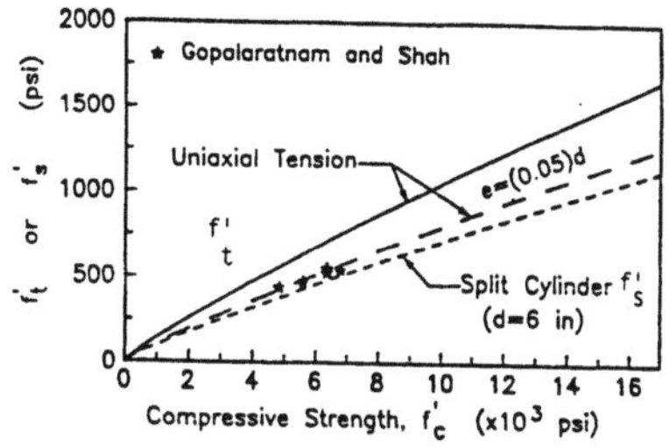


Fig. 15. Variation of tensile strength with compressive strength [John and Shah].

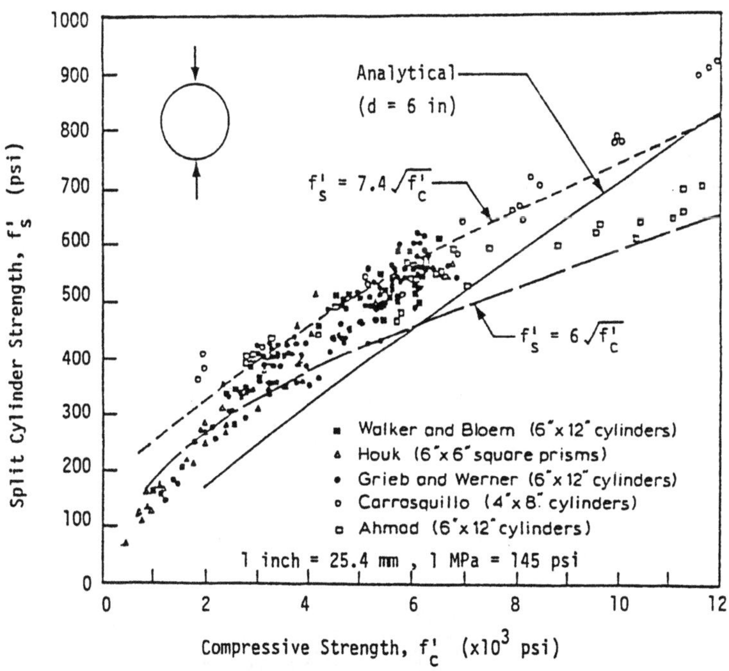


Fig. 16. Experimental and predicted relationship between split cylinder strength and compressive strength of concrete [John and Shah].

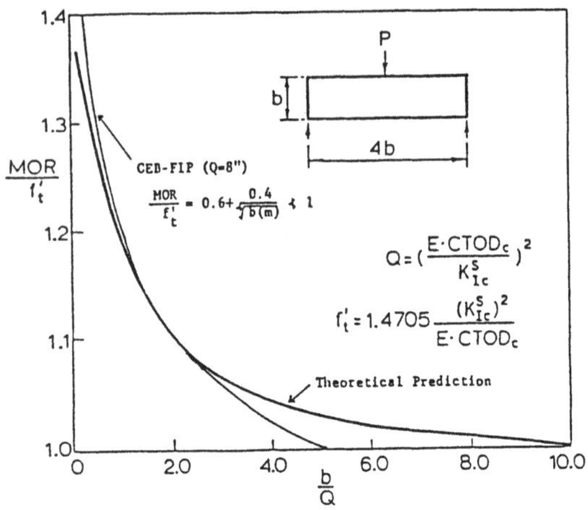


Fig. 17. Size effect on modulus of rupture (MOR) of concrete [Jenq and Shah].

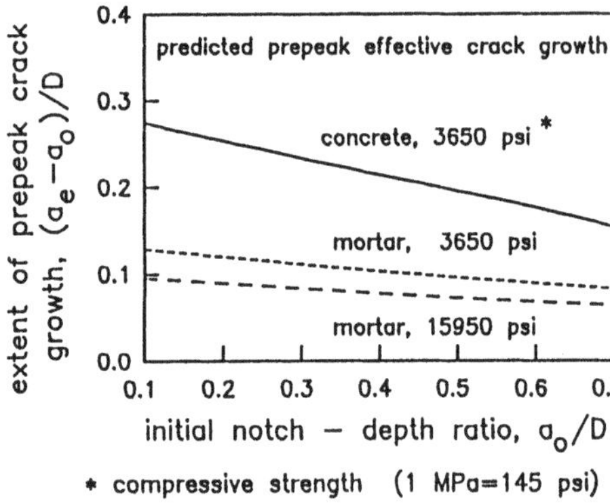


Fig. 18. Predicted variation of extent of prepeak crack growth with initial notch depth ratio .

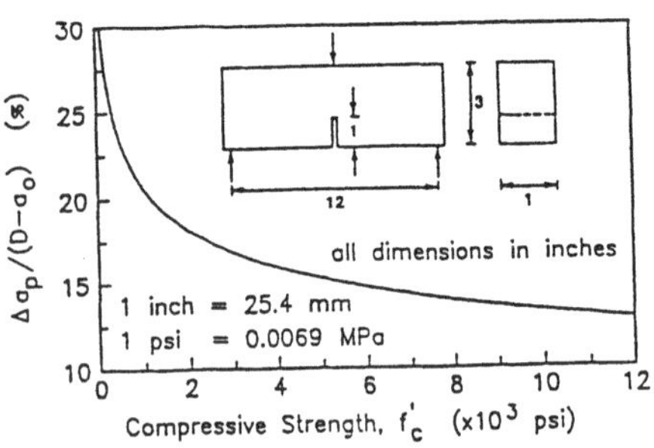


Fig. 19. Decrease of nonlinearity (extent of prepeak crack growth) with increasing compressive strength [John and Shah].

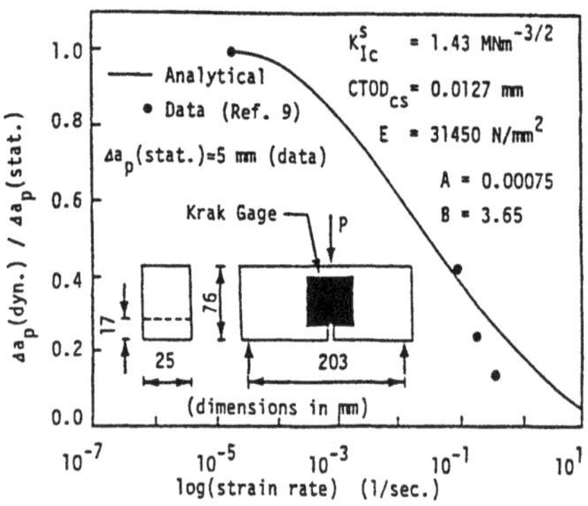


Fig. 20. Decrease of nonlinearity (extent of prepeak crack growth) with increasing strain rate (rate of loading) [John and Shah].

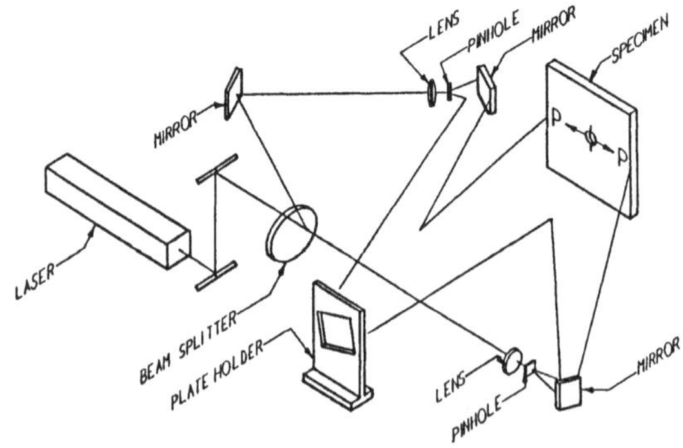


Fig. 21. Schematic of laser holography test set-up [Miller, Shah and Bjelkhagen].

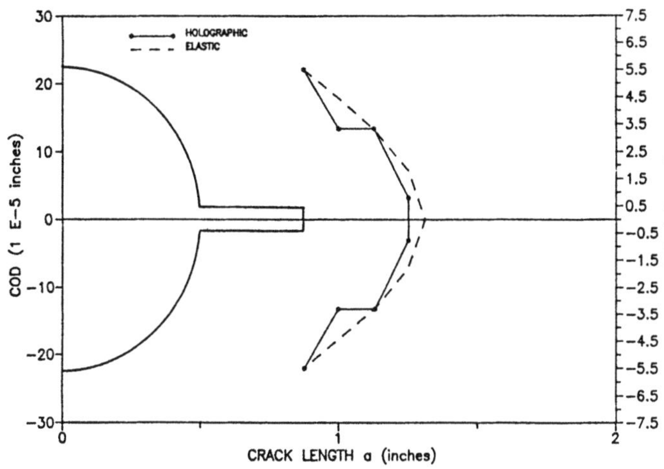


Fig. 22. Comparison of measured and elastic crack profiles at about 50% of peak load.

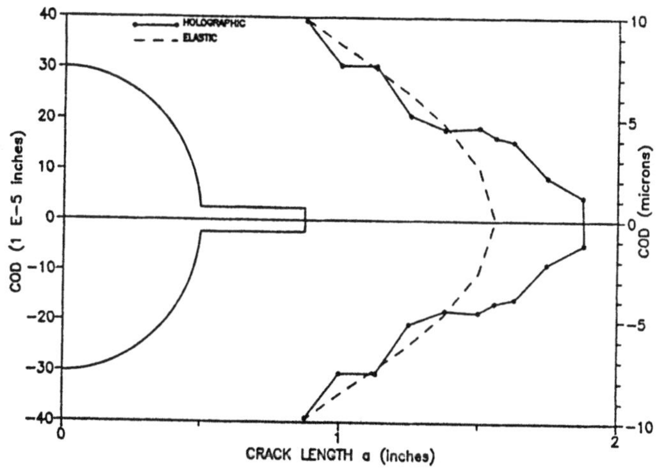


Fig. 23. Comparison of measured and elastic crack profiles at about 75% of peak load.

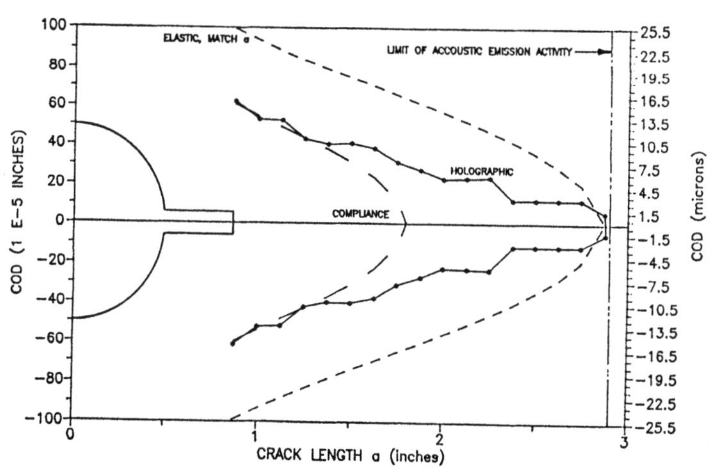


Fig. 24. Comparison of measured and elastic crack profiles and limit of acoustic emission activity at about the peak load.

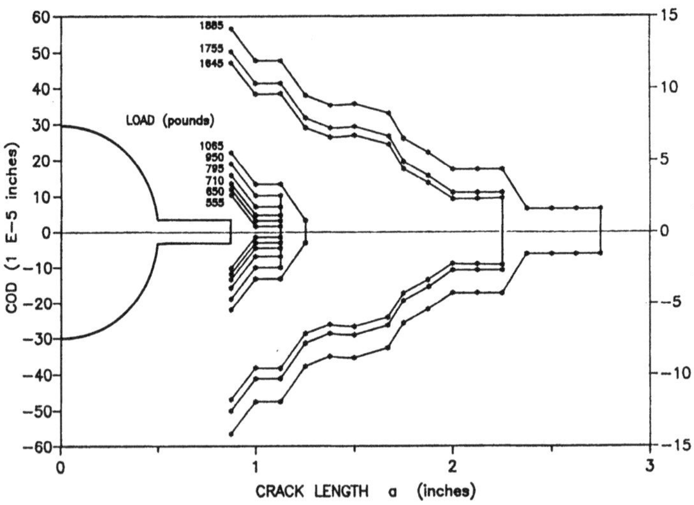


Fig. 25. Measured crack profiles at various loads.

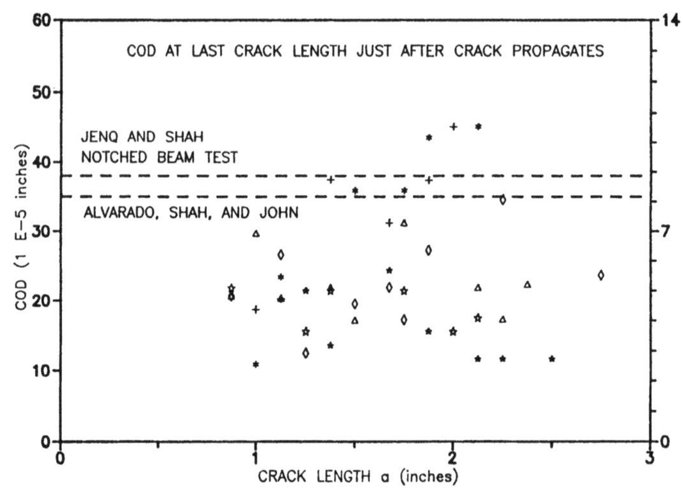


Fig. 26. Comparison of measured crack opening displacement (COD) at onset of propagation with critical crack tip opening displacement reported by Jenq and Shah, and Alvarado, Shah and John.

SEM/RILEM International Conference
on Fracture of Concrete and Rock
Houston, Texas, June 1987
Editors: S.P. Shah, S.E. Swartz

# FRACTURE TOUGHNESS DETERMINATION OF ROCKS WITH CORE-BASED SPECIMENS

K. P. Chong, M. D. Kuruppu and J. S. Kuszmaul

**ABSTRACT**
Rock and geomaterial specimens are usually extracted in circular cores. $K_{Ic}$ derived from various core specimens is discussed. A newly developed fracture toughness test specimen, semi-circular in shape and contains an edge-crack, is elaborated. The fracture load and the fracture energy of a typical layered rock are measured with static tests, and the fracture toughness is determined using a stress intensity factor method, a compliance method and a J-integral based method. A minimum dimensional requirement is proposed for plane strain fracture toughness. Application of $K_{Ic}$ in fragmentation modeling is illustrated. Mixed mode fracture specimen is also presented.

## INTRODUCTION

ASTM standards are available for determining plane strain fracture toughness of metallic materials [1]. However, most rocks and geomaterials are anisotropic, and the nature of fracture processes in rock is different from that in metals [2]. Thus, the direct application of such standards to rock is unacceptable. No standard has as yet been set for fracture toughness determination of rock material, though some are being drafted [3]. Such a test method should, in addition to conforming to specific material behavior, should utilize the geometry of a typical rock core for the test and require a minimum amount of specimen machining. Since geomaterials are weak in tension, tests should preferably be done with bending or compressive loadings where tensile fractures are induced. Short rod specimen developed by Barker [4], burst cylinder specimen developed by Clifton et al. [5], round bend bar specimen introduced by Ouchterlony [6,7], and semi-circular bend (SCB) specimen proposed by Chong [8] and shown in Fig. 1, satisfy most of those requirements. Hopefully, a future standard method for determining fracture toughness of rock material may make use of these types of specimens cored from a single block of material, as illustrated in Fig. 3, for a complete characterization in the materials principal directions.

Oil shale, a typical layered rock, transversely isotropic [9], containing recoverable crude oil in the form of an organic material called kerogen, is found in abundance in every continent of the world [9]. With the depletion of known petroleum reserves as a result of increasing consumption of natural crude oil, it would be necessary to develop synthetic crude oil such as that from oil shale to meet the future demand for energy supplies. Oil shale is a hard rock having a small grain size and a low permeability. The process of extracting oil from shale <u>in-situ</u> relies on the fracturing of tight underground formations to increase the permeability and to burn it in place, which is known as retorting. Both hydraulic fracturing [9, 10] and explosive fragmentation [11-13] have been used for this purpose. The mechanics of the fracture processes in oil shale are important for the optimization of retorting to make the large-scale extraction of oil economically viable. The elastic fracture toughness, $K_{Ic}$, has been found to be a useful parameter for the characterization of intact rock with respect to its resistance to crack propagation and as an index of fragmentation processes in the analysis of explosive or hydraulic fracturing [10,14-17]. Application of $K_{Ic}$ in rock fragmentation modeling is discussed at the end of this paper.

---

K. P. Chong (SEM Member) is Professor and Chairman of Structures/Solid Mechanics, Department of Civil Engineering, University of Wyoming, Laramie, WY, 82071; M. D. Kuruppu is Post-doctor, Department of Civil Engineering, University of Wyoming; and J. S. Kuszmaul is Member of Technical Staff, In Situ Technonologies, Sandia National Laboratoreis, Albuquerque, NM.

Most investigations in linear elastic fracture mechanics have been carried out under mode I conditions in which the crack opens and crack initiation occurs when the stress intensity factor reaches a critical value. It is the creation and propagation of fractures that are desired in many fracture applications such as explosive and dynamic loads on structures, rock fragmentation and massive hydrofracturing. Most of the applications are associated with combined mode failure [18], and it is important to study mode II fracture toughness and the combined mode fracture criterion.

## EXPERIMENTAL PROCEDURE

The experimental method adopted some of the important features of the ASTM standard method E399 for the plane strain fracture toughness testing of metals [1]. A newly developed semi-circular bend specimen was used instead of a standard fracture toughness test specimen for both modes I and II fracture.

### Specimen Type and Preparation

The SCB specimen has a single edge notch of length "a" and loaded in three-point bending as shown in Fig. 1. It can be prepared from typical cores of rock, concrete, asphalt, or other geomaterials with very little machining. It is especially suitable for applications requiring duplicate samples having similar composition since a circular disk could be cut into two halves to make two duplicate specimens. For mixed-mode fracture studies, the edge crack is cut at an angle (Fig. 2).

Layered rocks consist of horizontal layers of beddings and their properties may change substantially across the bedding planes. However, the beddings are consistently uniform in the horizontal direction resulting in nearly isotropic properties in that direction. Therefore, layered rocks can be adequately characterized as transversely isotropic materials [9,21]. Specimens are usually made such that the crack is aligned with one of three principal orientations known as the divider, the arrester, and the short transverse. Specimens of divider orientation yield average and representative properties of many different layers unlike those of the other two orientations. Hence, the present investigation was confined to measuring the fracture toughness in the divider orientation.

Deposits of Colorado oil shale containing about 250 ml/kg (60 gal/ton) are considered to be "rich" whereas those containing about 42 ml/kg (10 gal/ton) are considered to be "lean". The specific gravity of oil shale depends on the organic material content and, therefore, the volume of organic material in the rock can be determined by measuring the specific gravity [22]. Six different sizes of specimens were prepared for testing to determine any size effect on fracture toughness measurement. Specimens were machined using a diamond-impregnated wire saw to minimize any damage to the specimen introduced by machining operation. Clean water was used as a coolant. All specimens were to have a span-to-radius ratio (s/R) of 0.8. The largest specimen used has a radius of 76.2 mm (3.0 in) and a thickness of 35.6 mm (1.4 in). The smallest specimen has a radius of 22.4 mm (0.88 in) and a thickness of 11.43 mm (0.45 in).

### Testing

A closed-loop, servo-hydraulic testing machine was used to load specimens. Calibration of the load cell and the crack mouth opening displacement (CMOD) gauge produces a linear response in each of the signals. A fixture was designed to load the specimen in three point bending, minimizing frictional effects by allowing the support rollers to rotate and move apart slightly as the specimen was loaded, thus permitting roller contact. The rollers were initially positively positioned against stops that set the span length and were held in place by low-tension rubber bands. This fixture helps to achieve the proper alignment in the load transfer system. The CMOD gauge was attached to the specimen using knife edges positioned across the mouth of the notch using adhesives. These knife edges support the gauge arms and serve as the displacement reference points. Each specimen was fatigue cracked at 5 Hz by cycling between fixed limits of approximately 60 percent of the estimated failure load until the crack length to radius ratio (a/R) was approximately 0.5. Using the same fixtures for final failure, specimens were subsequently loaded under displacement control at a constant rate of about $3.4 \times 10^{-3}$ mm/sec until the specimen could sustain no further increase in load. The load and the CMOD were recorded on a x-y plotter. After testing, specimens were broken open, and the initial crack length "a" up to the fatigue crack front was determined as an average of three direct measurements.

## ANALYSES

Three different methods were used to calculate the fracture toughness based on experimental observations. They are the elastic stress intensity factor method, the compliance method, and the J-integral method. The J-integral method requires a measurement of area under load - load line displacement (LLD) curve to evaluate the fracture energy. This was achieved using the relationship between LLD and CMOD derived by Chong and Kuruppu [8].

A typical load versus LLD plot is shown in Fig. 4. Following ASTM standard procedure E399 [1], the load corresponding to 2 percent apparent increment of crack extension was established by a 5 percent deviation from the linear portion of the record. A secant line through the origin of the test record with slope (P/q) = 0.95 $(P/q)_o$ where $(P/q)_o$ was the slope of the tangent QA to initial linear part of the record was drawn. Load $P_Q$ was defined at the point of intersection of the secant line and the curve. If $P_{max}$ is taken as the maximum load achieved, the ratio $P_{max}/P_Q$ is found to be about 1.05 in most cases, but less than 1.1 in all tests satisfying one of the requirements for plane strain fracture toughness as given in the ASTM standard E399. Area A under the curve up to point $P_Q$ was measured.

Stress Intensity Factor Method

The SCB specimen was modeled under loading using the finite element method. The normalized stress intensity factor determined by strain energy method [23] and elliptic displacement of crack surface method [24] is shown in Fig. 5. The apparent fracture toughness $K_Q$ is determined as

$$K_Q = \bar{K}_I \sigma_u \sqrt{\pi a} \tag{1}$$

where, $\sigma_u = P_Q / (2Rt)$; $\bar{K}_I = K_I / (\sigma_o \sqrt{\pi a})$; $\sigma_o = P / (2Rt)$

and $\bar{K}_I$ is the normalized stress intensity factor, $P_Q$ is the load at 2 percent crack extension, R is the specimen radius, t is the thickness, and a is the crack length.

Compliance Method

Considering the energy balance of a cracked body, the energy release rate G can be expressed in terms of the compliance C (defined as the ratio of LLD and load), the applied load P, the thickness t, and the crack length a as

$$G = \frac{P^2}{2t} \frac{\partial C}{\partial a} \tag{2}$$

Normalized compliance is assumed to be a function of normalized crack length.

Therefore, ECt = F (a/R). (3)

By using eq.(3), an evaluation can be made of eq.(2) yielding

$$G = \frac{P^2}{2Et^2R} \frac{\partial F(a/R)}{\partial (a/R)} \tag{4}$$

$K_Q$ can then be related to the critical energy release rate $G_Q$ [25] as

$$K_Q = \left(\frac{EG_Q}{1-\nu^2}\right)^{1/2} = \frac{P_Q}{t\sqrt{2R(1-\nu^2)}} \left[\frac{\partial F(a/R)}{\partial (a/R)}\right]^{1/2} \tag{5}$$

The Young's modulus, E, and Poisson's ratio, $\nu$, of Colorado oil shale were computed using the evaluation formulae [21, 26] based on the oil yield and the stress level.

J-integral Method

J-integral [27, 28] is a valid parameter for anisotropic materials, and it can be used as a fracture parameter for nonlinear materials as well. In most cases, $K_Q$ is applicable only for isotropic elastic materials. Therefore, J-integral can be expected to yield a more accurate value for fracture toughness with anisotropic materials. It was computed from the load - LLD curve using the relationship between fracture energy, A, and J-integral, $J_Q$, derived by Rice et al. [29] as

$$J_Q = \frac{2A}{(R-a)t} \tag{6}$$

where A is the area under load - LLD curve up to the point corresponding to a 2 percent crack growth determined according to the method given in ASTM standard procedure E399 [1]. This simpler procedure was adopted instead of using the J-integral-based resistance curve [30], as the latter method could yield inaccurate results due to the difficulty in measuring the crack extension using the change in specimen compliance. The J-values were converted to the equivalent $K_Q$ values assuming linear elastic material properties.

Mixed-Mode Analysis

Using conic-section simulation, Westergaard's stress function and finite element analysis, mixed-mode stress intensity factors can be computed [18]. Fig. 6 shows the normalized modes I and II stress intensity factors of the SCB specimen subjected to three point bend loading with the crack orientation inclined to the loading axis. Crack length to radius ratio, a/R, and half span to radius ratio, s/R, were assigned values of 0.5 and 0.67, respectively. As would be expected, the mode I stress intensity factor has a maximum value when the direction of crack is parallel to that of loadings, and decreases with increasing crack angle, $\alpha$. The mode II stress intensity factor has a zero value for $\alpha$ of 0° and 90°, and a maximum value for $\alpha$ of approximately 32°. $\bar{K}_I$ reaches zero at a crack angle of 63° which corresponds to the pure mode II loading position. Therefore, the SCB specimen can be used to determine both pure modes I and II fracture behavior.

## RESULTS AND DISCUSSION

Static mode I fracture toughness values are determined using the stress intensity, the compliance, and the J-integral based methods. Data of a total of 55 tests are given. All three methods yield agreeable results, and the average values of the three methods are displayed in Fig. 7. Using regression analysis on the average results, the following relationship between the apparent fracture toughness and the specific gravity was obtained (plotted as solid line in Fig. 7).

$$K_Q = 94.294 - 50.05D + 8.4369\, D^2 \tag{7}$$

$$\text{and,} \quad r^2 = 0.7238;\ N = 55;\ S = 1.66\ \text{N/mm}^{3/2}$$

where $K_Q$ is the fracture toughness ($N/mm^{3/2}$), D is the specific gravity, N is the number of data, $r^2$ is the correlation coefficient and S is the standard error of estimate. Oil yield (organic material content) has an inverse variation with specific gravity [22]. Therefore, present results show that organic-rich oil shale has a higher fracture toughness than leaner material. Physically, rich oil shale has a higher ductility and requires more energy for the development of the fracture process zone than a leaner material. The results are comparable to the limited data obtained by Costin [30], who investigated one grade of oil shale using different specimen types, and to those of Young, Patti, and Trent [31]. Fracture surfaces appeared flat in all but a few of the samples indicating the absence of any shear or plane stress failure. In the remaining specimens, oblique fractures were observed having inclinations of not more than $5^o$ to the plane of notch.

Tests carried out with different sizes of specimens yield agreeable results as shown in Fig. 7. Even the smallest size specimens of radius 22.4 mm (0.88 in) and crack length 11.43 mm (0.45 in) yield results that agree well with those of larger sizes. Consequently, results of the smallest size specimens could be considered to yield the true fracture toughness, $K_{Ic}$. Fracture toughness of five specimens having approximately the same kerogen content characterized by their specific gravity are given in Table 1. The fifth column gives the tensile strength of Colorado oil shale as determined by Chong, et al. [19]. The last column shows the normalized smallest in-plane dimension of each specimen. These results suggest that the minimum size requirement for plane strain fracture toughness is $2.0\ (K_{Ic}\ /\ \sigma_t)^2$ where $\sigma_t$ is the tensile strength. However, more tests with other rock materials are necessary for further verification. Recent investigations by Li [32], using the tension-softening behavior on cementitious composites, may be a viable method of getting around size requirements. The tension-softening curve (derivable from the J-integral) can then be used to find $K_{Ic}$ according to Jenq and Shah [33].

## APPLICATION OF $K_{Ic}$ IN FRAGMENTATION MODELING

Fracture toughness has been identified as an important material property in determining the results of rock fragmentation. Grady [34] identified relationships that use $K_{Ic}$ and other material properties to relate fragment size to strain rate at fragmentation and to determine strain-rate dependent fracture stress. These relationships have been incorporated into a constitutive model of brittle rock response to dynamic loading [35]. An important feature of this material model is that it relies only on physically measurable material constants. When the model is incorporated into a finite-element code, it can be used to simulate explosive fragmentation of layered rocks as oil shale.

Therefore, this determination of $K_{Ic}$ values for various grades of oil shale permits a rock blasting simulation involving a wide range of grades of oil shale layers. The geometry of this blasting simulation and the grades of the 28 layers in this configuration are shown in Fig. 8. The design involved 2.5m of explosive (aluminized, emulsified ammonium nitrate and fuel oil mixture) in a 0.162-m-diameter hole beneath 2.5m of stemming. This problem was simulated using the two-dimensional, dynamic, finite-element code PRONTO [36]. The properties of the explosive used in the simulation are given in Table 2 [37]; the material properties of the oil shale layers are shown in Table 3.

The results of the simulation (given in Figure 9) use an interpretation of fragmentation that is based on percolation theory [38]. This provides a fragmentation threshold that represents the critical degree of fracturing necessary to achieve fragmentation. The constitutive model actually calculates the extent of fracturing, but with this extension, an estimation of the extent of fragmentation is made by this simulation. Also shown in Fig. 9 is the excavated crater profile obtained for a blasting experiment corresponding to this simulation [16]. This experiment took place in oil shale of the Anvil Points Mine near Rifle, Colorado. The experiment's crater extent is similar to the computer simulation of fragmentation except that the simulation shows deep fragmentation that may have been impossible to detect by excavation of the crater.

## SUMMARY AND CONCLUSIONS

Static fracture toughness tests were performed on different grades of Colorado oil shale, a typical layered rock, using a newly developed, core based, semi-circular bend specimen. The specimen is specially tailored for the needs of rock fracture testing. Specimens are tested in three point bending while measuring the load and the CMOD. Fracture toughness results were computed using a stress intensity factor method, a compliance method, and a J-integral based method. From the data shown in Figs. 6, 7 and 9, it can be stated that

1. Fracture toughness data determined by the stress intensity factor method, the compliance method, and the J-integral based method agree well with each other.

2. An average static fracture toughness for many layers of oil shale can be measured by choosing the divider orientation. Results do not depend on the particular position of the crack tip with respect to bedding planes unlike those of arrester and short transverse orientations.

3. Methods of fracture toughness evaluation based on linear elastic fracture mechanics, as well as J-integral based method, are applicable for anisotropic rock such as oil shale.

4. Static fracture toughness of organic-rich oil shale is higher than that of lean material. There is for example, about 20 percent decrease in fracture toughness for a change in organic volume from 208 ml/kg (50 gal/ton) to 83 ml/kg (20 gal/ton). Fracture toughness data compare favorably with the limited published data.

5. All sizes of specimens tested satisfy the minimum dimensional requirements for plane strain fracture. Based on the smallest specimen size, the minimum dimensional requirement for SCB specimen can be given as $2.0\ (K_{Ic}/\sigma)_t^2$.

6. Semi-circular bend specimens can be used for modes I, II and mixed mode fracture testing depending on the angle of the edge crack.

7. Fragmentation simulation using $K_{Ic}$ yields a similar crater profile.

In conclusion, the SCB test specimen has been shown to possess certain merits as a fracture toughness test specimen for rocks and other materials. It will hopefully compliment other specimens such as short rod and bend bar to be included in a standard procedure for rock fracture toughness testing that will be drafted in future. The SCB specimen is applicable for layered or non-layered rocks and other core-based materials.

Acknowledgements: This research was supported by the Sandia National Laboratories under contract 21-0981, project managers Theodore C. Bartke (METC) and Paul J. Hommert. Discussions with Victor C. Li of MIT and speedy typing by Barbara Powell and Connie Froman are also gratefully acknowledged.

**REFERENCES**


[1] Standard Method of Test for Plane Strain Fracture Toughness in Metallic Materials, ASTM E399 - 81, Am. Soc. for Testing and Materials, New York (1981).

[2] M. Wecharatana and S. P. Shah, Prediction of nonlinear fracture process zone in concrete, Journal of EMD of ASCE 109, 1231 - 1246 (1983).

[3] F. Ouchterlony, co-ordinator, Suggested Methods for Determining Fracture Toughness of Rock Materials, Int. Soc. Rock Mech. Commi. Testing Methods, 4th draft, June (1986).

[4] L. M. Barker, A simplified method for measuring plane strain fracture toughness, Engng. Fracture Mech. 9, 361 - 369 (1977).

[5] R. J. Clifton, E .R. Simonson, A. H. Jones, and S. J. Green, Determination of the critical stress intensity factor $K_{Ic}$ from internally -Pressurized thick-walled vessels, Exper. Mech. 16, 233-238 (1976).

[6] F. Ouchterlony, A new core specimen for the fracture toughness testing of rock, Swedish Detonic Research Foundation Rep., DS 1980: 17, Stockholm, Sweden (1980).

[7] F. Ouchterlony and Sun Zongqi, New methods of measuring fracture toughness on rock cores, Swedish Detonic Research Foundation Rep, DS 1983: 10, Stockholm, Sweden (1983).

[8] K. P. Chong and M. D. Kuruppu, New specimen for fracture toughness determination of rock and other materials, Int. J. Fract. 26, R 59 - R 62 (1984).

[9] K. P. Chong and J. W. Smith, eds, Mechanics of Oil Shale, Elsevier (1984).

[10] E. R. Simonson, A. S. Abou-Sayed and R. J. Clifton, Containment of massive hydraulic fractures, SPE paper No. 6089, 51st Annual Fall Meeting of SPE of AIME (1976).

[11] H. P. Rossmanith, Rock Fracture Mechanics, Springer - Verlag, Wien, New York (1983).

[12] F. Ouchterlony, Fracture Mechanics applied to rock blasting, Proceedings 3rd Int. Congress of the ISRM, Denver, Colorado, 2-B, 1377-1383 (1974).

[13] K. P. Chong, P.M. Hoyt, J. W. Smith and B.Y. Paulsen, Effects of strain rate on oil shale fracturing, Int. J. Rock Mech. Min. Sci 17, 35-43 (1980).

[14] R A. Schmidt, Fracture toughness of limestone, Exper. Mech. 16, 161-167 (1976).

[15] P.J Hommert, ed., Oil Shale Program Quarterly Reports, October, 1983 September, 1984, Sandia National Laboratories Reports, SAND84-1561 and SAND85-1170 Albuquerque, August (1985).

[16] J. S. Kuszmaul, Numerical modeling of oil shale fragmentation experiments, Proceedings 11th Annual Society of Explosives Engineers Conf., C. S. Konya, ed., San Diego, California (1985).

[17] M. E. Kipp, D .E. Grady and E. P. Chen, Strain-rate dependent fracture initation, Int. J. Fract. 16, 471-478 (1980).

[18] M. D. Kuruppu and K . P. Chong, New specimens for modes I and II fracture investigations of geomaterials, Proceedings Soc. Exp. Mech, Spring Conf., New Orleans, U. S. A., 31-38, June (1986).

[19] K. P. Chong, J. W. Smith, E. S. Borgman, Tensile strengths of Colorado and Utah oil shales, J. of Energy, AIAA, Vol. 6, No. 2, 81-85, March/Apr. (1982).

[20] K. P. Chong, J. P. Turner and G. F. Dana, Strain-rate effects on the mechanical properties of Utah oil shale, Proceedings Int. Symp. Mechanical Behavior Structural Media, Ottawa, Canada, 431-446 (1981).

[21] K. P. Chong, K. Uenishi, J W. Smith and A. C. Munari, Non-linear three dimensional mechanical characterization of Colorado oil shale, Int. J. Rock Mech. Min. Sci 17, 339-347 (1980).

[22] J. W. Smith, Relationship between rock density and volume of organic matter in oil shales, US DOE report ERDA/LERC/RI-76/6. Laramie, Wyoming (1976).

[23] J. R. Dixon and J. S. Strannigan, Determination of energy release rates and stress intensity factors by the finite element method, J. Strain Analysis 7, 125-131 (1972).

[24] C. W. Woo and M. D. Kuruppu, Use of finite element method for determining stress intensity factors with a conic-section simulation model of crack surface, Int. J. Fract. 20, 163-178 (1982).

[25] P. C. Paris and G. C. Sih, Stress analysis of cracks, in Symposium on Fracture Toughness Testing and Its Application, ASTM STP 381, 30-38 (1970).

[26] J. B. Sellers, G. R. Haworth and P. G. Zambas, Rock mechanics research on oil shale mining, Transactions Soc. Min. Engineers of AIME 252, 222-232 (1972).

[27] J. R. Rice, A path-independent integral and the approximate analysis of strain concentration by notches and cracks, Transactions ASME J. App. Mech 35, 379-386 (1968).

[28] C. P. Cherepanov, Crack propagation in continuous media, J. App. Math. Mech (PMM) 31, 476-488 (1967).

[29] J. R. Rice, P. C. Paris and J. G. Merkle, Some further results of J-integral analysis and estimates, in Progress in Flaw Growth and Fracture Toughness Testing, ASTM STP 536, 231-245 (1973).

[30] L. S. Costin, Static and dynamic fracture behavior of oil shale, in Fracture Mechanics Methods for Ceramics, Rocks and Concrete, ASTM STP 745, 169-184 (1981).

[31] C. Young, N. C. Patti and B. C. Trent, Stratigraphic variations in oil shale fracture properties, US DOE Report of Investigation, DOE/LC/RI-82-5, Laramie, Wyoming (1982).

[32] V. C. Li, Fracture resistance parameters for cementitious materials and their experimental determinations, Applications of Fracture Mechanics to Cementitious Composites, S. P. Shah, ed., Martinus Nijhoff Publ., 431-449 (1985).

[33] Y. S. Jenq and S. P. Shah, Nonlinear fracture parameters for cement based composites: theory and experiments, Applications of Fracture Mechanics to Cementitious Composites, S. P. Shah, ed., Martinus Nijhoff Publ. (1985).

[34] D. Grady, The mechanics of fracture under high-rate stress loading, in Mechanics of Geomaterials, John Wiley & Sons Ltd., (1985).

[35] L. M. Taylor, E. P. Chen, and J. S. Kuszmaul, Microcrack-induced damage accumulation in brittle rock under dynamic loading, Computer Methods in Applied Mechanics and Engineering, Vol. 55, No. 3, 301-320, (1986).

[36] L. M. Taylor and D. P. Flanagan, PRONTO2D--A Two-Dimensional Transient Solid Dynamics Program, in preparation.

[37] E. Gorham-Bergeron, private communication, (1985).

[38] J. S. Kuszmaul, "A Technique for Predicting Fragmentation and Fragment Sizes Resulting from Rock Blasting", in preparation.

**TABLE 1. Minimum Specimen Dimensional Requirements for Plane Strain Fracture Toughness**

| Specimen No. | Smallest In Plane Dimension, x (mm) | Specific Gravity | Average Fracture Toughness, $K_{Ic}$ ($N/mm^{3/2}$) | Tensile Strength, $\sigma_t$ ($N/mm^2$) | $\frac{x}{(K_{Ic}/\sigma_t)^2}$ |
|---|---|---|---|---|---|
| 48 | 10.80 | 1.866 | 29.12 | 13.58 | 2.35 |
| 51 | 10.29 | 1.934 | 29.63 | 13.60 | 2.17 |
| 52 | 10.41 | 1.913 | 28.97 | 13.59 | 2.29 |
| 53 | 10.67 | 1.823 | 32.24 | 13.56 | 1.89 |
| 55 | 10.50 | 1.877 | 31.06 | 13.61 | 2.02 |

**TABLE 2. Explosive Properties for simulation with PRONTO**

| | |
|---|---|
| Density | 1160 $kg/m^3$ |
| Detonation Velocity | 5650 m/sec |
| JWL Equation-of-State Parameters: | |
| A | 47.6 GPa |
| B | 0.524 GPa |
| Energy | 4.26 GPa |
| $R_1$ | 3.5 |
| $R_2$ | 0.9 |
| $\omega$ | 1.005 |

**Table 3. Oil Shale Properties for layered simulation with PRONTO**

| Grade (ml/kg) | Density (kg/m$^3$) | Young's Modulus | Poisson's Ratio | Yield Stress (MPa) (GPa) | m | k ($m^{-3}$) | $K_{Ic}$ ($N\,m^{-3/2}$) |
|---|---|---|---|---|---|---|---|
| 36.3 | 2523 | 30.23 | 0.259 | 102 | 6 | $2.97x10^{24}$ | 687000 |
| 42.2 | 2493 | 28.82 | 0.260 | 103 | 6 | $2.11x10^{24}$ | 694000 |
| 47.6 | 2465 | 27.51 | 0.261 | 103 | 6 | $1.51x10^{24}$ | 702000 |
| 56.3 | 2420 | 25.39 | 0.264 | 104 | 6 | $8.82x10^{23}$ | 714000 |
| 65.9 | 2370 | 23.07 | 0.266 | 106 | 6 | $4.46x10^{23}$ | 729000 |
| 72.2 | 2338 | 21.56 | 0.267 | 107 | 6 | $2.73x10^{23}$ | 740000 |
| 88.5 | 2261 | 17.83 | 0.271 | 106 | 6 | $7.20x10^{22}$ | 767000 |
| 100 | 2216 | 15.46 | 0.274 | 102 | 6 | $2.69x10^{22}$ | 785000 |
| 106 | 2192 | 14.20 | 0.276 | 100 | 6 | $1.51x10^{22}$ | 794000 |
| 121 | 2136 | 11.24 | 0.279 | 95 | 6 | $2.93x10^{21}$ | 818000 |
| 161 | 2008 | 7.03 | 0.289 | 78 | 6 | $1.18x10^{20}$ | 879000 |
| 213 | 1870 | 4.34 | 0.301 | 70 | 6 | $4.52x10^{18}$ | 955000 |

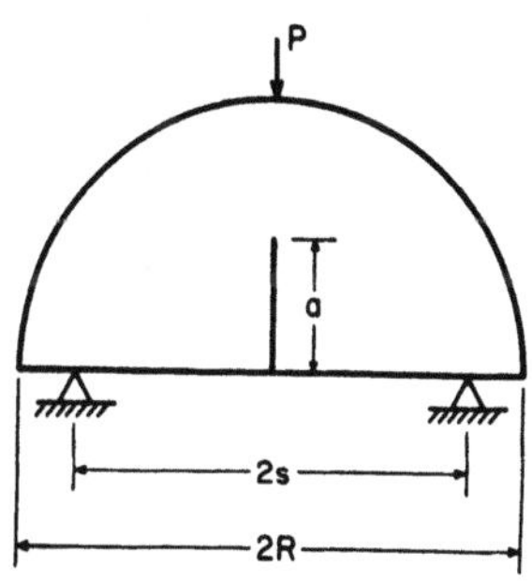


Fig. 1. Semi-circular bend (SCB) mode I test specimen.

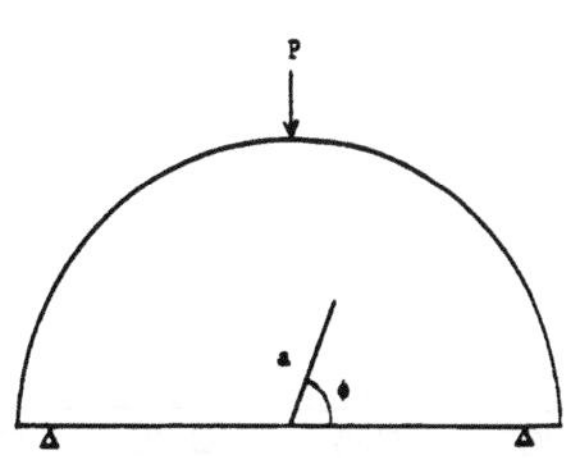


Fig. 2. Semi-circular bend mixed mode test specimen.

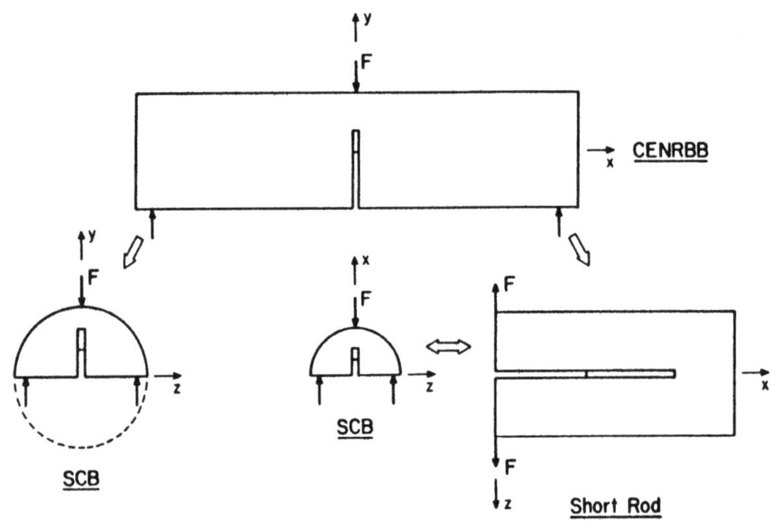


Fig. 3. Complete fracture toughness characterization from a single core.

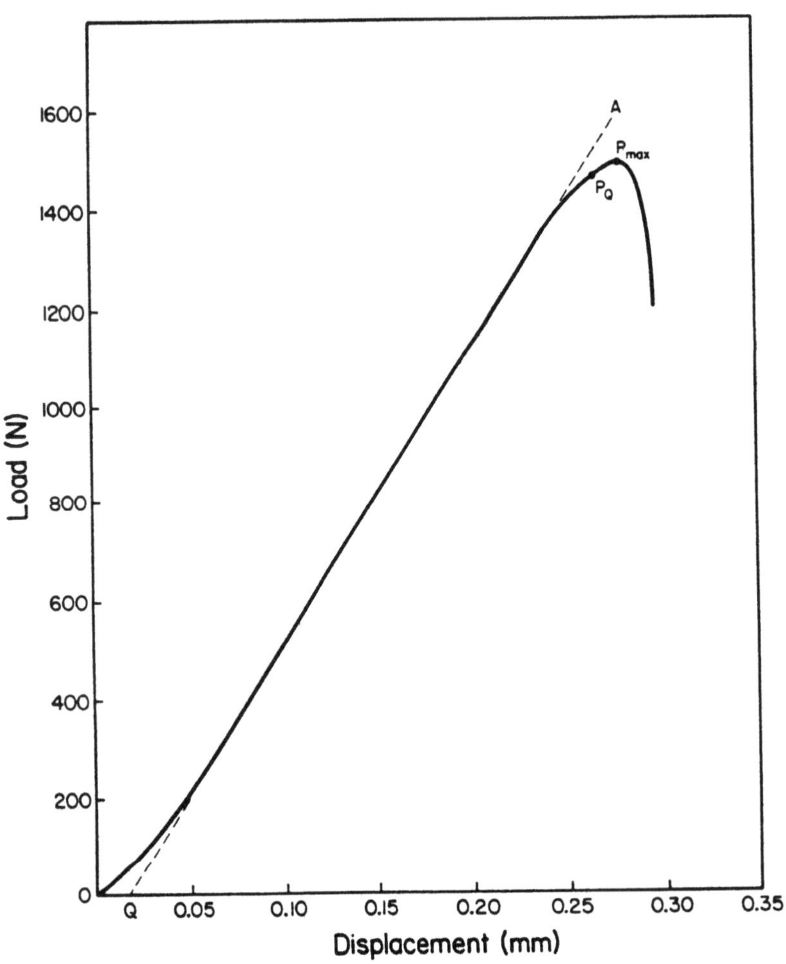


Fig. 4. A typical load versus load line displacement record.

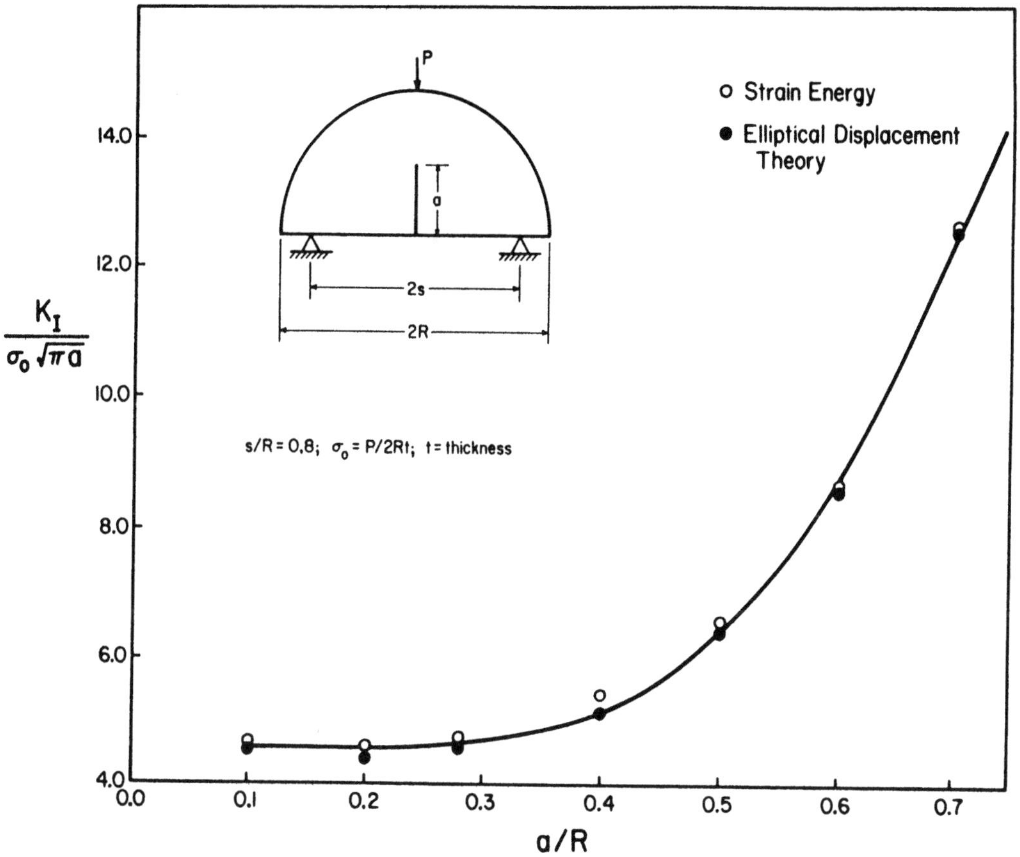


Fig. 5 Mode I stress intensity factor for the semi-circular bend specimen under three point bend loading.

Fig. 6. Modes I and II stress intensity factors for mixed mode specimen.

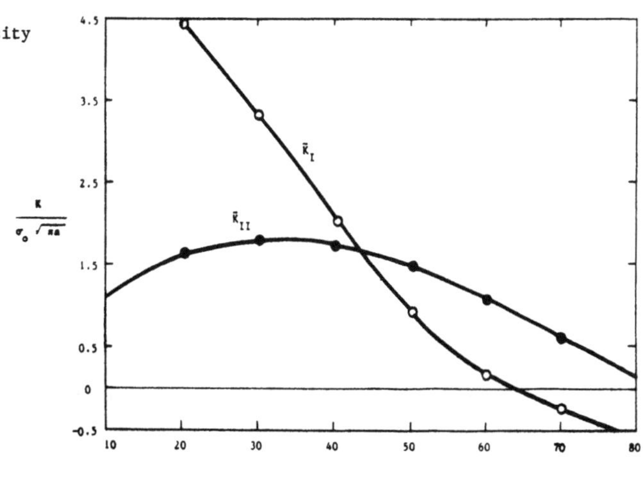

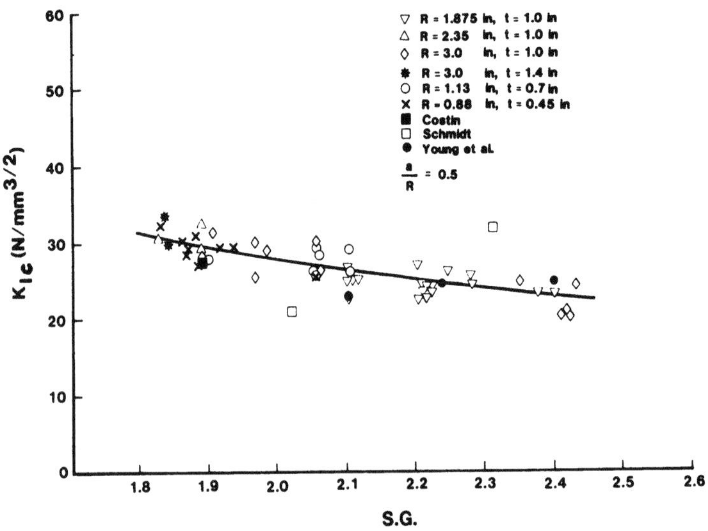


Fig. 7 Fracture toughness versus specific gravity using average of stress intensity factor, compliance and J-integral methods.

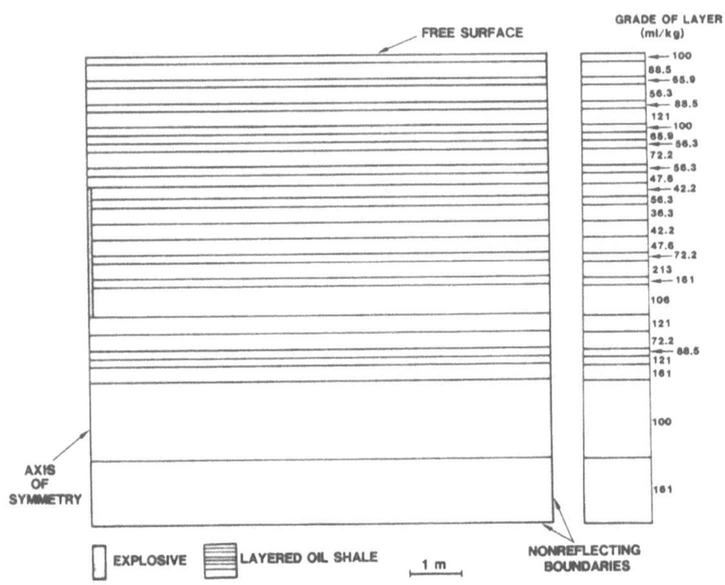


Fig. 8 Blasting simulation geometry and grades of the layers.

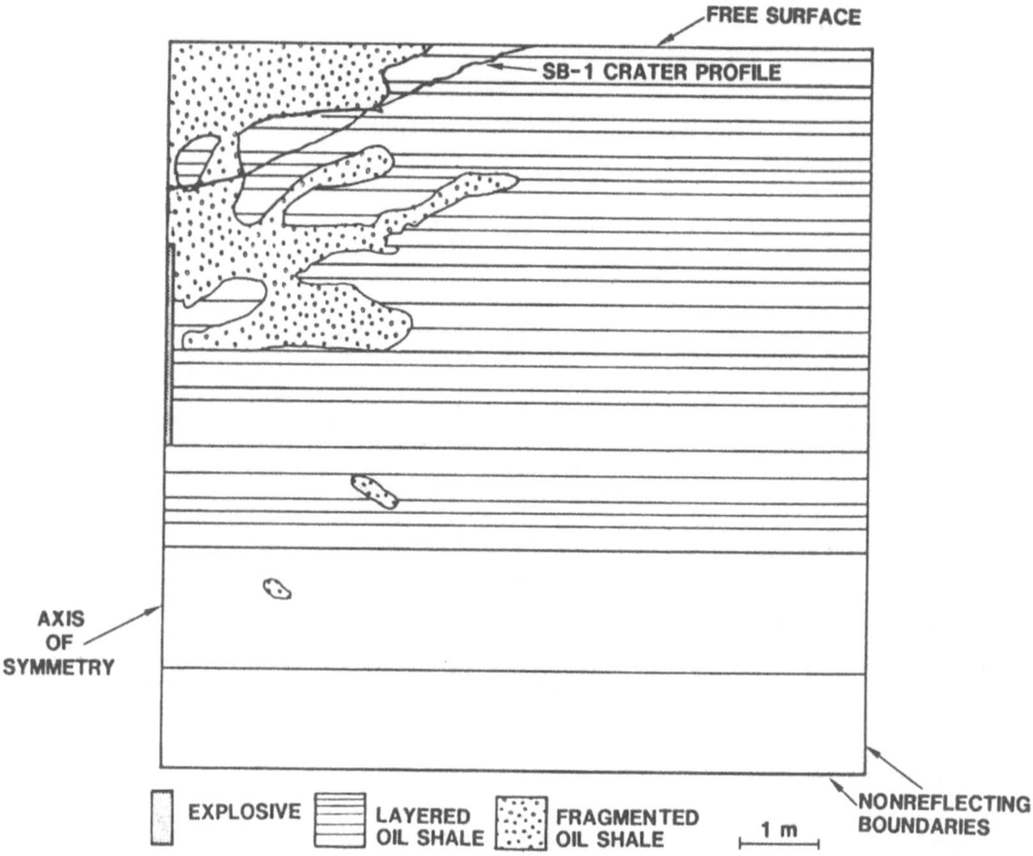


Fig. 9 Fragmented rock calculated by blasting simulation and experimental excavated crater profile.

SEM/RILEM International Conference
on Fracture of Concrete and Rock
Houston, Texas, June 1987
Editors: S.P. Shah, S.E. Swartz



# ENERGY BALANCE IN INSTRUMENTED IMPACT TESTS ON PLAIN CONCRETE BEAMS

N.P. Banthia, Sidney Mindess and Arnon Bentur

## ABSTRACT

Impact tests were carried out on plain concrete beams, 100 mm × 125 mm in cross-section and 1400 mm long, using an instrumented drop weight impact machine. The machine, capable of dropping a 345 kg mass through heights of up to 2.3 m, had strain gauges attached to the striking end of the hammer, and also to one of the support anvils. In addition, in order to record the beam response during the impact, three accelerometers were mounted along the length of the beam. Normal strength and high strength concretes were tested. For three different drop heights, the energy lost by the hammer was compared to the energy gained by the beam in various forms. It was found that up to the peak load, the energy gained by the beam was only a small percentage of the energy lost by the hammer. However, by the end of the impact event, most of the energy lost by the hammer could be accounted for.

## INTRODUCTION

There are numerous situations in which a concrete structure, or some of its structural elements, may be subjected to dynamic loading. In most of these situations, a great deal of energy is suddenly imparted to the structure. If the structural material is capable of absorbing this energy, failure may be avoided. But, if the material is brittle, like plain concrete, collapse may occur. Therefore, concrete is generally used in conjunction with other reinforcing materials, such as steel. However, an understanding of the composite behaviour of reinforced concrete at high stress rates requires a knowledge of the behaviour of its individual components at high stress rates. Although the static behaviour of concrete is fairly well understood, relatively little is known about its dynamic behaviour.

Many studies (1-5) have demonstrated the strain rate sensitivity of concrete. However, our knowledge of concrete behaviour at high stress rates still remains largely empirical. Part of the reason for this has been the inability to compare the results from different investigations, in the absence of any standard testing technique. The results of a particular investigation depend largely on the test method used in that particular investigation, because of the different energy losses associated with various testing machines, and different methods of analysis. The concept of energy balance to analyze impact data has been developed by Lueth (6) and Abe et al. (7). In the present work, the energy balance, or the various forms in which the hammer energy in a drop weight type impact machine is dissipated, has been applied to tests on concrete beams.

## EXPERIMENTAL PROCEDURES

### 1. The Drop Weight Impact Machine

The drop weight impact machine (Fig. 1), and a phenomenological description of the behaviour of concrete under this type of loading, have been described previously (8,9). Briefly, the machine is capable of dropping a 345 kg mass hammer through heights of up to 2.3 m on simply supported beams, on a test span of 960 mm. In this study the beam dimensions were (length × width × depth) 1400 mm × 100 mm × 125 mm. Two types of concretes were tested: normal strength (NS) concrete with a compressive strength of 42 MPa, and high strength (HS) concrete, with a compressive strength of 82 MPa. Details of the mix design and specimen preparation are given in (10).

N.P. Banthia and S. Mindess are Graduate Student and Professor, respectively, Department of Civil Engineering, University of British Columbia, Vancouver, B.C., Canada, V6T 1W5. Arnon Bentur is an Associate Professor, Building Research Station, Department of Civil Engineering, Technion - Israel Institute of Technology, Haifa 32000, Israel.

The instrumentation consists of strain gauges in the striking end of the hammer (called the 'tup') to measure the impact load, and in one of the support anvils to record the reaction at the supports. In addition, three accelerometer are mounted along the length of the beam (Fig. 2(a)) in order to record the beam response during an impact event. The data were acquired using a 5-channel data acquisition system based upon an IBM PC. All five channels were read simultaneously at 0.2 ms intervals. The duration of the entire impact event was about 10-15 ms.

## 2. Analysis of the Results

Typical tup load vs. time traces are shown in Fig. 3, for three different drop heights of the hammer. The three drop heights of 0.5 m, 0.25 m, and 0.15 m corresponded to hammer impact velocities of 3.13 m/sec, 2.21 m/sec, and 1.71 m/sec, respectively. [Due to the friction between the columns of the machine and the hammer, the downward acceleration of the hammer was found to be less than g (9.81 m/sec ). In the present study, downward accelerations of 0.95 g were obtained.] The three different starting times for the load vs. time pulses in Fig. 3 result from the fact that the data acquisition system was triggered by the hammer at a fixed height above the beam. Thus a faster hammer (0.50 m drop) took less time to strike the beam after triggering the system.

It has been pointed out by many investigators (e.g., 11,12) that the load, $P_t(t)$, as recorded by the tup is not the actual bending load on the beam because of the inertial loading effects. Part of the observed tup load consists of the inertial reaction of the beam. Moreover, in the case of brittle materials like concrete, since the time to failure is very small, the entire mechanical response of the beam up to failure may occur while the inertial reaction of the beam is still present. Accelerations in a beam undergoing dynamic testing are an inherent characteristic of such a test, and hence cannot be eliminated. However, knowing the accelerations along the length of the bean, a proper inertial correction can be applied.

On the basis of many tests carried out by the authors (8), it was assumed that for plain concrete, at any instant of time, the accelerations (Fig. 2(b)), and the displacements (Fig. 2(c)) are linear along the length of the beam. Let

$\rho$ = mass density of the beam material

$P_t(t)$ = observed tup load

$P_b(t)$ = equivalent bending load

$P_i(t)$ = generalized inertial load

$u(x,t)$ = displacements in the span of the beam between the supports

$u(y,t)$ = displacements in the portion of the beam overhanging the supports

$u_0(t)$ = displacement at midspan

$\delta u_0$ = virtual displacement at the centre compatible with the constraints (Fig. 2(c))

$\ddot{u}_0(t)$ = extrapolated value of the acceleration at midspan (Fig. 2(b))

$\dot{u}_0(t)$ = velocity at midspan

$l$ = span of the test beam

$h$ = length of the overhang

On the basis of the linear displacement assumption (Fig. 2(c)),

$$u(x,t) = \frac{2u_0(t)x}{\ell} \quad \text{in the supported span} \tag{1}$$

$$u(y,t) = \frac{-2u_0(t)y}{\ell} \quad \text{in the overhanging portion} \tag{2}$$

If the distributed inertial load along the length of the beam is to be replaced by a generalized inertial load, $P_i(t)$, then the virtual work done by all of the distributed inertial loads (Fig. 2(c)) should be equal to the virtual work done by $P_i(t)$ alone:

$$P_i(t)\,\delta u_0 = 2\int_0^{\ell/2} \rho A\left(\frac{2\ddot{u}_0(t)x}{\ell}\right)\left(\frac{2\delta u_0 x}{\ell}\right)dx + 2\int_0^{h} \rho A\left(\frac{-2\ddot{u}_0(t)y}{\ell}\right)\left(\frac{-2\delta u_0 y}{\ell}\right)dy \tag{3}$$

For a prismatic, homogeneous beam, Equation 3 can be further simplified to

$$P_i(t) = \rho A\ddot{u}_0(t)\left[\frac{\ell}{3} + \frac{8}{3}\frac{h^3}{\ell^2}\right] \tag{4}$$

The beam can now be modelled as a single degree of freedom system, and the equivalent bending load, $P_b(t)$, can be obtained from the equation of dynamic equilibrium:

$$P_b(t) = P_t(t) - P_i(t) \quad (5)$$

Once $P_b(t)$ has been found, the dynamically loaded beam with $P_t(t)$ applied at the centre and the inertial reaction distributed along the whole body of the beam (Fig. 3(a)) can be replaced by a beam loaded only by $P_b(t)$ at the centre (Fig. 3(b)). $P_b(t)$, therefore, can also be called the "equivalent static load". The values of bending moment, etc., obtained by using the static formula on such a beam will be the same as the ones obtained by doing a dynamic analysis of the beam shown in Fig. 3(a).

The velocity and the displacement at the centre can be obtained by integration,

$$\dot{u}_0(t) = \int_o^t \ddot{u}_0(t)dt \quad (6)$$

$$u_0(t) = \int_o^t \dot{u}_0(t)dt \quad (7)$$

Once the equivalent bending load, $P_b(t)$, at the centre (Equation 5) and the displacements at the centre (Equation 7) are known, the bending energy in the beam, $E_b(t)$, can be obtained from the area under the load vs. displacement plot,

$$E_b(t) = \int_o^t P_b(t)du_0 \quad (8)$$

As will be seen later, the energy given by Equation 8 is made up of the strain energy in the beam and the energy due to the work of fracture.

THE ENERGY BALANCE

The concept of "energy balance", which has its basis in the principle of the conservation of energy, compares the energy lost by the hammer, at any time during the impact, and the energy gained by the specimen. Theoretically, if any losses can be ignored, the law of conservation of energy would predict the two energies to be equal. In practice, however, the losses in the system cannot be ignored, and the energy gained by the specimen is, in general, less than the energy lost by the hammer.

The falling hammer, on striking the beam, suffers a loss of momentum. This loss of momentum, according to the impulse-momentum relationship, is equal to the impulse acting on the hammer. Using this principle, the loss in kinetic energy of the hammer ($\Delta E(t)$) can be evaluated:

$$\Delta E(t) = 1/2\ m_h[2a_h h - (\sqrt{2a_h h} - \frac{1}{m_h} \int_o^t P_t(t)dt)^2] \quad (9)$$

where

$m_h$ = mass of the hammer

$a_h$ = acceleration of the hammer

$h$ = height of hammer drop

$\int_o^t P_t(t)dt$ = the impulse

The energy lost by the hammer (Equation 9) may be transferred to the beam in various forms. This transfer of energy can best be studied at the following two points (Figure 4):

(1) Energy balance at the peak load ($t = t_p$)

(2) Energy balance at the end of impact event ($t = t_t$)

1. Energy Balance at the Peak Load ($t = t_p$)

At the peak load, the energy balance equation can be written as

$$\Delta E(t_p) = E_m(t_p) + E_{ker}(t_p) + E_b(t_p) \quad (10)$$

where

$E_m(t_p)$ = energy lost to the various machine parts in the form of vibrations or elastic strain energy

$E_{ker}(t_p)$ = rotational kinetic energy of the specimen

$E_b(t_p)$ = bending energy in the specimen

In Equation 10, the translational kinetic energy, and the vibrational energy in the specimen, have been ignored (6). The bending energy, given by the area under the load vs. load point displacement plot (Equation 8), comprises the elastic strain energy, $E_{se}(t_p)$, and the work of fracture, $E_{wof}(t_p)$, as indicated in Fig. 5.

$$E_b(t_p) = E_{se}(t_p) + E_{wof}(t_p) \tag{11}$$

From a load vs. centre point displacement plot, the elastic strain energy, $E_{se}(t_p)$, can reasonably be approximated by taking the secant modulus at 60% of the peak load (Figure 5):

$$E_{se}(t_p) = 0.5\ P_b(t_p)u_{0e} \tag{12}$$

where $u_{0e}$ is the elastic part of the midspan displacement. The secant modulus at 60% of the peak load was chosen because this is about the point at which significant matrix cracking is expected to begin. Prior to this point, most of the nonlinearity is due only to the development of bond cracks. The work of fracture, $E_{wof}(t_p)$, can then be obtained by subtracting the strain energy, $E_{se}(t_p)$, from the bending energy, $E_b(t_p)$, (Equation 11).

Knowing the velocity at the centre of the beam, and assuming that the velocity distribution is linear along the length of the beam, the rotational kinetic energy, $E_{ker}(t_p)$, of the specimen can be obtained by integrating over its length,

$$E_{ker}(t_p) = \frac{8\rho A\dot{u}_0(t_p)}{\ell_2}\left[\frac{\ell^2}{24} + \frac{h^3}{3}\right] \tag{13}$$

The energy lost to the machine, $E_m(t_p)$, can be obtained by subtracting the beam energies $E_b(t_p)$ and $E_{ker}(t_p)$ from the hammer energy $\Delta E(t_p)$, as given by Equation 10.

2. Energy Balance at the End of the Impact Event (at $t = t_t$)

At the end of the impact event, the external load, $P_t(t)$, is reduced to zero and the broken halves of the beam swing clear of the striking tup. At this instant, the energy balance can be written as

$$\Delta E(t_t) = E_m(t_t) + E_b(t_t) + E_{ker}(t_t) \tag{14}$$

Once again the energy $E_b(t_t)$ obtained from Equation 8 measures the work of fracture and the strain energy in the beam. Since the strain energy can be assumed to be negligible in the broken halves of the beam, all of the energy $E_b(t_t)$ represents the work of fracture, or the fracture energy. Equation 13 can then be used at $t=t_t$ to determine the rotational kinetic energy $E_{ker}(t_t)$. Once $E_{ker}(t_t)$ and $E_b(t_t)$ are known, Equation 14 can be used to determine $E_m(t_t)$.

## RESULTS

Tables 1 to IV and Figs. 6 to 9 present the experimental results. Tables I and III correspond to normal strength (NS) concrete and Tables II and IV correspond to high strength (HS) concrete. Tables I and II present the experimental results at the peak load ($t=t_p$) for normal strength and high strength concrete, respectively. Tables III and IV present the results at the end of the impact event ($t=t_t$). Data for three different heights of hammer drop are presented. Since Tables I through IV show substantial scatter in results, only the corresponding mean values are plotted in Figs. 6 to 9.

Fig. 6 shows the maximum values of $P_b(t)$, which occur at $t=t_p$. This peak equivalent bending load will be referred to here as the impact strength. It can be seen from Fig. 6 that HS concrete, which is stronger in static situations, is also stronger in impact (10). Moreover, the apparent strength increases with an increase in the drop height, that is, with an increase in the stress rate.

Fig. 7(a) shows the energy balance for NS and HS concretes at the peak load ($t=t_p$). At the peak load, the energy lost by the hammer, $\Delta E(t_p)$, is 2 to 4 times the energy gained by the beam, $E_s(t_p)$. The remainder of the energy is assumed to be absorbed in the machine itself, in the form of vibrations and elastic energy. The energy gained by the beam by virtue of its deformed shape, $E_b(t_p)$, is found to be much smaller than its kinetic energy, $E_{ker}(t_p)$. Also, the consistently lower values of energies $E_s(t_p)$, $E_{ker}(t_p)$, and $E_b(t_p)$ for HS concrete compared to NS concrete should be noted.

Fig. 7(b) presents the components of the bending energy, $E_b(t_p)$, at the peak load, i.e., the work of fracture, $E_{wof}(t_p)$, and the elstic strain energy, $E_{se}(t_p)$. Most of the energy consumed by the beam up to the peak bending load appears as the work of fracture. Both the work of fracture and the strain energy seem to increase with an increase in the hammer drop height; the work of fracture increases at a higher rate than does the strain energy. HS concrete, when compared to NS concrete, appears to have a reduced strain energy.

Fig. 8 presents the energy balance at the end of the impact event ($t=t_p$). Here, most of the energy lost by the hammer, $\Delta E(t_t)$, is gained by the beam, $E_s(t_t)$. The energy gained by the beam consists of the kinetic energy of the broken halves, $E_{ker}(t_t)$, and the bending energy, $E_b(t_t)$. Since, by the end of the impact event, the specimen has little or no strain energy, the bending energy represents only the fracture energy.

If the difference between the energy lost by the hammer and the energy absorbed by the beam can be assumed to be the energy lost to the machine, then the "machine loss" can be calculated at the peak load ($t=t_p$) and at the end of the event ($t=t_t$). Fig. 9 presents this machine energy calculated as a percentage of the total energy lost by the hammer ($E_m/\Delta E \times 100\%$). As can be seen from Fig. 9, at the peak load, 60 to 80 percent of the energy lost by the hammer is stored in the machine. However, by the end of the impact event ($t=t_t$), over 90 percent of the energy lost by the hammer appears as specimen energy. It can also be noted from Fig. 9 that HS concrete, being a stronger and stiffer material compared to NS concrete, showed higher machine losses.

## DISCUSSION

High stress rate testing of cementitious materials requires sophisticated testing equipment. Knowledge of the various modes in which energy can be lost during a test is essential to a proper analysis of the test results. In the absence of this knowledge, the results can be grossly misleading. The energy lost by the hammer cannot be assumed to be the energy consumed by the beam. Even if the machine losses can be assumed to be constant for a given drop height for a given machine, the energy gained by the specimen still has to be corrected for its kinetic energy.

The amount of energy lost to the machine seemed to depend upon the strength and stiffness of the material tested. Energy losses were found to be higher when HS concrete, which is stronger and stiffer than NS concrete, was tested. It is also likely that the energy lost to the machine depends upon the amount of energy available. In this study, the mass of the hammer was about eight times as large as that of the beams tested, and as a consequence the available hammer energy was very high compared to the energy absorption capacities of the specimens. It is possible that the use of a smaller impact hammer would lead to smaller energy losses to the machine.

HS concrete, although stronger in static as well as dynamic situations, was found to be more brittle than NS concrete. This restricts, at least to some extent, its use in situations in which it may be called upon to resist impact loading. Increased microcracking at higher stress rates is the probable cause for the increase in fracture energy requirements at higher stress rates (10); the energy suddenly imparted to the beam is dissipated in the form of microcracking at the aggregate-paste interface throughout the body of the specimen. As a result, the high strength concrete, which exhibits a better cement-aggregate bond, could not absorb as much energy as the normal strength concrete.

If it can be assumed that, after the beam fails, the broken halves of the beam have little or no strain energy, then in the post-peak load region, most or all of the strain energy stored in the beam is used in propagating the crack. Since the stored strtain energy at the peak load is much less than the overall fracture energy required, it seems possible that the crack propagates while the beam continues to absorb energy from the hammer and the various other machine parts.

## CONCLUSIONS

1. At the time of peak external load, only a fraction of the energy lost by the hammer is absorbed by the beam.

2. By the end of the impact event, over 90 percent of the energy lost by the hammer appears as the specimen kinetic energy and fracture energy.

3. For the correct estimation of the energies, in addition to the instrumentation in the tup, supplementary instrumentation (e.g., accelerometers) is essential.

4. The energy lost in the machine seems to be higher for a stronger and stiffer specimen.

5. High strength concrete was found to be more brittle than normal strength concrete.

6. The higher energy requirement in concrete at higher stress rates is probably due to the increased microcracking at higher stress rates.

ACKNOWLEDGEMENTS

The authors wish to thank Mr. Glenn D. Jolly and Mr. Rod B. Nussbaumer for designing and constructing the data acquisition system. This research was supported in part by a grant from the Natural Sciences and Engineering Research Council of Canada, and in part by a contract from Elkem Chemicals Inc., Pittsburgh, Pennsylvania.

REFERENCES

1. D.A. Abrams, "Effect of Rate of Application of Load on the Compressive Strength of Concrete," Proceedings, ASTM, V. 17, part 2, 1917, pp. 364-367.

2. D. Watstein, "Effect of Straining Rate on the Compressive Strength and Elastic Properties of Concrete," Journal of the American Concrete Institute, V. 49, No. 8, 1953, pp. 729-756.

3. B.P. Hughes and R. Gregory, "Concrete Subjected to High Rates of Loading in Compression," Magazine of Concrete Research, V. 24, No. 78, 1972, pp. 25-36.

4. W. Suaris and S.P. Shah, "Properties of Concrete Subjected to Impact," Journal of the Structural Division, ASCE, V. 109, No. 7, 1983, pp. 1727-1741.

5. A.J. Zielinsky and H.W. Reinhardt, "Stress Strain Behaviour of Concrete and Mortar at High Rates of Tensile Loading," Cement and Concrete Reserach, V. 12, No. 3, 1982, pp. 309-319.

6. R.C. Lueth, "An Analysis of Charpy Impact Testing as Applied to Cemented Carbide," Instrumented Impact Testing, ASTM STP 563, American Society for Testing and Materials, 1974, pp. 166-179.

7. H. Abe, H.C. Chandan and R.C. Bradt, "Low Blow Charpy Impact of Silicon Carbides," Bulletin of the American Ceramic Society, V. 57, No. 6, 1978, pp. 587-595.

8. A. Bentur, S. Mindess and N. Banthia, "The Behaviour of Concrete Under Impact Loading: Experimental Procedures and Method of Analysis," Materials and Structures (RILEM), in press.

9. S. Mindess and A. Bentur, "A Preliminary Study of the Fracture of Concrete Beams Under Impact Loading," Cement and Concrete Research, V. 15, No. 3, 1985, pp. 474-484.

10. N. Banthia, S. Mindess and A. Bentur, "Impact Behaviuor of Concrete Beams," Materials and Structures (RILEM), accepted for publication.

11. W.L. Server, "Impact Three Point Bend Testing for Notched and Pre-Cracked Specimens," Journal of Testing and Evaluation, V. 6, No. 1, 1978, pp. 29-34.

12. W. Suaris and S.P. Shah, "Inertial Effects in Instrumented Impact Testing of Cementitious Composites," Journal of Cement, Concrete and Aggregate, 1982, pp. 78-83.

TABLE I

NORMAL STRENGTH CONCRETE - RESULTS AT $t=t_p$

| | HEIGHT OF HAMMER DROP | | | | | | | | | | | |
|---|---|---|---|---|---|---|---|---|---|---|---|---|
| | 0.15 m (6)* | | | | 0.25 m (6)* | | | | 0.50 m (7)* | | | |
| | MAXIMUM | MINIMUM | MEAN | STANDARD DEVIATION | MAXIMUM | MINIMUM | MEAN | STANDARD DEVIATION | MAXIMUM | MINIMUM | MEAN | STANDARD DEVIATION |
| $P_b(t_t)$ (N) | 9440 | 7782 | 8460 | 604 | 14668 | 9178 | 12183 | 2401 | 17727 | 16452 | 16932 | 428 |
| $\Delta E(t_p)$ (N.m) | 25.78 | 19.75 | 22.86 | 2.37 | 42.06 | 33.36 | 37.24 | 3.21 | 71.60 | 60.89 | 64.64 | 3.80 |
| $E_{ker}(t_p)$ (N.m) | 7.52 | 4.60 | 6.20 | 1.27 | 10.00 | 6.12 | 7.98 | 1.37 | 16.96 | 10.96 | 12.23 | 2.00 |
| $E_b(t_p)$ (N.m) | 3.50 | 1.53 | 2.53 | 0.71 | 3.73 | 2.74 | 3.01 | 0.43 | 9.07 | 2.21 | 6.42 | 2.51 |
| $E_{se}(t_p)$ (N.m) | 0.64 | 0.32 | 0.49 | 0.12 | 0.82 | 0.52 | 0.63 | 0.11 | 1.35 | 0.78 | 1.14 | 0.19 |
| $E_{wof}(t_p)$ (N.m) | 2.85 | 1.20 | 2.05 | 0.58 | 2.90 | 2.04 | 2.38 | 0.32 | 7.90 | 1.42 | 5.27 | 2.41 |
| $E_s(t_p)$ (N.m) | 10.12 | 6.13 | 8.72 | 1.53 | 13.47 | 8.86 | 10.95 | 1.64 | 23.56 | 14.63 | 19.60 | 3.03 |
| $E_m(t_p)$ (N.m) | 15.66 | 12.69 | 14.12 | 1.10 | 31.46 | 24.45 | 26.32 | 2.97 | 48.39 | 40.47 | 45.03 | 3.49 |
| $\frac{E_m(t_p)}{E(t_p)} \times 100\%$ | 69 | 59 | 62 | 4.02 | 74 | 64 | 69 | 3.77 | 77 | 67 | 70.00 | 4.18 |

* Number of Specimens Tested

TABLE II

HIGH STRENGTH CONCRETE - RESULTS AT $t=t_p$

| | HEIGHT OF HAMMER DROP | | | | | | | | | | | |
|---|---|---|---|---|---|---|---|---|---|---|---|---|
| | 0.15 m (6)* | | | | 0.25 m (5)* | | | | 0.50 m (6)* | | | |
| | MAXIMUM | MINIMUM | MEAN | STANDARD DEVIATION | MAXIMUM | MINIMUM | MEAN | STANDARD DEVIATION | MAXIMUM | MINIMUM | MEAN | STANDARD DEVIATION |
| $P_b(t_t)$ (N) | 11694 | 8388 | 9906 | 1183 | 18759 | 10573 | 13371 | 2991 | 19206 | 18314 | 18760 | 446 |
| $\Delta E(t_p)$ (N.m) | 25.21 | 19.02 | 20.28 | 3.54 | 38.45 | 33.40 | 32.20 | 3.98 | 76.33 | 58.82 | 66.13 | 7.43 |
| $E_{ker}(t_p)$ (N.m) | 6.40 | 2.68 | 3.94 | 1.44 | 9.61 | 4.47 | 6.58 | 1.67 | 16.00 | 4.35 | 10.13 | 4.75 |
| $E_b(t_p)$ (N.m) | 2.92 | 1.79 | 2.37 | 0.54 | 2.96 | 1.86 | 2.55 | 0.376 | 5.41 | 3.80 | 4.64 | 0.659 |
| $E_{se}(t_p)$ (N.m) | 0.51 | 0.33 | 0.43 | 0.07 | 0.63 | 0.54 | 0.60 | 0.05 | 1.34 | 1.00 | 1.14 | 0.06 |
| $E_{wof}(t_p)$ (N.m) | 2.43 | 1.40 | 1.93 | 0.47 | 2.40 | 1.28 | 1.95 | 0.37 | 4.23 | 2.86 | 3.52 | 0.56 |
| $E_s(t_p)$ (N.m) | 9.32 | 4.47 | 6.31 | 1.83 | 11.47 | 7.19 | 9.14 | 1.39 | 19.80 | 1.08 | 14.78 | 4.40 |
| $E_m(t_p)$ (N.m) | 25.21 | 19.02 | 20.28 | 3.54 | 38.54 | 33.40 | 32.20 | 3.98 | 76.33 | 58.82 | 66.13 | 7.43 |
| $\frac{E_m(t_p)}{E(t_p)} \times 100\%$ | 75 | 63 | 69 | 4.47 | 72 | 70 | 72.0 | 0.80 | 84 | 74 | 78 | 4.96 |

* Number of Specimens Tested

TABLE III

NORMAL STRENGTH CONCRETE - RESULTS AT $t=t_t$

| | HEIGHT OF HAMMER DROP | | | | | | | | | | | |
|---|---|---|---|---|---|---|---|---|---|---|---|---|
| | 0.15 m (6)* | | | | 0.25 m (6)* | | | | 0.50 m (7)* | | | |
| | MAXIMUM | MINIMUM | MEAN | STANDARD DEVIATION | MAXIMUM | MINIMUM | MEAN | STANDARD DEVIATION | MAXIMUM | MINIMUM | MEAN | STANDARD DEVIATION |
| $\Delta E(t_t)$ (N.m) | 76 | 64 | 72 | 4.7 | 131 | 94 | 116 | 15.3 | 249 | 231 | 240 | 5.8 |
| $E_{ker}(t_t)$ (N.m) | 45 | 42 | 44 | 1.2 | 80 | 47 | 66 | 12.0 | 150 | 139 | 145 | 4.7 |
| $E_b(t_t)$ (N.m) | 31 | 19 | 25 | 4.3 | 60 | 27 | 43 | 12.4 | 100 | 87 | 90 | 6.4 |
| $E_s(t_t)$ (N.m) | 73 | 63 | 69 | 3.9 | 130 | 74 | 109 | 23.0 | 248 | 230 | 235 | 7.31 |
| $E_m(t_t)$ (N.m) | 3 | 1 | 2 | 1.0 | 20 | 1 | 7 | 7.7 | 12 | 0 | 5 | 4.0 |
| $\frac{E_m}{E} \times 100\%$ | 4 | 1.6 | 2.8 | 1.0 | 21 | 6.8 | 6.0 | 8.0 | 5.0 | 0 | 2.1 | 1.0 |

* Number of Specimens Tested

TABLE IV

HIGH STRENGTH CONCRETE - RESULTS AT $t=t_t$

| | HEIGHT OF HAMMER DROP | | | | | | | | | | | |
|---|---|---|---|---|---|---|---|---|---|---|---|---|
| | 0.15 m (6)* | | | | 0.25 m (5)* | | | | 0.50 m (6)* | | | |
| | MAXIMUM | MINIMUM | MEAN | STANDARD DEVIATION | MAXIMUM | MINIMUM | MEAN | STANDARD DEVIATION | MAXIMUM | MINIMUM | MEAN | STANDARD DEVIATION |
| $\Delta E(t_t)$ (N.m) | 89 | 43 | 68 | 16.7 | 133 | 100 | 109 | 12.2 | 238 | 214 | 223 | 10.5 |
| $E_{ker}(t_t)$ (N.m) | 54 | 13 | 36 | 16.8 | 89 | 54 | 64 | 13.6 | 127 | 116 | 121 | 5.5 |
| $E_b(t_t)$ (N.m) | 33 | 21 | 25 | 5.0 | 43 | 31 | 35 | 4.6 | 100 | 57 | 75 | 18.6 |
| $E_s(t_t)$ (N.m) | 89 | 34 | 61 | 21.0 | 133 | 86 | 99 | 18.0 | 213 | 193 | 196 | 10.0 |
| $E_m(t_t)$ (N.m) | 14 | 0 | 7 | 5.9 | 15 | 0 | 10 | 6.3 | 30 | 25 | 27 | 2.7 |
| $\frac{E_m}{E} \times 100\%$ | 21.1 | 0 | 10.3 | 5.4 | 10.8 | 0 | 9.2 | 3.7 | 13.7 | 10.2 | 12.1 | 1.5 |

* Number of Specimens Tested

Fig. 1. Overall view of the instrumented impact machine. The height of the machine is about 4.5 m.

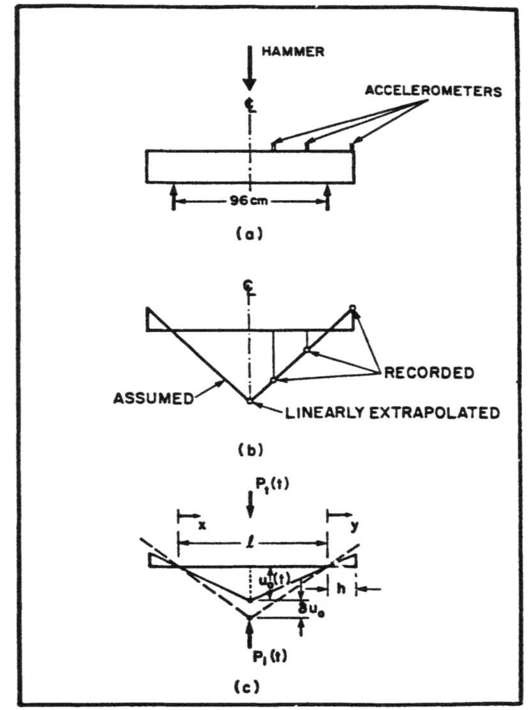


Fig. 2. Beam accelerations and deformations
(a) Positions of the accelerometers
(b) Acceleration distribution along the beam
(c) Displacements along the beam

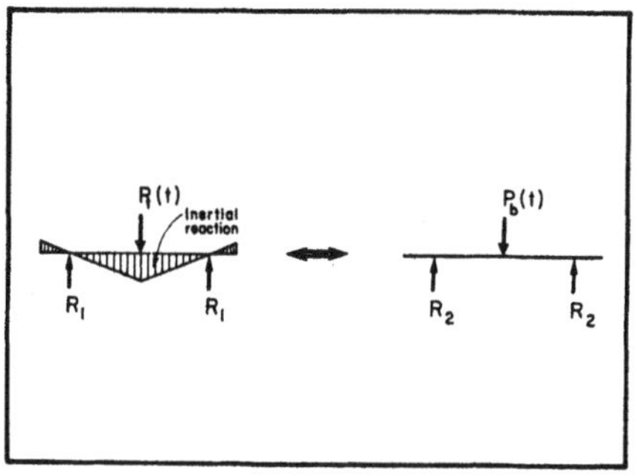


Fig. 3. Loading on the beam undergoing impact
(a) General dynamic loading
(b) Equivalent static loading

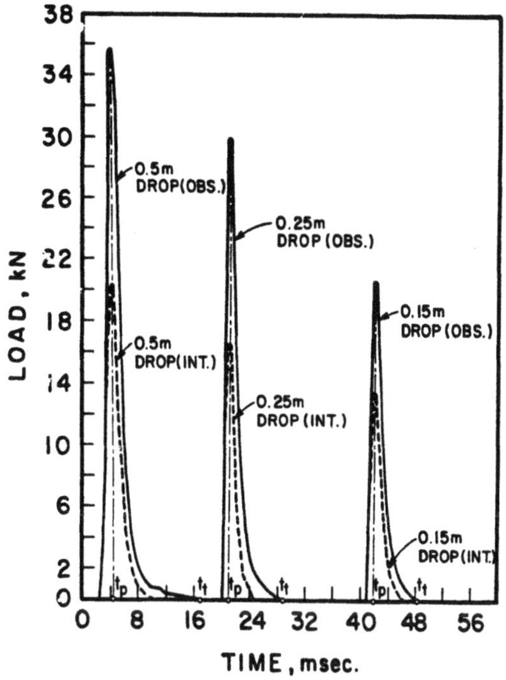


Fig. 4. Typical tup load vs. time traces for three different drop heights.

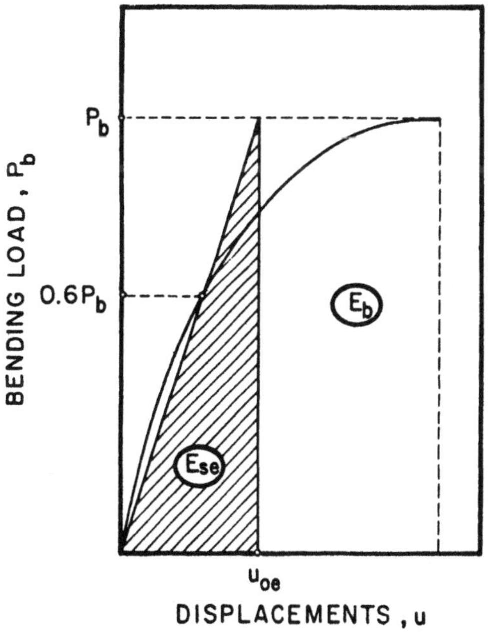


Fig. 5. Schematic load vs. displacement plot for the calcuations of strain energy and the work of fracture.

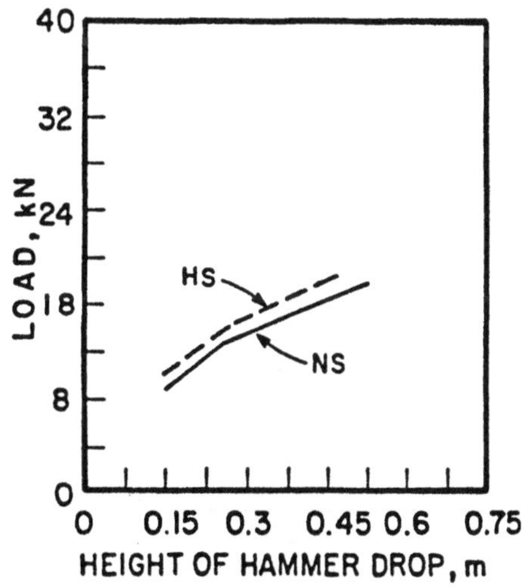


Fig. 6. Variation of impact strength with drop height for NS and HS concretes.

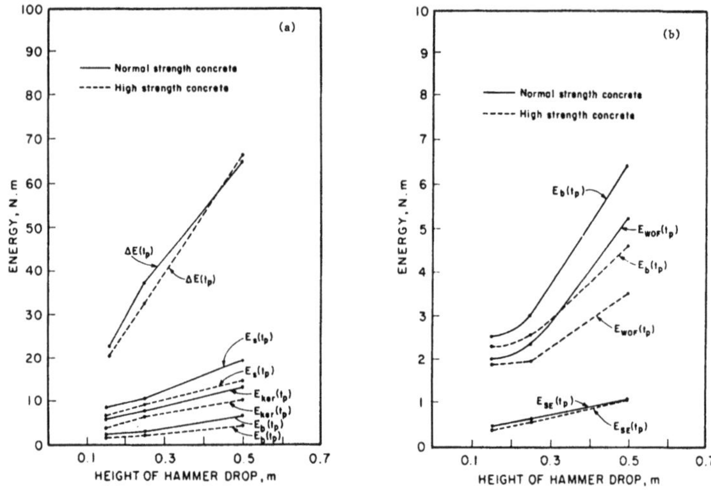


Fig.7. Energy balance at the peak load: (a) Overall energy balance; and (b) Energy distribution in the beam

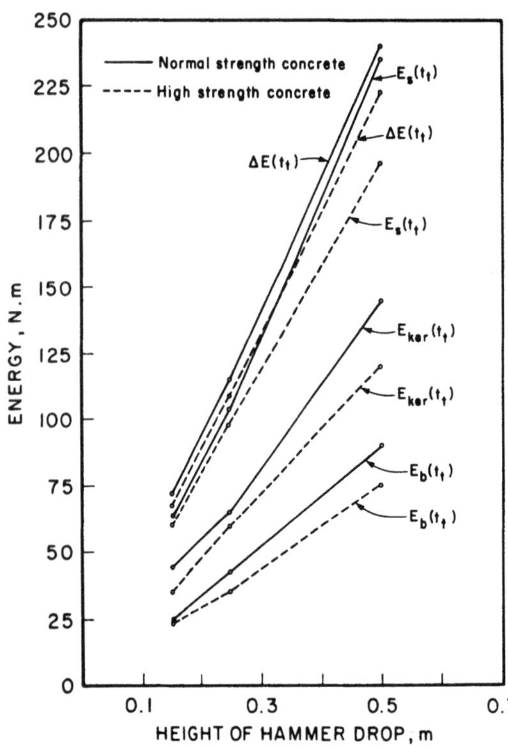


Fig.8. Energy balance at the end of the impact event.

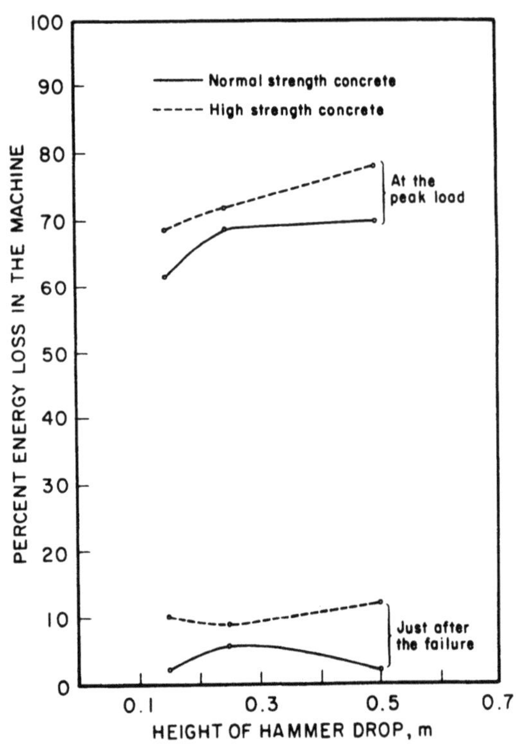


Fig. 9. Percent energy lost to the machine.

SEM/RILEM International Conference
on Fracture of Concrete and Rock
Houston, Texas, June 1987
Editors: S.P. Shah, S.E. Swartz

# A STATISTICAL THEORY OF TIME-DEPENDENT FRACTURE FOR CEMENTITIOUS MATERIALS

Xiao-Zhi Hu, Yiu-Wing Mai, Brian Cotterell

## ABSTRACT

By considering pre-existing flaw size distribution and the slow crack growth characteristics of these flaws a statistical theory is developed for the time-dependent strength behaviour of cementitious materials. Theoretical fracture strength and lifetime predictions in pure and three-point bend cement paste specimens subjected to both constant stress rates and sustained stresses are presented. In addition, creep curves under sustained stresses can also be predicted from the statistical fracture theory.

## INTRODUCTION

It is well recognised that the strength of cementitious matrices such as cement paste, mortar and concrete, like those of many other brittle materials, is dependent upon the loading rate as well as upon the time under which a sustained load is applied [1-3]. Two important factors affect these time-dependent strength characteristics. One is the flaw statistics with respect to density and flaw size distribution; the other is the slow crack growth process as determined by the influence of the chemical reactive species at the flaw tips [2-8]. The conventional approach to predict time dependence of strength of brittle materials ignores flaw statistics and only considers the growth of a single crack from some sub-critical flaw size to a critical flaw size in accordance with the basic crack growth law for the particular material [9]. Such an approach is inappropriate for cementitious materials which are known to contain a distribution of pre-existing flaws whose sizes are not constant. In these cases flaw statistics must be considered because the flaws within the stressed volume of the sample grow according to the particular crack growth law and alter the original flaw size distribution function. The significance of the statistical fracture approach and its difference to the single crack theory are discussed later in the paper.

In previous work [10] the authors presented a two-phase statistical fracture theory relevant to concretes and mortars; but time-dependent crack growth was not considered in the model. The authors have since extended that work [10] by incorporating slow crack growth in both single and two-phase materials. In this paper only the statistical time-dependent fracture theory for single phase materials like cement pastes and some brittle ceramics is presented. The theoretical fracture strength and time-to-failure predictions for specimens subjected to constant stress rates and sustained stresses are derived. Creep curves due to the growth of pre-existing cracks under sustained stresses are also obtained.

## A STATISTICAL THEORY OF TIME-DEPENDENT FRACTURE

Since brittle materials are more easily tested in bending than in tension the statistical fracture theory developed in this section is for a rectangular cross-section beam subjected to either three-point or four-point bending. In three-point bend specimens the span to depth ratio is considered large so that shear stresses are neglected. It is assumed that the specimen of volume V contains a certain distribution of flaw sizes q(a), where a is the equivalent length of the Griffith flaw lying normal to the maximum principal stress. The density of these pre-existing randomly distributed flaws $\rho_f$ is assumed to be small so that thre is no interaction between flaws.

---

Z.-Z. Hu, Y.-W. Mai and B. Cotterell are, respectively, PhD research student, Associate Professor and Reader, Department of Mechanical Engineering, University of Sydney, Sydney, N.S.W. 2006, Australia.

The weakest link theory due to Weibull [11,12] can be developed from the flaw size distribution function q(a). Hunt and McCartney [13] have shown that the failure probability at stress $\sigma_a$ is given by:

$$F(\sigma_a) = 1 - \exp\{- \int_V \int_{a(\sigma_{xy})}^{\infty} q(a)da\, dV\} \tag{1}$$

where $\sigma_{xy}$ is a function of $\sigma_a$ and is dependent on its position (x,y) within the stressed volume of the material. From Griffith's theory:

$$a(\sigma_{xy}) = [K_{Ic}/Y\sigma_{xy}]^2 \tag{2}$$

where $K_{Ic}$ is the fracture toughness and Y is the geometry correction factor which is equal to $\sqrt{2/\pi}$ for a penny-shaped crack. If q(a) is described by a Pareto function,

$$q(a) = \begin{cases} (\rho_f m/2a_o)(a_o/a)^{(m+2)/2}; & a_o \leq a \quad (3a) \\ 0 \quad ; & a < a_o \quad (3b) \end{cases}$$

where $a_o$ is the smallest flaw size, integrating Eq. (1) over V gives the Weibull equation:

$$F(\sigma_a) = 1 - \exp[- (\sigma_a/\sigma*)^m] \; . \tag{4}$$

Here m is the Weibull modulus and σ* is a normalising parameter where

$$\sigma* = \begin{cases} \sigma_o[2(m+1)/V\rho_f]^{1/m} \text{ (pure bend)} & (5) \\ \sigma_o[2(m+1)^2/V\rho_f]^{1/m} \text{ (three-point bend)} & (5b) \end{cases}$$

and $\sigma_o$ is given by Eq. (2) where $a = a_o$. Both m and σ* can be determined from "inert strength" experiments in which environmental effects are absent.

In the statistical treatment it is realised that the flaws grow with time (t) under a given applied stress ($\sigma_a$). The simplest equation used to describe slow crack growth in brittle materials is [9]:

$$da/dt = AK_a^n = A(\sigma_a Y \sqrt{a})^n \tag{6}$$

where da/dt is the crack velocity, $K_a$ is the applied crack tip stress intensity factor, and (A,n) are numerical constants depending on the particular material-environment system. There are two cases of time-dependent strength behaviour of interest. In the first case the applied stress is maintained constant (i.e. $\sigma_a$ = constant) and it is required to determine the corresponding time-to-fialure ($t_f$). In the second case the applied stress rate is constant (i.e. $\dot{\sigma}_a$ = constant) and it is required to evaluate the variation of the fracture strength ($\sigma_f$) with $\dot{\sigma}_a$.

Case 1: Constant applied stress ($\sigma_a$ = constant)

The growth of a flaw from a reference size $a_r$ at t = 0 to size a at time t is obtained from Eq. (6) where

$$\int_{a_r}^{a} a^{-n/2} da = \int_0^t A(\sigma_a Y)^n\, dt \tag{6a}$$

so that

$$a_r(\sigma_a,a,t) = [1 + (\tfrac{n-2}{2}) A(\sigma_o Y)^n\, t\, a^{(n-2)/2}]^{(\frac{2}{2-n})}\, a \; . \tag{7}$$

Alternatively, a at time t can be expressed as a function of $a_r$ at t = 0 i.e.

$$a(\sigma_a,a_r,t) = [1 - (\tfrac{n-2}{2}) A(\sigma_a Y)^n\, t\, a_r^{(n-2)/2}]^{(\frac{2}{2-n})}\, a_r \; . \tag{8}$$

The growth of flaws within the stressed volume V alters the flaw size distribution function from q(a,o) [= q(a)] to q(a,t) and the failure probability equation at time t is given by:

$$F(\sigma_a,t) = 1 - \exp\{- \int_V \int_{a(\sigma_{xy})}^{\infty} q(a,t)da\ dV. \tag{9}$$

Now, it can be shown from the definition of the two integrals that:

$$\int_{a(\sigma_{xy})}^{\infty} q(a,t)da = \int_{a_r(\sigma_{xy},t)}^{\infty} q(a)\ da. \tag{10}$$

The left hand side integral is the probability of finding a flaw which is bigger than a at time t and the right-hand side is that of finding a flaw bigger than $a_r$ at $t = 0$. From Eq. (7) it follows that any flaw size bigger than $a_r$ at $t = 0$ is bigger than a at time t. Hence the equality of Eq. (10). The failure probability equation becomes:

$$F(\sigma_a,t) = 1 - \exp\{- \int_V \int_{a_r(\sigma_{xy},t)}^{\infty} q(a)da\ dV\}. \tag{11}$$

From Eqs. (3) and (7) and noting that $\sigma_{xy}$ only varies linearly across the beam depth in pure bending but also varies along the length of the span between supports in three-point bending, Eq. (11) can be integrated to give:

$$F(\sigma_a,t_f) = 1 - \exp\{- (m+1)(\sigma_a/\sigma_*)^m \int_0^1 u^m[1+(n/2-1)AY^2 K_{Ic}^{n-2} \sigma_a^2 u^2 t_f]^{\frac{m}{n-2}} du\} \tag{12}$$

for the pure bending case; and

$$F(\sigma_a,t_f) = 1 - \exp\{- (m+1)^2(\sigma_a/\sigma_*)^m \int_0^1\int_0^1 (1-v)^m u^m[1+(n/2-1)AY^2 K_{Ic}^{n-2} \sigma_a^2(1-v^2)u^2 t_f]^{\frac{m}{n-2}} du\ dv\} \tag{13}$$

for the three-point bending case. When $t_f \to 0$, i.e. the effect of slow crack growth is insignificant, Eqs. (12) and (13) become Eqs. (4) and (5).

The lifetime predictions under a sustained stress $\sigma_a$ can now be obtained from Eqs. (12) and (13). For $\sigma_a^2 t_f >> 1$, these equations are reduced respectively to:

$$t_f\sigma_a^n = \bar{\lambda}_s = \frac{2}{(n-2)AY^2K_{Ic}^{n-2}} \{\sigma_*[\ln(\frac{1}{1-F})]^{1/m}[\frac{mn + n-2}{(m+1)(n-2)}]^{1/m}\}^{n-2} \tag{14}$$

and

$$t_f\sigma_a^n = \bar{\lambda}_s = \frac{2}{(n-2)AY^2K_{Ic}^{n-2}} \{\sigma_*\ln(\frac{1}{1-F})]^{1/m}[\frac{mn + n-2}{(m+1)(n-2)}]^{2/m}\}^{n-2}. \tag{15}$$

Case 2: Constant stress rate ($\dot{\sigma}_a$ = constant)

From Eq. (6a) and noting $\sigma_a = \dot{\sigma}_a t$ it can be easily shown that

$$a_r(a,\dot{\sigma}_a,t) = [1 + (\frac{n-2}{2})A(\dot{\sigma}_a Yt)^n\ t a^{(n-2)/2}]^{\frac{2}{2-n}} a\ . \tag{16}$$

Following the same procedure as detailed in the constant stress case the probability of failure at $\dot{\sigma}_a$ for the pure bending case is:

$$F(\dot{\sigma}_a,t_f) = 1-\exp\{-(m+1)(\dot{\sigma}_a t_f/\sigma_*)^m \int_0^1 u^m[1+ (\frac{n-2}{n+1}) \frac{AY^2}{2\dot{\sigma}_a} K_{Ic}^{n-2}(\dot{\sigma}_a t_f)^3 u^2]^{\frac{m}{n-2}} du\} \tag{17}$$

and for the three-point bending case this is

$$F(\dot{\sigma}_a,t_f) = 1-\exp\{-(m+1)^2(\dot{\sigma}_a t_f/\sigma_*)^m \int_0^1\int_0^1 (1-v)^m u^m[1 + (\frac{n-2}{n+1}) \frac{AY^2}{2\dot{\sigma}_a} K_{Ic}^{n-2}(\dot{\sigma}_a t_f)^3(1-v)^2u^2]^{\frac{m}{n-2}} du\ dv\}\ . \tag{18}$$

Again if slow crack growth is insignificant, that is $t_f \to 0$, Eqs. (17) and (18) become Eq. (4) of the Weibull theory. On the other hand, if slow crack growth is significant i.e. $\dot{\sigma}_a$ is small, it can be shown that the fracture strength $\sigma_f$ is related to $\dot{\sigma}_a$ by:

$$\sigma_f^{n+1} = \dot{\sigma}_a\bar{\lambda}_d = [2\dot{\sigma}_a(n+1)/(n-2)AY^2K_{Ic}^{n-2}]\ \{\sigma*[\ln(\frac{1}{1-F})/(m+1)]^{1/m}\}^{n-2}\ [\frac{mn+n-2}{n-2}]^{\frac{n-2}{m}} \tag{19}$$

for pure bending and

$$\sigma_f^{n+1} = \dot{\sigma}_a\bar{\lambda}_d = [2\dot{\sigma}_a(n+1)/(n-2)AY^2K_{Ic}^{n-2}]\ \{\sigma*[\ln(\frac{1}{1-F})]^{1/m}\ [\frac{mn+n-2}{(m+1)(n-2)}]^{2/m}\}^{n-2} \tag{20}$$

for three-point bending.

## LIFETIME AND FRACTURE STRENGTH PREDICTIONS FOR A CEMENT PASTE

In order to predict fracture strength ($\sigma_f$) as a function of stress rate $\dot{\sigma}_a$ using Eqs. (17) and (18) and to obtain lifetime distributions from Eqs. (12) and (13) it is required to determine five parameters: (A,n) of Eq. (6), (m,$\sigma$*) of Eq. (4) and $K_{Ic}$. (A,n) may be obtained from stable crack growth experiments in which the crack velocity (da/dt) is determined as a function of $K_a$ and then plotted in accordance with Eq. (6). Such experiments are usually carried out in the double torsion (DT) geometry, despite some reservations about its suitability [7], because $K_a$ is independent of crack length [14-16] and only depends on the applied load. Crack velocities can also be obtained from the slope of the load relaxation plot [7,14]. Using this technique Nadeau et al [2] have obtained the following $\dot{a}$-$K_a$ relationship for a particular cement paste tested in water (w/c = 0.5 and aged 3-4 months):

$$\log \dot{a} = 36 \log K_a - 200 \ (\text{m/s, Pa}\sqrt{\text{m}}) \tag{21}$$

and $K_{Ic}$ is estimated to be 0.34 MPa$\sqrt{m}$ [2]. (m,$\sigma$*) can only be derived from inert strength experiments appropriate to that particular specimen configuration (i.e. either four or three-point bending) and in order to avoid slow crack growth effects the tests must be done at a fast loading rate. Assuming the 50% failure probability modulus of rupture to be 12.7 MPa [2] and a typical Weibull modulus m = 10 [10], $\sigma$* can be calculated from Eq. (4). Although n is shown to depend on strain rates in impact fracture experiments [3,34] it only varies considerably when $\dot{\varepsilon} > 10^{-1}$ ($s^{-1}$) and is reasonably constant at lower strain rates. Since in the present work we only consider low strain rates n can be assumed to be approximately constant with a value of 36.

Now, with the values of the five parameters determined for the cement paste it is possible to construct the failure probability (F) versus log $t_f$ curves for any given applied stress $\sigma_a$ as well as the F versus $\sigma_f$ curves for different stress rates $\dot{\sigma}_a$. These results are shown in Figs. 1 and 2 according to Eqs. (12) and (17) for the four-point bending case. Alternatively, the results may be replotted in the more conventional form where log $\sigma_a$ - log $t_f$ and log $\sigma_f$ - log $\dot{\sigma}_a$ curves are constructed for a given failure probability F. Figures 3 and 4 show these two plots for F = 0.1, 0.5 and 0.9 and it is seen that linear straight lines are obtained when $t_f \geq 10^{-2}$s and $\dot{\sigma}_a \leq 10^2$ MPa$\sqrt{m}$/s. These conditions are of course implied in deriving Eqs. (14) and (19). Unfortunately, there are no published data for static fatigue and constant stress rate fracture strength for this cement paste to compare with the theoretical results in Figs. 3 and 4.

The values of (A,n) of the slow crack growth law may be determined from Eqs. (14),(15),(19) and (20) by ploting log $t_f$ against log $\sigma_a$ and log $\sigma_f$ against log $\dot{\sigma}_a$. n can be calculated from the slope of the straight line and A from the intercept if $K_{Ic}$, m, $\sigma$*, Y and F are given. In principle n determined in this way should agree with that obtained from the DT experiments, but in practice they are often very different to each other even for the same cement paste [1,7]. The values of n from DT tests are usually larger than from constant stress rate experiments, e.g. n = 36 compared to n = 17.7 [1]. In DT experiments Baldie and Pratt [7] observed that if repeated load relaxation measurements were taken at increasing maximum loads in a single testpiece this gave a shift of the da/dt versus $K_a$ curves along the $K_a$-axis giving different n and A values for each curve. The non-uniqueness of n and A poses a serious problem in the accuracy of lifetime

There is some evidence of a fracture process zone ahead of the visible crack tip [17-19] in cement pastes and the development of which gives rise to an increasing crack growth resistance curve. The relationship between fracture toughness and crack growth resistance curve has been studied recently by Mai and Lawn [20] and using this concept Cotterell and Mai [21] are able to explain the size-dependent fracture toughness results obtained by Higgins and Bailey [17] for cement pastes. It is plausible that the crack growth resistance curve and hence the development of the fracture process zone which is testpiece geometry dependent [21] may be responsible for causing the constants (n,A) of the slow crack growth law to be different for different geometries. Alternatively, the DT results of Baldie and Pratt [7] may be explained in terms of crack-interface bridging behind the crack tip because the effective stress intensity factor at the advancing tip is less than the nominally applied value so that there is a horizontal shift of the slow crack growth curves along the $K_a$-axis. Such a model has been given by Mai and Lawn [22] which successfully explains the toughening of non-transforming ceramics. Certainly, bulk material creeping of the two arms can also contribute to this shift [7]. Another problem with the DT test is that crack velocities are not directly measured. In order to fully understand this problem direct crack velocity and fracture process zone size measurements should be obtained in perhaps a few different specimen geometries other than the DT configuration.

From Eqs. (14) and (19) as well as (15) and (20) it can be easily proven that:

$$\bar{\lambda}_d = \bar{\lambda}_s(n+1). \tag{22}$$

This relationship provides a simple method to predict lifetimes of cementitious materials under sustained stresses. It is simple to conduct constant stress rate experiments in displacement-controlled machines so that a log $\sigma_f$ - log $\dot{\sigma}_a$ plot can be constructed to give n and $\bar{\lambda}_d$ from Eqs. (19) and (20). Now $\bar{\lambda}_s$ can be obtained from Eq. (22) and lifetimes for different applied stresses calculated from Eqs. (14) and (15). This technique for lifetime predictions is popular amongst the ceramics community [23,24]. It has the advantage of not having to determine the constants A,n,m and $\sigma_*$ from other independent experiments. It also appears that any crack growth resistance curve effect is reflected in $\bar{\lambda}_d$ and n so that its effect on lifetime predictions under static stresses will have been already assessed by $\bar{\lambda}_s$ provided the same test piece geometry and loading configuration for the two types of experiments are used.

## LIFETIME ESTIMATION-STATISTICAL FRACTURE VERSUS SINGLE CRACK THEORIES

The single crack theory of lifetime prediction subjected to static stresses assumes a single crack to grow from some initial flaw size ($a_i$) to a critical flaw size ($a_c = (K_{Ic}/Y\sigma_a)^2$) whose rate is governed by the slow crack growth law of Eq. (6). It is easy to show that the lifetime to applied stress relationship is given by [25]

$$t_f\sigma_a^n = \lambda_s = 2a_i^{1-n/2}/[(n-2)Y^nA] \tag{23}$$

provided $(a_i/a_c)^{n/2-1} << 1$. To account for the variation of $a_i$ inert strength measurements according to Eq. (4) with $\sigma_a = \sigma_c$ $(= K_{Ic}/Y\sqrt{a_i})$ can be performed. Thus, combining Eqs. (4) and (23) gives:

$$t_f\sigma_a^n = \lambda_s = 2\sigma_*^{n-2}[\ln(\frac{1}{1-F})^{\frac{n-2}{m}}/[A(n-2)Y^2K_{Ic}^{n-2}]. \tag{24}$$

Similarly, for the constant stress rate experiments, the fracture strength is given by:

$$\sigma_f^{n+1} = \dot{\sigma}_a\lambda_d = 2\dot{\sigma}_a(n+1)\sigma_*^{n-2}[\ln(\frac{1}{1-F})]^{\frac{n-2}{m}}/[A(n-2)Y^2K_{Ic}^{n-2}]. \tag{25}$$

It is noted that $\lambda_d$ and $\lambda_s$ are again related by Eq. (22). Thus, provided that value of $\lambda$ ($\lambda_s$ or $\bar{\lambda}_s$) determined from the constant stress rate tests is used to predict lifetimes for the same specimen geometry and loading arrangement, there is no need to distinguish whether the ($\sigma_f, \dot{\sigma}_a$) data should be described by the single crack theory or the statistical fracture approach. However, if it is required to predict lifetimes under static loads in other specimen geometries and different loading configurations, the precise values of (A,n) from the constant stress rate data have to be evaluated. It is now required to decide whether Eq. (19) or (25) is used to fit the ($\sigma_f, \dot{\sigma}_a$) data. Although n can be uniquely determined from the slope of the straight line log $\sigma_f$ versus log $\dot{\sigma}_a$ plot, the estimated value of A will be different depending on which equation is used. Inaccuracies in lifetime predictions can therefore result. This uncertainty has given much weight to the argument of using controlled flaws such as surface indentation cracks of known sizes, (which is equivalent to predetermining $a_i$), in the experimental studies of the time-dependent strength behaviour of ceramic materials [26] so that the single crack approach can be employed to determine A and n unambiguously.

In the single crack theory Eqs. (24) and (25) are applicable to uniform tensile stress, three-point bending and pure bending cases. This inability to distinguish the effect of non-uniform stress fields across the specimen section is a shortcoming of the theory when compared to the statistical fracture approach. It is well worth noting that the largest flaw is not necessary subjected to the largest stress in a non-uniform stress field and hence the deficiency of the single crack approach. The ratio of the predicted lifetimes based on these two approaches is:

$$\frac{t_f(\text{statistical fracture})}{t_f(\text{single crack})} = \left[\frac{mn + n-2}{(m+1)(n-2)}\right]^{(n-2)/m} \tag{26}$$

for pure bending and

$$\frac{t_f(\text{statistical fracture})}{t_f(\text{single crack})} = \left[\frac{mn + n-2}{(m+1)(n-2)}\right]^{2(n-2)/m} \tag{27}$$

for three-point bending. Table 1 shows how this ratio varies with m and n. For cementitious materials with $m \leq 10$ and $n \leq 80$ the error in lifetimes can be quite significant in three-point bending and the single crack theory is inaccurate. This is even more so in the case of cyclic bending fatigue where the difference can be an order of magnitude [27,33]. Such a prediction can be supported by the work of Gurney and Pearson [28] on glasses and Krohn and Hasselman [29] on a polycrystalline alumina subjected to both static and cyclic fatigue.

## PREDICTION OF CREEP

Creep in cementitious materials increases as the applied stress is increased. There are two components to creep at any time t and applied stress $\sigma_a$ or stress rate $\dot{\sigma}_a$: one due to homogeneous creep of the bulk material and the other due to the crack mouth opening displacement of the pre-existing flaws which grow according to Eq. (6). For high $\dot{\sigma}_a$ it can be assumed that creep is largely due to the opening up of the flaws, but for slow and medium $\dot{\sigma}_a$ the creep elongation comes from both components [30] and in addition the crack tip stresses are relaxed due to localised creep deformation. The same comments applied respectively to high applied stress and medium/low applied stresses. A full solution for creep can be obtained if time-dependnet deformation of the bulk material is superposed in the analysis provided that the increase in process zone size due to creep is small compared to the crack growth. In the following only that creep component due to the propagation of pre-existing flaws is considered. The analytical results obtained give some insight as to the role of fracture mechanics in determining creep in cementitious materials. It is noted that this method of analysis has also been applied to creep in polymeric materials immersed in organic solvents [31]. Here the creep elongation is due to the opening up of the surface crazes which grow according to certain time-dependent craze growth equation.

Consider a beam of length $\ell$ and cross-sectional area hw subjected to constant applied moments $M_a$ at the ends. Only the flaws situated in that half of the beam under tension will propagate and hence contribute to creep. For simplicity, penny-shaped cracks are assumed so that $Y = \sqrt{2/\pi}$ in Eq. (2). Creep strains can be determined from energy balance considerations in a similar way to that used by the authors in calculating the stress-strain relationship for cementitious materials in a previous paper [32].

At time t the amount of energy stored in an elemental volume $\ell$hdx at a distance x from the neutral axis due to cracks between a and (a + $\delta$a) is given by:

$$\delta U_c = \ell hdx.q(a,t)\delta a.\int_0^a \frac{G\pi a}{2}\,da$$

$$= \ell hq(a,t).\ \frac{(1-\nu^2)}{3E}\ .\ \sigma_{xy}^2\ a^3\delta a \tag{28}$$

where $\sigma_{xy}$ is the uniform stress acting over the elemental volume of material and varies along the beam depth (= $2\sigma_a x/W$); G is the strain energy release rate (= $2\sigma_{xy}^2(1-\nu^2)a/\pi E$) and $\nu$ is the Poisson's ratio. Considering all cracks from a $(a_o,\sigma_a,t)$ to infinity the total stored energy is:

$$U_c = \int_o^{w/2} (2\sigma_a x/w)^2 \ell h(1-\nu^2)dx/3E\int_{a(a_o,\sigma_a,t)}^{\infty} q(a,t)a^3 da \tag{29}$$

where a at time t is given by Eq. (8) except $a_r$ now becomes $a_o$. q(a,t) in Eq. (29) has to be replaced by q(a) as shown in Eq. (10). Thus

$$\int_{a(a_o,\sigma_a,t)}^{\infty} q(a,t)da = \int_{a(a_o,\sigma_a,t)}^{a_c(\sigma_a)} q(a,t)da + \int_{a_c(\sigma_a)}^{\infty} q(a,t)da \tag{30a}$$

$$= \int_{a_o}^{a_r(\sigma_a,t)} q(a)da + \int_{a_r(\sigma_a,t)}^{a_c(\sigma_a)} q(a)da + \int_{a_c(\sigma_a)}^{\infty} q(a)da \tag{30b}$$

$$= \int_{a_o}^{\infty} q(a)da \tag{30c}$$

where $a_c$ is given by Eq. (2) with $\sigma_a = \sigma_{xy}$ and is related to $a_r(\sigma_a,t)$ by Eq. (7). In the right hand side of Eq. (30b) the first integral refers to those flaws which are smaller than $a_c(\sigma_a)$ and therefore extend according to the slow crack growth law of Eq. (6); the second integral refers to those flaws whose size is smaller than $a_c(\sigma_a)$ at t = 0 but equals to $a_c(\sigma_a)$ at t when fracture takes place; the third integral corresponds to those flaws bigger than $a_c(\sigma_a)$ and therefore the sample breaks at t = 0. To exclude those samples that fail at t = 0 this last integral can be neglected in Eq. (30b). Thus, Eq. (29) becomes:

$$U_c = [\sigma_a^2 \ell hw(1-\nu^2)/6E]\int_0^1 u^2\left(\int_{a_0}^{a_r(\sigma_a u,t)} q(a)[1-(n-2)A(\sigma_a u\sqrt{2a/\pi})^n\, t/2a]^{\frac{6}{2-n}}\, a^3 da\right.$$

$$\left. + \int_{a_r(\sigma_a u,t)}^{a_{cu}(\sigma_a u)} a_{cu}^3(\sigma_a u)q(a)da + \int_{a_{cu}(\sigma_a u)}^{\infty} a^3 q(a)da\right)du \qquad (31)$$

where $u = 2x/w$ and $a_{cu} = \pi K_{Ic}^2/2\sigma_a^2 u^2$.

The strain energy stored in the beam is

$$U_s = 2\int_0^{w/2} \sigma_{xy}^2 h\ell dx/2E = \int_0^{w/2} \frac{4\sigma_a^2 x^2 h\ell dx}{Ew^2}$$

$$= \sigma_a^2 \ell hw/6E \ . \qquad (32)$$

The external work done by the bending moment $M_a$ is

$$W_e = \frac{1}{2} M_a d\theta = \frac{1}{2}\left(\frac{hw^2\sigma_a}{6}\right)\left(\frac{2\varepsilon}{w}\right)$$

$$= hw\ell\sigma_a\varepsilon/6 \qquad (33)$$

where $\theta$ is the angular displacement and $\varepsilon(= \Delta\ell/\ell)$ is the strain at the outermost fiber of the beam. From work and energy balance considerations it can be shown that [32]

$$W_e = U_c + U_s$$

i.e.

$$\varepsilon(\sigma_a,t) = \frac{\sigma_a}{E}\{1 + (1-\nu^2)\int_0^1 u^2\left(\int_{a_0}^{a_r} q(a)[1-(n-2)A(\sigma u\sqrt{2a/\pi})^n\, t/2a]^{\frac{6}{2-n}}\, a^3 da\right.$$

$$\left. + \int_{a_r}^{a_{cu}} a_{cu}^3 q(a)da + \int_{a_{cu}}^{\infty} a^3 q(a)da\right)du\} \ . \qquad (34)$$

The first term within the bracket, i.e. $\sigma_a/E$, represents the instantaneous strain due to $\sigma_a$ at $t = 0$; the other terms are contributions due to pre-existing cracks. A numerical integration procedure is now required to solve for $\varepsilon(\sigma_a,t)$. For the same cement paste as discussed in the previous sections, the values of A,n,m and $\sigma*$ are already known; $E = 17.4$ GPa [2] and $V = 10x10x228$ $mm^3$ [1] so that $\rho_f$ can be calculated from Eq. (5a) to be ~100 $mm^{-3}$ if $a_o$ is assumed to be some 100 µm. Figure 5 shows the creep strain curves for three applied stresses. As expected the initial strain is larger for larger applied stresses. The predicted creep curves are, however, rather flat indicating the insignificant contribution to creep by the growth of pre-existing flaws. This is largely due to the large slow crack growth law exponent $n = 36$ for this cement paste. If n is small, say $n = 8$ for a polymer-modified cement paste, the contribution to creep by crack growth is significant. The creep strain at failure at low $\sigma_a$ even exceeds that at high $\sigma_a$ because of the longer lengths the crack can grow to at low $\sigma_a$ before final fracture instability [35].

Flaw size distributions as a function of $\sigma_a$ and t, i.e. q(a,t), can also be easily computed from this statistical fracture theory. Details of this will be given elsewhere.

## CONCLUSIONS

A theoretical analysis is presented for the time-dependent strength behaviour of single phase materials like cement pastes and some brittle ceramics. Fracture mechanics and flaw statistics are combined to give predictions for fracture strengths under constant stress rates and for lifetimes and their distributions under sustained stresses. Although the predicted results for a cement paste appear to be qualitatively correct there are no existing data for quantitative comparison. The statistical fracture theory also allows creep strains to be estimated by considering the growth of pre-existing flaws.

ACKNOWLEDGEMENT

The authors wish to thank the Australian Research Grants Scheme for the support of this work which is part of a larger continuing project: Fracture in Cementitious Materials.

REFERENCES

1. Mindess, S. and Nadeau, J.S., "Effect of loading rate on the flexural strength of cement and mortar", Bull. Am. Ceram. Soc., 56, 429-430 (1977).

2. Nadeau, J.S., Mindess, S. and Hay, J.M., "Slow crack growth in cement paste", J.Am.Ceram.Soc., 57, 51-54 (1974).

3. Mindess, S., "Rate of loading effects on the fracture of cementious materials", in Application of Fracture Mechanics to Cementitious Composites, S.P. Shah ed., Martinus Nijhoff Publ., The Netherlands, 617-636 (1985).

4. Wittman, F.H.., "Influence of time on crack formation and failure of concrete", in Application of Fracture Mechanics to Cementitious Composites, S.P. Shah ed., Martinus Nijhoff Publ., The Netherlands, 593-615 (1985).

5. Beaudoin, J.J., "Mechanisms of subcritical crack growth in Portland cement paste", in Proc.Int.Conf. on Fracture Mechanics of Concrete, F.H. Wittmann ed., Lausanne, Switzerland, 1, 3-11 (1985).

6. Beaudoin, J.J., "Stress corrosion and sub-critical crack growth in Portland cement paste", Cement and Concrete Research, to be published.

7. Baldie, K.D. and Pratt, P.L., "Crack growth in hardened cement paste", to be published.

8. Tait, R.B., "Fatigue and fracture of cement mortar", PhD Thesis, University of Cape Town, 1984.

9. Atkins, A.G. and Mai, Y.-W., Elastic and Plastic Fracture: metals, polymers, ceramics, composites, biological materials, Ellis Horwood/John Wiley, Chichester, UK, 1985, Chapter 5.

10. Hu, X.-Z., Cotterell, B., and Mai, Y.-W., "A statistical theory of fracture in a two-phase brittle material", Proc. R. Soc., Lond. A401, 251-265 (1985).

11. Weibull, W., "A statistical theory of the strength of materials", Proc. 151, Ing. Veterskaps. Akad. Hanal, 1939.

12. Weibull, W., "A statistical distribution function of wide applicability", J.Appl.Mech., 18, 293-297 (1951).

13. Hunt, R.A. and McCartney, L.N., "A new approach to Weibull's statistical theory of brittle fracture", Int.J.Fract., 15, 365-375 (1979).

14. Evans, A.G., "Method for evaluating the time-dependent failure characteristics of brittle materials - and its application to polycrystalline alumina", J.Mater.Sci., 7, 1137-1146 (1972).

15. Mai, Y.-W. and Gurney, C., "Crack growth in mineral glass", Phys. and Chem. of Glasses, 16, 70-72 (1975).

16. Williams, D.P. and Evans, A.G., "Simple method for studying slow crack growth", J.Test.Eval., 1, 264-270 (1973).

17. Higgins, D.D. and Bailey, J.E., "Fracture measurements on cement paste", J.Mater.Sci., 11, 1995-2003 (1976).

18. Eden, N.B. and Bailey, J.E., "On the factors affecting strength of Portland cement", J.Mater.Sci., 19, 150-158 (1984).

19. Diamond, S. and Mindess, S., "Scanning electron microscopic observations of cracking in Portland cement paste" in Proc. 7th Int. Congress on the Chemistry of Cement, Vol. 3, pp.VI-114-119, 1980, Paris.

20. Mai, Y.-W. and Lawn, B.R., "Crack stability and toughness in brittle materials", Ann. Review Mater.Sci., 16, 415-439 (1986).

21. Cotterell, B. and Mai, Y.-W., "Crack growth resistance curve and size effect in fracture of cement paste", J.Mater.Sci. (1987), in press.

22. Mai, Y.-W. and Lawn, B.R., "Crack-interface grain bridging as a fracture resistance mechanism in ceramics: II. Theoretical fracture mechanics model", J.Am.Ceram.Soc., 70,289-294 (1987).

23. Fuller, E.R., Lawn, B.R. and Cook, R.F., "Theory of fatigue for brittle flaws originating from residual stress concentrations", J.Am.Ceram.Soc., 66, 3214-321 (1983).

24. Gonzalez, A.C., Multhopp, H., Cook, R.F., Lawn, B.R. and Freiman, S.W., "Fatigue properties of ceramics with natural and controlled flaws: a study on alumina", in ASTM STP 844, 43-56 (1984).

25. Hu, X.-Z., Mai, Y.-W. and Cotterell, B., "Lifetime prediction of ceramic materials subjected to static loads"< J.Mater.Sci.Lett., 6, 462-464 (1987).

26. Lawn, B.R., "The indentation crack as a model surface flaw", Fracture Mechanics of Ceramics, R.C. Bradt et al eds., Plenum Press, N.Y., 5, 1-25 (1983).

27. Hu, X.-Z., Mai, Y.-W. and Cotterell, B., "A statistical fracture theory for cyclic fatigue of brittle materials", to be published.

28. Gurney, C. and Pearson, S., "Fatigue of mineral glass under static and cyclic loading", Proc. R. Soc., Lond. A192, 538-544 (1948).

29. Krohn, D.A. and Hasselman, D.P.H., "Static and cyclic fatigue behaviour of a polycrystalline alumina", J.Am.Ceram.Soc., 55, 208-211 (1972).

30. Newman, K., "Concrete control tests as measures of the properties of concrete", Proc. Symp. Concrete Quality, Cement and Concrete Association, London, 120-138 (1964).

31. Weidman, G.W. and Williams, J.G., "Crazing and the creep behaviour of PMMA in methanol", Polymer, 16, 921-924 (1975).

32. Hu, X.-Z., Cotterell, B. and Mai, Y.-W., "Computer simulation models of fracture in concrete", in Proc. Int. Conf. on Fracture Mechanics of Concrete, F.H. Wittmann ed., Lausanne, Switzerland, 73-82 (1985).

33. Hu, X.-Z., Mai, Y.-W. and Cotterell, B., "A statistical time-dependent fracture theory for brittle materials", submitted to J. Mech. Phys. & Solids.

34. Shah, S.P., "Fracture toughness of concrete", presented at the Int. Conf. on Fracture of Concrete and Rock, SEM-RILEM, June 17-19, 1987, Houston, Texas.

35. Kendall, K. Private communication, 1986.

TABLE 1 Ratio of predicted lifetimes from statistical fracture and single crack theory for a beam of rectangular cross-section (After Hu et al [25]).

| | | $t_f$(statistical fracture)/$t_f$(single crack) | |
|---|---|---|---|
| | | 3-point bend | pure bend |
| 6 | 20 | 1.73 | 1.31 |
| 6 | 40 | 1.75 | 1.32 |
| 6 | 80 | 1.76 | 1.33 |
| 10 | 20 | 1.41 | 1.19 |
| 10 | 40 | 1.43 | 1.19 |
| 10 | 80 | 1.43 | 1.20 |
| 14 | 20 | 1.29 | 1.14 |
| 14 | 40 | 1.30 | 1.14 |
| 14 | 80 | 1.30 | 1.14 |

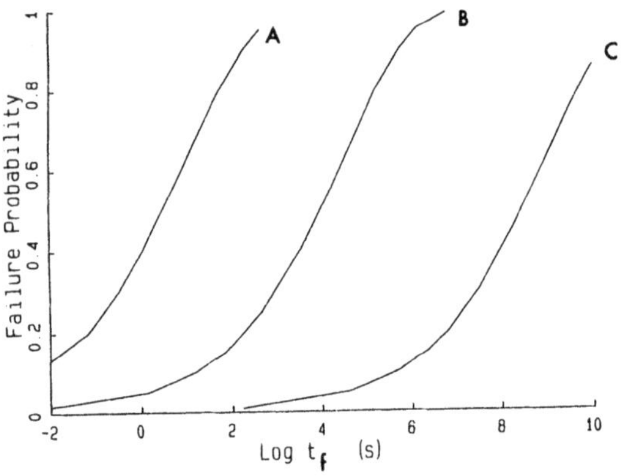


Fig. 1. Failure probability against time-to-failure ($t_f$) for (A) $\sigma_a$ = 10 MPa, (B) $\sigma_a$ = 8 MPa, and (C) $\sigma_a$ = 6 MPa.

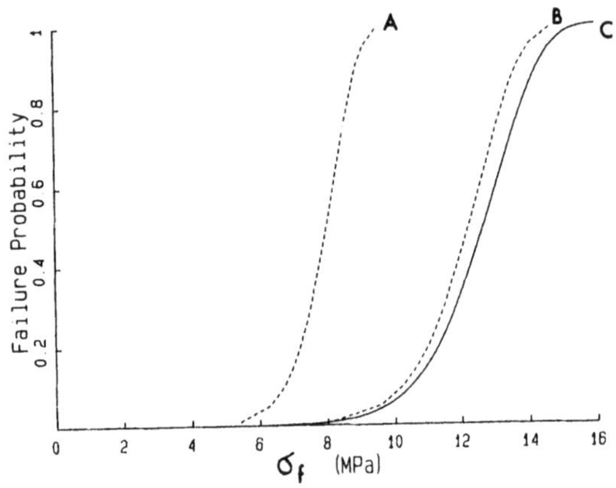


Fig. 2. Failure probability against strength ($\sigma_f$) for (A) $\dot{\sigma}_a$ = $10^{-5}$ MPa/s, (B) $\dot{\sigma}_a$ = $10^2$ MPa/s and (C) $\dot{\sigma}_a \to \infty$. i.e. Eq. (4) with m = 10 and $\sigma_* = 13.17$ MPa.

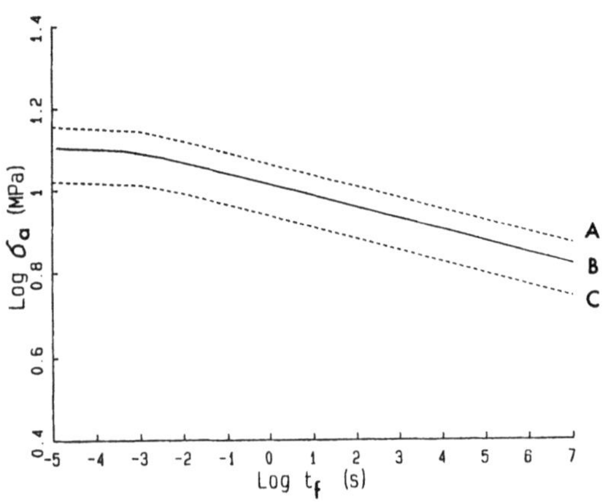


Fig. 3. Relationship between applied stress ($\sigma_a$) and time-to-failure ($t_f$) for (A) F = 0.9, (B) F = 0.5, and (C) F = 0.1.

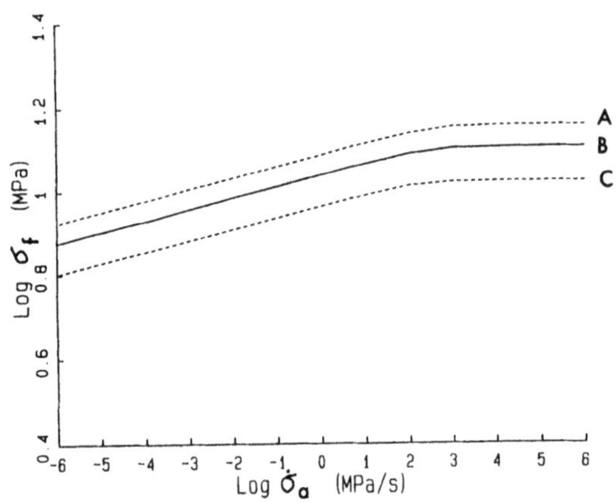


Fig. 4. Relationship between strength ($\sigma_f$) and stress rate ($\dot{\sigma}_a$) for (A) F = 0.9 and (B) F = 0.5 and (C) F = 0.1.

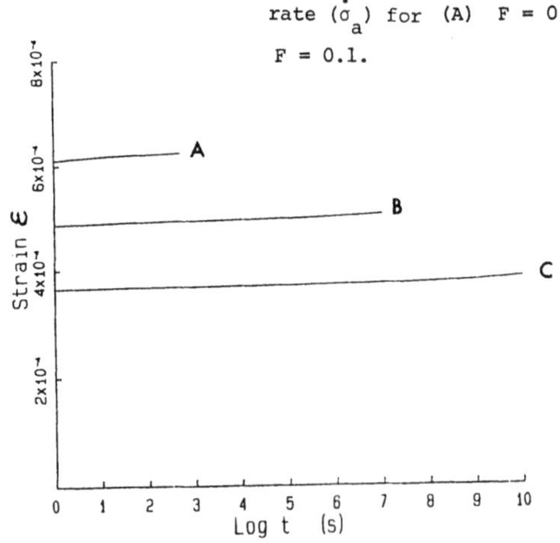


Fig. 5. Creep strain ($\varepsilon$) as a function of elapsed time (t) for (A) $\sigma_a$ = 10 MPa, (B) $\sigma_a$ = 8 MPa and (C) $\sigma_a$ = 6 MPa.

SEM/RILEM International Conference
on Fracture of Concrete and Rock
Houston, Texas, June 1987
Editors: S.P. Shah, S.E. Swartz

# THE DESIGN METHOD FOR THE DEMOLITION OF CONCRETE WITH EXPANSIVE DEMOLITION AGENTS

Tetsuo HARADA Takashi IDEMITSU

Akira WATANABE Shun-ichi TAKAYAMA

## ABSTRACT

Recently, some chemical expansive agents have a tendency to be used for demolition. With these agents, demolition work can be done in safety almost without any pollution, since high expansive pressure is obtained gradually by only mixing the agents with water and pouring the slurry into boreholes. To estimate the time for demolition works, it is necessary to clarify the amount of the expansive pressure acting on the inner surface of the boreholes and the mechanism of fracture of materials to be demolished. The authors carried out experimental and theoretical studies to clarify the problems mentioned above, and proposed a demolition design method.

## INTRODUCTION

Recently in Japan, some concrete structures, which still maintain their own strength enough, have a tendency to be demolished intentionally to utilize the spaces of the cities widely and effectively. In these demolition works, explosives such as dynamite or heavy machines have been used, but sounds, vibrations and some other pollution are also caused. Thus, it is required to consider the safety and the prevention of pollution. In order to solve the problems in these situations, some chemical expansive materials called "expansive demolition agents" have been developed and used in the demolition works of concrete structures and rocks.

With these agents, demolition works can be done in safety without sounds, vibrations, fly rocks, gases, and considerable reduction of dust and some other pollution, since high expansive pressure is obtained gradually by only mixing the agents with water and pouring the slurry into boreholes, and after several hours, cracks occur statically between each of two boreholes [1]. From the above points, the method using these agents is the most suitable one for demolition works. However, practical demolition works with these agents depend mainly on laborers' experiences because there are still many unknown points regarding fundamental matters such as estimation method of expansive pressure and the mechanism of demolition by expansive pressure which plays a principal role in demolition.

In order to estimate a rational design method using these agents, the relation between the spacing of the boreholes to inject the slurry and the demolition time are required [2]. Therefore, the following items should be clarified:

1) Demolition criterion of concrete like materials subjected to the internal pressure
2) Mechanism of demolition
3) Characteristics of expansive pressure of the agent

In this paper, the authors clarify items 1) and 2) experimentally and theoretically, and indicate the important and fundamental data for the design of the demolition. Moreover, the authors present the estimation method of demolition time.

Studies on the item 3) will be discussed in another paper in this conference as "THE CHARACTERISTICS OF EXPANSIVE DEMOLITION AGENTS AND DEVELOPMENT OF PRESSURE TRANSDUCERS".

## DEMOLITION OF HOLLOW CONCRETE CYLINDERS

In order to clarify the demolition criteria, it is necessary to measure what amount of expansive pressure is required for the demolition. In practical demolition works with these agents, boreholes drilled with hand hammer or crawler drill were mainly used to inject the slurry of these agents. It is fundamental and important to investigate the demolition criteria of one borehole. So, hollow concrete or mortar cylinders were used as specimens. Physical properties of these specimens are shown in Table 1. After the concrete hardened, a

T. HARADA is a research associate, at the Department of Structural Engineering Nagasaki University, Nagasaki, Japan; T. IDEMITSU, A. WATANABE and S. TAKAYAMA are an associate professor, a professor and a research associate, respectively, at the Department of Civil Engineering, Kyushu Institute of Technology, Kitakyushu, Japan.

borehole was made by pulling out a vinyl chloride pipe which was set previously.

Although the definition of demolition is different from that of fracture , the authors regard demolition as the state when the first crack occurs from the inside surface of the borehole to the outside one. Therefore, in this paper, both definitions are considered equal.

The expansive pressure at fracture was measured by the pressure transducers newly developed. And also some instantaneous tests simulating the demolition were conducted. In these tests oil-hydraulic pressure was applied instead of the agent, since it was considered that the pressure at fracture was not influenced by the pressure medium. All experiments were carried out in a room of constant temperature and moisture.

Measurement of the Pressure at Demolition when Using Oil-hydraulic Pressure

A testing apparatus using oil pressure is shown in Fig. 1 [3] . The oil-hydraulic pressure was increased at the rate of 8∿10 kgf/cm$^2$ a minute. The pressure was kept constant by bolting the circular plates with O-rings. Since the axial strain of specimen caused by bolting was only $5\sim10\times10^{-6}$, its influence to the pressure at fracture was neglected. It is necessary to coat the upper and lower ends and inner surface of the specimen in order to prevent the permeation of pressurized oil. According to preliminary tests, it was found that the most suitable materials to coat the specimen were soft-type vinyl tube and epoxy adhesive. Therefore, in all tests with hydraulic pressure, specimens were coated with materials mentioned previously. Hollow cylindrical specimens made of concrete or mortar have various dimensions: 7.5∿30 cm in outer diameter, 0.9∿3.2 cm in inner one and 15 cm in height. The fracture occurred when the pressure got the maximum value.

Measurement of the Pressure at Demolition when Using the Demolition Agent

In order to measure the expansive pressure at fracture actually and directly, two kinds of embedding type of pressure transducers were newly developed. One is made of a steel pipe and orthogonal strain gauges are attached to its inner surface as shown in Fig. 2(a). This transducer can be located in the center of a borehole filled with the slurry and the external expansive pressure acting on the transducer can be directly measured in the boreholes. The expansive pressure p is calculated from Eq.(1) using the strains $\varepsilon_\theta$ and $\varepsilon_z$ measured by this transducer. The measuring method with this transducer is named "inner pipe method" [4]. The other transducer consists of two thin steel plates that are joined with the epoxy adhesive as shown in Fig. 2(b). At the middle of one of these plates a hole is cut so that a circular diaphragm is made and a strain gauge is attached on the inner surface of the diaphragm. With this transducer, the expansive pressure acting on the inner surface of the boreholes as well as ditches made by a concrete cutter can be accurately measured. This method is named "diaphragm method" [5].

$$p = \frac{-E_s(k^2-1)}{2(1-\nu_s{}^2)k^2}(\varepsilon_\theta + \nu_s\varepsilon_z) \tag{1}$$

where, Es:Young's modulus of steel pipe
$\nu_s$:Poisson's ratio of steel pipe
k :Ratio of outer diameter to inner one

In addition to these methods, there is an indirect measuring method named "outer pipe method". This is a practical and simple method to measure the fundamental characteristics of the expansive pressure. As shown in Fig. 2(c), a steel pipe on the outer surface of which orthogonal strain gauges were attached was treated as a pressure transducer. After the slurry was poured into the pipe, the circumferential and axial strains at the outer surface were measured successively and the corresponding internal expansive pressures were calculated from the following Eq.(2).

$$p = \frac{E_s(k^2-1)}{2(1-\nu_s{}^2)}(\varepsilon_\theta + \nu_s\varepsilon_z) \tag{2}$$

Test specimens for demolition have 15 cm in height, 2.2∿4.8 cm in inner diameter and 7.5∿60 cm in outer diameter.

Fig. 3 shows the time-dependent change of the expansive pressure measured by the diaphragm method. At the maximum point, the pressure starts to decrease suddenly. The authors regard this maximum point as the pressure at fracture. In fact, at this time, a crack occurs from the inside surface of a borehole to the outside one.

Experimental Results and Discussion

In Figs. 4, 5 and 6, ⓐ indicates the cracks which occurred first. The crack ⓐ was caused by a circumferential stress $\sigma_\theta$ since the crack mode of the outside surface occurred parallel to the longitudinal direction as shown in Fig. 6. In the tests with hydraulic pressure, the crack a occurred suddenly and the crack width at the outer surface of the specimens was about 5 mm (Fig. 4). In the tests with the demolition agent, two kinds of fracture mode were observed at the moment of fracture according to the ratio k of the hollow cylinder. One was a successive cracking following the first crack which did not reach the outer surface when k>5 (Fig. 7(a)) and the other was an instantaneous straight cracking from the inside surface to the outside one when k<5 (Fig. 7(b)). In the case of Fig. 7(a), the crack propagation could be clearly observed by rubbing cotton soaked with acetone on the surface of the specimen. The crack width was so small that it could not even be measured.

Figs. 8 and 9 show the relation between $p_u/\sigma_t$ and the ratio k, where $p_u$ is the internal pressure at fracture and $\sigma_t$ is the splitting tensile strength. Sato et al.[6] reported that the results of the fracture tests with brittle materials are plotted in the range between lines (I) and (II) and explained that the fracture is influenced by the ductility of the materials, that is, when the material is more ductile, the value of $p_u/\sigma_t$ is plotted near the straight line (I), and when the material is more brittle, the value of $p_u/\sigma_t$ is plotted near the curve (II). It was found that the authors' results are almost plotted in the range of (I) and (II).

When k<5, many values exist near the straight line (I) in Figs. 8 and 9. It is considered that the fracture is caused when the mean value of the circumferential stress reaches the tensile strength of the specimen. That is,

$$\bar{\sigma}_\theta = \sigma_t \tag{3}$$

From the circumferential equilibrium at t-t section, as shown in Fig. 10,

$$p_i \cdot r_1 = \bar{\sigma}_\theta (r_2 - r_1) \tag{4}$$

$$\bar{\sigma}_\theta = p_i/(k-1) \qquad (k=r_2/r_1) \tag{5}$$

The internal pressure at fracture $p_u$ is described as

$$p_u/\sigma_t = k - 1 \tag{6}$$

thus, Eq.(6) agrees with curve (I). Also, Eq.(6) represents the fracture criterion of a hollow cylinder regarded as a thin-walled or perfectly plastic cylinder.

When k<5, even if a circumferential stress at the inner surface of a borehole reaches the tensile strength of the specimen, its fracture does not occur because the outer part which has small circumferential stress and large volume prevents cracks to occur in the inner part. Thus, fracture occurs when the plastic zone of the surrounding of a borehole spreads up to the outside surface.

When k>5, in both pressure media, values of $p_u/\sigma_t$ deviate from the straight line (I) as k increases. In the case of k>5 and with demolition agent, the first crack occurs at the surrounding of a borehole before the average value of circumferential stress reaches the tensile strength of the specimen, although the first crack does not cause the instantaneous fracture. It is necessary to increase the expansive pressure for the propagation of the first crack up to occurrence of fracture. However, since the expansive pressure does not increase until the average value of the circumferential stress reaches the tensile strength, values of $p_u/\sigma_t$ deviate from the line (I). On the other hand, in the case of oil-hydraulic pressure, the coating of the specimen ruptures as soon as the first crack appears, and the fracture of the specimen occurs instantaneously since the hydraulic pressure acts on the crack tip. Thus, the values of $p_u/\sigma_t$ in the case of hydraulic pressure are smaller than those in the case of the demolition agent.

The above phenomena of demolition can be explained considering the stress intensity factor $K_I$ pressurized thick-walled vessels, as R.J.Clifton et al.[7] reported. Let us consider the demolition using the agent and the case of a single radial crack. As shown in Fig. 11, when k<5, $K_I$ is a monotonically increasing function of crack length, and the fracture occurs suddenly, what does not happen for k>5. When k>5, from point A, the $K_I$-crack length curve has negative slope segments, and the stable crack growth with increasing pressure is expected. When using the oil-hydraulic pressure, $K_I$ is always a monotonically increasing function of crack length no matter the value of k. The relation between stress intensity factor($K_I$) and expansive pressure at fracture is going to be investigated in details by the authors.

When using the demolition agent, the tensile axial stress $\sigma_z$ occurs but its influence to the fracture is so small that it can be neglected. And as the ratio of $\bar{\sigma}_z$ (average value of axial stress) to $\bar{\sigma}_\theta$ (average value of circumferential stress) is 1/(k+1) , then $\bar{\sigma}_z$ is much less than $\bar{\sigma}_\theta$ .

Based on the above results, the practical demolition criterion is written as

$$\bar{\sigma}_\theta/\sigma_t = \alpha \tag{7}$$

When k<5, Eq.(6) is obtained by adopting $\alpha$=1 in Eq.(7). When k>5, the values estimated from the experimental results varied from 0.5 to 0.8 as in Fig. 9. If the value of $\alpha$=0.8 is adopted, the estimated demolition time is longer than the practical demolition time, so the design of demolition using the agent is safe. Then the following equation is proposed.

$$p_u/\sigma_t = 0.8\ k \tag{8}$$

Equations (6) and (8) that describe the demolition criterion give the same value of $pu/\sigma t$ for k=5, so the two lines defined by these equations are connected smoothly.

Behavior of Circumferential Strain at the Inner Surface of Boreholes or around them

Since the strain at fracture can be adopted for the design of demolition instead of the expansive pressure, it is important to verify the relation between the expansive pressure and the circumferential strain Then, the behavior of the circumferential strain at the inner surface of hollow concrete or mortar cylinders on which strain gauges are attached at the position of 3~4 cm from the top surface of the specimen is investigated by using oil-hydraulic pressure. Also, to investigate the behavior of the circumferential strain at the position of 2 cm from the inner surface of the hole of hollow concrete or mortar cylinders, strain gauges are embedded in these specimens. The tests are conducted using the expansive pressure of the agent.

Figs. 12 and 13 show the behavior of the circumferential strain up to fracture at the inner surface of mortar specimens and at the position 2 cm from the hole of concrete cylinders , respectively . The lines curve slightly at the point where strain is $200\sim 300\times10^{-6}$ and this is the so-called limit value of extensibility. However, the change of strain is so small that the behavior of the strain until fracture is indicated by an almost straight line. The values of tensile strain at fracture are about $400\sim750\times10^{-6}$. These values are two or three times larger than those obtained from the splitting test. The $p$-$\varepsilon_\theta$ relations calculated from Eq.(9) are also shown in Fig. 12.

$$\varepsilon_u = \frac{p\{(k^2+1) + \nu_c(k^2-1)\}}{E_c(k^2-1)} \qquad (9)$$

In computation, Young's modulus Ec and Poisson's ratio $\nu c$ obtained from the compression tests are used. The measured value of strain is larger than the calculated one. It is considered that the properties of deformation under the compressive stress differ from those under the tensile stress.

Mechanism of Demolition

Two kinds of fracture modes when using hydraulic pressure or demolition agent are shown in Figs. 4 and 5, respectively. Let us now compare the mechanism of demolition and the fracture mode of hollow cylinders. The first-occurred cracks ⓐ are similar, but the other cracks differ from each other. While the mode of cracking of specimens under the oil-hydraulic pressure is that the specimen is separated into two blocks (Fig. 4), in the case of the expansive pressure of the agent, the specimen is separated into three fan-shaped parts at vertical angles of about 120° (Fig. 5). As the internal pressure does not become zero just after crack ⓐ occurs, crack ⓑ is caused by the bending moment due to the internal pressure. The point on the inner surface that has the maximum bending moment as well as the principal stress under the hydraulic pressure is opposite to crack ⓐ. In case of the demolition agent, two points of maximum bending moment originate because of the friction between the surface of the borehole and the hardened agent. Therefore, the point of origin of crack ⓑ changes according to the conditions of the inner surface of boreholes.

In order to confirm where the maximum stress is caused by the bending moment, a photoelastic method was used. After an uniform expansive pressure was applied to the inner surface, a slit was made simulating the first-occurred crack of concrete. Figs. 14 and 15 show the photoelastic fringe pattern and the stress trajectories, respectively. The principal stress distributions along A-A', B-B' and C-C' sections are shown in Fig. 16. By analyzing this figure, it is considered that the second cracks are more probable to occur at section C-C'.

The typical fracture mode (Fig. 5), when using the demolition agent, can be applied to make the arrangement of boreholes, that is, the hexagonal arrangement. The demolition of concrete slabs with the hexagonal arrangement of boreholes will be described in the following section.

## DESIGN OF DEMOLITION AND ESTIMATION OF DEMOLITION TIME

Basic Concepts

Considering two boreholes in an infinite plate, the internal pressure acts on the inner surfaces as shown in Fig. 17. The average of the circumferential stress is given by the circumferential equilibrium at t-t section and is expressed in Eq.(10), where $p_i$ is the internal pressure and $\ell$ is the distance between a borehole and another one.

$$\sigma_\theta = \frac{p_i \cdot r_1}{\ell/2 - r_1} = \frac{p_i}{k-1} \qquad (10)$$

where, $k=(\ell/2)/r_1$ .

By using the demolition criterion $\bar{\sigma}_\theta/\sigma_t = 1$, $p_u$ is written as

$$p_u/\sigma_t = k - 1$$

This equation is the same as Eq.(6). In the case of two boreholes, it can be treated that there are two hollow cylinders whose inner and outer diameters are $2r_1$ and $\ell$, respectively. The tensile strength $\sigma t$ of the materials of the object to be demolished and k are used in Eq.(6) or Eq.(8) to compute the expansive pressure up required for demolition. By using this expansive pressure, it becomes possible to estimate the demolition time from Fig. 18. As shown in Fig. 18, the solid and broken curves represent the regression curves of time-dependent change of expansive pressure measured by the inner pipe method and the outer pipe method, respectively. The regression curves computed using the least squares method are given by the following equations.

$$p_i = 418 - 560 \exp(-0.0460\ t) \quad \text{(by the inner pipe method)} \quad (11)$$

$$p_0 = 553 - 672 \exp(-0.0328\ t) \quad \text{(by the outer pipe method)} \quad (12)$$

Fig. 19 shows the relation between demolition time tu and k calculated by using tensile strength as the parameter, where the solid and broken curves are calculated by using Eq.(6),(8),(11), and Eq.(6),(12), respectively. Table 2 shows the comparison of calculated and measured values of expansive pressure. It was found that the calculated values are in good agreement with the measured ones.

In the practical demolition work, if the curves shown in Figs. 18 and 19 are obtained by using the inner pipe method, the design of demolition can be done more accurately. However, the difference between the demolition time estimated by the solid and broken curves in Fig. 19 is only 10% for the range of $\sigma_t < 30$ kgf/cm$^2$ and $k<15$. When the external conditions during the experimental tests and execution of demolition works are the same, estimation of the demolition time by the outer pipe method is also effective.

Demolition Time and Demolition Mode

The demolition time would be changed owing to the arrangement, the length and any other conditions of boreholes. Thus, the fundamental experiments were carried out in order to do the demolition works more effectively and economically.

The demolition tests are carried out using high strength concrete slabs with the hexagonal and square arrangements and the same number of boreholes. The results are shown in Fig. 20. In the hexagonal arrangement, cracks occurred between a borehole and another one from 10.5 to 12.5 hours after mixing and divided the specimen plate into hexagonal blocks just as expected. But in the square arrangement, cracks occurred a little later and there were a few cracks that did not reach the next borehole. It was found that the hexagonal arrangement of the boreholes is advantageous for the same number and diameter of boreholes as compared to the square arrangement.

Making free surfaces of the object to be demolished is advantageous to demolish effectively. Fig. 21 shows the relation between failure time and the distance from a borehole to the free end. Test specimens have various spacings "$\ell$" between boreholes and distances "a" from the borehole to the free end. When the distance from the borehole to the free end is short, the crack on the free end occurs earlier but the extension of the cracks between boreholes that induce fracture is slow. From this figure, it is found that when the distance from the borehole to the free end is half the spacing of boreholes, the demolition time is the shortest and the demolition can be done effectively.

In the previous experiments, the effect of demolition in two dimensions was investigated with concrete slab specimens. The following experiments were carried out to investigate the demolition of mass concrete. The test specimens were concrete blocks with the following dimensions 60x70x40 cm$^3$ and the hexagonal arrangement was adopted. In order to examine how deep the boreholes should be drilled to demolish effectively, various depths of boreholes were made in the specimens. When the depth of a borehole is deeper, the crack occurs earlier and the crack width increases remarkably. Fig. 22 shows experimental results. The cracking time was 9 hours when h/H was 1 or 3/4, 13 hours when h/H was 1/2, and 18 hours when h/H was 1/4. The demolished shape of concrete block indicated hexagonal column when h/H was 1 or 3/4. When h/H was 1/2, the lower half part of the specimen did not divide as expected and large blocks remained indivisible. When h/H was 1/4, the lower part of the specimen also did not demolish perfectly. The demolition time is the shortest when the borehole is drilled from top to bottom of the specimen. But as economical aspects should be considered, it is not advised to drill such deep boreholes. Based on the above results, the demolition can be done easily when the depth of the borehole to be drilled is about three quarters of total height of the block. When the depth of borehole is larger than half of total height, although demolition can be done, the demilition time is longer.

The phenomenon of demolition using ditches made by concrete cutters was investigated. As shown in Fig. 23 concrete blocks with ditches having width of 9.5 mm and depth of h were used as specimens. Table 3 shows the relation between the demolition time and the depth h of the ditch. When the depth of ditch was deeper than a quarter of the total height, the decrease of demolition time was very small.

Demolition experiments were also carried out to investigate the relation between the shape of boreholes and the cracking time. The test specimen was a rectangular concrete plate with two boreholes at 10 and 20 cm from the free end and it was reinforced by reinforcing bars at the other three sides. Fig. 24 shows the behavior of strain at the free end for various shapes of boreholes. In the case of acute angled hole, as the degree of stress concentration was higher, crack appeared more promptly than that of a circular one. In particular, the cracking time of the borehole with a vertical angle of 30° is less than half of that of a circular one. When the borehole is not circular, the direction of crack propagation is easier to control. The effect of the borehole with acute angle on shortening the demolition time was clearly observed.

In order to examine the relation between the expansive pressure at fracture and the stress concentration factor, the same experiments were carried out with elliptic hole specimens having various radii of curvatures. As a result, it has been confirmed that the relation between the expansive pressure at fracture and the inverse of the stress concentration factor $\beta$ is almost linear, as shown in Fig. 25. Therefore, it was suggested that in the design of demolition using the stress concentration of the elliptic or acute angled boreholes, the inverse of the stress concentration factor $\beta$ should be multiplied to the right term of the equation of the demolition criterion (Eq.7).

## CONCLUSIONS

The results from these studies are summarized below.

1) The first crack that appeared on concrete hollow cylinders subjected to the expansive pressure was caused by a circumferential stress. According to the ratio k of the outer to the inner diameter of the hollow cylinder, two kinds of cracking modes were observed at the moment of fracture. One was an instantaneous straight cracking from the inside surface to the outside one when k<5, and the other one consisted of successive crackings following the first crack which did not reach the other surface when k>5.

2) The mode of cracking of the specimens under the oil-hydraulic pressure is that the specimen is separated into two blocks. Under the expansive pressure of the agent, the specimen is separated into three blocks and its cross section has three fan-shaped parts at vertical angles of about 120°.

3) The demolition criterion is defined by this equation $\bar{\sigma}_\theta/\bar{\sigma}_t=1$. When k<5, $\alpha=1$ is taken and the equation is written as $p_u/\sigma_t=k-1$. When k>5, $\alpha=0.8$ is taken, and the equation is written as $p_u/\sigma_t=0.8k$.

4) The circumferential strain $\varepsilon_\theta$ at the inner surface of the specimen increases monotonically up to fracture and the extensibility is two or three times larger than the value obtained from the splitting test.

5) The demolition time can be estimated accurately by using the equation of the time-dependent changes of expansive pressure measured by the inner pipe method or the diaphragm method and by using the equation of the demolition criterion with the tensile strength as a parameter. When the estimation of demolition time is done by the outer pipe method, if the surrounding conditions are considered, the range of error is not more than 10%, so it is also effective.

6) The ordinary shape of a borehole is circular, but with an acute angled hole or a ditch made by a concrete cutter the direction of crack propagation is easier to control.

We must recognize that not a few concrete structures are not eternal monuments and someday they will be demolished in a cycle of construction-demolition due to public requirements. In future, the demolition works using demolition agents will increase from the viewpoint of economy and prevention of pollution. Therefore, when new concrete structures are constructed, it is recommendable to provide pipes or holes to inject the slurry of the demolition agent considering the strength of the structure and the workability of concreting in order to do the demolition work easily later.

## REFERENCES

[1] Kawano,T., "Non-Explosive Demolishing Agent," Gypsum & Lime, No. 176, pp.41-48, 1982 (in Japanese).
[2] Watanabe,A., T.Goto and H.Matsuda, "Experimental Study on Demolition of Concrete Structures with Non-Explosive Demolition Agent," Review of the 36th General Meeting, The Cement Association of Japan, pp.183-186, 1982 (in Japanese)
[3] Harada,T., A.Watanabe, H.Kanetake and M.Hamada, "Experimental Study on Rupture of Hollow Cylinders Concrete-like Materials Subjected to Internal Pressure," Reports of The Faculty of Engineering, Nagasaki University, Vol. 13, No. 21, pp.179-185, 1983 (in Japanese).
[4] Idemitsu,T. and N.Takeda, "Study on Demolition of Concrete with Non-Explosive Demolition Agent - Measurement of Expansive Pressure and Estimation of Demolition Time ," Proceedings of the 38th Annual Conference of the Japan Society of Civil Engineers (5), pp.539-540, 1983 (in Japanese).
[5] Harada.T., T.Idemitsu and A.Watanabe, "The Method of the Measurement on the Expansive Pressure of the Agent and Characteristics of the Expansive Pressure," Proceedings of the 41th Annual Conference of the Japan Society of Civil Engineers (5), pp.449-540, 1986 (in Japanese).
[6] Sato,Y., T.Shigemura and F.Nagai, "Influence of Ductility of Brittle Materials upon its Rupture of Cylinder under Internal Pressure," Trans. of JSME, Series A, Vol. 45, No. 391, pp.220-226, 1979 (in Japanese).
[7] Clifton,R.J., E.R.Simonson, A.H.Jones and S.J.Green, "Determination of the Critical-stress-intensity Factor KIc from Internally Pressurized Thick-walled Vessels," Experimental Mechanics, pp.233-238, June 1976.

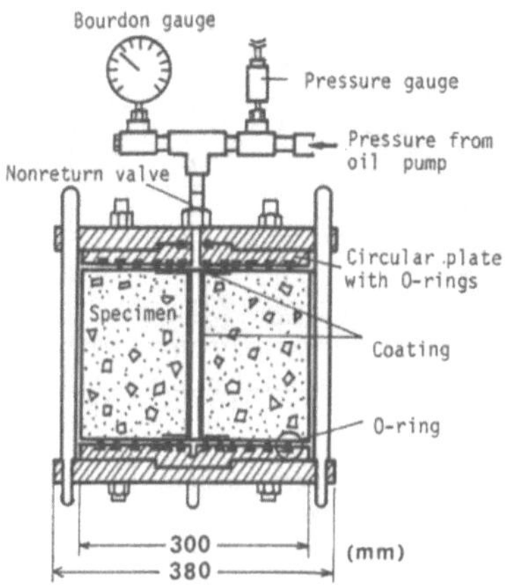


Fig. 1 Specimen and device for oil-hydraulic pressure test.

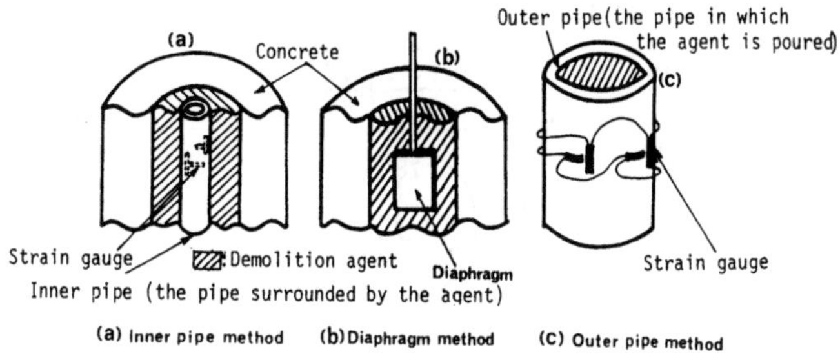


Fig. 2 Methods of measurement of expansive pressure.

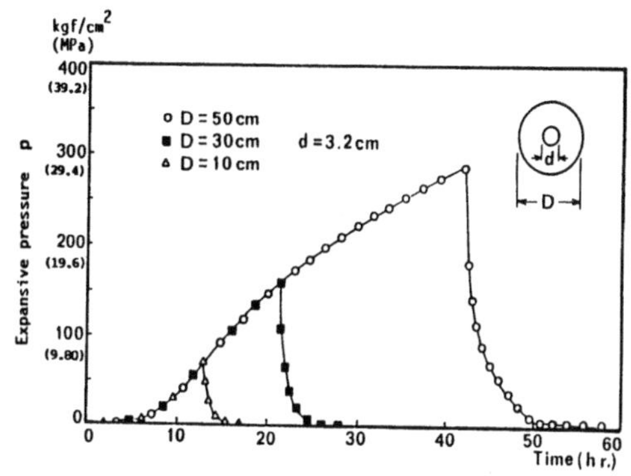


Fig. 3 Time-dependent change of the expansive pressure up to fracture.

Table 1 Physical properties of concrete and mortar.

| Kinds of concrete | Compressive strength $kgf/cm^2$ (MPa) | Tensile strength $kgf/cm^2$ (MPa) | Young's modulus $\times 10^5 kgf/cm^2$ (GPa) |
|---|---|---|---|
| Plain concrete | 305 ~ 456 (29.9 ~ 44.7) | 23.8 ~ 32.8 (2.33 ~ 3.21) | 2.77 ~ 3.63 (27.1 ~ 35.6) |
| Mortar | 382 ~ 490 (37.4 ~ 48.0) | 22.7 ~ 29.8 (2.22 ~ 2.92) | 2.68 ~ 3.70 (26.3 ~ 36.3) |
| High strength concrete | 616 (60.4) | 43 (4.21) | 4.00 (39.2) |

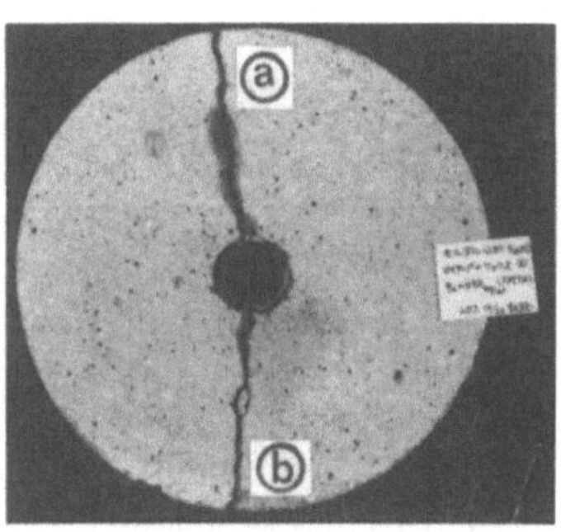


Fig. 4 Fracture mode of the cylinder demolished by oil-hydraulic pressure. (D=15, d=2.2 cm)

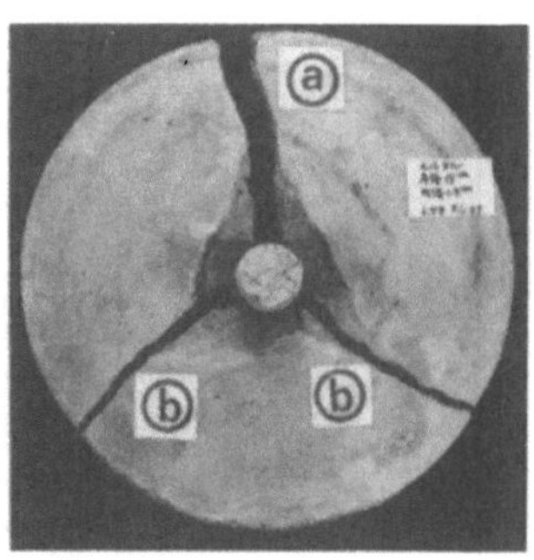


Fig. 5 Fracture mode of the cylinder demolished by the agent. (D=15, d=1.5 cm)

Fig. 6 The crack mode of the outside surface.

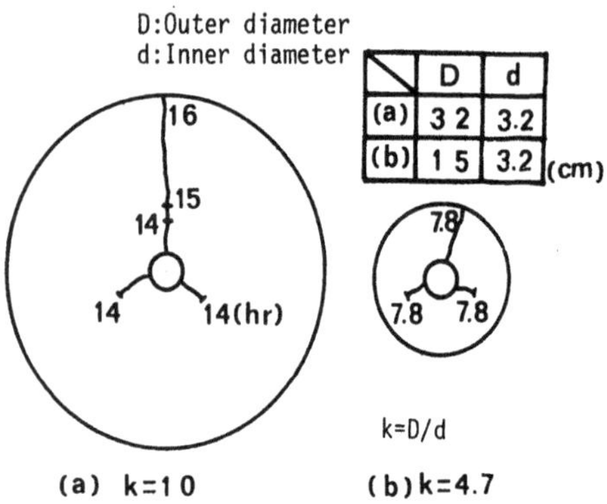


Numerals indicate the time of crack propagation.

Fig. 7 Difference of the crack mode due to k.

(a) k>5, (b) k<5

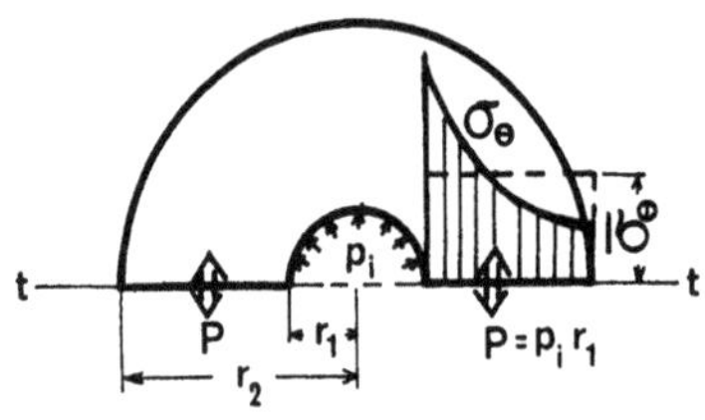


Fig. 10 Circumferential equilibrium at t-t section in the case of one borehole.

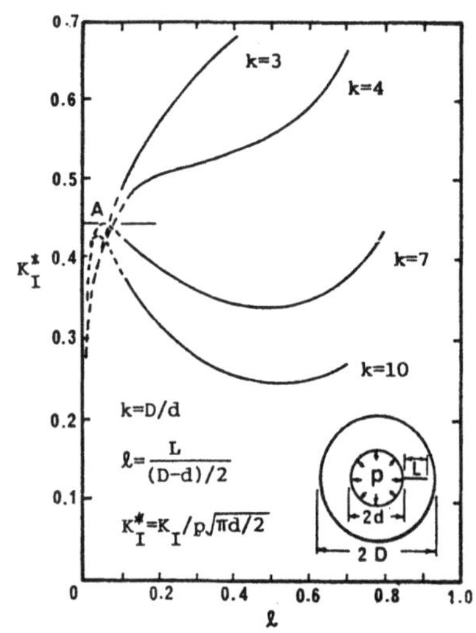


Fig. 11 Stress-intensity factor for jacketed cylinder with one radial crack. (R.J. Clifton et al. 1976 [7])

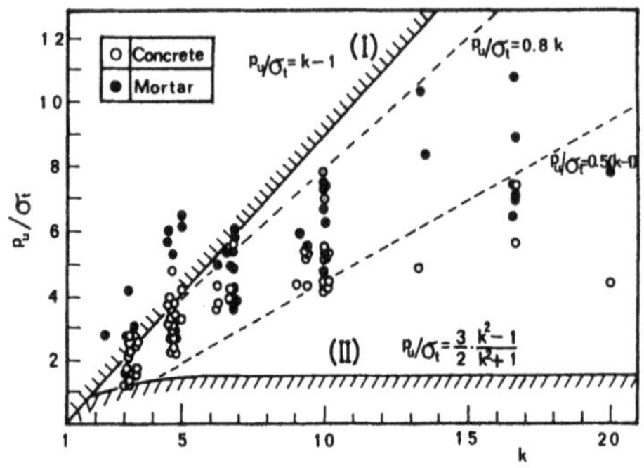


Fig. 8 Relationship between pu/st and k when using oil-hydraulic pressure.

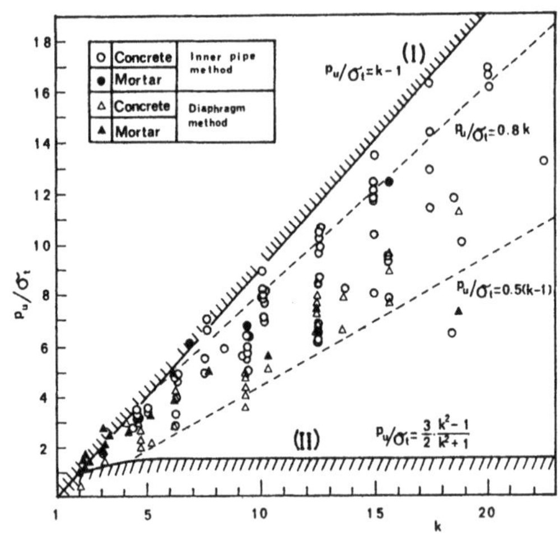


Fig. 9 Relationship between pu/st and k when using the demolition agent.

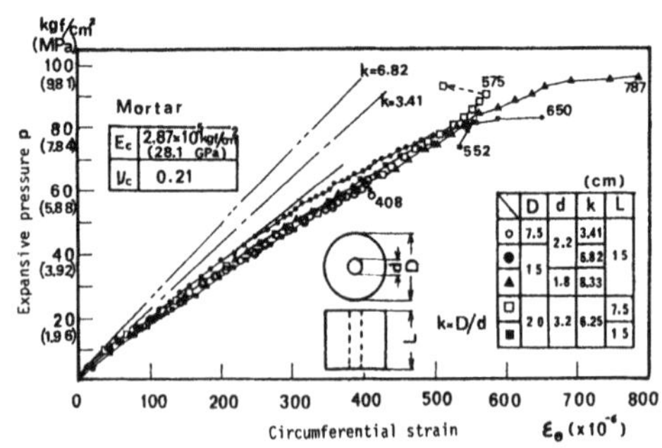


Fig. 12 Behavior of the circumferential strain at the inner surface of mortar cylinder subjected to the oil-hydraulic pressure up to rupture.

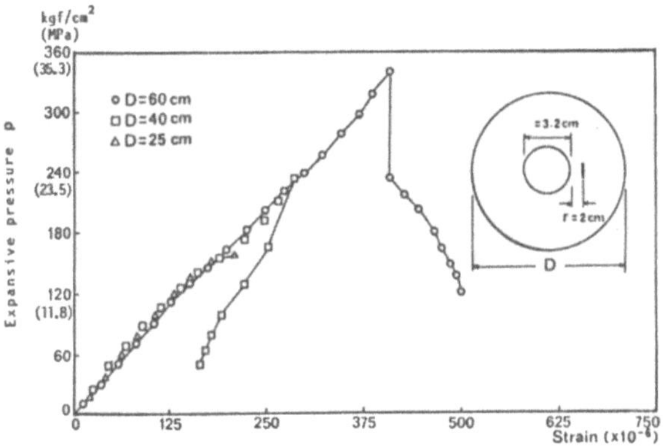


Fig. 13 Behavior of the circumferential strain at the positionof 2 cm from the inner surface of concrete cylinder subjected to the expansive pressure up to rupture.

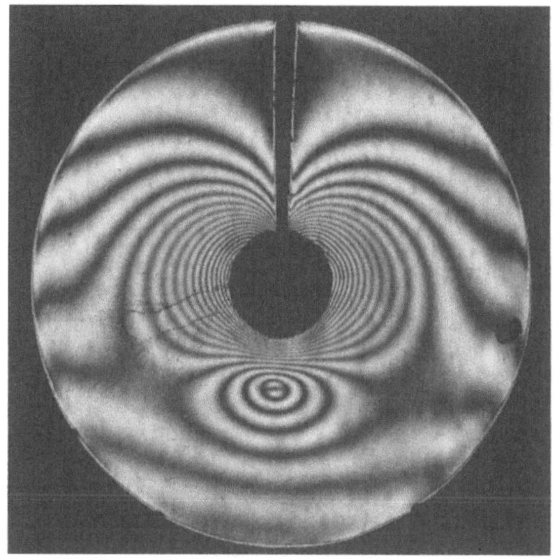

Fig. 14 Photoelastic fringe pattern.

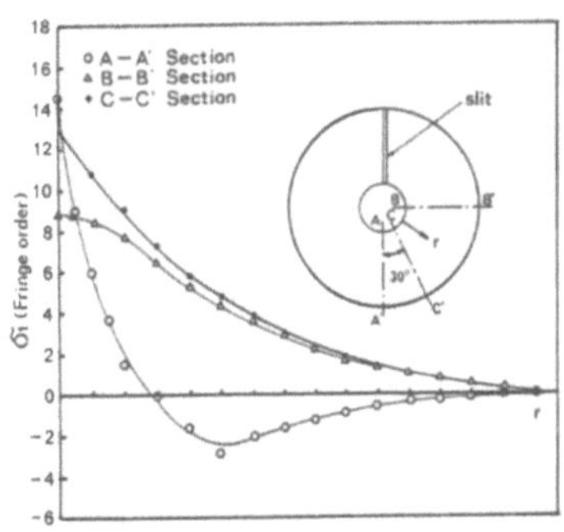


Fig. 16 Principal stress distributions along A-A', B-B' and C-C' sections.

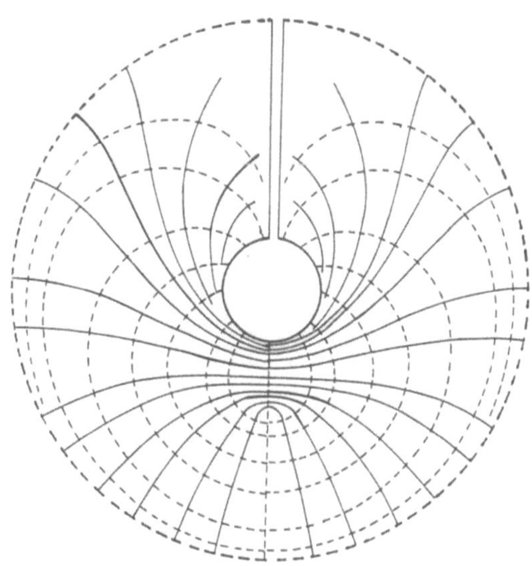

Fig. 15 Stress trajectories.

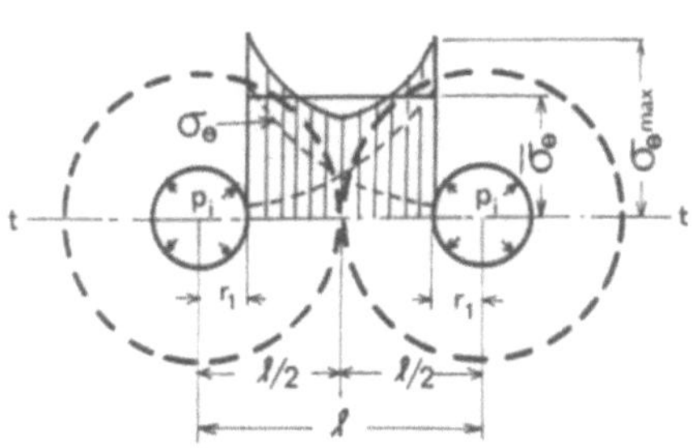


Fig. 17 Circumferential stress and its average value at t-t section in the case of two boreholes.

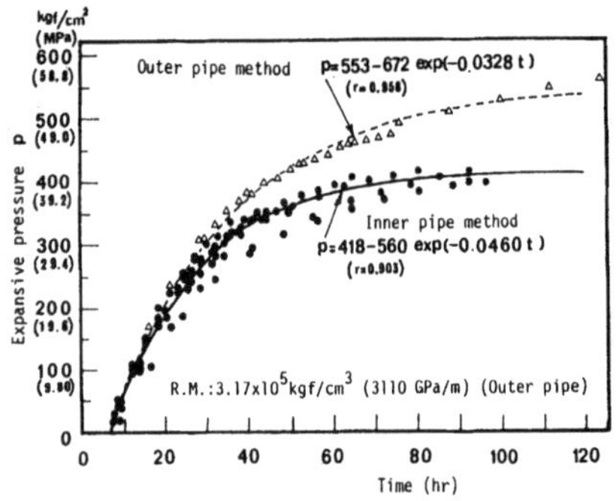


Fig. 18 Regression curves of expansive pressures.

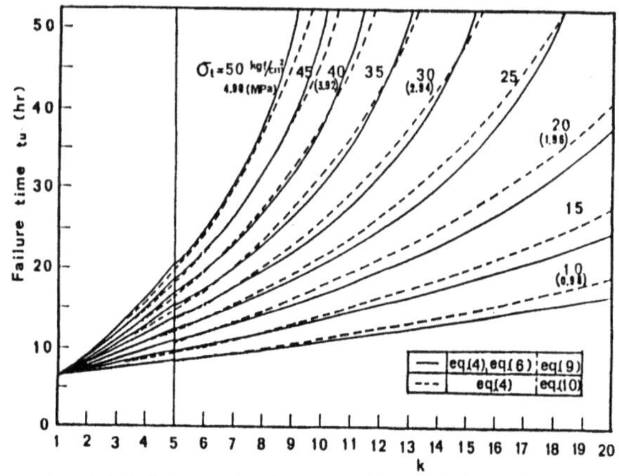


Fig. 19 Relationship between failure time and k for various tensile strengths of concrete.

Table 2 Comparison of the calculated values with measured ones.

| | | $k^{*}$ | Expansive pressure at failure kgf/cm² (MPa) | | Failure time (hr) | |
|---|---|---|---|---|---|---|
| | | | Calculated | Measured | Calculated | Measured |
| A | i) | 1 | — | — | — | — |
| | ii) | 5.91 | 126 (12.3) | — | 14.1 | 12.8 |
| B | i) | 1.88 | 23 (2.25) | 57 (5.59) | 9.0 | 8.5 |
| | ii) | 5.63 | 120 (11.8) | — | 13.7 | 13.5 |
| C | i) | 3.13 | 57 (5.59) | 57 (5.59) | 8.0 | 8.5 |
| | ii) | 5.22 | 111 (10.9) | — | 13.1 | 12.0 |
| D | i) | 4.69 | 98 (9.60) | 71 (6.96) | 10.4 | 9.5 |
| | ii) | 4.69 | 98 (9.60) | 88 8.62 | 10.4 | 10.5 |
| E | i) | 6.25 | 133 (13.0) | 90 (8.82) | 14.7 | 11.0 |
| | ii) | 4.16 | 84 (8.23) | — | 9.6 | 11.5 |
| F | i) | 7.5 | 160 (15.7) | 90 (8.82) | 16.8 | 11.0 |
| | ii) | 3.75 | 73 (7.15) | 90 (8.82) | 8.9 | 11.0 |

$k^{*}$ i) (Distance from a borehole to a free end / Radius of a borehole)

ii) (Distance between boreholes/ Diameter of a borehole)

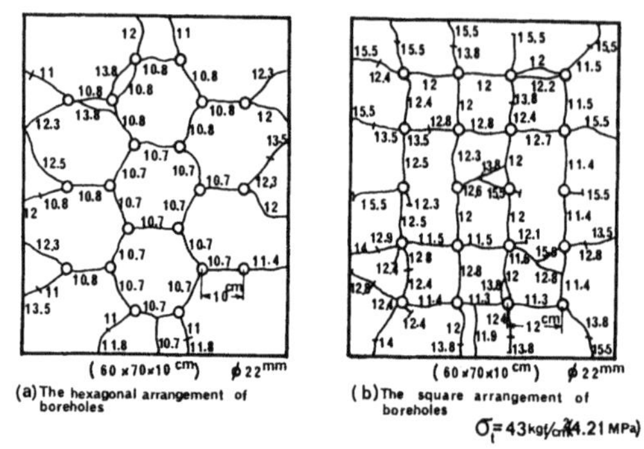


Fig. 20 The crack modes of concrete slabs. (Numerals indicate the cracking time.)

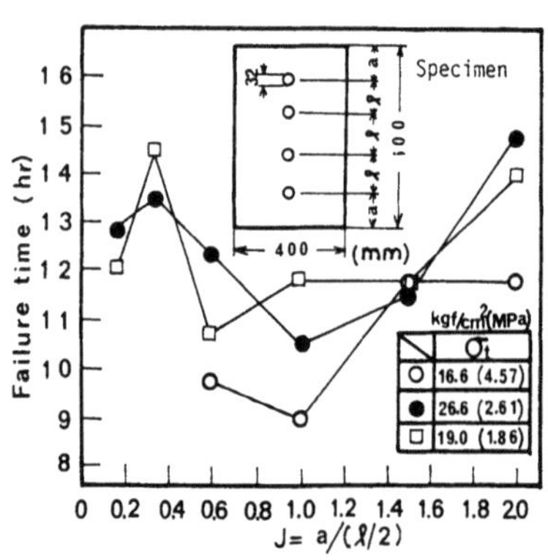


Fig. 21 Relationship between failure time and J.

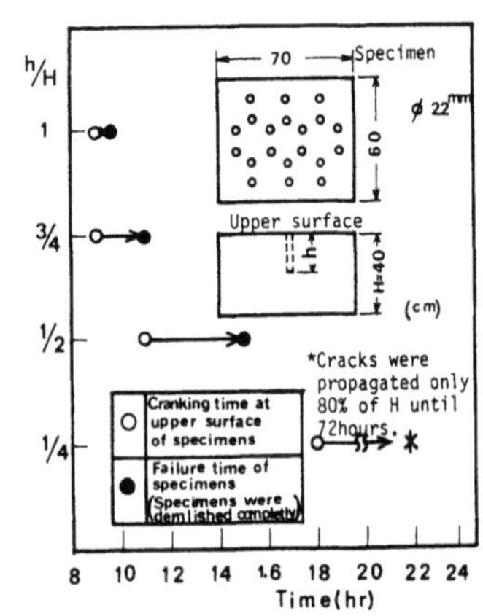


Fig. 22 Relationship between h/H and failure time.

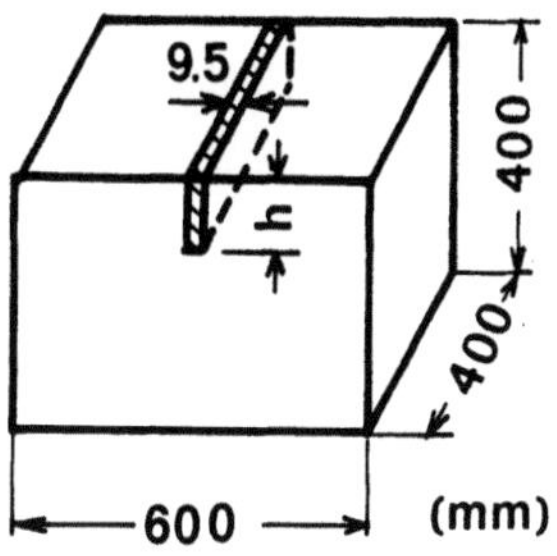


Fig. 23 Concrete specimen containing a ditch.

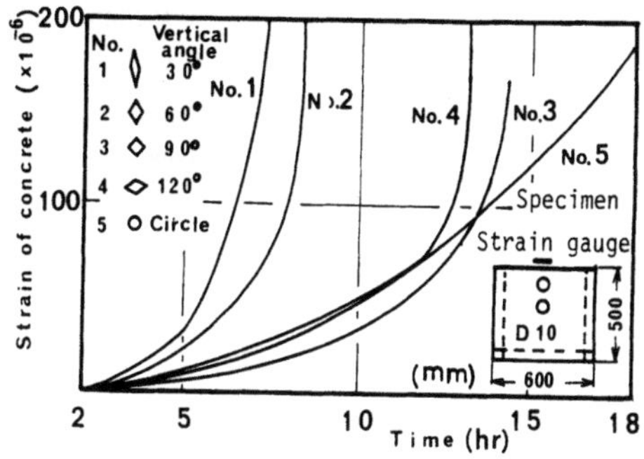


Fig. 24 The cracking time versus the shape of borehole.

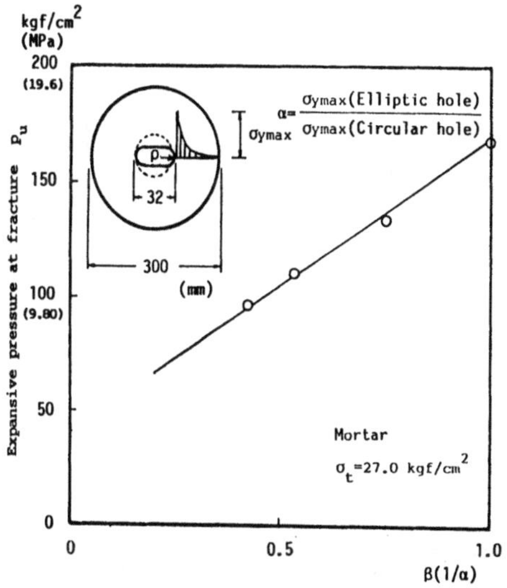


Fig. 25 Relationship between the expansive pressure at fracture and the inverse of the stress concentration factor.

Table 3 Relationship between h and failure time.

| h (mm) | 50 | 100 | 150 | 200 |
|---|---|---|---|---|
| Failure time (hr) | 16 | 9 | 7 | 6 |

$\sigma_t$=30 kgf/cm² (2.94 MPa)

SEM/RILEM International Conference
on Fracture of Concrete and Rock
Houston, Texas, June 1987
Editors: S.P. Shah, S.E. Swartz



# FATIGUE RUPTURE OF PLAIN CONCRETE ANALYSED BY FRACTURE MECHANICS

Zhang Binsheng Zhu Zhaohong Wu Keru

ABSTRACT

The fatigue damage mechanisms of concrete under repeated loads are studied. Based on the results of fatigue tests of concrete beam specimens, new fatigue phenomena and process are found and a computing formula is put forward. The physical meanings of the formula are explained from the standpoint of mesoscopic fracture mechanics and damage mechanics. This formula could be used to predict the fatigue life or remaining life of concrete constructions.

## INTRODUCTION

The fatigue rupture of concrete is the dominant failure form in some concrete constructions such as concrete pavements and airport runways. Although traditional fatigue strength theorems have been used to evoluate the fatigue fracture characteristics of plain concrete, these are rather limited. In traditional stress fatigue ($N_f = C\sigma^m$)(1) and strain fatigue ($N_f = C\varepsilon^m$)(2) theorems, initial defects and inperfect of concrete materials and constructions are usually neglected so that a great difference between the fatigue life predicted by experimental formula and practical life would be produced. The linear hypothesis $\Sigma \frac{n_i}{N_i} = 1$ proposed by Miner(3) is not enough sufficient. In this paper, the method of fracture mechanics is introduced to study the fatigue mechanisms of concrete under repeated loads.

According to the viewpoint of energy equilibrium, Griffith(4) first researched the cracking propagation of britte materials such as glass, ceramics etc., establishing the basis of fracture mechanics. Many later researchers enriched and improved this theory further. It was Paris ($\frac{da}{dN} = C\Delta K^m$)(5) and Forman ($\frac{da}{dN} = \frac{C\Delta K^m}{K_{Ic}(1-R)-\Delta K}$)(6) who started to study fatigue fracture theory, which has been replenished and modified by later researchers. Kaplan(7) first applied fracture mechanics to the study of fracture characteristics of concrete. Since then, experts have been controverting the above problem, but after all the concrete fracture mechanics has great vitality.

The main purposes of our work are to investigate the fatigue properties of concrete from fracture mechanics, to find out a fatigue fracture law more suitable to concrete and to analyse the suitability of the traditional fatigue theorems.

## FATIGUE FRACTURE TESTS

Concrete beams with single edge notch in three-point bending are loaded in a 100 KN program control high frequency testing machine which works by nuclear magnetic resonant principle at working frequencies of 80 to 120 Hz and in the form of sinusoidal load-time function. In testing, we control stress intensity factor amplitude $\Delta K = K_{max} - K_{min}$ and cycle characteristic coefficient $R = K_{min}/K_{max} = 0.2$. Loads are measured by a load sensor and displacement by an extensometer. The signals are dynamically received by a dynamic strain indicator and recorded on an oscilloscope. The whole testing apparatus is shown in Fig. 1.

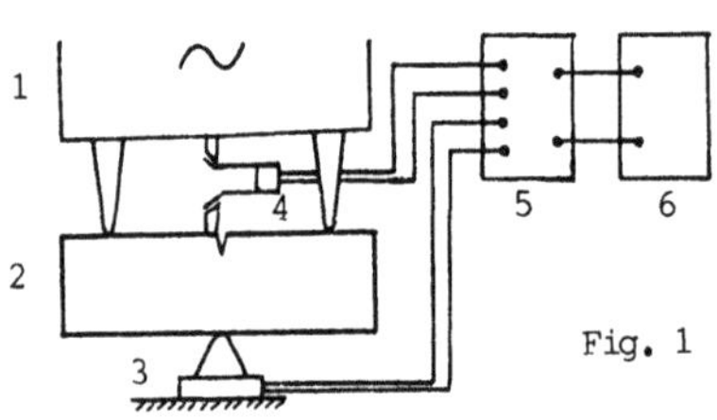


1 - Testing machine
2 - Specimen
3 - Load sensor
4 - Extensometer
5 - Dynamic strain indicator
6 - Oscilloscope

Fig. 1 General sketch of fatigue testing apparatus

Zhang Binsheng, Research Assistant; Zhu Zhaohong, Professor of Highway Engineering; Wu Keru, Professor of Concrete Science; Tongji University, Shanghai, China.

Beam specimens of 10 × 10 × 30 cm are made of fine aggregate concrete, whose mix ratio is as follows: water:cement:sand:stone = 0.48:1:1.1:2.48, maintained in wetting conditions for more than 28 days. The relative depths a/h vary from 0.1 to 0.75, most of which are 0.25 and whose radius is 0.5 mm.

The standard compressive strength of these specimens at 28 days is $f'_c = 30$ MN/m², the flexural strength $f_t = 5$ MN/m², elastic modulus $E = 3.01 \times 10^4$ MN/m², fracture toughness $K_{Ic} = 0.84$ MN/m$^{3/2}$ and dynamic fracture toughness $K_{Icd} = 1.16$ MN/m$^{3/2}$ measured in the testing machine, dynamic coefficient being 1.37.

PHENOMENA OF FATIGUE TESTS

In the fatigue testing, the following phenomena are found:

1. When $\Delta K$ is smaller than some critical value $\Delta K_{th}$, which is called threshold value, the elastic modulus keeps constant with the increase of loading cycles. For the specimens used in testing, $\Delta K_{th}$=0.47MN/m$^{3/2}$, being about 0.5 $K_{Icd}(1-R)$.

2. When $\Delta K$ is greater than $\Delta K_{th}$, macroscopically cracking does not occur on the specimens for about 95% of the fatigue life. Only near rupture do cracks propagate macroscopically, and even do some specimens break suddenly, before which no rupture information appears. The life by which the above critical rupture phenomenon occurs is called linear life, expressed as $N_L$. For our testing, $N_L$ is about 97% of the total fatigue life.

3. Within the linear life N , dynamic elastic modulus $E_d$ of concrete decreases linearly with increase of cycle number N. Only when $N > N_L$, does the dynamic elastic modulus change nonlinearly and rapidly so that in a very short time, it would decrease to very small value and specimens would fail. The dynamic elastic modulus at which N reaches $N_L$ is named the linear limit of dynamic elastic modulus, $E_{dL}$. From our testing results, $E_{dL}$ is 75 ~ 80% of initial dynamic elastic modulus (Fig. 2 and 3). Because dynamic measure is adopted in the whole testing, the measured elastic modulus is dynamic, which is about 1.3 times as much as static elastic modulus.

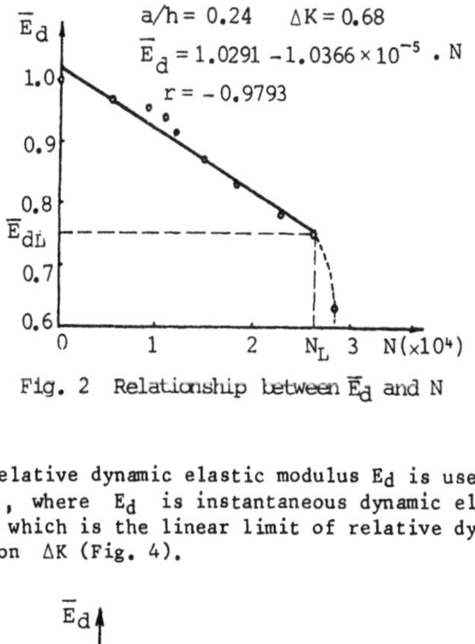


Fig. 2 Relationship between $\bar{E}_d$ and N

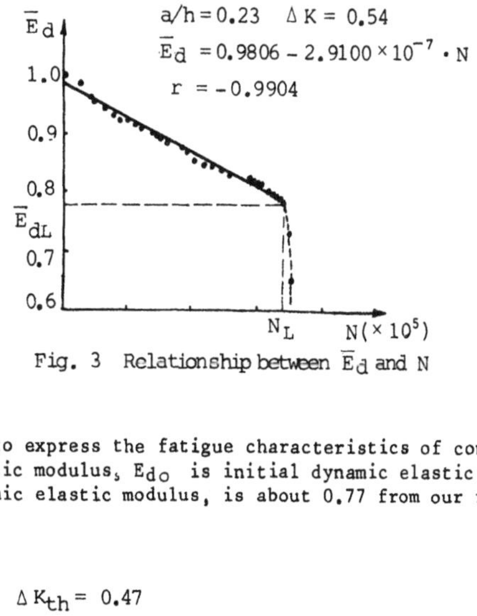


Fig. 3 Relationship between $\bar{E}_d$ and N

4. If relative dynamic elastic modulus $\bar{E}_d$ is used to express the fatigue characteristics of concrete, $\bar{E}_d = E_d / E_{do}$, where $E_d$ is instantaneous dynamic elastic modulus, $E_{do}$ is initial dynamic elastic modulus, then $\bar{E}_{dL}$, which is the linear limit of relative dynamic elastic modulus, is about 0.77 from our results and independent on $\Delta K$ (Fig. 4).

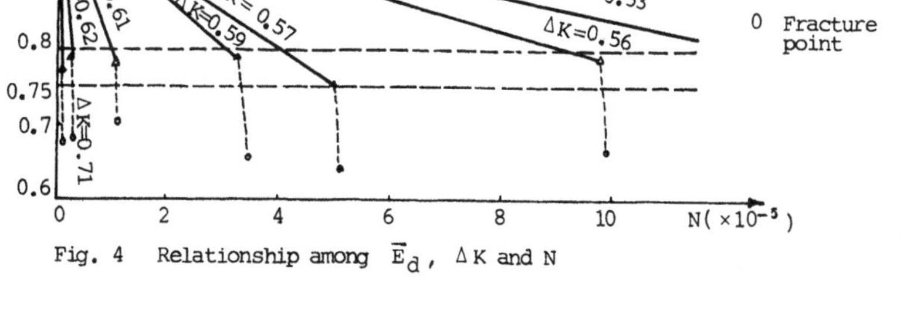


Fig. 4 Relationship among $\bar{E}_d$, $\Delta K$ and N

5. When several stress levels are exerted to the same specimen, it is found that the dynamic elastic modulus respectively decreases linearly corresponding to different stress level (Fig. 5 and 6), the relative linear limit $\bar{E}_{dL}$ being still unchanged. (In Fig. 2, 3, 5, 6 and 7, r is linear correlation coefficient.)

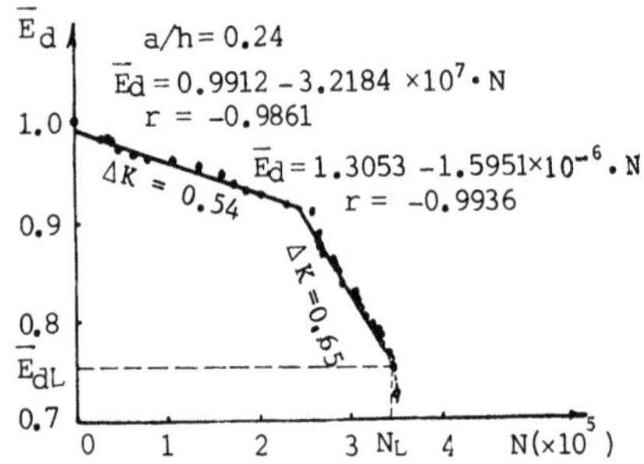


Fig.5 Relationship between $\bar{E}_d$ and N under different $\Delta K_S$

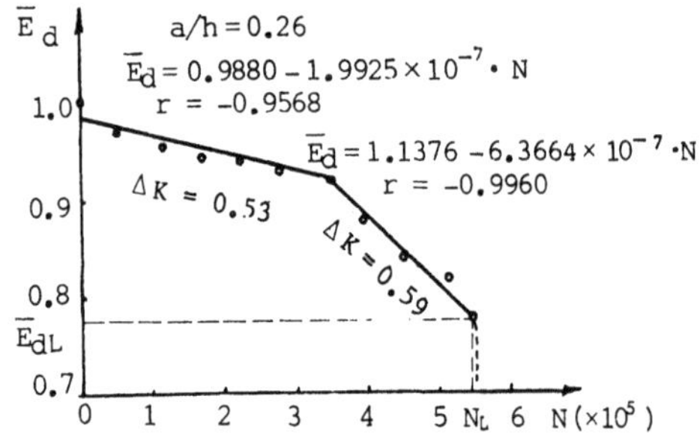


Fig.6 Relationship between $\bar{E}_d$ and N under different $\Delta K_S$

FATIGUE FRACTURE REGULARITY

From our testing results (Fig. 7, 8), considering the critical conditions which should be satisfied, i.e.

i. when $\Delta K < \Delta K_{th}$, $\frac{d\bar{E}_d}{dN} = 0$ and $\frac{d(d\bar{E}_d/dN)}{d(\Delta K)} = 0$

ii. when $\Delta K = K_{Ic}(1-R)$, $\frac{d\bar{E}_d}{dN} = \infty$

the fatigue formula is put forward:

$$\frac{d\bar{E}_d}{dN} = \frac{\alpha < \Delta K - \Delta K_{th} >^n}{K_{Icd}(1-R) - \Delta K}$$

where $\Delta K_{th}$ = threshold value of stress intensity factor, $K_{Icd}$ = I type dynamic fracture toughness of concrete, R = cycle characteristic coefficient, $R = K_{min}/K_{max}$, $\alpha$, n = material constants, < > = jump operator $<x> = \begin{cases} 0 & x<0 \\ x & x>0 \end{cases}$.

In practice, K is always greater than $K_{th}$, so the above formula can be transformed into another form:

$$\frac{d\bar{E}_d}{dN} = \frac{\alpha(\Delta K - \Delta K_{th})^n}{K_{Icd}(1-R) - \Delta K}$$

By statistical treatment of testing results, we obtain

$$\frac{d\bar{E}_d}{dN} = \frac{-1.9168 \times 10^{-12} (\Delta K - \Delta K_{th})^{3.0683}}{K_{Icd}(1-R) - \Delta K}$$

i.e. $\alpha = -1.9168 \times 10^{-12}$, $n = 3.0683$.

Some explanations about the above formula are given as follows:

1. Even though $\Delta K$ keeps constant, $\frac{d\bar{E}_d}{dN}$ changes with $K_{max}$. In the above formula, the denominator term expresses the effect of average stress.

2. Practical concrete constructions are subjected to a loading spectrum consisting of various loads, in which, loads with high or low amplitudes occur alternatively and randomly. From our results, it is found that if $\Delta K$ is smaller than $K_{Ic}(1-R)$, the quantity and order of exerted loads do not effect the fatigue regularity of concrete.

3. In general, environmental conditions such as temperature and humidity effect fatigue life of concrete to some degree. Temperature produces a temperature field in practical constructions so that temperature stress intensity factor $K_t$ would be produced. In predicting fatigue life, $K_t$ should be added to the stress intensity factor K due to loading, so the practical constructions are subjected to a complex stress intensity factor $K' = K + K_t$. Because $K_t$ keeps constant in certain time, so $R' = \frac{K_{min} + K_t}{K_{max} + K_t}$ is used to modify cycle characteristic coefficient R. Humidity also affects fatigue life of concrete. The strength and elastic modulus of concrete decrease with relative humidity [8]. The effect of humidity may be indeed modified by probable distribution.

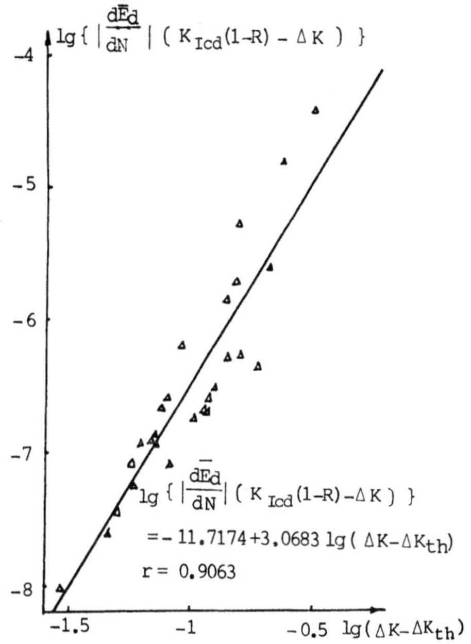


Fig.7 Logarithmic relationship between $\left|\frac{d\bar{E}_d}{dN}\right|(K_{Icd}(1-R)-\Delta K)$ and $(\Delta K-K_{th})$

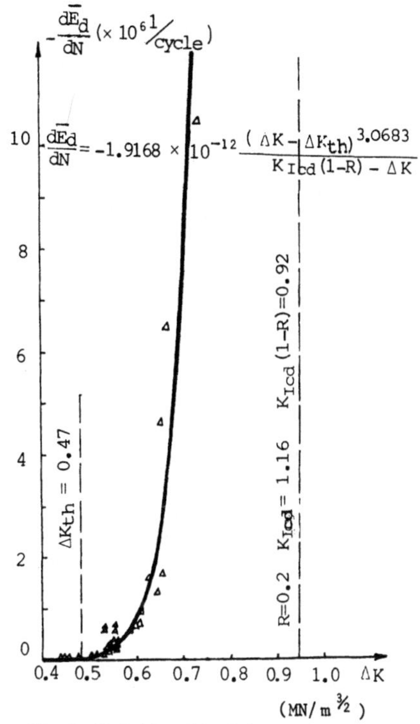


Fig.8 Relationship between $\frac{d\bar{E}_d}{dN}$ and $\Delta K$

4. Elastic modulus of concrete increases with age, so the aging effect should be considered. According to the data concerned, after three months, elastic modulus almost does not increase. Here modifying coefficient is regarded as about 1.26.

## DISCUSSION OF FATIGUE FORMULA

Static failure of a plain concrete specimen contains the following several stages. Before loading, there exist initial defects and microcracks on the interface between matrix and aggregate, which distribute randomly. In the initial loading stage, stress concentration takes place at the tip of interface cracks, which causes cracks to propagate along the weakest direction on the interface. These microcracks only propagate mesoscopically with increasing loading, but no macrocracks form. In the last loading stage, where the load is added to its ultimate value, cracks propagate rapidly so that at least a macrocrack through the specimen forms. At this time, the specimen fails.

For a concrete specimen subjected to repeated loading, correspondingly, when $\Delta K < \Delta K_{th}$, the initial cracks existing on the interface between matrix and aggregate of the specimen are in stable state without propagation, so $\Delta K$ may be regarded as a critical parameter to judge whether the cracks propagate or not. When $\Delta K$ continues to increase and exceed $\Delta K_{th}$, the cracks will mesoscopically propagate so that the relative displacement produced on the interface may cause elastic modulus to decrease. When the cracks do not run through the specimen, elastic modulus linearly decreases with the increase of cycle number of loading. Even if the cracks penetrate the specimen, due to the mutual engagement and friction of cracks surfaces, the specimen still has certain capability to transform loads, so that when $N_L < N < N_f$, elastic modulus decreases nonlinearly. Benouch[9] combined the change of elastic modulus with damage degree D of concrete. He proposed $E = E_0(1-D)$, where D = damage degree of concrete materials, $E_0$ = initial elastic modulus, E = instanteneous elastic modulus. For non-damaged materials, D = 0, while for wholly failed materials, D = 1.

If the term of relative dynamic elastic modulus $\bar{E}_d$ is used, then $\bar{E}_d = E_d / E_{d0} = 1 - D$. As mentioned above, under certain stress intensity factor amplitude, $E_d$ decreases with the increase of N, which means D increases. When $N \le N_L$, $\bar{E}_d$ or D is the linear function of N, while when $N > N_L$, $\bar{E}_d$ or D changes nonlinearly. When $N = N_f$, $\bar{E}_d = 0$ and $D = 1$. Keep it in mind that when $\Delta K < \Delta K_{th}$, $\bar{E}_d = 1$ and $D = 0$. So we use the change of damage degree to express fatigue fracture process as follows:

$$D = \begin{cases} 0 & \Delta K < \Delta K_{th} & \text{without damage} \\ \alpha N & \Delta K_{th} < \Delta K < K_{Icd}(1-R),\ N < N_L & \text{linear damage} \\ f(N)_{non-linear} & \Delta K_{th} < \Delta K < K_{Icd}(1-R),\ N_L < N < N_f & \text{nonlinear damage} \\ 1 & \Delta K = K_{Icd}(1-R) \ \text{or} \ N = N_f & \text{failure} \end{cases}$$

where $K_{Icd}(1-R)$, similar to $K_{Ic}$, is critical fatigue fracture toughness amplitude.

In the following part, Miner's hypothesis will be evoluated.

Miner's hypothesis is simply expressed as follows: if at some stress level, at which the fatigue life is $N_i$, cycle number is $n_i$, then after $n_i$, the damage $\frac{Dn_i}{N_i}$ is reached. The total damage is D, $D = \Sigma \frac{Dn_i}{N_i}$, i.e. $\Sigma\frac{n_i}{N_i} = 1$. From our results, we find:

1. Miner's hypothesis contains alternative fatigue effects of different stress levels, so it is convenient to use and easily acceptable.

2. It can not cover initial cracks and defects in concrete constructions.

3. It neglects the nondamage effect of low stress. In our testing, when $\Delta K < \Delta K_{th}$, such very low stress level will increase $\Delta K_{th}$ without damaging the constructions. This means smaller $\Delta K$ may drill the constructions, so that initial propagation could be delayed.

4. Damage D in Miner's hypothesis has no clear physical meanings.

5. The hypothesis neglects the interaction of high and low stresses. As mentioned above, low stress may increase $\Delta K_{th}$, but it is not very clear how the final life is. We find for either single or complex loads, if $\bar{E}_{dL}$ is used as evoluating standard, $N_L$ is regular. But the cycle number N between $N_L$ and $N_f$ is disorganized because $\Delta N = N_f - N_L$ is 8% or more of $N_f$. When $N > N_L$, macrocracks form, but the constructions behave random rupture: sometimes very low stress may make them fail at once, while sometimes very high stress cannot cause them to fail completely. This means the error produced by Miner's hypothesis is more than 8%. But the fatigue fracture regularity under different stress levels are similar. If $N_L$ is used as fatigue damage life, the error will be minimized.

Hence, how approximate Miner's hypothesis is depends on what is used as the fatigue damage standard. If dynamic elastic modulus is used, the hypothesis not only may be applied to concrete, but also has high precision. We restate Miner's hypothesis: $D = \Sigma D_i$, where $D_i$ is the damage under $i^{th}$ stress level. $D_i$ may be expressed with either $\bar{E}_{di}$ or $N_i$. If $\bar{E}_{di}$ is used, $\frac{1}{\Delta \bar{E}_{dL}} \Sigma(\Delta \bar{E}_{di}) = 1$, where $\Delta \bar{E}_{di} = \bar{E}_{di} - \bar{E}_{di0}$, $\Delta \bar{E}_{dL} = \bar{E}_{dL} - \bar{E}_{d0}$, $\bar{E}_{di}$ = the initial relative dynamic elastic modulus for $i^{th}$ stress level, $\bar{E}_{di}$ = the final relative dynamic elastic modulus for $i^{th}$ stress level, $\bar{E}_{dL}$ = linear limit of relative dynamic elastic modulus, and $\bar{E}_{d0}$ = initial relative dynamic elastic modulus, $\bar{E}_{d0} = 1$. So we get, $\bar{E}_{di} = f(\Delta K_i)$. If $N_i$ is used, $\Sigma\frac{n_i}{N_i} = 1$, where $n_i = f(\bar{E}_{di}, \Delta K_i)$, $N_i = f(\Delta K_i)$. Notice that the mentioned fatigue formula different from Miner's hypothesis lies in that $N_L$ is scientifically considered as $N_f$ in our work.

PREDICTION OF FATIGUE LIFE

From fatigue fracture regularity $\frac{d\bar{E}_d}{dN} = \frac{\alpha(\Delta K - \Delta K_{th})^n}{K_{Icd}(1-R) - \Delta K} = f(\Delta K)$, we get $dN = \frac{1}{f(\Delta K)} d\bar{E}_d$. Integrating the differential formula, we get

$$\int_{N_1}^{N_2} dN = \int_{\bar{E}_{d2}}^{\bar{E}_{d1}} \frac{1}{f(\Delta K)} d\bar{E}_d = \frac{1}{f(\Delta K)}(\bar{E}_{d2} - \bar{E}_{d1}), \text{ i.e.}$$

$$N_2 - N_1 = \frac{1}{f(\Delta K)}(\bar{E}_{d2} - \bar{E}_{d1}).$$

1. Before loading, $N_1 = 0$, $\bar{E}_{d1} = \bar{E}_{d0} = 1$, so $N_2 = \frac{1}{f(\Delta K)}(\bar{E}_{d2} - 1)$.

When fatigue rupture on concrete constructions occur, $N_2 = N_f$, where $N_f$ is fatigue life. If $N_L$ is regarded as $N_f$, as discussed above, the predicted fatigue life is more safe and rational. Now let $N_2 = N_L = N_f$, $\bar{E}_{d2} = \bar{E}_{dL}$, so we get fatigue life $N_f$:

$$N_f = \frac{1}{f(\Delta K)}(\bar{E}_{dL} - 1).$$

2. If $\bar{E}_{d1}$ and $\bar{E}_{d2}$ are the relative dynamic elastic module at two arbitrary moments, the cycle number between the moments $\Delta N = N_2 - N_1$ and $\Delta \bar{E}_d = \bar{E}_{d2} - \bar{E}_{d1}$. So we get, $\Delta N = \frac{1}{f(\Delta K)} \Delta \bar{E}_d$. If $\Delta K$ is known, $\Delta N$ causing damage $\Delta \bar{E}_d$ can be determined.

3. When constructions fail, $N_2 = N_L = N_f$ and $\bar{E}_{d2} = \bar{E}_{dL}$. We get

$$N_f - N_1 = \frac{1}{f(\Delta K)}(\bar{E}_{dL} - \bar{E}_{d1}).$$

Using the above formula, the residual life after $N_1$ can be determined.

4. Under multiple stress level, the damage caused by every level can be accumlated. Once the relative dynamic modulus of concrete constructions is smaller than $\bar{E}_{dL}$, the constructions may be considered in rupture state.

CONCLUSIONS

1. Fatigue failure of concrete consists of four stages: formation of initial microcracks, mesoscopic propagation and macroscopic propagation of the cracks, and failure, in which mesoscopic propagation of microcracks contains 95% of the whole fatigue life. This is actually a gradual damage process, which can be expressed by damage degree D.

2. When $\Delta K$ is smaller than $\Delta K_{th}$, the cracks do not propagate, which is equivalent to nondamage state where D = 0.

3. When $\Delta K > \Delta K_{th}$, damage begins, whose macroscopic expression is the decrease of dynamic elastic modulus. Within 95% of fatigue life, dynamic elastic modulus decreases linearly with the increase of cycle number. When it decreases to between 75 and 80% of its initial value, it decreases nonlinearly so that concrete constructions fail.

4. The formula $\frac{d\bar{E}_d}{dN} = \frac{\alpha(\Delta K - \Delta K_{th})^n}{K_{Icd}(1-R) - \Delta K}$ proposed in this paper can be used to predict the fatigue life of concrete constructions.

5. Miner's hypothesis is only an approximate method. If the linear limit of fatigue life $N_L$ is regarded as the moment when concrete constructions fail, the suitability of Miner's hypothesis to these constructions will be improved greatly.

REFERENCES


(1) C.L. Monismith, "Asphalt Mixture Behavior in Repeated Fracture" IER Report TE 66-6 University of California, Berkeley, 1966.

(2) C.L. Monismith, "Flexibility Characteristics of Asphalt Paving Mixture", Proc. Asso. of Asphalt Paving Tech., Vol. 27, 1958, pp. 74-106.

(3) M.A. Miner, "Cumulative Damage in Fatigue", Trans. ASME, Vol. 67, A159, 1945.

(4) A.A. Griffith, "The Phenomenon of Rupture and Flow in Solids", Phil. Trans. Royal Soc., London 221A, 1920-1921, pp. 163-198.

(5) P.C. Paris, "The Fracture Mechanics Approach to Fatigue", Proc. of 10th Sagamore Army Mete. Rese. Conf., (Aug. 1963), Syracuse Uni. Press.

(6) R.G. Forman, V.E. Kearny and R.M. Engle (1967), "Numerical Analysis of Crack Propagation in Cycle-Loaded Structures", Jnl. Bas. Engng., ASME Trans. Series D. 89, 459.

(7) M.F. Kaplan, "Crack Propagation and Fracture of Concrete", Jour. of ACI, Proc. Vol. 58, No.5, Nov. 1961

(8) V.S. Ramachandran, R.F. Feldmen and J.J. Beaudoin, CONCRETE SCIENCE — Treatise on Current Research, pp. 83.

(9) S. Benouche, "Modélization de l'endommagement du béton hydraulic par micro fissuration en compression", Thèse 3ème cycle (Paris VI - ENSET), 1980.

SEM/RILEM International Conference
on Fracture of Concrete and Rock
Houston, Texas, June 1987
Editors: S.P. Shah, S.E. Swartz

# KINETICS OF CRACK GROWTH IN PLAIN CONCRETE

P.C. Perdikaris, A.M. Calomino

## ABSTRACT

The number of load cycles, N, load-point displacement (LPD) and crack- mouth-opening displacement (CMOD) compliance were measured in a series of fatigue tests on single-edge-notched concrete beams (SENB) under 4-point bending to investigate the kinetics of crack propagation in plain concrete. The prenotched beams were subjected to either a constant or variable pulsating load up to a maximum load level of about 75% of the static ultimate strength. Typically, the crack growth rate, $d\ell/dN$ decreased for the first 8 mm of crack extension. The crack growth rate and the strain energy release rate, $G_I$, are plotted versus the crack length to beam depth ratio, which is determined from the CMOD compliance measurements. The crack speed varied considerably along the crack path but increasing strain energy release rates produced on the average an increase in the crack speed. Finally, kinetic data from three beams subjected to a constant amplitude repeated loading is compared to the Paris model.

## INTRODUCTION

Several controlled crack growth studies have been carried out for cementitious materials on double torsion specimens [ 1 to 4 ]. In these studies, the average value of the exponent $\alpha$ in the Paris model [5]

$$d\ell/dN = \beta \ \Delta K_I^{\alpha} \quad , \tag{1}$$

often used to describe the subcritical crack growth (slow crack growth) in brittle materials, is 30 in the case of hydrated cement paste and cement mortar [6]. Also, although a significant scatter in the value of $\alpha$ was observed, it was found that polymer impregnation [2] or fiber reinforcement [4] did not have any great effect on its value.

It has been reported that crack speeds vary considerably along the crack path, while in general they tend to increase as the crack propagates [7]. This may be attributed to the inherent heterogeneity of concrete. Due to its random strength field and various crack arresting mechanisms (aggregate, inclusions etc.) crack growth could be stopped resulting in extensive crack branching [8] or at least slowed down considerably [9,10].

The experimental data on the kinetics of crack growth in plain concrete presented in this paper are based on a series of tests on prenotched SEN concrete beams subjected to pulsating load (4-point bending). These tests were part of a research study on the effect of fatigue loading on the fracture toughness of plain concrete [11,12]. The crack speed, $d\ell/dN$, and energy release rate, $G_I$, are determined as a function of crack extension to depth ratio, $\ell/d$, based on the CMOD compliance measurements. Finally, comparison of the results of three beams subjected to a constant maximum fatigue load level is performed based on the Paris model of crack growth.

## EXPERIMENTAL PROGRAM

An initial midspan precast notch of $\ell/d = 0.20$ was cast in all fatigued SENB specimens. The dimensions of the specimens are 80 mm (width) x 160 mm (depth) x 960 mm (length), as shown in Fig. 1. These dimensions were selected so that the minimum dimension of the beam would be at least 10 times the maximum characteristic aggregate size of about 7 mm. The precast notch was obtained with a 5 mm thick brass plate with a wedge of 20 degrees at its tip. This brass plate was inserted into the specimen mold prior to casting and then removed prior to testing when the concrete had hardened.

Two types of repeated loading of a constant and variable load amplitude were used to qualitatively

---

P.C. Perdikaris (SEM member) is Associate Professor, Department of Civil Engineering, Case Western Reserve University, Cleveland, Ohio 44106; A.M. Calomino is Materials Engineer, NASA Lewis Research Center, Cleveland, Ohio 44106 (he is a former graduate student of Civil Engineering, Case Western Reserve University).

investigate the effect of loading history on the crack propagation rate. The beams fatigued under a constant load amplitude are labeled L (low level) along with their specimen number (specimens FT4L, FT6L, FT7L, FT9L and 1FT10L). The maximum fatigue load amplitude for these beams varied between 2.72 and 3.11 kN. The beams fatigued under a variable load amplitude are labeled H (high level), such as specimens FT6H and FT7H. Typically, they were initially subjected to relatively high loads of up to 3.3 kN stepwise decreasing to about 2.9 kN. It is important to note here that the average static ultimate strength of prenotched beams with the same initial $\ell/d = 0.20$ is 3.9 kN [11]. The loading frequency varied between 0.5 and 1.0 Hz. A minimum applied force of 0.02 kN was always maintained in order to keep the load fixture and specimen in continuous contact. Instead of fatiguing the specimens to failure, the fatigue crack propagation was terminated after the crack length to beam depth ratio, $\ell/d$, reached a value of about 0.50, so that the beam could be failed statically while a dye-penetrant was injected into the root of the precast notch.

The concrete mix used has aggregate/sand/cement/water weight proportions of 2.0/1.5/1.0/0.5. The average cylinder compressive strength (75 x 150 mm cylinder), $f_c'$, and elastic modulus, $E_c$, of the concrete mix are 51 MPa and 33 GPa, respectively.

The experimental setup is shown in Fig. 2. Both the CMOD and load-point displacement (LPD) were measured and all specimens were tested upside down to minimize the effect of self-weight. The crack length based on CMOD compliance measurements and the corresponding number of load cycles, N, were recorded when changes in the CMOD compliance were noticed. Also, the theoretical strain energy release rate, $G_I$, based on LEFM is calculated as a function of the crack length, $\ell$, which is taken as the average of the two crack lengths used to determine crack speed, $d\ell/dN$.

The following formulas for the CMOD compliance, $C_{CMOD}$, stress intensity factor, $K_I$, and energy release rate, $G_I$, are used [13]:

$$C_{CMOD} = \frac{72\ell}{E_c d^2}\left[0.8 - 1.7\frac{\ell}{d} + 2.4\frac{\ell^2}{d^2} + \frac{.66}{(1-\ell/d)^2}\right], \tag{2}$$

$$K_I = \frac{18F\sqrt{\pi\ell}}{d^2}\left\{\left(\frac{2d}{\pi\ell}\tan\frac{\pi\ell}{2d}\right)^{1/2}\left[\frac{.923 + .199(1-\sin\pi\ell/2d)^4}{\cos\pi\ell/2d}\right]\right\} \tag{3}$$

$$\text{and}\quad G_I = \frac{K_I^2}{E_c}(1-\nu^2) \tag{4}$$

where $\ell$ is the crack length, F the applied force, d the depth of the beam and $\nu = 0.18$ the Poisson's ratio for concrete.

## KINETICS OF CRACK PROPAGATION

The crack propagation rate, $d\ell/dN$, calculated as the change in the predicted crack length ($d\ell$) divided by the change in the number of load cycles (dN) is plotted together with the energy release rate, $G_I$, versus $\ell/d$ for specimens FT4L, FT6L, FT6H, FT7H, FT7L and FT9L in Figs. 3(a), 3(b), 3(c), 3(d), 3(e) and 3(f), respectively. The load history for each specimen is also included in these figures. Review of the crack propagation rate plots reveals some interesting characteristics.

In almost all fatigued beams, except for specimen FT6H, crack speeds decreased for the first 8 mm of crack extension. In the case of specimen FT6H, the initial crack speed was based on a somewhat larger than usual change in crack length and may not be representative. The change in CMOD compliance which produced the initial crack propagation rate occurred rapidly while additional changes typically required a large number of additional load cycles. This large number of load cycles required for crack growth is reflected in the decreasing crack speeds.

In general, crack propagation rates did correspond to changes in the strain energy release rate. For the beams subjected to variable load amplitude, increasing or decreasing maximum fatigue load levels resulted in increasing or decreasing crack propagation rates, respectively. Although typically the average behavior was that increasing crack lengths resulted in increasing crack speeds and energy release rates, periodic deceleration in the crack growth was observed. This is probably due to the random nature of the strength field of concrete and the possible development of microcracking. Deceleration of crack propagation has been also observed in other materials and is usually associated with the presence of a damage zone near the crack tip [14].

The kinetics for specimens FT4L, FT6L and FT7L, which were subjected to load histories of constant load amplitude were investigated further. The $d\ell/dN$ values for these three specimens are plotted versus $\Delta K_I$ in Fig. 4. The crack growth data for specimen FT9L is not included since its overall propagated crack length was relatively small and the available data was limited. A first-order least square fit of the data with a coefficient of correlation greater than 0.96 produced the solid straight lines shown in Fig. 4. Using a log-log scale the coefficients of the Paris model of crack growth (Eq. 1) are determined. The value of the exponent $\alpha$, as shown in Table 1, varies very little with an average standard deviation $\sigma_\alpha = 0.33$ and an average value of about 14. If this type of crack growth model is used to predict crack speeds, however, at least a 100% possible error in the predicted value of crack speed is possible. Therefore, significant errors can result in predicting life expectancy in the case that instability is described by a critical crack length. The latter is one of the shortcomings of the Paris model. Also, this model does not allow for the natural phenomenon of crack deceleration.

Fatigue loading was interrupted at an estimated crack length, $\ell$, of about 0.5d. The crack propagation was not allowed to become unstable and the specimens were failed under a monotomically increasing static loading. Crack growth data in the region of crack instability, however, would have been very helpful in improving our understanding of the nature of crack propagation in plain concrete.

## CONCLUSIONS

1. Decreasing crack speeds observed at the initiation of fatigue loading may be due to the development of a damage zone preceding crack extension.

2. Increasing strain energy release rates produced on the average increasing crack speeds, although there is significant scatter in the determined values of crack speed.

3. Periodic deceleration of crack propagation observed under fatigue may be attributed to the random nature of the strength field of concrete and possibly the development of microcracking.

4. Under the fatigue loading of variable load amplitude for the SEN beam specimens tested, the crack speeds tended to increase or decrease for increasing or decreasing load amplitudes, respectively.

5. The use of the Paris expression describing crack growth in plain concrete can result in significant errors of at least 100% for the predicted values of crack speed or life expectancy.

6. Further testing of prenotched or unnotched concrete beams under repeated loading is needed to study the kinetics in both the slow crack growth and instability regions. This, together with some input on the stochastic nature of crack propagation derived from the topography of the fracture surface [15], would lead to a better understanding of the fracture process in plain concrete.

## ACKNOWLEDGMENTS

The research reported here was supported by a grant from the National Science Foundation (Grant No. CEE-8307937).

## REFERENCES

1. Mindess, S., Nadeau, J.S. and Hay, J.M., "Effects of Different Curing Conditions on Slow Crack Growth in Cement Paste", Cement and Concrete Research, 4, 953-965 (1974).

2. Evans, A.G., Clifton, J.R. and Anderson, E., "The Fracture Mechanics of Mortar", Cement and Concrete Research, 6, 535-548 (1976).

3. Wecharatana, M. and Shah, S.P., "Double Torsion Tests for Studying Slow Crack Growth of Portland Cement Mortar", Cement and Concrete Research, 10, 833-844 (1980).

4. Yam, A.S.-T and Mindess, S., "The Effects of Fibre Reinforcement on Crack Propagation in Concrete", International Journal of Cement Composites and Lightweight Concrete, 4, 83-93 (1982).

5. Paris, P.C., Gomez, M.P. and Anderson, N.E., "A Rational Analytic Theory of Fatigue", The Trend of Engineering, 13, 9-14 (1961).

6. Mindess, S., "Rate of Loading Effects on the Fracture of Cementitious Materials", NATO Advanced Research Workshop, Applications of Fracture Mechanics to Cementitious Composites, Ed. S.P. Shah, Northwestern University (Sept. 1984).

7. Alford, N.McN., "Dynamic Considerations of Fracture in Mortars", Materials Science and Engineering, 56, 279-287 (1982).

8. Mindess, S., and Diamond, S., "A Preliminary SEM Study of Crack Pro- pagation in Mortar", Cement and Concrete Research, 10, 509-519 (1980).

9. Mindess, S. and Diamond, S., "The Cracking and Fracture of Mortar", Materiaux et Constructions, 15, 107-113 (1982).

10. Wittmann, F.H., "Influence of Time on Crack Formation and Failure of Concrete", NATO Advanced Research Workshop, Application of Fracture Mechanics to Cementitious Composites, Ed. S.P. Shah, Northwestern University (Sept. 1984).

11. Perdikaris, P.C., Calomino, A.M., and Chudnovsky, A., "Effect of Fatigue on the Fracture Toughness of Concrete", Journal of Engineering Mechanics, ASCE, 112 (8), 776-791 (Ang. 1986).

12. Perdikaris, P.C., and Calomino, A.M., "Load History Effects on the Fracture Properties of Plain Concrete", Report to NSF (Res. Grant CEE-83-07937), Dept. of Civil Engrg., Case Western Reserve University (Feb. 1986).

13. Tada, H., Paris, P.C. and Irwin, G.R., The Stress Analysis of Cracks Handbook, Del Research Corporation (1973).

14. Chudnovsky, A., Moet, S.A. and Bankert, M.T., "Effect of Damage Dissemination on Crack Propagation in Polypropylene", Journal of Applied Physics, 54 (10), 5562-5567 (Oct. 1983).

15. Perdikaris, P.C. and Chudnovsky, A., "New Approach to the Fracture Toughness of Concrete-Probabilistic Model", Proceedings of the 6th International Conference on Fracture, 4, New Delhi, India, 2769-2775 (Dec. 1984).

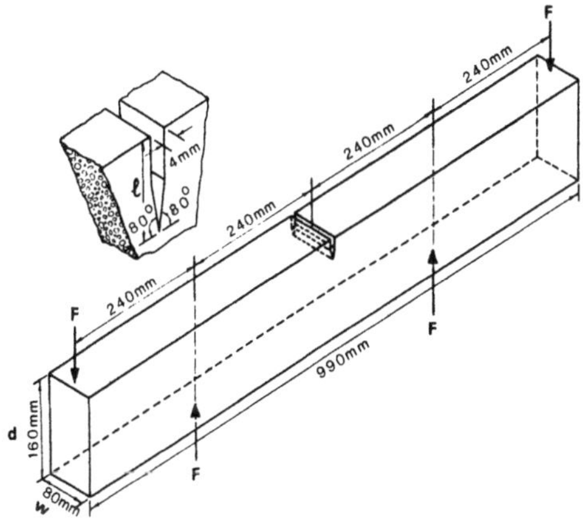


Fig. 1 - Prenotched SEN beam specimen subjected to 4-point bending.

Fig. 2 - Experimental setup.

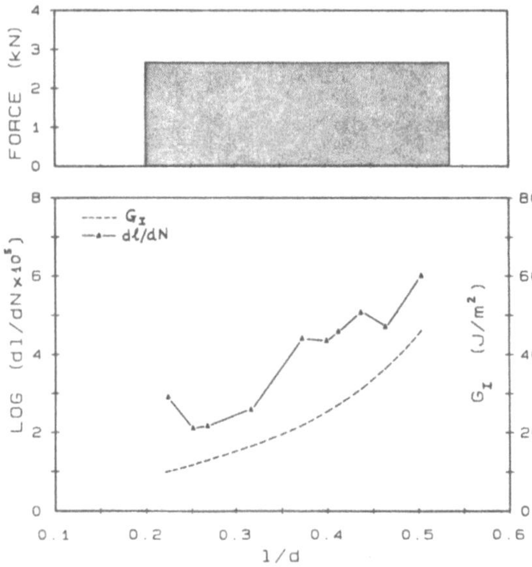


a) Specimen FT4L

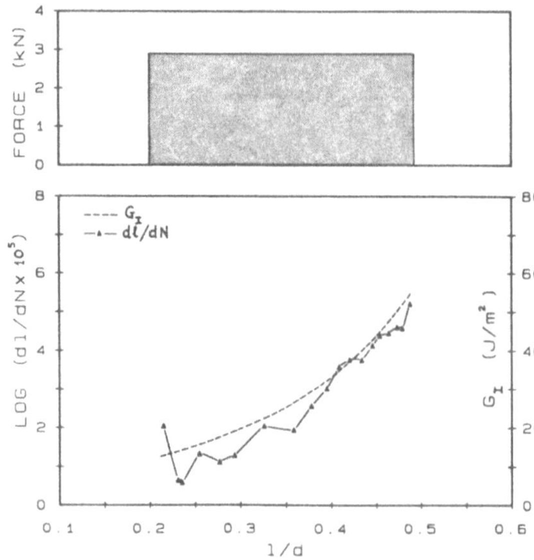


b) Specimen FT6L

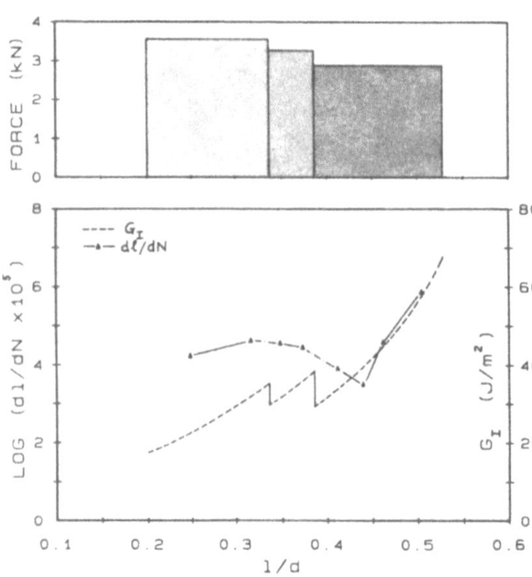


c) Specimen FT6H

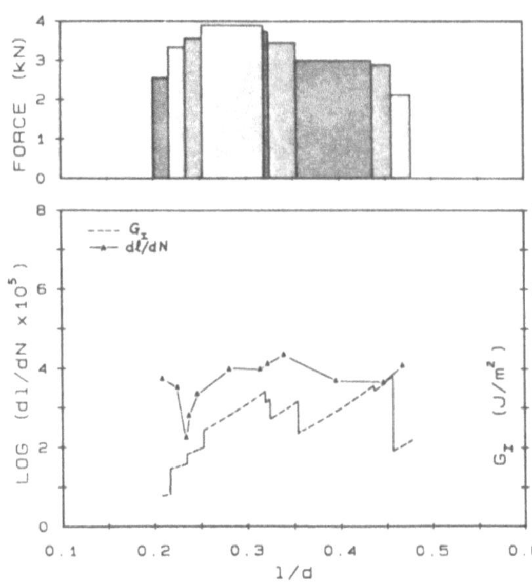


d) Specimen FT7H

Fig. 3 - Crack speed, dℓ/dN, energy release rate, $G_I$, and load history for fatigue loading versus crack length to beam depth ratio, ℓ/d.

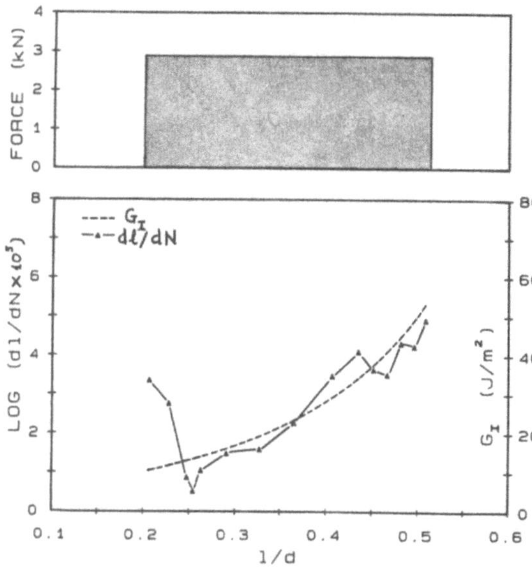


e) Specimen FT7L

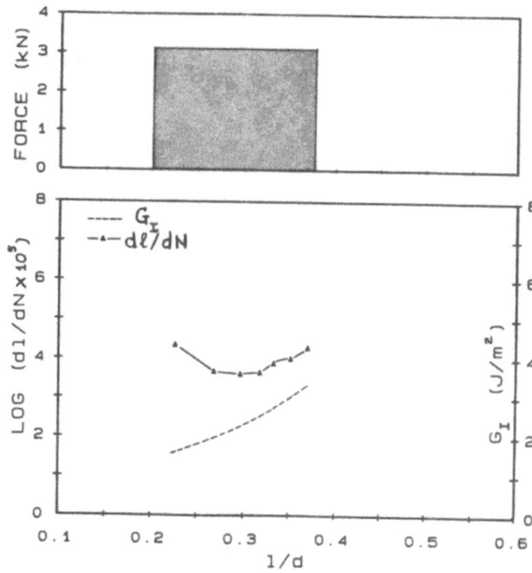


f) Specimen FT9L

Fig. 3 (cont'd)

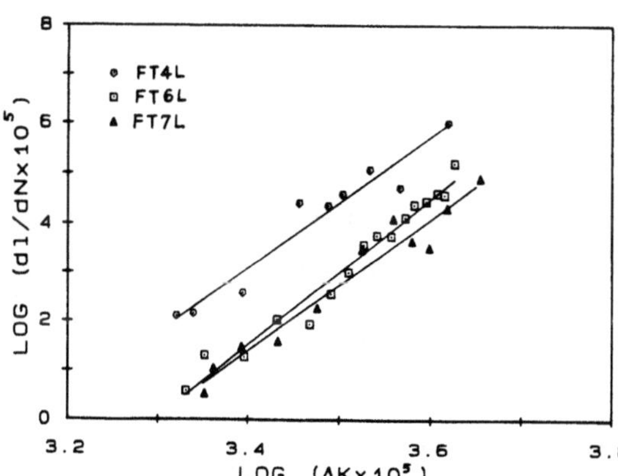


Fig. 4 - Crack speed, dℓ/dN, versus $\Delta K_I$ for specimens FT4L, FT6L and FT7L.

Table 1 - Paris model for crack propagation for specimens FT4L, FT6L and FT7L

| Beam Specimen | Exponent $\alpha$ | Correlation Coefficient | Standard deviation, $\sigma_\alpha$ |
|---|---|---|---|
| FT4L | 13.2 | 0.97 | 0.37 |
| FT6L | 14.9 | 0.99 | 0.26 |
| FT7L | 13.1 | 0.98 | 0.28 |

SEM/RILEM International Conference
on Fracture of Concrete and Rock
Houston, Texas, June 1987
Editors: S.P. Shah, S.E. Swartz

# ON THE ESTIMATING METHOD OF FATIGUE STRENGTH OF THE BRITTLE MATERIALS UNDER REPEATED IMPULSIVE BENDING LOAD

Y. MATSUFUJI, T. OHKUBO

ABSTRACT

In the case that reinforced concrete structures are subjected to repeated impulsive bending load, it is important to clarify the fatigue strength and the fatigue life of R/C members under repeated impulsive bending load. In this paper, a method estimating the fatigue life of the brittle materials, such as concrete and rock, is shown on the basis of the test results of cement-mortar beams under repeated impulsive bending load.

## 1. INTRODUCTION

A resistant power of the repeated impact load is a characteristic value of the strength of each material. And, it is important to clarify the fatigue life of materials under repeated impulsive bending load.

The strength of metals under repeated impulsive load has been investigated by various authors from defferent points of view, for example the studies from the point of impact energy by Kawamoto[1] and Hartbower [2] and the studies of impact stress by wornock[3] and Cotterell[4]. However, a study on brittle materials such as concrete and rock under repeated impulsive bending load have not been carried out so much. The method to obtain the fatigue life of brittle materials has not been established yet because the test method and the estimating method have not been standardized. Furthermore, most of previous studies were carried out with the test method that the repeated impact level was defined by changing the velocity of a impact hammer, so the effect of impact velocity was disregarded. We have recognized that the resistant power of brittle materials under impulsive bending load[5] depends on the impact velocity. Thereby, it is reasonable that the fatigue life of materials under repeated impulsive load changes with the impact velocity. In this paper, a method to estimate the fatigue life of the brittle materials is shown on the basis of the test results of cement-mortar beams under repeated impulsive bending load.

## 2. EVALUATING METHOD AND FATIGUE LIFE

### 2.1 EVALUATING METHOD

In this study, the impact hammer repeatedly strikes on the middle of beam type specimen which is simply supported.

In most of previous experiments[6,7] which were carried out with using above-mentioned test method, the impact level (S) is selected by changing the velocity of the impact hammer ($V_0$) of which mass is constant. And, the strength of materials depending on the impulsive velocity is neglected in this method.

Nakahara[8] has been reported that the strength of materials under impulsive load is affected by not only the magnitude of the impulsive stress but also by the time of impulsive load applied. When the magnitude of impulsive stress is constant and the longer the time in applying the impulsive load, the shorter the fatigue life of the specimen takes place.

Because of the changes in time of the impulsive load applied, the velocity of impact hammer can not be neglected in the repeated impact bending test. Therefore in this study, the experiments will be carried out with employing two methods described as follows.

[Experiment I]
The impact level (S) is changed by the velocity ($V_0$) of the impact hammer of which mass is constant.

[Experiment II]
The impact level (S) is changed by the mass (M) of the impact hammer of which velocity is constant.

In the method of [Experiment I], the effect of impulsive velocity is neglected, and this is a same method with the previous experiments. In the method of [Experiment II] the effect of impulsive velocity is considered on a basic principle that the strength of materials using the impulsive load depends on the hammer velocity.

Y. MATSUFUJI is a Associate Professor and T. OHKUBO is a Research Associate, Department of Architecture, Faculty of Engineering, Kyushu University, Fukuoka, Japan

Rearranging the results of both experiments, the fatigue life of specimen can be estimated under repeated impulsive load.

## 2.2 FATIGUE LIFE

Fatigue strength of materials is evaluated as the relation between the magnitude of stress and the number of strokes applied until the specimen is broken. In this study, "fatigue life" indicates the number of hammer strokes until the specimen is broken under repeated impulsive load. Generally, in fatigue test, the fatigue lifes of specimens vary widely even if a repeated constant load is applied to the specimens made of the same material. The main causes as described above are a lack of uniformity and an original lack of specimen, therefore, in order to obtain results accurately it is needed to prepare more specimens for each impact level to estimate the fatigue life of material under repeated impulsive load.

Fig.1, shows an example of frequency distribution of fatigue life of specimens in the impact level. The broken probability $P(N_L)$ of specimen at the number of strokes $N_L$ is given as

$$P(N_L) = \int_0^{N_L} Y(N)\, dN \,/\, (n+1) \qquad (1)$$

where n is total number of specimens and N is the number of hammer strokes.

Using Eg.(1) in evaluating the fatigue life of specimens with the basis of the broken probability $P(N_L)$ may obtain the results accurately, however it is needed so many specimens to calculate $P(N_L)$, and in practice may be difficult and imposible to examine them. Therefore, another way is to be used in this study. There are two assumptions of Freudenthal's test results[9] can be used for calculating $P(N_L)$ in Eg.(1) for an experiment limiting the number of specimens. The two assumptions are as follows,

[Assomption I]: The frequency distribution of fatigue life of specimens under repeated impulsive bending load can be treated as a logarithmic normal distribution — when Y(N) represents the number of specimens broken at the number of strokes N, the relation between Y(N) and logarithm of $\overline{N}$ can be treated as a normal distribution.(See Fig.2)

[Assumption II]: For each impact level, the broken probability $P(N_E)$ of specimens at the number of strokes $N_E$ can be obtained on the basis of orderly statistical quantity theory. By this theory, $P(N_E)$ is given by

$$P(N_E) = \frac{r}{n+1} \times 100\ (\%) \qquad (2)$$

where n is total number of specimens and r is the increasing order of the number of fatigue lifes of specimens.

In this paper, the fatigue life $\overline{N}$ that is the representatine value of specimens of the impact level is obtained as follows, the relation of $P(N_E)$-Log $N_E$ is drawn on normal probability chart and then fatigue life $\overline{N}$ is given as the number of strokes at P(N)=50% obtained from the line ($P(N_E)$-Log $N_E$). An outline of the method to obtain $\overline{N}$ is shown in Fig.3.

## 2.3 IMPACT LEVEL

Most of previous studies of the impulsive repeated test, the physical unit of impact level was expressed with the unit of stress[3 ,10] or the unit of energy[11]. When the impulsive strength of materials is expressed with the unit of stress, the strength changes with the shape of the hammer. In the studies of the method expressing the impact level with the unit of energy, the impact level is a kinetic energy of hammer before striking. But this impact level is inadequate as a measure to express the magnitude of impact level to the specimen.

In this paper, the impact level is expressed by using an impulse as a vector so that it is possible to consider the direction of striking and rebounding, the magnitude of load and the impulsive load time. This impact level is represented by S as follows,

$$S = M\,(V_0 - V_1) \qquad (3)$$

where M is the mass of hammer, $V_0$ is the velocity of hammer before striking and $V_1$ is the velocity after rebounding.

Impulse S is a characteristic value obtained from the acceleration curve of the hammer stroke and also as a measure to express the magnitude damage of a specimen caused by one stroke of a hammer. Generally known that the impact level obtained from acceleration curve of the hammer stroke varies according to the kind of material specimen even if a hammer strikes the specimen with constant velocity. In the previous results, although the other researchers have reported about the impact level of all kinds of materials, the effect of free-vibration of the specimen after striking is still neglected. (See Fig.9 a,b)

# 3. TESTING METHOD AND SPECIMENS

## 3.1 THE TESTING MACHINE FOR REPEATED IMPACT BENDING TEST

In this study, a testing machine which utilized the principle of crank movement is used. Fig.4 is its design drawing and Fig.5 shows the mechanism of the hammer strikes the specimen repeatedly.

In Fig.4, ⑨ indicates an inertia wheel that drives around by a motor (200V-400W). ③ indicates the holder of the hammer, which is connected to ⑨ by a connection rod ⑧. When ⑨ rotates, ③ moves horizontally along the guide rail ⑦. In Fig.5, the position of the hammer changes from (STATE I) to (STATE III) with rotating the inertia wheel ⑨. The impact hammer ② that has a protrusion entered to the hole of hammer holder and glides back together on the rail ④ [(STATE Ⅲ)→(STATE Ⅳ)→(STATE I)]. The hammer ② strikes the specimen ⑥ in the state of a free collision and then rebounds from the specimen [(STATE Ⅱ)]. By the repetition of these movements, the specimen is being given the repeated impulsive strokes. The rotational frequency of the inertia wheel can be changed in the range of 25～75 r.p.m. The shape of the impact hammer is square and it has a curved edge at the striking side.

### 3.2 TESTING METHOD AND SPECIMEN

Table.1 shows the static strength and dynamic modulus of elasticity of the mortar specimens. The moulds were removed 24 hours after compaction. The specimens were cured in 20 ± 1 ℃ water for a month, and then they were stored in the curing room until the experiment done, where temperature is controlled at 20 ± 2 ℃ and humidity at about 70%. The age of mortar specimen in this experiment was 3 months after compaction.

The specimens were 6x6x30(cm) prisms. In the testing machine,a specimen was simply supported by a 18 cm of bending span and girdled firmly by rubber bands. The hammer repeatedly struck at the middle of specimen with a given velocity. In this study, 6 kinds of velocity ($V_0$) were selected in the range of 0.4 ～ 1.2 m/sec and 4 kinds of hammer (the mass of 2 kg, 3 kg, 4 kg and 5 kg) were used. There were 24 kinds of impact level in this experiment, and five specimens were prepared for each impact level.

To measure the acceleration of impact hammer, a acceleration transducer ( capacity:1000 g, Frequency response:0～20 kHz ) was attached at the central axis of the hammer. The strain of the specimen was measured on the reverse side stricken of the specimen by wire strain gages.

Fig.6 shows the measuring system. The responses of acceleration and strain were recorded in a data recorder ( 0～20 kHz ) through dynamic strain amplifiers ( 0～50 kHz ). The recorded data was stored once in the micro computer and then to be calculated.

## 4. TEST RESULTS

### 4.1 ACCELERATION CURVE OF THE HAMMER AND STRAIN OF THE SPECIMEN

Fig.7 is an example of acceleration curves and strain curves measured in the experiments. This example was measured in the case that the 2 kg hammer repeatedly strikes the mortar specimen A with velocity 1.2 m/sec.

Property of the change in time of the acceleration of hammer almost coincides with that of the specimen strain so that the acceleration wave of hammer could be employed as a measure to express the change in time of impulsive loading of the specimen.

The acceleration curves of hammer measured when 3 kg hammer repeateadly struck the mortar specimen A at various velocities are shown in Fig.8. Since the impact velocity $V_0$ of the hammer was so slower than the velocity when the specimen vibrated freely each acceleration curves were complex shape as shown in Fig.8. When the impact velocity $V_0$ was speeded up, the peak of acceleration curve became higher therewith the time of impulsive load also became shorter. Therefore, it has become clear that it is necessary to carry out experiment shown in pragraph 2.1. As described before, it was impossible to consider the effect of free-vibration of the specimen after striking by the impact level suggested in this paper. There are many methods have considered the effect of free-vibration, such as the theories by miner and Newmark.

Fig.9 shows the measured strain wave of cement mortar specimens A which were given various impact levels. M was the mass of hammer and $V_0$ represents the impact velocity of hammer. In Fig.9 could be seen that the specimen was still vibrating freely for a while after being struck. However, the magnitude of strain wave — it so-called as "first strain"— during the hammer touched the specimen was bigger than that of free-vibration. When mass M was lighter and velocity $V_0$ was slower, then the rate of the magnitude of free-vibration to first strain became bigger.

When the specimen was given high magnitude of impulsive load, such as the case at the mass M heavier or the velocity V faster, the magnitude of free-vibration strain was so smaller than that of first strain in which its effect to the fatigue life of specimen could be neglected. This was caused by the microscopic cracking of specimen during being striked when the impact level was higher. When the magnitude of impact level was lower, such as the case of the lighter mass M or the slower velocity $V_0$ , the specimen was not all cracked by one stroke and conseguently the longer time of free-vibration would take place. Since the free-vibration strains of specimen were in the limit of elastic strain which were so smaller than first strain, in this case, the specimen did not reach a failure point by the repeated strokes.

In both the cases of above-described, the strain waves of free-vibration differed so much from the first wave depend on their magnitudes and periods.

Therefore in this study, the effect of free-vibration was disregarded in calculating the impact level because of its strain is not regarded as the impulsive strain.

### 4.2 FATIGUE LIFE

Table.2 shows the fatigue life of mortar specimen obtained from the experiment with using 3 kg hammer. The experimental values of fatigue life $N_E$ were very scattered. In Fig.10, the relation between $P(N_E)$ by Eg. (2) and Log $N_E$ was drawn on normal probability chart, as the values shown in Table 2. It became clear that the relation of $P(N_E)$-Log $N_E$ was expressed as a linear equation in each impact velocity $V_0$. Therefore, the two assumptions made up in Paragraph 2.2 has been right as a method to obtain the fatigue life $\overline{N}$ from the experiment with limiting the number of specimens.

The fatigue life N, as shown in Table 2, was the representatible value of specimen for each impact level, where the fatigue life N was given as the number of strokes at P(N )=50 % in Fig 10. It could be obtained obviously that the fatigue life N greatly increasing with the decreasing of velocity V .

### 4.3 IMPACT LEVEL

In this paper, the impact level S was calculated by Eg.(3) in the repeated impulsive bending tests. It has been reported that the strain wave changed with the number of strokes of the specimen under repeated impulsive bending load. As described above, the impact level S changed with the specimen. This paragraph will be considered whether the impact level S changes with the number of strokes or not.

The relation between impact level S and the number of strokes was shown in Fig.11, which was the test results of the 2 kg hammer repeatedly striked the mortar specimen A at various velocities. The values of impact levels were almost unchanged in each impact velocity without reference to the number of strokes N. But the values at a stroke number N=1 were scattered and larger than the constant value as described above due to the striked part of the specimen was not stable. And the impact level S was settled in the range of N>3, so the effect of impact level at N=1 to the fatigue life of specimen was small so that it was neglected.

From these results, the impact level S was obtained from the acceleration curve measured in the range of N>3 with exception that the specimen was broken before N=3.

Fig 12 shows the impact levels of mortar specimen A, B and C in Table 1. The static strength and the modulus of elasticity became bigger in the order C, A, B. As shown in Fig 12, when the strength of specimen was lower, then the magnitude of impact level became larger. The magnitudes of damage of the specimens by one stroke were smaller in their order of magnitudes of strength. So the impact level S has been obtained for each specimen.

Fig.13 (a), (b) and (c) show the relation between the impact level S and the impulsive velocity $V_0$ for specimen A, B and C.

In every case, impact level S is in direct proportion to the velocity $V_0$ and their relation could be approximated by

$$S = \alpha V_0 \qquad (4)$$

The value of coefficient $\alpha$ increased with the mass of hammer and became smaller in the order of the magnitude of strength of the specimen. Coefficient $\alpha$ of each specimen calculated with the method of least squares is shown in Table 3.

In the next paragraph, the value of impact level S is obtained from Eg.(4) by using $\alpha$ in Table 3.

## 4.4 RELATION OF S - $\overline{N}$ BY [EXPERIMENT I]

In this paragraph, it was considered that the results of [Experiment I] of which experimental method was described in paragraph 2.1. Fig.14 shows the relation between the impact level S and fatigue life $\overline{N}$ obtained from the case of the mortar specimen A and B, in Table 1, were repeatedly striked by 5 kg or 2 kg hammer.

The relation of S - $\overline{N}$ for the result of each specimen was approximated by a line which had a bent point as in Fig.14. Comparing the test result of specimen A and B, the fatigue life $\overline{N}$ of A was longer than one of B at the same impact level S. This was caused by the impact level of specimen B was lower than that of specimen A when they were repeatedly striked by a hammer at a same velocity.

If the momentum $MV_0$ of the hammer before striking was used as the impact level, the fatigue life $\overline{N}$ of specimen B was longer than $\overline{N}$ of specimen A at the same impact level. But from the point of view for examining the strength of the material under impulsive load, it was proper to use the impulse ($MV_0 - MV_1$) as the impact level. Because the magnitude damage of the specimen by impulsive load was expressed by impulse as described above.

At the same time, as explanation from Fig.14, the fatigue life $\overline{N}$ changed longer in the order magnitude of hammer's mass. This result as follows : when the impact velocity $V_0$ was faster, the fatigue life $\overline{N}$ of specimen became shorter even if in the same impact level.

## 4.5 RELATION OF S - $\overline{N}$ BY [EXPERIMENT II]

In the [Experiment II], the impact level is defined by changing the hammer's mass at a fixed velocity.

Fig.15 shows an example of the relation between impact level S and fatigue life N of specimen A obtained from the [Experiment II]. In this example, the impact velocity has been fixed in three steps, 0.6 m/sec, 0.8 m/sec and 1.0 m/sec. The relation of S - $\overline{N}$ for each impact velocity $V_0$ showed their correlation respectively and differed with the result of [Experiment I] such as the relation in Fig.14. By these results were obtained that the fatigue life $\overline{N}$ changed under the influence of impact velocity $V_0$, thereby it was evident that the fatigue strength of specimen under repeated impulsive load also depended on the impact velocity V .

Fig 16 is a typical result of the resistant power of mortar specimen with using impulsive load, which has been obtained from the single impact test. The range in which the impact level was slow, the impulse-expressed resistant power greatly changed with the impact velocity. Generally, the static strength of materials changes with the velocity load applied in static test, so that it is reasonable that the fatigue life $\overline{N}$ changes when a change in impact velocity is done as shown in Fig 15.

## 4.6 RELATION OF S - $\overline{N}$ - $V_0$

Rearranging both of the results of [Experiment I] and [Experiment II], they will became possible to estimate the fatigue life $\overline{N}$ of specimen A under repeated impulsive bending load.

Fig 17 shows the fatigue life $\overline{N}$ of specimen A obtained from both of the results of [Experiment I] and [Experiment II], and so-called as "S - $\overline{N}$ - $V_0$ curves". In Fig.17, the fatigue life $\overline{N}$ changed with the mass of hammer and impact velocity $V_0$. It seems that these results had a contradiction because of some fatigue lifes was found in the same impact level. However, for that matter these results had no contradiction because the impulsive phenomenon had some factors that could be expressed with impulse or energy, such as a propagation of stress wave. And it was assumed enough to show the test results of impulse for estimating the fatigue lifes of specimens. From the S - $\overline{N}$ - $V_0$ curves could be obtained the fatigue life $\overline{N}$ of specimen for a specific mass and specific velocity of the hammer. In connection with these curves, the relation between S and $\overline{N}$ of the [Experiment I], shown with dotted line in Fig 17, the areas ranging before or behind of the bent point were calculated as Eg.(5)

$$S = c - d \text{ Log } N \qquad (5)$$

Fig.18 shows the relation of coefficient c, d in Eg.(5) and the mass of hammer M. In Fig.18, the correlation was observed between coefficient c, d and the mass M that the relation of S - $\overline{N}$ could be obtained with altering Eg.(5) as follows :

$$\text{(before bent point)} \quad S = (2.45 + 0.76\ M)-(1.13 + 0.22\ M) \text{ Log } \overline{N}$$

$$\text{(behind bent point)} \quad S = (1.73 + 0.67\ M)-(0.15 + 0.11\ M) \text{ LOg } \overline{N} \qquad (6)$$

Furthermore, the relations of S - $\overline{N}$ obtained from the results of the [Experiment II], shown with full lines in Fig.17, were expressed as

$$S = e\ \overline{N}^{f} \qquad (7)$$

Fig.19 shows the relation between coefficient e, f and impact velocity $V_0$ of mortar specimen A. The coefficient e showed the constant value ( -2.3) without reference to the velocity $V_0$ and the coefficient f which was expressed with an experimental equation,

$$f = -0.02-0.0016\ V_0^2 \qquad (8)$$

As concerns mattar specimen A in relation of S - $\overline{N}$ of Eg.(8) obtained from [Experiment II] is expressed as the following equation,

$$S = -2.3\ \overline{N}^{(-0.02-0.0016V_0^2)} \qquad (9)$$

Eg.(6) of [Experement I] in which M was a variable, but in Eg.(9) of [Experiment II],V was a variable. When the mortar specimen A was repeatedly stricken by a specific mass M of hammer at a specific velocity $V_0$ , the fatigue life $\overline{N}$ was obtained by, solving the simultaneous equations of Eg.(6) and Eg.(9) after substituting the $M_x$ and $V_{0x}$ .

## 5. CONCLUSION

In this paper, a method to estimate the fatigue life of brittle materials had been shown on the basis of the test results of cement-mortar specimen under repeated impulsive bending load. In this test, the impact level is expressed by using an impulse which is a measure to express the magnitude of the damage of a specimen by one stroke of a hammer in repeated strokes.

The main results which have been obtained in this repeated impulsive bending tests are as follows,

(1) Fatigue life $\overline{N}$ can satisfactorily be obtained from the experiment by limiting the number of specimens using two assumption in paragraph 2.2.

(2) The impact level can properly be expressed with impulse obtained from the measured acceleration wave of hammer because the change in time of acceleration almost coincides with that of the strain of specimen.

(3) From S - $\overline{N}$ - $V_0$ curves, can be obtained the fatigue life of specimen under repeated impact bending load for a specific mass and specific velocity of the hammer.

The purpose of this study is to establish the test method for estimating the fatigue life of brittle materials under repeated impulsive bending load. As described in this paper, when the impact level is expressed with the magnitude of impulse, then the fatigue life can be obtained by using a specific mass and specific velocity. We recognize that this method is also available in the case of impact compressive repeat test. We will study further concerning of this method.

ACKNOWLEDGEMENT

The authors would like to acknowledge the continuing guidence and encouragement of Dr. T. SAJI. The authors also thank Mr. KATSUKI and Mr. KANDA for their help in this experiment.

REFERENCES

[1] Kawamoto, 1968, Journal of Mechanical Engineering, Vol.34,257
[2] Hartbower, 1955, ASTM STP 176
[3] Wornock, 1949, Proc. Inst. Mech. Engr.
[4] Cotterell, 1962, Brit. Weld. J. no.83
[5] Matsufuji, 1985, Journal of Structural and Construction Engineering, no.358
[6] Wright, 1952, Magazine of Concrete Research
[7] Durtron, 1930, Premier Congress de La route
[8] Nakahara, 1984, Journal of the J. S. M. E. Vol.73. no.622
[9] Freudenthal, 1951, Symposium on Statistical Aspects, ASTM STP 121
[10] Tanaka, 1963, Journal of Basic Engineering
[11] Saji, 1984, Proc. of the 27 th Japan Congress on Materials Research

Table 1. Mechanical properties of cement-mortar specimen

| Specimen | compressive strength (Pa)x$10^5$ | Flexural strength (Pa)x$10^5$ | Static modulus of elasticity (Pa)x$10^8$ | Dynamic modulus of elasticity (Pa)x$10^8$ |
|---|---|---|---|---|
| A | 520 | 102 | 248 | 279 |
| B | 807 | 125 | 377 | 411 |
| C | 440 | 87 | 208 | 239 |

Table 2. Fatigue life of mortar specimen A

| | Impact velocity $V_0$ (m/s) | Number of hammer strokes until specimen was broken $N_E$ | Fatigue life LogN̄ | N̄ |
|---|---|---|---|---|
| Specimen A | 1.2 | 1 1 1 2 2 | 0.120 | 1.3 |
| | 1.0 | 2 2 2 2 3 | 0.336 | 2.2 |
| | 0.8 | 4 6 10 11 13 | 0.908 | 8.1 |
| | 0.7 | 12 20 43 58 1261 | 1.775 | 59.6 |
| | 0.6 | 31 338 1019 2367 3000 | 2.775 | 595.7 |
| | 0.4 | 9576 13163 20056 33686 39421 | 4.306 | 20230.0 |

Table 3. The Coefficient α

(*S=αx$V_0$)

| Specimen | Hammer's mass (kg) | Coefficient α |
|---|---|---|
| A | 2 | 2.88 |
| | 3 | 3.97 |
| | 4 | 5.57 |
| | 5 | 7.53 |
| B | 2 | 2.81 |
| | 3 | 3.47 |
| | 4 | 4.79 |
| | 5 | 5.72 |
| C | 2 | 2.91 |
| | 3 | 4.31 |
| | 4 | 5.82 |
| | 5 | 7.70 |

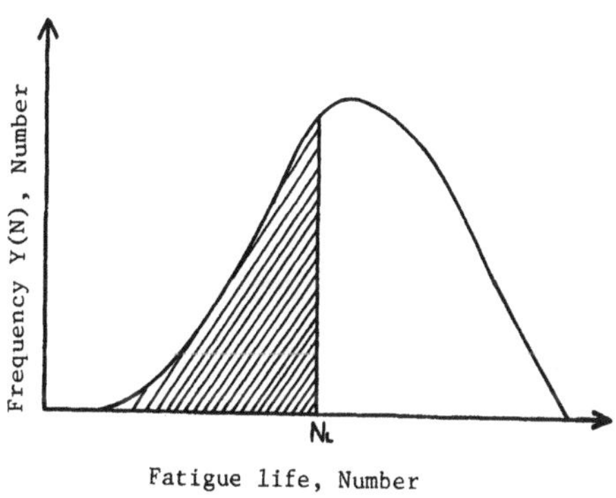


Fig.1 Frequency distribution of fatigue li-

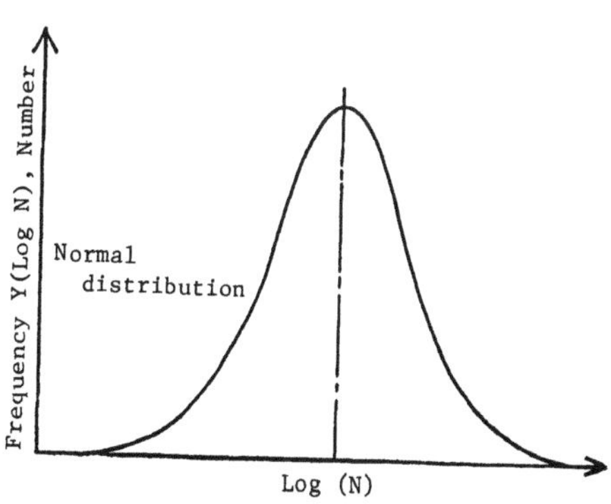


Fig.2 Frequency distribution of fatigue life, the relation of Y(N)-Log N.

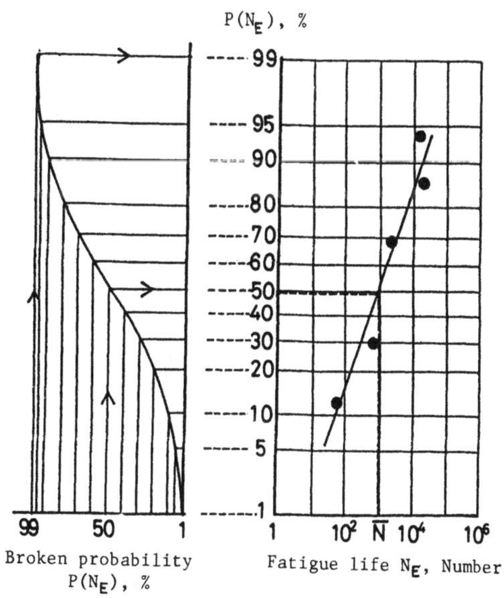


Fig.3 The relation between broken probability P($N_E$) and fatigue life $N_E$.

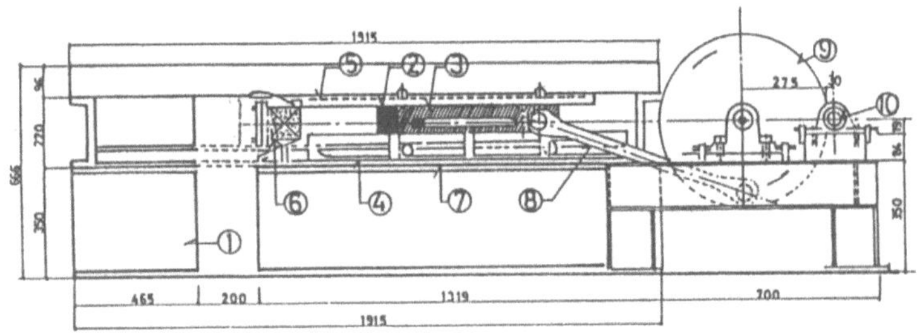


| | | | | |
|---|---|---|---|---|
| ① projection weight | ② impact hammer | ③ hammer holder | ④ down side rail | ⑤ up side rail |
| ⑥ specimen | ⑦ guide rail of hammer holder | ⑧ connection rod | ⑨ inertia wheel | ⑩ rolling adjuster |

Fig.4 Side view of testing machine

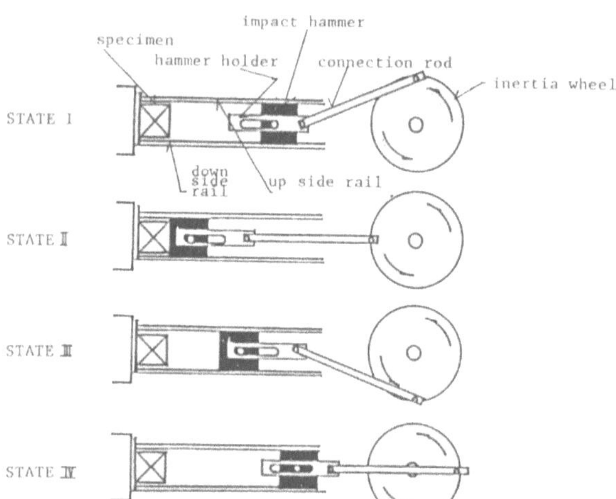


Fig.5 The mechanism of the hammer strikes the specimen repeatedly.

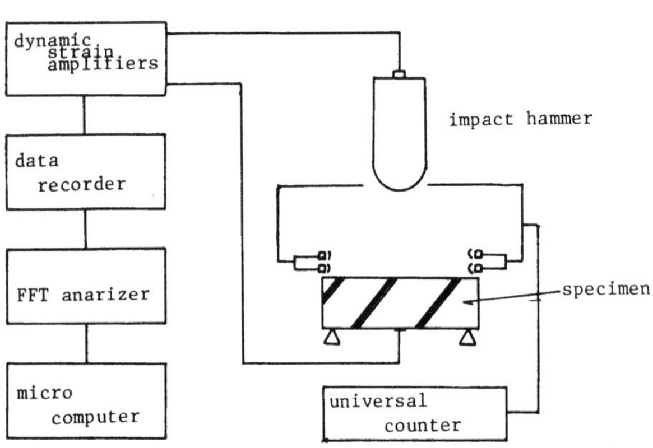


Fig.6 Measuring system

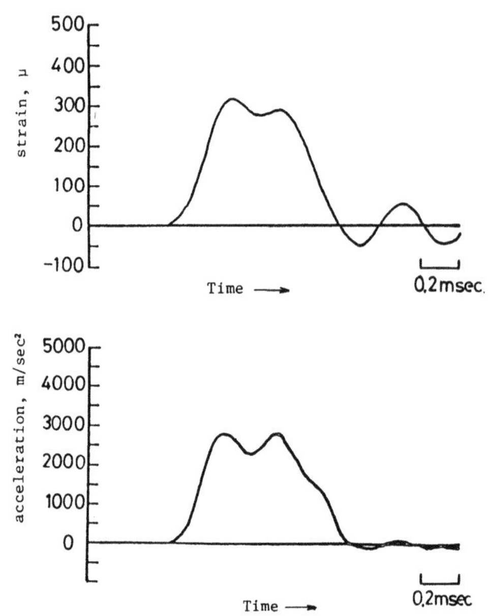


Fig.7 An example of acceleration curves and strain curves

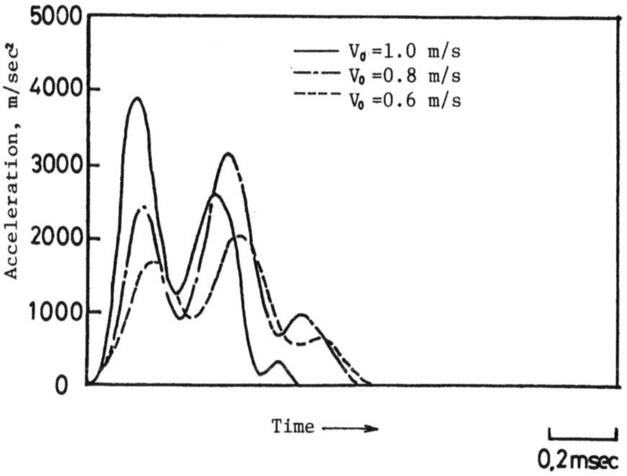


Fig.8 The acceleration curves of hammer measured when 3 kg hammer struck the mortar specimen A.

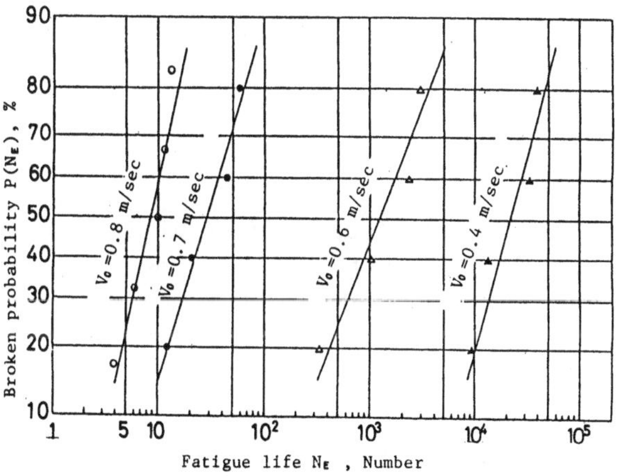


Fig.10 The relation between P($N_E$) by Eg.(2) and $N_E$ obtained by experiments

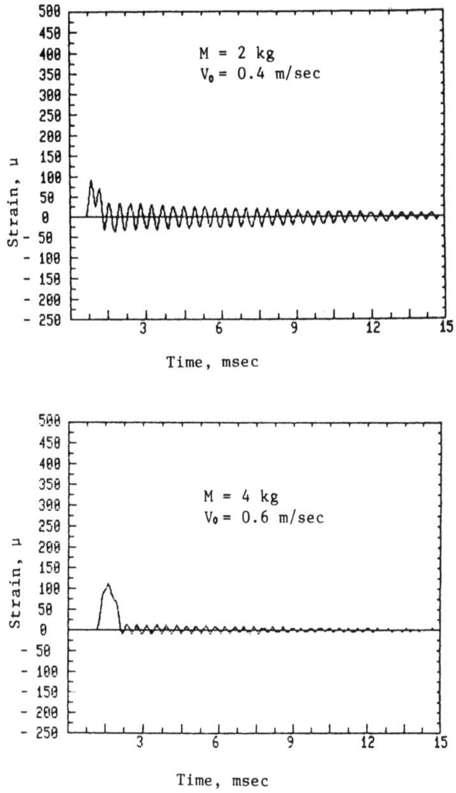


Fig.9 Measured strain waves of cement-mortar specimen A

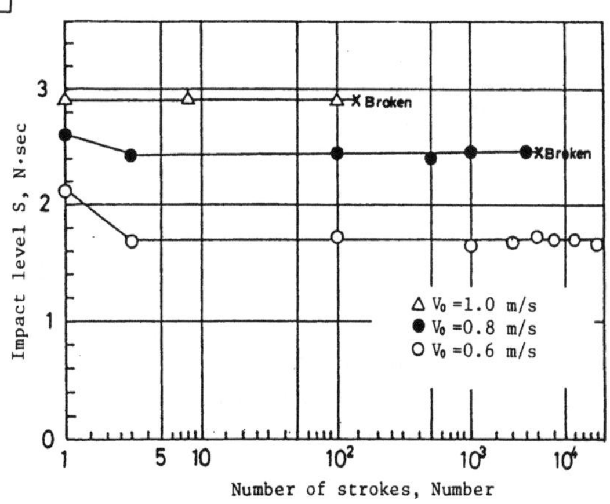


Fig.11 The relation between impact level S and the number of strokes.

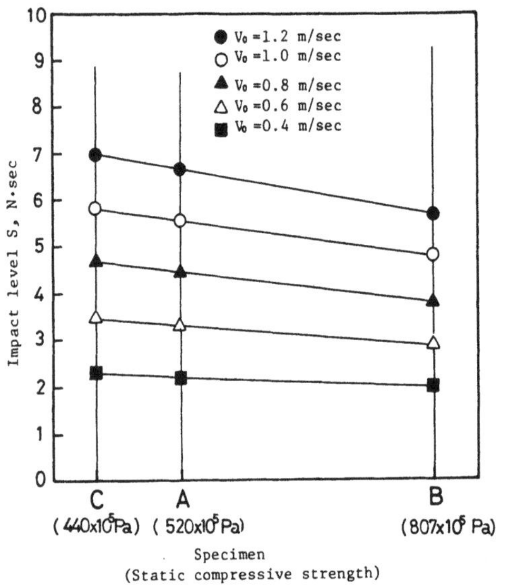


Fig.12 Impact levels of mortar specimen A, B and C

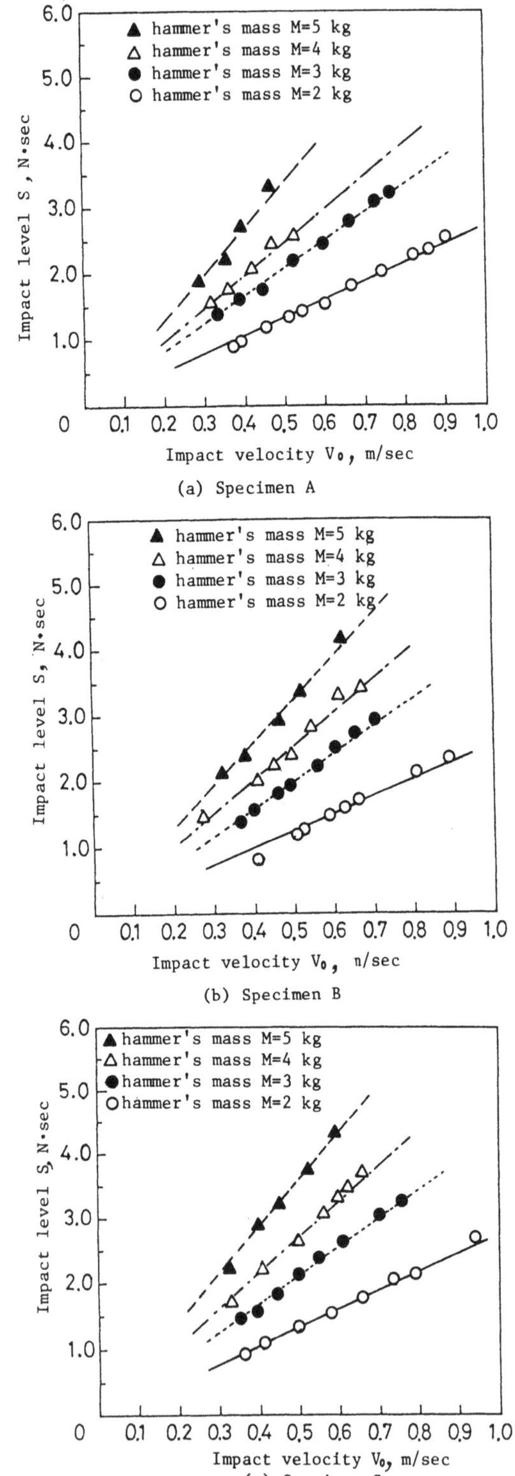


Fig.13 The relations between impact level S and impact velocity $V_0$ for specimen A, B and C

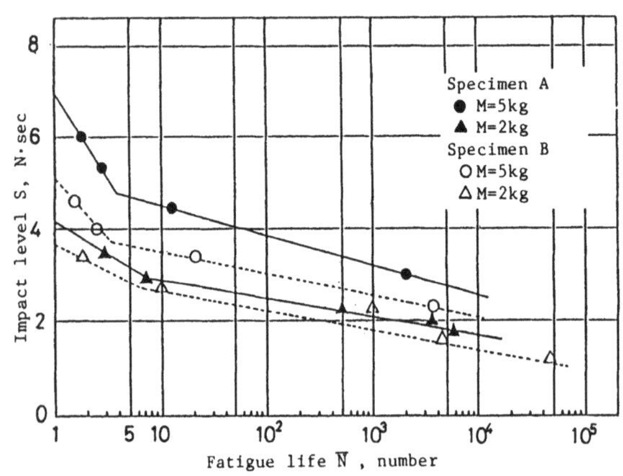


Fig.14 Relation of S-$\overline{N}$ by [Experiment I]

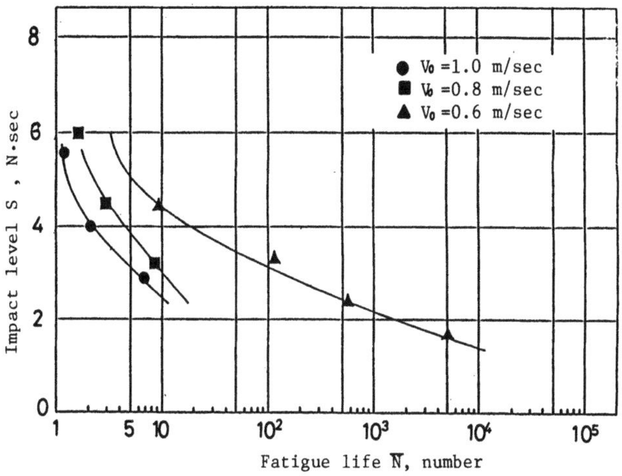


Fig.15 Relation of S-$\overline{N}$ by [Experiment II]

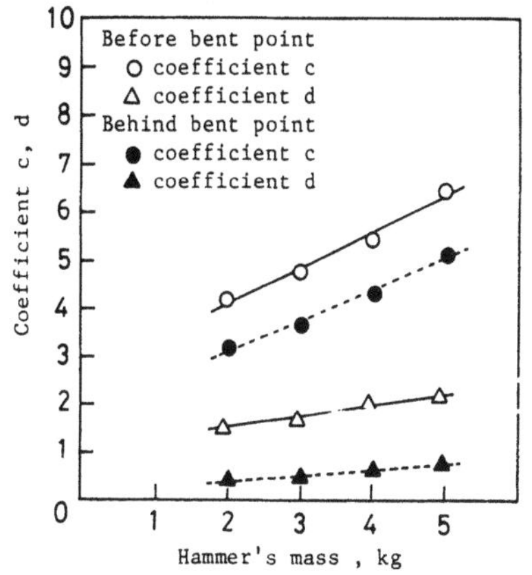


Fig.18 Relation of coefficient c, d in Eg.(5) and the mass of hammer

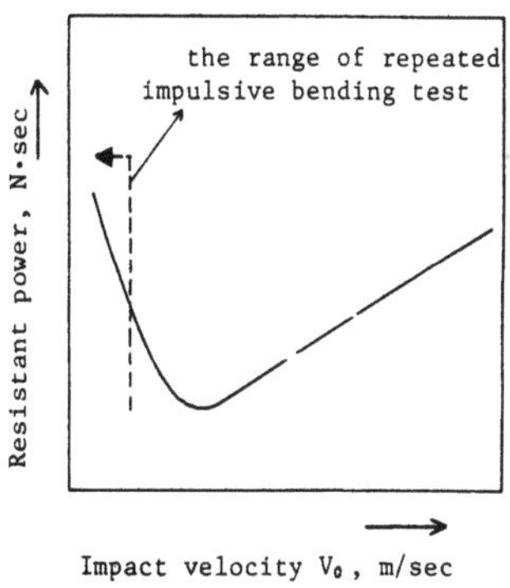


Fig.16 Typical result of the resistant power of mortar specimen under single impact bending load.

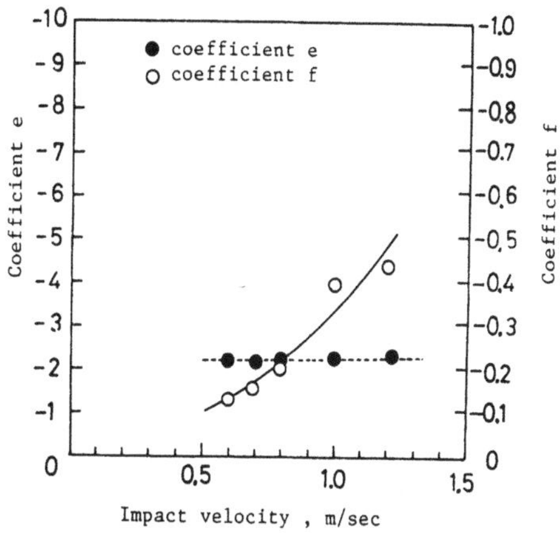


Fig.19 Relation between coefficient e, f in Eg.(7) and impact velocity

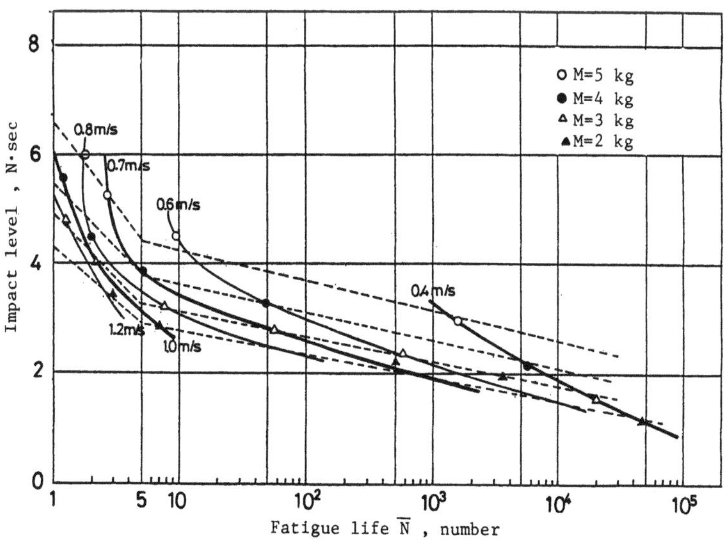


Fig.17 S - $\overline{N}$ - $V_0$ curves

SEM/RILEM International Conference
on Fracture of Concrete and Rock
Houston, Texas, June 1987
Editors: S.P. Shah, S.E. Swartz

# FATIGUE CRACK PROPAGATION IN PLAIN CONCRETE

M. H. Baluch, A. B. Qureshy and A. K. Azad

## ABSTRACT

The work in this investigation is aimed at establishing whether the empirical Paris' Law for crack propagation in metals and rocks given by $da/dN = C(\Delta K)^m$ is valid for use in crack propagation in plain concrete. Compliance and fatigue testing was carried out on single edged notched beam (SENB) specimens in 3 point bending (Mode I), and results indicate that Paris' Law may be applicable for crack growth in plain concrete, with m being a constant and C dependent on the stress cycle ratio R.

## INTRODUCTION

Since fatigue phenomenon was first observed in structures, much of the research has been limited to the investigation of its influence on metals. The study of fatigue crack propagation in metals has been done on an exhaustive level and the laws governing the said propagation have been well established [1,2]. Better analysis methods and more accurate information about the material properties of concrete have made possible the construction of increasingly slender concrete structures. These structures are prone to damage caused by fatigue. The present investigation is aimed at understanding the phenomenon of fatigue crack growth in plain concrete.

The simplest model used to describe crack growth in metals is given by :

$$da/dN = C(\Delta K)^m \tag{1}$$

where $\Delta K$ = differential stress intensity factor at the tip of a crack and C,m are experimentally determined material constants. Equation (1) is normally referred to as Paris' Law [3], and although somewhat modified on occasions, has gained general acceptance for use in metals. In addition, Kim and Mubeen [4] have reported its validity for Westerly granite, although C was found to depend on the stress cycle ratio R. The objective of this paper is to see if Paris' Law is applicable for crack propagation in plain concrete.

The work in this paper uses principles of linear elastic fracture mechanics to describe the stress intensity for any crack size. The existence of a nonlinear process or damage zone preceding the main crack is well known [5,6], but this present work does not attempt to incorporate this effect. However, a recent paper by Perdikaris et al. [7] indicates that compliance behavior of beam specimens may be predicted accurately from linear elastic theory for crack lengths upto 60% of the beam's depth.

## MATERIALS

The beam specimens were made using Type I Portland cement and locally available coarse and fine aggregates. The curing time in water was 28 days. Relative quantities of the materials used, by weight, were as follows: coarse aggregate 45%; fine aggregate 27%; cement 18.5% and water 9.5%. Gradation of coarse aggregate used was :

| Sieve Size (mm) | % Passing |
|---|---|
| 12.7 | 100 |
| 9.5 | 60 |
| 4.75 | 40 |
| 2.36 | 20 |
| 1.18 | 5 |

M. H. Baluch and A. K. Azad are Associate Professors in the Department of Civil Engineering, University of Petroleum and Minerals, UPM Box 57, Dhahran 31261, S.Arabia; A.B.Qureshy is presently a graduate student in the Department of Civil Engineering at Texas A&M University, College Station, Texas.

Gradation of fine aggregate used was :

| Sieve Size (mm) | % Passing |
|---|---|
| 2.36 | 100 |
| 1.18 | 63 |
| 600μm | 45 |
| 300μm | 20 |

Compressive strength as obtained from testing of 76.2 mm $\times$ 152.4 mm cylinders was 27.6 MPa.

SPECIMEN TYPE

The first step in the experimental program was the determination of specimen type, shape and size. There were a number of constraints regulating this selection, including (i) the crack growth had to be stable through most part of the duration of the experiment (ii) the stiffness of the specimen had to be effectively sensitive to small crack increment (iii) the method of measuring change in stiffness due to crack propagation had to be simple, effective and reliable and (iv) the specimen had to conform with the available testing equipment. A simply supported three point beam specimen was found to be the best choice under the constraints mentioned. The final dimensions were arrived at after a few trial runs.

The criterion that had to be met under all conditions was stable crack growth. It was observed that under a given loading condition, an increase in specimen depth and width and a decrease in supported span andinitial notch depth contributed towards the instability of the crack growth, once it had initiated. The objective, therefore, was to minimize the first two dimensions and maximize the latter two as far as the requisites of a stable crack growth were concerned.

A minimum value of the width of the specimen was dictated by the aggregate size used as well as by enough strength and structural stability required to withstand handling during, and in between casting, curing, notch-cutting, and testing. The same factors applied to the case of the gross depth of the beam. An additional constraint against a minimum depth was the available length for crack growth. For a given notch depth, a decrease in gross depth meant less length for crack growth and consequently fewer data points.

A maximum span size is, obviously, restricted by the handling difficulties and limitations of the testing equipment. For the notch depth, a state of dichotomy existed inasmuch as fracture is unstable for small notch depths, whereas stable large notch depths reduce the number of possible data points. A compromise was felt to be achieved by using a notch to beam depth ratio of 0.5.

The eventual beam specimen dimensions for compliance calibration and fatigue testing were chosen to be 51 mm wide * 152 mm deep * 1360 mm supported span length and a 3 mm notch width. The total specimen length was 1524 mm. This provided an 82 mm overhang over each support. A notch was provided at mid-span and was cut using an electric masonry saw. The notch depth was varied. Specimens are referred to as single edged notched beam (SENB) specimens.

TESTING EQUIPMENT

Testing was conducted on Instron 1196, an electrically driven model. The machine has a maximum capacity of 500 kN. The crack mouth opening displacement (CMOD) measurements were made through the machine's built-in strain data unit via the Instron built extensometer. The load measurements were recorded via the built-in load cell. The plot of load versus CMOD was obtained from the built-in x-y recorder.

Instron G-51-14-M-A extensometer was used to obtain crack mouth opening displacement (CMOD) measurements. An assembly of brackets and connecting arms was designed to attach the extensometer to the specimen. The brackets were glued to the specimen surface, on either side of the notch, using a five minute epoxy adhesive. Details of assembly are provided in [8].

TESTING SCHEDULE

The testing schedule is best described by Figs. 1 and 2, and can be divided into two parts. The first part comprises of compliance calibration as illustrated in Fig. 1. The objective of the calibration test was to get a relation between crack length and compliance of the specimen. This facilitates in the monitoring of the propagated crack length during the fatigue testing phase of the experiment. The specimen was loaded at the rate of 0.5 mm/min. The load and the CMOD were plotted on the x-y recorder of the Instron 1196. The load was increased until a maximum value was reached (further downward movement of the crosshead did not result in an increase in load). At this point the crack length 'a' was measured using a hand held scope. The specimen was unloaded and then reloaded for a new maximum load and a corresponding crack length measurement. This procedure was repeated,yielding a number of compliance curves (load vs. CMOD) and corresponding measured crack lengths. These two were then plotted against each other to give a crack length versus compliance curve. This curve (data) was fitted on to a regression equation (third order in this case). This crack length versus compliance curve is referred to as the compliance calibration curve.

The second part of the testing schedule was the fatigue testing; Fig. 2. Before starting, a cycle ratio of $R = \sigma_{min}/\sigma_{max}$ was chosen. A compliance (i.e. load vs. CMOD) measurement was taken as was done for the compliance calibration part of the work. Fatigue testing was then started at the selected cycle ratio and continued for a predetermined number of cycles. At this point, testing was interrupted for another compliance measurement. The process was repeated until failure, yielding a number of compliance values at different numbers of cycles N.

These values of compliance, when compared to the compliance calibration curve from the first part of the testing, yield corresponding crack lengths incurred at the respective number of cycles. These two parameters were then plotted to yield a crack length versus number of cycles curve (a vs. N).

From this latter plot, the slope of the curve (da/dN) was calculated at discrete intervals. The corresponding values of crack length 'a' were used to calculate differential stress intensity factor ($\Delta K$) values. The data for checking the validity of Paris' Law was now available in the form of these two final parameters - da/dN was plotted against $\Delta K$ on a log-log scale.

The fatigue testing testing procedure was then repeated for a different cycle ratio, a different notch depth, a different load range or a combination of these three parameters for a clearer picture of the behaviour of fatigue crack propagation in plain concrete.

The results of these two types of tests (compliance calibration and fatigue testing) are presented in the next section.

## COMPLIANCE CALIBRATION TESTS [9]

An initial notch-depth of 75 mm was cut in the specimen and it was loaded in 3-point bending at the rate of 0.5 mm/min. The load and the CMOD (crack mouth opening displacement) were plotted on the x-y recorder of the Instron 1196. The load was increased until a maximum value was reached. At this point the crack length 'a' was measured using a hand held scope on both sides of the beam, and an average value used. The specimen was unloaded and then reloaded for a new maximum load and a corresponding crack length measurement. This procedure was repeated yielding a number of compliance curves (load vs. CMOD) and corresponding measured crack lengths. Compliance and corresponding crack length were then plotted against each other to give a crack length versus compliance curve. This data was fitted on to a third degree polynomial regression equation, resulting in the compliance calibration curve. For the present work two compliance calibration tests were run. The results from both these tests were very close. The initial notch depth was 75 mm in each case. The load versus CMOD curve for one specimen is shown in Fig. 3. The compliances were measured as CMOD/load. Plotting compliances against the corresponding crack lengths, gave the compliance calibration curves for the specimen. Fig. 4 shows a typical compliance calibration curve.

A third order regression equation was obtained next. Denoting compliance C and crack length by a, with units of a and C in millimeters and meters/Newton, respectively, the following functional representation was obtained for the compliance :

$$C = 2.3978(10^{-6})Z^3 + 1.7843(10^{-4})Z^2 + 8.5271(10^{-3})Z + 0.3049 \qquad (2)$$

where $Z = a - 113.44$ (3)

## FATIGUE TESTING

Fatigue testing was done using three different stress cycle ratios, R = 0.1, 0.2 and 0.3. It was found that a maximum cyclic load of less than about one fourth of the static load capacity was not enough to cause crack growth. On the other hand, maximum cyclic load values exceeding half the static load capacity led to very fast crack growth. For a specimen with an initial notch depth of 75 mm, the static load capacity was about 900 Newtons. For all fatigue tests, the cycling rate was about 0.5 Hz. For the specimen used, the non-hydraulic Instron 1196 could not provide a higher frequency. The cycling was load controlled for all tests. A total of fourteen fatigue tests were run. These were divided into three categories and designated as Type (A), Type (B) and Type (C) fatigue tests. All tests were conducted utilizing 3-point bending single edge notched beam (SENB) specimens, which were of identical size and concrete mix as these used in the compliance calibration tests.

## TYPE (A) FATIGUE TESTS

A total of six tests were conducted; two per cycle ratios R = 0.1, 0.2 and 0.3 each. Initial notch depth was 75 mm and the maximum load was 300 Newtons for each test. This load level at 75 mm notch depth was equivalent to an initial maximum stress intensity of $3.49 * 10^5$ $N/M^{3/2}$. Compliance records were taken after every 10,000 cycles. These readings were used to yield corresponding crack lengths from compliance calibration equation (2). The a-N plots (number of cycles versus crack length) for cycle ratio R = 0.1, 0.2 and 0.3 were obtained and a typical plot is shown in Fig. 5 for R = 0.1. The a-N plots were used to calculate the rate of crack growth (da/dN), i.e. the slope of a versus N curves for the three cycle ratios at different crack lengths. The values of crack lengths were used in the following equation to yield corresponding differential stress intensity factor ΔK:

$$\Delta K = (\sigma_{max} - \sigma_{min})\, Y a^{\frac{1}{2}} \tag{4}$$

where Y is the calibration factor for a single edged notched beam and given by [10]

$$Y = 1.93 - 3.07(a/d) + 14.53(a/d)^2 - 25.11(a/d)^3 + 25.8(a/d)^4 \tag{5}$$

where a is the crack length and d is the total depth of the beam. In equation (4), $\sigma_{max}$ and $\sigma_{min}$ are the maximum and minimum gross stresses corresponding to the maximum and minimum loads, respectively. These stresses are given by

$$\sigma_{max} = 3P_{max}\ell/2bd^2 \tag{6}$$

and

$$\sigma_{min} = 3P_{min}\ell/2bd^2 \tag{7}$$

where $P_{max}$ and $P_{min}$ are the maximum and minimum loads, respectively, in a cycle and b is the width of the beam specimen.

These values of ΔK were plotted against respective da/dN values on a log-log scale. Figs. 6 and 7 show the typical ΔK-da/dN plots for two tests corresponding to R = 0.3. From these plots it can be seen that the da/dN vs. ΔK data does fall on a straight line as predicted by equation (1). The last few data points show a gradual change in slope.

The values of ΔK in Type (A) tests did not exceed $7.0 * 10^5$ $N\text{-}m^{3/2}$. In order to obtain da/dN vs. ΔK data points for higher ΔK values, the maximum load was increased. These tests are presented next as Type (B) tests.

TYPE (B) FATIGUE TESTS

A total of four tests were run in this category, two each per cycle ratios R = 0.2 and 0.3. The tests could not be run for R = 0.1 because of equipment limitations encountered in this range. At this cycle ratio, Instron 1196 was unable to consistently maintain the minimum load level. This led to an occasional complete seperation of the crosshead from the specimen, introducing an impact in the following cycle. This eventually led to a premature failure.

The notch depth was again chosen to be 75 mm as in Type (A) tests. The maximum load was increased to 400 Newtons from 300 Newtons in Type (A) tests. This load level, with 75 mm notch depth, was equivalent to an initial maximum stress intensity of $4.65 * 10^5$ $N/m^{3/2}$. As specimen life to failure was expected to be less than in Type (A) tests due to higher load range, compliance records were taken after every 2,000 cycles. The a-N plots for the two cycle ratios R = 0.2 and 0.3 were obtained, with two plots per cycle ratio.

From the slopes of these curves and using equation (4), the da/dN vs. ΔK plots were subsequently obtained. A slight S-shaped behaviour is apparent when the data from the two types of tests, (A) and (B), is superimposed. Such a shape is also documented for metals [11]: however, the slope is constant in the working range life of most cracked structures. This is shown in Fig. 8. Both Type (A) and Type (B) tests were conducted using an initial notch depth of 75 mm. In order to investigate any effects of a change in initial notch depth, Type (C) tests were run. These are presented next.

TYPE (C) FATIGUE TESTS

A total of four tests were run in this category, with two tests per cycle ratios R = 0.2 and 0.3. Tests for R = 0.1 could not be run, again, due to equipment limitations. Impact, as observed in Type (B) tests, was again the cause of premature failure of the specimens at R = 0.1. Initial notch depth was increased to 85 mm from 75 mm as used in Type (A) and Type (B) tests. The load was decreased to 250 Newtons from 300 Newtons for Type (A) tests. This load level at 85 mm notch depth was equivalent to an initial maximum stress intensity of $3.62 * 10^5$ $N/m^{3/2}$. The compliance records were taken after every 10,000 cycles. The a-N plots for the two cycle ratios R = 0.2 and 0.3 were obtained and corresponding da/dN vs. ΔK plotted (Fig. 9). The slope of the straight line in this plot is taken from Type (A) tests for comparative purposes.

DISCUSSION

The results of the tests suggest that the rate of fatigue crack propagation in plain concrete is proportional to the differential stress intensity factor, ΔK. The relationship can be successfully represented by equation (1), i.e. Paris' Law. Fig. 10 shows the straight lines, correlating da/dN to ΔK, for all three cycle ratios used. These lines are found to be almost parallel. The calculated values of the constant m in equation (1) for the cycle ratios R=0.1,0.2,0.3 are 3.12, 3.12 and 3.15, respectively. The values of the constant 'C' are $7.71 * 10^{-25}$, $5.78 * 10^{-24}$ and $1.72 * 10^{-24}$, respectively. It may, therefore, be concluded that the m is independent of R while C may be a function of R. A regression model for C versus R could be developed, but the number of data points is inadequate.

A possible explanation for the shift of da/dN - ΔK plots to the left for increasing R may be hypothesized. For the same ΔK, since the differential stress ($\sigma_{max} - \sigma_{min}$) is greater for R = 0.1 than that for R = 0.2, the crack length for the latter (R = 0.2) would have to be greater than that of the former (R = 0.1). As indicated by plots of type shown in Fig. 5, the slope da/dN increases for increasing values of crack length a. This may explain the leftward shift. However, for metals, the parameters C and m are constant and independent of R. Kim and Mubeen [4] in their work on Westerly granite reported a trend similar to the one found in plain concrete in this work.

All observations of crack propagation in this work were within the $K_{Imax}$ range of 40 to 70 percent of $K_{Ic}$. The value of $K_{Ic}$ was taken as $1.16 * 10^6$ $N/m^{3/2}$ for this work. When using a max $K_I$ above 80% of $K_{Ic}$, the crack propagation rate was very high and specimen failed in very few cycles. Data for this range of $K_I$ may be obtained using displacement control mode instead of the load control mode used in the present work.

Foreman et al. [12] proposed a modified form of Paris' Law in the form

$$\frac{da}{dN} = A(\Delta K)^P/[(1-R)K_{IC} - \Delta K] \qquad (8)$$

where A, P are material constants to be determined experimentally. Equation (8) has greater flexibility than Paris' Law, inasmuch as the influence of R-variation can be taken into account.

The data from the present work was checked against equation(8), but the validity of that expression was not ascertained. The values of the constant 'p' for the three cycles, R = 0.1, 0.2, 0.3 were 2.31, 2.41 and 2.51, respectively and the values of 'A' were $1.56 * 10^{-14}$, $6.55 * 10^{-15}$ and $2.62 * 10^{-15}$, respectively. In view of this scatter of values, it has to be concluded that the Foreman model cannot take into account the effect of different cycle ratios for plain concrete.

CONCLUSIONS

1. The relationship between log da/dN and log ΔK for plain concrete is linear.

2. The parameter m in equation $da/dN = C(\Delta K)^m$ is independent of the cycle stress ratio R. The parameter C, however, appears dependent upon R.

3. To correlate C and R successfully, additional information about log da/dN versus log ΔK is required for more values of cycle ratio R.

4. Modified form of Paris' Law proposed by Foreman et al. [12] does not appear to be valid for concrete.

ACKNOWLEDGEMENTS

The support of the University of Petroleum and Minerals in the pursuit of this work is acknowledged.

REFERENCES

[1] Knott, J.F., Fundamentals of Fracture Mechanics, Butterworth, Sevenoaks (1973).

[2] Rolfe, S.T. and Barsom, J.M., Fracture and Fatigue Control in Structures, Prentice Hall, New Jersey, 1977.

[3] Paris, P.C. and Erdogan, F., "A Critical Analysis of Crack Propagation Laws", Transactions of ASME, Journal of Basic Engineering, V. 85, pp. 528-534, 1963.

[4] Kim, Kunsoo and Mobeen, Abdul, "Relationship Between Differential Stress Intensity Factor and Crack Growth Rate in Cyclic Tension in Westerly Granite", Fracture Mecahnics Methods for Ceramics, Rocks and Concrete, ASTM STP 745, S.W. Frieman and E.R. Foller, jr., Eds., American Society for Testing and Materials, 1981, pp. 157-168.

[5] Hillerborg, A., Modeer, M. and Peterson, R.E., "Analysis of Crack Formation and Growth in Concrete by Means of Fracture Mechanics and Finite Elements", Cement and Concrete Research, V.6, No.6, Nov. 1976, pp.773-782.

[6] Bazant, Z.P. and Oh, B.H., "Crack Band Theory for Fracture of Concrete", Materials and Structures, Vol.16, No.93, May-June 1983, pp. 155-178.

[7] Perdikaris, P.C., Calomino, A.M. and Chudnovsky, A., "Effect of Fatigue on Fracture Toughness of Concrete", Journal of the Engineering Mechanics Division, ASCE, Vol.112, No.8, Aug. 1986, pp.776-791.

[8] Qureshy, A.B., "Fatigue of Concrete in the Presence of a Crack", Master of Science Thesis, Department of Civil Engineering, University of Petroleum and Minerals, Dhahran 31261, Saudi Arabia.

[9] Swartz, S.E., Hu, K.K. and Jones, G.L., "Compliance Monitoring of Crack Growth in Concrete", Journal of the Engineering Mechanics Division, ASCE, Vol.104, No.EM4, Aug. 1978, pp. 789-800.

[10] Brown, J.R. and Srawley, J.E., "Plane Strain Crack Toughness Testing of High Strength Metallic Materials", ASTM STP 410, ASTM, 1966.

[11] Parker, A.P., The Mechanics of Fracture and Fatigue, E.&F.N. Spon Ltd., London, 1981.

[12] Foreman, R.G., Kearny, V.E. and Englee, R.M., "Numerical Analysis of Crack Propagation in a Cyclic Loaded Structure", ASME Transactions, Journal of Basic Engineering, 89D, p.459 (1967).

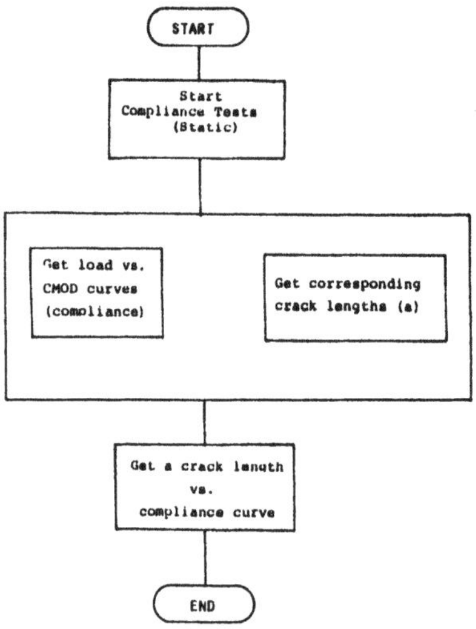


FIG-1 : Compliance Calibration Flow Chart

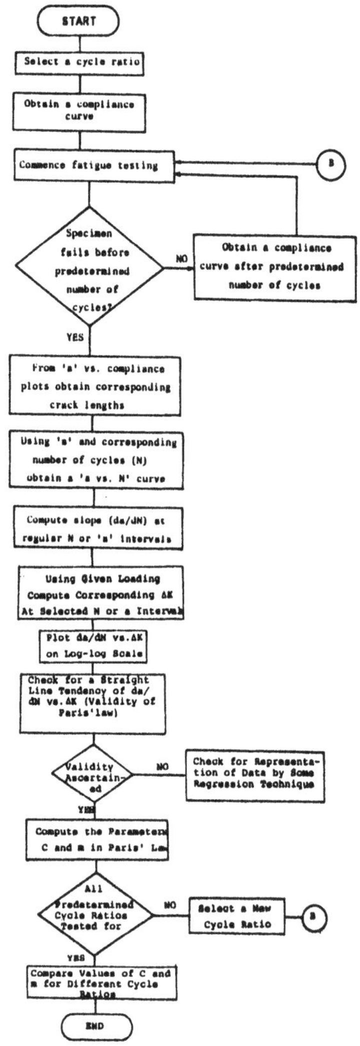


FIG. - 2 : Fatigue Testing Flow Chart

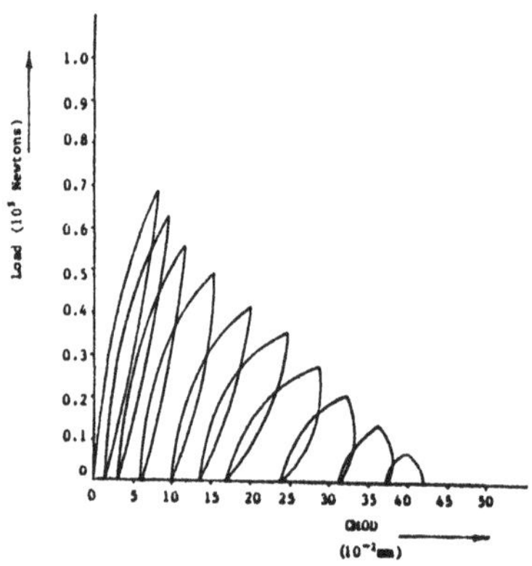


Fig- 3 : . Compliance (load vs. CMOD) curves.
Specimen # 1

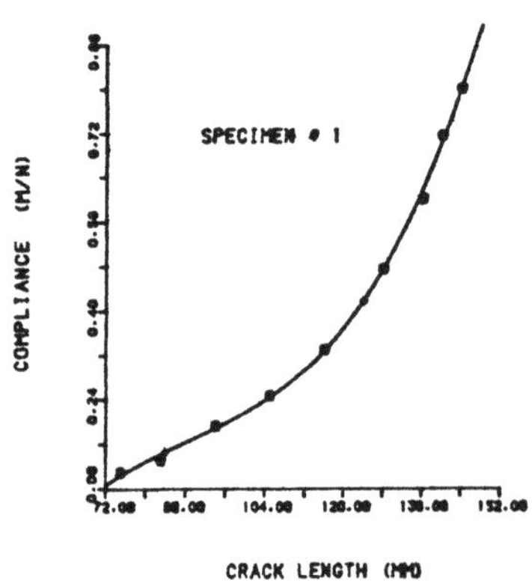


FIG. 4 : COMPLIANCE CALIBRATION CURVE

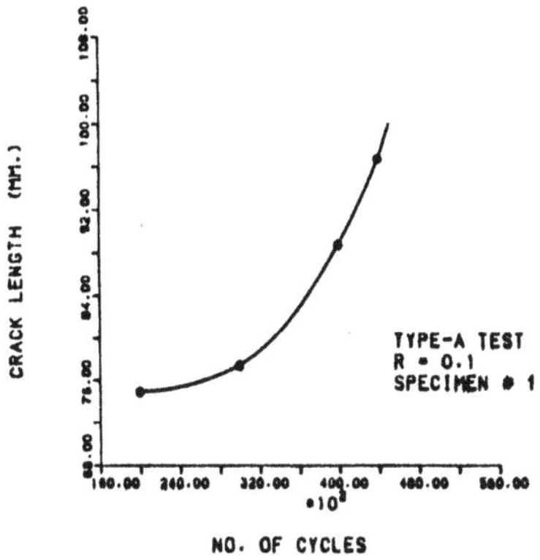


FIG. 5 : NO. OF CYCLES VS. CRACK LENGTH

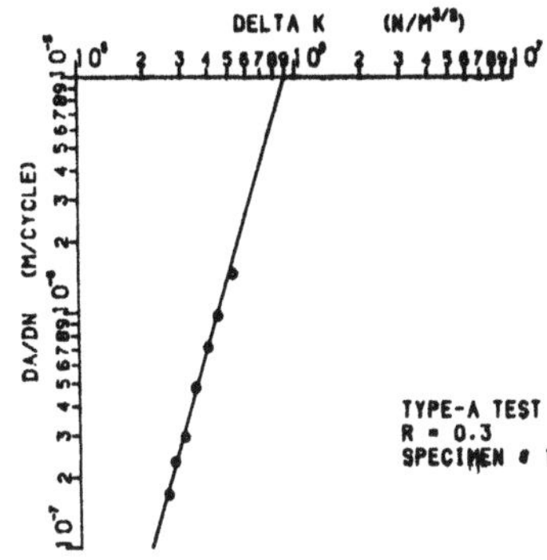


FIG. 6 : LOG DA/DN VS. LOG DELTA K

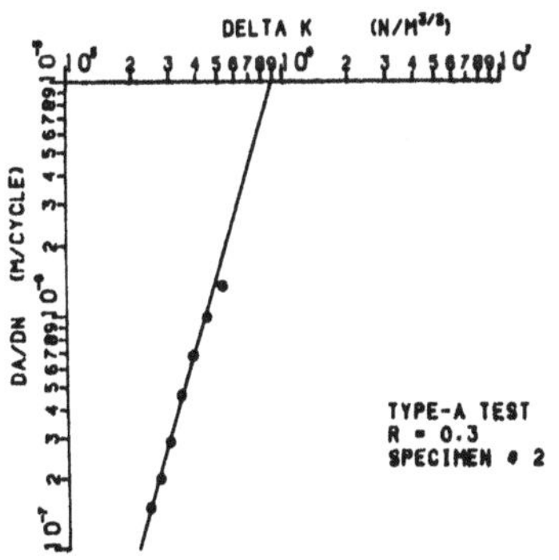


FIG. 7 : LOG DA/DN VS. LOG DELTA K

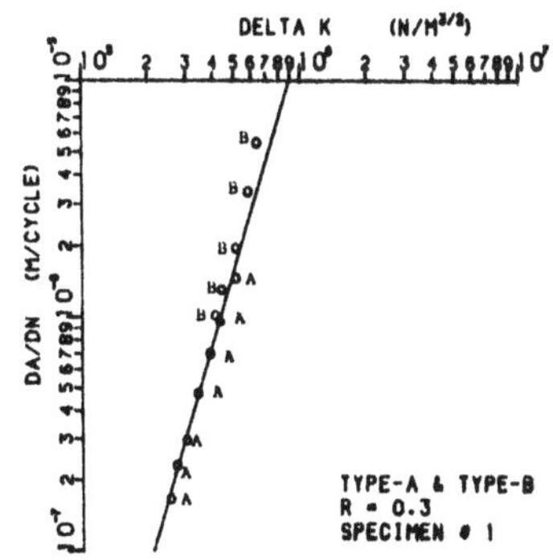


FIG. 8 : LOG DA/DN VS. LOG DELTA K

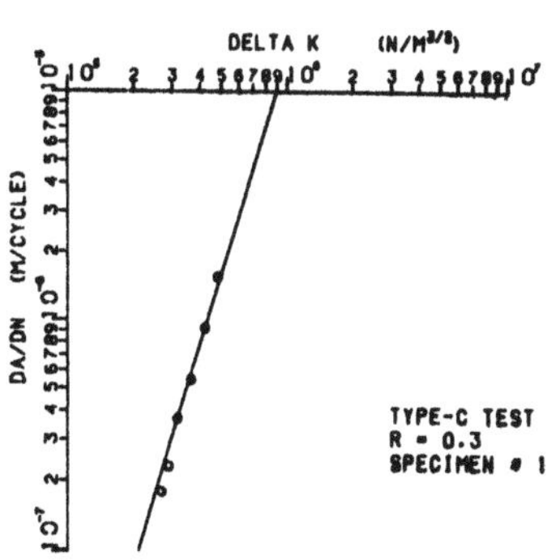


FIG. 9 : LOG DA/DN VS. LOG DELTA K

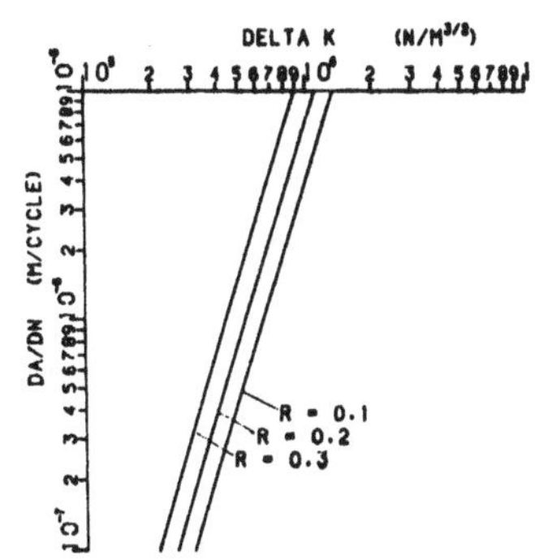


FIG.10 : LOG DA/DN VS. LOG DELTA K
FOR DIFFERENT VALUES OF R.

SEM/RILEM International Conference
on Fracture of Concrete and Rock
Houston, Texas, June 1987
Editors: S.P. Shah, S.E. Swartz

# FATIGUE DAMAGE OF CONCRETE

A. ALLICHE and D. FRANCOIS *

ABSTRACT

Cyclic flexural load tests of simply supported beams under central concentrated charge were performed, for two different concrete Young modulus. The strain at the most tensioned fiber was measured during experiments. A model for cumulative damage which is proportional to the strain was used to present the results.

KEY-WORDS: Concrete, Damaging, Fatige, Experiments, Modelling.

## 1- INTRODUCTION.

Fatigue of concrete was studied by various authors [1 to 8] who usualy gave empirical results not resulting in a fatigue life prediction model.

Damage mechanics [9 to 12] look like a promising way to deal with this problem.

The fatigue life of concrete displays a wide scatter when expressed as a function of the load amplitude. However this scatter is strongly reduced when the fatigue life is expressed as a function of the deformation per cycle. It seems therefore that the strain amplitude is responsible for concrete damage. This was our conclusion when studying the cement paste fatigue [1]. In order to extend this to concrete we undertook experiments, in which we measured strains on three points bent beams. The different steps are the following:

1) Identification of damage mechanisms on the basis of experimental observations [1 to 4].

2) Damage modelling in a simple case of repeated uniaxial loading of constant amplitude.

3) Always in this simple case, the elaboration of the laws connecting the various parameters involved in a fatigue experiment.

## 2- EXPERIMENT.

### 2.1- Materials.

Two concrete types were used. The type A concrete was made with a CPA HP55 cement and the type B with a CPJ 45 one. The type A showed highest strength and stiffness than the type B concrete.

* Ecole Centrale des Arts et Manufactures. Paris

Table 1-Weight composition per m3.

| Components | Concrete A ( Kg ) | Concrete B ( Kg ) |
|---|---|---|
| Gravel 5/12 mm | 1103 | 1650 |
| Sand 0/4 mm | 706 | 1264 |
| Cement | CPA HP 55 MPa | CPJ 45 MPa |
| Water | 201 | 314 |

After 24 hours the demolded samples were covered by a plastic layer to avoid the humidity exchanges with the environment air.

2.2- Samples and characterisation.

The samples were of a prism shaped with dimensions of 140 x 140 x 560 mm. Three points bend tests were used since they allow damage concentration and are easy to reaize.

We measured the longitudinal strains on the tensioned face at two location, the mid span deflection and the load.

Three months after demolding we measured the static strength for 6 samples of each type of concrete. The average fracture loads for type A concrete was Prupt =29.2 kN ( $E_A$ = 37700 MPa ), and for type B: Prupt = 24.30 kN ( $E_B$ = 32700 MPa ).

2.3- Fatigue experiments.

A triangular shaped cyclic loading was applied. The frequency was 1 Hz for concrete type A and 3 Hz for concrete type B. The maximum load was chosen as a function of the average static strengths.The experiments were carried out at seven loading levels: 0.90, 0.85, 0.80, 0.75, 0.70, 0.65, 0.60 of the static strength

Two values of the ratio R, characterizing the mean load (R= Pmin/Pmax) were used : 0.1 and 0.4.

2.4- Data acquisition.

We used a micro computer supplied with analogic digital and timing interfaces for the continual acquisition of the four measure refered above. This acquisition was stored each time the maximum mid span deflection value ( at time t ) exceeded its previous maximum value (at time t-1) by a certain amount, fixed in advance.

3- RESULTS.

3.1- S.N curve.

Many publications on the fatigue of concrete used the S.N curve [5,6,7]. All these studies showed a large scatter of the results and a large incertainty regarding the previsions of the life time based on this diagram.

### 3.2- Evolution of the maximum strain for the extreme tensioned fiber.

Maximum strain evolution of the extreme tensioned fiber ($\varepsilon_t$) could be visualized as consisting of three stages. Stage I would extend over the first 10% of the total life. It is characterized by a quick increase of $\varepsilon_t$ as a function of the number of cycles. Stage II would correspond to a stabilization of the evolution of $\varepsilon_t$ with an increase per cycle that many authors consider as being constant [2,3]. This longest stage would extend from 10 to 70- 80 % of the life time. Finally during the third stage the accelerated increase of $\varepsilon_t$ would lead to the final fracture.

It can be observed that if in stage II the evolution is considered to be linear. Its slope is rather higher when the life time is shorter.

## 4- DISCUSSION and MODELLING.

### 4.1- S. N Diagram Analysis.

Despite the above mentioned scatter, the S.N curves concerning concretes of type A and B allow to stress some tendancies

It can be seen in fig 1 that the life times of concrete type B samples are larger than those of type A, and this for the same ratio Pmax/ Prupt.

It can be understood noticing that the static fracture stain for type A concrete was larger than for type B, so that for the same load ratio, type A concrete was submitted to a larger strain amplitude than type B.

This observation was reported in many previous studies [1,6,8]. Furthermore the diagrams show the influence of the R ratio. The life time increases when the loading amplitude decreases.

### 4.2- Analysis of the Curves $\varepsilon_t = f\,(N)$.

This paragraph concerns the interpretation of the three stages described above.

The material before any loading contains imperfections due to manufacture and different phenomena like shrinkage and swelling. These structur defects are of random dimensions, orientation and distribution. A given applied load induces a material adaptation to the strain field involving the defects which are perpendicular to the maximum main extension direction and releasing the others. This accounts for stage I.

Stage II relates to the slow evolution of the activated defects. The rate of their development and their multiplication accelerating, inducing in the end an instability.

Stage III would not involve any new processes. It should thus be included as part of stage II during which the strain per cycle would keep increasing, however hardly perceptible at the beginning.

### 4.3- Model.

This model is based on damage mechanics which suppose that the stress acting in the material is an effective stress ( $\sigma_{eff}$) which is higher than the apparent stress owing the damage D.

$$\sigma_{eff} = \sigma/(1 - D) \qquad (1)$$

We introduce the fundamental hypothesis that the damage parameter per cycle, dD/dN, is a scalar function of the maximum strain $\varepsilon_t$ and of the ratio R. We therefore neglect for simplification purpose, the damage anisotropy even though it results from the observed evolution in the stage I. Our analysis concerns only the stage II.

In the case of a fatigue experiment with repeated tension and stress amplitude $\Delta\sigma$, this hypothesis allows write:

$$\Delta\sigma = E\left[1 - \int_0^N \frac{dD}{dN}(\varepsilon, R)\,dN\right].\Delta\varepsilon \qquad 2$$

E being the Young modulus of concrete at the beginning of stage II.

This equation allows to find out the maximum strain evolution as a function of N, for an assumed damage law. For instance if we choose the following law:

$$\frac{dD}{dN} = K(R, \varepsilon_0) . \varepsilon^n \quad \ldots\ldots 3$$

we find that the strain increases to infinity (and that D tends towards 1), when the number of cycles reaches $N_F$ given by the relation:

$$N_F = [\, \varepsilon_0 / (n+1) \,] . \left( \frac{d\varepsilon}{dN} \right)_0 \quad \ldots\ldots 4$$

$\varepsilon_0$ being the maximum strain, and $(d\varepsilon / dN)_0$ its increase per cycle at the beginning of stage I.

For a beam under bending, a damage law such as law (3) leads to very difficult calculations if n is not equal to 1. In this case when considering the hypothesis of plane sections, we find that the maximum strain of the most tensioned fiber $\varepsilon_t$ and the number of cycles N vary as function of the neutral fiber reduced abcisse $\zeta_n$ (fig 3) according to the following relations.

-$\varepsilon_t$ expression:

$$\varepsilon_t = \frac{4\zeta_n}{6\zeta_n^2 - 15\zeta_n + 8} . (\varepsilon_t)_0 \quad \ldots\ldots 5$$

-N expression:

$$N = \frac{1}{K(R).(d\varepsilon / dN)_0 . (\varepsilon_t)_0} \left[ -\frac{9}{2} \ln(2\zeta_n) - \frac{63}{4\zeta_n} + \frac{69}{8\zeta_n^2} - \frac{2}{\zeta_n^3} + 13 \right] \quad \ldots\ldots 6$$

The total life time is given by the value of $\zeta_n$ in equation (5) which makes $\varepsilon_t$ infinite (D = 1), hence $\zeta_n$ = 0.77.

$$N_F = \frac{0.77}{K(R).(\frac{d\varepsilon}{dN})_0 . (\varepsilon_t)_0} \quad \ldots\ldots 7$$

Relations (5) and (6) allows to draw the curves $\varepsilon_t = f(N)$, by using the parameter $\zeta_n$.

The phenomenon trend is well reproduced but the observed final acceleration is generally larger than the computed one.

Table 2 summarizes the experimental results and the total life time value furnished by the model.

| N° samp. | Conc | R | $\frac{P\ max}{P\ rupt}$ | $(\varepsilon_t)_0$ $10^4$ | $(d\varepsilon/dN)_0$ $10^9$ | $N_F$ calcul D=1 | $N_F$ exp | $D_C$ | $N_F$ calcul D=$D_C$ |
|---|---|---|---|---|---|---|---|---|---|
| OMD 8 | A | 0.1 | 0.70 | 0.81 | 1.065 | 32062 | 19575 | 00.34 | 19824 |
| OMD 9 | A | 0.1 | 0.70 | 0.82 | 0.75 | 40176 | 18792 | 0.34 | 28427 |
| OMD11 | A | 0.1 | 0.75 | 0.68 | 1.26 | 22333 | 11166 | 0.34 | 14094 |
| OMD18 | A | 0.1 | 0.75 | 0.72 | 1.67 | 18132 | 11400 | 0.34 | 11225 |
| OMD13 | A | 0.1 | 0.80 | 0.84 | 26.2 | 1324 | 863 | 0.34 | 831 |
| OMD23 | A | 0.1 | 0.85 | 0.89 | 6 | 6180 | 3236 | 0.34 | 3856 |
| OMD17 | A | 0.4 | 0.75 | 0.80 | 2.1 | 15999 | 9704 | 0.34 | 9905 |
| OMD16 | A | 0.4 | 0.80 | 0.60 | 2.8 | 8552 | 7014 | 0.34 | 5578 |
| OMD21 | A | 0.4 | 0.85 | 0.90 | 64 | 538 | 300 | 0.34 | 365 |
| OMD22 | A | 0.4 | 0.85 | 0.77 | 2.4 | 13475 | 4898 | 0.34 | 8342 |
| OMD25 | A | 0.4 | 0.85 | 0.76 | 0.15 | 212800 | 77187 | 0.34 | 131737 |
| OMD37 | B | 0.1 | 0.75 | 0.75 | 0.2 | 158000 | 136000 | 0.52 | 128100 |
| OMD50 | B | 0.1 | 0.75 | 0.76 | 0.9 | 35466 | 29600 | 0.52 | 28771 |
| OMD31 | B | 0.1 | 0.80 | 0.68 | 0.47 | 60636 | 61145 | 0.52 | 49190 |
| OMD32 | B | 0.1 | 0.80 | 0.69 | 1.7 | 17047 | 19700 | 0.52 | 13882 |
| OMD46 | B | 0.1 | 0.80 | 0.68 | 6.75 | 4231 | 2600 | 0.52 | 3432 |
| OMD54 | B | 0.1 | 0.80 | 0.59 | 3.57 | 6941 | 6825 | 0.52 | 5583 |
| OMD38 | B | 0.1 | 0.85 | 0.85 | 7.3 | 4990 | 5100 | 0.52 | 3990 |
| OMD48 | B | 0.1 | 0.85 | 0.74 | 0.9 | 34533 | 17800 | 0.52 | 28014 |
| OMD39 | B | 0.4 | 0.75 | 0.66 | 0.3 | 92400 | 68000 | 0.52 | 74957 |
| OMD51 | B | 0.4 | 0.75 | 0.61 | 0.36 | 71666 | 43375 | 0.52 | 57732 |
| OMD44 | B | 0.4 | 0.80 | 0.79 | 0.21 | 158000 | 102300 | 0.52 | 128781 |
| OMD 47 | B | 0.4 | 0.80 | 0.80 | 0.48 | 70000 | 52800 | 0.52 | 56431 |
| OMD 43 | B | 0.4 | 0.85 | 0.76 | 5.2 | 6138 | 4440 | 0.52 | 5012 |

We note that calculated NF (D = 1) is always larger than the experimental NF.
The calculation were done again by introducting in the model the experimental values of NF which allowing to determine $\zeta_n$ at rupture:

*Concrete A, $\zeta_n = 0.56$, Dc = 0.33

*Concrete B, $\zeta_n = 0.60$, Dc = 0.50

The results are shown on fig 6.

In this approach we used the simple Bernouilli hypothesis. It would be necessary to undertake numerical calculations in order to show more correctly the damage localisation phenomenon.
In the other hand if the damage law is not linear, as shown in equation (3), the life time should be shorter ($n > 1$).
The main advantage of such a model would be to allow the prediction of damage cumulation in the case of variable amplitude loadings. The verification remains to be done.

5- CONCLUSION.

We have studied the strain evolution in concrete beams loaded in fatigue by repeated bending, as well as their life time. We found a life time larger for a concrete with a Young modulus of 32 700 MPa, than for a concrete of a higher modulus: 37 700 MPa.
The strain of the most tensioned fiber increases rapidly at the beginning of the experiment, during stage I which is related to the material adaptation.
It then increases slowly at each cycle but with an increasing rate, leading to fracture by instability without visible cracking.

The life time is well correlated with the strain increase per cycle at the beginning of stage II.
The relation is a function of the R ratio. A damage mechanical model with a damage parameter increasing at each cycle proportionaly to the strain, gives a good approximation of the experimental results.

REFERENCES.

1. Alliche A., François D. _ Fatigue Behaviour of Hardened Cement Paste.Cement and Concrete Researche, Vol.16(1986).

2. Lenshow R.- Resistance against Fatigue of Concrete in Offshore Structures,with Special Emphasis on Uncracked Concrete. 1er Col. ARBEM, Publ. CNEXO ,Brest (1981).

3. Holmen J.O.- Fatigue of Concrete by Constante and Variable Amplitude Loading. Norwegian Institute of technology FCB , Trondheim (1979).

4. Cornelissen H.A.W., Reinhard H.W.- Uniaxial Tensile Fatigue Failure of Concrete under Constante and Programme Loading . Magazine of Concrete Research , Vol. 36 - 126(Dec . 1984).

5. Nordby G.N.- Fatigue of Concrete - A Review of Researche . A.C.I. V.30 _ 2(Aug. 1958).

6. Antrim J.D. - The Mechanism of Fatige in Cement Past and PLain of Concrete. Highway Researche Record , Vol. 210(1968).

7. Tepfers R. and Kutti T. - Fatigue Strength of Plain , Ordinary and Lightweight Concrete . J.A.C.I., Vol. 76 - 5(May 1979).

8. Sparks p.R. - The Influence of Rate of Loading and Material Variability in the Fatigue Characteritics of Concrete. Fatigue of Concrete Structures . Detroit. A.C.I. Publication SP 75, Detroit (1982).

9. Chaboche J.L. - These d' Etat , Univ. P. & M. Curie (Paris 6) , Juin 1978.

10. Chaboche J.L. - Une loi différentiele d'endommagement avec cumul non linéaire. Revue Française de Mécanique . 50-51 ,1974

11. Lemaitre J., Mazars J. - Application de la théorie de la mécanique non linéaire de l'endommagement à la rupture du béton de structure. Annales de l'I.T.B.T.P., N° 401(1982).

12. Lemaitre J. - Damage Modelling for Prediction of Plastic or Creep Fatigue Failure in Structures . 5th Conference on Structurale Mechanics in Reactor Technology, SMIRT 5(1979).

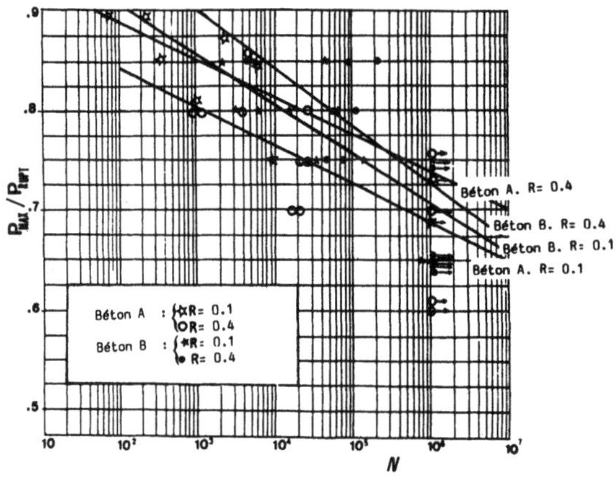


Fig . 1 - Results of the four experimental types using Wöhler curve

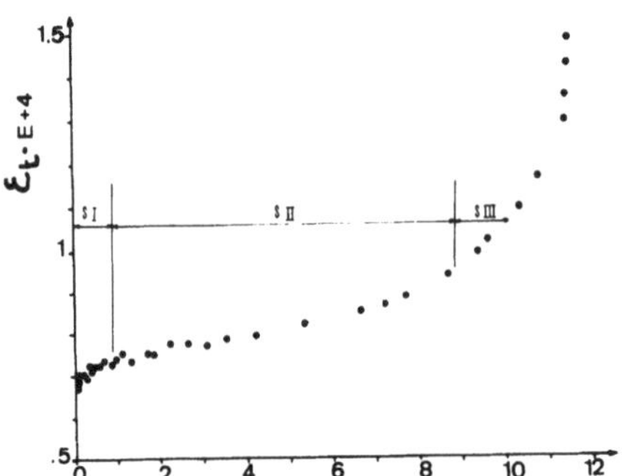


Fig. 2 - $\varepsilon_t$ evolution in function of the number of cycles.

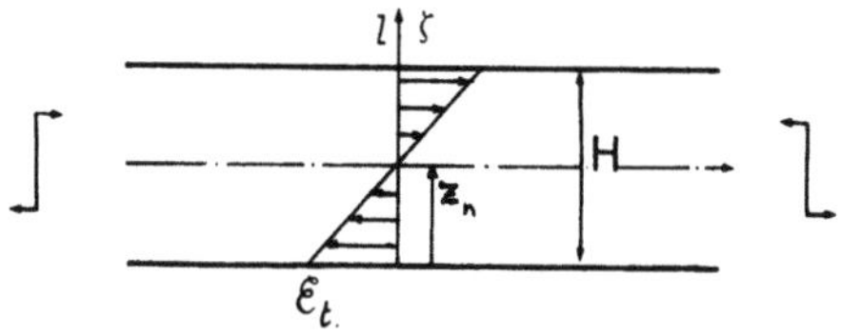


Fig. 3 - Calculation Parameters.

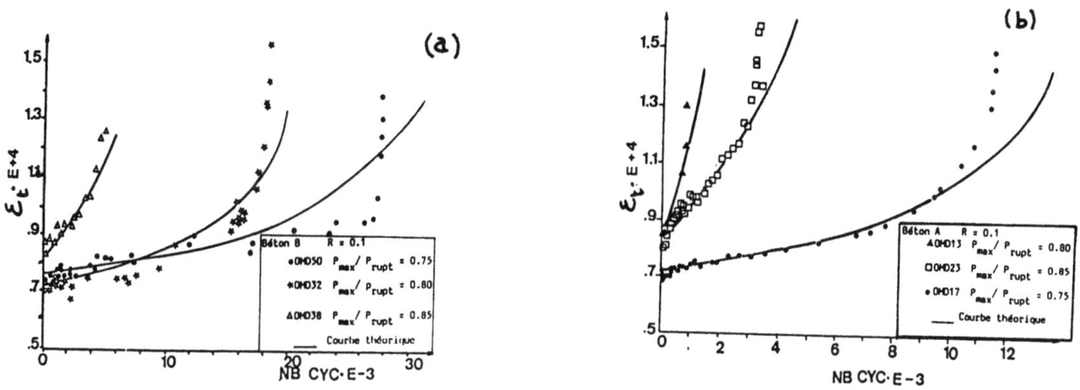


Fig. 4 - Strain evolution in function of the number of cycles.

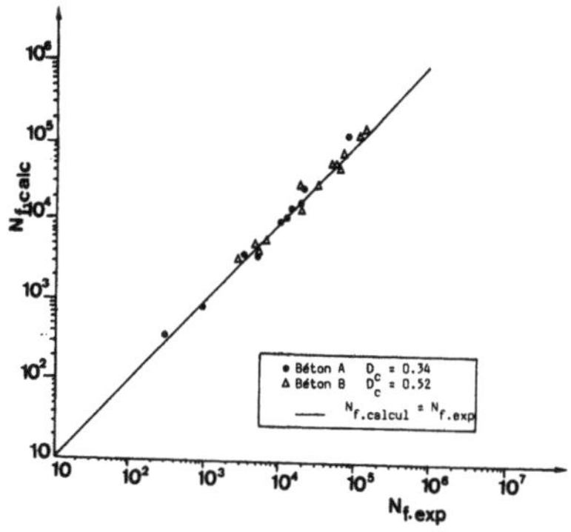


Fig. 5 - Calculated and experimental

$N_F$ comparison.

SEM/RILEM International Conference
on Fracture of Concrete and Rock
Houston, Texas, June 1987
Editors: S.P. Shah, S.E. Swartz



# DO PLAIN AND FIBRE-REINFORCED CONCRETES HAVE AN R-CURVE BEHAVIOUR ?

B.L. Karihaloo

## ABSTRACT

The fracture toughness of plain and (fibre) reinforced concretes has been determined from axial compression splitting tests. The results suggest the existence of R-Curve behaviour. The $K_c$ value for a given water/cement ratio and mix proportions would seem to be roughly three times the initiation value.

## INTRODUCTION

The fracture toughness of plain and (fibre) reinforced concretes is commonly determined from tension, compact tension and three-point bend tests [1-6]. It has been consistently noticed that the fracture toughness ($K_{Ic}$, $G_{Ic}$ or $G_F$) so determined varies with the specimen dimensions and geometry [7-9]. Several modifications of the above fracture parameters have been proposed [10-12] to render them independent of the specimen size.

It has also been observed that cementitious materials exhibit crack growth resistance behaviour [13-14], which could be empirically established from $G_F$ using the size effect law [15]. An alternative procedure for determining both the initiation value $K_i$ and the plane stress value $K_c$ from a single test in compression has been recently proposed [16]. In this procedure, the specimens are formed with a pre-crack and are compressed in a similar manner to standard compression strength tests, but are loaded through a narrow platen. Providing the loading platen is rather narrow and the pre-crack is sufficiently deep, the specimen will fail by axial splitting, allowing the calculation of both $K_i$ and $K_c$. The main advantage of the compression splitting test is the relative ease of specimen preparation and testing under load control.

Our investigation would seem to confirm the existence of an R-curve behaviour in plain and (fibre) reinforced concretes.

## COMPRESSION SPLITTING MODEL

Several forms of compression splitting tests have been in use for a long time. The best known of these are the Brazilian test and the indentation test. They are also called the indirect tensile tests, as a crack forms along the axis of compression due to the tensile stresses developed in the specimens. The problem with such compression splitting tests is that crack formation and growth are extremely rapid giving a catastrophic failure and inhibiting detailed examination of the failure process. Kendall [17] noted that the rapid crack growth occurred because of the large (compressive) force necessary to create the initial crack. He therefore proposed the axial splitting model, in which a sharp starting crack is introduced into the specimen along the loading axis, providing an initial flaw from which the compression crack can easily grow.

The geometry of the axial splitting specimens used in this work is shown in Fig. 1. This geometry was chosen because of simplicity, but there is no reason why other geometries and/or sizes may not be adopted. In fact, several different geometries and sizes have been previously tried [18-19] with no noticeable effect on the resulting fracture toughness value. However, in order to ensure failure by axial splitting rather than by yielding under the platen or by buckling it is important that $c/d \geq 3/2$ and $w/d \leq 1/2$ [18-19].

---

B.L. Karihaloo is Professor of Civil Engineering, University of Sydney, N.S.W. 2006, AUSTRALIA.

When axial splitting occurs the pre-crack extends along the axis of compression, creating more or less two identical struts which bend under eccentric compression. Using the energy balance principle, it has been shown [18] that the Griffith surface energy per unit area R is given by

$$R = \frac{F^2e^2}{32C} (2-kc \tan kc/2) \sec^2 kc/2 \quad (1)$$

where F is the compressive force corresponding to the central gap length c, $k^2 = F/(2EI)$ and E is Young's modulus of the material. For rectangular specimens, $e = (d-w)/4$, $I = bd^3/96$ and $C = EIb$. For circular specimens of diameter d, $e = [(2d/3\pi) - w/4]$, $I = d^4(9\pi^2-64)/(1152\pi)$ and $C = EId$. The relation (1) is only valid when no "plastic" deformation takes place either under the platen or at the propagating crack front. The surface energy R may be expressed in terms of the plane stress intensity factor $K_I^2 = ER$ [18].

The replacement of R with the opening mode stress intensity factor $K_I$ implies that the pre-crack growth results from tensile stresses developed at the crack front by the bending of the struts. The existence of such stresses has been confirmed in previous studies [16, 18, 19] by the use of elastic finite element and boundary element calculations and is again confirmed by elastic finite element calculations, performed for all the tests reported here.

SPECIMEN PREPARATION

The mix proportions, water/cement ratios and the type of coarse aggregate and of fibre used in the thirteen mixes are listed in Table 1. Of these mixes, six contained either steel or polypropylene fibres.

Table 1 Mix proportion, kg

| Batch No. | 20 mm Crushed Stone | 10 mm Crushed Stone | Beach Sand | River | Cement | Water | Fibre Type | Fibre % of Cement |
|---|---|---|---|---|---|---|---|---|
| 1 | 30.0 | 30.0 | 15.00 | 15.00 | 21.24 | 8.52 | Steel | 1.0 |
| 2 | 30.0 | 30.0 | 15.00 | 15.00 | 21.24 | 8.52 | Steel | 2.0 |
| 3 | 30.0 | 30.0 | 15.00 | 15.00 | 21.24 | 9.56 | Steel | 1.0 |
| 4 | 30.0 | 30.0 | 15.00 | 15.00 | 21.24 | 9.56 | Steel | 2.0 |
| 5 | 30.0 | 30.0 | 15.00 | 15.00 | 21.24 | 9.56 | Poly** | 1.0 |
| 6 | 30.0 | 30.0 | 15.00 | 15.00 | 21.24 | 9.56 | Poly** | 0.5 |
| 7 | 37.5* | 37.5 | 23.00 | 23.00 | 18.20 | 10.00 | None | 0.0 |
| 8 | 37.5 | 37.5 | 23.00 | 23.00 | 18.20 | 10.00 | None | 0.0 |
| 9 | 37.5 | 37.5 | 20.75 | 23.75 | 16.60 | 10.00 | None | 0.0 |
| 10 | 37.5* | 37.5 | 20.75 | 23.75 | 16.60 | 10.00 | None | 0.0 |
| 11 | 37.5 | 37.5 | 23.75 | 20.75 | 16.60 | 10.00 | None | 0.0 |
| 12 | 37.5 | 37.5 | 23.00 | 23.00 | 18.20 | 10.00 | None | 0.0 |
| 13 | 37.5 | 37.5 | 18.75 | 18.75 | 26.55 | 10.65 | None | 0.0 |

* Uncrushed stone
** Polypropylene

In each experimental factorial group of tests, five specimens and three cylinders generally formed the sample. Both the fibre types are commercially manufactured and are readily available. The steel fibre used is 19mm 'Fibresteel'. This fibre has enlarged ends giving it a slight resemblance to a dumbbell. The enlarged ends help the fibre bond to the concrete and avoid the workability problems encountered by constant diameter fibres which need long lengths to develop bond strength. The polypropylene fibre used is grade 19 'Fibremesh'. It is a collated fibrillated mesh which when mixed with concrete opens up into a birdmesh-like lattice.

The test specimens were cast in steel moulds with steel bases. The geometrical properties of the test specimens are given in Table 2. The pre-crack (notch) was introduced using a wedge made from 5mm steel plate milled to a sharp edge. The edge was remilled as often as necessary to retain its sharpness. Each batch was prepared in the mixer with the fibre being added last, to achieve an even distribution of fibres.

Each specimen was thoroughly vibrated to remove air bubbles. The moulds were stripped after twenty-four hours and the wedge gently removed. The specimens were placed in the fog chamber for curing. Three standard cylinders (diameter 150mm, height 300mm) were also cast from each mix and cured together with the test specimens. These were used to determine the elastic modulus (E) appearing in the formula (1).

Table 2 Geometrical and selected mix properties of specimens

| Batch | Pre-crack Length mm | Width, d mm | Depth, b mm | Platen width, w mm | Water/ Cement |
|---|---|---|---|---|---|
| 1 | 175 | 95 | 100 | 38 | 0.40 |
| 2 | 175 | 95 | 100 | 38 | 0.40 |
| 3 | 175 | 95 | 100 | 38 | 0.45 |
| 4 | 175 | 95 | 100 | 38 | 0.45 |
| 5 | 175 | 95 | 100 | 38 | 0.45 |
| 6 | 175 | 95 | 100 | 38 | 0.45 |
| 7 | 145 | 95 | 100 | 38 | 0.55 |
| 8 | 145 | 95 | 100 | 38 | 0.55 |
| 9 | 145 | 95 | 100 | 50 | 0.60 |
| 10 | 145 | 95 | 100 | 50 | 0.60 |
| 11 | 175 | 95 | 100 | 38 | 0.60 |
| 12 | 175 | 95 | 100 | 38 | 0.55 |
| 13 | 175 | 95 | 100 | 38 | 0.40 |

On the twenty-eight day after mixing, the specimens were removed from the fog room. The cylindrical specimens were placed under a sun lamp until the ends were dry and were then sulphur capped. On the other hand, the ends of the axial splitting specimens were made perfectly flat by diamond grinding. Likewise, the faces containing the pre-crack were cleaned with a grinding stone until no calcium hydroxide was visible on them. After grinding, the test specimens were placed in the sun to dry. They were then spray painted with a thin layer of paint which would dry quickly. Finally, any debris which had accumulated in the pre-crack was removed. A fine piano wire was held taut and drawn through the pre-crack front with a sawing action. This left a sharp crack tip visible on both faces of the specimen.

TESTING PROCEDURE

The axial splitting specimens were tested in a standard hydraulic-ram testing machine. The loading platen was carefully placed on the specimen and aligned. The painted faces of the specimen were illuminated by powerful spotlights in order to monitor the progress of the crack front which showed as a dark line on the painted surface. The force was applied at a constant rate of approximately 1 kN/sec. The load was applied in 10 kN increments and held constant after each increment while the crack front was located on each face. The crack tip was marked and the corresponding force level indicated. The path of the crack was also traced to allow easy measurement of the crack extension. A typical crack growth pattern highlighting the location of crack tip at various load levels is shown in Fig. 2. The procedure of incrementing the load and of marking the position of the crack tip was continued until either unstable crack growth occurred or extensive crushing took place under the loading platen. Extensive crushing indicated that local yielding was beginning to take place under the platen. In weak mixes this can happen even before the growth of pre-crack, because the local compressive stress under the narrow platen exceeds the compressive strength. To remedy this situation a wider platen may be used (cf. mixes 9 and 10, Table 2). The same problem may also be resolved by increasing the depth of central cut (cf. mixes 11 and 12, Table 2). This problem is however unlikely to arise with stronger mixes.

The (curved) crack extension was measured by approximating it with a series of straight lines. It was seldom necessary to use more than eight straight line segments to determine the extension caused by any load increment. This measurement technique assumes that the crack front is straight.

To gain an idea of the shape of the crack front and hence an indication of the accuracy of the crack extension measurements, ultrasound tests were performed. Ultrasound is transmitted through the specimen only when there is no discontinuity. Any discontinuities (e.g. a crack) reflect most of the ultrasound resulting in no (or weak) transmission. It was therefore possible to get an estimate of the location of crack front by locating the boundary of full and no (partial) transmission. For good acoustic coupling, the two faces of the specimen not containing the pre-crack were heavily smeared with vaseline. Only three specimens were used for ultrasound examination. They gave similar results to the effect that the crack front is somewhat convex with a large radius of curvature.

Measurements of the extension in crack length are listed in Table 3 for mixes 1-6 and 13. For other mixes, reference should be made to [16]. Table 3 shows the mean crack extension and deviation as obtained from recordings over the two faces of each of the five speciments tested from a batch. For each mix (batch) the entries in Table 3 begin at no crack extension, i.e. at the load level corresponding to the onset of pre-crack growth, and end at the load level corresponding to unstable crack growth.

Table 3 Compressive force and crack extension

| FORCE (kN) | BATCH 1 | 2 | 3 | 4 | 5 | 6 | 13 |
|---|---|---|---|---|---|---|---|
| 70 | 0 | 0 | 0 | 0 | 0 | 0 | 0 |
| 80 | $2^{+13}_{-2}$ | 0 | 0 | 0 | 0 | $2^{+9}_{-2}$ | 0 |
| 90 | $5^{+27}_{-5}$ | $2^{+7}_{-2}$ | $1^{+3}_{-1}$ | 0 | $2^{+10}_{-2}$ | $4^{+14}_{-4}$ | 0 |
| 100 | $6^{+10}_{-3}$ | $7^{+5}_{-13}$ | $8^{+10}_{-8}$ | 0 | $8^{+12}_{-8}$ | $6^{+18}_{-6}$ | 0 |
| 110 | $15^{+27}_{-9}$ | $16^{+8}_{-3}$ | $14^{+12}_{-5}$ | $1^{+6}_{-1}$ | $13^{+14}_{-7}$ | $10^{+7}_{-8}$ | 0 |
| 120 | $23^{+34}_{-17}$ | $26^{+13}_{-5}$ | $24^{+8}_{-12}$ | $6^{+2}_{-3}$ | $23^{+5}_{-11}$ | $15^{+11}_{-9}$ | $6^{+8}_{-6}$ |
| 130 | $35^{+45}_{-23}$ | $38^{+13}_{-11}$ | $39^{+9}_{-7}$ | $18^{+7}_{-9}$ | $30^{+10}_{-13}$ | $25^{+6}_{-4}$ | $11^{+10}_{-9}$ |
| 140 | $52^{+48}_{-29}$ | $50^{+19}_{-8}$ | $44^{+14}_{-16}$ | $30^{+10}_{-14}$ | $39^{+14}_{-15}$ | $35^{+9}_{-9}$ | $21^{+18}_{-16}$ |
| 150 | $61^{+58}_{-28}$ | $64^{+16}_{-12}$ | $65^{+15}_{-22}$ | $46^{+9}_{-14}$ | $48^{+20}_{-19}$ | $35^{+3}_{-6}$ | $32^{+28}_{-13}$ |
| 160 | $72^{+62}_{-50}$ | $76^{+17}_{-16}$ | $59^{+18}_{-14}$ | $71^{+27}_{-22}$ | $59^{+21}_{-19}$ | $43^{+5}_{-7}$ | $45^{+17}_{-12}$ |
| 170 | $97^{+49}_{-42}$ | $99^{+27}_{-21}$ | $79^{+9}_{-15}$ | $93^{+23}_{-18}$ | $72^{+13}_{-12}$ | $60^{+17}_{-7}$ | $75^{+32}_{-31}$ |
| 180 | 110 | $121^{+24}_{-34}$ | $77^{+8}_{-6}$ | $126^{+28}_{-30}$ | $87^{+22}_{-19}$ | $66^{+8}_{-6}$ | $110^{+36}_{-37}$ |
| 190 | $152^{+51}_{-54}$ | $134^{+25}_{-24}$ | $99^{+14}_{-12}$ | $152^{+28}_{-23}$ | $103^{+16}_{-21}$ | $80^{+21}_{-21}$ | $129^{+18}_{-25}$ |
| 200 | | $154^{+17}_{-29}$ | | 195 | $106^{+17}_{-13}$ | $99^{+22}_{-25}$ | $162^{+42}_{-23}$ |
| 210 | | $187^{+14}_{-14}$ | | | | | |

The elastic modulus (E) and compressive strength ($\sigma_c$) of each mix were determined from separate cylinder tests. Note however that $\sigma_c$ is not required in the calculation of $K_I$ from the compression splitting tests.

## RESULTS AND DISCUSSION

Eq. (1) and the relation $K_I^2 = ER$ were used to calculate the values of $K_I$ from the onset of pre-crack growth to the moment of unstable crack growth. The initiation value $K_i$ was calculated from the initial pre-crack length and the compressive force at the onset of crack growth. The values of $K_I$ during the stable crack growth were calculated using the actual load level and the augmented crack length (equal to the sum of pre-crack length and mean crack extension). This is necessarily an approximation because, as already noted the crack front is not straight, nor does it advance exactly along the middle of the specimen. However, Eq. (1) is still valid even at the full crack extension because the corresponding buckling load of the struts is well in excess of the load at the onset of instability.

The results of these calculations for all of the mixes used in this study are shown in Fig. 3a-m. For each mix the R-curve behaviour is clearly visible, allowing the estimation of the plane stress value, $K_c$. The fracture toughness values obtained from finite element analysis are given in Tables 4 and 5 for several mixes and are compared with the results of the compression splitting model.

Table 4 Comparison of stress intensity values ($MPa.m^{\frac{1}{2}}$) obtained by experimental (compression splitting model) and finite element analyses, for three batches with a water/cement ratio of 0.40

| Crack Length (mm) | Batch 1 1% Steel | | Batch 2 2% Steel | | Batch 13 No Fibre | |
|---|---|---|---|---|---|---|
| | $K_{IFE}$ | $K_{IEXP}$ | $K_{IFE}$ | $K_{IEXP}$ | $K_{IFE}$ | $K_{IEXP}$ |
| 174.0 | 0.86 | 0.94 | 0.97 | 1.07 | 1.27 | 1.40 |
| 182.5 | 1.11 | 1.22 | 1.16 | 1.27 | 1.43 | 1.57 |
| 191.0 | 1.29 | 1.41 | 1.27 | 1.40 | 1.55 | 1.70 |
| 199.5 | 1.40 | 1.54 | 1.38 | 1.51 | 1.64 | 1.80 |
| 212.5 | 1.88 | 1.69 | 1.69 | 1.66 | 2.14 | 1.91 |
| 225.5 | 1.64 | 1.81 | 1.51 | 1.77 | 1.83 | 2.01 |
| 242.5 | 1.76 | 1.95 | 1.74 | 1.94 | 1.91 | 2.12 |
| 259.5 | 1.86 | 2.07 | 1.86 | 2.07 | 1.99 | 2.21 |
| 276.5 | 1.93 | 2.17 | 1.95 | 2.20 | 2.04 | 2.30 |
| 325.5 | 2.49 | 2.40 | 2.56 | 2.46 | 2.56 | 2.46 |
| 335.5 | - | - | 2.61 | 2.51 | 2.60 | 2.50 |
| 362.5 | - | - | 2.62 | 2.61 | - | - |

Table 5 Comparison of stress intensity values ($MPa.m^{\frac{1}{2}}$) obtained by experimental (compression splitting model) and finite element analyses, for four batches with a water/cement ratio of 0.45

| Crack Length (mm) | Batch 3 1% Steel | | Batch 4 2% Steel | | Batch 5 1% Polyprop. | | Batch 6 0.5% Polyprop. | |
|---|---|---|---|---|---|---|---|---|
| | $K_{IFE}$ | $K_{IEXP}$ | $K_{IFE}$ | $K_{IEXP}$ | $K_{IFE}$ | $K_{IEXP}$ | $K_{IFE}$ | $K_{IEXP}$ |
| 144.5 | - | - | - | - | - | - | 1.06 | 0.96 |
| 153.0 | - | - | - | - | - | - | 1.41 | 1.27 |
| 161.5 | - | - | - | - | - | - | 1.69 | 1.51 |
| 174.0 | 1.00 | 1.10 | 1.24 | 1.40 | 0.98 | 1.08 | 1.60 | 1.76 |
| 182.5 | 1.15 | 1.26 | 1.39 | 1.52 | 1.15 | 1.26 | 1.72 | 1.89 |
| 191.0 | 1.29 | 1.41 | 1.49 | 1.64 | 1.31 | 1.43 | 1.85 | 2.02 |
| 199.5 | 1.39 | 1.52 | 1.56 | 1.71 | 1.42 | 1.56 | 1.94 | 2.13 |
| 212.5 | 1.91 | 1.71 | 2.02 | 1.81 | 1.94 | 1.74 | 2.38 | 2.26 |
| 225.5 | 1.70 | 1.87 | 1.72 | 1.90 | 1.73 | 1.91 | 2.18 | 2.40 |
| 242.5 | 1.87 | 2.07 | 1.80 | 2.00 | 1.89 | 2.10 | 2.30 | 2.55 |
| 259.5 | 2.07 | 2.30 | 1.87 | 2.09 | 2.02 | 2.25 | - | - |
| 276.5 | 2.14 | 2.41 | 1.92 | 2.17 | 2.11 | 2.38 | - | - |
| 280.5 | - | - | - | - | 2.53 | 2.42 | - | - |
| 325.5 | - | - | 2.46 | 2.36 | - | - | - | - |
| 335.5 | - | - | 2.49 | 2.40 | - | - | - | - |
| 362.5 | - | - | 2.47 | 2.46 | - | - | - | - |
| 371.5 | - | - | 2.41 | 2.48 | - | - | - | - |

A comparison of $K_i$ and $K_c$ obtained by the compression splitting model (Eq. 1) and finite element analysis is of particular interest. This has been done in Table 6 for several mixes. The results are in good agreement, the maximum variation being only 0.18 $MPa.m^{\frac{1}{2}}$ or 7%. This proves that the compression splitting model is an accurate working model.

Table 6 Values of stress intensity factor $K_i$ ($MPa.m^{\frac{1}{2}}$) and the plane stress toughness $K_c$ ($MPa.m^{\frac{1}{2}}$) determined by both the experimental (compression splitting model and finite element analyses.

| BATCH | W/C | Fibre | Finite element | | Experimental | |
|---|---|---|---|---|---|---|
| | | | $K_i$ | $K_c$ | $K_i$ | $K_c$ |
| 1 | 0.40 | 1% steel | 0.86 | 2.49 | 0.94 | 2.40 |
| 2 | 0.40 | 2% steel | 0.97 | 2.62 | 1.07 | 2.61 |
| 3 | 0.45 | 1% steel | 1.00 | 2.14 | 1.10 | 2.41 |
| 4 | 0.45 | 2% steel | 1.24 | 2.41 | 1.40 | 2.48 |
| 5 | 0.45 | 1% polyp. | 0.98 | 2.53 | 1.08 | 2.41 |
| 6 | 0.45 | 0.5 polyp. | 1.06 | 2.30 | 0.96 | 2.55 |
| 13 | 0.40 | none | 1.27 | 2.60 | 1.40 | 2.50 |

## CONCLUSIONS

1. The results of the present investigation confirm the existence of an R-curve behaviour in plain and (fibre) reinforced concrete. Whether or not this behaviour is also size dependent is currently under investigation.

2. The results of the compression splitting test are generally independent of the shape and size of the test specimens. However, to ensure failure by axial splitting the width of the loading platen and the depth of the pre-crack have to be within certain limits.

3. The limited test results on fibre reinforced concrete show that $K_i$ is marginally lower, and $K_c$ is marginally greater, than the corresponding plain concrete. The primary role of the fibre reinforcement would seem to be in the intermediate range between $K_i$ and $K_c$ in which it reduces crack extension.

## ACKNOWLEDGEMENT

Batches 1-6 and 13 were tested by Nick Johnston and Greg Stigter.

## REFERENCES

1. Nallathambi, P., Karihaloo, B.L. and Heaton, B.S. "Effect of specimen and crack sizes, water-cement ratio and coarse aggregate texture upon fracture toughness of concrete", Magazine of Concrete Research, Vol. 36, No. 129, December 1984, pp. 227-36.

2. Nallathambi, P., Karihaloo, B.L. and Heaton, B.S. "Various size effects in fracture of concrete", Cement and Concrete Research, Vol. 15, No. 1, February 1985, pp. 117-26.

3. Legendre, D. and Mazars, J. "Damage and fracture mechanics for concrete (a combined approach)", Advances in the Fracture Research: Proceedings of the 6th International Conference on Fracture (ICF 6), Pergamon Press, Editors: S.R. Valluri et al., Vol. 4, 1984, pp. 2841-8.

4. Petersson, P.E. "Fracture energy of concrete: method of determination", Cement and Concrete Research, Vol. 10, No. 1, February 1980, pp. 78-89.

5. Rossi, P., Acker, P. and Francois, D. "Measurements of fracture toughness $K_{Ic}$ of concrete" Advances in Fracture Research: Proceedings of the 6th International Conference on Fracture (ICF 6), Pergamon Press, Editors: S.R. Valluri et al., Vol. 4, 1984, pp. 2833-40.

6. Barker, D.B., Hawkins, N.M., Jeang, F.L., Cho, K.Z. and Kobayashi, A.S. "Concrete fracture in a CWL-DCB specimen", Journal of Engineering Mechanics, Proceedings, American Society of Civil Engineers, Vol. 111, No. EM5, May 1985, pp. 623-38.

7. Hillsdorf, H.K. and Brameshuber, W. "Size effects in the experimental determination of fracture mechanics parameters", NATO Advanced Research Workshop: Application of fracture mechanics to cementitious materials, 4-7 September 1984. Editor: S.P. Shah, Northwestern University, Evanston, U.S.A., pp. 213-54.

8. Bazant, Z.P., Kim, J.K. and Pfeiffer, P. "Determination of nonlinear fracture parameters from size effect tests", ibid, pp. 143-70.

9. Hillerborg, A. "Results of three comparative test series for determining the fracture energy $G_F$ of concrete", Materiaux et Constructions, Vol. 18, No. 107, September-October 1985, pp. 407-13.

10. Wecharatana, M. and Shah, S.P. "Predictions of nonlinear fracture process zone in concrete", Journal of Engineering Mechanics, Proceedings, American Society of Civil Engineers, Vol. 109, No. EM5, May 1983, pp. 1231-46.

11. Cho, K.A., Kobayashi, A.S., Hawkins, N.M., Barker, D.B. and Jeang, F.L. "Fracture process zone of concrete cracks", Journal of Engineering Mechanics, Proceedings, American Society of Civil Engineers, Vol. 109, EM8, August 1984, pp. 1174-84.

12. Nallathambi, P. and Karihaloo, B.L. "Determination of specimen-size independent fracture toughness of plain concrete", Magazine of Concrete Research, Vol. 38, No. 135, June 1986, pp. 67-76.

13. Wecharatana, M. and Shah, S.P. "Slow crack growth in cement composites", Journal of Structural Division, Proceedings, American Society of Civil Engineers, Vol. 108, No. ST6, June 1982, pp. 1400-13.

14. Wecharatana, M. and Shah, S.P. "A model for predicting fracture resistance of fibre reinforced concrete", Cement and Concrete Research, Vol. 13, No. 6, December 1983, pp. 819-29.

15. Bazant, Z.P. "Size effect in blunt fracture: concrete, rock, metal", Journal of Engineering Mechanics, Proceedings, American Society of Civil Engineers, Vol. 110, No. EM4, April 1984, pp. 518-535.

16. Karihaloo, B.L. "Fracture toughness of plain concrete from compression splitting tests", Int. Journal of Cement Composites and Lightweight Concrete, Vol. 8, No. 4, November 1986, pp.251-9.

17. Kendall, K. "Complexities of compression failure", Proceedings of Royal Society London, Vol. A361, 1978, pp. 245-63.

18. Karihaloo, B.L. "Compressive failure of brittle materials", ibid, Vol. A396, 1984, pp. 297-314.

19. Guest, B.W. and Karihaloo, B.L. "Fracture of brittle materials", Journal of Materials Science Letters, Vol. 4, No. 10, October 1985, pp. 1285-9.

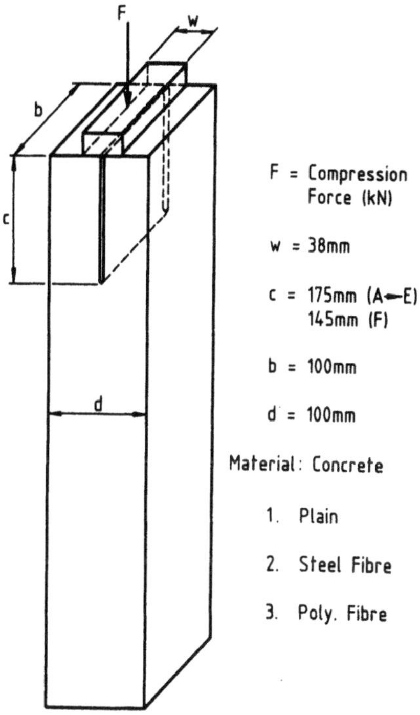


FIG. 1 Geometry of test specimen

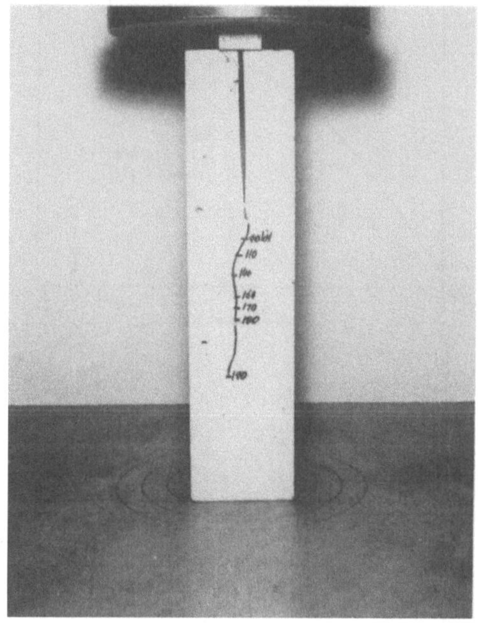

FIG. 2 Location of crack tip at various load levels

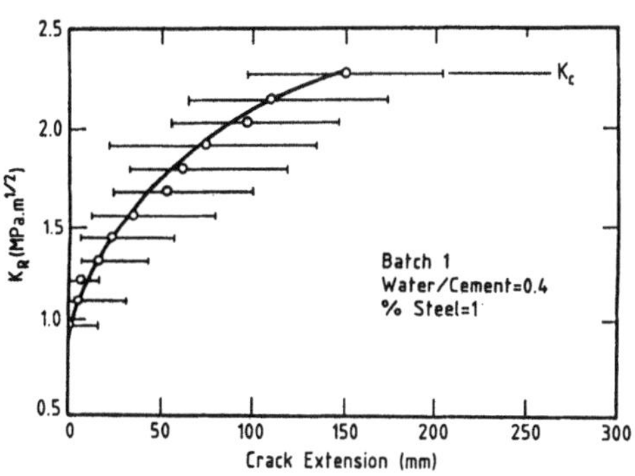


FIG. 3a

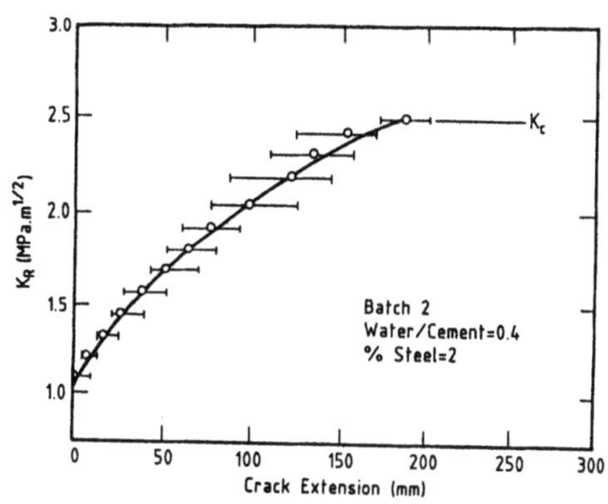


FIG. 3b

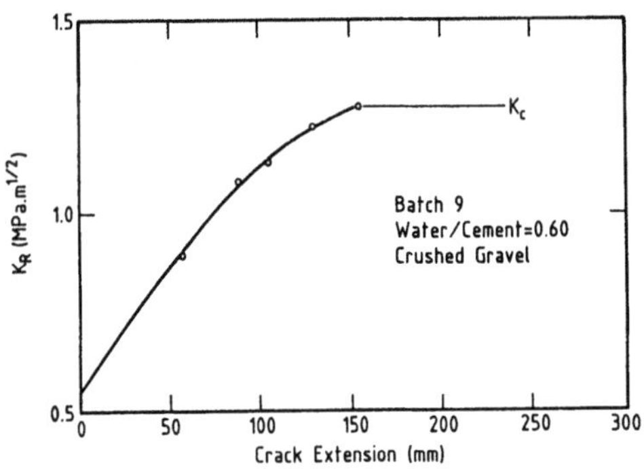


FIG. 3i

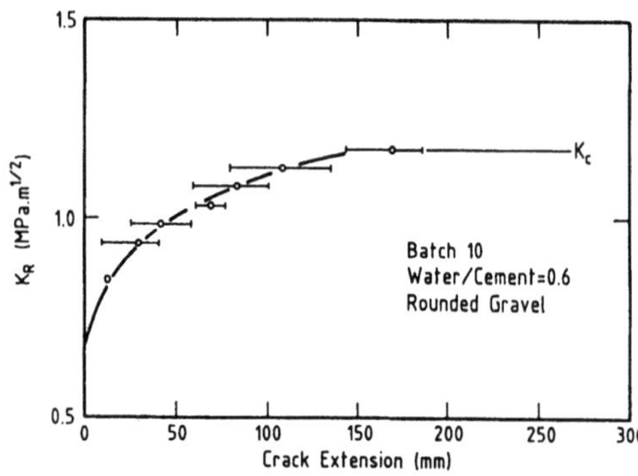


FIG. 3j

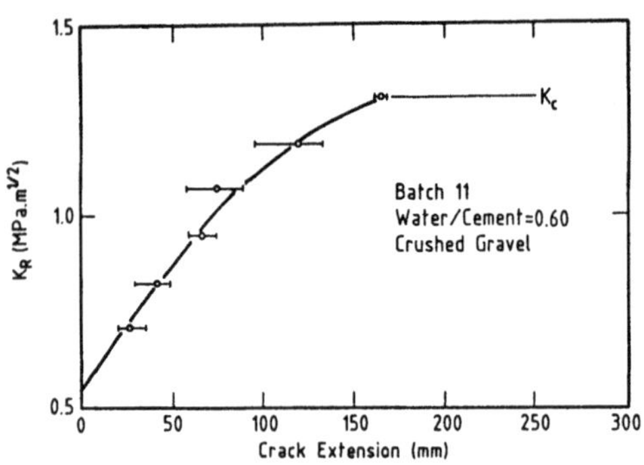


FIG. 3k

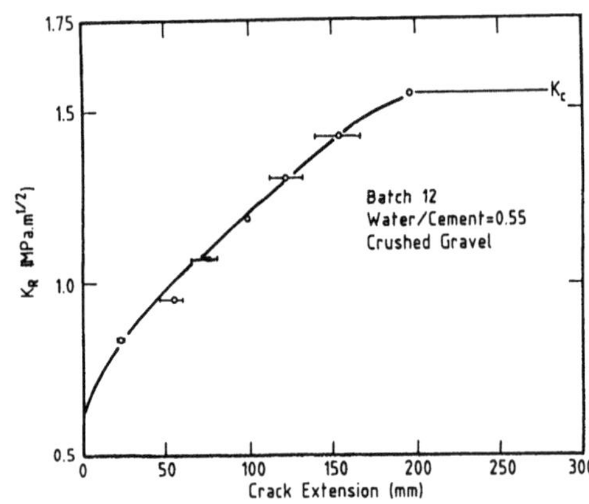


FIG. 3ℓ

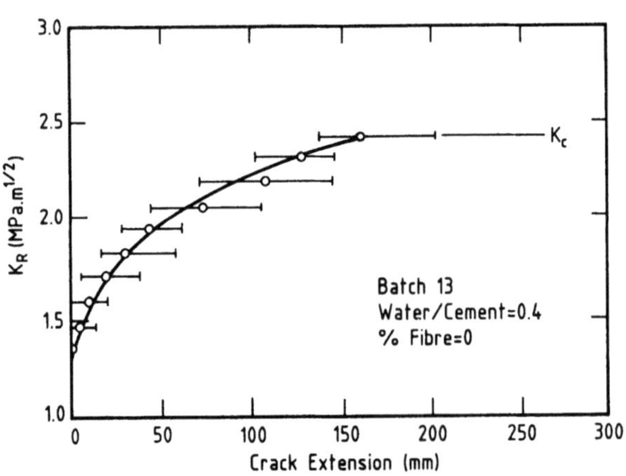


FIG. 3m  Increase in $K_R$ with crack extension up to unstable crack growth ($K_R = K_c$)

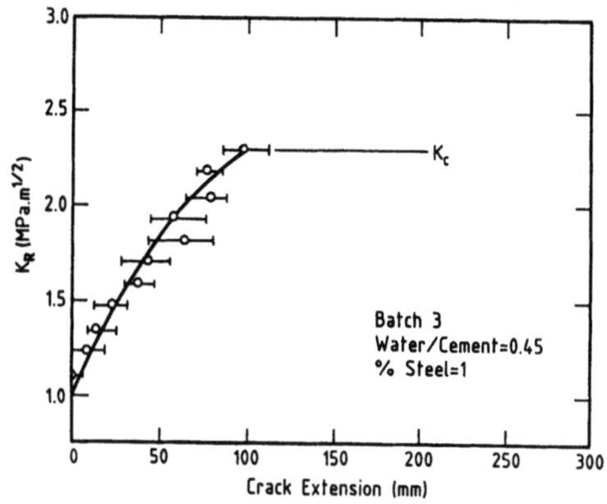


FIG. 3c

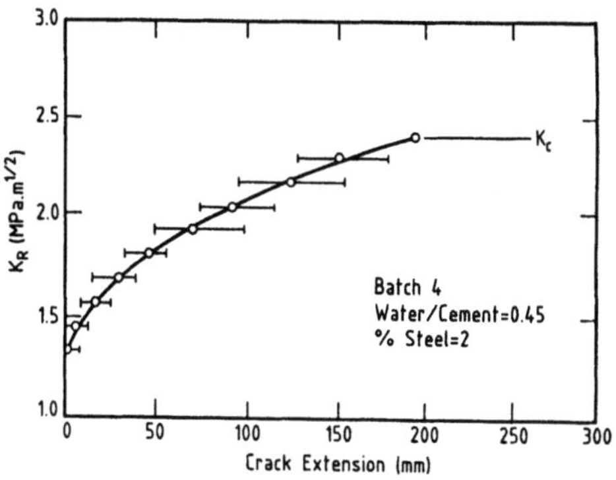


FIG. 3d

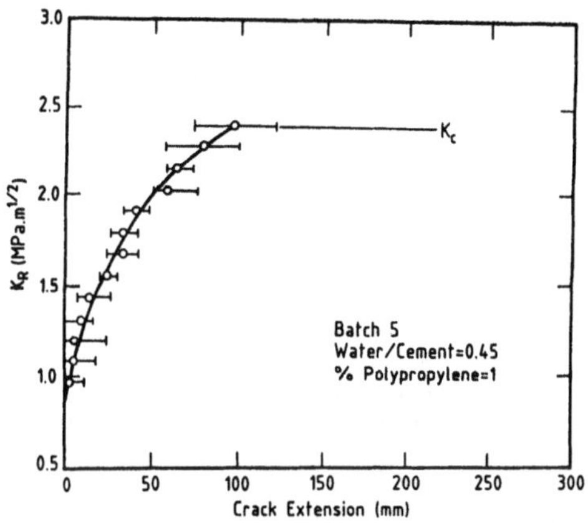


FIG. 3e

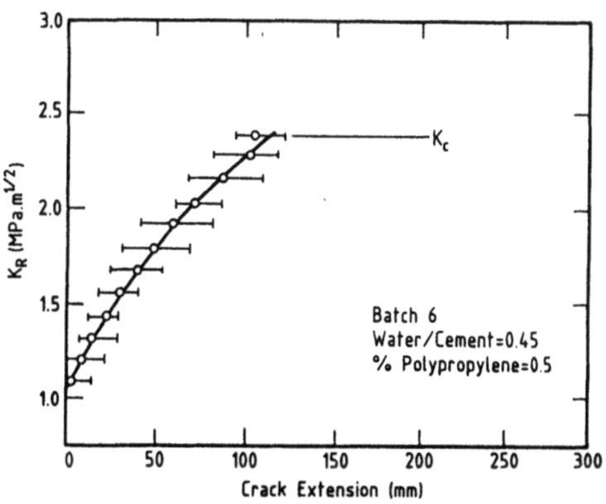


FIG. 3f

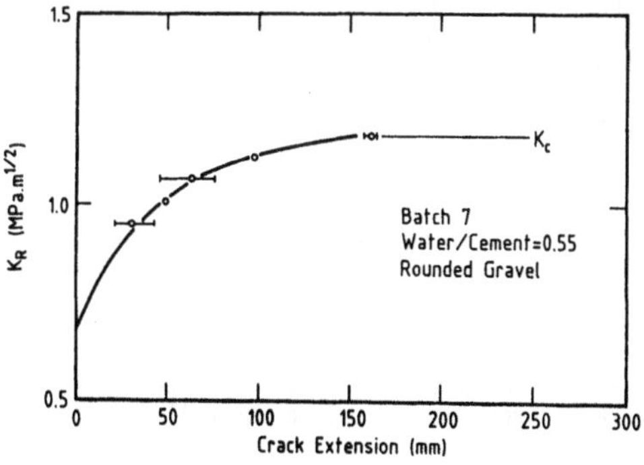


FIG. 3g

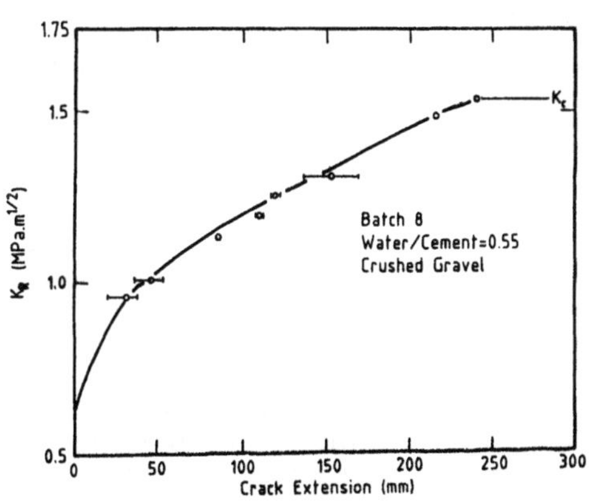


FIG. 3h

SEM/RILEM International Conference
on Fracture of Concrete and Rock
Houston, Texas, June 1987
Editors: S.P. Shah, S.E. Swartz

# FRACTURE OF CONCRETE AT CRYOGENIC TEMPERATURES

M. Elices, J. Planas and P. Maturana
Dept. of Materials Science. Polytechnical University of Madrid

ABSTRACT

Stable three point bend tests on notched concrete specimens have been done at -170 C and at room temperature. Specimens were water saturated and four sizes have been tested at each temperature to ascertain the influence of size on measured parameters. Problems related with concrete cracking due to temperature non uniformity during cooling have been considered.

## 1. INTRODUCTION

Reinforced concrete at low temperatures is currently used for land-based storage of liquefied gases and for barge hulls. Other applications of prestressed concrete at low temperatures devised in the Artic include mobile exploratory drilling vessels, floating terminals and floating industrial plants. This interest is reflected in several international conferences, held in the last years, on this subject, see references in [1].

At low temperature, plain concrete exhibits many desirable features for the design engineer. Most of the available information on cryogenic concrete can be found in the proceedings of two recent International Conferences [2, 3], a FIP State of the Art Report [4] and references there in.

The compressive strength of concrete increases markedly with reduction of temperature, the main influence upon this increase is exerted by the moisture content. Water saturated concrete may attain a relative strength at -170 C of 400%, or more, in comparison to the strength at 20 C. The dependence of the tensile strength on temperature -as determined by splitting, bending or double punch tests- is similar to the compressive strength, though less pronounced. The stress-strain diagram of axially compressed concrete is also markedly influenced by the temperature and moisture content of concrete. At -170 C the stress-strain record is linear elastic with brittle failure, in the engineering sense.

To the authors knowledge no fracture tests at cryogenic temperatures have been reported, with the exception of [5] , where short rods of concrete mortar were tested. Applying linear Elastic Fracture Mechanics techniques, these tests show that there was a substantial increase in $K_{IC}$, as temperature was lowered. Test results for 22 C, -78 C and -197 C were, respectively, 0.9, 2.2 and 1.9 MPa $m^{1/2}$.

In this paper, fracture tests done in accordance with a RILEM proposal [6] have been performed at room temperature and at -170 C, and the fracture energy measured. Experimental set-up as well as preliminary results are reported in the next sections.

## 2. MATERIALS AND SAMPLE PREPARATION

CEMENT

Rapid hardening Portland cement (ASTM type III) has been used.

AGGREGATES

Natural rounded aggregates, classified as siliceous, have been used. Table 1 shows gradding of aggregate.

TABLE 1. GRADING OF AGGREGATES (Cumulative percentage passing)

| ASTM number<br>Metric size (mm) | 3/8<br>10 | 4<br>5 | 8<br>2.5 | 16<br>1.25 | 30<br>0.63 | 50<br>0.32 | 100<br>0.16 | 200<br>0.08 |
|---|---|---|---|---|---|---|---|---|
| Fine aggregate | 100 | 99 | 93 | 52 | 27 | 10 | 3 | 1 |
| Coarse aggregate | 100 | 44 | 1 | 0 | 0 | 0 | 0 | 0 |

CONCRETE

Concrete used in this research is standard RILEM concrete, as described in [7]. Table 2 summarizes the characteristics of concrete mix and table 3 shows some concrete properties measured according to ASTM standards. Grading of concrete is shown in Fig. 1.

TABLE 2. PROPORTIONAL MIXING OF CONCRETE BY WEIGHT

| Cement* | Coarse aggregate | Fine aggregate | Water |
|---|---|---|---|
| 1 | 1.35 | 3.02 | 0.55 |

* Cement content 400 kg/m$^3$.

TABLE 3. CONCRETE PROPERTIES

| Slump (cm) | 28 day strength (MPa) | | Tangent Modulus 28 days (GPa) |
|---|---|---|---|
| | compresive | tensile | |
| 5 | 33.10 | 2.80 | 26.64 |

CONCRETE SAMPLES

Test specimens for fracture measurement were concrete notched beams. The geometry is sketched in Fig. 2.

Notched beams have been instrumented with five thermocouples, as sketched in Fig. 2, in order to ascertain the temperature distribution near the notched section during cooling and warming.

Also, standard cylindrical specimens of 150 mm length and 75 mm diameter have been moulded to measure tensile strength as well as modulus of elasticity.

All the specimens were cast from a single batch in steel moulds and compacted by means of a vibrating table. After demoulding, samples were stored under lime saturated water at 20±3 C until testing time.

Inmediately after the specimens were taken out of the water, and to avoid moisture losses during handling, they were wrapped in a plastic sheet covering all, but the future load bearing areas. These areas were then wet-ground and cleaned to get a good contact with the supports.

The bases of tension specimens were wet-ground and brushed, surface dried and bonded to steel plates with fast-curing water-insensitive epoxy resin. After a short period of curing (about 3 hours at room temperature) the specimens were re-inmersed in water until testing time (minimum 12 hours).

## 3. EXPERIMENTAL PROCEDURE

Fracture tests were stable three point bend tests and were performed following, essentially, the tentative RILEM recommendation for the determination of the fracture energy of concrete [6]. The special features of the used procedure are presented below:

### LOW TEMPERATURE EQUIPMENT

Two low temperature chambers differing only in their respective volumes have been used throughout the experimentation. Fig. 3 shows the block diagram description of either chamber.

The temperature into the chamber is lowered at a constant rate by spraying a mist of liquid nitrogen controlled by an electromagnetic valve. The intended cooling ramp is input to the temperature controller via a digital function generator and the feed-back signal is provided by a Platinum resistor sensing the gas coolant temperature. The controller automatically operates the electrovalve through a PID control algorithm. The stability of the gas temperature achieved with this system is ±1 C.

The gas coolant temperature is continuously recorded during the cooling process. The temperature of the five type K thermocouples attached to the specimen are measured via an Automatic Data Acquisition System (ADAS) Hewlett Packard 9825A and a multiple temperature vs. time plot is output in real time. The temperature stability at the specimen points is better than 0.1 C.

The cooling rates used ranged from 20 C/hour, for the largest specimens, to 28 C/hour, for the smallest specimens.

### TENSION TESTS

Tension tests were performed in a 500 kN servohydraulic testing machine INSTRON 1270, run in displacement control mode. Loads were measured by a 250 kN load cell with a resolution of 12.5 N and 0.5 percent accuracy. Strains were measured by two specially built low temperature extensometers, 75 mm in gauge length, located on oposite sides of the specimen. Resolution was 0.025 µm and accuracy ±0.5 µm.

In these tests, load cell and extensometer outputs were continuously recorded and, simultaneously, readings were taken by the ADAS and stored as a 700 x 3 matrix. Stress-strain curves and deformation modulus were obtained by processing the stored data.

All the tests were conducted at the same displacement rate, leading to mean strain rates of 1 µm/(ms) and 0.6 µm/(ms) for tests at 20 and -170 C respectively.

### STABLE BENDING TESTS ON NOTCHED BEAMS

Fig. 4 shows a sketch of the experimental set-up for three point bend tests.

The tests were performed in a 1 MN servohydraulic testing machine INSTRON 1275, run in CMOD control mode. Loads were measured by a 25/50 kN load cell with a resolution of 1.25/2.5 N and 0.5 percent accuracy. CMOD was measured by a clip-on gauge MTS 632.03C-51, with 0.2 µm resolution and ±2 µm accuracy.

Deflection was measured as the relative displacement of the central loading head and the line defined by the points on the upper surface of the specimen located on the verticals of the lower supports. Such line was materialized by a rigid frame and a cross bar (a short bar crossing through the loading head), as depicted in Fig. 4. The displacement was measured by an extensometer located into a transverse hole in the loading head. The accuracy of the extensometer was better than 5 µm.

In all tests weight compensation has been used. This was automatically accomplished by using specimens twice as long as the loading span for the two smallest sizes. Prestressed springs on both sides of the notch provided the load compensation for the larger specimens.

Load cell and extensometer outputs were continuously recorded and, simultaneously, readings were taken by the ADAS and stored as a 700 x 3 matrix. Force-deflection and force-CMOD curves and related fracture parameters were obtained by processing the stored data.

All tests were run at the same rate of CMOD/depth ratio, namely 7 µm/(m s). With this condition the maximum load was reached in between 30 and 60 s in all cases.

## 4. EXPERIMENTAL RESULTS

TENSILE TESTS

The main results of direct tensile tests on saturated cylindrical specimens are summarized in table 4, where values in brackets are standard deviations.

TABLE 4. MECHANICAL PROPERTIES AT ROOM TEMPERATURE AND -170 C

| TEMPERATURE ( C) | AGE (days) | TENSILE STRENGTH (MPa) | ELASTIC MODULUS (GPa) |
|---|---|---|---|
| 20 | 90 | 3.14 (0.05) | 25.4 (0.35) |
| -170 | 90 | 4.01 (0.12) | 47.9 (3.70) |

During the cooling process the difference of temperature in a cross section of the specimen was less than 2,5 C.

STABLE BENDING TESTS ON NOTCHED BEAMS

These tests were performed on notched beams, as sketched in figure 2, and the main results are summarized in table 5, where values in brackets are standard deviations.

TABLE 5. FRACTURE TESTS

| SPECIMEN | d(m) | TEMPERATURE ( C) | $P_u$(kN) | $f_N$(MPa) | RILEM $G_F$(N/m) |
|---|---|---|---|---|---|
| FT-1 | 0.05 | 20 | 2.89 (0.07) | 4.31 | 57 (3) |
| FT-2 | 0.10 | 20 | 5.21 (0.09) | 4.15 | 75 (18) |
| FT-3 | 0.20 | 20 | 9.37 (0.09) | 3.92 | 82 (3) |
| FT-4 | 0.30 | 20 | 11.25 (0.82) | 3.08 | 94 (7) |
| FT-1 L.T. | 0.05 | -170 | 6.36 (0.02) | 10.84 | 175 (24) |
| FT-2 L.T. | 0.10 | -170 | 13.94 (0.87) | 11.42 | 213 (3) |
| FT-3 L.T. | 0.20 | -170 | 26.70 (3.12) | 11.03 | 289 (7) |
| FT-4 L.T. | 0.30 | -170 | 35.80 | 9.76 | 300 |

a, b, d, l as in figure 2. P is the maximum load and $f_N = 1.5 P_u l/b(d-a)^2$.

The maximum temperature difference between two points in a cross section during the cooling process is summarized in Table 6 for each specimen size.

TABLE 6. MAXIMUM DIFFERENCE OF TEMPERATURE DURING THE COOLING PROCESS

| SPECIMEN | MAXIMUM DIFFERENCE OF TEMPERATURE ( C) |
|---|---|
| FT-1 L.T. | 6 |
| FT-2 L.T. | 9 |
| FT-3 L.T. | 10 |
| FT-4 L.T. | 11 |

## 5. DISCUSSION ON EXPERIMENTAL PROCEDURES

Testing of concrete at very low temperatures is, in general, a difficult matter for quite different reasons:

One of the most usual source of disturbances comes from devices which work very well at room temperature but fail to work properly at low temperatures due to either icing or to mismatched dimensional changes. In this experimental set, mechanical apparatus have been excluded and the chamber was sealed to prevent moisture inflow and frost formation. The only remaining potential source of error comes from the roller supports. Gripping of the rollers is very difficult but it can not be excluded. Rollers efficiency was not directly checked but as load deflection curves do not display a stick-slip pattern we disregard this effect.

When dealing with moist concrete, additional difficulties arise from the appearance of internal stresses due to non uniform temperature distribution during cooling. Cooling rates must be small enough to prevent both extensive bulk microcracking and extensive development of a fracture process zone at the notch tip.

In order to assess the extent of damage during cooling, a simplified model has been developed and is included as an APPENDIX. Far field maximum tensile stress and stress intensity factor during cooling have been estimated for the different specimens using experimental data quoted in Table 6 and the theoretical results depicted in Figs. A1 and A4. For this evaluation, values of deformation modulus E and dilatation coefficient $\alpha$, were taken as

$E$ = 30 GPa
$\alpha$ = 10 μm/mK in the contraction range
$\alpha$ = -15 μm/mK in the expansion range

where this later value was taken from previous dilatometric tests on similar concretes [8, 9].

The relative fracture process zone extension was also evaluated, in the frame of a cohesive crack model, using the zeroth order asymptotic approach described in [10], as:

$$FPZ = (\pi/8)(K_{IN}/f_t)^2 \tag{1}$$

where $K_{IN}$ is the nominal stress intensity factor and $f_t$ the tensile strength.

The results are summarized in table 7. In these specimens, the maximum tensile stress always occured during the expansion range and reaches about 75 percent of the room temperature strength, which may be considered acceptable, but in the very limit of acceptability. On the contrary, the induced stress intensity factors are very small and the FPZ size is less than 1 percent of the initial ligament size.

TABLE 7. ESTIMATION OF THERMALLY INDUCED MAXIMUM TENSILE STRESSES, STRESS INTENSITY FACTORS AND FRACTURE PROCESS ZONE SIZES

| SPECIMEN | $\sigma_{max}$ (MPa) | $K_{IN}$ (MPa m)$^{1/2}$ | FPZ (mm) | FPZ/(d-a) percent |
|---|---|---|---|---|
| FT1-LT | 1.4 | 0.056 | 0.14 | 0.42 |
| FT2-LT | 2.1 | 0.102 | 0.45 | 0.67 |
| FT3-LT | 2.3 | 0.134 | 0.78 | 0.58 |
| FT4-LT | 2.3 | 0.145 | 0.92 | 0.46 |

It must be emphasized that the proposed simplified model can not take into account localized crack growth at the corners of the crack, the description of which needs a three dimensional treatment.

If moisture losses are not prevented, shrinkage and non uniform temperatures will add their effects, and may cause severe damage. In this research, moisture losses were prevented by plastic sheet wrapping.

## 6. DISCUSSION ON EXPERIMENTAL RESULTS

The deformation modulus at low temperature is nearly twice that at room temperature. This is in agreement with published results for moduli measured in compression tests on saturated concrete [4].

Tensile strength from direct tensile testing increases only about 30 percent on lowering temperature down to -170 C. This is a smaller value than those reported for indirect tensile strength measurements [2-4]. Neither tensile tests or correlations between direct and indirect tensile strengths at low temperatures have been found in the literature for saturated concrete. Since the values of the characteristic length found in this work are larger than the usual values at room temperature it may be expected that differences between direct and splitting strength are also larger for low temperature situations. Hence the present 30 percent increase may be a consistent value.

Examination of $G_F$ values in Table 5 show that there is an apparent size effect for this parameter, both at 20 and -170 C. This trend was already found for one of the previously tested concrete mixes, and was reported in [11]. This effect is nearly the same, in relative terms, for both temperatures.

$G_F$ values increase, roughly, in a 3/1 relation on lowering temperature. This means that saturated concrete is tougher at low temperature than in ordinary conditions. Moreover, when characteric lengths are calculated using average $G_F$ values, the following values are obtained for this concrete.

$l_{ch}$ (20) = 200 mm
$l_{ch}$(-170) = 730 mm

These values show that the structural brittleness for a given structural size is roughly 3.7 times less for low temperature situations than for ordinary conditions, which is a relatively unexpected behaviour since the compressive behaviour of saturated concrete at low temperatures is usually described [4] as linear elastic up to the occurrence of explosive, hence "brittle" failure.

**ACKNOWLEDGEMENTS**

The authors gratefully knowledge Dr. Félix Tamayo R. for helping in experimental work. Financial support under a Cooperative Research Project U.S.-Spain Joint Committee for Science and Technology 83/071 is also acknowledged.

**REFERENCES**

1. Elices, M., "Cryogenic prestressed concrete: Fracture aspects", Theoretical and Applied Fracture Mechanics, Vol. 7, No. 1 (1987).
2. The Concrete Society, "Cryogenic concrete", Proceedings of the 1st. International Conference, Newcastle upon Tyne, March 1981, Construction Press, pp. 336 (1982).
3. The Concrete Societies of the Netherlands and the U.K., "Cryogenic concrete", Proceedings of the 2nd. International Conference, Amsterdam, Netherlands Concrete Society preprints (1983).
4. Elices, M., Rostasy, F.S., Faas, W.M., Wiedemann, G., "Cryogenic behaviour of materials for prestressed concrete", FIP State of the Art Report, pp. 84 (1982).
5. Arvidson, J.M., Sparks, L.L., Steketee, E., "Mechanical properties of concrete mortar at low temperatures", NBS Report 82-1658 (1982).
6. 50-FMC RILEM Draft Recommendation, "Determination of the fracture energy of mortar and concrete by means of three-point bend tests on notched beams", Materials and Structures, Vol. 18, No. 106, pp. 285-224 (1974).
7. RILEM Standard, Materials and Structures, No. 39, pp. 207-224 (1974).
8. Planas, J., Corres, H., Elices, M., Chueca, R., "Thermal deformation of loaded concrete during thermal cycles from 20 C to -165 C", Cement and Concrete Research, 14, pp. 639-644, (1984)
9. Elices, M., Planas, J., Corres, H., "Thermal deformation of loaded concrete at low temperatures, 2: Transverse deformation", Cement and Concrete Research, 16, pp. 741-748, (1986).
10. Planas, J., Elices, M., "Asymptotic analysis of the development of a cohesive crack zone in mode I loading, for arbitrary softening curves", These proceedings.
11. Planas, J., Elices, M., "Towards a measure of $G_F$: An analysis of experimental results", Fracture Toughness and Fracture Energy of Concrete, F.H. Wittmann Ed., Elsevier Science Publishers, B.V., Amsterdam, pp. 381-390, (1986).
12. Tada, H., Paris, P.C., Irwin, G.R., "The stress analysis of cracks handbook", del Research Corporation, Hellertown, Pa, (1973)

## APPENDIX. ESTIMATION OF THERMALLY INDUCED FAR STRESS FIELD AND STRESS INTENSITY FACTOR IN A CONCRETE NOTCHED BEAM COOLED AT A CONSTANT RATE

The purpose of this analysis is to find a reasonable approximation for the stress field in a notched beam with rectangular cross section subject to steady cooling, from which limiting criteria for the cooling rate can be set.

Having in mind testing of concrete-like materials at low temperatures limiting criteria have to be set such that during the cooling process both extensive bulk microcracking and extensive crack growth are avoided.

The description of the actual behaviour would require a three dimensional dependent and highly nonlinear treatment. In this preliminary work, however, reasonable simplifications are introduced in order to reduce the problem to a two dimensional, time independent and linear one. It is assumed, moreover, that the fracture process zone, described as a cohesive crack, is limited to such small size that the Small Scale Yielding approach may be used so that, as shown in reference [10], the nominal stress intensity factor $K_{IN}$ fully characterizes the behaviour near the crack tip.

The analysis is developed in three steps. In the first one, a temperature distribution is obtained for what can be thought to be the worse situation in a cooling process. In a second step the far field stresses are estimated from the temperature distribution. Finally, the problem is reduced to a clasical two dimensional situation by averaging stresses through the thickness, and the thermally induced $K_I$ computed using a Green's function formulation.

ESTIMATION OF THE TEMPERATURE DISTRIBUTION

The following assumptions are made:

H1. The beam is long enough for the end efects to be negligible.
H2. The rate of heat transfer from the coolant surrounding gas through the crack (notch) faces is negligible.
H3. Heat capacity and thermal conductivity are both constant, and so is thermal difusivity $D_T$.

This reduces the problem of finding the temperature distribution to that of solving the classical bidimensional heat transfer equation over the cross section of the beam subject to adequate boundary conditions.

The actual boundary conditions are much dependent on the experimental equipment used (convection coeficients), and on the kind and degree of control achieved (gas or surface controlled cooling rate). We will consider here the case where the surface temperature $T_s$ is decreased at constant rate v from an initially uniform temperature $T_0$, i.e.: $T_s = T_0 - vt$.

In this situation it can be shown that the temperature differences across the specimen increase steadily with time and tend exponentially to an asymptotic time independent value. Taking, within a cross section, central axis x and y respectivelly parallel to the beam thickness, b, and beam depth, d, this may be written as:

$$T(x, y, t) \simeq T_s + T_1(x, y) \qquad \text{for} \qquad t >> l^2/D_T \quad \text{with} \quad l = \frac{bd}{2(b+d)} \tag{A-1}$$

$$T_1(x, y) = 0 \qquad \text{on the boundary} \tag{A-2}$$

Substitution of Eq. (1) into the heat conduction equation leads to the following quasi-harmonic equation for T :

$$\nabla^2 T_1(x, y) = -v/D_T \tag{A-3}$$

The solution of this equation for a rectangular section with boundary conditions (2) is well known and may be written as:

$$T_1(x, y) = vt_0 \; S(x/b, y/d; d/b) \tag{A-4}$$

where:

$$t_0 = b^2/(4D_T) \tag{A-5}$$

$$S(x^*, y^*; s) = \frac{16}{\pi^3} \sum_{m=0}^{\infty} \frac{(-1)^m \cos\left[(2m+1)\pi x^*\right]}{(2m+1)^3} \left[1 - \frac{\text{ch}\left[(2m+1)\pi s y^*\right]}{\text{ch}\left[\frac{(2m+1)\pi s}{2}\right]}\right] \tag{A-6}$$

The maximum temperature difference across the specimen $\Delta T$ is taken to be the easiest variable to measure during the test, and is given by $T_1$ (0, 0). Writing Eqn. (1) in terms of $\Delta T$, one finally gets for the asymptotic temperature distribution:

$$T(x, y, t) = T_s + \Delta T \;\; S(x/b, y/d; d/b)/S_0(d/b) \tag{A-7}$$

where $S_0$ (d/b) stands for S (0, 0; d/b).

This temperature distribution is taken as the "design distribution" since it gives upper bounds for the far field stresses.

ESTIMATION OF THE FAR FIELD STRESSES

The thermomechanical behaviour of moist concrete at low temperatures is known to be highly nonlinear [4, 8, 9]. Nevertheless since both the stress level and the temperature differences will be kept small in practice, this behaviour is considered as linear and time independent, being understood that this represents a tangent approach to the actual constitutive relation, and write, for the uniaxial case:

$$\varepsilon = \varepsilon(T_s) + \sigma/E + \alpha(T-T_s) \tag{A-8}$$

where E and $\alpha$ are, respectively, the instantaneous elastic modulus and the tangent coefficient of thermal expansion at the surface temperature.

Using Saint Venant's principle and neglecting Poisson effect, the stresses normal to the cross section are found to be:

$$\sigma(x, y) = \alpha E \ \{<T-T_s> - (T-T_s)\} \tag{A-9}$$

where < > denotes average value over the cross section.

Use of Eqn. (7) leads to the following axial stress distribution, where the dependency on the arguments is not explicity shown,

$$\text{''}) = \alpha E \ \Delta T \ (<S> - S)/S_0 \tag{A-10}$$

Moist concrete displays, on cooling, a strong expansion in the temperature range between -20 C and -60 C, and, accordingly, in this interval $\alpha$ becomes negative [4, 8] and can reach very high absolute values (30 μm/m C or greater). Taking this into account, the maximum tensile stress is given by:

$$\sigma_{max} = \alpha E \ \Delta T \ <S>/S_0 \qquad \text{for} \quad \alpha > 0$$

$$\sigma_{max} = |\alpha| E \ \Delta T\{1 - <S>/S_0\} \qquad \text{for} \quad \alpha < 0 \tag{A-11}$$

Equation (11) has been computed for different depth-to-thickness ratios. Averaging of S has been done by analytic term-to-term integration. The resulting series for <S>, as well as that for $S_0$, has been approximately summed by taking 101 terms. The results are shown in Fig. A.1.

ESTIMATION OF THE STRESS INTENSITY FACTOR

The elastic stress field taking into account the effect of the crack may be analysed, as shown in Fig. A-2, as the superposition of the remote stress field just calculated (state I) and the field generated by a stress distribution on the crack surface oposite to the far stress field (state II). Only state II gives rise to singular stresses, but it can not be analysed in a straightforward way because of through the thickness variations of the stresses.

This is overcome by reducing state II to the approximate biaxial case obtained by through the thickness averaging, as shown in Fig. A-3a. Denoting by $<\ >_x$ the averaged value over the thickness, the stress distribution can be written as

$$<\sigma>_x(y/d;\ d/b) = \alpha E \ \Delta T\{<S>(d/b) - <S>_x(y/d;\ d/b)\}/S_0(d/b) \tag{A-12}$$

Using the approximate Tada's solution [12] for $K_I$ in the point loading case depicted in Fig. A-3b, given by:

$$K_I = 2P(\pi a)^{-1/2}F(u/a;\ a/d) \tag{A-13}$$

where the influence function F is given by:

$$F(u^*;\ a^*) = \frac{3.52(1-u^*)}{(1-a^*)^{3/2}} - \frac{4.35-5.28u^*}{(1-a^*)^{1/2}} + \left[\frac{1.30-0.3u^{*3/2}}{(1-u^{*2})^{1/2}} + 0.83-1.76u^*\right]\left[1-(1-u^*)a^*\right] \tag{A-14}$$

one gets the following Green's formulation for the stress intensity factor:

$$K_I = 1.128\sqrt{a} \int_0^1 <\sigma>_x(0.5 - u^*\frac{a}{d};\ \frac{d}{b})\ F(u^*;\ \frac{a}{d})\ du^* \tag{A-15}$$

The stress intensity factor has been computed numerically for a number of depth to thickness and notch length to depth ratios, using Gauss-Chebyshev quadrature (101 integration points over the (0, 1) interval). The average values of S appearing implicity in the integrand have been evaluated by analytic term to term averaging of Eq. (6) followed by summation of 101 terms of the series. The results have been plotted in dimensionless form in Fig. A-4.

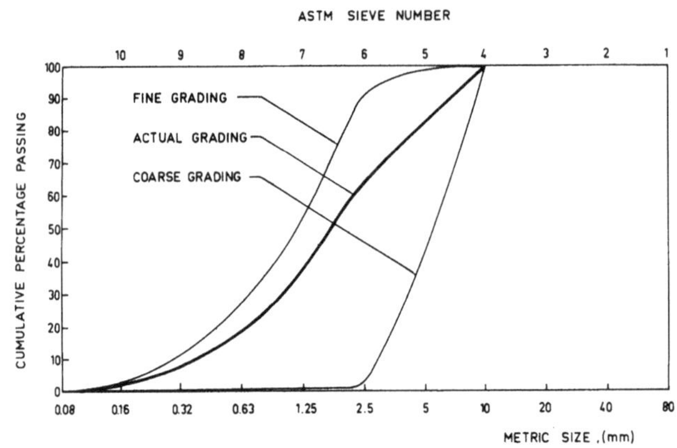


Fig. 1.- Grading of concrete

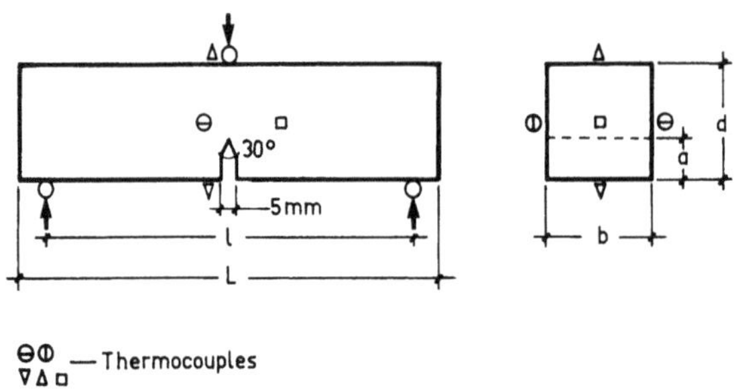


| L | l | d | a | b |
|---|---|---|---|---|
| 300 | 125 | 50 | 17 | 100 |
| 550 | 250 | 100 | 33 | 100 |
| 550 | 500 | 200 | 68 | 100 |
| 800 | 750 | 300 | 100 | 100 |

dimensions in m.m.

Fig. 2.- Notched beam geometry and thermocouples arrangement

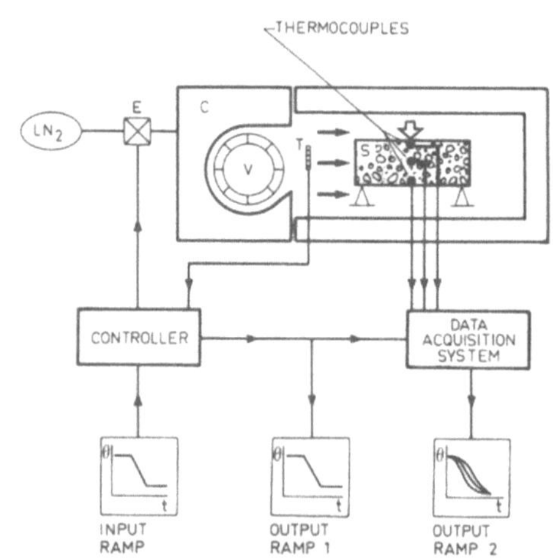


Fig. 3.- Cooling system. Block diagram

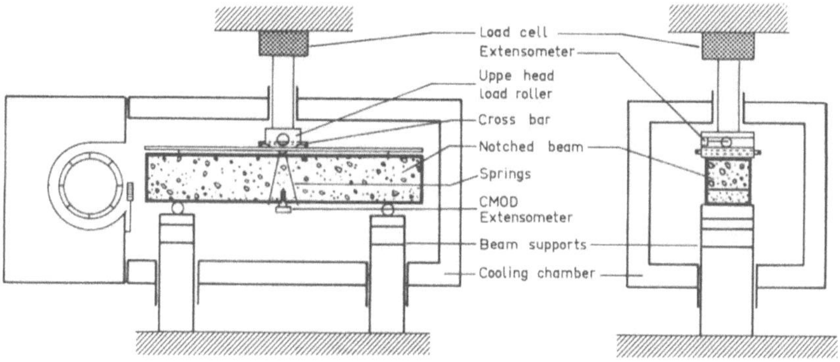


Fig. 4.- Beam testing set-up

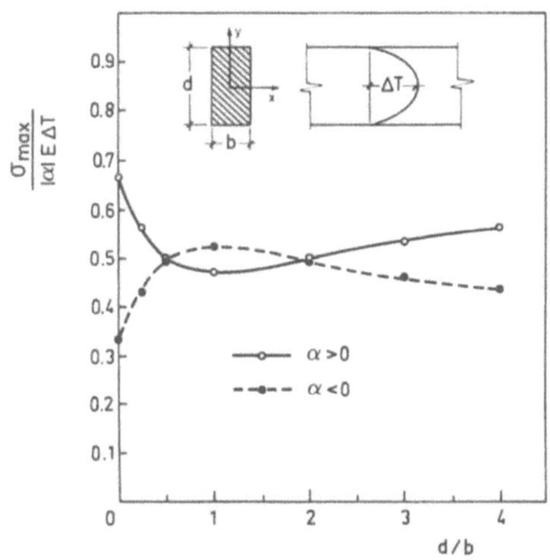


Fig. A-1.- Values of $\sigma_{max}$ across the specimen section

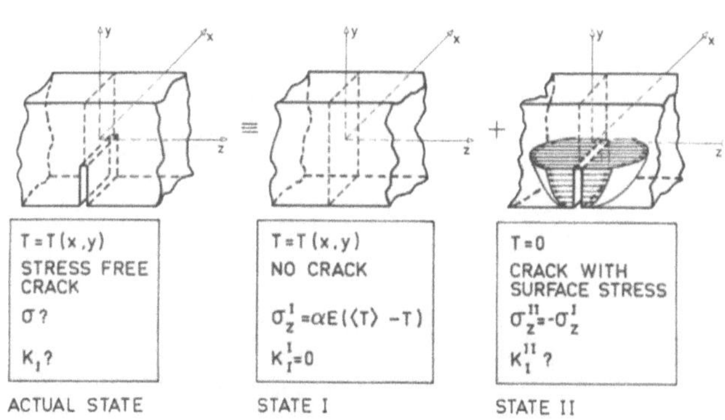


Fig. A-2.- Estimation of the stress intensity factor by the superposition method

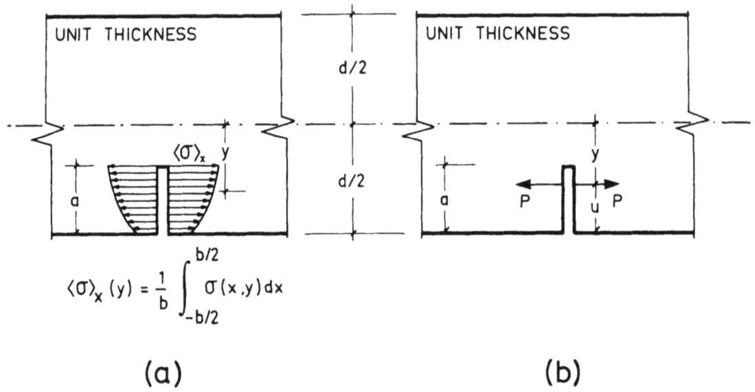


Fig. A-3.- Through the thickness averaging procedure

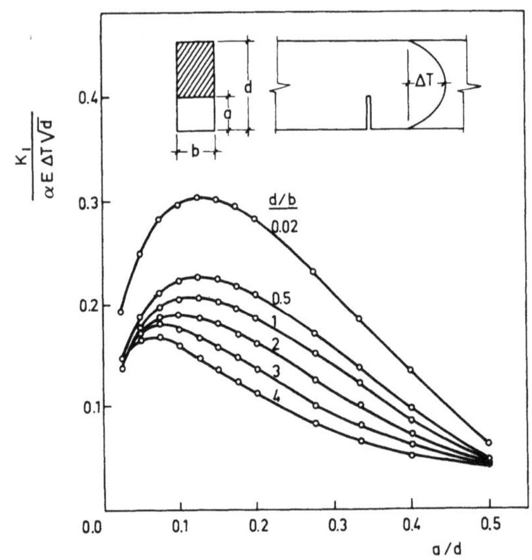


Fig. A-4.- Stress intensity factor for a notched beam subjected to non uniform temperature distribution during cooling

SEM/RILEM International Conference
on Fracture of Concrete and Rock
Houston, Texas, June 1987
Editors: S.P. Shah, S.E. Swartz

# MIXED MODE FRACTURE TESTS ON CONCRETE

H.W. Reinhardt, H.A.W. Cornelissen, D.A. Hordijk

ABSTRACT

Mixed mode fracture testing should provide a more comprehensive knowledge of concrete in the softening range, which is needed for a complete modelling in finite element codes. After a survey of some mixed mode testing methods which have been applied so far it is shown that all of these are not capable to supply necessary data in a physically clear and independent way. This is the reason why a new equipment has been designed and built which uses plane quadratic push-off type specimens. Loading is controlled in two independent directions by displacement control. The main features of the equipment are described and the results of a few preliminary tests are presented. It is concluded that the equipment fulfills the requirements of stable testing. Further experiments are announced.

## 1. INTRODUCTION

During the last decade, cracking of concrete has received considerable attention. This is true for experimental investigations using displacement controlled loading devices which enable the determination of the complete force-displacement relation of a specimen in uniaxial tension. It could be shown that this relation consists of an almost linear part up to about 80% of the tensile strength, a top value and, after that value, a steep decay. This decay is due to the occurance of a single crack which spreads into the specimen. After complete formation of the crack no force can be transmitted anymore.

This force-displacement relation can be transposed to a stress-strain relation before cracking. After cracking a stress-crack opening displacement can be defined. The relevant quantities describing the material are Young's modulus, tensile strength, stress free crack opening displacement and shape of the descending branch of the stress-crack opening relation. The total area under this relation is the specific fracture energy $G_f$.

Material models have been derived on the basis of these experimental findings. In terms of fracture mechanics, they belong to mode I crack opening. They have been incorporated in finite element codes which lead to good agreement between numerical prediction and experimental results of structural tests.

The shortcomings of these models are that they are confined to mode I. Many structural problems, however, ask for more generally applicable models since the behaviour of a structure is only seldom confined to mode I. An example can illustrate this. Let us consider a simply supported reinforced concrete beam in four point bending. The first vertical cracks occur due to tensile stresses in the most stressed fibres. The cracks in the pure bending zone will keep propagating vertically and will open in a mode I sense. In the shear zone, however, the cracks tend to bend which is a consequence of the inclination of the principal stresses. The cracks are formed perpendicularly to the largest principle stress. In this stage the crack is a mode I and opens perpendicularly to the direction of the crack tip. During further loading, the crack opens more and is loaded by shear simultaneously. Tests on reinforced beams have clearly demonstrated [1] that the opening of the crack is no longer perpendicular to the crack but is influenced by a mode II shear movement. The transition from pure mode I to mixed mode I and II is due to the change of the structural system after an inclined crack has formed. Tensile forces are taken by the reinforcement while shear is taken by the partly cracked concrete web and the compressive zone.

In most finite element codes, the criterion for a crack is that the largest principle stress reaches the tensile strength. The crack direction is perpendicular to this principle stress. Whereas this point is quite clear the question arises with respect to the available fracture energy. In mode I, the whole fracture energy

H.W. Reinhardt is Professor of Structural Engineering at Stevin Laboratory, Delft University of Technology, The Netherlands, and Darmstadt University, West Germany. H.A.W. Cornelissen is Senior Research Engineer in the Civil and Architectural Engineering Department of N.V. KEMA, Arnhem, The Netherlands. D.A. Hordijk is a Research Engineer at Stevin Laboratory, Delft University of Technology, The Netherlands.

is dissipated by the crack opening normal to the crack. What happens if, in subsequent loading steps, the opening changes to mixed mode I and II? How much of the energy may be attributed to mode I and how much to mode II, or is fracture energy as determined in a mode I test, the total attainable energy or only a part of it? Besides energy, stiffness (or compliance) should be known for simultaneous mode I and mode II occurrence

There are theoretical models derived and there are test results available on cracks under mode II loading. The tests, however, are carried out on precracked specimens with crack width of 0.05 mm and larger. What is still missing is the understanding of the continuous transition from uncracked to cracked concrete under displacement controlled circumstances.

The intention of a current research programme at Delft University of Technology is to carry out relevant tests to analyse these in terms of fracture mechanics. The paper will deal with the experimental set up and will show some preliminary results. Before this will be done various test methods for mixed mode testing will be discussed.

## 2. MIXED MODE TESTING METHODS

During the past, many test configurations have been conceived for mode II and mixed mode testing. In pure mode II testing, it is tried to create a shear force without simultaneous bending and without tensile or compressive normal force. Fig. 1 shows a few examples. The idea of Fig. 1a is often applied to the testing of joints in timber structures, of joints in masonry or joints between prefabricated concrete slabs. Fig. 1b shows a specimen shape which has extensively been used in testing the shear transfer in cracks in concrete and is suggested in [2] for determining the critical stress intensity factor in mode II loading of concrete. Finite element analyses have shown that the ratio a/W should be less than 0.25 in order to reach a smooth shear stress distribution [3].

The notched beam was used in two ways: Arrea/Ingraffea [4] used a beam with one notch whereas Bažant/Pfeiffer [5] applied two edge notches. The first authors observed a crack at the tip of the notch which propagated under a certain angle into the material and proceeded subsequently more or less vertically and ended under the top force. Arrea/Ingraffea concluded that shear failure as such does not exist, but cracks are caused by the largest principal stress and are perpendicular to the direction of that stress. The tests have been analysed by several researchers using finite element codes [6,7,8]. They showed good agreement between test and analysis although the shear model for concrete was rather crude and not verified by tests. This means that an appropriate crack softening model for mode I would be sufficient.

Bažant refused the idea of the non-existence of shear failure strongly. He argued that the crack path as observed by Ingraffea is due to the wide zone of shear force. In order to support this argumentation Bažant carried out tests on beams with two edge notches and a narrow zone of shear force, Fig. 2b. In these tests, two cracks started to propagate at the tip of the notches and joined each other about half way the beam depth. Bažant concluded from this behaviour that shear fracture exists, i.e. the crack can propagate in mode II. On the other hand, he says that "the shear fracture is likely to form as a zone of inclined tensile microcracks which only later connect by shearing". This means, it is a question of level of observation how shear failure would have to be modelled. The fracture propagated vertically in the tests by Bažant and remained in the highest stressed zone of the beam. This suggests that crack propagation direction is governed by the criterion of the maximum energy release rate rather than by the largest principal stress.

From the tests and by use of the size effect law the fracture energy in mode II was calculated which lead to values of the order of $G_f^{II} = 1000\ J/m^2$ or about ten times the value of mode I. However $G_f^{II}$ is mainly due to compressive failure of struts between tensile microcracks and is no material property as such.

The idea to use an inclined crack in a plate where a combination of mode I and mode II occurs is common practice in linear fracture mechanics. Formulae exist which allow the computation of $K_I$ and $K_{II}$ and the angle of crack propagation [1]. The theory is extended to anisotropic solids which is of particular interest to rock [10]. From uniaxial mode I testing it is known that concrete is a crack softening material and that critical stress intensity factors may apply only to very large specimens (and structures). On the other hand, there will be always two stress intensity factors at the tip of an inclined crack as shown by Fig. 3a and it is very probable that mode I governs crack extension because $K_{Ic}$ of concrete is rather low. Therefore the test according to Fig. 3a is not suited for supplying the knowledge on concrete behaviour.

The idea of the inclined crack has been further developed [11]. A compressive force is applied which causes a negative stress intensity in mode I, i.e. no fracture due to crack opening can occur and mode II can be studied. This loading configuration is used by Izumi et al. for determining $K_{IIc}$, but no attempt has been made to establish $G_f^{II}$.

All tests described so far with the exception of the beam test (Figs. 1c, 1d) originate from linear elastic fracture mechanics (LEFM) and are conceived for the determination of a critical stress intensity factor in mode II or mixed mode. Since concrete does not obey the conditions of LEFM (at least not in usual specimen sizes) these tests are not suited for an appropriate investigation of concrete. The beam tests are tools for studying the cracking behaviour but they do not allow to establish material properties and influences on that properties as discussed in Chapter 1. Therefore, new testing methods should be developed which enable a separate control of mode I and mode II displacement in a mixed mode test.

Chapter 3 will provide information on the testing method developed in the Stevin Laboratory of Delft University of Technology. Before this will be done, the testing method of Darmstadt University of Technology will be discussed shortly [12]. This method uses a hollow cylinder which is tested uniaxially in a deformation controlled way until the descending branch of the stress displacement diagram. Then, torsion is applied which leads to almost uniform shear stress distribution, i.e. strain softening is combined with mode II loading. This testing method uses the same idea and will have similar capabilities as the one which is described in the following chapter although the design and mechanical functioning are completely different.

## 3. PLANE BIAXIAL LOADINC EQUIPMENT

### 3.1 Motives

At Delft University of Technology, much experience is gained in performing deformation controlled uniaxial tensile tests as illustrated by Fig. 4a. It turned out that deformation control was possible by use of a rather stiff loading frame and a sufficiently fast electro-hydraulic circuit. Having this experience it was a simple idea to add a second circuit perpendicularly to the first one which should provide shear loading (Fig. 4b). As the specimen geometry and the loading configuration is concerned two requirements had to be fulfilled:

a) The stress distribution in the critical cross-section should be as uniform as possible with respect to normal stress and shear stress.

b) Forces should be applied as direct as possible in order to avoid loss of stiffness and also as independent as possible in order to be able to control normal and shear stress separately.

Requirement a) leads to a certain length to width ratio of the specimen while requirement b) asks for a rather small specimen and for two independently operating closed-loop control systems. The advantage of such a testing method is that each combination of mixed mode can be realized.

### 3.2 Specimen

The requirements as stated above lead to a specimen type as indicated by Fig. 5. It has already been discussed that the distance between the saw cuts should be chosen in such a way that a smooth fracture plane in pure shear develops [3]. However, since most of the tests which will be performed, will be loaded in shear only after the softening has started in the axial loading direction the ratio a/W is thought to be less important. Softening redistribution of stresses will occur in the crack plane rather than elastic stress concentrations.

The shape of the specimen is quadratic with an edge length of 200 mm, the thickness is 50 mm. The saw cuts are 5 mm wide and 100 mm long with the distance a being 50 mm. In order to force the crack to propagate in a straight line some specimens are grooved additionally as illustrated by Fig. 6.

### 3.3 Testing rig

Main points for the design of the testing rig were that tensile as well as compressive stresses could be applied both in normal and transverse (shear) direction which should result in pure normal and/or shear stresses in the fracture zone. This required stiff loading frames in order to prevent eccentricities during non-uniform fracture.

Fig. 7 shows the stiff square loading frames, an inner one for the normal direction and because of symmetry two coupled frames for the transverse direction. As can be seen, plate springs are mounted which prevent the frames from rotation. Because these springs have a limited (linear) bending capacity an extra guiding system was necessary in order to facilitate mounting of the specimen (see Fig. 8). This guiding system is blocked during the execution of the test, so only the plate springs follow the deformations. The resistance of the springs can be calculated from strain gauge measurements (1 mm bending equals 50 N) and be subtracted from the force which is read from the load cells. The readings from the plate springs can be used for checking the centricity during the test. The load cells are situated outside the loading frame in order not to impair the stiffness of the equipment.

Fig. 7b shows that the hydraulic actuators act along the centre lines of the specimen in both directions. The capacity of the actuators is 100 kN in tension and in compression. The closed-loop hydraulic systems can operate under load as well as under displacement control.

Two LVDTs are attached to the specimen which measure the relative displacement over the fracture zone in normal direction. The signals are averaged and compared with the external ramp generator control signal. The usual constant rate of deformation is 0.08 μm/s. The signal for the load in the transverse direction (shear) is supplied by a simple internal generator.

Electro-mechanical extensometers of 35 and 110 mm gauge length are used for the deformation readings. Since the equipment is still in the stage of development the type, place and number of extensometers is not yet fixed. Fig. 9 shows two different arrangements. The data from the measurements are sent to a micro computer and stored on a floppy disk. Later on the results are processed and presented in a graphical form. Fig. 10 gives a schematic view of the total testing set-up.

## 4. PRELIMINARY TESTS AND RESULTS

### 4.1 General

At the moment that the preliminary tests were performed which will be described here, the data aquisition system was not yet complete. This is the reason for the fact that the readings from the spring platens are not available and that the correct force could not be displayed. However, it is assumed that the basic idea and behaviour of the specimen can be presented despite of this shortcoming.

### 4.2 Concrete mix and specimen preparation

Table 1 shows the mix of the concrete with 8 mm maximum aggregate size. The specimens were sawn from 300mmx300mmx50mm plates which were cast vertically in a battery mould. After two days the plates were demoulded and cured under water for two weeks whereafter sawing took place. After sawing, the specimens were stored in the laboratory at 20°C and 60% relative humidity until the instant of testing. The testing age of the specimen varied between 28 and 60 days.

Standard mechanical properties of the concrete at 28 days are also given by Table 1.

| Mix proportions | ($kg/m^3$) |
|---|---|
| Portland cement | 375 |
| sand 0-2 mm | 905 |
| sand 2-4 mm | 363 |
| gravel 4-8 mm | 540 |
| water | 187.5 |
| fresh density | 2366 |
| **Mechanical properties** | **($N/mm^2$)** |
| cube compressive strength | 48.4 (2.6%) *) |
| cube splitting strength | 3.12 (5.6%) |

*) coefficient of variation

Table 1. Mix proportions and 28 day mechanical properties.

### 4.3 Mounting of the specimen

Before glueing the extra guiding system was unlocked and the two loading frames were positioned in such a way that maximum space between the glue platens was available. Glue was supplied at the appropriate edges of the specimen while it was mounted in an auxiliary lifting and positioning device. Due to this device, it was not complicated to position the 4.5 kg weighing specimen accurately and smoothly between the loading frames which were then moved to the specimen edges which completed the mounting procedure.

### 4.4 Types of tests

All tests so far were preliminary tests, i.e. the aim of the tests was rather to check and improve the equipment and the testing method than to investigate basic and well defined properties of a material. The loading combinations and the procedure of the tests was therefore more or less arbitrarily chosen.

First, some uniaxial tensile tests were performed the results of which could be compared with earlier tests. For these tests, the extensometers were chosen as shown in Fig. 9a. Then, tests were carried out in which a softening zone was created by uniaxial tension first up to a deformation of about 15 to 20 μm. Then the uniaxial deformation was fixed and a shear force was applied gradually up to a certain value. This value was kept constant when the normal deformation was activated again at the same constant rate as before.

Subsequently, two tests were performed in which a crack opening was created by uniaxial tension. Then, cyclic shear was applied with an upper shear stress level of 2 $N/mm^2$ and a lower one of zero. After three manually controlled cycles the shear stress was increased till 8 $N/mm^2$ and decreased to zero again. Then the specimen was unloaded completely. In these tests, the measuring devices were arranged according to Fig. 9b on a specimen as shown by Fig. 6.

Finally a shear test was carried out on a virgin specimen which was not precracked in tension. This test should give the elastic shear modulus and served also as a check for the loading and measuring devices.

4.5 Test results

The deformation distribution of a specimen subjected to a uniaxial tensile test in the mixed mode testing equipment, is shown in Fig. 11. The upper part of the figure is composed of four measurements which are the means of the corresponding readings of the extensometers at the front and the rear side of the specimen. The lower part shows a cross-section with the mean of all measurements of the front side at one edge and the mean of all measurements of the rear side at the other edge. This gives the out-of-plane rotation. It can be seen that the non-uniform crack opening is of minor importance and that an out-of-plane rotation occurs at about 5 to 15 μm deformation. Later on the crack opens parallel again. It can be concluded that the loading equipments meets the requirements as stated in chapter 3.

Stress-deformation diagrams of two mixed mode tests are shown by Fig. 12. In these tests, a softening zone was created first which had a normal deformation of about 15 μm in test number 02-2 and about 20 μm in test number 02-4. Then a shear force was applied gradually which was 1 kN in the test 02-2 and 2 kN in test 02-4, corresponding to average shear stresses of 0.4 $N/mm^2$ and 0.8 $N/mm^2$, respectively. Then, loading in the normal direction was resumed.

Fig. 12 shows that the normal stress decreases at constant crack opening when a shear force is applied. There are two mechanisms which might be responsible for that behaviour: true shear strain or relaxation. When the loading in normal direction is resumed while the shear stress is still active, the normal stress increases again. It looks like the stress-deformation relation were following a uniaxial envelope curve. Whether this speculation is true can only be answered by future experiments.

The following two sets of figures give the results of cyclic shear loading at constant normal deformation. Fig. 13a shows the normal stress versus normal deformation relation during the complete test. Uniaxial loading is applied first until the normal deformation reaches the value of 58 μm. Then a shear stress is applied at the normal deformation which was intended to be constant. However it can be seen that normal deformation increased a little and reached a value of 63 μm after four cycles. However, it should be kept in mind how small the absolute and relative displacements are.

Fig. 13b shows the shear stress versus shear displacement relation for the same test. First, three cycles were applied up to 2 $N/mm^2$. It can be seen that the shear stiffness varies from the first to the second cycle, but that the second and third cycle are almost the same. Obviously, there is energy dissipation which seems to be partly due to friction and partly to damage. The following cycle with higher stress shows an almost linear relation between stress and deformation. Unloading is similar to the first small cycle, i.e. large energy dissipation occurs.

Fig. 13c gives the normal stress versus shear deformation relation for the four cycles. It can be seen that a small normal stress develops during the first small shear cycles with negligible energy dissipation. The last cycle however shows a large degradation of the material.

Fig. 14 is a comparable set of figures with the difference that the normal deformation during cycling is 107 μm. Fig. 14a gives the normal stress versus normal deformation relation, Fig. 14b is the shear stress versus shear deformation plot. The features are the same as in Fig. 13b but it is even more clear that energy dissipation is mainly due to a frictional process rather than a damaging process. Fig. 14c shows once more the normal stress versus shear deformation relation.

In a last test, the shear modulus should be determined on a virgin specimen. It turned out that the shear modulus $G_0$ was much higher than what have been expected from calculating by

$$G_0 = \frac{E_0}{2(1+\nu)}$$

with Young's modulus $E_0$ and Poisson's ratio $\nu$. After re-examining the test set-up it appeared that the measurements of the shear gages consist of two parts: one is the translating movement, the second is the rotation due to shear. What was intended to measure was the translation only. The two parts have opposite sign irrespective of the direction of the external shear force. So, the final measuring result must be too small and lead to a shear modulus which is too high.

Before another slip gage will be available the results are corrected by adding an elastic shear deformation. This has been done for the results of Figs. 13 and 14. The result is shown in Fig. 15 together with the result of the uncracked specimen. Now it can be seen how previous precracking leads to less a shear stiffness and how normal stress develops during shearing. Normal stiffness decreases also with larger precracking. The shear retention factor $\beta$ which is a commonly used factor accounting for shear degradation in cracks can be estimated from Fig. 15. $\beta$ is 0.38 for an axial deformation of 58 μm, and 0.25 for a deformation of 107 μm. This is considerably more than has been proposed so far [13,14] which is 0.15 and 0.08 for comparable normal strain.

The moment being, the tests have not been evaluated further. A testing program will start soon which will provide more systematic data.

## 5. CONCLUSIONS

A few conclusions can be drawn:

- There is a need for the thorough knowledge of concrete behaviour under mixed mode loading.
- There are several testing methods suggested or applied which pretend to investigate mixed mode behaviour.
- It should be defined in advance whether crack propagation under mixed mode loading should be investigated or whether a crack which is due to tensile stress (maximum stress failure criterion) should be studied under subsequent shear force.
- A test set-up with two independent deformation controlled electro-hydraulic circuits has been developed at Delft and is described. This set-up allows testing on precracked or non-precracked specimens.
- First results shows that the equipment operates according to the requirements.
- It has been demonstrated how shear stiffness depends on precracking. However, the test results need more evaluation.
- A testing series is recommended which will start soon to produce more systematic results.

## 6. ACKNOWLEDGEMENTS

The authors are very indebted to Mr. G. Timmers for his encouragement and substantial assistance during testing and to the Measurement and Instrumentation Group of the Stevin Laboratory for their assistance in designing the loading frame.

## 7. REFERENCES

[1] Walraven, J.C. The influence of depth on the shear strength of lightweight concrete beams without shear reinforcement. Stevin report 5-78-4, Delft University of Technology, Delft 1978

[2] Liu, K., Barr, B.I.G., Watkins, J. Mode II fracture of fibre reinforced concrete materials. Intern. J. Cement Composites and Lightweight Concrete 7 (1985), no. 2, pp 93-101

[3] Watkins, J., Liu, K.L.W. A finite element study of the short beam test specimen under mode II loading. Intern. J. Cement Composites and Lightweight Concrete 7 (1985), no. 1, pp 39-47

[4] Arrea, M., Ingraffea, A.R. Mixed-mode crack propagation in mortar and concrete. Report No. 81-13, Dept. Struct. Eng., Cornell University, Ithaca, NY, 1982

[5] Bažant, Z.P., Pfeiffer, P.A. Shear fracture tests of concrete. RILEM Materials and Structures 19 (1986), no. 110, pp 111-121

[6] Rots, J.G. Strain softening analysis of concrete fracture specimens. Proc. Int. Conf. on Fracture Mechanics of Concrete. (ed. F.H. Wittmann), Lausanne 1985, pp 137-148

[7] De Borst, R. Non-linear analysis of frictional materials. Diss. Delft University of Technology, Delft 1986

[8] Ohlsson, U., Gylltoft, K. Finite element analysis of mixed mode fracture using a discrete crack approach. Report 86-15, Div. Struct. Eng., University of Lulea 1986

[9] Erdogan, F., Sih, G.C. On the crack extension in plates under plane loading and transverse shear. Trans. ASME. J. Basic Engineering 85 (1963), no. 4, pp 519-527

[10] Saouma, V.E., Ayari, M.L., Leavell, D.A. Mixed mode crack propagation in homogeneous anisotropic solids. Eng. Fracture Mechanics (1986)

[11] Izumi, M., Mihashi, H., Nomura, N. Fracture toughness of concrete for mode II. Intern. Conf. on Fracture Mechanics of Concrete. ed. F.H. Wittmann, Lausanne 1985, pp 347-354

[12] Keuser, W. Kornverzahnung bei Zugbeanspruchung. Forschungskolloquium DAfStb, Darmstadt 1986, pp 13-18

[13] Bažant, Z.P., Gambarova, P. Rough cracks in reinforced concrete. J. Struct. Div. ASCE 106 (1980), pp 819-842

[14] Rots, J.G., Kusters, G.M.A., Nauta, P. Variable reduction factor for the shear resistance of cracked concrete (in Dutch). TNO-IBBC Report BI-84-33, Delft 1984

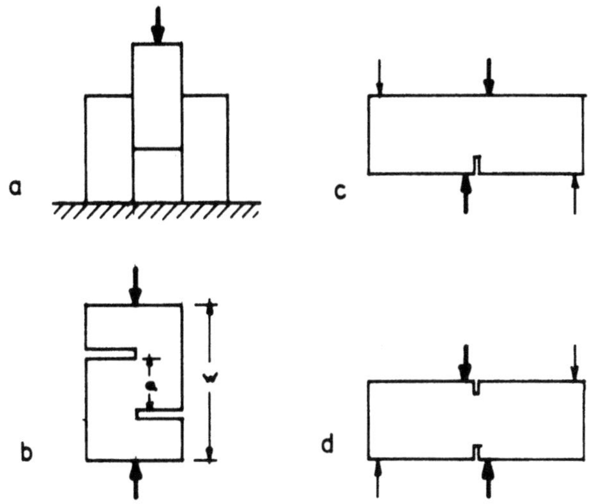


Fig. 1 Various types of mode II tests.

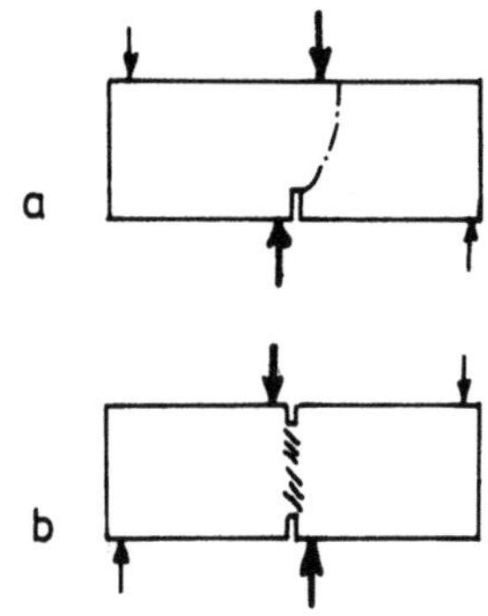


Fig. 2 a) Crack development in a wide shear zone.
b) Crack development in a narrow shear zone.

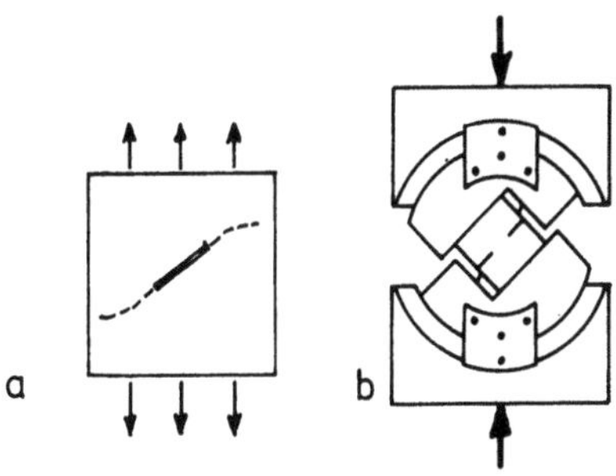


Fig. 3 a) Specimen with inclined crack for mixed mode testing.
b) Mode II testing.

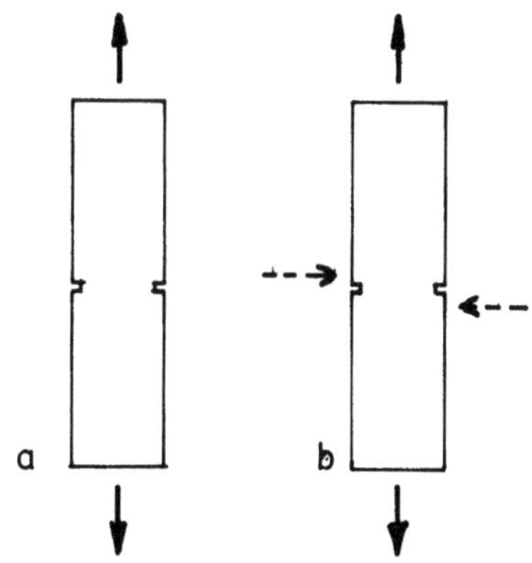


Fig. 4 a) Uniaxial tensile test.
b) Basic idea of mixed mode testing.

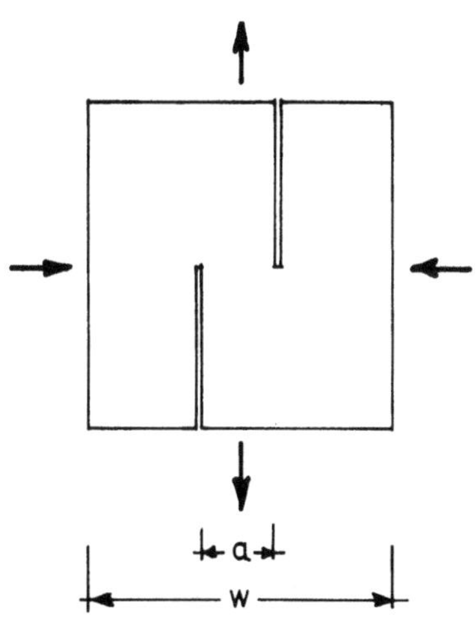


Fig. 5 Push-off type specimen for plane biaxial testing.

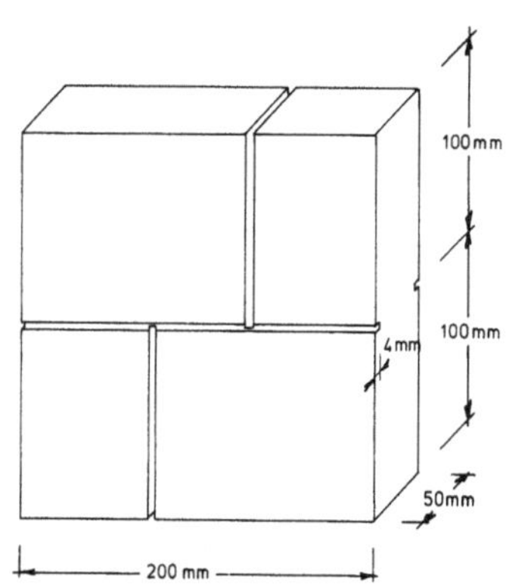


Fig. 6 Improved specimen with additional side groove.

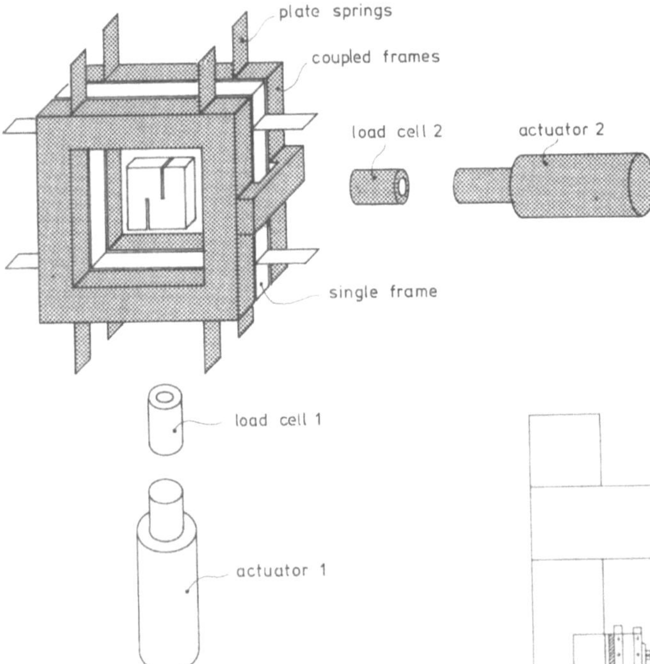


a

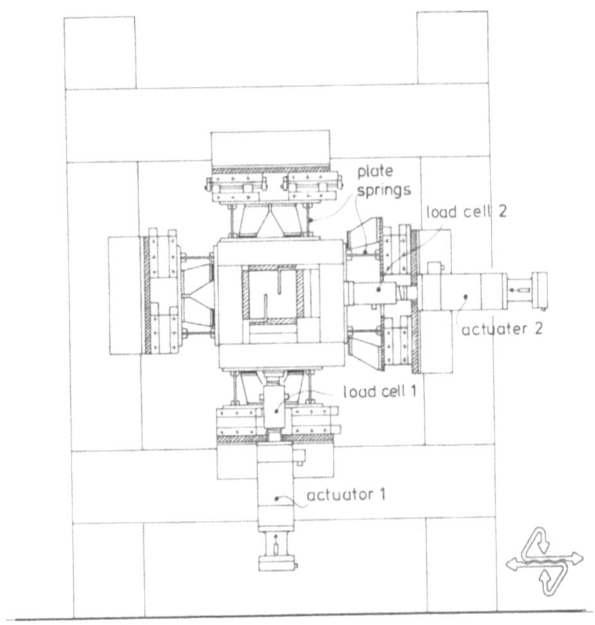


b

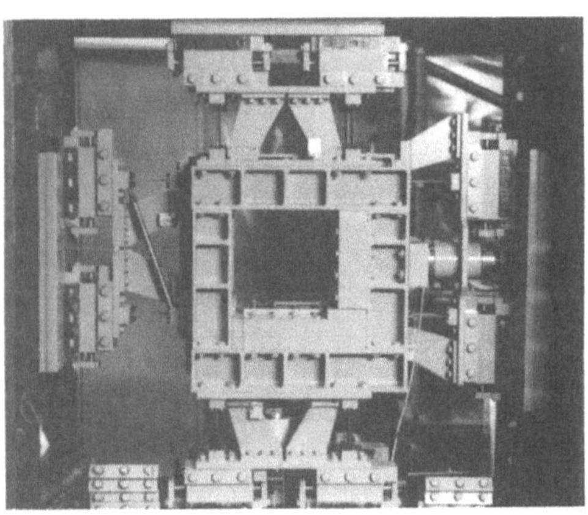

c

Fig. 7 a) Loading frames with specimen and loading devices.
b) Loading frames within support frame and guiding devices.
c) Photograph of loading equipment.

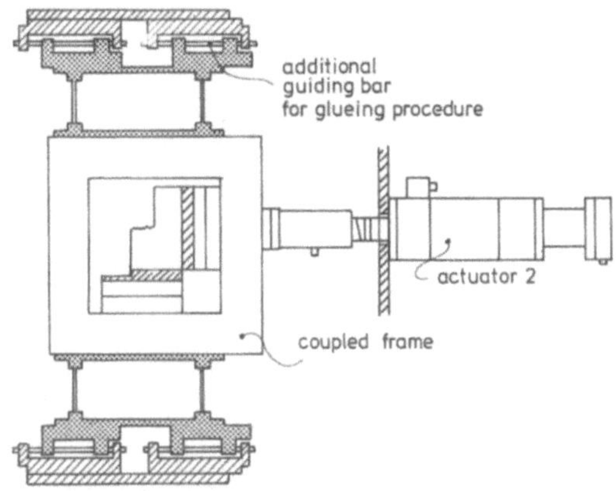


Fig. 8 Loading frame stiffed along additional guiding bar during glueing procedure.

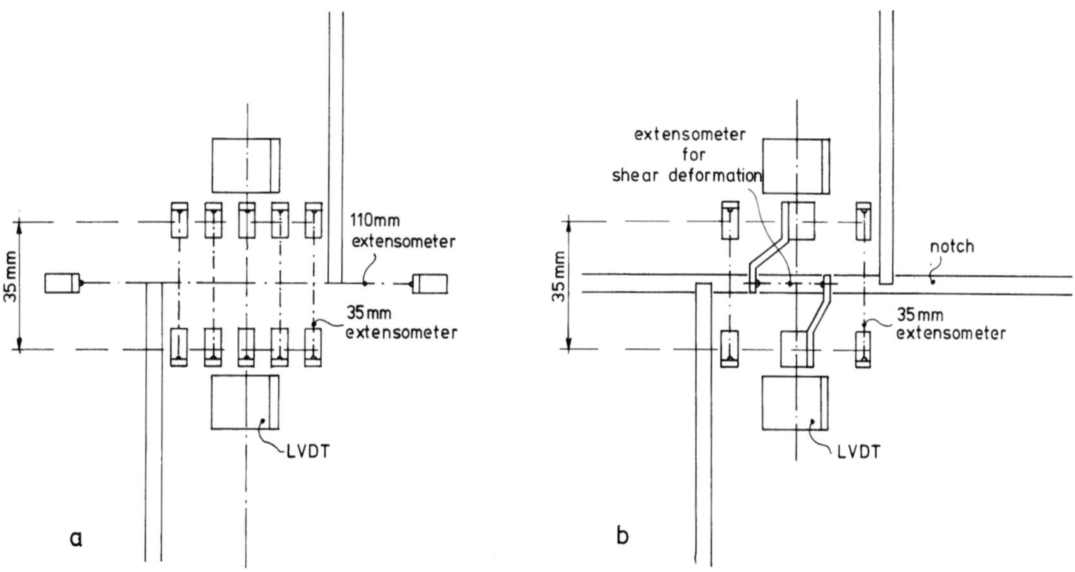

c

Fig. 9 a) and b) Various measuring arrangements.
c) Photograph of instrumented specimen.

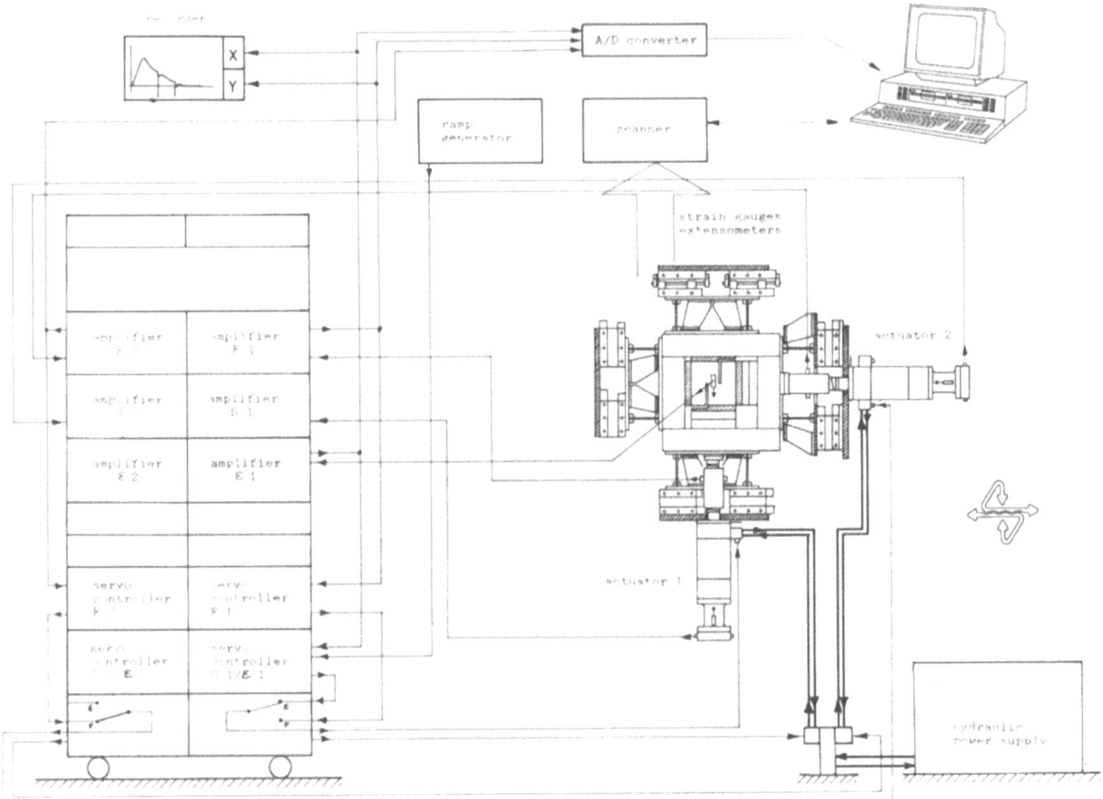


a

b

Fig. 10 a) Total schematic view of testing equipment.
b) Photographic impression of total testing equipment.

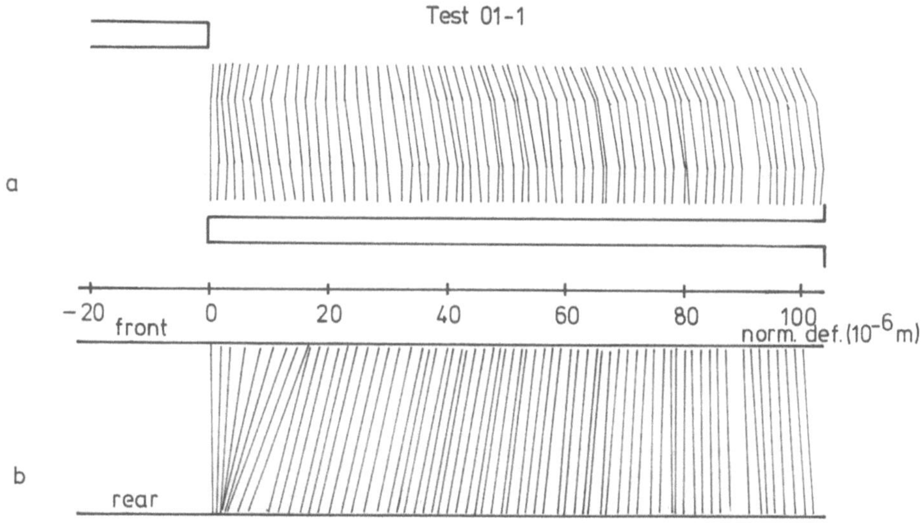


Fig. 11 a) In-plane deformation distribution at subsequent loading steps.
b) Out-of-plane rotation at corresponding loading steps.

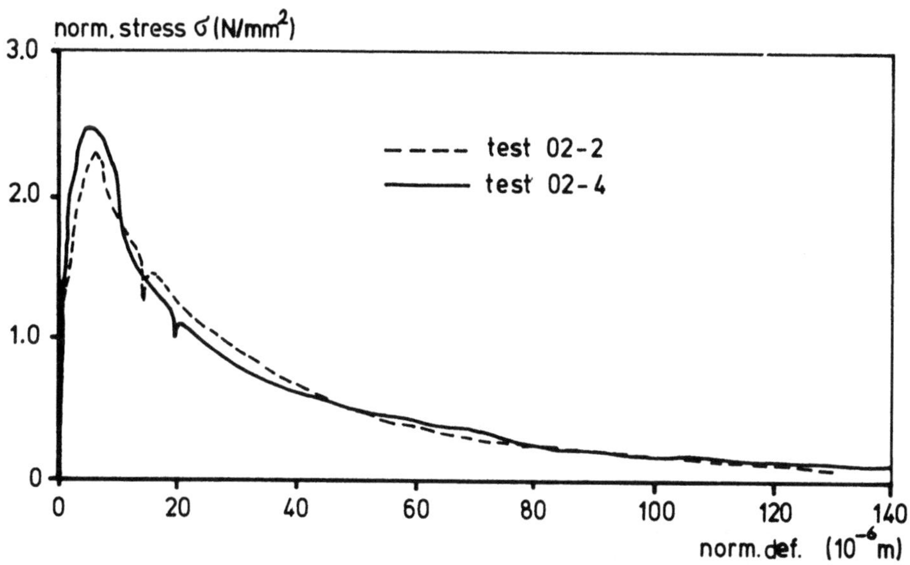


Fig. 12 Stress versus normal deformation of two tests.

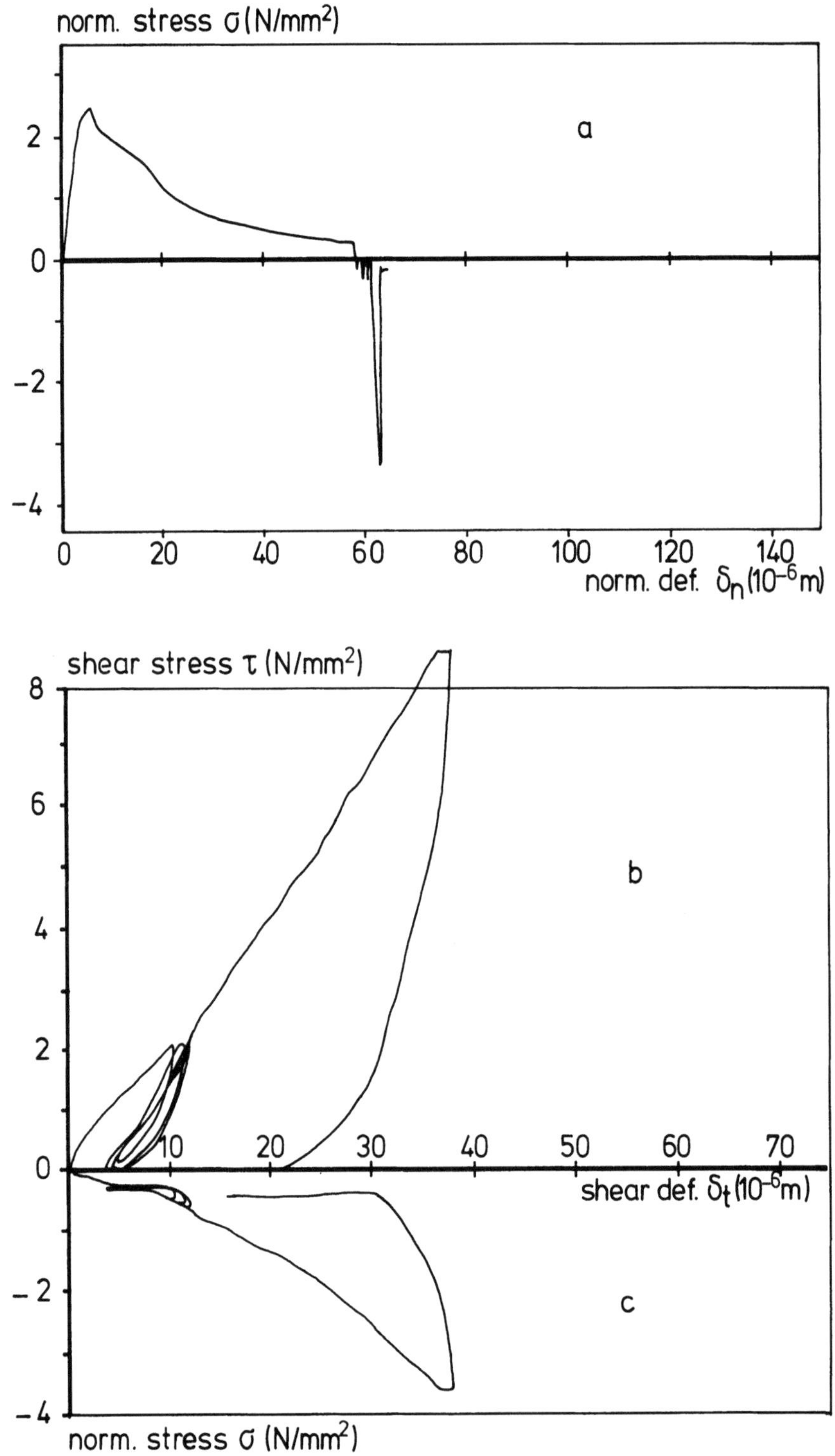


Fig. 13 a) Normal stress versus normal deformation.
b) Shear stress, and
c) Normal stress versus shear deformation during cycling at constant normal deformation of 58 μm.

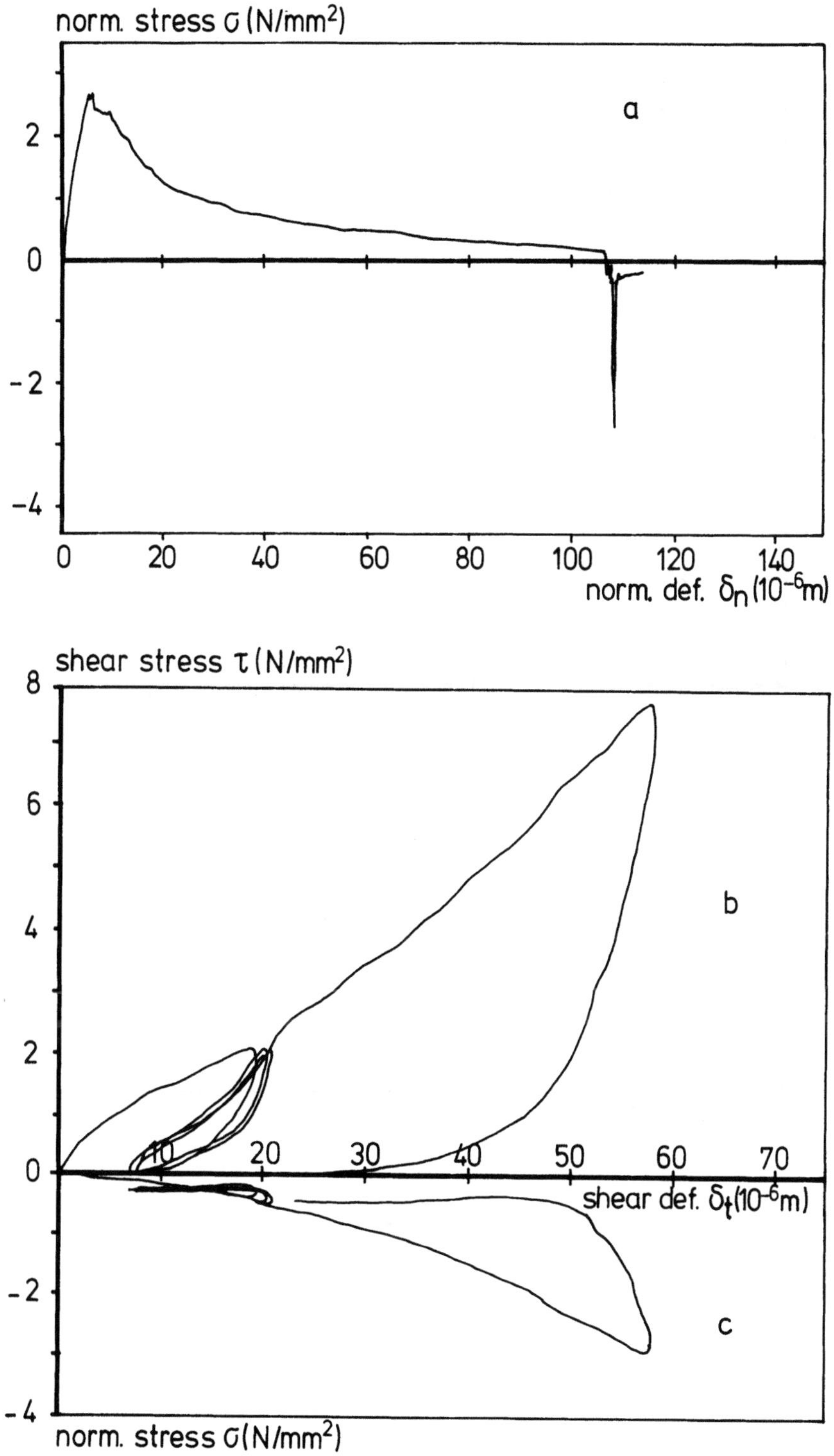


Fig. 14 Same as Fig. 13, but for normal deformation of 107 μm.

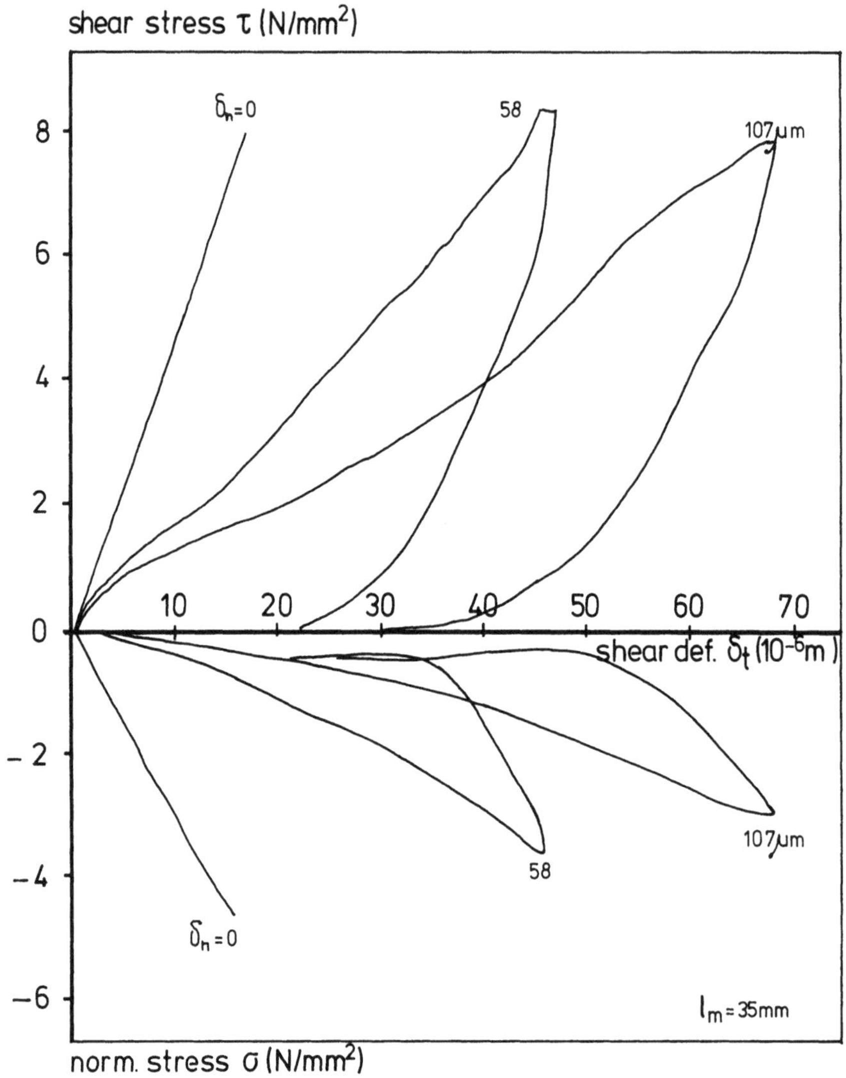


Fig. 15 Complete plot of four interrelated quantities: shear stress, normal stress, shear deformation and normal deformation.

SEM/RILEM International Conference
on Fracture of Concrete and Rock
Houston, Texas, June 1987
Editors: S.P. Shah, S.E. Swartz



# NUMERICAL SIMULATION OF CONCRETE FRACTURE THROUGH A BILINEAR SOFTENING STRESS-CRACK OPENING DISPLACEMENT LAW

A. Carpinteri, G. Colombo, G. Ferrara, G. Giuseppetti

## ABSTRACT

The crack growth process in concrete is analyzed by a cohesive crack model. A bilinear descending stress-crack opening displacement diagram is considered to simulate the flexural behaviour of pre-cracked specimens. Although the shape of the diagram affects the theoretical value of the maximum load, such a dependence is smaller and smaller by decreasing the material fracture energy and/or increasing the specimen size.

## INTRODUCTION

A three point bending test according to the R.I.L.E.M. Recommendation [1] is numerically simulated. The crack growth process is analyzed by a cohesive crack model with cohesive forces at the crack tip [2-5]. The tensile strength $f_t$ of concrete is obtained by a direct test, whereas the Young's modulus E and Poisson ratio $\nu$ are assumed as the average values between tension and compression. The fracture energy $G_F$ is then considered to be equal to the area under the experimental load-deflection curve, divided by the initial ligament area. Besides the linear descending stress $\sigma$-crack opening displacement w law, other twenty-five (25) bilinear $\sigma$-w diagrams with the same area $G_F$ are considered, the discontinuity point (knee) being located in 25 different positions. The numerical load-deflection response appears to be very sensitive to the $\sigma$-w shape. Only one of the twenty-five theoretical curves reproduces the experimental curve with sufficient accuracy.

The numerical stability and the graphycal resolution of the load-deflection curves are investigated by varying the knee location and the brittleness number $s_E = G_F/f_t b$ (b = beam depth) [6,7]. For each value of the dimensionless number $s_E$, there is a set of discontinuity points on the plane stress-crack opening displacement, which provide numerically stable load-deflection curves. It is proved that such an area of numerical stability shrinks by decreasing the number $s_E$.

The ratio of the maximum loading capacity obtained by the cohesive crack model to the maximum load according to Linear Elastic Fracture Mechanics (LEFM) is equal to the ratio of the fictitious fracture toughness to the true one $K_{IC}$. Such a ratio is demonstrated to tend to unity when $s_E \to 0$, for the linear as well as for the bilinear $\sigma$-w constitutive softening relations. This means that, for large structure sizes and/or low fracture energies, the bifurcation of the load-deflection curve [8,9] tends to describe the classical LEFM-instability. Therefore, it is possible to assert that, <u>although the shape of the curve $\sigma$-w affects the theoretical value of the maximum load, such a dependence is smaller and smaller by decreasing the number</u> $s_E$.
As a limit-case, for $s_E \to 0$, the maximum load derives from the well-known LEFM-relation: $K_I = K_{IC} = \sqrt{G_F E}$, and is depending only on the area $G_F$ under the assumed $\sigma$-w diagram and not on the shape of this diagram.

---

A. Carpinteri is Full Professor at the Department of Structural Engineering of the Politecnico di Torino, 10129 Torino, Italy; G. Colombo, G. Ferrara and G. Giuseppetti are Researchers at the Hydraulics and Structural Engineering Research Center of the Italian Power Agency (ENEL-CRIS), 20162 Milano, Italy.

## COHESIVE CRACK MODEL

The cohesive crack model is based on the following assumptions [2-5]:

(1) The cohesive fracture zone (plastic or process zone) begins to develop when the maximum principal stress achieves the ultimate tensile strength $f_t$ (Fig. 1-a).

(2) The material in the process zone is partially damaged but still able to transfer stress. Such a stress is dependent on the crack opening displacement w. A bilinear σ-w law is assumed as in Fig. 1-b.

The real crack tip is defined as the point where the distance between the crack surfaces is equal to the critical value of crack opening displacement $w_c$ and the normal stress vanishes (Fig. 2-a). On the other hand, the fictitious crack tip is defined as the point where the normal stress attains the maximum value $f_t$ and the crack opening vanishes (Fig. 2-a).

The closing stresses acting on the crack surfaces (Fig. 2-a) can be replaced by nodal forces (Fig. 2-b). The intensity of these forces depends on the opening of the fictitious crack, w, according to the σ-w constitutive law of the material (Fig. 1-b). When the tensile strength $f_t$ is achieved at the fictitious crack tip (Fig. 2-b), the top node is opened and a cohesive force starts acting across the crack, while the fictitious crack tip moves to the next node.

With reference to the three point bending test (TPBT) geometry in Fig. 3, the nodes are distributed along the potential fracture line. The coefficients of influence in terms of node openings and deflection are computed by a finite element analysis where the fictitious structure in Fig. 3 is subjected to (m + 1) different loading conditions. Consider the TPBT in Fig. 4-a with the initial crack tip in the node k. The crack opening displacements at the m fracture nodes may be expressed as follows:

$$\underset{\sim}{w} = \underset{\sim}{K}\,\underset{\sim}{F} + \underset{\sim}{C}\,P + \underset{\sim}{\Gamma}, \tag{1}$$

being:

$\underset{\sim}{w}$ = vector of the crack opening displacements,
$\underset{\sim}{K}$ = matrix of the coefficients of influence (nodal forces),
$\underset{\sim}{F}$ = vector of the nodal forces,
$\underset{\sim}{C}$ = vector of the coefficients of influence (external load),
P = external load,
$\underset{\sim}{\Gamma}$ = vector of the crack opening displacements due to the specimen weight.

On the other hand, the initial crack is stress-free and therefore:

$$F_i = 0, \quad \text{for} \quad i = 1, 2, \ldots, (k - 1), \tag{2-a}$$

while at the ligament there is no displacement discontinuity:

$$w_i = 0, \quad \text{for} \quad i = k, (k + 1), \ldots, m. \tag{2-b}$$

Eqs (1) and (2) constitute a linear algebraical system of 2m equations and 2m unknowns, i.e., the elements of vectors $\underset{\sim}{w}$ and $\underset{\sim}{F}$. If load P and vector $\underset{\sim}{F}$ are known, it is possible to compute the beam deflection, δ:

$$\delta = \underset{\sim}{C}^T \underset{\sim}{F} + D_P\,P + D_\gamma, \tag{3}$$

where $D_P$ is the deflection for P = 1 and $D_\gamma$ is the deflection due to the specimen weight.

After the first step, a cohesive zone forms in front of the real crack tip (Fig. 4-b), say between nodes j and n. Then Eqs (2) are replaced by:

$$F_i = 0, \quad \text{for} \quad i = 1,2,\ldots, (j - 1), \tag{4-a}$$

$$F_i = F_t^* \left(1 - \frac{w_i}{w_c}\right), \quad \text{for} \quad i = j,(j + 1)),\ldots,\ell, \tag{4-b}$$

$$F_i = F_t \left(1 - \frac{w_i}{w_c^*}\right), \quad \text{for} \quad i = (\ell + 1), (\ell + 2),\ldots, n, \tag{4-c}$$

$$w_i = 0 \qquad \text{for} \quad i = n, (n+1), \ldots, m, \qquad (4\text{-d})$$

where $F_t$ is the ultimate strength nodal force:

$$F_t = \frac{f_t\, b}{(m+1)}, \qquad (5)$$

and analogously: $F_t^* = f_t^* b/m+1$ (the extrapolated values $f_t^*$ and $w_c^*$ are shown in Fig. 1-b).
Eqs (1) and (4) constitute a linear algebraical system of (2m + 1) equations and (2m + 1) unknowns, i.e., the elements of vectors $\underset{\sim}{w}$ and $\underset{\sim}{F}$ and the external load P.

At the first step, the cohesive zone is missing (n = j = k) and the load $P_1$ producing the ultimate strength nodal force $F_t$ at the initial crack tip (node k) is computed. Such a value $P_1$, together with the related deflection $\delta_1$ computed through Eq. (3), gives the first point of the P-$\delta$ curve. At the second step, the cohesive zone is between the nodes k and (k + 1), and the load $P_2$ producing the force $F_t$ at the second fictitious crack tip (node k + 1) is computed. Eq. (3) then provides the deflection $\delta_2$. At the third step, the fictitious crack tip is in the node (k + 2), and so on. The present numerical program simulates a loading process where the controlling parameter is the fictitious crack depth. On the other hand, real (stress-free) crack depth, external load and deflection are obtained at each step after an iterative procedure.

The program stops with the untieing of the node m and, consequently, with the determination of the last couple of values $F_m$ and $\delta_m$. In this way, the complete load-deflection curve is automatically plotted by the computer.

## NUMERICAL SIMULATION OF AN EXPERIMENTAL TEST

The numerical simulation of a three point bending test according to the R.I.L.E.M. Recommendation [1] is considered, through the application of the cohesive crack model previously introduced. The numerical procedure takes into account the weight of the specimen (P), which is 80 cm long (L) and 10 cm deep (B) and thick (D), with a crack of depth 5 cm (A0/B = 0.5).

The concrete aggregate is crushed and the material properties are the following ones:

| | | | |
|---|---|---|---|
| Ultimate strength | $f_t$ = 43.13 | kg/cm² | (FT), |
| Young's modulus | E = 289.000 | kg/cm² | (E), |
| Poisson ratio | $\nu$ = 0.20 | | (V), |
| Fracture energy | $G_F$ = 0.06 | kg/cm | (GF). |

The tensile strength $f_t$ is obtained by a direct test, whereas the Young's modulus E and Poisson ratio $\nu$ are the average values between tension and compression. The fracture energy $G_F$ is assumed to be equal to the area under the experimental load-deflection curve, divided by the initial ligament area.

Besides the linear descending stress ($\sigma$)-crack opening displacement (w) law, other twenty-five (25) bilinear $\sigma$-w diagrams with the same area $G_F$ (Fig. 1-b) are considered, the discontinuity point (knee) being located in twenty-five (25) different positions (Fig. 5). Practically, the grid-points are given by the intersections of respectively five (5) slanting lines departing from the limit point (0, $f_t$) and five (5) horizontal lines equally spaced (Fig. 5).

The numerical load-deflection response appears to be very sensitive to the $\sigma$-w shape (Figs 6-a to e) [10]. Only one of the twenty-five (25) theoretical curves reproduces the experimental curve with sufficient accuracy (Fig. 6-e, curve E5).

Some of the numerical curves reported in Fig. 6-a to e do not present a sufficient degree of numerical accuracy and graphycal resolution. As will be clearer in the next section, the number of finite elements on the center line (80), see Fig. 7-a, is too low if related with the britteness number ($s_E$ = 0.06/(43.13 x 10) = 13.9 x $10^{-5}$) in those cases.

The slanting column providing the best approximation to the experimental curve is the 5th (points A5, B5, C5, D5, E5) in Fig. 6-e. The best approximation point of the twenty-five considered is E5 (Fig. 6-e). On the other hand, point F5 in Fig. 5 reproduces the experimental curve even better, as shown in Fig. 8.

## NUMERICAL STABILITY AND GRAPHYCAL RESOLUTION BY VARYING THE BRITTLENESS NUMBER

The numerical stability and the graphycal resolution of the load-deflection curves is investigated by varying the knee location and the brittleness number $s_E = G_F/f_t b$.

A span to depth ratio $\ell/b$ equal to 4 and an initial crack depth $a_0/b$ equal to 0.1 are considered. The weight of the specimen is neglected and 40 elements adjacent to the center line are utilized (Fig. 7-b).

For each value of the dimensionless number $s_E$, there is a set of discontinuity points on the plane $\sigma/f_t$ versus $w/w_{eq}$ ($w_{eq} = 2\, G_F/f_t$) which provide numerically stable load-deflection curves. As the diagrams in Fig. 9 make evident, the area of numerical stability shrinks by decreasing the number $s_E$.

It is important to observe that a steep slope of the $\sigma$-$w$ diagram in the first stage ($0 \leq w \leq w_{knee}$) tendentially produces numerically unstable results.

## DIMENSIONAL TRANSITION TOWARDS LEFM

Let us consider a three point bending test specimen with geometrical ratios $\ell = 4b$ and $t = b$ (Fig. 3) and a central crack of depth $a_0/b = 0.1$, and a material with ultimate strain $\varepsilon_t = 0.87 \times 10^{-4}$ and Poisson ratio $\nu = 0.1$. The $\sigma$-$w$ diagram is assumed with the knee at the point of coordinates $w = 2/9\, w_{eq}$, $\sigma = 1/3\, f_t$. The dimensionless load-deflection diagrams are represented in Fig. 10 by varying the brittleness number $s_E$.

The ratio of the maximum loading capacity obtained by the cohesive crack model to the maximum load according to LEFM is equal to the ratio of the fictitious fracture toughness to the true one $K_{IC}$. Such a ratio is demonstrated to tend to unity when $s_E \to 0$, for the linear as well as for the bilinear $\sigma$-$w$ constitutive softening relations (Fig. 11). This means that, for large structure sizes and/or low fracture energies, the bifurcation of the load-deflection curve tends to describe the classical LEFM-instability.
Therefore, it is possible to conclude that, <u>although the $\sigma$-$w$ shape affects the theoretical value of the maximum load, such a dependence is smaller and smaller by decreasing the number</u> $s_E$. As a limit-case, for $s_E \to 0$, the maximum load derives from the well-known LEFM-relation: $K_I = K_{IC} = \sqrt{G_F\, E}$, and is depending only on the area $G_F$ under the assumed $\sigma$-$w$ diagram, and not on the shape of this diagram.

## ACKNOWLEDGEMENTS

The numerical results reported in the present paper were obtained in a joint research program between ENEL-CRIS-Milano and the University of Bologna.

## REFERENCES

[1] Determination of the fracture energy of mortar and concrete by means of three-point bend tests on notched beams, R.I.L.E.M. Draft Recommendation, 50-FMC Committee Fracture Mechanics of Concrete (Chairman: Prof. F.H. Wittmann).

[2] A. Hillerborg, M. Modeer and P.E. Petersson, Analysis of crack formation and crack growth in concrete by means of fracture mechanics and finite elements, Cement and Concrete Research 6, 773-782 (1976).

[3] P.E. Petersson, Crack growth and development of fracture zones in plain concrete and similar materials, Report TVBM-1006, Division of Building Materials, Lund Institute of Technology (1981).

[4] G. Colombo and E. Limido, Un metodo numerico per l'analisi di prove TPBT stabili: confronto con alcuni dati sperimentali, XI Convegno Nazionale dell'Associazione Italiana per l'Analisi delle Sollecitazioni, Torino, 233-243 (1983).

[5] A. Carpinteri, A. Di Tommaso and M. Fanelli, Influence of material parameters and geometry on cohesive crack propagation, International Conference on Fracture Mechanics of Concrete, Lausanne, October 1-3, 1985, Vol. I, 95-113.

[6] A. Carpinteri, G. Colombo and G. Giuseppetti, Accuracy of the numerical description of cohesive crack propagation, International Conference on Fracture Mechanics of Concrete, Lausanne, October 1-3, 1985, Vol. II, 25-31.

[7] A. Carpinteri, Softening instability and equilibrium branching in cohesive solids, Physical-Numerical Modelling in Non-Linear Fracture Mechanics,

Euromech Colloquium No. 198, September 11-12, 1985, Stuttgart.
[8] A. Carpinteri, Interpretation of the Griffith instability as a bifurcation of the global equilibrium, Application of Fracture Mechanics to Cementitious Composites, NATO Advanced Research Workshop, September 4-7, 1984, Northwestern University, edited by S.P. Shah, Martinus Nijhoff Publishers, 287-316 (1985).
[9] A. Carpinteri, Limit analysis for elastic-softening structures: scale and slenderness influence on the global brittleness, Structure and Crack Propagation in Brittle Matrix Composite Materials, Euromech Colloquium No. 204, November 12-15, 1985, Jablonna (Warsaw), edited by A.M. Brandt, Elsevier Applied Science Publishers, in press.
[10] P.E. Roelfstra and F.H. Wittmann, Numerical method to link strain softening with failure of concrete, International Conference on Fracture Mechanics of Concrete, October 1-3, 1985, Lausanne, Vol. I, 127-139.

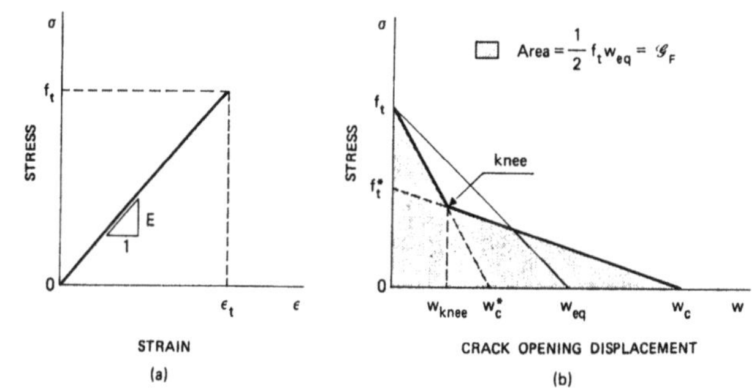


Fig. 1. Double constitutive law of concrete : (a) stress-strain and (b) stress-crack opening displacement .

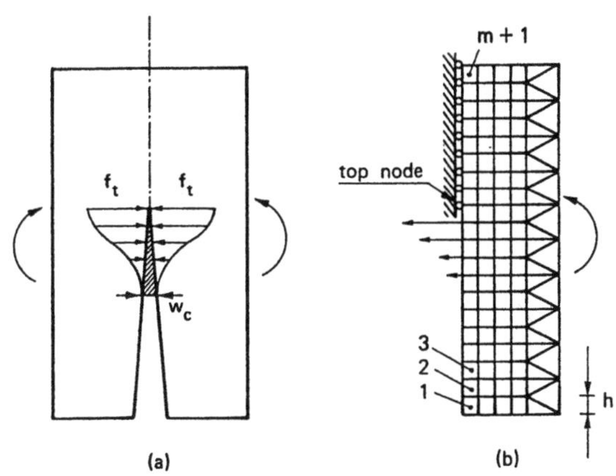


Fig. 2. Cohesive crack model : (a) cohesive force distribution at the crack tip ; (b) finite elements discretization .

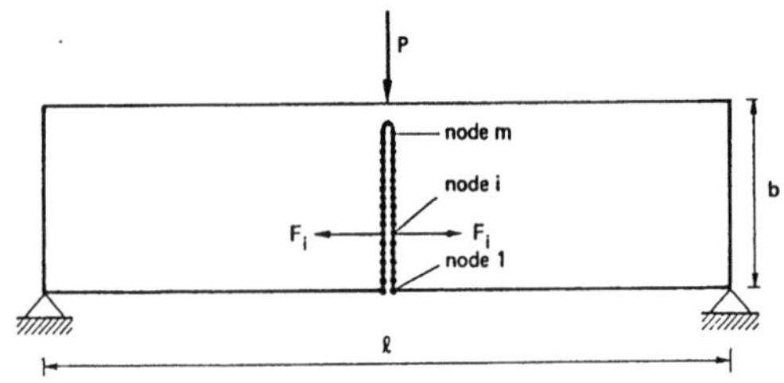


Fig. 3. Three point bending test with the potential fracture line on the center-line .

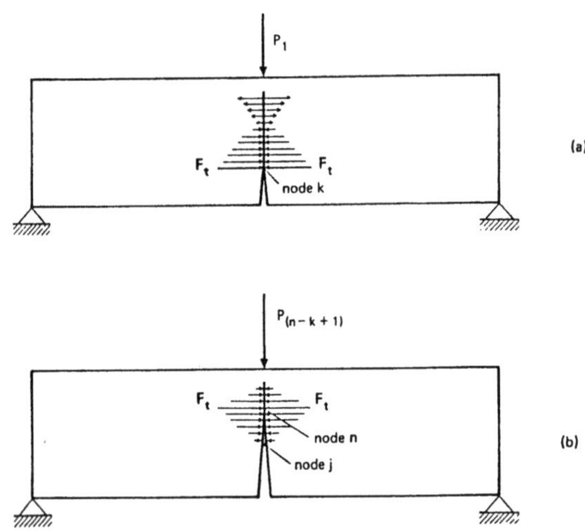


Fig. 4. Cohesive crack configurations at the first (a) and (n-k+1)-th (b) crack growth increment .

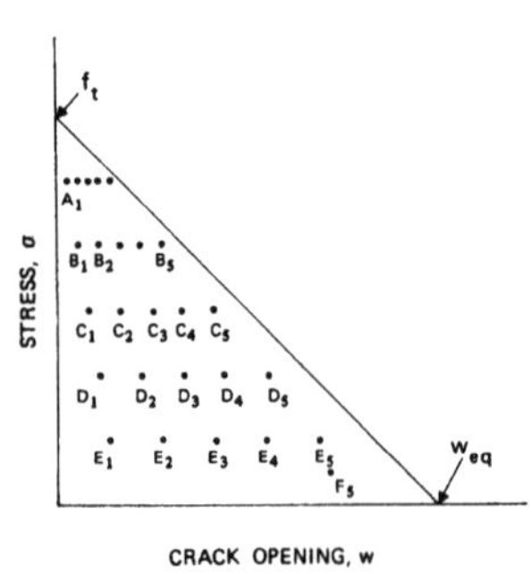


Fig. 5. Locations of the knee-points in the stress-crack opening displacement diagram of Figure 1-b .

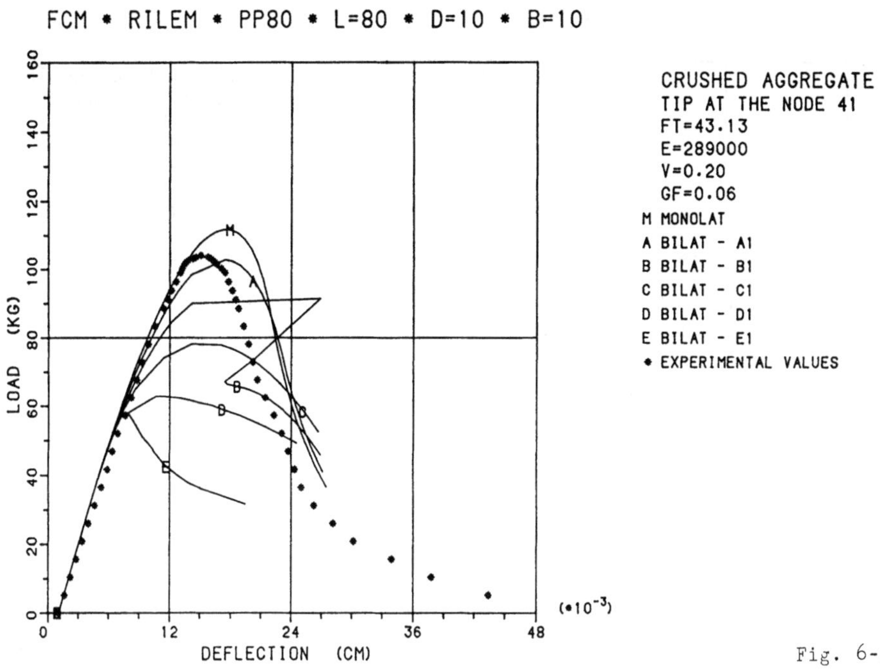
FCM * RILEM * PP80 * L=80 * D=10 * B=10
LOAD (KG)
DEFLECTION (CM)
(*10^-3)
0
20
40
60
80
100
120
140
160
12
24
36
48
CRUSHED AGGREGATE
TIP AT THE NODE 41
FT=43.13
E=289000
V=0.20
GF=0.06
M MONOLAT
A BILAT - A1
B BILAT - B1
C BILAT - C1
D BILAT - D1
E BILAT - E1
* EXPERIMENTAL VALUES

Fig. 6-a

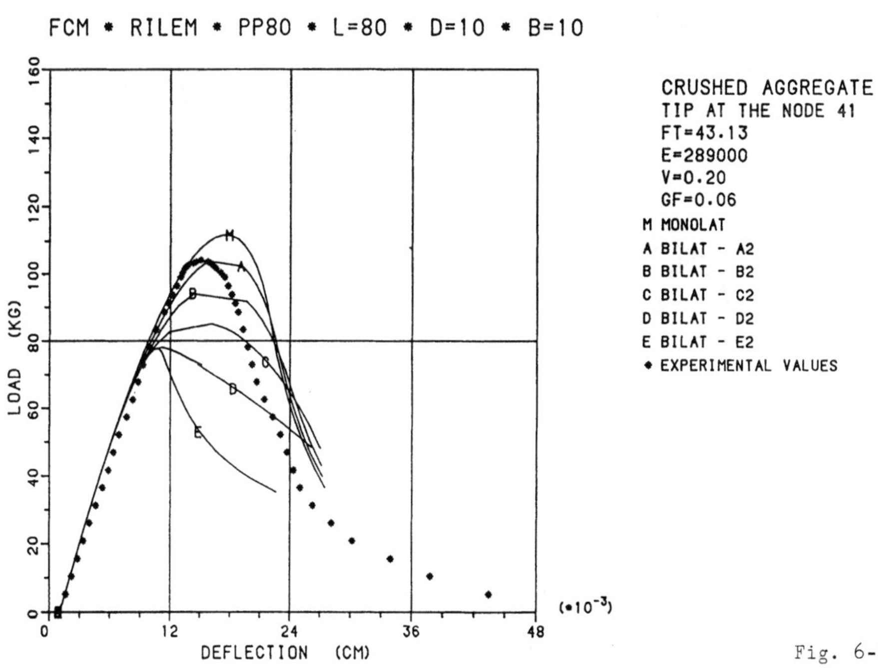
FCM * RILEM * PP80 * L=80 * D=10 * B=10
LOAD (KG)
DEFLECTION (CM)
(*10^-3)
0
20
40
60
80
100
120
140
160
12
24
36
48
CRUSHED AGGREGATE
TIP AT THE NODE 41
FT=43.13
E=289000
V=0.20
GF=0.06
M MONOLAT
A BILAT - A2
B BILAT - B2
C BILAT - C2
D BILAT - D2
E BILAT - E2
* EXPERIMENTAL VALUES

Fig. 6-b

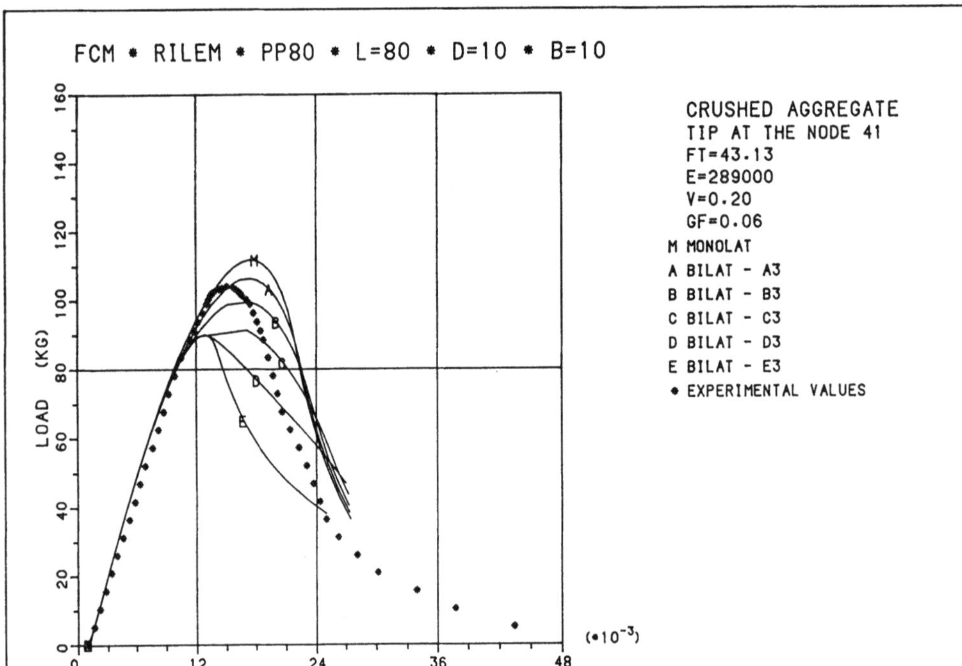
FCM * RILEM * PP80 * L=80 * D=10 * B=10
CRUSHED AGGREGATE
TIP AT THE NODE 41
FT=43.13
E=289000
V=0.20
GF=0.06
M MONOLAT
A BILAT - A3
B BILAT - B3
C BILAT - C3
D BILAT - D3
E BILAT - E3
* EXPERIMENTAL VALUES
LOAD (KG)
DEFLECTION (CM)
(*10^-3)
0
12
24
36
48
20
40
60
80
100
120
140
160

Fig. 6-c

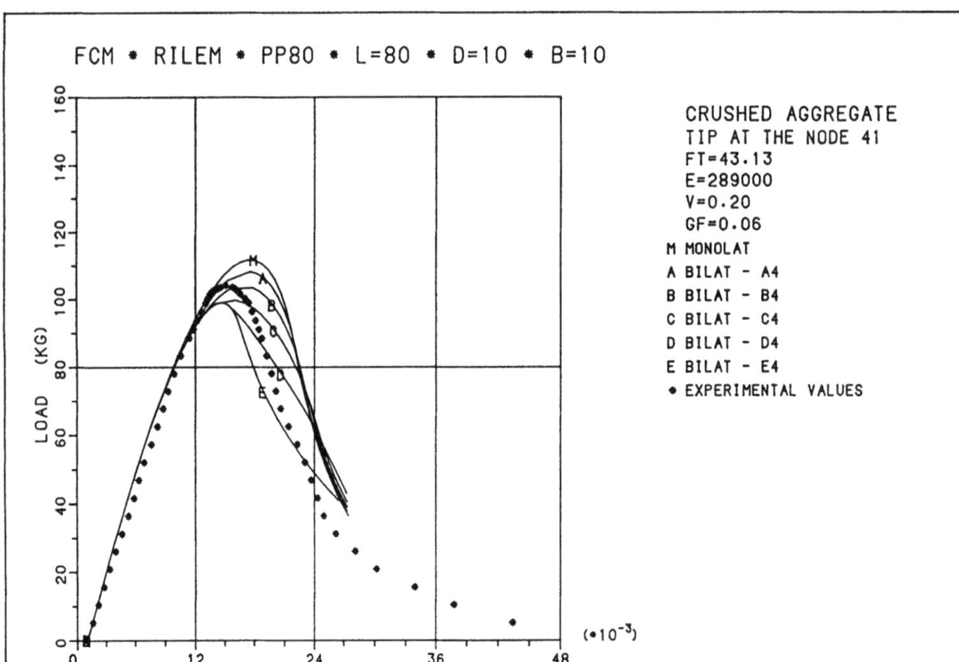
FCM * RILEM * PP80 * L=80 * D=10 * B=10
CRUSHED AGGREGATE
TIP AT THE NODE 41
FT=43.13
E=289000
V=0.20
GF=0.06
M MONOLAT
A BILAT - A4
B BILAT - B4
C BILAT - C4
D BILAT - D4
E BILAT - E4
* EXPERIMENTAL VALUES
LOAD (KG)
DEFLECTION (CM)
(*10^-3)
0
12
24
36
48
20
40
60
80
100
120
140
160

Fig. 6-d

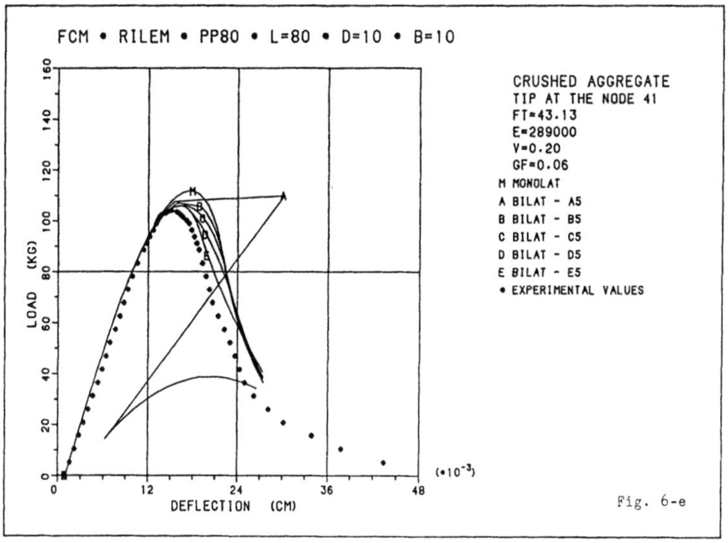


Fig. 6. Numerical simulation of the load-deflection response by varying the location of the knee-point (see Figure 5) .

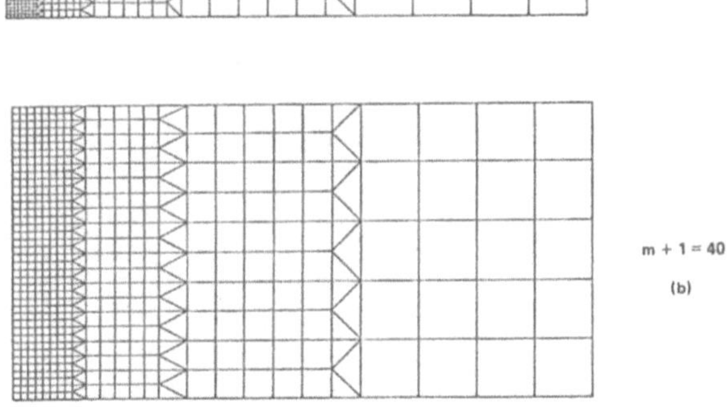


Fig. 7. Refined (a) and coarse (b) finite element meshes .

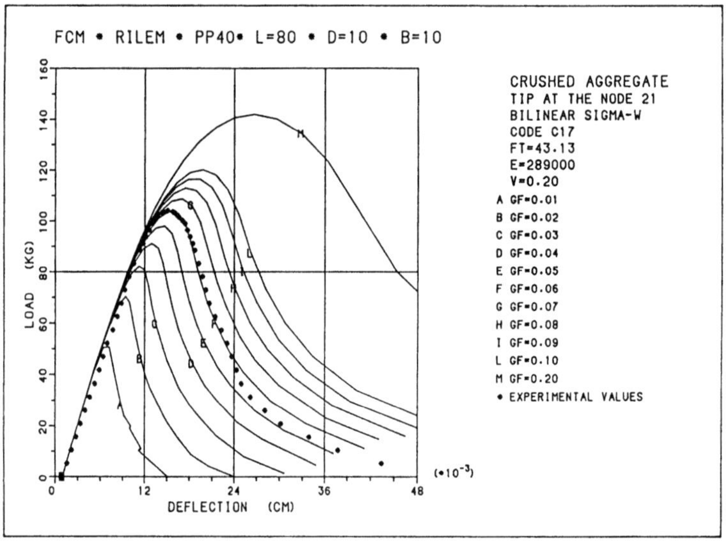


Fig. 8. Numerical simulation of the load-deflection response when the knee-point coincides with point F5 in Figure 5.

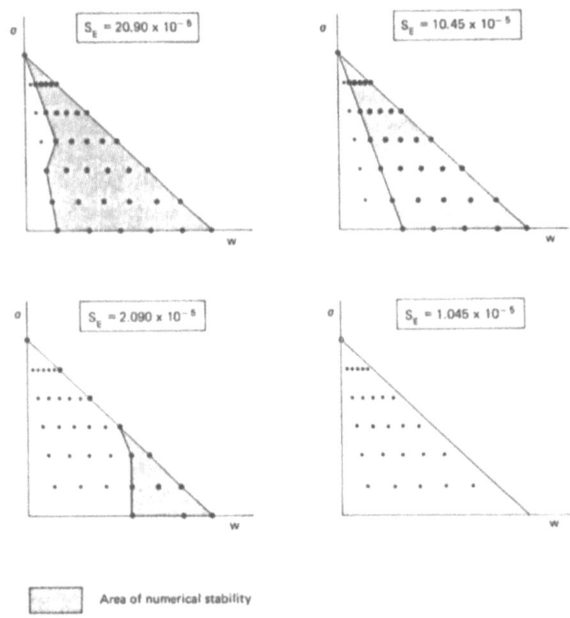


Fig. 9. Knee-point areas of numerical stability by varying the brittleness number $s_E$ .

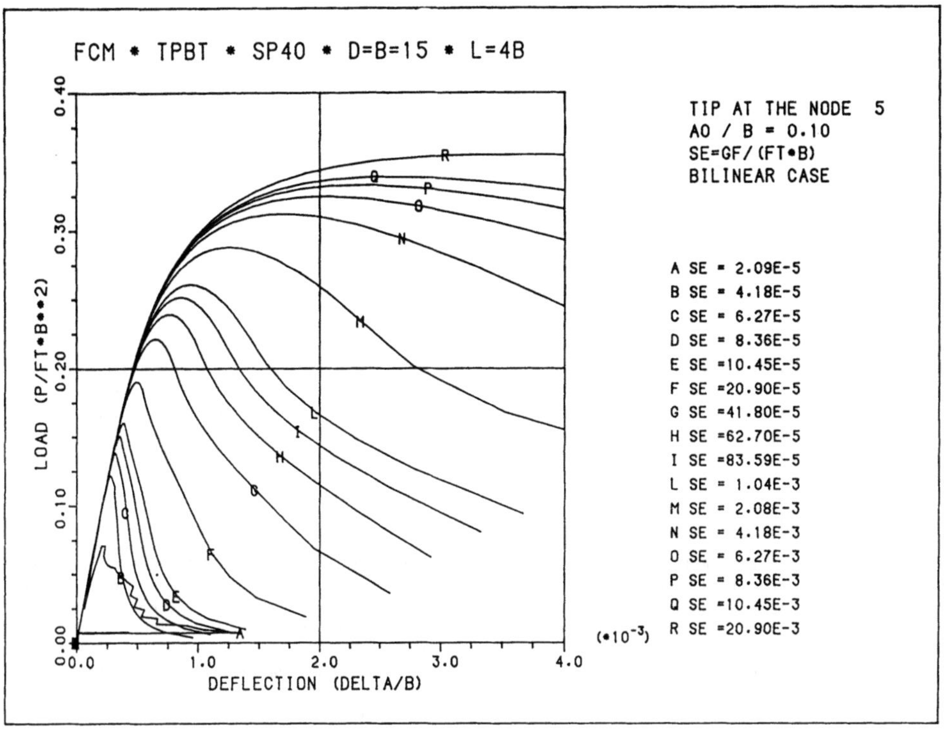


Fig. 10. Dimensionless load-deflection diagrams by varying the brittleness number $s_E$ .

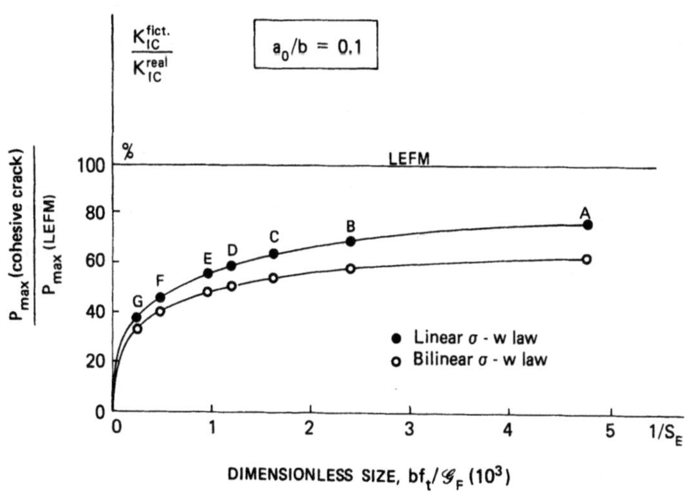


Fig. 11. Transition towards LEFM : fictitious fracture toughness vs. dimensionless size .

SEM/RILEM International Conference
on Fracture of Concrete and Rock
Houston, Texas, June 1987
Editors: S.P. Shah, S.E. Swartz

# FUNDAMENTAL ISSUES OF SMEARED CRACK MODELS

Kaspar Willam*, Eddy Pramono** and Stein Sture*

ABSTRACT

For numerical simulation of fracture in concrete and rock the "smeared crack approach" is receiving increasing attention. On one hand renewed attempts in terms of the fixed and rotating crack models resort to fracture mechanics in order to refine the traditional orthotropic crack formulation. Along this approach the original concept involving Mode I type cracking is being broadened to include mixed mode fracture interpretation of the shear retention factor, if the crack memory is fully retained. On the other hand, fracture energy-based plasticity models are advocated by the authors, as well as other investigators, which describe the degradation of strength due to tensile cracking and decohesion in shear in terms of isotropic and anisotropic strain-softening concepts.

The paper addresses the settled differences between elasticity and plasticity-based concepts which are currently used to describe localized and distributed failure due to tensile cracking and decohesion. In particular, the issues concerning the shear retention factor and rotation of principal directions will be scrutinized in view of recent discussions related to fixed versus rotating crack models. In the context of fracture energy-based plasticity the prevalent issues are also related to the problems of isotropic versus anisotropic strain-softening and the proper choice of appropriate directional fracture process variables.

The treatise concludes with the analysis of two elementary model problems involving fixed and rotating principal coordinates in order to illustrate the subtle constitutive arguments pertaining to progressive fracture under Mode I and mixed mode fracture.

## INTRODUCTION

Recent refinements of the original orthotropic crack model by Rashid [1] include several contributions beyond the early concepts of tension stiffening and shear retention. These extensions were originally introduced to account for the composite action of reinforced concrete due to interface bond between primary cracks and due to dowel action in the crack region. The reader is referred to the original proposals by Scanlon and Murray [2], Cervenka and Gerstle [3] and Lin and Scordelis [4]. Recently, strain-softening and residual shear stiffness are also attributed to plain, fractured concrete because of interface asperities and aggregate interlock of the tortuous crack surface, see e.g. Hillerborg, Modeer and Petersson [5] as well as Millard and Johnson [6]. The current developments include fracture energy concepts to formulate "equivalent" softening relations in tension and shear as advocated by Bazant and Oh [7], de Borst [8] and Rots and de Borst [9]. In addition, the rotation of principal stresses and strains during the fracture process has rejuvenated the debate involving fixed and rotating crack models [10] where the previous crack directions are erased from memory altogether, see e.g. Gupta and Akbar [11], Milford and Schnobrich [12] and Cope [13].

Strain-softening plasticity has been utilized within the context of thermo-plasticity and cyclic plasticity in the past. However, it was not until recently that uniqueness and stability were addressed in the context of non-associated plastic flow for pressure sensitive solids, see e.g. Rudnicki and Rice [14] and Maier and Hueckl [15]. The intriguing aspects of localization were examined recently by Prevost and Hughes [16], by Pietruszczak and Mroz [17] and by the authors in refs [18, 19, 20]. The earlier discussions were confined to isotropic softening formulations based on hypothetical material properties. The identification of softening properties was addressed by the authors in refs. [21, 22], when fracture energy release concepts were included in the definition of the plastic softening modulus in order to account for surface-driven macro-defects which are primarily responsible for the overall degradation of strength and stiffness in the post-critical response regime, see also Nilsson and Oldenburg [23]. Very little work has been done to include the anisotropy induced during the course of the actual degradation process, except for the early plane stress plasticity model by Murray et. al. in ref. [24]. In fact a systematic investigation is entirely missing which correlates the

* Kaspar Willam and Stein Sture are Professors of Civil Engineering at the University of Colorado, Boulder, Colorado 80309-0428.

** Eddy Pramono is a Research Graduate Assistant who is completing his Ph.D. degree in Civil Engineering at the University of Colorado, Boulder, Colorado 80309-0428.

strain-softening plasticity formulation to the fixed crack approach based on orthotropic concepts and the rotating crack model based on nonlinear elasticity relations in principal coordinates.

The objective of this paper is to compare the two strategies for constructing "equivalent" softening continuum models and to illustrate the underlying differences of the resulting tangential modulus operators when neutral loading is considered. Particular attention will be devoted to a rational analysis of the shear retention factor and the aspects of crack memory, when the directions of the principal stress/strain coordinates rotate during the actual fracture process. Two elementary model problems will be used to illustrate the performance of the elasticity - and plasticity-based formulations for smeared crack analysis of Mode I and mixed mode fracture.

## ELASTICITY-BASED CRACK MODELS

The original format of the orthotropic crack model is essentially a fixed crack approach in which the principal axes of orthotropy are aligned with the principal stress direction at incipient cracking. The underlying assumption is that the Rankine criterion of maximum normal stress governs the initial formation of tensile cracks that are responsible for the sudden transition from isotropic into transversely-isotropic and orthotropic behavior with fixed crack directions. Subsequent cracking is thus exclusively monitored along the three fixed principal axes of orthotropy irrespective of the magnitude of maximum principal stress along different directions. Normally a shear retention factor is introduced which describes the residual shear stiffness of the crack due to aggregate interlock that diminishes with increasing crack-opening or rather tensile straining normal to the smeared crack surface. In addition, the early concepts of brittle stress release of the "residual" stress components are being extended in order to include continuous strain-softening, because of the tortuous debonding process in the matrix-aggregate composition of concrete.

### Fixed Crack Approach - Orthotropic Model

For plane stress conditions, the fixed crack approach results invariably in a constitutive relation for the "equivalent" continuum along the principal n-t axes of orthotropy normal and tangential to the critical crack direction.

$$\begin{bmatrix} \sigma_n \\ \sigma_t \\ \tau_{nt} \end{bmatrix} = \begin{bmatrix} E_{nn} & E_{nt} & 0 \\ E_{tn} & E_{tt} & 0 \\ 0 & 0 & G_{nt} \end{bmatrix} \begin{bmatrix} \varepsilon_n \\ \varepsilon_t \\ \gamma_{nt} \end{bmatrix} \qquad (1)$$

The format of the stress-strain relation in Eq. 1 has its origin in anisotropic elasticity where the compliance relation for plane stress is defined along the principal axes of orthotropy as

$$\begin{bmatrix} \varepsilon_n \\ \varepsilon_t \\ \gamma_{nt} \end{bmatrix} = \begin{bmatrix} \frac{1}{E_n} & -\frac{\nu_{nt}}{E_t} & 0 \\ -\frac{\nu_{tn}}{E_n} & \frac{1}{E_t} & 0 \\ 0 & 0 & \frac{1}{G_{nt}} \end{bmatrix} \begin{bmatrix} \sigma_n \\ \sigma_t \\ \tau_{nt} \end{bmatrix} \qquad (2)$$

with the out-of-plane strain component being defined as

$$\varepsilon_s = -\left(\frac{\nu_{sn}}{E_n}\sigma_n + \frac{\nu_{st}}{E_t}\sigma_t\right)$$

For symmetry $\nu_{nt}/E_t = \nu_{tn}/E_n$, thus the Poisson's ratios are in general not symmetric in their indices, i.e., $\nu_{nt} \neq \nu_{tn}$ when $E_t \neq E_n$. Therefore the individual terms of the orthotropic stiffness relation in Eq. 1 involve in general four independent material moduli for the in-plane behavior which may be expressed in terms of $E_n$, $E_t$, $\nu = \nu_{nt}$ and $G_{nt}$ such that

$$E_{nn} = \frac{E_n}{1-\nu_{nt}\nu_{tn}} = \frac{E_n}{1-\nu^2 E_n/E_t}$$

$$E_{tt} = \frac{E_t}{1-\nu_{nt}\nu_{tn}} = \frac{E_t}{1-\nu^2 E_n/E_t} \qquad (3)$$

$$E_{tn} = \frac{\nu_{tn} E_t}{1-\nu_{nt}\nu_{tn}} = \frac{\nu_{nt} E_n}{1-\nu_{nt}\nu_{tn}} = \frac{\nu E_n}{1-\nu^2 E_n/E_t}$$

Very often the orthotropic shear modulus $G_{nt}$ is expressed in terms of the other elastic moduli while assuming isotropic shear behavior for the case of pure shear where $\sigma_n = -\sigma_t$ and $\varepsilon_n = -\varepsilon_t$. In this case the orthotropic stiffness tensor can be fully characterized in terms of three moduli which reduce Eq. 1 to

$$\begin{bmatrix}\sigma_n\\ \sigma_t\\ \tau_{nt}\end{bmatrix} = \frac{E_n}{1-\nu^2 E_n/E_t}\begin{bmatrix}1 & \nu & 0\\ \nu & E_t/E_n & 0\\ 0 & 0 & G^*\end{bmatrix}\begin{bmatrix}\varepsilon_n\\ \varepsilon_t\\ \gamma_{nt}\end{bmatrix} \qquad (4)$$

where

$$G^* = (1-\nu^2 E_n/E_t)\frac{E_t}{E_n + E_t + 2\nu E_n}$$

This expression was e.g. derived in detail by Link [25] assuming invariance of the "isotropic" shear modulus with regard to arbitrary states of stress and strain which differ from the state of pure shear. Enforcing isotropy for the out-of-plane deformation, the transverse Poisson ratios can be expressed in terms of the in-plane ratio's, i.e. $\nu_{sn} = \nu_{tn}$ and $\nu_{st} = \nu_{nt}$ such that

$$\varepsilon_s = -(\sigma_n + \sigma_t)\frac{\nu}{E_t} \qquad (5)$$

As a result of these simplifying assumptions three independent moduli fully determine the orthotropic properties for two-dimensional plane stress behavior. Remark: In spite of the isotropic shear stiffness the stress and strain tensors are in general non-coaxial, i.e. the principal axes of stress do not coincide with the principal axes of strain as long as $E_t \neq E_n$.

In the fixed crack approach the Poisson effect is usually neglected after cracking has initiated assuming $\nu = 0$. In addition the ill-defined shear stiffness is normally expressed in terms of the retention factor $\beta_s$ which expresses the effective shear stiffness due to asperities and aggregate interlock as a fraction of the initial shear modulus $G = E/\ 2(1+\nu)$ for isotropic conditions. Therefore initial cracking due to excessive tension in the n-direction transforms the isotropic solid into an equivalent "orthotropic" solid which is characterized by the diagonalized crack modulus tensor below.

$$\begin{bmatrix}\sigma_n\\ \sigma_t\\ \tau_{nt}\end{bmatrix} = \begin{bmatrix}E_n & 0 & 0\\ 0 & E_t & 0\\ 0 & 0 & \beta_s G\end{bmatrix}\begin{bmatrix}\varepsilon_n\\ \varepsilon_t\\ \gamma_{nt}\end{bmatrix} \qquad (6)$$

It is important to realize that the orthotropic format of the fixed crack model in Eq. 6 does not maintain coaxiality between stress and strain, i.e. the principal axes of stress does not coincide in general with that of strain. This basic orthotropy argument can be readily verified, if we consider the angles of principal stresses and strains with regard to the fixed Cartesian x-y reference frame

$$\tan 2\alpha_\sigma = \frac{2\sigma_{xy}}{\sigma_{xx} - \sigma_{yy}} \quad \text{and} \quad \tan 2\alpha_\varepsilon = \frac{2\varepsilon_{xy}}{\varepsilon_{xx} - \varepsilon_{yy}} \qquad (7)$$

If we align the principal axes of orthotropy with the Cartesian reference system then $\alpha_\sigma \neq \alpha_\varepsilon$ for orthotropic behavior as long as $E_n \neq E_t$ and $2\beta_s \neq E/G$. Moreover, frame-invariance is assured when the crack-orientation remains fixed during the active fracture process i.e. the principal axes of orthotropy subtend an angle $\theta$ with the Cartesian reference frame which remains invariant during the entire fracture process, where

$$\tan 2\theta_\sigma = \left.\frac{2\sigma_{xy}}{\sigma_{xx} - \sigma_{yy}}\right|_{t=t_{cr}} \quad \text{and} \quad \tan 2\theta_\varepsilon = \left.\frac{2\varepsilon_{xy}}{\varepsilon_{xx} - \varepsilon_{yy}}\right|_{t=t_{cr}} \qquad (8)$$

For initially isotropic solids the initial crack-orientation is thus the same regardless whether the maximum stress or strain criterion is utilized to monitor crack initiation. Thus $\theta_\sigma = \theta_\varepsilon$ and it is not necessary to make a distinction between the stress and strain formulations for initial cracking.

The post-crack behavior is described by the orthotropic expression in Eq. 6 where the individual material moduli represent secant moduli that are usually a function of the tensile strains normal to the two crack surfaces in the n- and t-direction, e.g.

$$\begin{aligned}E_n &= E_n(\varepsilon_n)\\ E_t &= E_t(\varepsilon_t)\\ \beta_s &= \beta(\varepsilon_n,\varepsilon_t)\end{aligned} \qquad (9)$$

The resulting strain-softening description is recently cast in terms of the fracture energy release rate for Mode I type cracking which introduces a characteristic length for transforming the crack-opening displacement of a single crack into an equivalent tensile strain between adjacent primary cracks. As a result we may express e.g. the softening behavior due to cracking by the differential expression of Eq. 6

$$\begin{bmatrix} \dot\sigma_n \\ \dot\sigma_t \\ \dot\tau_{nt} \end{bmatrix} = \begin{bmatrix} E_n^T & 0 & 0 \\ 0 & E_t^T & 0 \\ 0 & 0 & \beta_s^T G \end{bmatrix} \begin{bmatrix} \dot\varepsilon_n \\ \dot\varepsilon_t \\ \dot\gamma_{nt} \end{bmatrix} \quad \text{where} \quad \begin{aligned} E_n^T &= \frac{d\sigma_n}{d\varepsilon_n} = E_n + \frac{dE_n}{d\varepsilon_n}\varepsilon_n \\ E_t^T &= \frac{d\sigma_t}{d\varepsilon_t} = E_t + \frac{dE_t}{d\varepsilon_t}\varepsilon_t \\ \beta_s^T &= \beta_s + \frac{\partial\beta}{\partial\varepsilon_n}\varepsilon_n + \frac{\partial\beta}{\partial\varepsilon_t}\varepsilon_t \end{aligned} \qquad (10)$$

Note that the change of secant moduli characterizes the fracture energy release rate during the separation process. For cracking in the n-direction, an "equivalence" argument relates the strain energy release rate during softening in the elementary volume dV to the fracture energy release rate for propagating the crack by dA

$$\int \dot\sigma_n \varepsilon_n \, dV = \int \dot G_f \, dA \qquad \text{with } \dot G_f = \sigma_n \, du_n \quad \text{and } \dot\sigma_n = \dot E_n \varepsilon_n \qquad (11)$$

or in an equivalent manner

$$\frac{d\sigma_n}{du_n} h = \frac{\sigma_n}{\varepsilon_n} \qquad \text{where } h = dV/dA \qquad (12)$$

Therefore the strain-softening relationship involves a characteristic length "h" which represents the width of the equivalent strain-softening zone participating in the crack formation. The rate of fracture energy release is herewith defined by the fracture modulus

$$E_n = E_d/h \qquad \text{where } E_d = d\sigma_n/du_n = d^2G_f/du_n^2 \quad \text{and } E_n = \sigma_n/\varepsilon_n \qquad (13)$$

Fig. 1 illustrates the basic relationships that relate strain-softening due to tensile cracking to the fracture energy release rate via $E_d$ and h. In the fixed crack approach, tensile cracking is monitored independently in the fixed orthogonal material n-t coordinates. It is therefore safe to assume that there is little interaction between the fracture energy release rates which are required to develop two arrays of orthogonal cracks. On the other hand, their effect on the shear retention factor is less than obvious, particularly when multiple cracking is considered as suggested recently by de Borst [8]. In fact, in a further extension Rots and de Borst [9] did resort to Mode II type fracture interpretation of the shear retention factor such that $\beta_s = \beta(G_f^{II})$. However, at this stage it is unclear, whether there exists a critical fracture energy release rate for Mode II type fracture that can be be determined independently from Mode I type tensile cracking. Therefore, in practical applications the shear retention factor is assumed to be a constant value in the range $0 \leq \beta_s \leq 0.5$, or a monotonically decreasing function of tensile cracking strain, i.e. $\beta_s = \beta(\varepsilon_n)$ since the cohesive shear stiffness rapidly diminishes with crack opening to the residual strength level attributed to interface aggregate interlock, see Fig. 2. For simultaneous cracking in the n- and t-directions a rational approach is still missing which quantifies the shear retention factor in terms of separate tensile strains, $\beta_s = \beta(\varepsilon_n, \varepsilon_t)$. This is even more so when multiple cracking is considered on crack planes which differ from 90° [8,9].

Rotating Crack Approach - Isotropic Model

In contrast to the permanent memory of the fixed crack approach the rotating crack approach follows essentially the concepts of nonlinear elasticity in which the stress is expressed as a nonlinear tensor function in terms of strain. In the case of frame-invariance this leads to isotropic representations of stress in terms of invariants or equivalent principal values of strains, i.e.

$$\boldsymbol\sigma = \mathbb{F}(\boldsymbol\varepsilon) \rightarrow \mathbb{F}(I_\varepsilon, II_\varepsilon, III_\varepsilon) \rightarrow \mathbb{F}(\varepsilon_1, \varepsilon_2, \varepsilon_3) \qquad (14)$$

For the sake of argument let us adopt the direct notation where the second order tensors of stress and strains are written in matrix form. Then spectral representation of stress and strain leads to the following expressions

$$\boldsymbol\sigma = \sum_{i=1}^{3} \sigma_i \, \mathbf{e}_i^\sigma \otimes \mathbf{e}_i^\sigma \quad \text{and} \quad \boldsymbol\varepsilon = \sum_{i=1}^{3} \varepsilon_i \, \mathbf{e}_i^\varepsilon \otimes \mathbf{e}_i^\varepsilon \qquad (15)$$

where the principal directions of stress and strains are spanned by the orthonormal base vectors $\mathbf{e}_i^\sigma$ and $\mathbf{e}_i^\varepsilon$.

The usual isotropy argument is based on the coaxiality argument of stress and strain, i.e., $\mathbf{e}_i^\sigma = \mathbf{e}_i^\varepsilon = \mathbf{e}_i$ and as a result the nonlinear elasticity relations can be solely expressed in terms of principal values of stress and strain such that

$$\boldsymbol\sigma\boldsymbol\varepsilon = \boldsymbol\varepsilon\boldsymbol\sigma \qquad (16)$$

Therefore, the most general format of a nonlinear elastic solid may be expressed as

$$\boldsymbol{\sigma} = \phi_0 \mathbf{II} + \phi_1 \boldsymbol{\epsilon} + \phi_2 \boldsymbol{\epsilon}^2 \qquad (17)$$

where the invariant response functions $\phi_i$ are cyclic symmetric functions of principal strains, i.e. $\phi_i = \phi_i(\varepsilon_1, \varepsilon_2, \varepsilon_3)$. Explicity, this infers that

$$\begin{bmatrix} \sigma_1 & 0 & 0 \\ 0 & \sigma_2 & 0 \\ 0 & 0 & \sigma_3 \end{bmatrix} = \begin{bmatrix} \phi_0 + \phi_1\varepsilon_1 + \phi_2\varepsilon_1^2 & 0 & 0 \\ 0 & \phi_0 + \phi_1\varepsilon_2 + \phi_3\varepsilon_2^2 & 0 \\ 0 & 0 & \phi_0 + \phi_1\varepsilon_3 + \phi_2\varepsilon_3^2 \end{bmatrix} \qquad (18)$$

The equivalent vector form leads to the following secant relationship for the stress-strain law

$$\begin{bmatrix} \sigma_1 \\ \sigma_2 \\ \sigma_3 \end{bmatrix} = \begin{bmatrix} E_{11} & E_{21} & E_{23} \\ E_{21} & E_{22} & E_{23} \\ E_{31} & E_{32} & E_{33} \end{bmatrix} \begin{bmatrix} \varepsilon_1 \\ \varepsilon_2 \\ \varepsilon_3 \end{bmatrix} \qquad (19)$$

The individual secant moduli are in general nonlinear functions of strain exhibiting "strain-induced anisotropy" where

$$\begin{aligned} E_{ii} &= \phi_0/\varepsilon_i + \phi_1 + \phi_2\varepsilon_i \\ E_{ij} &= \phi_0/\varepsilon_j + \phi_1\varepsilon_i/\varepsilon_j + \phi_2\varepsilon_i^2/\varepsilon_j \end{aligned} \qquad (20)$$

In fact, it has been shown by Ogden [26] that the nonlinear elastic representation in Eq. 17 can be expressed by a single scalar response function of the three principal strains

$$\sigma_i = f(\varepsilon_i, \varepsilon_j, \varepsilon_k) = f(\varepsilon_i, \varepsilon_k, \varepsilon_j) \qquad (21)$$

that is only restricted by the cyclic symmetry argument in Eq. 21. In this case, the secant moduli in Eq. 19 are simply defined as

$$\begin{aligned} E_{ii} &= f(\varepsilon_i, \varepsilon_j, \varepsilon_k)/\varepsilon_i \\ E_{ij} &= f(\varepsilon_i, \varepsilon_j, \varepsilon_k)/\varepsilon_j \end{aligned} \qquad (22)$$

The "uniaxial" character of this particular form of the nonlinear elastic representation in principal coordinates has been the key argument for the so-called "orthotropic" concrete models [27, 28]. Unfortunately, this misnomer has been widely accepted in the literature, see e.g. ref [10], although the basis of Eq. 10 is an isotropic tensor representation with inherent coaxiality between stress and strain.

It remains to reduce the triaxial nonlinear elastic relations to the case of cracking under plane stress conditions. In this case incipient cracking in the direction of the major principal stress formally leads to

$$\begin{bmatrix} \sigma_1 \\ \sigma_2 \end{bmatrix} = \begin{bmatrix} E_{11} & E_{12} \\ E_{21} & E_{22} \end{bmatrix} \begin{bmatrix} \varepsilon_1 \\ \varepsilon_2 \end{bmatrix} \qquad (23)$$

In fact, alignment of the principal 1-2 axes of stress/strain with the principal n-t axes of orthotropy furnishes the fundamental differences between the fixed and the rotating crack models. In the orthotropic case of Eq. 1 we encounter an explicit statement of shear, whereby the shear stiffness $\beta_s G$ is in general independent of the normal stiffness moduli $E_n$ and $E_t$. As a result we observe that the stress and strain tensors are no longer coaxial except for the isotropic condition when $E_n = E_t$ and $\beta_s = E/2G$. In the nonlinear elastic case of Eq. 23 we do not encounter any statement with regard to shear because of the underlying hypothesis of isotropy and coaxiality which allows us to express the nonlinear stress-strain relations exclusively in terms of their principal values. In fact the inherent symmetry of the single response function f in Eq. 21 imposes definite restrictions on the representation of the response moduli in Eq. 22 which go beyond the usual stability constraints of the orthotropic moduli in order to maintain positive definitive behavior. In the case of diagonal behavior, if we neglect the Poisson effect the secant moduli in Eq. 23 are only functions of the associated normal strain, i.e.

$$\begin{bmatrix} \sigma_1 \\ \sigma_2 \end{bmatrix} = \begin{bmatrix} E_1 & 0 \\ 0 & E_2 \end{bmatrix} \cdot \begin{bmatrix} \varepsilon_1 \\ \varepsilon_2 \end{bmatrix} \quad \text{where} \quad \begin{aligned} E_{11} &= f(\varepsilon_1)/\varepsilon_1 = E_1(\varepsilon_1) \\ E_{22} &= f(\varepsilon_2)/\varepsilon_2 = E_2(\varepsilon_2) \end{aligned} \qquad (24)$$

$$\text{and} \quad \sigma_i = f(\varepsilon_i) \qquad E_{12} = 0$$

Therefore, the canonical forms of the fixed crack and the rotating crack models in Eqs. 6 and 24 coincide in this case except for the shear term and the underlying difference in coaxiality of stress and strain.

The fundamental differences are further obviated if we compare the tangential moduli tensors of the fixed crack and the rotating crack model. In the case of the orthotropic crack model differentiation of Eq. 4 with regard to fixed axes or orthotropy led to the incremental expression in Eq. 10. In the case of the nonlinear elastic description in terms of principal coordinates differentiation of Eq. 19 or Eqs. 23, 24 is complicated by the fact that the base vectors $\mathbf{e}$ of the principal stresses and strains rotate with the underlying changes of stress and strain. In the past this has led to the popular belief that the so-called orthotropic models are restricted to proportional load histories during which the axes of principal stress and strain remain fixed [10]. This is, however, not necessarily true if we include the spin, i.e. the rate of change of direction of the principal coordinates in the derivation of the tangential modulus operator. In fact, this is directly related to classic issue of defining "objective" stress rates which properly account for the rotation of the base vectors $\dot{\mathbf{e}} \neq 0$.

Let us start with relating the current principal directions of stress and strain $\mathbf{e}$ to the fixed Cartesian reference frame with the orthonormal basis $\mathbf{i}$

$$\mathbf{e} = \mathbf{Q}\,\mathbf{i} \tag{25}$$

where $\mathbf{Q}$ is the usual orthogonal transformation defining the rotation of the base vectors. The rate of change of the principal stress axes is thus simply defined by the spin tensor $\dot{\boldsymbol{\omega}}$ as

$$\dot{\mathbf{e}} = \dot{\mathbf{Q}}\,\mathbf{i} = \dot{\boldsymbol{\omega}}\,\mathbf{e} \qquad \text{where} \qquad \dot{\boldsymbol{\omega}} = \mathbf{Q}^t\,\dot{\mathbf{Q}} \qquad \text{or} \qquad \dot{\boldsymbol{\omega}} = \sum_{i=1}^{3} \dot{\mathbf{e}}_i \otimes \mathbf{e} \tag{26}$$

The chain rule of differentiation of the spectral representation of stress and strain in Eq. 15 leads to the following differential expressions for

$$\dot{\boldsymbol{\sigma}} = \sum_{i=1}^{3} \dot{\sigma}_i\, \mathbf{e}_i \otimes \mathbf{e}_i + \sum_{i=1}^{3} \sigma_i\, \dot{\mathbf{e}}_i \otimes \mathbf{e}_i + \sum_{i=1}^{3} \sigma_i\, \mathbf{e}_i \otimes \dot{\mathbf{e}}_i$$
$$\dot{\boldsymbol{\varepsilon}} = \sum_{i=1}^{3} \dot{\varepsilon}_i\, \mathbf{e}_i \otimes \mathbf{e}_i + \sum_{i=1}^{3} \varepsilon_i\, \dot{\mathbf{e}}_i \otimes \mathbf{e}_i + \sum_{i=1}^{3} \varepsilon_i\, \mathbf{e}_i \otimes \dot{\mathbf{e}}_i \tag{27}$$

Using the antisymmetric property of the spin tensor $\dot{\omega}_{ij} = - \dot{\omega}_{ji}$ we can rewrite Eq. 27 in terms of

$$\dot{\boldsymbol{\sigma}} = \sum_{i=1}^{3} \dot{\sigma}_i\, \mathbf{e}_i \otimes \mathbf{e}_i + \sum_{i\neq j}^{3} \dot{\omega}_{ij} (\sigma_j - \sigma_i)\, \mathbf{e}_i \otimes \mathbf{e}_i$$
$$\dot{\boldsymbol{\varepsilon}} = \sum_{i=1}^{3} \dot{\varepsilon}_i\, \mathbf{e}_i \otimes \mathbf{e}_i + \sum_{i\neq j}^{3} \dot{\omega}_{ij} (\varepsilon_j - \varepsilon_i)\, \mathbf{e}_i \otimes \mathbf{e}_i \tag{28}$$

We recall that coaxiality of stress and strain has been assumed by the isotropic tensor representation leading to Eq. 19. Therefore, the base vectors of principal stresses and strains coincide and no distinction needs to be made between the spins of the stress axes and the strain axes. Differentiation of the principal stress-strain relation in Eq. 19 and substitution into Eq. 28 leads to

$$\dot{\boldsymbol{\sigma}} = \sum_{i=1}^{3} \sum_{j=1}^{3} \frac{\partial \sigma_i}{\partial \varepsilon_j}\, \dot{\varepsilon}_j\, \mathbf{e}_i \otimes \mathbf{e}_i + \sum_{i\neq j}^{3} \dot{\omega}_{ij} (\varepsilon_j - \varepsilon_i)\, \frac{1(\sigma_j - \sigma_i)}{2(\varepsilon_j - \varepsilon_i)}\, \mathbf{e}_i \otimes \mathbf{e}_i \tag{29}$$

As a result of coaxiality we can extract the tangent modulus tensor and express it directly in terms of the rates of change of principal stress and strain where the first term on the right hand side of Eq. 29 yields

$$\begin{bmatrix} \dot{\sigma}_1 \\ \dot{\sigma}_2 \\ \dot{\sigma}_3 \end{bmatrix} = \begin{bmatrix} \frac{\partial \sigma_1}{\partial \varepsilon_1} & \frac{\partial \sigma_1}{\partial \varepsilon_2} & \frac{\partial \sigma_1}{\partial \varepsilon_3} \\ \frac{\partial \sigma_2}{\partial \varepsilon_1} & \frac{\partial \sigma_2}{\partial \varepsilon_2} & \frac{\partial \sigma_2}{\partial \varepsilon_3} \\ \frac{\partial \sigma_3}{\partial \varepsilon_1} & \frac{\partial \sigma_3}{\partial \varepsilon_2} & \frac{\partial \sigma_3}{\partial \varepsilon_3} \end{bmatrix} \begin{bmatrix} \dot{\varepsilon}_1 \\ \dot{\varepsilon}_2 \\ \dot{\varepsilon}_3 \end{bmatrix} \tag{30}$$

while the second term introduces a shear stiffness which is associated with the spin of the principal axes

$$\begin{bmatrix} \dot{\tau}_{12} \\ \dot{\tau}_{23} \\ \dot{\tau}_{31} \end{bmatrix} = \begin{bmatrix} \frac{1}{2}\frac{\sigma_1 - \sigma_2}{\varepsilon_1 - \varepsilon_2} & 0 & 0 \\ 0 & \frac{1}{2}\frac{\sigma_2 - \sigma_3}{\varepsilon_2 - \varepsilon_3} & 0 \\ 0 & 0 & \frac{1}{2}\frac{\sigma_3 - \sigma_1}{\varepsilon_3 - \varepsilon_1} \end{bmatrix} \begin{bmatrix} \dot{\gamma}_{12} \\ \dot{\gamma}_{23} \\ \dot{\gamma}_{31} \end{bmatrix} \tag{31}$$

As a result of coaxiality between stress and strain the resulting tangent modulus tensor is now comprised of two contributions; one represents the "normal" stiffness and the other the "shear" stiffness associated with the difference of principal stresses due to the rotation of principal axes. Note that for strain states where $\varepsilon_i = \varepsilon_j$ the difference format of the shear stiffness has to be interpreted as differential operation including $d(\sigma_i - \sigma_j)/d\varepsilon_i$.

For the two-dimensional case the tangential stiffness moduli tensor associated with the secant relation in Eq. 23 reduces to

$$\begin{bmatrix} \dot{\sigma}_1 \\ \dot{\sigma}_2 \\ \dot{\tau}_{12} \end{bmatrix} = \begin{bmatrix} \frac{\partial \sigma_1}{\partial \varepsilon_1} & \frac{\partial \sigma_1}{\partial \varepsilon_2} & 0 \\ \frac{\partial \sigma_2}{\partial \varepsilon_1} & \frac{\partial \sigma_2}{\partial \varepsilon_2} & 0 \\ 0 & 0 & \frac{1}{2}\frac{\sigma_1 - \sigma_2}{\varepsilon_1 - \varepsilon_2} \end{bmatrix} \begin{bmatrix} \dot{\varepsilon}_1 \\ \dot{\varepsilon}_2 \\ \dot{\gamma}_{12} \end{bmatrix} \tag{32}$$

whereby the off-diagonal coupling terms vanish, when we neglect the Poisson effect according to the secant format expressed in Eq. 24

In this case, the tangential moduli in the principal coordinates play the equivalent role of the diagonal terms in the orthotropic fixed crack expression of Eq. 10

$$\begin{bmatrix} \dot{\sigma}_1 \\ \dot{\sigma}_2 \\ \dot{\tau}_{12} \end{bmatrix} = \begin{bmatrix} \frac{\partial \sigma_1}{\partial \varepsilon_1} & 0 & 0 \\ 0 & \frac{\partial \sigma_2}{\partial \varepsilon_2} & 0 \\ 0 & 0 & \frac{1}{2}\frac{\sigma_1 - \sigma_2}{\varepsilon_1 - \varepsilon_2} \end{bmatrix} \begin{bmatrix} \dot{\varepsilon}_1 \\ \dot{\varepsilon}_2 \\ \dot{\gamma}_{12} \end{bmatrix} \tag{33}$$

except for the shear term, which is here related to the difference of principal stresses and strains and is not an independent function of shear or normal strains.

Remarks: It is intriguing that the shear stiffness explicitly appears when we recast the secant relation of nonlinear elasticity into tangential form. This shear stiffness arises, however, only due to rotation of the principal coordinates, whereby coaxiality of stress and strain underlies the entire postulate of the rotating crack approach.

Note that the same tangential expression in Eq. 32 for the rotating crack model was recently developed independently by Adeghe and Collins starting from a description in a fixed Cartesian reference frame. On the other hand the rotating crack model advocated recently in refs. [11, 12] does not fully coincide with the formulation above, since it was derived for reinforced panels using limit analysis concepts and does not enforce coaxialty between stress and strain.

## PLASTICITY-BASED CRACK MODELS

In analogy to the previous treatment we distinguish between isotropic and anisotropic strain-softening plasticity models. The former are associated with the rotating crack formulation without memory of the crack direction while the latter are connected to the fixed crack approach with permanent crack memory.

Strain-softening plasticity formulations have been used in the past primarily in the context of isotropic $J_2$-theories. Little attention has been directed towards anisotropic softening and the underlying issues of fixed versus rotating cracking with the associated question of crack memory. Therefore we will develop in the following an isotropic strain-softening plasticity model without directionality which will be compared with an anisotropic strain-softening plasticity formulation in which the yield condition and flow rules are expressed in terms of local material coordinates associated with the initial crack direction. These two approaches follow essentially the basic concepts of the rotating and fixed crack models of the isotropic and orthotropic elasticity-based formulations above.

### Fixed Crack Approach - Orthotropic Plasticity Model

In this case we adopt the previous concept that the material has experienced tensile cracking in a plane defined by the normal n. Therefore, the subsequent yield condition involves the normal and shear stress $\sigma_n$-$\tau_{nt}$ in the crack plane. For biaxial conditions the elliptic yield function depicted in Fig. 3 restricts the crack stresses according to

$$F(\boldsymbol{\sigma}, \varepsilon_{np}) = \sigma_n^2 + \frac{\tau_{nt}^2}{\alpha^2} - f_t^2 = 0 \tag{34}$$

where it is assumed that the normal stress component $\sigma_t$ in the crack plane is not affected by the formation of the primary crack normal to the n-direction.

For associated plastic flow the evolution of plastic strain rates is controlled by

$$\dot{\boldsymbol{\varepsilon}}_p = \dot{\lambda}\, \mathbf{m} \qquad \text{where } \dot{\boldsymbol{\varepsilon}}_p = [\dot{\varepsilon}_{np}, \dot{\gamma}_p]^t \quad \text{and } \mathbf{m} = \partial F/\partial \boldsymbol{\sigma} = [2\sigma_n, \frac{2\tau_{nt}}{\alpha^2}]^t \tag{35}$$

The plastic multiplier is extracted from the consistency condition which is expressed in terms of $\boldsymbol{\varepsilon}$ for strain control as follows.

$$\dot{\lambda} = \frac{\mathbf{n}^t \mathbb{E} \dot{\boldsymbol{\varepsilon}}}{E_p + \mathbf{n}^t \mathbb{E} \mathbf{n}} \qquad \text{where} \quad E_p = - \frac{\partial F}{\partial f_t} \frac{\partial f_t}{\partial \varepsilon_{np}} \left(\frac{\partial \varepsilon_{np}}{\partial \boldsymbol{\varepsilon}_p}\right)^t \frac{\partial F}{\partial \boldsymbol{\sigma}} \tag{36}$$

$$= 4 f_t \sigma_n \frac{\partial f_t}{\partial \varepsilon_{np}}$$

$$\text{and} \quad \mathbf{n}^t \mathbb{E} \mathbf{n} = 4\sigma_n^2 E_n + 4\tau_{nt}^2 G/\alpha^4$$

Following earlier arguments in refs. [19, 21] for defining strain-softening in terms of an equivalent fracture modulus $E_f = d\sigma_t/du_n$ which is related to the fracture energy release rate $G_f{}^I$ for Mode I type cracking we can recast the plastic modulus in terms of $E_f$ and the characteristic length "h" in Eq. 12 as

$$E_p = 4 f_t \sigma_n h E_f \qquad \text{where } h = dV/dA \qquad \text{and } E_f = d\sigma_n/du_n \tag{37}$$

The elastic-plastic stress-strain rate expression results in the familiar relationship

$$\boxed{\dot{\boldsymbol{\sigma}} = \mathbb{E}_{ep} \dot{\boldsymbol{\varepsilon}}} \qquad \text{where } \mathbb{E}_{ep} = \mathbb{E} - \frac{\mathbb{E} \mathbf{n} \mathbf{n}^t \mathbb{E}}{E_p + \mathbf{n}^t \mathbb{E} \mathbf{n}} \tag{38}$$

The elastic plastic modulus tensor $\mathbb{E}_{ep}$ is expanded below for plane stress conditions with $\dot{\boldsymbol{\sigma}} = [\dot{\sigma}_n, \dot{\sigma}_t, \dot{\tau}_{nt}]^t$ and $\dot{\boldsymbol{\varepsilon}} = [\dot{\varepsilon}_n, \dot{\varepsilon}_t, \dot{\gamma}_{nt}]^t$

$$\mathbb{E}_{ep} = \begin{bmatrix} E_{11} & E_{12} & 0 \\ E_{21} & E_{11} & 0 \\ 0 & 0 & G \end{bmatrix} - 4c \begin{bmatrix} \sigma_n^2 E_{11}^2 & \sigma_n^2 E_{11} E_{12} & \sigma_n \tau_{nt} \frac{G E_{11}}{\alpha^2} \\ \sigma_n^2 E_{11} E_{12} & \sigma_n^2 E_{12}^2 & \sigma_n \tau_{nt} \frac{G E_{12}}{\alpha^2} \\ \sigma_n \tau_{nt} \frac{G E_{11}}{\alpha^2} & \sigma_n \tau_{nt} \frac{G E_{12}}{\alpha^2} & \frac{\tau_{nt}^2 G^2}{\alpha^4} \end{bmatrix} \tag{39}$$

where $c = 1/(E_p + \mathbf{n}^t \mathbb{E} \mathbf{n})$ and where the elastic moduli $E_{11}$ and $E_{12}$ denote the usual terms of isotropic linear elasticity which are defined for plane stress as

$$E_{11} = \frac{E}{1-\nu^2} \qquad E_{12} = E_{21} = \frac{\nu E}{1-\nu^2} \tag{40}$$

Comparison with the analogous expression of the orthotropic elasticity model in Eq. 10 for fixed crack direction indicates that we have now considerably more coupling between the shear and normal stiffness terms. Even if we neglect the Poisson effect, i.e $E_{12} = 0$, which was responsible for the diagonal format of Eq. 10 the tangent modulus stiffness is still comprised of

$$\begin{bmatrix} \dot{\sigma}_n \\ \dot{\sigma}_t \\ \dot{\tau}_{nt} \end{bmatrix} = \begin{bmatrix} E_{11}(1-4E_{11}c\sigma_n^2) & 0 & -4cE_{11}G\sigma_n\tau_{nt}/\alpha^2 \\ 0 & E_{11} & 0 \\ -4cE_{11}G\sigma_n\tau_{nt}/\alpha^2 & 0 & G(1-4cG\tau_{nt}^2/\alpha^4) \end{bmatrix} \begin{bmatrix} \dot{\varepsilon}_n \\ \dot{\varepsilon}_t \\ \dot{\gamma}_{nt} \end{bmatrix} \tag{41}$$

We observe that the dependence of the yield condition on the shear stress $\tau_{nt}$ introduces not only the term of the shear stiffness but also the coupling term between the normal and shear components. It is intriguing that isotropic softening which is here entirely controlled by the plastic normal strain component $\varepsilon_p = \varepsilon_{np}$ does introduce an effect which is analogous to the shear retention factor in addition to the $\sigma_n$-$\tau_{nt}$ interaction effect which is characteristic for frictional materials.

## Rotating Crack Approach - Isotropic Softening Model

In the case of no permanent crack-memory the plasticity formulation is entirely phrased in terms of principal quantities without consideration of the actual crack direction.

The yield condition for tensile cracking is in this case simply the Rankine criterion depicted in Fig. 4.

$$F(\boldsymbol{\sigma}, \varepsilon_{1p}) = \sigma_1 - f_t = 0 \tag{42}$$

where $f_t$ denotes the current value of tensile strength. Upon cracking the irreversible fracture strain is controlled by the "plastic flow rule" as

$$\dot{\boldsymbol{\epsilon}}_p = \dot{\lambda}\, \mathbf{m} \qquad \text{where } \dot{\boldsymbol{\epsilon}}_p = [\dot{\epsilon}_{1p}, \dot{\epsilon}_{2p}]^t \quad \text{and } \mathbf{m} = \frac{\partial F}{\partial \boldsymbol{\sigma}} = [1, 0]^t \tag{43}$$

The plastic multiplier $\dot{\lambda}$ results from the consistency condition as

$$\dot{\lambda} = \frac{\mathbf{m}^t\, \mathbb{E}\, \dot{\boldsymbol{\epsilon}}}{E_p + \mathbf{m}^t\, \mathbb{E}\, \mathbf{m}} \qquad \text{where } E_p = -\frac{\partial f_t}{\partial \epsilon_{1p}} \left(\frac{\partial \epsilon_{1p}}{\partial \boldsymbol{\epsilon}_p}\right)^t \mathbf{m} = -\frac{\partial f_t}{\partial \epsilon_{1p}} \quad \text{and } \mathbf{m}^t \mathbb{E}\, \mathbf{m} = E_{11} \tag{44}$$

The resulting tangent modulus relation for isotropic strain-softening plasticity relates the two principal components by

$$\mathbb{E}_{ep} = \begin{bmatrix} E_{11} & E_{12} \\ E_{21} & E_{22} \end{bmatrix} - c \begin{bmatrix} E_{11}^2 & E_{11}E_{12} \\ E_{11}E_{12} & E_{12}^2 \end{bmatrix} \tag{45}$$

where $\qquad c = 1/(E_p + \mathbf{m}^t\, \mathbb{E}\, \mathbf{m})$

In the absence of Poisson's effect, the coupling $E_{12} \rightarrow 0$ vanishes and Eq. 45 reduces to

$$\begin{bmatrix} \dot{\sigma}_1 \\ \dot{\sigma}_2 \end{bmatrix} = \begin{bmatrix} E_{11}(1 - c\, E_{11}) & 0 \\ 0 & E_{22} \end{bmatrix} \begin{bmatrix} \dot{\epsilon}_1 \\ \dot{\epsilon}_2 \end{bmatrix} \tag{46}$$

Comparison with the analogous expression of the fixed crack approach in Eq. 41 reveals that there is close agreement except for the shear contribution which arises due to the elliptic yield condition in Eq. 34. One could argue that in the tension-compression region the maximum stress criterion should be augmented by a more refined yield condition e.g. along the line of the "Leon-Model" adopted earlier by the authors in ref. [21] for general triaxial loading conditions. However, since the forthcoming example problems will be restricted to tension-shear problems the isotropic strain-softening formulation is based on the simple maximum stress condition, whereby the fracture concept above is incorporated along the earlier developments in [19, 20] in order to express the strain-softening modulus in terms of the fundamental postulate of constant fracture energy release rates.

## MODEL PROBLEMS FOR SMEARED CRACKING

In the previous sections two families of fixed and rotating crack models were developed. One was based on concepts of nonlinear isotropic and anisotropic elasticity and the other on strain-softening plasticity. In both cases, the fracture postulate of invariant fracture energy release rate introduced a characteristic length in the constitutive relations which correlates the governing crack strains to the underlying crack-opening and crack-sliding displacements controlling the actual fracture process. The four groups of smeared cracking feature the characteristic crack interface behavior in the form of stiffness and strength degradation in the normal and tangential components. The elasticity-based models effectively reduce the strength through the degradation of stiffness which is accompanied by fully reversible deformations. In contradistinction the plasticity-based models monitor directly the degradation of strength through softening yield conditions without concomitant change of the elastic stiffness properties. Thus the fracture process is in this case accompanied by irreversible deformations.

The underlying constitutive relations of the four crack models are specified in Eqs. 10, 33, 41 and 46 for plane stress conditions assuming that the Poisson effect can be neglected in the post-crack regime. The pertinent features are summarized in Table 1 for the sake of comparison.

In the rotating crack approach there is no memory of previous crack directions. Therefore, the current states of principal stress and strain entirely suffice to control the fracture process. The main disadvantage is in this case the isotropic format which does not account for the rotation of principal coordinates, except via the spin-induced shear stiffness in Eq. 33. Even for the case of proportional loading with fixed principal directions, there remains a distinct difference because of the orthotropic format of the fixed crack approach and the explicit consideration of shear stiffness. In fact, the rotating nonlinear elasticity model enforces coaxiality between stress and strain because of the underlying isotropy argument, while all other formulations maintain this property neither in the incremental nor total sense.

In order to illustrate the performance of the four groups of smeared crack formulations we will examine below two different model problems which exhibit both invariant as well as rotating principal axes of strain

| SHEARED CRACK MODEL | ELASTICITY-BASED | PLASTICITY-BASED |
|---|---|---|
| FIXED CRACK AXES: n-t<br>$\begin{bmatrix}\sigma_n\\ \tau_{nt}\end{bmatrix}$ and $\begin{bmatrix}\varepsilon_n\\ \gamma_{nt}\end{bmatrix}$ | Orthotropic Stiffness Representation<br>Permanent Crack Direction<br>Shear Retention Factor $\beta_s(\varepsilon_n)$<br>Misalignment with $e_1$-$e_2$ Axes<br>No Control of Maximum Tensile Stress<br>Reversible Crack Deformations | Anisotropic Strength Representation<br>Permanent Crack Direction<br>Combined $\sigma_n$-$\tau$ Yield Condition<br>Misalignment with $e_1$-$e_2$ Axes<br>No Control of Maximum Tensile Stress<br>Irreversible Crack Deformations |
| ROTATING CRACK AXES: $e_1$-$e_2$<br>$\begin{bmatrix}\sigma_1\\ \sigma_2\end{bmatrix}$ and $\begin{bmatrix}\varepsilon_1\\ \varepsilon_2\end{bmatrix}$ | Isotropic Stiffness Representation<br>No Memory of Crack Direction<br>Spin-Induced Shear Stiffness<br>Full Alignment with $e_1$-$e_2$ Axes<br>Control of Maximum Tensile Stress<br>Reversible Crack Deformations<br>(for proportional loading only) | Isotropic Strength Representation<br>No Memory of Crack Direction<br>No Shear Stiffness in $e_1$-$e_2$<br>Full Alignment with $e_1$-$e_2$ Axes<br>Control of Maximum Tensile Stress<br>Irreversible Crack Deformations<br>(for proportional loading only) |

Remarks: In the fixed crack approach the initial crack direction is memorized permanently. Thus the principal axes of orthotropy n-5 are normally misaligned with the principal axes of stress and strain $e_1$-$e_2$. The major defect is in this case that the fracture process is controlled exclusively by the stress-strain components along the n-5 axes of orthotropy irrespective of the current state of principal stress and strain.

after crack initiation. In order to minimize the bias only tension-shear problems are considered in which the constitutive formulation in compression is not mobilized. For simplicity, the elementary forms of linear strain-softening are adopted as depicted in Fig. 5. In the orthotropic crack model the tangential strain-softening modulus $E_n = d\sigma_n/d\varepsilon_n$ is augmented by a linear degradation of the shear retention factor $\beta_s^T = d\beta/d\varepsilon_n$. In the plasticity-based models the plastic strain-softening modulus $E_p = d\sigma_n/d\varepsilon_p$ is related to $E_n$ via additive decomposition of strains into elastic and plastic components. For fixed crack plasticity approach the combined state of tension-shear stress is confined by the elliptic yield criterion in Eq. 34 which is assumed to soften isotropically according to the plastic normal strain component $\varepsilon_{np}$.

The model problems were implemented in strain control in order to stabilize the softening computations. It is thereby understood that the smeared crack analysis involves an elementary volume of finite dimensions even though the solution of actual boundary value problems was reduced to the integration of the underlying incremental constitutive relations assuming homogeneous deformations.

Simple Shear Problem

It is this tensile fracture example which has recently found wide attention because of the extensive experiments by Vecchio and Collins on reinforced concrete panels [30]. In the absence of normal load components, the simple shear problem leads to principal strain coordinates which remain fixed during the entire crack process.

Fig. 6 depicts the lay-out of the model problem which is subjected to monotonically increasing shear strain. Clearly, the ensuing tensile principal stress reaches the cracking limit when $\sigma_1 = f_t'$, after which the tensile separation progress leadsto progressive debonding down to zero residual strength when $\varepsilon_1 > \varepsilon_r$.

Fig. 7 summarizes the predictions of the four crack models in terms of nominal stress-strain relations in the normal and tangential directions. The four $\sigma_n$ - $\varepsilon_n$ predictions essentially coincide and reproduce the input softening relationship in Fig. 5. The shear response predictions however illustrate the variation of the remaining shear stiffness beyond cracking. There is relatively close agreement between the four formulations because the monotonic increase of the shear strain maintains a fixed crack direction during the entire loading history with n-t being coaxial with $e_1$-$e_2$.

Combined Tension and Simple Shear

In order to induce rotation of the principal stress/strain directions and therefore misalignment with the material axes of orthotropy associated with the initial crack direction we consider the following strain-driven load history. In the pre-crack regime the concrete specimen is stretched in the y-direction according to $\dot\varepsilon_1 = \dot\varepsilon_n = \dot\varepsilon_y > 0$ up to incipient cracking when $\sigma_1 = \sigma_n = \sigma_y = f_t'$. During this loading phase the lateral contraction $\dot\varepsilon_2 = \dot\varepsilon_t = \dot\varepsilon_x = -\nu\dot\varepsilon_y < 0$ corresponds to the unconstrained deformation associated with Poisson's effect when an isotropic solid is subjected to uniaxial tension in the y-direction. Upon cracking the concrete specimen is loaded under simultaneous shearing $\dot\gamma_{xy} > 0$ and extension $\dot\varepsilon_y > 0$, $\dot\varepsilon_x > 0$ such that the axes of minor principal strain rotates continuously between $0 \leq \alpha_\varepsilon \leq 30°$ where $\tan 2\alpha_\varepsilon = \gamma_{xy}/(\varepsilon_x - \varepsilon_y)$. In the rotating crack models the associated crack direction swings correspondingly between $0 \leq \theta_c \leq 30°$. In the fixed crack models the rotation of principal strain axes with the initial crack direction introduces an increasing misalignment with the fixed crack direction, i.e., $\alpha_\varepsilon \neq \theta_c$. Fig. 8 illustrates the two basic crack approaches which correspond to material and spatial descriptions of crack directions with and without full crack memory.

The predictions of the four crack models are summarized in Figs. 9, 10, and 11 which depict the response behavior in terms of nominal stress-strain components in the Cartesian reference configuration $\sigma_y - \varepsilon_y$, $\sigma_x - \varepsilon_x$ and $\tau_{xy} - \gamma_{xy}$. In addition Fig. 12 depicts the coaxially between the resulting stress states and the strain input in the form of cross plots of the angle $\alpha_\sigma$ and $\alpha_\varepsilon$ which subtend the minor principal axes of strain and stress with the Cartesian x-axis.

The results of the normal $\sigma_n - \varepsilon_n$ components across the initial crack direction are shown in Fig. 9. They exhibit full agreement in the pre-peak regime as long as $\sigma_n < f_t' = 400$ psi. In the post-peak regime the linear softening relation is followed identically by the orthotropic fixed crack model which monitors the degradation of tensile stress directly in terms of the strain-softening relationship $\sigma_n = E_n^T \varepsilon_n$ where $\sigma_n = \sigma_y$ and $\varepsilon_n = \varepsilon_y$ throughout. Since the governing Eq. 10 does not exhibit coupling between normal and shear components the increasing shear deformation does not affect the normal stress-strain response across the crack. In contrast the fixed crack plasticity model does exhibit this coupling and thus introduces a more rapid degradation of strength in the post-peak regime. It is intriguing that both rotating crack models also exhibit this coupling because of the rotation of principal axes in spite of the fact that the incremental constitutive relations in Eqs. 33 and 46 do not exhibit coupling in the principal coordinates.

The associated normal stress-strain response tangential to the initial crack direction is shown in Fig. 10. It illustrates the principal difference between fixed and rotating crack models. Again the orthotropic fixed crack models do not induce degradation of the normal stiffness because $\sigma_t < f_t'$ and because of zero coupling between the two normal stiffness components since $E_{12}=0$. On the other hand, the rotating crack models exhibit increasing degradation of stiffness and strength, whereby the isotropic elasticity approach introduces a discontinuity of incipient cracking because of the sudden release of the minor principal stress component associated with $E_{12} \to 0$.

The corresponding shear response predictions are depicted in Fig. 11. It clearly illustrates the basic difference between the fixed and rotating crack models. In the case of the fixed crack formulations the initial shear stiffness degrades gradually due to the decrease of the shear retention factor with increasing normal strain i.e., $\beta_s = \beta_s(\varepsilon_n)$ and due to strain-softening of the elliptic yield surface with increasing normal plastic strain $\varepsilon_{np}$. Note that for constant values of the shear retention factor $\beta_s$ the shear stiffness would remain invariant and the shear strength of the fixed orthotropic model would grow without bound. In the case of of the rotating crack models the shear strength decreases implicitly by the rotation of principal axes and the strain-softening control along the major axes of principal strain $\varepsilon_1$ and $\varepsilon_{1p}$.

Fig. 12 illustrates the rotation of the principal axes of strain and stress. Up to peak all the principal directions coincide and are aligned with the global Cartesian reference coordinates because of initial isotropy of the elastic response. After initial cracking when $\sigma_y = f_t'$ the strain history was designed such that the $e_2$-axes of minor principal strain increases monotonically between $0 \leq \alpha_\varepsilon \leq 30°$. The corresponding axis of minor principal stress $\alpha_\sigma$ remains coaxial only for the rotating elasticity-based crack model because of the underlying isotropy argument. Clearly, both fixed crack models lead to increasing deviation of the minor principal stress axis because of the underlying orthotropy postulate. Moreover, the isotropic plasticity model for rotating cracking does also exhibit misalignment of the principal stress and strain axes because of stress-induced anisotropy of the incremental constitutive relation in Eq. 46.

In summary, the different constitutive formulations for smeared cracking exhibit pronounced differences when the principal coordinates rotate in the post-crack regime. This results in substantial discrepancies of the residual shear strength predictions as shown in Fig. 11. The orthotropic format of the fixed crack models leads to considerably higher strength and ductility estimates of the shear resistance than the isotropic concepts behind the two rotating crack formulations.

## ACKNOWLEDGMENT

The liaison efforts of Dean Norman and the financial support by the U.S. Army Waterways Experiment Station under Contract DACA 39-86-K-0006 are gratefully acknowledged.

## REFERENCES

[1] Rashid, Y.R., "Analysis of Prestressed Concrete Pressure Vessels," Nucl. Eng. Design, Vol. 7 (1968), pp. 334- 344.

[2] Scanlon A. and Murray, D.W., "Time Dependent Reinforced Concrete Slab Deflection," ASCE J. Structural Div, Vol. 100, ST9 (1974)

[3] Cervenka, V. and Gerstle, K.H., "Inelastic Analysis of Reinforced Concrete Panels", Int. Assoc. of Bridge and Structural Eng. Publications, Part I, Vol. 31-11 (1971), pp. 31-45 and Part II, Vol. 32-11 (1972), pp. 25-39.

[4] Lin, C-S and Sordelis, A.C., "Nonlinear Analysis of RC Shells of General Form," ASCE J. Structl Div., Vol. 101, ST3 (1975), pp. 523-538.

[5] Hillerborg, A., Modeér, M. and Petersson, P.E., "Analysis of Crack Formation and Crack Growth in Concrete by Means of Fracture Mechanics and Finite Elements," Cement and Concrete Research, Vol. 6, No. 6 (1978), pp. 773-782.

[6] Millard, S.G. and Johnson, R.P., "Shear Transfer Across Cracks in Reinforced Concrete Due to Aggregate Interlock and Dowel Action," Mag. Concrete Research, Vol. 36, No. 126 (1984), pp. 9-21.

[7] Bazant, Z.P. and Oh, B.H., "Crack Band Theory for Fracture of Concrete," Materials and Structures, RILEM, Vol. 16 (1983), pp. 155-177.

[8] de Borst, R., "Computational Aspects of Smeared Crack Analysis," Chapter 2 in Computational Modelling of Reinforced Concrete Structures (E. Hinton and R. Owen eds.) Pineridge Press, Swansea, 1986, pp. 44-83.

[9] Rots, J.G. and de Borst, R., "Analysis of Mixed-Mode Fracture in Concrete," paper submitted to ASCE J. EMD, 1986.

[10] ASCE Task Committee on Finite Element Analysis of Reinforced Concrete Structures: A State-of-the-Art Report on Finite Element Analysis of Reinforced Concrete Structures, ASCE Spec. Publ, New York, 1981.

[11] Gupta, A.K. and Akbar, H., "Cracking in Reinforced Concrete Analysis," ASCE J. Structl Div., Vol. 110, No. 8 (1984), pp. 1735-1746.

[12] Milford, R.V. and Schnobrich, W.C., "Application of the Rotating Crack Model to R/C Shells," Computer and Structures, Vol. 20 (1985), pp. 225-239.

[13] Cope, R.J., "Nonlinear Analysis of Reinforced Concrete Slabs," Chapter 1 in Computational Modelling of Reinforced Concrete Structures (E. Hinton and R. Owen eds.), Pineridge Press, Swansea, 1986, pp. 3-43.

[14] Rudnicki, J.W. and Rice, J.R., "Conditions of the Localization of the Deformation in Pressure-Sensitive Materials," J. Mech. Phys. Solids, Vol. 23, (1975), pp. 371-394.

[15] Maier, G. and Hueckl, T., "Nonassociated and Coupled Flow Rules of Elastoplasticity for Rock-Like Materials," Int. J. Rock Mech. Min. Sci., Geomech. Abstr. Vol. 16 (1979), pp. 77-92.

[16] Prevost, J.H. and Hughes, T.J.R., "Finite Element Solution of Elastic-Plastic Boundary Value Problems," J. Appl. Mech, ASME, Vol. 48 (1981), pp. 69-74.

[17] Pietruszczak, St. and Mroz, Z., "Finite Element Analysis of Deformation of Strain-Softening Materials," Int. J. Num. Methods Eng., Vol. 17 (1981), pp. 327-334.

[18] Willam, K. Bicanic, N. and Sture, S., "Constitutive and Computational Aspects of Strain-Softening and Localization in Solids," ASME/WAM '84 meeting, New Orleans, Dec 1984, Symposium on Constitutive Equations, Macro and Computational Aspects, ASME Vol. G00274 (K. Willam ed.), New York (1984), pp. 233-252.

[19] Willam, K., Pramono, E., Sture, S., "Stability and Uniqueness of Strain-Softening Computation," Europe-US Symposium on Finite Element Methods for Nonlinear Problems (P. Bergan, W. Wunderlich and K-J. Bathe eds), Springer Verlag, Berlin Heidelberg (1986), pp. 119-142.

[20] Willam, K., Bicanic, N. Pramono, E. and Sture, S., "Composite Fracture Model for Strain-Softening Computations of Concrete," Proc. Fracture Toughness and Fracture Energy of Concrete (F.H. Wittmann, ed.) Elsevier Science Publ. Amsterdam (1986), pp. 149-162.

[21] Willam, K. Hurlbut and Sture, S., "Experimental and Constitutive Aspects of Concrete Failure," Proc. Seminar on Finite Element Analysis of Reinforced Concrete Structures, Tokyo May 21-24, 1985, ASCE Special Publication (1986), pp. 226-254.

[22] Willam, K. and Montgomery, K., "Fracture Energy-Based Softening Plasticity Model for Shear Failure," Structures Research Series 87-04, University of Colorado, Boulder, 1987.

[23] Nilsson, L. and Oldenburg, M., "On the Numerical Simulation of Tensile Fracture," Europe-US Symposium on Finite Element Methods for Nonlinear Problems (P. Bergan, W. Wunderlich and K-J. Bathe, eds.), Springer Verlag, Berlin Heidelberg (1986), pp. 103-117.

[24] Murray, D.W., Chitnuyanondh, L., Rijub-Aghola K.Y. and Wong, C., "A Concrete Plasticity Theory for Biaxial Stress Analysis," ASCE J. Eng. Mech. Dev., Vol. 105, EM6 (1979), pp. 989-1106.

[25] Link, J., "Eine Formulierung des Zweiaxialen Verformungs und Bruchverhaltens von Beton," Deutscher Ausschuss für Stahlbeton, Issue 270, Berlin, 1976.

[26] Ogden, R.W, "Non-linear Elastic Deformations," Ellis Harwood Ltd., Chichester, 1984.

[27] Liu, T.C.Y., Nilson, A.H. and Slate, F.O., "Biaxial Stress-Strain Relations for Concrete," ASCE J. Structl. Div., Vol. 98, ST5 (1972), pp. 1025-1034.

[28] Darwin, D. and Pecknold, D.A., "Nonlinear Biaxial Stress-Strain Law for Concrete," ASCE J. Eng. Mech. Div., Vol. 103, EM2, (1977), pp. 229-241.

[29] Adeghe, L.N. and Collins, M.P., "A Finite Element Model for Studying Reinforced Concrete Detailing Problems," Department of Civil Engineering Publication No. 86-12, University of Toronto, 1986.

[30] Vecchio, F. and Collins, M.P., "The Response of Reinforced Concrete to In-Plane and Normal Stress", Department of Civil Engineering, Publication No. 82-03, University of Toronto, 1983.

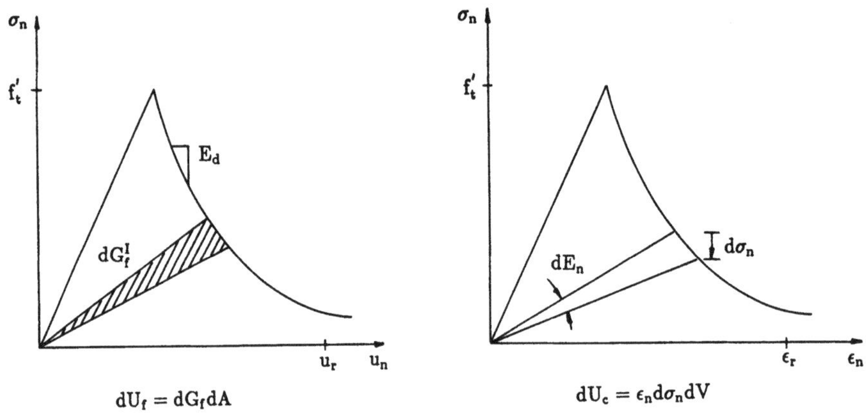


Fig. 1 Fracture Energy Equivalent Strain - Softening

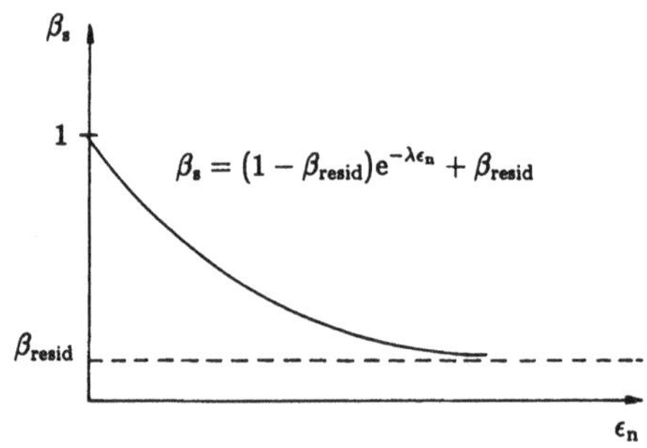


Fig. 2 Shear Retention Factor $G^* = \beta_s G$

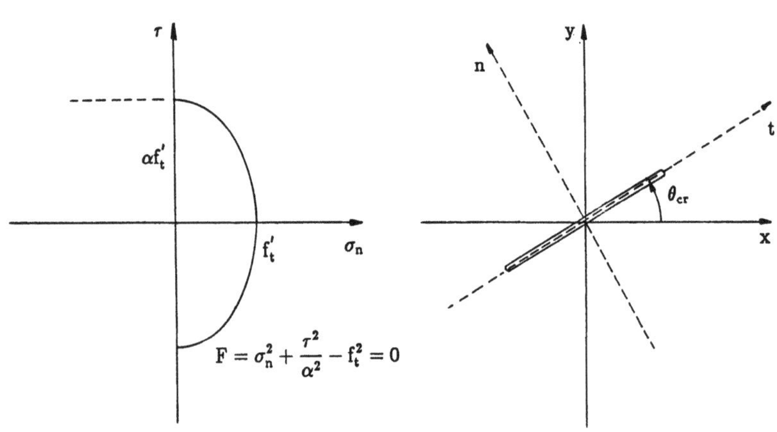


Fig. 3 Plasticity - Based Fixed Crack Approach

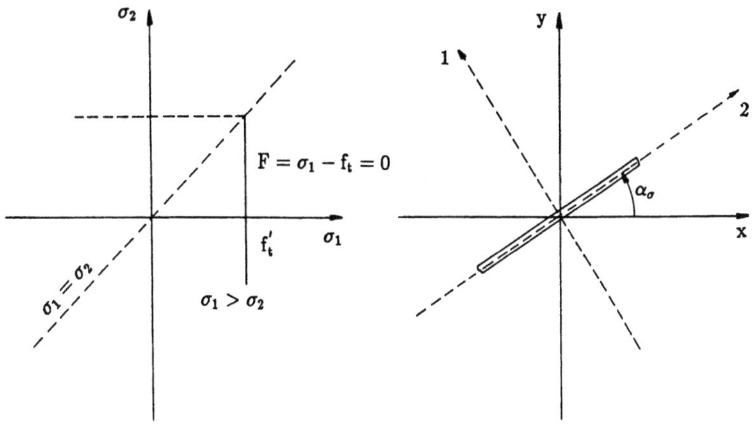


Fig. 4 Plasticity - Based Rotating Crack Approach

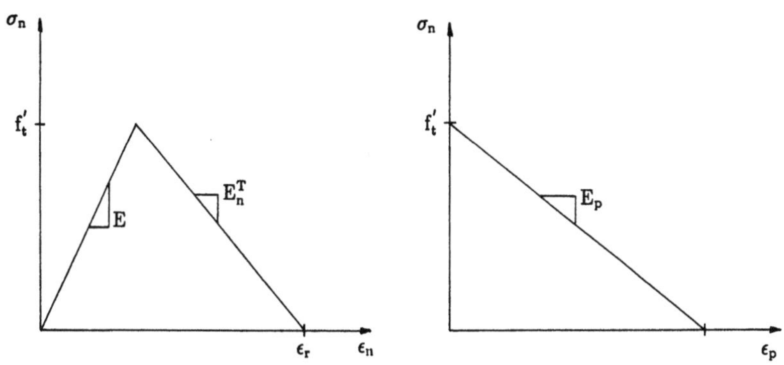


$E = 2800\text{Ksi}$ $\quad$ $E^T_n = -2000\text{Ksi}$

$\nu = 0.2$ $\quad$ $f'_t = 0.4\text{Ksi}$

Fig. 5a Tensile Strain - Softening Properties

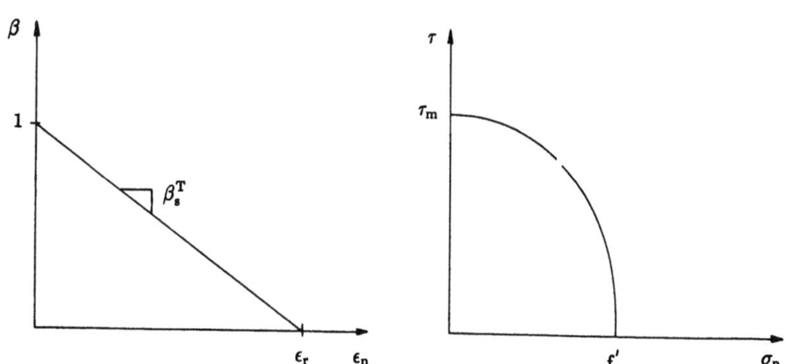


$G^* = \beta^T_s G$ $\quad$ $\tau_m = \alpha f'_t$

$\beta^T_s = 2917$ $\quad$ $\alpha = 2.06$

Fig. 5b Shear - Softening Properties

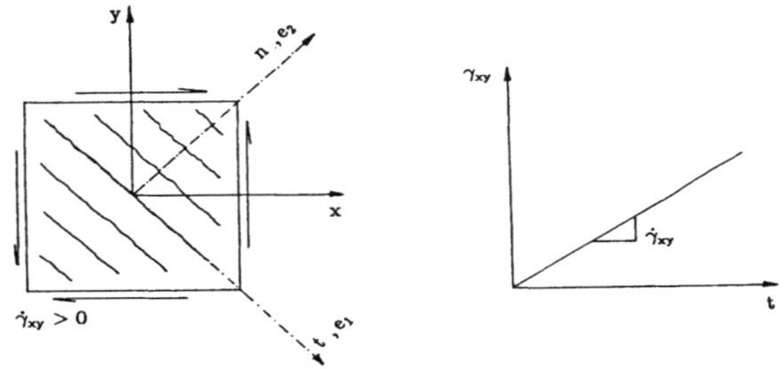


Fixed Principal Strain Axes $e_1 - e_2$

Fig. 6 Pure Shear Model Problem

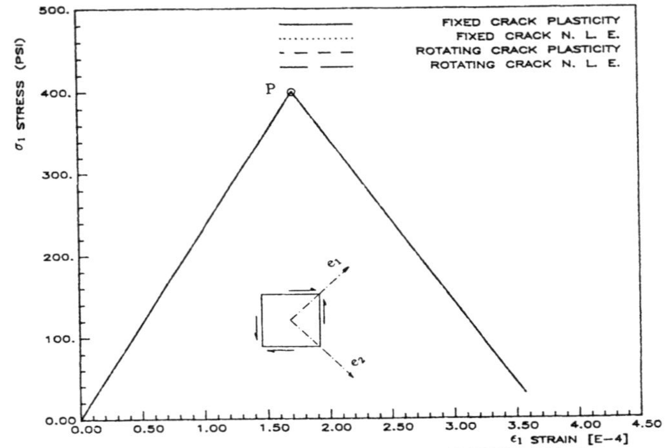


Fig. 7a STRESS – STRAIN RESPONSE IN MAJOR PRINCIPAL DIRECTION

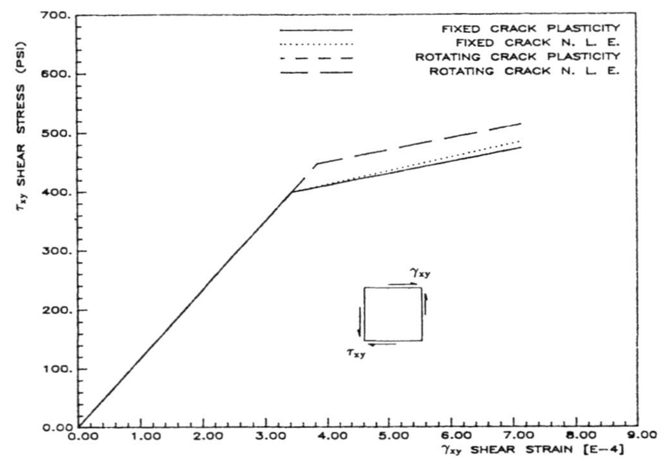


Fig. 7b SHEAR STRESS–STRAIN RESPONSE FOR SIMPLE SHEAR PROBLEM

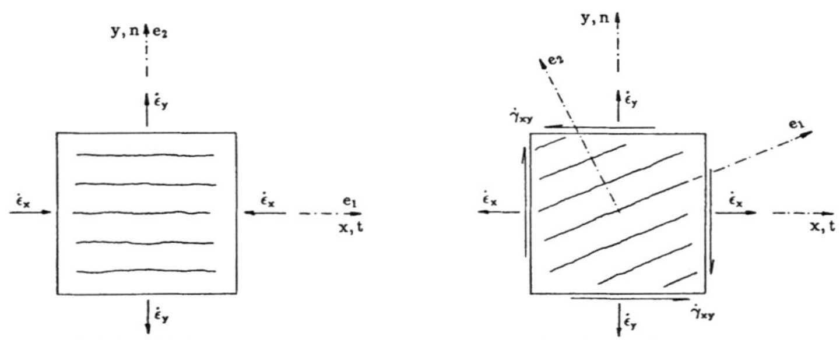


Fixed Principal Strain $e_1 - e_2$

Rotating Principal Strain Axes $e_1 - e_2$

Fig. 8 Tension - Shear Model Problem

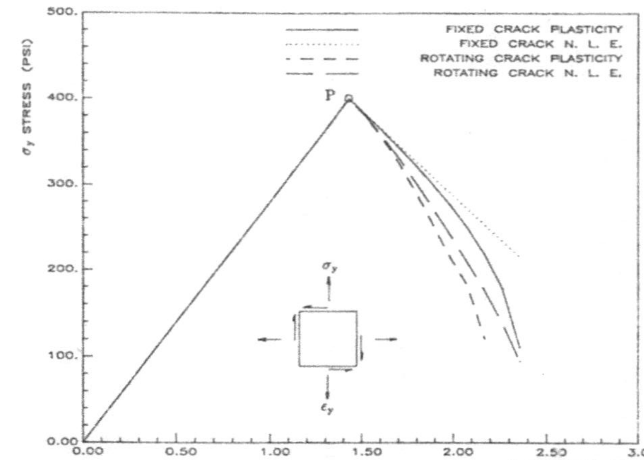


Fig. 9 NOMINAL STRESS-STRAIN RESPONSE IN Y - DIRECTION

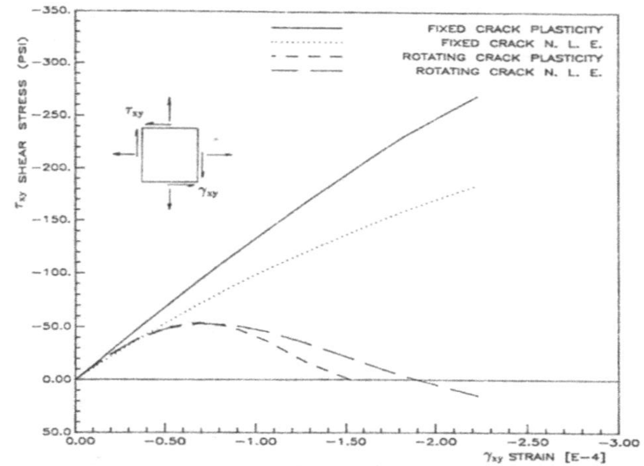


Fig. 11 NOMINAL SHEAR RESPONSE IN X - Y DIRECTION

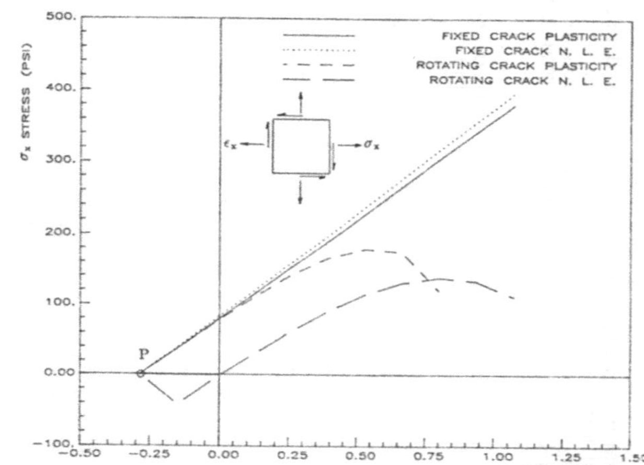


Fig. 10 NOMINAL STRESS-STRAIN RESPONSE IN X - DIRECTION

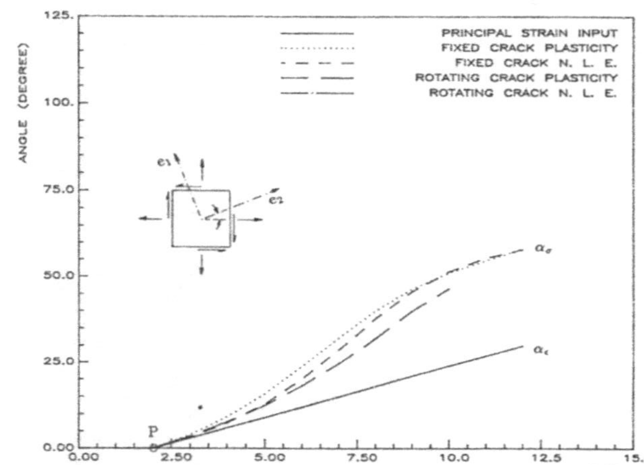


Fig. 12 ROTATION OF MINOR PRINCIPAL STRESS - STRAIN AXES

SEM/RILEM International Conference
on Fracture of Concrete and Rock
Houston, Texas, June 1987
Editors: S.P. Shah, S.E. Swartz

# MATHEMATICAL MODELLING OF DAMAGE EVOLUTION IN CONCRETE AND FRC-MATERIALS

Henrik Stang

## ABSTRACT

The present paper deals with constitutive modelling of materials whose behaviour is governed by the nucleation and growth of microcracks. The modelling is based on composite material theory and continuum damage mechanics. The theoretical predictions for fibre reinforced cementitious materials are discused and compared with experimental findings.

## INTRODUCTION

It is by now a well-established fact that failure of materials such as concrete and other cementitious materials is governed by the nucleation and growth of microcracks rather than by the growth of one single crack [1].

It is also acknowledged that the evolution of microcrack fields is responsible for the major part of the nonlinearities observed in the mechanical response of concrete and related materials [2]. Note that these observations apply not only in the post peek but also in the pre peek regime of the uniaxial stress/strain curve.

In the field of cementitious composite materials the evolution of microcrack fields is especially important since these materials are designed with the aim of controlling the microcrack growth in the brittle cementitious matrix material.

Depending on the design of the composite material, the mechanical behaviour can range from a concrete like behaviour with a catastrophic growth of the microcracks to a steel like behaviour with a "yield stress" at which nucleation of stable microcracks takes place.

The macroscopic modelling of the distributed cracking which takes place in concrete and related materials has been the subject of discussion for more than a decade and still is. Some of the keywords in this discussion are strain softening, strain localization, mesh independent constitutive modelling for FEM-analysis, and smeared crack contra descrete crack modelling.

I will not go into this discussion here especially since a comprehensive review article by Bazant has appeared recently [1].

I will, however, deal with the constitutive modelling of the microcrack fields and the modelling of microcrack growth and nucleation in homogeneous as well as inhomogeneous materials.

The modelling will be based on continuum damage mechanics in the sense that it will be assumed that it is always possible, in any state of material degeneration, to produce a representative volume element containing many microcracks.

When dealing with cementitious (fibre reinforced) composite materials this assumption does not cause any problems, since the structure of the material usually ensures sufficient distribution of the microcracking. In the case of more brittle materials, e.g. concrete with small aggregates it is necessary to assume that a possible strain localization takes place in a band of finite width.

---

Henrik Stang is a Research Associate at the Department of Structural Engineering, Technical University of Denmark, Building 118, DK-2800 Lyngby, Denmark.

The basic tool used in the constitutive modelling is a general linear elastic composite material theory which is able to model a composite material containing different inclusion types in non-dilute concentrations.

## COMPOSITE MATERIAL THEORY

The linear elastic composite material theories which are based on the analysis of a representative volume element (RVE) and which assumes that the composite material consists of a number of inclusions embedded in an otherwise continuous matrix are well established by now [3], [4], [5] and need no detailed introduction. Here we will just summarize the basic results.

Prescribing either a macroscopically homogeneous stress or strain field on the RVE by means of the boundary conditions

$$\sigma_{ij} n_j = \Sigma_{ij} n_j \qquad \text{on } \delta(V) \tag{1}$$

or

$$u_i = E_{ij} x_j \qquad \text{on } \delta(V) \tag{2}$$

the stiffness, L, and the compliance, M, of the composite material are defined by

$$\bar{\sigma}_{ij} = L_{ijkl} \bar{\varepsilon}_{kl} \qquad \text{given (1) or (2)} \tag{3}$$

and

$$\bar{\varepsilon}_{ij} = M_{ijkl} \bar{\sigma}_{kl} \qquad \text{given (1) or (2)} \tag{4}$$

where

$$\bar{\sigma}_{ij} = \frac{1}{V} \int_V \sigma_{ij} \, dV \tag{5}$$

and

$$\bar{\varepsilon}_{ij} = \frac{1}{2V} \int_V (u_{i,j} + u_{j,i}) dV. \tag{6}$$

Here $\sigma$ denotes the microscopic stress field, $\Sigma$ a homogeneous stress tensor, and n the outward unit normal to the surface $\delta(V)$ of the RVE with volume V. Furthermore, u denotes the microscopic displacement field, E a homogeneous strain tensor, and x denotes the coordinates.

The volume average of the strain energy, $\bar{W}$, in the RVE is given by

$$\bar{W} = \frac{1}{V} \int_V \frac{1}{2} \sigma_{ij} \varepsilon_{ij} \, dV = \frac{1}{2} \bar{\sigma}_{ij} \bar{\varepsilon}_{ij} \tag{7}$$

provided that either (1) or (2) is prescribed. Finally, we give expressions for the free energy density, $\psi$, defined by

$$\psi = \bar{W} - \frac{1}{V} \int_{\delta(V)_p} u_i \sigma_{ij} n_j \, dA \tag{8}$$

where $\delta(V)$, subscript p denotes the part of $\delta(V)$ where surface loads are prescribed. It can be shown by means of (7) and the divergens theorem that

$$\psi = -\frac{1}{2} \Sigma_{ij} M_{ijkl} \Sigma_{kl} \qquad \text{given (1)} \tag{9}$$

and

$$\psi = \frac{1}{2} E_{ij} L_{ijkl} E_{kl} \qquad \text{given (2).} \tag{10}$$

For the purpose of deriving solutions for L or M we apply a composite material theory derived by Levin [6] which in an approximate way takes into account the effect of nondilute concentrations of the inclusions which are assumed to be ellipsoidal.

## DAMAGE IN A HOMOGENEOUS MATERIAL

First we consider the case of some sort of microdamage in an otherwise homogeneous and isotropic material. Using Levins method it can be shown that the solution for the compliance can be written in the following compact form

$$M_{ijkl} = M^m_{ijkl} + \beta\, H_{ijmn}\, M^m_{mnkl} \tag{11}$$

where $\beta$ is a scalar which describes the damage density, and where M with superscript m denotes the compliance of the matrix material.

Equation (11) covers the case of randomized penny-shaped cracks, aligned penny-shaped cracks, and spherical porosity.

In the case of spherical porosity $\beta$ is given by

$$\beta = \frac{c}{1 - c} \tag{12}$$

where c denotes volume concentration. The H-tensor is isotropic and given by

$$H = \left(\frac{3}{2}\,\frac{(1 - \nu^m)}{(1 - 2\nu^m)}\,,\ 15\,\frac{(1 - \nu^m)}{(7 - 5\nu^m)}\right) \tag{13}$$

where $\nu$, superscript m denotes Poisson's ratio for the matrix material.

In the case of randomly distributed penny-shaped cracks $\beta$ is given by

$$\beta = \frac{4\pi}{3}\, n(a)^3 \tag{14}$$

where n is the number of cracks per volume and a is the crack radius. Again H is isotropic and given by

$$H = \left(\frac{4}{3\pi}\,\frac{(1 - (\nu^m)^2)}{(1 - 2\nu^m)}\,,\ \frac{8}{15\pi}\,\frac{(1 - \nu^m)(5 - \nu^m)}{(2 - \nu^m)}\right). \tag{15}$$

Finally, in the case of aligned penny-shaped cracks $\beta$ is given by eq. (14), but now H is transversely isotropic and given by

$$H = \frac{4}{\pi}\left(0,\ \frac{(1 - \nu^m)^2}{(1 - 2\nu^m)}\,,\ 0,\ \frac{(1 - \nu^m)}{(2 - \nu^m)}\,,\ \frac{(1 - \nu^m)\,\nu^m}{(1 - 2\nu^m)}\,,\ 0\right). \tag{16}$$

In the equations (13), (15), and (16) we have adopted the decomposition technique for fourth order tensors developed by Hill and Walpole (see e.g. Walpole's paper, [3]). Thus we have given only the invariants instead of writing out all the relevant components of the fourth order H-tensor.

With the composite material compliance given by eq. (11) the corresponding stiffness is given by

$$L_{ijkl} = L^m_{ijmn}\, T_{mnkl} \tag{17}$$

with

$$T^{-1}_{ijkl} = I_{ijkl} + \beta\, H_{ijkl} \tag{18}$$

where $I_{ijkl}$ denotes the fourth order identity tensor.

## DAMAGE IN AN INHOMOGENEOUS MATERIAL

Here in this section we will concentrate on describing microdamage in a composite material which consists of an isotropic matrix material which is reinforced with long uniaxial fibres. The fibres are assumed to be circular cylindrical, and the matrix material is assumed to contain some sort of microscopic damage.

One way of deriving the stiffness or compliance of the composite material would be to apply the Levin-solution for a material consisting of uniaxial fibres in an otherwise homogeneous, possibly anisotropic matrix material:

$$L = L^* + c^f(L^f - L^*)[I + (1 - c^f)P^f(L^f - L^*)]^{-1} \tag{19}$$

or

$$M = M^* + c^f(M^f - M^*)[I + (1 - c^f)Q^f(M^f - M^*)]^{-1} \tag{20}$$

where the tensor indices have been omitted for the sake of brevity, and where a notation similar to that of e.g. Walpole's [3] has been used. Here superscript f relates the quantity in question to the fibres. I denotes the fourth order identity tensor, while P, superscript f denotes the socalled polarization tensor corresponding to circular cylindrical inclusions, and Q, superscript f is given by

$$Q^f = L^* - L^* P^f L^* . \qquad (21)$$

In L (M), superscript *'s place we would usually see the stiffness (compliance) of the matrix material, but here we prescribe that

$$M^* = (I + \beta H)M^m \qquad (22)$$

or

$$L^* = L^m(I + \beta H)^{-1} \qquad (23)$$

according to eqs. (11), (17), and (18).

Thus the influence of the damage in the matrix on the overall stiffness of the composite material is described via the equations (22) and (23).

Since L (M), superscript * might not be isotropic we cannot use the simple solution for the P-tensor which is based on Eshelby's results [7]. More general expressions for the P-tensor has been derived by e.g. Laws [8], [9]. On the basis of Laws' expressions it can be shown that in the case of a transversely isotropic matrix material with stiffness L, superscript *, and with a symmetry axis which coincides with the fibre axis, the P-tensor for a long fibre with circular cross section is given by

$$P^f = \left( \frac{1}{\ell^3 + \ell^1} , 0, \frac{\ell^3 + \frac{1}{2}\ell^1}{\ell^3(\ell^3 + \ell^1)} , \frac{1}{2\ell^4} , 0, 0 \right) \qquad (24)$$

where

$$L^* = (\ell^1, \ell^2, \ell^3, \ell^4, \ell^5, \ell^6). \qquad (25)$$

See also Laws and McLaughlin [10].

Again Walpole's decomposition technique has been used. Note that the decomposition (25) can be used both for transversely isotropic and isotropic tensors [11].

The fibre reinforced material with matrix cracks can alternatively be modelled by a more direct approach. In [11], Stang has extended Levin's method to cover the case of a composite material containing more than one inclusion type. In this approach each inclusion type is treated equally with respect to its mathematical modelling. In [11] it is shown that with this approach the compliance and stiffness of a composite material containing fibres and aligned penny-shaped cracks can be written as

$$M = [I + c^f R^{-1} + \beta\, U H]\, M^m \qquad (26)$$

and

$$L = M^{-1} \qquad (27)$$

with

$$U = I + c^f[(I - S^f)^{-1} M^m (M^f - M^m)^{-1} + (1 - c^f) I]^{-1} \qquad (28)$$

and

$$H = \frac{4}{\pi} \left( 0, \frac{(1 - \nu^m)^2}{(1 - 2\nu^m)} , 0, \frac{(1 - \nu^m)}{(2 - \nu^m)} , \frac{(1 - \nu^m)\nu^m}{(1 - 2\nu^m)} , 0 \right) \qquad (29)$$

and

$$R = M^m (M^f - M^m)^{-1} + (1 - c^f)(I - S^f) \qquad (30)$$

where S, superscript f denotes the usual Eshelby S-tensor, [7], for the fibre inclusion embedded in the isotropic matrix material. Expressions for the invariants of the S-tensor is given in [12].

In eqs. (26)-(30) it is assumed that the penny-shaped cracks are perpendicular to the fibre axis thus the situation modelled by these equations are similar to the situation modelled by eqs. (19)-(23) with β and H given by eqs. (14) and (16). However, the situations are not identical; the geometry modelled by the two sets of equations are outlined in Fig. 1.

## DAMAGE EVOLUTION

In order to model the evolution of the material damage described above, we apply the principles of continuum damage mechanics as described in e.g. [13], [14].

First, we note that in all the examples given above, the degree of material damage is described by one non-dimensional damage parameter $\beta$. This parameter can be split into two non-dimensional damage parameters $\alpha$. In the case of penny-shaped cracks we get

$$\beta = \alpha_1\alpha_2 \tag{31}$$

and in the case of spherical voids we can write

$$\beta = \frac{\alpha_1\alpha_2}{1-\alpha_1\alpha_2}\,, \qquad \alpha_1\alpha_2 < 1 \tag{32}$$

where in both cases

$$\alpha_1 = (k)^3 n\,, \qquad \alpha_2 = \frac{4\pi}{3}\left(\frac{a}{k}\right)^3 \tag{33),(34}$$

where k is a scalar with dimension length, and a is the radius of either crack or spherical void.

Assuming that the $\alpha$ s are independent internal variables, a thermodynamic force A can be defined as

$$A_\gamma = -\frac{\partial\psi}{\partial\alpha_\gamma}\,, \qquad \gamma = 1,\ 2 \tag{35}$$

where $\psi$ is given by eqs. (9) or (10) depending on the prescribed boundary condition.

Next, we assume that for a given admissible damage state, $\alpha$, the admissible stress and strain states can be described by means of a positive scalar function, f, so that

$$f(A_\gamma) \leqq 1\,. \tag{36}$$

Now, assume that the RVE is loaded with an incremental stress or strain which would violate eq. (36). Then the damage state would have to be adjusted according to the following damage evolution law, which is basicly a postulate:

$$d\alpha_\gamma = (0,0) \qquad \text{when} \quad f(A_\gamma) < 1 \tag{37a}$$

$$d\alpha_\gamma = (0,0) \qquad \text{when} \quad \begin{cases} f(A_\gamma) = 1 & \text{and} \\ df \leqq 0 \end{cases} \tag{37b}$$

$$d\alpha_\gamma = \delta\,\frac{\partial f}{\partial A_\gamma} \qquad \text{when} \quad \begin{cases} f(A_\gamma) = 1 & \text{and} \\ df > 0\,. \end{cases} \tag{37c}$$

When the scalar $\delta$ is given by

$$\delta = \left(\frac{\partial f}{\partial\alpha_\gamma}\,\frac{\partial f}{\partial A_\gamma}\right)^{-1}\left(-\frac{\partial f}{\partial\Sigma_{ij}}\,d\Sigma_{ij}\right), \tag{38a}$$

or

$$\delta = \left(\frac{\partial f}{\partial\alpha_\gamma}\,\frac{\partial f}{\partial A_\gamma}\right)^{-1}\left(-\frac{\partial f}{\partial E_{ij}}\,dE_{ij}\right) \tag{38b}$$

depending on the prescribed boundary condition (either (1) of (2)), it is assured that eq. (36) is not violated.

In general the f-function depends on $\alpha$ and possibly other micromechanical parameters, but here we will only investigate the simplest possible situation where f is independent of all the internal parameters involved.

## THE ALGORITHM

Results can now be obtained from the equations above by using a numerical scheme. Let a load history be given in, say, the macroscopic strain tensor, and let also an initial damage state $\alpha$, superscript o be given.

The load history is now modelled by incremental steps. In each step the value of f is checked and the damage state $\alpha$ is adjusted according to eqs. (37)-(38) if necessary. Thus in each step $\alpha$ is known and by applying the solutions for the material stiffness, the stiffness and thus the stress are known.

In this way the numerical scheme determines both the stress/strain curve and the damage evolution for an arbitrary load history.

SOME EXPERIMENTAL OBSERVATIONS

In a series of tests with polypropylene fibre reinforced cement paste, a family of uniaxial stress/strain curves were observed. The varying parameter in the tests was the fibre volume concentration, while all other parameters were unaltered.

The fibre reinforcement consisted of uniaxial, infinitely long pp-fibres of the KRENIT-type which is a low modulus, high bond fibre made from modified polypropylene and developed at the Department of Structural Engineering at the Technical University of Denmark [15].

This FRC-material is interesting because the reinforcing fibres do not contribute to the overall stiffness of the composite material, actually the fibres have a slight stiffness reducing effect on the composite material. Thus the crack arresting effect of the fibres becomes very apparent and it can be assumed that any changes in mechanical behaviour of the composite material as compared to the behaviour of the pure matrix material are due to changes in microcracking behaviour.

In relation to the resulting uniaxial stress/strain curves which were observed, see fig. 2, it should be noted that even at an axial strain of 1% no cracks were visible, i.e. the material was perfectly homogeneous. Thus it is reasonable to assume that the physical process which is responsible for the creation of a stress plateau on the stress/strain curve is crack or void formation rather than crack or void growth.

If this assumption holds, it is also reasonable to assume that a registration of the accumulated acoustic emission during the load history is a registration of the total number of microcracks. A typical accumulated acoustic emission curve is shown in fig. 3.

MODELLING OF DAMAGE GROWTH

According to the experimental observations and assumptions made in the previous section the results in figs. 2 and 3 should be modelled by means of a f-function of the following form

$$f(A_1, A_2) = \frac{A_1}{A^{cr}} . \tag{39}$$

By choosing this form we prescribe damage growth only in one direction in the $\alpha$ space:

$$d\alpha_\gamma = \delta\cdot\left(\frac{1}{A^{cr}} , 0\right) \tag{40}$$

which is consistent with the assumptions made above and eqs. (33) and (34).

Let us now turn to the different models which could be applied in the modelling of the results in figs. 2 and 3. In fig. 4 we have outlined the predicting ability of the four relevant models which have been described above (eqs. (19) (23) (12) (13), (19) (23) (14) (15), (19) (23) (14) (16), and (26)-(30)). We will refer to these models as model 1, 2, 3, and 4 respectively. As an example we have chosen a fibre volume concentration of 0.12.

The load history under consideration is a monotonically increasing strain prescribed in such a way that the stress is always uniaxial.

It is evident that model 1, which describes nucleation of spherical voids, does not describe the experiments very well. Models 2 and 3 predict an almost bilinear curve with an initial slope which depends on the initial damage, the fibre volume concentration, and the stiffness of the matrix and the fibres. The slope of the curve after the damage process has started is very close to the volume concentration of the fibres multiplied with their Young's modulus.

Model 4 also predicts an almost bilinear curve. It predicts approximately the same "bend-over-point" as model 3 but hereafter it continues with a nearly horizontal curve.

Note that these characteristics are found also for other fibre volume concentrations, the fibre volume concentration of 0.12 is chosen merely as an example.

By analyzing the experiments it turns out that the experimental curves always fall between the predictions of models 3 and 4, and in general we can say that the larger the fibre volume concentration the more relevant is model 4, and the less the fibre volume concentration the more relevant is model 3.

In other words, taking the fibre volume concentration into account the specimens with low fibre content tend to be stiffer (in the microcracking regime) than specimens with a higher fibre content.

It is difficult on the basis of the experimental data and the theoretical predictions for the uniaxial stress/axial strain curves to say anything about whether model 2 or model 3 is the more relevant to use in the modelling of the experimental data with relatively small fibre volume concentrations. However, if we turn to a plot of uniaxial stress vs. transverse strain there is a clear distinction between models 2 and 3 due of course to the difference in crack orientation. Model 2 predicts much larger transverse strains than model 3 does, see fig. 5. Unfortunately recordings of the transverse strain were not made in the experiments referred to here, however, such recordings will be made in future experiments.

In fig. 6 we have plotted the damage evolution, in terms of $\beta$, vs. axial strain as predicted by the four models. We can see that apart from model 1, in which beta never exeeds 1, all models give an almost linear damage-axial strain relationship. Such a linear relationship does not agree directly with the accumulated acoustic emission curves observed in the experiments. This might simply reflect the fact that some cracks are easier to nucleate than others. Thus we would typically get a steeper AAE-axial strain curve in the begining just after the crack nucleation has begun than later on when the more "difficult" cracks are formed.

This effect could be modelled by prescribing a f-function which depends on $\alpha$, however, here we still stick to the simple assumption that f is independent of $\alpha$ and any other internal variable or parameter.

If no fibres were present in the material under consideration the material stiffness and compliance should be determined according to the three models, eqs. (11)-(18), depending on damage shape and orientation. Furthermore, the prescribed f-function should allow both nucleation and growth. As an example let us prescribe a f-function in the following way:

$$f(A_1, A_2) = \left(\frac{A_1}{A_1^{cr}}\right)^q + \left(\frac{A_2}{A_2^{cr}}\right)^q \tag{41}$$

and write the critical values of A as

$$(A_1^{cr}, A_2^{cr}) = A^{cr}(1, g) \tag{42}$$

thus

$$g = \frac{A_2^{cr}}{A_1^{cr}} , \quad A^{cr} = A_1^{cr} . \tag{43a, b}$$

For finite values of g the three models all predict some sort of strain softening. As an example we have chosen q in eq. (41) to be 3. The load history that we consider is again a monotonically increasing strain prescribed in such a way that the stress is uniaxial. Given a damage orientation and geometry and an initial damage state ($\alpha^o = (0.1, 1)$) different values of g result in a family of stress/strain curves as shown in fig. 7, where we have considered the case of aligned penny-shaped cracks. The cracks are oriented so that the crack surfaces are perpendicular to the uniaxial stress. Note, how the behaviour changes from a "plastic" (large g) to a very brittle one (small g). Thus, if damage growth is difficult relative to damage nucleation the material behaves in a "plastic" manner and if damage growth is easy relative to damage nucleation then the material behaves in a brittle way. (Note in this connexion that the word "plastic" is not formally correct since unloading branches from any part on the stress-strain curve are always linear through (0, 0)).

Finally, we have outlined the effect of the initial damage state. In fig. 8 we have changed the damage density for a given f-function (g = 0.5) and for a given direction in the damage space (given by the initial damage state used above). It is evident how a large damage density produces a less brittle behaviour while a smaller damage density produces a more distinctly brittle behaviour. This is consistent with experimental observations in e.g. ceramics [16].

## FUTURE DEVELOPMENTS

The most obvious shortcoming of the present model is its missing ability to deal with crack closing effects, i.e. we cannot expect that the model is able to describe damage growth under compression. In order to design a model which can describe tension as well as compression we must firstly introduce a crack description which works differently during crack opening and crack closing. Secondly, we must introduce frictional forces on the crack surfaces. Work in this direction has been done by Horii and Nemat-Nasser [17], though, to my knowledge such solutions have not yet been introduced in models for damage evolution.

Furthermore, it would be very interesting to couple the present model with a material instability investigation in order to deal with the strain localization problem. In this connection it should be noted that the strain softening curves shown in figs. 7 and 8 refer to one material point, and thus it has no meaning to compare with experimental data obtained from test specimens with strain localization. However, if it was possible to combine the present model with an instability investigation then interesting results could be derived concerning the influence of e.g. composite material composition and initial damage on the instability of the material in question.

Finally, it should be mentioned that time dependent phenomena can be dealt with by introducing a viscoelastic composite material description in the continuum damage mechanics model. The paper by Laws and McLaughlin [10] is an example of a work in this direction. However, it must be noted that often a linear viscoelastic description is insufficient to describe the materials used in composite materials, especially the polymers. This of course makes the problem much more difficult but if a proper solution can be obtained it must be expected that it can describe material response to long term loading and make life time predictions.

REFERENCES

[1] Z.P. Bažant. Mechanics of distributed cracking. Appl. Mech. Rev., 39, 675-705, (1986).

[2] S. Ziegeldorf. Phenomenological aspects of the fracture of concrete. In Fracture Mechanics of Concrete, ed. F.H. Wittmann. Elsevier Science Publishers, Amsterdam. Pp. 31-41, (1983).

[3] L.J. Walpole. Elastic behavior of composite materials: Theoretical foundations. Adv. Appl. Mech., 21, 169-242, (1981).

[4] Z. Hashin. Analysis of Composite Materials - A Survey. J. Appl. Mech., 50, 481-505, (1983).

[5] N. Laws, G.J. Dvorak, and M. Hejazi. Stiffness changes in unidirectional composites caused by crack systems. Mech. of Mater., 2, 123-137, (1983).

[6] V.M. Levin. Determination of effective elastic moduli of composite materials. Dokl. Akad. Nauk. SSSR, 220, 1042-1045, (1975). (Sov. Phys. Dokl., 20, 147-148, (1975)).

[7] J.D. Eshelby. The determination of the elastic field of an ellipsoidal inclusion and related problems. Pro. R. Soc. Lond. A. 241, 376-396, (1957).

[8] N. Laws. A note on interaction energies associated with cracks in anisotropic media. Phil. Mag., 36, 367-372, (1977).

[9] N. Laws. The determination of stress and strain concentrations at an ellipsoidal inclusion in an anisotropic material. J. Elasticity, 7, 91-97, (1977).

[10] N. Laws and R. McLaughlin. Self-consistent estimates for the viscoelastic creep compliances of composite materials. Proc. R. Soc. Lond. A. 359, 251-273, (1978).

[11] H. Stang. Strength of composite materials with small cracks in the matrix. Int. J. Solids Structures, 22, 1259-1277, (1986).

[12] H. Stang. A Double inclusion model for microcrack arrest in fibre reinforced brittle materials. J. Mech. Phys. Solids, 35, 325-342, (1987).

[13] D. Krajcinovic. Continuous damage mechanics revisited: Basic concepts and definitions. J. Appl. Mech., 52, 829-834, (1985).

[14] J.J. Marigo. Modelling of brittle and fatigue damage for elastic material by growth of microvoids. Eng. Frac. Mech., 21, 861-875, (1985).

[15] H. Krenchel and S.P. Shah. Synthetic fibres for tough and durable concrete. In Developments in fibre reinforced cement and concrete, Rilem Symposium, FRC 86, Vol. 1, (1986).

[16] A.G. Evans and R.M. Cannon. Toughening of brittle solids by martensitic transformations. Acta Metall., 34, 761-800, (1986).

[17] H. Horii and S. Nemat-Nasser. Overall moduli of solids with microcracks: load-induced anisotropy. J. Mech. Phys. Solids, 31, 155-171, (1983).

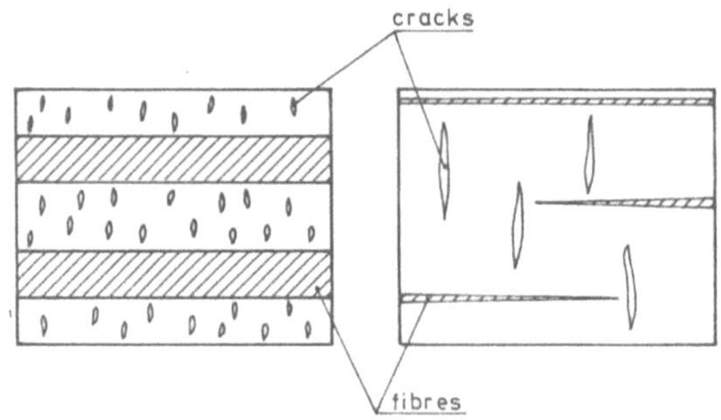


Fig. 1. The two geometrical models described by eqs. (19) + (23) + (14) + (16) (the left model) and eqs. (26)-(30) (the right model).

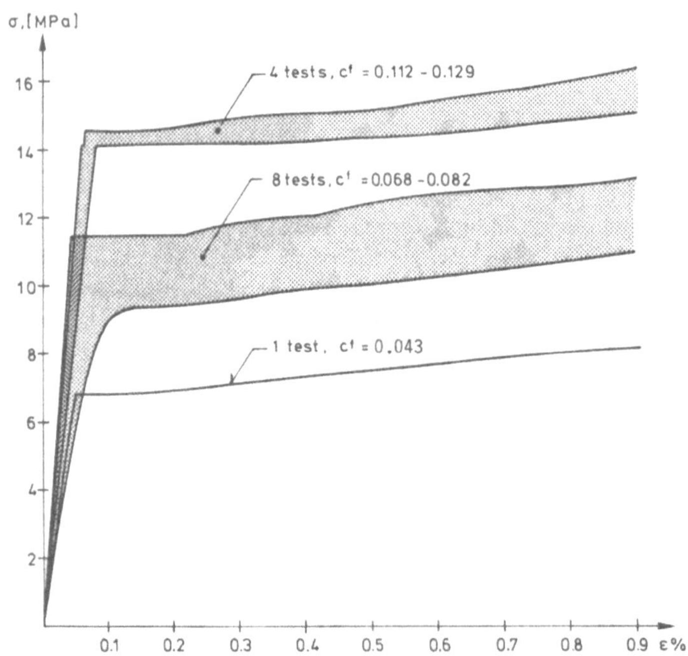


Fig. 2. The uniaxial stress/strain tests with pp-fibre reinforced cement paste specimens.

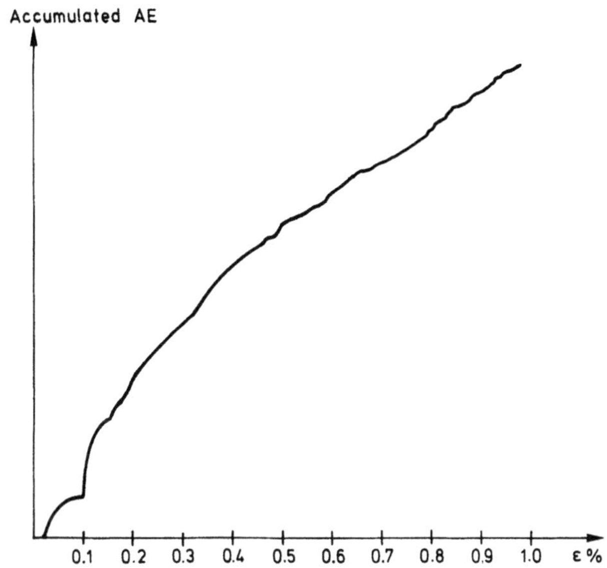


Fig. 3. A typical registration of the accumulated acoustic emission in a pp-fibre reinforced cement paste specimen tested in uniaxial stress ($c^f = 0.043$).

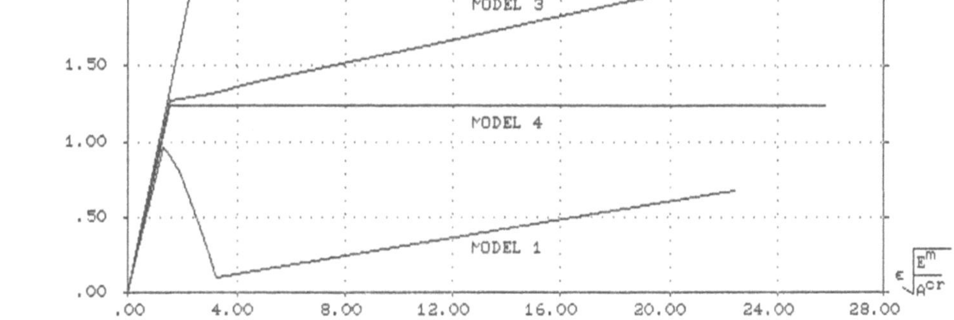


Fig. 4. The uniaxial stress/axial strain curves predicted by model 1-4. The material parameters used are:

$E^m/E^f = 32/8$ , $\nu^m = 0.25$ , $\nu^f = 0.35$ , $c^f = 0.12$

where E and $\nu$, superscript m or f denotes Young's modulus and Poisson's ratio for the matrix and the fibres respectively. The initial damage state is given by:

$\alpha^o = (0.1, 1)$.

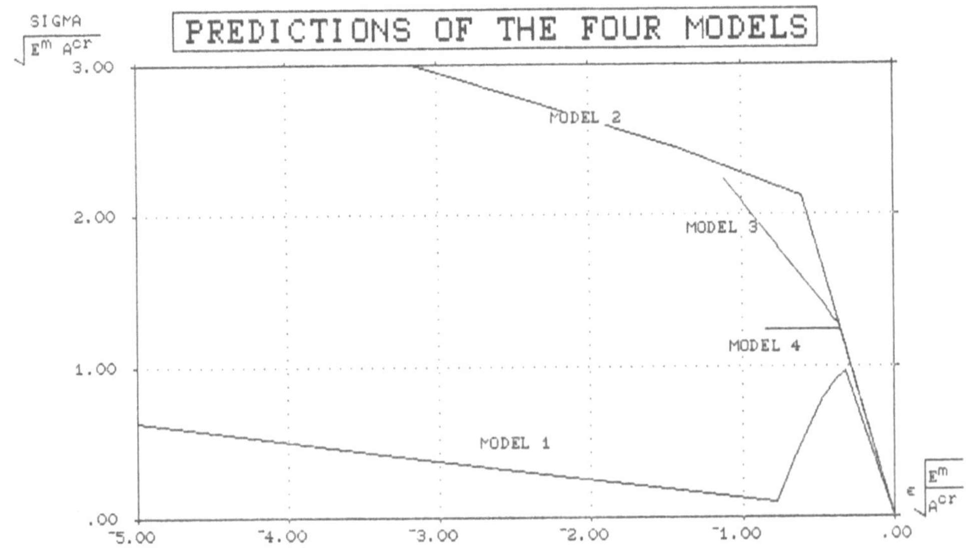


Fig. 5. The uniaxial stress/transverse strain curves predicted by model 1-4. The material parameters used are the same as in fig. 4. Also, the initial damage state is the same.

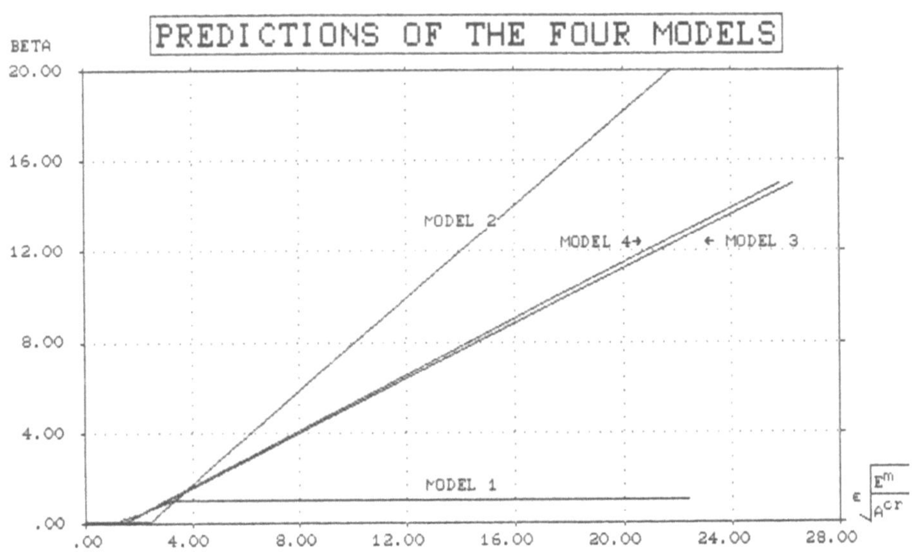


Fig. 6. The damage (in terms of $\beta$) vs. axial strain curves predicted by the four models. The material parameters and the initial damage state are all the same as in fig. 4 and 5.

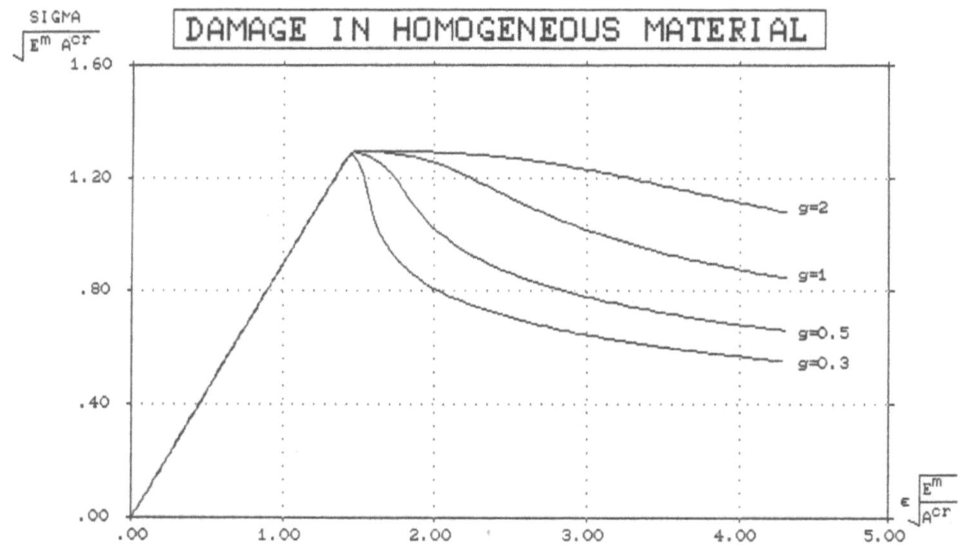


Fig. 7. Uniaxial stress vs. axial strain curves based on nucleation and growth of aligned penny-shaped cracks in a homogeneous material for different values of g. The parameters chosen were:
$\alpha^o = (0.1,\ 1)$ , $\nu^m = 0.25$.

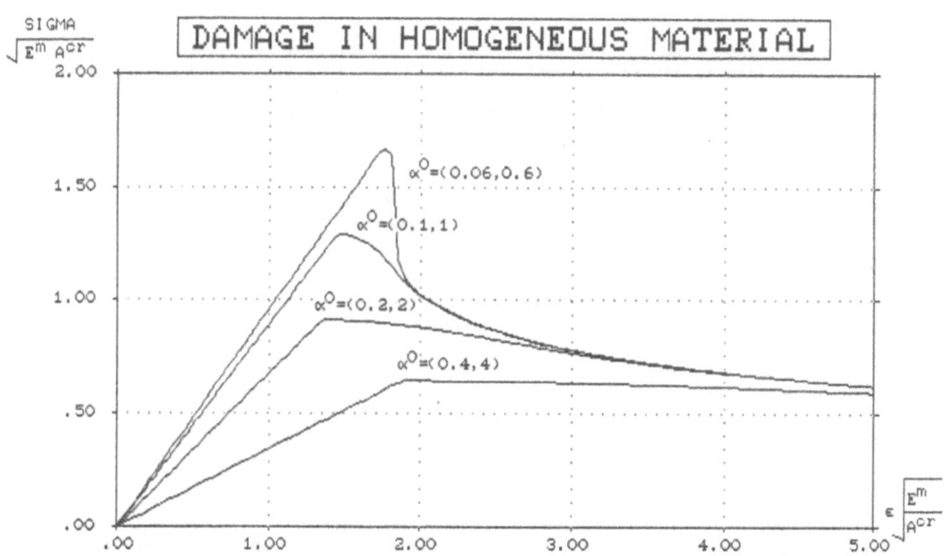


Fig. 8. The influence of the initial damage state on the uniaxial stress vs. axial strain curve. The damage geometry and the parameters are the same as in fig. 7.

SEM/RILEM International Conference
on Fracture of Concrete and Rock
Houston, Texas, June 1987
Editors: S.P. Shah, S.E. Swartz

# A HYPOTHESIS CONCERNING THE EFFECTS OF MACRO-POROSITY ON MECHANICAL PROPERTIES OF CONCRETE

Sandor Popovics

ABSTRACT

A hypothetical mechanism is offered for the observation that the porosity has a greater effect on the compressive strength of a concrete than on its flexural strength which effect is again greater than that on pulse velocity. The explanation is based on Griffith's theory of fracture of brittle materials which states that even uniaxial loadings, such as uniaxial compression, produce not only a uniaxial stress field but also tensile stress concentrations around the pore tips in the brittle material. These, in turn, crack the material progressively until it fails. The greater the magnitude of the uniform stress field, the greater the stress concentrations, thus the greater the damage caused by the pores. This is why, say, 1% increase in porosity causes greater reduction in the compressive strength of a concrete than in its flexural strength or in other mechanical properties. A quantitative form of this mechanism is also presented.

## INTRODUCTION

It has been reported already by Abrams (1) that changes in the water-cement ratio of a concrete affect its compressive strength more than its flexural strength. The recognition came much later that the air content (macro-porosity) in a concrete causes similar difference in the two strengths as well as in other mechanical properties of a hardened concrete. This is illustrated for four properties by an example based on Kaplan's experimental data (2) in Figure 1.

An explanation is offered for these differences in this paper which is based on the dissimilarities in tensile stress concentrations around the pore tips. A quantitative form of this hypothesis is also derived that provides the relative influence of porosity on a given mechanical property of a hardened concrete as a function of the magnitude of the prevailing uniaxial stress field.

It is not claimed that the presented formulas, especially their parameters in the formulas are final. After all very little experimental evidence is available for the reliable establishment of these parameters. It seems more important that (a) a mechanism is offered for the varying effects of porosity on various mechanical properties of the concrete; and (b) a direction is shown for the development of a quantitative form for this mechanism which can be utilized in the future for more refined formulas whenever more pertinent experimental data will be available.

Before the presentation of the new results, two pertinent basic concepts will be discussed briefly, namely, porosity and nature of strength in a hardened cement paste.

## DISCUSSION OF POROSITY

The term "Porosity" can be defined as a dispersed phase the strength of which is negligible as compared to the strength of the matrix. In other words, porosity means voids filled with air, liquid, or other strengthless material. Assuming that the porosity in the aggregate particles does not influence the concrete strength, the effective porosity means voids in the hardened cement paste which is the matrix portion of the concrete. Porosity in cement paste includes gel pores, capillary pores and macro-porosity or large (air) voids. The gel pores are an intrinsic feature of the hydrated portland cement that one can not influence. Therefore, it may be omitted from any further discussion here. The capillary porosity is related to the water-cement ratio, curing and age of the paste, whereas the macro-porosity is the results of incomplete consolidation, or entrained air, or both (3).

The size of the effect of capillary pores on concrete strength seems to be larger than that of the macro-porosity at low cement contents whereas at high cement contents these effects are reversed; the effect of macro-porosity will be greater. (4) Only the effects of macro-porosity are discussed further in this paper. A similar analysis on the effects of capillary pores is planned for another publication.

S. Popovics is the Samuel S. Baxter Professor of Civil Engineering, Drexel University, Philadelphia, PA 19104.

ORIGIN OF STRENGTH IN CEMENT PASTES

Since mechanical properties, that is the behavior under stress, of a concrete are related in one way or another to the strength of the hardened cement paste, a brief discussion of the nature of the strength is also presented here.

The strength of a hardened cement paste is determined primarily by three factors; (a) by the number and intensity of the interpartical bonds in the hydration products, as the source of the strength; (b) by the relative volume of the solid material; and (c) by the nature and magnitude of porosity of the cement paste as a strength-reducing factor. However, these factors are not independent from each other which causes an ambiguity. In the hardening cement paste new bonds are produced by the developing hydration products which, however, simultaneously increase the volume of solids, thus reduce the capillary porosity by filling these pores. At present we do not know what percentage of an observed strength increase of a cement paste is due to the increased number of bonds, what percentage to the increased volume of load carrying solids, and what to the reduction of porosity. It has been shown many times, however, that the strength of a given concrete shows good correlation with its porosity without any consideration of the bonds. A possible reason for the goodness of this approximation is the interdependence of the three factors mentioned as influencing the concrete strength. Note also that it is easier to measure porosity than bonds. For these practical reasons, it has become customary to attribute practically any strength increase solely to a reduction in the paste porosity.

It would follow from all this logically that the magnitude of porosity can be expressed more meaningfully in terms of the volume of matrix rather than that of the concrete. Nevertheless, only the latter is used in the practice almost exclusively, probably because it is simpler and still provides acceptable correlation with concrete properties.

EFFECT OF POROSITY ON MECHANICAL PROPERTIES OF CONCRETE

The weaking effect of the presence of pores is not surprising. After all a 1% increase in porosity means 1% less solid, load-carrying material in the same volume. It has been noticed, however, that the strength reduction by porosity is much greater than what would follow solely from the reduction of the solid material. This is also true for most other mechanical properties, although the rates of reduction are different. For instance, Figure 1 shows that an increase of 5% in the macro-porosity produces about 33% recution in the compressive strength, about 25% reduction in the flexural strength, and only in the case of pulse velocity is the reduction 5% which is approximately equal to the relative reduction of the solid content.

This rapidity of the strength reduction by porosity is due to the fact that not only the decrease of solid material causes the strength reduction but also the reduction of the number of bonds and, most importantly, tensile stress concentrations developing around the pore tips during loading. The greater the stress field in the specimen during testing, the greater the stress concentrations. (5) Thus, the greatest stress concentrations develop in a concrete in the standard compression strength test since the uniaxial average stress goes up to 30 to 50 MPa or more (several thousand psi). On the other hand, stress concentrations are practically missing when a pulse travels through a concrete because here the created remote stresses are not higher than about 0.01 MPa (1 psi). Since a larger stress concentration causes more cracking damage in the concrete, the porosity produces the largest reduction in the uniaxial compressive strength and the least in the pulse velocity. The reductions in other mechanical properties are between these two extremes according to the magnitudes of the prevailing uniaxial stress fields.

THE MATHEMATICAL MODEL

On the basis of the qualitative discussion above, a mathematical model is offered for the quantitative description of the role of macro-porosity in the behavior of hardened concrete under stress. This new mathematical model is the following:

$$f_{rel} = \left(1 - \frac{a}{a_{cr}}\right) \cdot \left(1 + 10^{-\gamma a/a_{cr}}\right) \qquad 1)$$

where

$f_{rel}$ = relative value of a mechanical property of a porous hardened concrete or some other brittle material as fraction of the same property of the pore free material, %/100

$a$ = relative volume of macro-porosity (air content or large pores) of the material, %/100

$a_{cr}$ = critical porosity, that is, the macro-porosity at which the strength or the characteristics of other mechanical properties become zero, %/100

$\gamma$ = experimental parameter which is a function of the type of mechanical property in question.

The first term of the model in Eq. 1 reflects the reduction in the quantity of solid material in the specimen, thus it is independent of the mechanical property in question, or the magnitude of the stress field produced by the related test. The second term represents the additional effects of macro-porosity, primarily the effect of the tensile stress concentrations around the pore tips. The model also reflects the fact that the strength or a characteristic of a mechanical property of a hardened concrete can be zero when the porosity reaches a critical value $a_{cr}$ which is less than 100%. $a_{cr}$ is probably a function of the mechanical property in question; more precisely, it may increase with a decrease in the magnitude of the stress field in the concrete. For instance, in the case of compressive strength it is around 60%, in the case of pulse velocity it can be greater than 100%. In this paper the constant value of 66.7% is assumed for concrete for the sake of simplicity.

Eq. 1 can be approximated quite well, at least for values of $a \leq 30\%$, with the following simpler form:

$$f_{rel} = \left(1 - \frac{a}{a_{cr}}\right) \times 10^{-\gamma a/a_{cr}} \qquad 2)$$

and even further as

$$f_{rel} = 10^{-\gamma a} \qquad 3)$$

where the symbols are the same as in Eq. 1.

The suggestion of using two terms for the description of the effect of macro-porosity on compressive strength, namely one for the reduction of the solid content and the other for the effect of porosity, came first probably from Millard (6) for coal, although he used a different formula, a power function for stress concentrations. The exponential formula of Eq. 3 was first recommended probably by Ryskewitsch for ceramics (7), and calibrated for concrete by Popovics (8).

CALIBRATION OF EQ. 2

The parameter $\gamma$ increases with increasing magnitude of the stress field. This means that at low stress concentrations, such as in pulse velocity measurements, the first term of Eqs. 1 and 2 controls the value of $f_{rel}$ whereas at high stress concentrations (compressive strength) the second term.

The numerical values of $\gamma$ can be obtained for each mechanical property from pertinent experimental data. For instance, several $\gamma$ values are shown in Fig. 1 for Eq. 3. Experimental data support Eq. 3 with these $\gamma$ values quite well as shown in Fig. 2 for compressive strength. Far fewer pertinent experimental data are available for other concrete properties.

The $\gamma$ values, of course, are different for Eq. 1 or 2. The following values were obtained for Eq. 2:

| | |
|---|---|
| compressive strength | 0.019 |
| flexural strength | 0.012 |
| pulse velocity | -0.0027 |

These values of $\gamma$ can be approximated with a simple continuous function. Assuming the value of 28 MPa (4000 psi) for the compressive strength, 2.8 MPa (400 psi) for the flexural strength and the stress field of 0.01 MPa (1 psi) for pulse velocity, such a function is the following:

$$\gamma = 0.006 \log \bar{f} - 0.0027 \qquad 4)$$

where $\bar{f}$ is the average uniaxial stress relatively far away from the pore causing the concentration of tensile stresses around the tip.

Eq. 4 is presented not so much for calculation purposes but rather to illustrate a method for the development of a mathematical model for a general description of the effects of macro-porosity on various mechanical properties of a concrete. Coefficients can be refined easily in the future when more pertinent experimental data are available.

It is also conceivable that further analysis of the $\gamma$ values will provide us a better picture of stress concentrations and their role in the behavior of concrete under load.

Substitution of Eq. 4 into Eq. 2 provides the following, again for $a_{cr}$ = 66.7%:

$$f_{rel} = \left(1 - \frac{a}{a_{cr}}\right) \left(\frac{1.0063}{\bar{f}^{0.006}}\right)^{100a/a_{cr}} \qquad 5)$$

where the symbols are the same as in Eq. 1 and 4.

Eq. 5 seems to indicate that the weaking effect of porosity is greater in high-strength concretes than in low-strength concretes.

Eq. 3 can also be written in the following form:

$$f_{rel} = 10^{-\gamma a} = \frac{1}{B^a} \qquad 6)$$

where $B = 10^{\gamma}$.

Thus, a similar form for Eq. 2 is, as follows:

$$f_{rel} = \frac{1 + \frac{a}{a_{cr}}}{B^a} \qquad 7)$$

which is the same form that the Abrams Formula has for the concrete strength vs. water-cement ratio relationship.

INTERPRETATION OF EQ. 3

It has been shown earlier that Eq. 3 can be written in the following form (8):

$$f = f_0 \times 10^{-\gamma a} = f_1^a \times f_0^{(1 - a)} \qquad 8)$$

where

$f$ = a numerical measure of the mechanical property of the porous concrete
$f_0$ = the same measure of the poreless concrete
$f_1$ = the same measure attributed to the pores.

The other symbols are the same as in Eq. 3

Eq. 8 shows that the numerical measure of a mechanical property of a concrete with macro-porosity is the weighted geometric average of the measure of property of the poreless material and that of the pores. In the case of pulse velocity $f_0$ is the velocity in the voidless material and $f_1$ is the velocity in the void. In the case of strength, the interpretation is not so clear since Eq. 8 attributes a strength, that is a load carrying capacity to the voids, as follows:

$$f_1 = f_0 \times 10^{-100\gamma} \qquad 9)$$

where the symbols are the same as in Eq. 8.

That is, if the compressive strength of the voidless concrete is 28 MPa (approximately 4000 psi) and $\gamma = 0.038$, than from Eq. 9:

$$f_1 = 28 \times 10^{-3.8} = 0.0044 \text{ MPa}$$

or 0.6 psi. It is conceivable that this apparent strength of macro-pores is the result of a stress-relieving effect of the pores. It is also possible, however, that the assumption of $a_{cr} = 0$ in Eq. 3 for compressive strength is an oversimplification that produces the error $f_1$.

CONCLUSIONS

The presented formulas, although approximate, appear to be suitable for the quantitative description of the varying effect of macro-porosity on mechanical properties of concrete. Such properties are: compressive and flexural strengths, pulse velocity, etc. The underlying mechanism of these formulas can be justified by Griffith's theory of fracture. More specifically, the weakening effect of porosity can be attributed to two sources: (a) the reduction in the quantity of solid material; and (b) the tensile stress concentrations around the tips of the pores in the brittle material under stress. It follows from this that the greater the stress that produces the failure of concrete, the greater the relative weakening effect of porosity.

The paper deals only with effects of macro-porosity in concrete, although the presented mechanism as well as the formulas may be generalized for micro-prosity as well as brittle materials other than concrete.

REFERENCES


1. Abrams, D. A., "Design of Concrete Mixtures," Bulletin 1 , Structural Materials Research Laboratory, Lewis Institute, Chicago, December, 1918, p. 22.

2. Kaplan, M. F., "Effects of Incomplete Consolidation on Compressive and Flexural Strength, Ultrasonic Pulse Velocity and Dynamic Modulus of Elasticity of Concrete," ACI Journal, Proc. Vol. 56, March 1960, pp. 853 - 867.

3. Popovics, S., Concrete Making Materials, McGraw-Hill Book Company, New York, etc. and Hemisphere Publishing Corporation, Washington, etc. 1979, p. 370.

4. Popovics, S., "New Formulas for the Prediction of the Effect of Porosity on Concrete Strength," ACI Journal, Proceedings V. 82, No. 2, Mar.-Apr. 1985, pp. 136-146.

5. Popovics, S., "The Fracture Mechanism in Concrete: How Much Do We Know?" Journal of the Engineering Mechanics Division, Proc. ASCE, EM 3, June 1969, pp. 531-544.

6. Millard, D. J., Discussion of the paper "Porosity and Strength of Brittle Solids," by K. K. Schiller in Walton, W. H. Mechanical Properties of Non-Mettalic Brittle Materials, Mechanical Properties of Non-metallic Brittle Materials. W. H. Walton, Editor, Interscience Publishers, Inc., New York, 1958. pp. 45-47.

7. Ryskewitsch, E., "Compressive Strength of Porous Sintered Alumina and Zirconia - 9th Communication to Ceramography," Journal of the American Ceramic Society, Vol. 36, No. 2, February, 1953. pp. 65-68.

8. Popovics, S., "Effect of Porosity on the Strength of Concrete," Journal of Materials, JMLSA, Vol. 4, No. 2, June, 1969. pp. 356-371.

9. Popovics, S., "Structural Model Approach to Two-Phase Composite Materials: State of the Art," American Ceramic Society Bulletin, Vol. 48, No. 11, November, 1969. pp. 1060-1064.


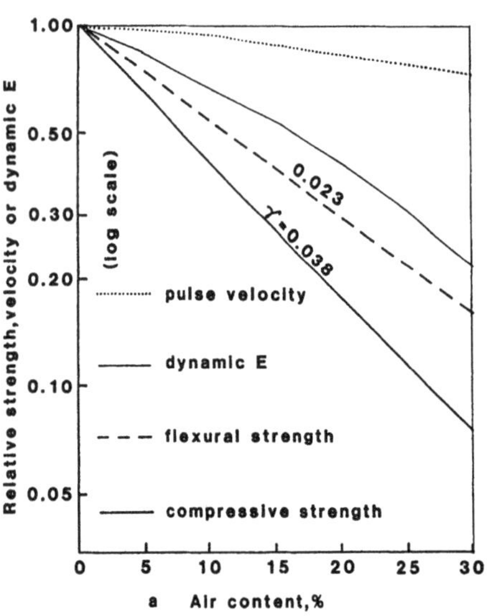


Figure 1- Relative effects of the macro-porosity on several mechanical properties of concrete (8) γ = slope.

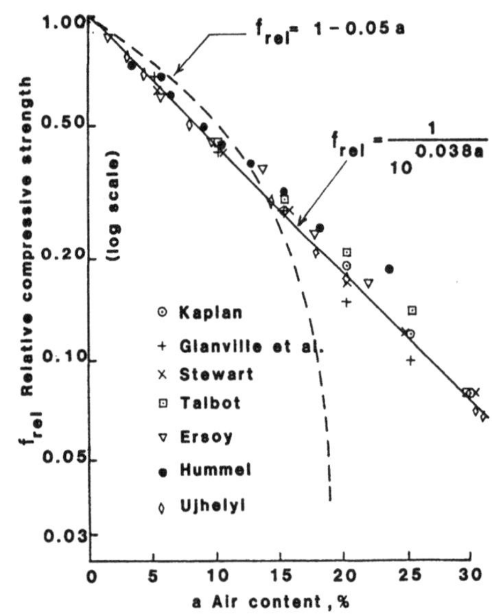


Figure 2- Comparison of the relative effects of macro-porosity on the compressive strength of concrete (8).

SEM/RILEM International Conference
on Fracture of Concrete and Rock
Houston, Texas, June 1987
Editors: S.P. Shah, S.E. Swartz

# STRAIN-SOFTENING SIMULATIONS OF MIXED-MODE CONCRETE FRACTURE

J.G.Rots[1], G.M.A.Kusters[2], J.Blaauwendraad[3]

ABSTRACT

Mixed-mode fracture in concrete is analysed using elastic-softening models and finite elements. Both a smeared crack concept and a discrete crack concept are deployed and emphasis is placed on the possibilities of these concepts for modelling stress rotations in fracture propagation problems. The example problems involve a CLWL-DCB specimen and a double-notched shear beam.

## INTRODUCTION

Fracture research is mostly set-up in such a way that the fracture occurs in the form of a straight separation band which only shows opening (mode-I). In finite element studies mode-I problems have become quite popular, not only because of the absence of crack shear displacements but also because the principal stresses in the cracked elements do not rotate. These studies have revealed that a great deal of experimental observations can be explained by using tensile-softening models (e.g. Hillerborg et al. [8], Bažant and Oh [1]).

In structural practice, however, the crack path will generally not be straight and the crack surfaces will undergo sliding displacements (mode-II) in addition to opening displacements. Such mixed-mode fracture problems pose strong demands on the constitutive model as we need a stress-displacement (or stress-strain) law which embodies a mode-I relation between the normal traction and the crack opening displacement, a mode-II relation between the shear traction and the crack sliding displacement, and probably also some interaction in the form of mixed-mode coupling terms. Furthermore, the fact that shear tractions are allowed to develop along the crack surfaces involves that the axes of principal stress may rotate after crack formation. Consequently, the finite elements must be able to capture these stress rotations accurately for otherwise the response may be much too stiff.

The purpose of this paper is to explore some of the possibilities and the limitations of softening models and finite elements for predicting mixed-mode fracture in heterogeneous materials. To this end, a mixed-mode elastic-softening model is deployed, which is formulated both within a smeared crack concept as well as a discrete crack concept. Subsequently, smeared crack studies as well as discrete crack studies will be presented for two mixed-mode fracture problems. We will concentrate on the computational results rather than discuss the constitutive aspects in detail, as these aspects have been treated before by Rots et al. [18] for the case of tensile softening and Rots & de Borst [20] for the case of tensile softening combined with shear softening.

Particular attention will be given to the provisions which are required for simulating stress redistributions and stress rotations in the neigbourhood of the fracture. With regard to this issue the merits and demerits of smeared versus discrete models will be discussed. The two example problems are a

1) Research Engineer, Delft Univ. of Technology, Civil Eng. Dept., Delft, The Netherlands.
2) Research Manager, TNO Inst. for Building Materials and Structures, DIANA Section, Delft.
3) Professor of Struct. Engrg., Delft Univ. of Technology, Civil Eng. Dept., Delft.

single-notched CLWL-DCB specimen and a double-notched shear beam, displaying rotating stress combinations in the tension-tension and tension-compression domain of the principal stress space respectively.

## SMEARED CRACK MODEL

For the smeared crack computations in this paper a framework has been employed which is based on a decomposition of the total strain increment $\Delta\boldsymbol{\varepsilon}$ into a concrete strain increment $\Delta\boldsymbol{\varepsilon}^{co}$ and a crack strain increment $\Delta\boldsymbol{\varepsilon}^{cr}$,

$$\Delta\boldsymbol{\varepsilon}=\Delta\boldsymbol{\varepsilon}^{co}+\Delta\boldsymbol{\varepsilon}^{cr} \tag{1}$$

While the crack strain vector $\Delta\boldsymbol{\varepsilon}^{cr}$ in Eq. (1) refers to the global axes, we may also define a local crack strain vector $\Delta\boldsymbol{e}^{cr}$ in the crack axes. The relation between global and local crack strains reads

$$\Delta\boldsymbol{\varepsilon}^{cr}=\boldsymbol{N}\Delta\boldsymbol{e}^{cr} \tag{3}$$

where $\boldsymbol{N}$ is a transformation matrix reflecting the crack orientation. For a plane-stress configuration $\Delta\boldsymbol{\varepsilon}^{cr}$ consists of three components and $\Delta\boldsymbol{e}^{cr}$ consists of two components (a crack normal strain and a crack shear strain), so that $\boldsymbol{N}$ is a 3*2 matrix. This matrix is kept unchanged and therefore the model belongs to the class of fixed-crack models.

In a similar way, a crack traction vector $\Delta\boldsymbol{t}^{cr}$ can be defined, which is related to the global stress tensor according to

$$\Delta\boldsymbol{t}^{cr}=\boldsymbol{N}^{t}\Delta\boldsymbol{\sigma} \tag{4}$$

Introducing a constitutive relation for the concrete,

$$\Delta\boldsymbol{\sigma}=\boldsymbol{D}^{co}\Delta\boldsymbol{\varepsilon}^{co} \tag{5}$$

and a constitutive relation for the crack,

$$\Delta\boldsymbol{t}^{cr}=\boldsymbol{D}^{cr}\Delta\boldsymbol{e}^{cr} \tag{6}$$

the final stress-strain relation for the cracked concrete may be derived to be [4,18]:

$$\Delta\boldsymbol{\sigma}=\left\{\boldsymbol{D}^{co}-\boldsymbol{D}^{co}\boldsymbol{N}\left[\boldsymbol{D}^{cr}+\boldsymbol{N}^{T}\boldsymbol{D}^{co}\boldsymbol{N}\right]^{-1}\boldsymbol{N}^{T}\boldsymbol{D}^{co}\right\}\Delta\boldsymbol{\varepsilon} \tag{7}$$

It is noted that not only the total strain increment, but also the crack strain increment may be decomposed in order to model multi-directional cracking. In that case the local crack strain vector, the crack traction vector, the transformation matrix and the crack constitutive matrix $\boldsymbol{D}^{cr}$ may be extended to include the contribution of the subsequent crack, each time a subsequent crack arises (Litton [12], de Borst & Nauta [4]).

## DISCRETE CRACK MODEL

The decomposition of strain into a concrete part and a crack part is also present if we employ the so-called discrete crack concept. Even stronger, the discrete crack concept completely separates the concrete and the crack since it represents these two components by separate finite elements. The uncracked concrete is represented by continuum elements with the same constitutive relation as mentioned above (Eq. 5), while the crack is modelled by means of interface elements which are characterized by crack displacements $\boldsymbol{u}^{cr}$ rather than crack strains. The constitutive behaviour of the discrete crack is then described by an incremental relation between the interface tractions and the crack displacements:

$$\Delta\boldsymbol{t}^{cr}=\boldsymbol{C}^{cr}\Delta\boldsymbol{u}^{cr} \tag{8}$$

Interface elements may be distinguished into two categories, viz. lumped interface elements (e.g. Ngo & Scordelis [15]) and smeared interface elements (e.g. Goodman et al. [7]). Upon testing the latter type of elements oscillating stress profiles were encountered in this study. This aspect is currently investigated and for the computations reported here we have resorted to the more simple, former type of elements which lump the interface behaviour to the nodes.

Furthermore, the discrete cracks were assumed to have a predefined orientation. Consequently, the interface elements could be incorporated within the original finite element mesh so that the topology of the mesh remained intact. The initial stiffness of the predefined elements was assigned a very large value in order to model the uncracked state, and only upon violating the crack initiation condition the stiffness was changed according to Eq. (8). Of course, the difficulty with the discrete crack approach is to predict the orientation of the discrete crack, which requires a change in the topology of the finite element mesh (Ingraffea & Saouma [10]) or other provisions (Blaauwendraad & Grootenboer [3]). However, this aspect has not been pursued here since we rather concentrate on the capability of the concept to simulate stress redistributions.

ELASTIC-SOFTENING MODEL

In this study a linearly elastic model was assumed for the concrete and a softening model for the crack, so that we arrive at an elastic-softening model for the cracked concrete. Consequently, $\boldsymbol{D}^{co}$ (Eq. 4) was assumed to be the elasticity matrix, while $\boldsymbol{D}^{cr}$ (Eq. 6) and $\boldsymbol{C}^{cr}$ (Eq. 8) were assumed to reflect the softening characteristics.

Considering the softening law for discrete cracks first (Eq. 8), it appears that for a 2D configuration a discrete crack only has a normal displacement $u_n^{cr}$ (mode-I) and a shear displacement $u_t^{cr}$ (mode-II). The particular form of Eq. (8), which relates these quantities to the normal traction $t_n^{cr}$ and the shear traction $t_t^{cr}$ across the crack, was assumed to be given by

$$\begin{Bmatrix} \Delta t_n^{cr} \\ \Delta t_t^{cr} \end{Bmatrix} = \begin{bmatrix} C^I & 0 \\ 0 & C^{II} \end{bmatrix} \begin{Bmatrix} \Delta u_n^{cr} \\ \Delta u_t^{cr} \end{Bmatrix} \tag{9}$$

where $C^I$ is the tensile softening modulus and $C^{II}$ is the crack shear modulus.

The tensile softening modulus $C^I$ may be related to three tensile softening parameters, viz. the direct tensile strength $f_{ct}$, the fracture energy $G_f^I$ and the shape of the tensile softening diagram. In this paper the tensile softening diagram proposed by Reinhardt et al. [17] has been adopted (Fig. 1). The shear softening modulus $C^{II}$ is generally assumed to be a constant, but recent research has shown that it may also be related to a set of shear softening parameters, such as an ultimate shear traction $\tau_u$, a mode-II fracture energy $G_f^{II}$ and a specific shape of the shear softening diagram [20]. In this paper numerical experiments have been undertaken using either a very low but constant $C^{II}$ or a bilinear shear softening diagram (Fig. 2). Unloading and reloading phenomena have been modelled using a secant approach, both for mode-I as well as mode-II as indicated in Figs. 1 and 2.

The smeared crack analogy of Eq. (9) reads

$$\begin{Bmatrix} \Delta t_n^{cr} \\ \Delta t_t^{cr} \end{Bmatrix} = \begin{bmatrix} D^I & 0 \\ 0 & D^{II} \end{bmatrix} \begin{Bmatrix} \Delta \varepsilon_{nn}^{cr} \\ \Delta \gamma_{nt}^{cr} \end{Bmatrix} \tag{10}$$

where $\varepsilon_{nn}^{cr}$ is the crack normal strain and $\gamma_{nt}^{cr}$ is the crack shear strain. Here, it should be kept in mind that these crack strains represent crack displacements which are smeared out over a certain crack band width $h$ (Bazant & Oh [1]). Assuming a constant crack strain distribution over $h$ (which is justified for lower order finite elements), we have $u_n^{cr}=h\varepsilon_{nn}^{cr}$ and $u_t^{cr}=h\gamma_{nt}^{cr}$. Consequently, $D^I$ and $D^{II}$ have been set equal to $hC^I$ and $hC^{II}$ respectively in order to achieve mesh-objective release of the mode-I and the mode-II fracture energy. The crack band width $h$ is related to the particular finite element configuration and may be estimated in advance.

It is noted that for the smeared crack computations in this paper the fracture was assumed to be initiated in mode-I, that is normal to the axis of the principal tensile stress. A direct consequence hereof is that the crack shear traction is zero at the onset of cracking, which explains why the crack shear diagram of Fig. 2 starts in the origin. Only upon subsequent rotation of the principal stress axes a shear stress may develop across the crack faces.

## SMEARED CRACK ANALYSIS OF CLWL-DCB SPECIMEN

The first specimen analysed is a single-notched Crack-Line-Wedge-Loaded Double-Cantilever-Beam (CLWL-DCB) which was tested by Kobayashi et al. [11]. The finite element mesh is shown in Fig. 3 and consists of four-noded bilinear finite elements which have been integrated using four-point Gauss quadrature. The specimen had a thickness of 50.8 mm and was subjected to a wedge load $F_1$ as well as a diagonal compression load $F_2$. The ratio of the diagonal force to the wedge force was kept constant at 0.6 until a predetermined diagonal force of 3.78 kN was reached, whereafter the diagonal force was kept constant and only the wedge force altered. This particular loading configuration involves that the fracture propagating from the notch will show opening as well as sliding (mixed-mode). Fig. 4 gives a typical experimental result, revealing a highly localized crack which propagates towards the upper-right corner.

The elastic properties of the concrete were assumed to be Young's modulus $E$=30000N/mm$^2$ and Poisson's ratio $\nu$=0.2. The mode-I softening properties of the crack were taken as tensile strength $f_{ct}$=3.0N/mm$^2$ and mode-I fracture energy $G_f^I$=75J/m$^2$. For this example we did not add shear softening, but assumed a very low constant value of $D^{II}$ equal to 0.1 percent of the elastic shear modulus, which corresponds to the use of a shear retention factor $\beta$ [22] equal to 0.001. The fracture was pre-assumed to be distributed over a crack band width equal to the side of one finite element, i.e. $h$=14.3mm. Similar to the experiment, the numerical analysis was performed under wedge opening displacement control. Furthermore, a full Newton-Raphson procedure was employed with the tangent stiffness matrix being updated before each iteration.

Upon incrementing the loads the analysis started to predict slow crack growth up to peak. At peak load the tangent stiffness matrix ceased to be positive definite and an eigenvalue analysis of the tangent stiffness matrix yielded one negative eigenvalue with the eigenmode being plotted in Fig. 5. This eigenmode shows that the fracture has developed only partially at peak, which is underscored by plotting the corresponding crack pattern (Fig. 6). This observation entirely parallels the experimental findings by Kobayashi et al. [11], indicating that at peak load the crack was "barely visible with the naked eye" and had reached only the first position indicated in Fig. 4.

After peak the load was decremented and the equilibrium solution could be continued until the eigenmode of Fig. 7 and the crack pattern of Fig. 8 were obtained. At this stage the solution was disturbed by zero-energy modes [20] and had to be terminated. The convergence problems were emphasized by the tendency of the crack to branch, which gave rise to additional negative eigenvalues and suggested the existance of alternative equilibrium states. Nevertheless, Figs. 7 and 8 shows that the final fracture orientation fairly resembles the experimental result, which indicates that the assumption of crack initiation in the direction normal to the principal tensile stress (i.e. in mode-I) was justified.

Fig. 7 is also of interest as it demonstrates that a square mesh of bilinear elements is capable of simulating a diagonal stretch band. The effective width of the band is relatively large because the fracture must zig-zag through the mesh, which involves interlocking of the finite elements. Consequently, the assumption of a crack band width $h$ equal to the side of one finite element was too conservative. A multiplication of $h$ by a factor $\sqrt{2}$ might be more realistic.

## COMPARISON WITH DISCRETE CRACK ANALYSIS

The CLWL-DCB specimen was re-analysed with the discrete crack model, using precisely the same material parameters as for the smeared crack analysis. A 30° inclination angle was assumed for the orientation of the predefined discrete crack. The discrete crack analysis progressed relatively easily and the complete post-peak regime could be traced. Figs. 9 and 10 show the incremental displacements at peak load and at a residual load level half way down the softening branch. It may be observed that the rate of crack extension parallels the smeared crack result and the experimental result.

Fig. 11 gives a comparison between load-CMOD curves. Here, the load is the wedge load $F_1$, while the CMOD is defined as the Crack Opening Displacement measured at the Mouth of the notch. The Figure exhibits too little post-peak softening for the smeared crack analysis, whereas it shows a nice and smooth softening curve for the discrete crack analysis. The trend of the experimental result is reproduced quite well by the discrete crack analysis, but it is noted that no attempt has been made to closely fit the experimental results by changing the parameters.

In conclusion we are left with two basic questions. First, the discrepancy between the smeared crack results and the discrete crack results seems to cast some doubt as to the objectiveness of "smearing out" the fracture energy over an assumed crack band width $h$ in case of a diagonal stretch band in a square mesh. However, it should be kept in mind that if the crack band width is pre-assumed to equal $\sqrt{2}$ times the element side instead of one times the element side, the input tensile softening diagram becomes steeper and, consequently, the peak load decreases and the smeared crack result comes closer to the discrete crack result. Hence, for smeared crack analyses it is very essential to pre-estimate the crack band width as accurate as possible.

A second, more serious question relates to the fact that the smeared crack analysis yields too little overall softening, which is a phenomenon that has been observed before for similar problems [16,18]. Again, this may be partly attributed to the underestimation of the crack band width, but different reasons can be put forward which provide more clarity. To illustrate this, the principal tensile stresses have been plotted for the smeared crack analysis as well as for the discrete crack analysis, both at peak load and for a residual post-peak load (Figs. 12 to 15).

The discrete crack results exhibit a smooth stress pattern and the stress rotations in the neighbourhood of the crack tip are captured with realism. In contrast, the smeared crack result displays a somewhat distorted stress pattern and does not properly represent the stress rotations. The explanation for this discrepancy lies in the fact that for the discrete crack analysis the stress rotations simply occur in the uncracked elastic finite elements, whereas for the smeared analysis the stress rotations must be modelled in cracked finite elements.

In the latter case, the most important parameter which controls the stress rotations is the mode-II crack shear modulus $D^{II}$. For the present smeared crack analysis an almost zero value was adopted for $D^{II}$, which implies an almost zero crack shear traction and, consequently, an almost neglegible rotation of the principal stress axes. It seems natural then to increase $D^{II}$ up to for instance 25 percent of the elastic shear modulus. Such an additional analysis was undertaken and the principal stresses indeed turned out to rotate much better, but the overall result was completely incorrect because the analysis did not produce any limit load at all. The intermediate option of an initial shear softening modulus up to an ultimate shear traction $\tau_u$ followed by shear softening (Fig. 2) was also tried out but did not furnish any positive effect. Obviously, for mixed-mode fracture problems like the present CLWL-DCB specimen, which are governed by tension-tension stress states, the best possible solution is obtained for a zero or almost zero value of $D^{II}$.

Interrelated to the above questions is the observation of extensive multi-directional cracking, as appears from the smeared crack pattern of Fig. 8. A considerable number of integration points displays double cracking and the lack of post-peak softening is partly explained from the fact that the fracture energy consumed in creation of the second crack was taken equal to the fracture energy consumed in creation of the first crack. A reduction of the fracture energy for subsequent cracks in an integration point is more realistic and improvements of the smeared crack model in this direction are currently worked out.

Other improvements are probably given by rotating-crack models (e.g. Milford & Schnobrich [14]) instead of the present fixed-crack model, and by the application of more flexible finite elements and/or single point integration schemes. In this respect it is noted that the use of a predefined discrete line-crack is identical to the use of a predefined smeared crack which is modelled by means of inclined elements with center-point integration [19,21]. In the latter case the question of zig-zag crack paths, interlocking and spurious crack shear effects is circumvented. In addition it is possible that more effort should be spent in developing remeshing techniques both for discrete cracks as well as smeared cracks, in order to align them with the true crack direction (Ingraffea et al. [10]).

## SMEARED CRACK ANALYSIS OF DOUBLE-NOTCHED SPECIMEN

In this Section a mixed-mode fracture problem is considered which relates to the tension-compression quadrant of the principal stress space rather than the tension-tension quadrant. It is a double-notched shear beam which is loaded anti-symmetrically and which was tested by Bažant & Pfeiffer [2]. Simulations of this fracture test using the above smeared crack model have been presented before by Rots & de Borst [20], and have shown that the use of shear softening is essential. The purpose of this Section is to briefly extend the results obtained in [20] with novel insight regarding the type of fracture in double-notched specimen geometries (van Mier [13], Hordijk et al. [9]).

The finite element mesh with anti-symmetric loading scheme is shown in Fig. 16. In contrast to the preceding analyses [20] we did not make use of the anti-symmetry, but modelled the complete specimen. The left side of the specimen was given a slight material imperfection in the form of one percent reduction of $G_f^I$, the importance of which will appear from the sequel. The specimen was analysed under indirect displacement control (de Borst [5,6]) of the CMOD over the bottom notch in Fig. 16.

The elastic concrete properties were assumed to be: Young's modulus $E$=25000N/mm$^2$ and Poisson's ratio $\nu$=0.2. The softening properties were taken as: tensile strength $f_{ct}$=3N/mm$^2$, mode-I fracture energy $G_f^I$=75J/m$^2$, initial crack shear modulus $D_0^{II}$ equal to 25 percent of the elastic shear modulus, ultimate crack shear traction $\tau_u$=0.5N/mm$^2$, mode-II fracture energy $G_f^{II}$=75J/m$^2$ and the crack band width was estimated as $h$=12.7mm.

The analysis progressed as shown in Figs. 17 to 21, which refer to key-events in the fracture localization process. Up to peak, anti-symmetric deformations were predicted with both notches being active (Fig. 17), but at peak load the tangent stiffness matrix ceased to be positive definite and the eigenmode corresponding to the negative eigenvalue resembles a strongly non-antisymmetric deformation (Fig. 18). Obviously, the bottom notch with the slight material imperfection suddenly becomes very active while the opposite side of the specimen drastically unloads. After peak the fracture continued propagating from the bottom notch right through the specimen (Fig. 19), and only at a certain stage half way down the post-peak regime the upper notch became active again (Fig. 20).

Fig. 21 shows the load-CMOD curves, both for the bottom notch and for the upper notch. Post-peak unloading for the upper notch is evident and it is noted that the load-CMSD response showed a similar trend. The anti-symmetric solution without the material imperfection has been added to Fig. 21 and lies in between the two curves obtained for the present computation.

The above described behaviour entirely parallels the numerical simulation of a double-notched direct tensile test (Rots et al. [21]), and reveals that "symmetry" or "anti-symmetry" is impossible in fracture testing of heterogeneous materials. Because of the heterogeneous nature of concrete, material imperfections occur in any specimen and, consequently, the post-peak response measured from double-notched fracture tests will always continue on an equilibrium path which is associated with a non-symmetric or a non-antisymmetric fracture mode. For symmetric tests on mode-I double-notched specimens this statement is supported by experimental evidence (Hordijk et al. ([9]), but for the present anti-symmetric mixed-mode test the phenomenon was not reported (Bažant & Pfeiffer [2]).

Further results of the analyses were very similar to the anti-symmetric results [20]. It was found that mode-I effects prevailed up to peak, while mode-II effects became gradually more important during the post-peak regime. A dissimilarity with the CLWL-DCB specimen is that for this notched shear beam cracking was accompanied by significant lateral compression, whereas the CLWL-DCB specimen was governed by tension-tension stress states. Hence, for the present specimen shear softening could be added, which differs from the CLWL-DCB analysis which required a zero or almost zero crack shear modulus.

## CONCLUDING REMARKS

Finite element studies of mixed-mode concrete fracture were undertaken using an elastic-softening model which was formulated within a smeared (fixed-) crack concept as well as a discrete crack concept. The smeared crack concept was demonstrated to be capable of simulating diagonal crack bands through a square mesh. In order to model the stress rotations involved in fracture propagation this concept requires provisions in the form of a very low crack shear modulus, or shear softening in addition to tensile softening. The discrete crack concept poses less strong demands on the constitutive model as it implies the stress rotations to be modelled in uncracked, elastic concrete elements adjacent to the discrete crack elements.

Numerical results for a double-notched shear beam cast doubt on the applicability of double-notched geometries for enforcing perfectly 'direct' mode-II tests. Here, the situation is similar to double-notched mode-I tests, which reveal that 'direct' tension does not exist [9,13,21].

## ACKNOWLEDGEMENTS

The research for this paper was supported financially by the Netherlands Technology Foundation (STW), by CUR-committee A26 "Concrete Mechanics" and by the Department of Public Works (RWS), Section Bouwspeurwerk. The computations have been performed with the DIANA * finite element code on the GOULD/SEL 9000 computer facilities of the TNO Institute for Building Materials and Structures (TNO-IBBC). Support from René de Borst, Pier Nauta and other colleagues from the DIANA-group of TNO-IBBC is gratefully acknowledged.

## REFERENCES

[1] Bažant, Z.P., and Oh, B.H., 1983, "Crack band theory for fracture of concrete," *Materiaux et Constructions*, RILEM, Vol. 16, pp. 155-177.

[2] Bažant, Z.P., and Pfeiffer, P.A., 1986, "Shear fracture tests of concrete," *Materiaux et Constructions*, RILEM, Vol. 19, pp. 111-121.

[3] Blaauwendraad, J., and Grootenboer, H.J., 1981, "Essentials for discrete crack analysis", Final Report IABSE Colloquium on "*Advanced Mechanics of Reinforced Concrete*", Delft University Press, Delft, The Netherlands, pp. 263-273.

[4] de Borst, R., and Nauta, P., 1985, "Non-orthogonal cracks in a smeared finite element model", *Engineering Computations* Vol. 2, pp. 35-46.

---

*) DIANA is a trademark of TNO-IBBC

[5] de Borst, R., 1986, "Non-linear analysis of frictional materials", *Dissertation*, Delft University of Technology, Delft, The Netherlands, 140 pp.

[6] de Borst, R., 1986, "Computation of post-bifurcation and post-failure behaviour of strain-softening solids", *Computers & Structures*.

[7] Goodman, R.E., Taylor, R.L., and Brekke, T.L., 1968, "A model for the mechanics of jointed rock", *Journal of the Soil Mechanics and Foundation Division*, ASCE, Vol. 94, No. SM3, pp. 637-659.

[8] Hillerborg, A., Modeer, M., and Petersson, P.E., 1976, "Analysis of crack formation and crack growth in concrete by means of fracture mechanics and finite elements," *Cement and Concrete Research*, Vol. 6, No. 6, pp. 773-782.

[9] Hordijk, D.A., Reinhardt, H.W., and Cornelissen, H.A.W., 1987, "Fracture mechanics parameters of concrete from uniaxial tensile tests as influenced by specimen length", *Proc. Int. Conf. on Fracture Mechanics of Concrete and Rock*, S.E. Swartz, S.P. Shah et al. Ed., to be held in Houston.

[10] Ingraffea, A.R., and Saouma, V., 1985, "Numerical modelling of discrete crack propagation in reinforced and plain concrete", Chapter 4 in *Fracture Mechanics of Concrete*, G.C. Sih, A. DiTomasso Eds., Martinus Nijhoff Publishers, Dordrecht, pp. 171-225.

[11] Kobayashi, A.S., Hawkins, M.N., Barker, D.B., Liaw, B.M., 1985, "Fracture process zone of concrete", *Application of Fracture Mechanics to Cementitious Composites*, S.P. Shah Ed., Martinus Nijhoff Publishers, Dordrecht, pp. 25-50.

[12] Litton, R.W., 1974, "A contribution to the analysis of concrete structures under cyclic loading," *Dissertation*, University of California, Berkeley.

[13] van Mier, J.G.M., 1986, "Fracture of concrete under complex stress", *HERON*, Vol. 31, No. 3, 90 pp.

[14] Milford, R.V., and Schnobrich, W.C., 1984, "Numerical model for cracked reinforced concrete", *Computer-Aided Analysis and Design of Concrete Structures*, F. Damjanic et al. Eds., Pineridge Press, Swansea, Part 1, pp. 261-273.

[15] Ngo, D., and Scordelis, A.C., 1967, "Finite element analysis of reinforced concrete beams, *Journal of the American Concrete Institute*, Vol. 64, No. 14, pp. 152-163.

[16] Oldenburg, M., 1985, "Finite element analysis of tensile fracturing structures", *Licentiate Thesis 1985:010L*, Div. of Comp. Aided Analysis and Design, Lulea University of Technology, Lulea, Sweden, 74 pp.

[17] Reinhardt, H.W., Cornelissen, H.A.W., and Hordijk D.A., 1986, "Tensile tests and failure analysis of concrete", *Journal of Structural Engineering*, Vol. 112, No. 11, pp. 2462-2477.

[18] Rots, J.G., Nauta, P., Kusters, G.M.A., and Blaauwendraad, J., 1985, "Smeared crack approach and fracture localization in concrete," *HERON*, Vol. 30, No. 1, 48 pp.

[19] Rots, J.G., 1986, "Strain-softening analysis of concrete fracture specimens", *Fracture Toughness and Fracture Energy of Concrete*, F.H. Wittmann Ed., Elsevier Science Publ., Amsterdam, pp. 137-148.

[20] Rots, J.G., and de Borst R., 1987, "Analysis of mixed-mode fracture in concrete", submitted for publication in *Journal of Engineering Mechanics*, ASCE.

[21] Rots, J.G., Hordijk, D.A., and de Borst, R., 1987, "Numerical simulation of concrete fracture in 'direct' tension", Proc. Fourth Int. Conf. on *Numerical Methods in Fracture Mechanics*, A.R. Luxmoore et al. Eds., Pineridge Press, Swansea.

[22] Suidan, M. and Schnobrich, W.C., 1973, "Finite element analysis of reinforced concrete," *Journal of the Structural Division*, ASCE, Vol. 99, No. 10, pp. 2109-2122.

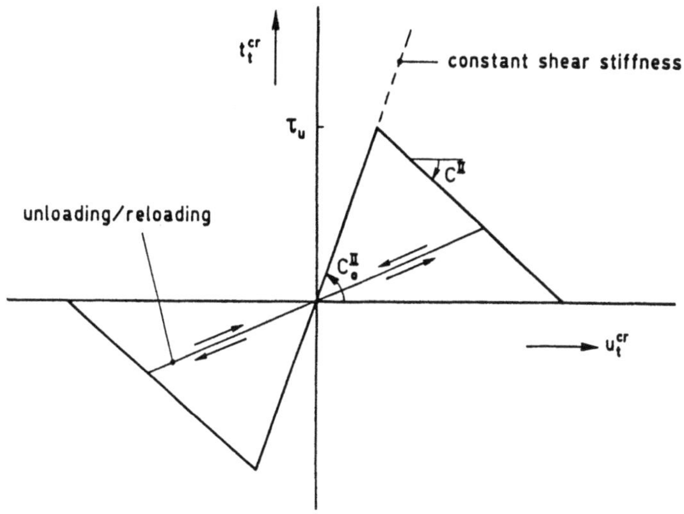


Fig. 2. Crack shear diagram.

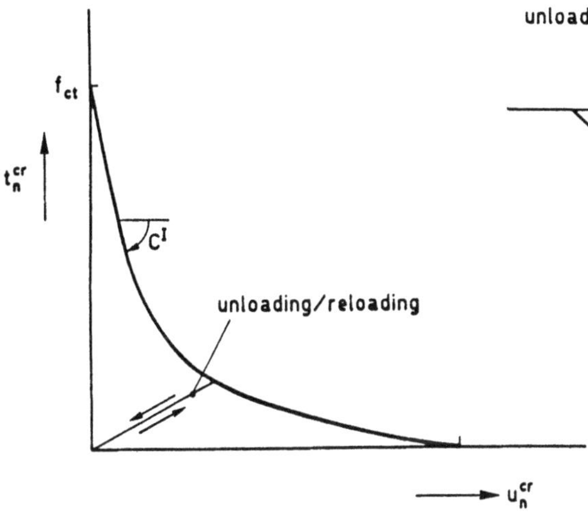


Fig. 1. Nonlinear tensile-softening diagram.

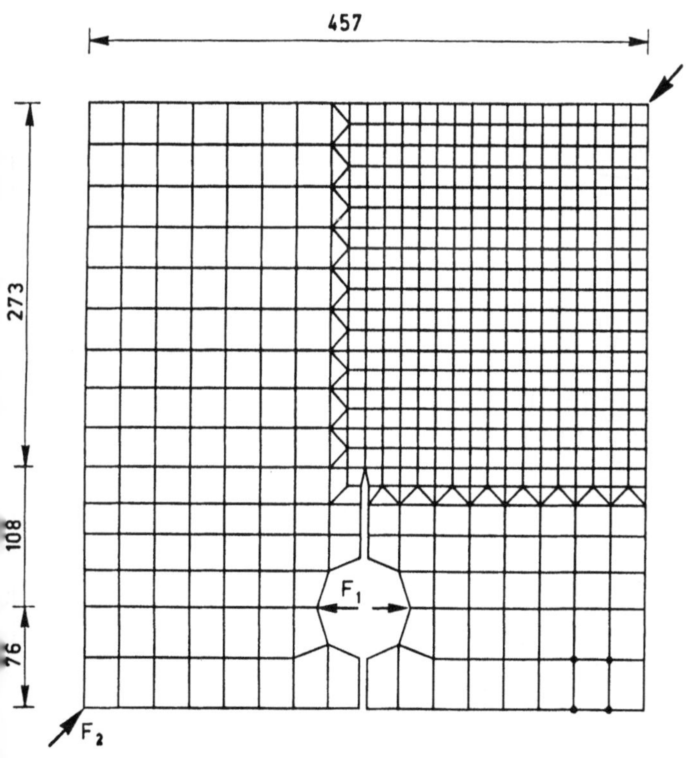


Fig. 3. Finite element mesh for CLWL-DCB specimen.

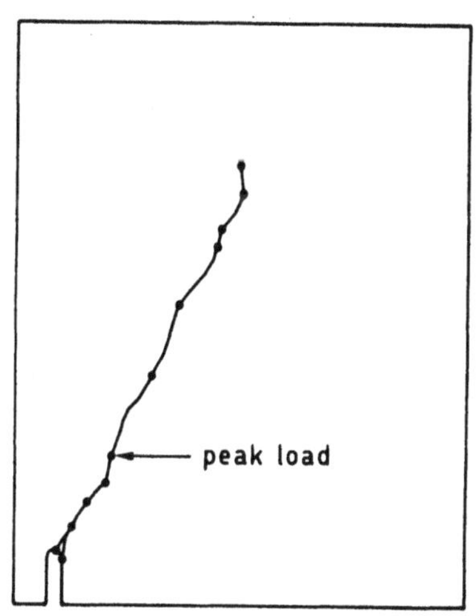


Fig. 4. Experimental crack (Kobayashi et al. [11]).

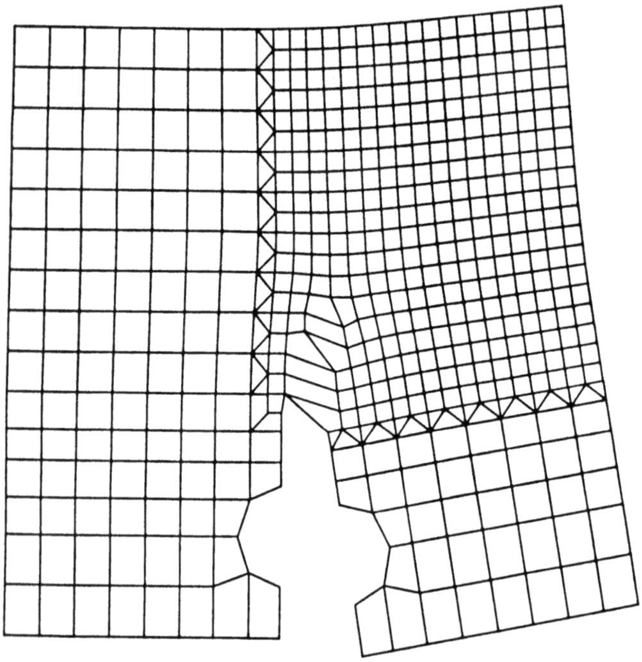

Fig. 5. Eigenmode at-peak ($F_2$=6.87kN).

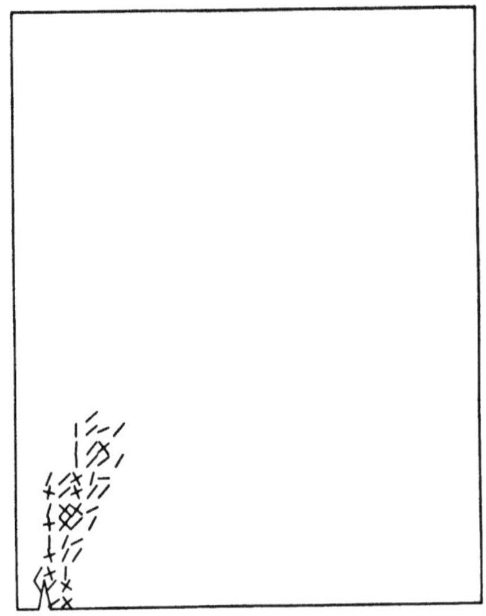

Fig. 6. Crack pattern at-peak.

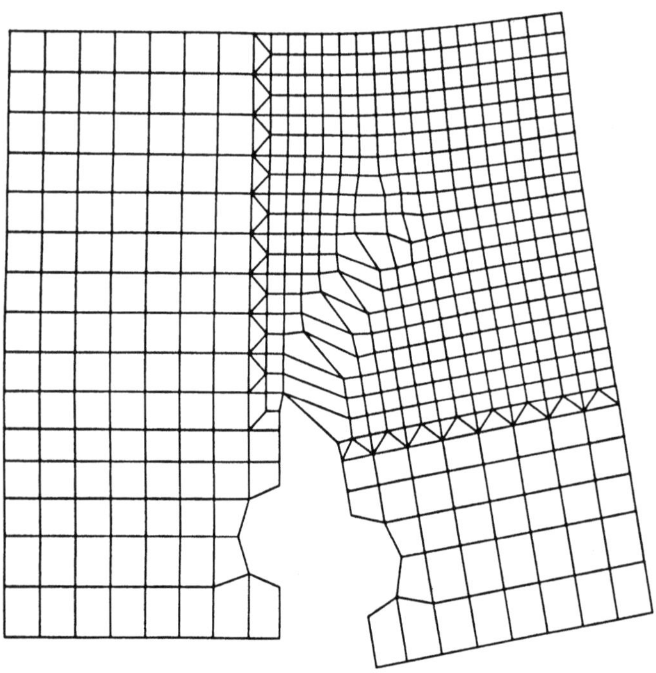

Fig. 7. Eigenmode post-peak ($F_2$=5.54kN).

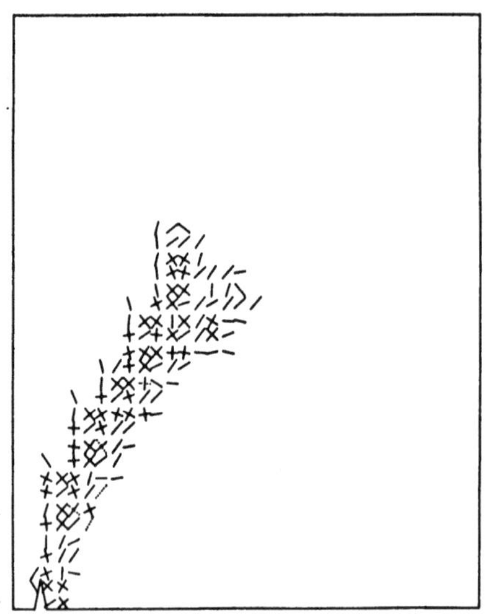

Fig. 8. Crack pattern post-peak.

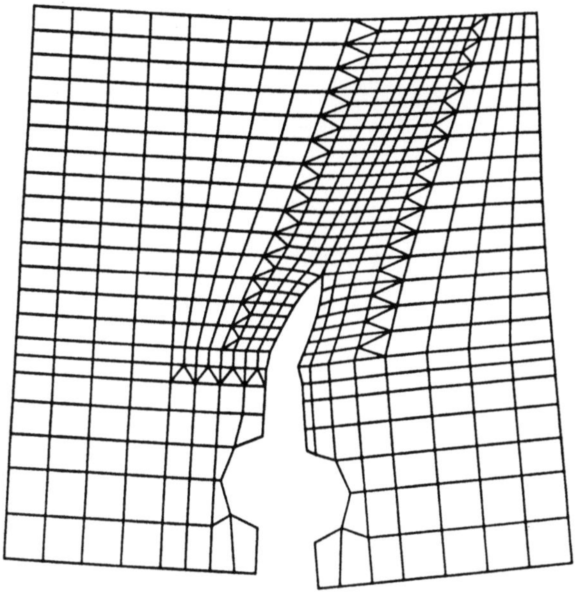

Fig. 9. Incremental displacements, at-peak, discrete crack analysis.

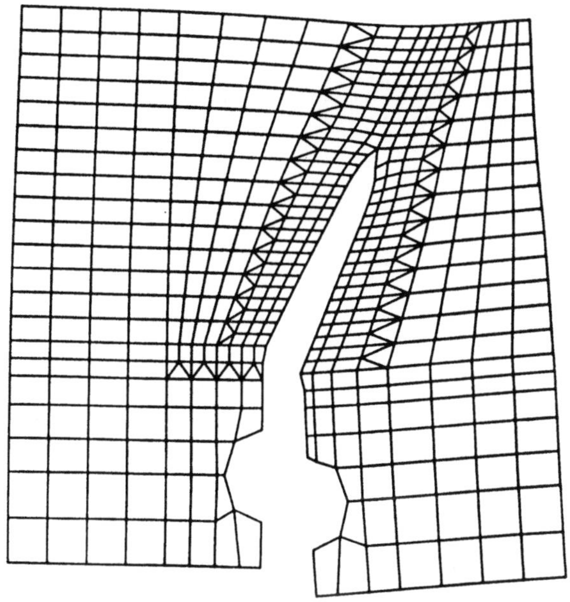

Fig. 10. Incremental displacements, post-peak, discrete crack analysis.

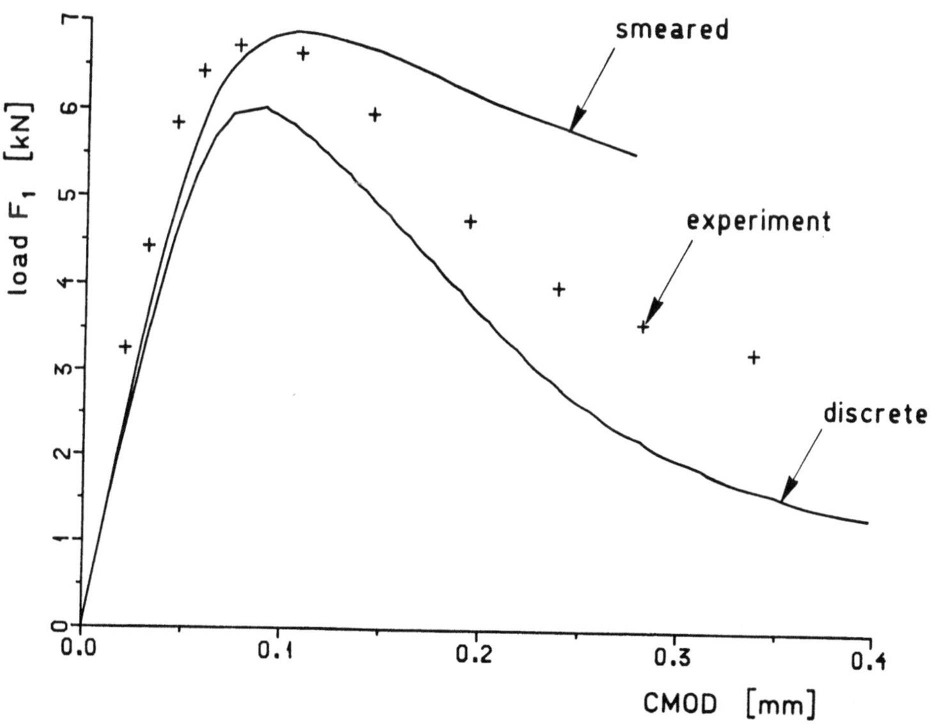


Fig. 11. Load-CMOD response.

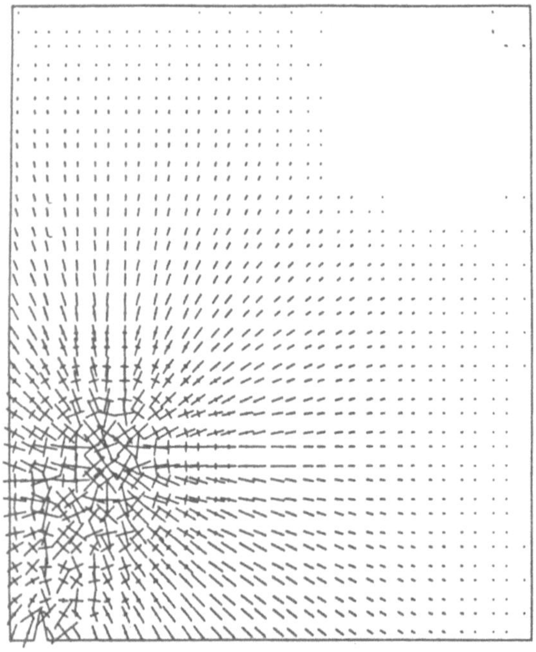

Fig. 12. Principal tensile stresses, at-peak, smeared crack analysis.

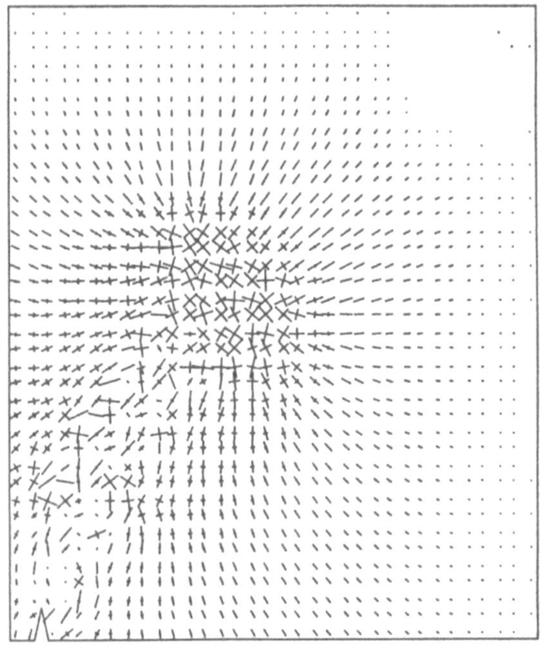

Fig. 13. Principal tensile stresses, post-peak, smeared crack analysis.

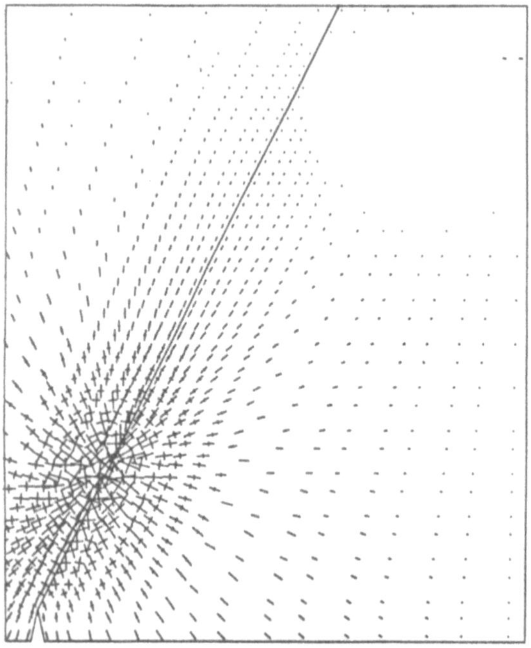

Fig. 14. Principal tensile stresses, at-peak, discrete crack analysis.

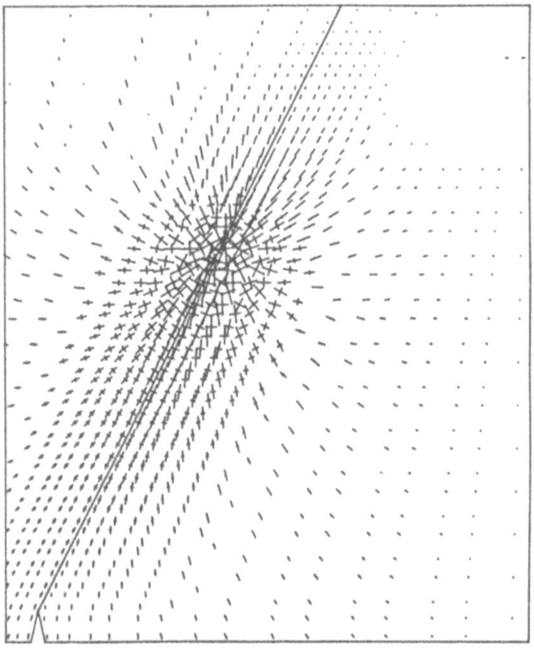

Fig. 15. Principal tensile stresses, post-peak, discrete crack analysis.

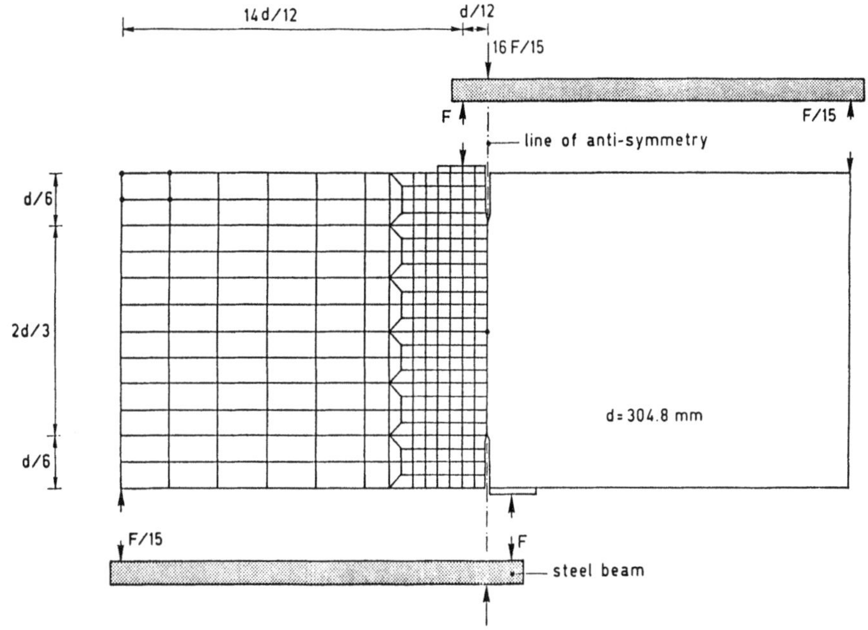


Fig. 16. Finite element mesh for double-notched shear beam.

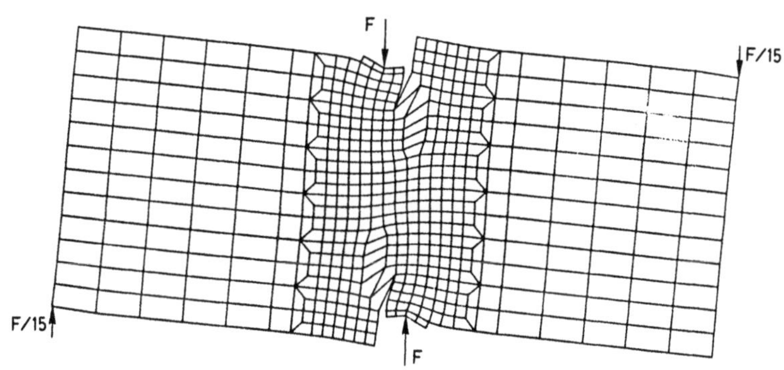


Fig. 17. Eigenmode, pre-peak, P=43.8kN.

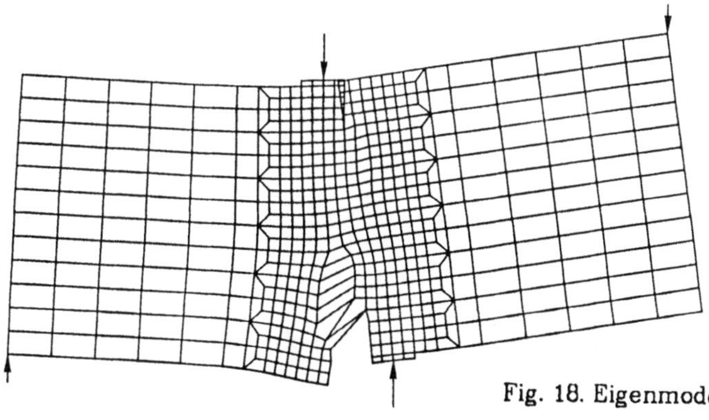

Fig. 18. Eigenmode, at-peak, P=44.1kN.

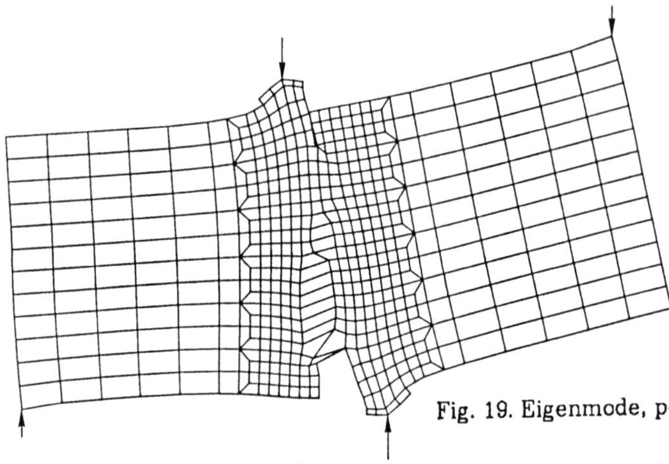

Fig. 19. Eigenmode, post-peak, P=39.8kN.

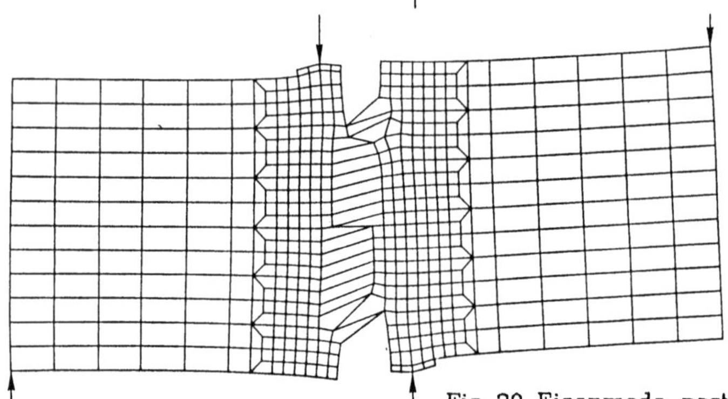

Fig. 20. Eigenmode, post-peak, P=22.0kN.

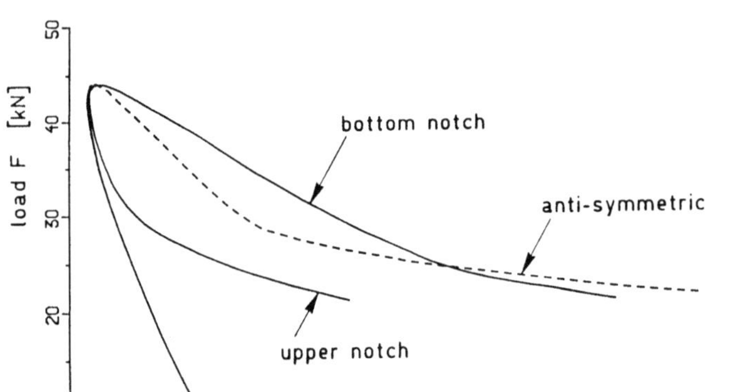


Fig. 21. Load-CMOD response for double-notched shear beam.

SEM/RILEM International Conference
on Fracture of Concrete and Rock
Houston, Texas, June 1987
Editors: S.P. Shah, S.E. Swartz

# MODELS OF QUASI-STATIC AND DYNAMIC FLUID-DRIVEN FRACTURING IN JOINTED ROCKS

R. J. Shaffer, F. E. Heuze, R. K. Thorpe, Lawrence Livermore National Laboratory, Livermore, California
A. R. Ingraffea, Cornell University, Ithaca, New York,
R. H. Nilson, S-CUBED, La Jolla, California

ABSTRACT: We describe the development and the applications of a numerical model to simulate fluid-driven fracturing in rock masses. It is a finite element computer program which couples solid mechanics, fracture mechanics and fluid mechanics. The fractures are driven either in a quasi-static fashion by conventional hydrofracturing liquids, or in a dynamic fashion by gases from burning solid propellants. Fracture propagation can be arbitrarily modeled in a mixed-mode, in media which already contain discontinuities. The code is applied to the analysis of stimulation of tight gas reservoirs such as the Gas Sands in the Western U.S. and the Coal Beds in the Southern U.S.

## 1. INTRODUCTION

Considerable quantities of natural gas are locked tightly into deep gas formations in the Western gas sands and the Eastern devonian shales of the United States. In the Western gas sands the gas is localized in sandstone lenses that are embedded in shale. In the Eastern Shales the gas is mostly drained from the natural fractures, which tend to be vertical and spaced a few metres apart. In addition, large quantities of methane can be extracted from coal beds, for energy value and to enhance mine safety.

To recover this gas, it is necessary to "stimulate" the rock reservoirs, that is, to produce large, man-made fractures that penetrate the reservoirs and drain the gas toward wells. Fractures can be created either by injecting fluids under presure (hydrofracturing) or by burning solid propellants in the wells. To use either technique effectively, one must be able to predict how the rock will fracture. Rock reservoirs already contain natural fractures and joints, making it difficult to predict the effects of induced fracturing. An analysis of fracturing in such a medium must include models for the joint systems, and for the preexisting cracks packed with infill material. The joints are subject to compressive and shear forces; if compression becomes large, the joints become very stiff, and if shearing becomes too large, slippage occurs. Consequently, in a useful model, joint properties must change with stress conditions. In the Western gas sands, interfaces are the regions where shales and sandstones meet, and these interfaces behave much like joints.

We have developed a computer code, FEFFLAP (Finite-Element Fracture and Flow Analysis Program), that has enabled us to make great progress in describing the complex physics of fluid-driven fractures propagating in jointed media [1,2]. The coupled FEFFLAP model includes solid mechanics, fracture mechanics, and fluid mechanics. From the quasi-static to the dynamic regime this has applications to fluid-flow in jointed rocks, hydraulic fracturing for hydrocarbon recovery, and comminution of rock masses. For dynamic analyses, the steady-state FEFFLAP was coupled to the FAST fluid dynamics module [3].

## 2. DESCRIPTION OF THE QUASI-STATIC FRACTURING MODEL

### 2.1 The Solid Fracture Model (FEFAP)

Our FEFFLAP code represents the coupling of the FEFAP discrete fracture propagation code [4] and of the JTFLO program, a LLNL-enhanced version of an earlier code for analysis of fluid flow in rock fractures [5]. FEFAP analyzes planar and axisymmetric structures for crack initiation and growth. The program combines linear and non-linear fracture mechanics theory, the use of interactive computer graphics, and a unique, automatic remeshing capability to allow the user to initiate and propagate up to ten discrete cracks simultaneously. The salient capabilities available in FEFAP were:

- o complete interactive-graphical execution of the program. Each analysis step is directed by the user with alphanumerical and graphical feedback of the results of this step. After any complete crack propagation step, the analysis can be terminated and restarted from the previous step. The emphasis in the program design is on providing versatility to the analyst. One is not locked into a batch-produced result via the initial data input.
- o automatic, discrete crack nucleation at arbitrary points and angles on an edge as specified by the analysis.

- o automatic, discrete crack propagation capability with optional interactive mesh adjustment along the propagating crack.
- o automatic model adjustment for singular elements, and direct, automatic extraction of the stress intensity factors.
- o automatic bandwidth minimization and nodal renumbering.

This automatic crack extension is an intricate process, the logic of which is unique to the FEFAP code.

FEFAP is built under the assumption of linear elastic fracture mechanics (LEFM); stress intensity factors control crack stability and trajectory. The logic of FEFAP is:

1. compute stress intensity factors, $K_I$ and $K_{II}$, for present structure and loading.
2. substitute $K_I$ and $K_{II}$ into any of three mixed-mode interaction formulas. Compute new crack direction and assess crack stability. If the crack is unstable, continue. If stable, go to step 4.
3. mesh for a selected increment of propagation. Repeat steps 1 through 3 until the crack is stable or fracture occurs.
4. if the crack is stable, raise the load level until instability is predicted by interaction formula. Continue with step 3.

### 2.2. The joint models (JPLAXD)

To apply FEFAP to the problem of fluid driven crack propagation in jointed rock, two major extensions of its capabilities were required. First a rezoning was implemented which can handle crack propagation into, across, or from discrete joints. Second, the joint constitutive models were enhanced to accommodate non-linear shear and normal behavior, such as tension cut-off, maximum closure, and shear-strain softening. The non-linear algorithms were those developed earlier in the JPLAXD code [6]. Then FEFAP was coupled to the flow model described next.

### 2.3 The flow model (JTFLO)

The fluid flow model in FEFFLAP is that of the JTFLO finite element program. JTFLO is a LLNL-enhanced version of a coupled stress and flow analysis model developed by Noorishad [5]. It provides for the steady-state solution of flow in parallel or tapered channel such as joints, cracks, and interfaces in rock. Both flow rate and pressure boundary conditions can be specified. The fluid conductivity of individual flow elements is described by

$$k_p = \rho g b^2/12\mu$$

where b is the mean element aperture, $\rho$ is the specific gravity of fluids, g the acceleration of gravity and $\mu$ the dynamic viscosity. Then, the areal permeability along a family of parallel such fractures is proportional to $b^3$. The merits and limitations of this so-called "cubic law" model have been discussed at length by others [7,8].

### 2.4 Coupling of JTFLOW with the enhanced FEFAP: The FEFFLAP code

FEFFLAP allows for flow and pressurization in existing and propagating cracks as well as pressurization and fluid flow in joints. In the general operation of the model one initiates or extends a crack, obtains the new pressure distribution due to the new crack geometry, and follows any non-linear joint behavior by secant iterations on interface element properties, as required. Since the joints can behave non-linearly, care must be exercised in selecting the sequence of events. Figure 1 shows the logic of the coupled structural, fracture, and flow analyses.

## 3. VERIFICATION OF THE QUASI-STATIC FRACTURING MODEL

### 3.1 Comparison of FEFFLAP Analyses with Hydrostone block experiments

Sixteen hydrostone block experiments were performed at LLNL to provide physical test data related to hydrofractures crossing interfaces [9]. The basic geometry is shown in Figure 2. The problem involved two types of hydrostone separated by an interface, and also included the steel platens that were used to load the block. Thus three different solid material types were used in the analysis. Four joint-interface types were required: (1) the interface between the two hydrostone materials, (2) the interfaces between steel platens and the hydrostone, (3) the joint elements that are inserted into the crack as it propagates, and (4) a set of joint elements around the interior of the borehole, which provides a convenient way to pressurize the hole. The last two joint types are necessary for the fluid flow part of the analysis.

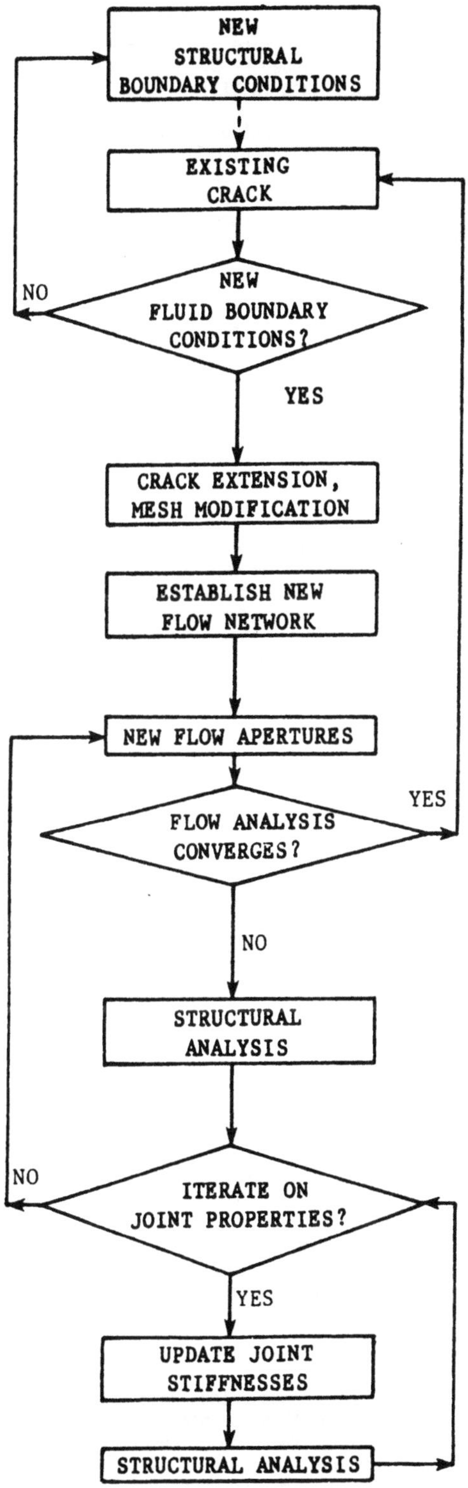


Figure 1. Logic of the FEFFLAP Code for Fluid-Driven Fracture Analysis.

In order to determine the adequacy of FEFFLAP, a 2-D code, to handle the 3-D geometry of Figure 4, the stresses in the mid-vertical section of the block were calculated both with a plane stress FEFFLAP solution, and with JROC3D, a 3-D jointed block code developed by C. St. John at Imperial College [10] and enhanced by F. Heuze, at LLNL. Results agreed to better than 1% [11].

Then, two of the tests were analyzed with FEFFLAP using the mesh shown in Figure 3. Figure 4 shows the results of a FEFFLAP analysis of one experiment in which the crack reinitiates from the interface; for this case the vertical and horizontal loads were 1800 and 750 psi respectively, and the peak borehole pressure was 3400 psi. The borehole pressure required in FEFFLAP to reinitiate the crack was higher than that in the experiment. We suspect the cracks crossed the interface dynamically in the experiments because the cracks went straight through the interfaces. In Figure 5 the FEFFLAP analysis of another experiment shows that a crack stopped at the interface; vertical and horizontal loading stresses were 700 and 100 psi, respectively, and the peak pressure in the borehole was 2800 psi.

### 3.2 Comparison of FEFFLAP with problems for which solutions are known

FEFFLAP was tested on a cracked borehole problem (standard hydrofrac geometry) by calculating Mode I stress intensity factors for two types of loading: a remote biaxial tensile stress, and uniformly pressurized borehole and cracks. The results were compared to established values [12] to obtain an estimate of the code's accuracy. For both types of loading each crack length was 1.5 times the borehole radius. The Mode I stress intensity factor calculated in FEFFLAP was 7 percent higher than the established value for both cases. These results are quite good when one considers the coarse finite element mesh. In addition, the mesh is truncated at 10 times the borehole radius while the established values correspond to an infinite medium.

The multicrack capability of FEFFLAP was verified against an analytical solution for six pressurized cracks emanating from a borehole. The analytic results are due to Ouchterlony [13]. The geometry of the cracks in half plane symmetry are shown in Figure 6; it has 6-fold symmetry. The borehole and the cracks are subject to a constant pressure P. Table 1 lists the values for the nondimensional Mode I stress intensity factors for each crack tip. The quantity $\mu$ is a multiplicative factor of the borehole radius and its value yields the distance from the center of the borehole to the crack tip. Thus $\mu$ is a measure of crack length + borehole radius. There is a slight variation in the values of $\mu$ for the three cracks. This is due to the fact that the crack tip locations are identified by positioning a cursor on the computer screen. Every crack tip should have the same value of stress intensity. Table 1 shows the FEFFLAP values, which vary by less than 4% from the analytical one.

Table 1. Comparison of FEFFLAP Results With Analytically Obtained Results of Ouchterlony.

| Crack Number | $\mu$ | $K_I/P\sqrt{\pi\mu R}$ |
|---|---|---|
| 1 | 2.679 | 0.779 |
| 2 | 2.678 | 0.781 |
| 3 | 2.676 | 0.775 |
| Ouchterlony [13] | $>$ 1.5 | 0.743 |

## 4. QUASI-STATIC APPLICATIONS

### 4.1 Fracturing of a jointed sandstone reservoir

Figure 7 shows the kind of realistic field problem FEFFLAP was designed to solve. In this case, there is a joint system (heavy lines) around a borehole, and the ends of the two discontinuous joints above the borehole are treated as points of possible crack extension. The sequence in Figure 7 illustrates successive steps in the analysis. In this example, the fluid flows into a preexisting joint system; in other cases, the induced crack may cross the first interface. This numerical calculation also showed, for the first time, a phenomenon that had been predicted only analytically: the advancing crack front tends to open the natural fracture by inducing tensile stresses ahead of itself. In Figure 7c the crack has proceeded from the borehole and intersected a joint. The flow can now go out into the joint network. For this particular problem the flow was apportioned as shown in Figure 8. Notice how lubrication (some flow) occurs in almost every joint. If the medium is under stress one could expect slippage from these joints and therefore some seismic noise.

Sandia National Laboratories, Albuquerque, is conducting the Multi-Well Experiment in the Western gas sand formations near Rifle, Colorado, at a depth of several thousand feet [14]. Three proximate wells have been drilled, and one has been hydraulically fractured. The progress of a hydrofracture in one well was monitored from a second well by passive seismic means as shown in Figure 9. Sandia found that the seismic noises originated from a zone between 6 and 24 m wide, considerably greater than the original fracture width (about 2.5 cm). FEFFLAP can demonstrate (see Fig. 8) how a large volume of rock may react to a single hydrofracture injection, and we are using it to help explain the above observations at the Multi-Well Experiment.

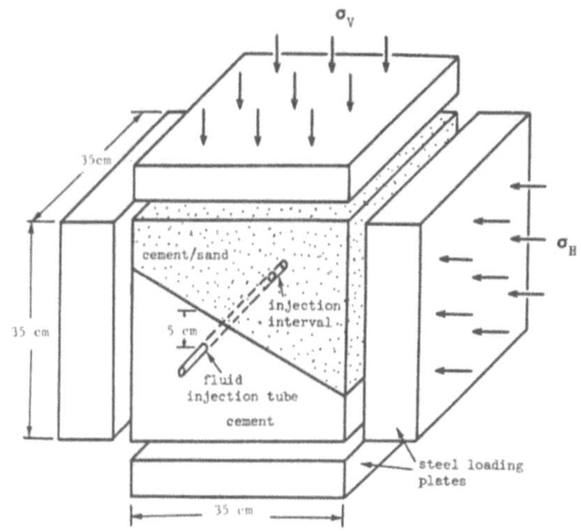


Fig. 2. Figure 2. Geometry of Hydrofracturing in Jointed Blocks Made of Two Different Materials.

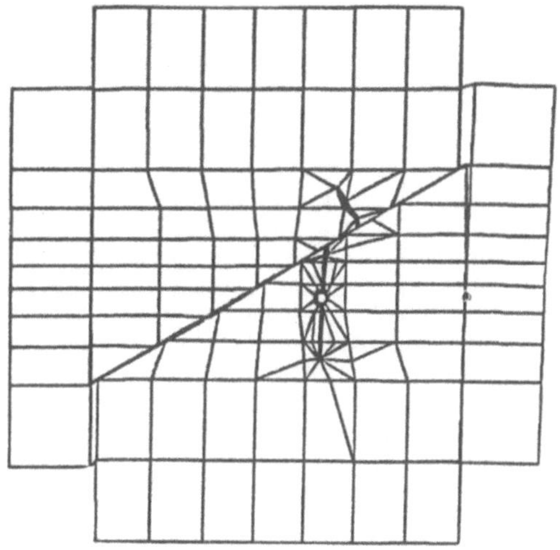

Figure 4. Fracture Crosses Interface, and Restarts Offset ($\sigma_V$ = 12.4 MPa, $\sigma_h$ = 5.2 MPa).

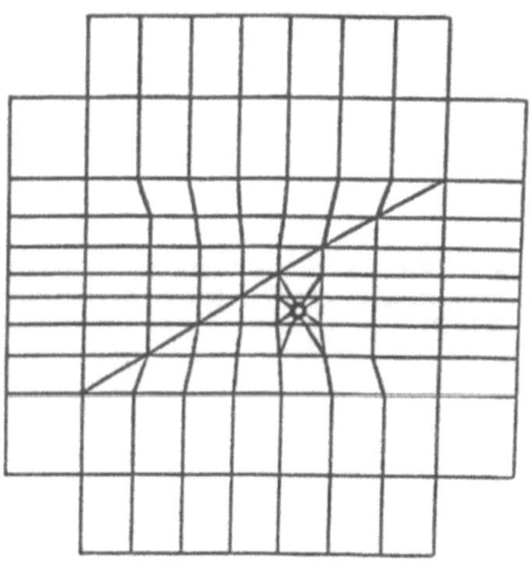

Figure 3. Finite Element Mesh for the Jointed Block Hydrofrac Analysis with FEFFLAP.

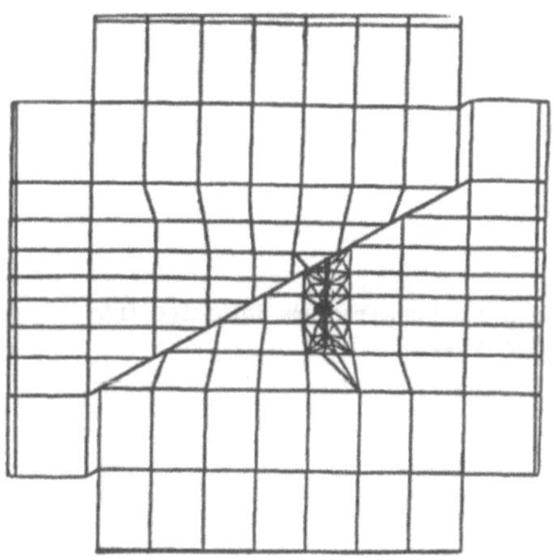

Figure 5. Fracture Does Not Cross the Interface ($\sigma_V$ = 4.8 MPa, $\sigma_h$ = 0.7 MPa).

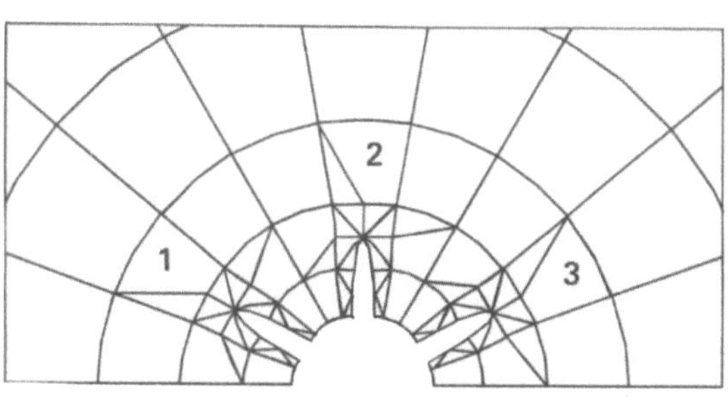


Figure 6. FEFFLAP Analysis of Multiple Crack Propagation, for Comparison with Ouchterlony's Analytical Results.

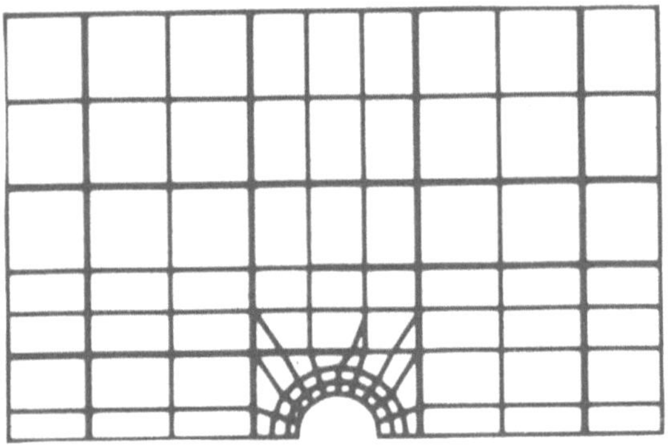

a) Pressurized Hole in Jointed Rock

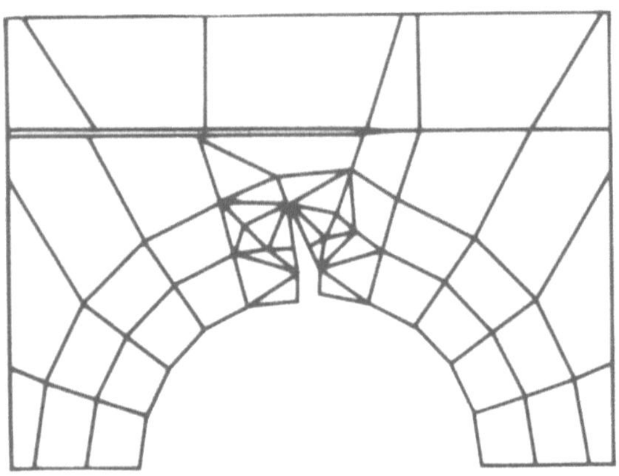

b) Fluid-Driven Fracture Initiated from the hole.

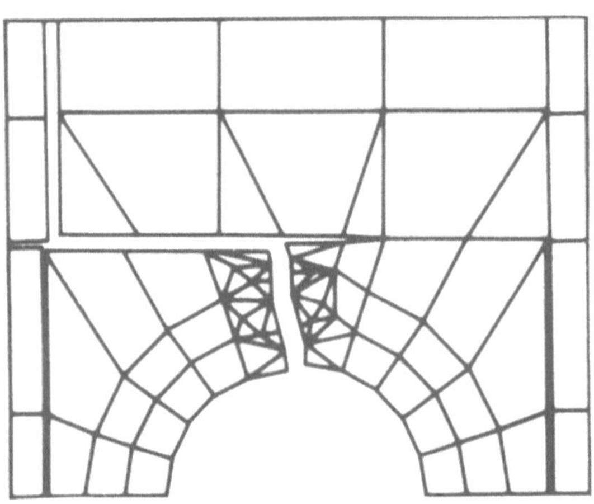

c) Fracture Intersects Natural Joint
c) Fracture Intersects Natural Joint

Figure 7: FEFFLAP Analysis of Fluid-Driven Jointed Block System.

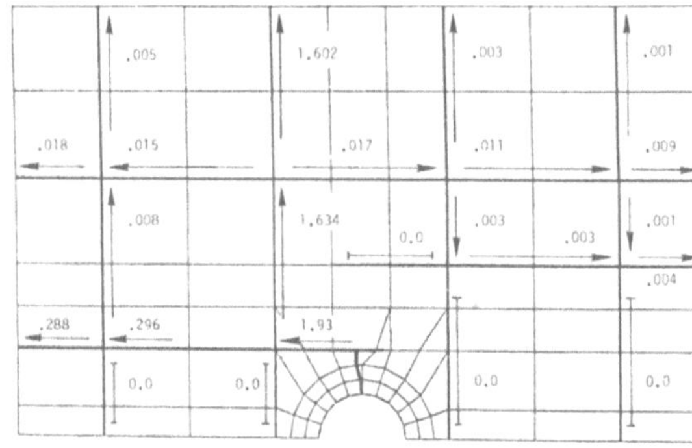


Figure 8: Coupled Fracturing and Flow Analysis in a Jointed Medium.

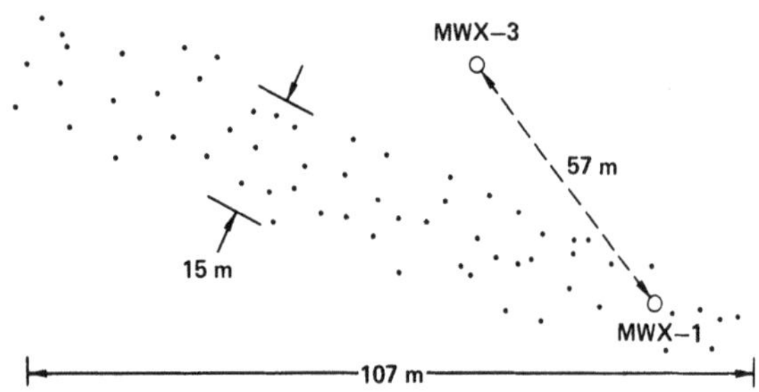


Figure 9: Microseismic Activity Due to Massive Hydrofracturing at MWX Well No. 1, as Monitored from Well 3.

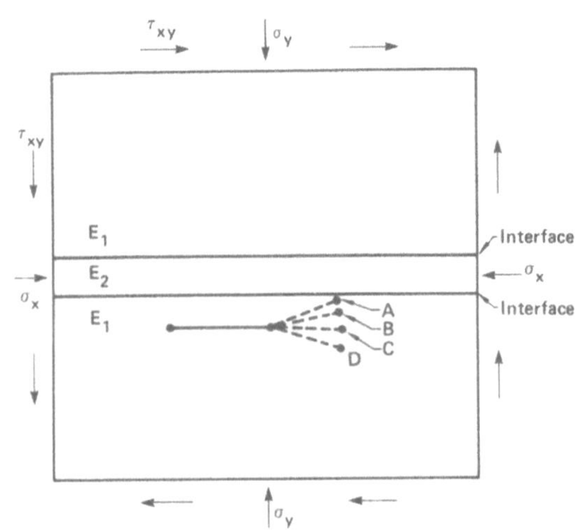


Figure 10: Behavior of a Hydrofracture Near a Rock Layer Interface.

### 4.2 Crack curvature near a lens

Figure 10 shows the analysis of a hydrofracture initiated parallel to a layer boundary to determine under what conditions the crack would turn toward or away from the interface. When the layer is twice as stiff as the surrounding medium the crack tends to curve toward the layer (Points A and B on Fig. 10). When the surrounding medium is twice as stiff as the layer, the crack tends to curve away from the layer (Points C and D). The amount of curvature also depends on the relative values of the in-situ stresses parallel and perpendicular to the layer boundary. When the stresses approach the hydrostatic state, the fractures tend to curve less than when the maximum stress is aligned with the initial crack direction. ($\sigma_x = 2.25\ \sigma_y$ for Points A and D, and $\sigma_x = \sigma_y$ for Points B and C). These results provide some insight in the stimulation mechanics of lenticular gas reservoirs.

### 4.3 Hydrofracturing in coal for methane recovery

Figure 11 shows a plan view of a hydrofracture in the Oak Grove Mine near Birmingham, Alabama. An analysis was done to determine why the fracture went into the roof rock [15]. The results indicate that no net tension would be created in the shale roof rock due to the coal hydrofracture. However, inflation of an existing flaw in the shale roof could occur if fluid got to it, and this new crack then could propagate if it were as small as a centimeter, to start.

The geometry of Figure 12 was constructed to model the Blue Creek coal seam. It also includes the one foot thick Marylee seam for completeness. There are five horizontal joint systems; all of them are coal-shale interfaces except for the coal-coal interface at the bottom of the hydrofrac. This geometry represents a cross section of about 12 m x 12 m. The hydrofracture in the coal is inflated. An analysis was done to obtain the stresses and determine if there was any nonlinear behavior of the slickensided joints near the hydrofracture. There was no joint inelastic slip.

Figure 13 represents a "blow up" of the hydrofracture - shale intersection. This grid corresponds to an area 60 cm wide by 52 cm high. This analysis was done for the same reasons as above, except here the details are fine enough to look at five face cleats and treat the fracture as being 5 cm wide. The boundary conditions for this analysis were obtained from the previous analysis, i.e., the stresses corresponding to the perimeter of Figure 13 were extracted from the results of the larger grid stress analysis. As before, there was no joint inelastic slip at these loads. Note that this slip, if it occurred, would deconcentrate stresses and for our particular problem it would tend to retard or diminish the creation of tensile stress across the interface in the shale. The stresses just across the interface did not show tension above the 5 cm crack. Indeed, there was compression, with the horizontal component in the 2.1 to 3.5 MPa range. What happened, then? We know from field observations that the shale had a dilated proppant-filled crack.

A possible explanation is the following: suppose that there is a flaw of some height h in the shale, that intersects the coal-shale interface. Suppose further that the 5 cm wide hydrofracture comes in contact with this flaw. If fluid can get into this crack while it is under compression the crack could extend. There is a way for this to occur - namely - a mismatch of the flaw's surfaces. Assuming inflation due to mismatch, the problem to solve with FEFFLAP is the minimum height h of the pre-existing crack to allow the fluid to extend it. Five calculations were run: crack lengths of 1/2 cm and 2 cm with the mesh of Fig 13, and crack lengths of 7, 12, and 62 cm with the mesh of Fig. 12. Figure 12, for example, shows the 62 cm crack in the shale overburden. Table 2 summarizes the results of the calculations in terms of stability. It shows the Mode I stress intensity factor $K_I$ and the load factor, F, for each calculation. A load factor F < 1 means crack instability. If the load factor is greater than 1, the crack will become unstable when the load vector is multiplied by F. The shale fracture toughness was taken as $K_{IC} = 1.2\ MPa\sqrt{m}$.

Table 2: Summary of 5 Crack Stability Calculations.

| Crack Length (cm) | $K_I(MPa\sqrt{m})$ | Load Factor, F |
|---|---|---|
| 0.5 | 0.63 | 1.065 |
| 2.0 | 1.16 | 1.003 |
| 7.0 | 1.26 | < 1 |
| 12.0 | 1.54 | < 1 |
| 62.0 | 2.47 | < 1 |

The calculations show that a crack as short as 2 cm is incipiently unstable if the 7.6 MPa fluid pressure can get to it. In fact, the shortest crack is also potentially unstable if the horizontal in-situ stresses were slightly less than the assumed value of 6 MPa, or if the fluid pressure were slightly in excess of 7.6 MPa. Additionally, it is well established that $K_{IC}$ decreases with very short crack length so that we may be over estimating the load factor, F, for the shorter cracks. Thus, the pressurizing of a small flaw in the roof shale is proposed as the most likely explanation for the observed hydrofrac vertical propagation.

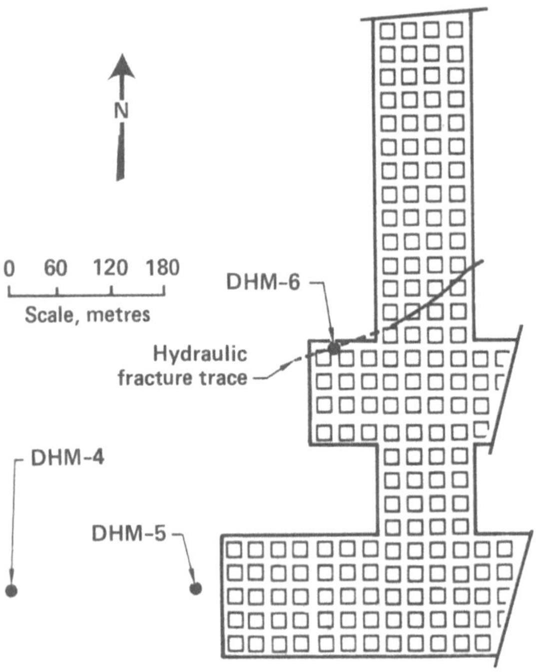


Figure 11: Hydrofracturing at the Oak Grove Mine, near Birmingham, Alabama, in the Blue Creek Coal Seam.

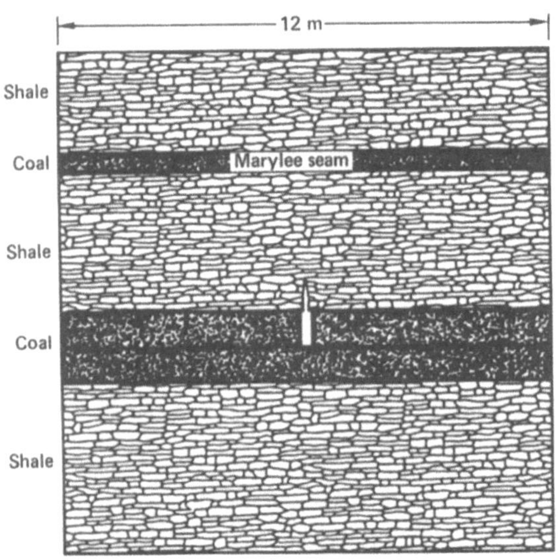


Figure 12: Model of the Hydrofracture in Coal, Extending in the Shale Roof.

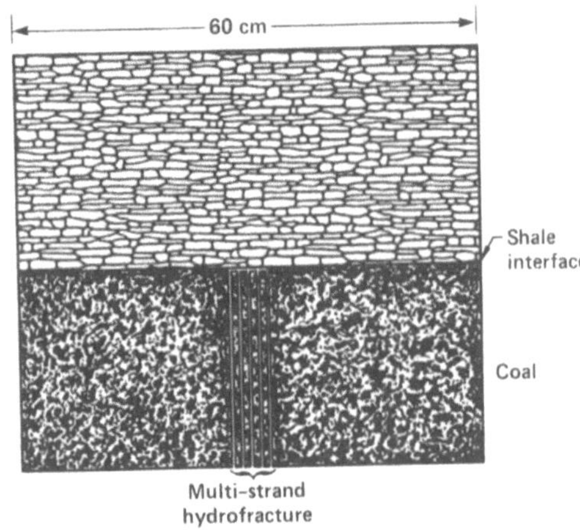


Figure 13: Detailed FEFFLAP Analysis of the Fracture at the Interface.

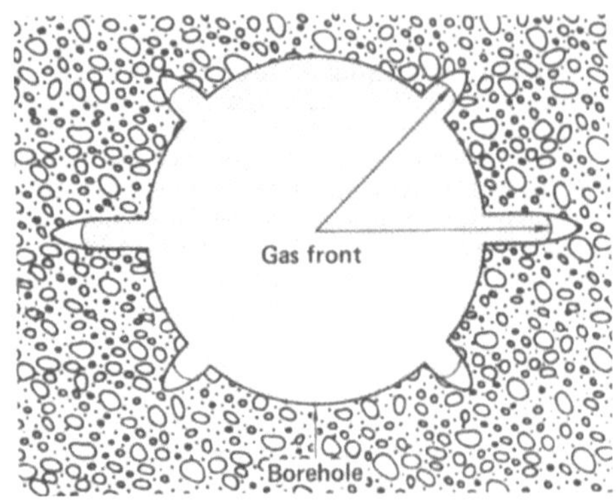


Figure 14: Geometry for FEFFLAP/FAST Analysis of Multiple Gas-Driven Fractures.

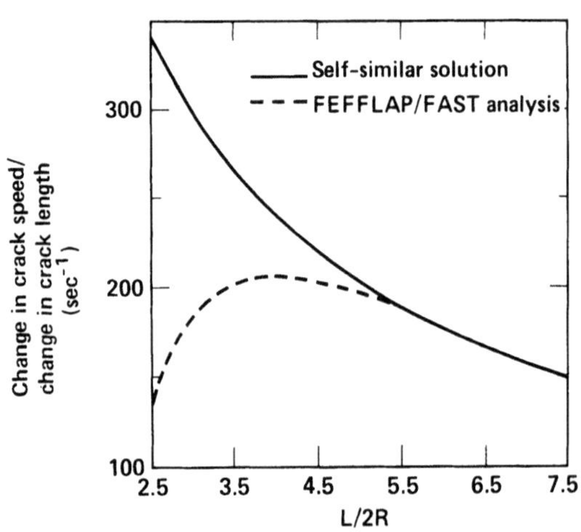


Figure 15: Comparison of FEFFLAP/FAST Results with the Self-Similar Solution.

## 5. DEVELOPMENT OF THE DYNAMIC FEFFLAP MODEL

### 5.1 The FAST/FEFFLAP coupled model

The new developments build upon the existing steady state FEFFLAP model and a fluid propagation model called FAST [16]. In the FAST model, the rate of fluid advance and the pressure variation along each individual crack is determined by solving the conservation equations for mass and momentum of the fluid. The one-dimensional form of the equations is appropriate, since the flow channels are very long and narrow. The friction factor is approximated by a simple additive expression, which includes both laminar and turbulent components. The lateral seepage velocity of fluid into the walls of the fracture is estimated with Darcy-flow analysis normal to the fracture plane. The temperature distribution along the crack is determined from conservation of energy.

These fluid/thermal transport equations are solved by numerical implementation of an integral method [3,16]. In brief, the continuity equation is satisfied in an integral sense over three regions of the flow including: the entire volume of the fracture, a small subvolume at the leading edge of the flow, and the borehole volume which feeds the fractures. The other equations, for momentum and energy conservation, are satisfied on a local basis, using finite-difference approximations to replace the derivative operators. The numerical procedure has been tested by comparison with analytical solutions and with experimental data from field experiments. The opening displacements and stress intensity factors were evaluated using closed form solutions for special geometries in linear elastic media.

In terms of coupling with FEFFLAP, the FAST module essentially replaces the steady state flow module. FEFFLAP provides crack and joint apertures to FAST and FAST returns pressure profiles due to those apertures and the flow boundary conditions.

### 5.2 Application problems and initial verification

Figure 14 typifies a geometry of interest for gas driven fractures from a borehole, where two fractures have begun to dominate. The interaction of these expanding cracks with joint systems (not shown) is also of interest. The example calculations reported here are based on the material properties and the process parameters outlined in Table 3.

Table 3: Parameters and Material Properties.

| | |
|---|---|
| borehole radius | 0.1 m |
| borehole pressure | 40 MPa |
| borehole temperature | 300°K |
| confining stress | 10 MPa |
| fracture toughness | 0 |
| shear modulus | 3 GPa |
| Poisson's ratio | 0.3 |
| surface roughness | 400 μm |
| rock porosity | 0.3 |
| gas viscosity | $10^{-5}$ Pa-S |
| molecular weight | 28 |
| ambient pore pressure | 0.1 MPa |

An "exact" similarity solution for fracture tip velocity is applicable under some restrictions. In comparison with the similarity solution we assumed that the rock permeability was negligible. In Figure 15 L is the half crack length and R is the borehole radius. As shown in Figure 15, the numerical results are in good agreement with the analytical results at larger crack lengths; this is consistent with the fact that the similarity solution requires the cracks to be long, compared to the borehole diameter.

REFERENCES

1. Ingraffea, A.R., Shaffer, R.J., and Heuze, F.E., (1985) "FEFFLAP: A Finite Element Program for Analysis of Fluid-Driven Fracture Propagation in Jointed Rock, Vol. 1: Theory and Programmer's Manual," Lawrence Livermore National Laboratory, UCID-20368, March.

2. Shaffer, R.J., Ingraffea, A.R., and Heuze, F.E., (1985) "FEFFLAP: A Finite Element Program for Analysis of Fluid-Driven Fracture Propagation in Jointed Rock, Vol. 2: User's Manual and Model Verification," Lawrence Livermore National Laboratory, UCID-20369, March.

3. Nilson, R. H., (1986) "An Integral Method for Predicting Hydraulic Fracture Propagation Driven by Gases and Liquids," Int. J. Num. and Analyt. Methods in Geomechanics, Vol. 10, pp 191-211.

4. Ingraffea, A. R., Saouma, V., (1984) "Numerical Modeling of Discrete Crack Propagation in Reinforced and Plain Concrete," Application of Fracture Mechanics to Concrete Structures, G. C. Sih and A. DiTommaso, editors, Martinus Nijhoff Publishers.

5. Noorishad, J., Witherspoon, P. A., and Brekke, T. L. (1971) "A Method for Coupled Stress and Flow Analysis of Fractured Rock Masses," Publ. No. 71-6, University of California, Berkeley, March.

6. Heuze, F. E. (1981) "JPLAXD: A Finite Element Program for Static, Plane and Axisymmetric Analysis of Structures in Jointed Rock," Lawrence Livermore National Laboratory, UCID-19047.

7. Witherspoon, P. A., Wang, J. S. Y., Iwai, K., and Gale, J. E. (1979) "Validity of Cubic Law for Fluid Flow in a Deformable Fracture," Lawrence Berkeley Laboratory, LBL-9557.

8. Gale, J. E. (1982) "The Effect of Fracture Type (Induced vs. Natural) on the Stress-Fracture Closure-Fracture Permeability Relationships," Proc. 23rd U.S. Symp. on Rock Mechanics, Berkeley, CA (AIME, Littleton, CO).

9. Thorpe, R. K., Heuze, F. E., and Shaffer, R. J., (1984) "An Experimental Study of Hydraulic Fracture-Interface Interaction," UCID-20114, Lawrence Livermore National Laboratory, Livermore, July.

10. St. John, C. M., (1971) "Three-Dimensional Analysis of Rock Slopes," Proc. Symp. Int. Soc. Rock Mech, Nancy, France, Sept.

11. Shaffer, R. J., Thorpe, R. K., Ingraffea, A. R., and Heuze, F. E., (1984) "Numerical and Physical Studies of Fluid-Driven Fracture Propagation in Jointed Rock," 25th Rock Mechanics Symposium, (SME, Littleton, CO), p 117-126.

12. Sih, G. C. (1973) Handbook of Stress Intensity Factors, Institute of Fracture and Solid Mechanics, Lehigh Univ., Bethlehem, PA, p 1.2.8.

13. Ouchterlony, F., (1982) "Analysis of Cracks Related to Rock Fragmentation," Part 1 of Lectures at the International Center for Mechanical Sciences (CISM), Udine, Italy, Report DS 1982:4.

14. Northrop, D. A., (1986) "Second Technical Poster Session for the Multiwell Experiment," SAND 86-0945, Sandia National Laboratories, Albuquerque. Also, Proc. SPE/DOE Symposium on Low Permeability Reservoirs, May, 1986. Denver, CO. (Soc. Pet. Eng., Richardson, TX).

15. Shaffer, R. J., Nilson, R. H., Heuze, F. E., and Swift, R. P. (1986) "Quasi-static and Dynamic Arbitrary Fracture Propagation in Jointed Rock," Proc. 27th U.S. Symposium on Rock Mechanics, University of Alabama, June 13-26. Also, Lawrence Livermore National Laboratory report UCRL-93429.

16. Nilson, R. H., and Peterka, D. L. (1986) "User's Manual for the FAST Hydrofracture Code," Report SSS-R-86-7590 to Lawrence Livermore National Laboratory, by S-CUBED, La Jolla, CA, August.

ACKNOWLEDGMENTS

This research was performed under the LLNL Unconventional Gas Program - F. Heuze, Manager. Program monitoring is performed by Morgantown Energy Technology Center under contract W-7405-ENG-48 with the U.S. Department of Energy. The careful typing of Lydia Grabowski is gratefully acknowledged.

SEM/RILEM International Conference
on Fracture of Concrete and Rock
Houston, Texas, June 1987
Editors: S.P. Shah, S.E. Swartz



# Fracture Process Zone of a Concrete Fracture Specimen

J.J. Du*, A.S. Kobayashi* and N.M. Hawkins**

## ABSTRACT

For concrete three-point bend specimens Moire interferometry was used to locate the micro-crack tip and to measure the crack opening displacement (COD) along the fracture process zone preceeding that tip. The COD data was used as input to a finite element model which was then executed in its generation mode, to yield crack closure forces along the fracture process zone. The crack closure forces versus COD relationship for the fracture process zone was consistent with that postulated previously based on the use of hybrid experimental-numerical techniques utilizing finite element methods in both their generation and propagation modes.

## INTRODUCTION

In earlier work [1,2], two of the authors and their colleagues used the double compliance method and a replica technique to identify the relation between the micro-crack tip for the fracture process zone and the effective crack length for concrete, crackline wedge-loaded double cantilever beam (CLWL-DCB) specimens. Crack closure stresses along the fracture process zone were then determined through a hybrid experimental-numerical analysis [3] for both Mode I and Mixed Modes I and II fractures in the CLWL-DCB specimens [4]. The resulting stress distribution is shown schematically in Figure 1. The distribution consists of three straight line segments, which can be characterized from the micro-crack tip backwards as a micro-crack coalescence zone in which the closure stress remains constant, a micro-crack opening zone in which the stress decreases linearly and rapidly with increasing crack opening displacements (COD), and a macro-crack opening zone where, due to aggregate bridging, the closure stress decreases linearly, but slowly, with increasing crack openings. That crack closure stress distribution was in qualitative agreement with the earlier results of Hillerborg et al. [5], Modeer [6], Petersson [7] and Wecharatana and Shah [8]. The three segment crack closure stress versus COD relation, obtained by analysis, was found to be specimen size independent but material property dependent. The constitutive relation between crack width and closure stress for the fracture process zone was concrete strength and aggregate characteristic dependent. A critical COD criterion was postulated where macro-crack extension occurred when the COD at the macro-crack tip reached the critical value of 250 micron for a fully developed fracture process zone. A critical strength criterion was used to advance the micro-crack tip when the maximum stress in the crack tip element reached the tensile strength of the concrete. This constitutive relationship was also used together with a finite element model of an impacted concrete beam [9], to make dynamic response predictions for that beam. Good agreement was obtained between the measured and computed average crack velocity and load point displacements [10,11].

Crack closure stresses cannot be measured directly and experimental evidence on COD distributions along the fracture process zone are few [12]. Therefore, despite extensive analyses such as those presented previously by two of the authors, there is a lack of consensus among researchers as to the uniqueness of the crack closure stress model, or the factors that might affect that uniqueness [13]. This paper reports direct experimental evidence on the variation in COD values along the fracture process zone and uses that evidence to determine crack closure stresses by a hybrid experimental-numerical procedure.

## EXPERIMENTAL PROCEDURE

The experimental procedure involved the use of white-light moire interferometry to determine the location of the micro-crack tip and to measure the crack opening displacement (COD) within the fracture process zone for a concrete three-point bend specimen.

### Specimen

Because this investigation was initially exploratory, and later equipment limited, small scale three-point concrete bend specimens were used. Those specimens had the proportions, and were loaded, as shown in

Figure 2. They were cut from undisturbed portions of the CLWL-DCB specimens tested previously [4] and a 20.3 mm deep crack initiator with a root radius of 1.6 mm was sawed into each specimen's edge. The average aggregate size for the specimens was 6.3 mm. Portions of one side of the specimen were polished so that an active grating could be attached to the remaining ligament ahead of the notch tip.

White Light Moire Interferometry

Fracture process zone characteristics were measured using the set-up shown in Fig. 3. White light moire interferometry, proposed originally by Post [14], uses an incandescent light source without filtering and thus provides adequate light intensity for low speed sequential photography. The required incoherent light source necessitates that the gap distance between the reference and active gratings, as shown on the right of Fig. 3, to be less than 1 mm. The achromatic light emerges from the compensator grating, as shown in the center of Fig. 3, as monochromatic light beams at different diffraction angles. However those beams generate the same moire interferometric pattern for each wave length. Thus, the camera records the scalar sum of the light intensities associated with each wave length. In the optical setup of Fig. 3, a fringe multiplication of two is accomplished by the use of two light beams, one of which is a zero order and the other of which is a first order beam. Those beams are distorted by the active grating and interfere with the reference grating. Thus, the compensator and active gratings are of half frequency, i.e., f/2, while the reference grating is of full frequency, i.e., f = 1200 lines/mm. The motor driven camera could record 5 frames per second during a period of rapid development of the fracture process zone.

Experimental Setup

The concrete three-point bend specimen was loaded in a displacement controlled loading fixture and the applied load was recorded on a strip chart recorder through a load cell. Figure 4 shows the loading fixture, specimens, and the white-light moire interferometry setup. Crack mouth opening displacement (CMOD) were also recorded in some specimens.

The test procedure consisted of applying an incremental increase in applied displacement and photographing the moire fringes for that load. The COD along the extending fracture process zone was small when the micro-crack tip was contained within the remaining ligament of the beam. Sudden increases in the number of moire fringes and hence the CODs were observed once the micro-crack tip penetrated through the entire depth of the remaining ligament. At this point, the applied load decreased with increased applied displacement thus exhibiting "strain softening effects" for this material. Much of the moire fringe recording was conducted in that strain softening regime.

EXPERIMENTAL RESULTS

A total of four specimens were loaded near to complete penetration of micro-crack tip of the fracture process zone or near the full depth of the notched beam.

Figures 5 and 6 show typical moire interferometric patterns for Specimen Nos. 1 and 4 at different load levels. The clearly identifiable micro-crack tip of the fracture process zone is marked on each figure. The fracture process zones in those two specimens extended from the tip of the machined notches at angles of $40^\circ$ and $30^\circ$, respectively, to the axes of those notches. This inclination of the angle of propagation of fracture process zone to axis of the notch was caused by aggregates which were located approximately 2-3 mm below the surface to which the active grating was attached. The locations of those aggregates are identified by dotted lines in Figures 5 and 6. The micro-crack tip location, after being projected onto the net section of the specimen, was used to determine the amount of crack extension for a given increase in load. The order of the Moire fringes was used to determine the COD variation along the fracture process zone. The data points in Figures 7 and 8 represent the COD variations along fracture process zones at different load levels for Specimen Nos. 1 and 4, respectively. Those loads are for increasing applied displacements and they occurred under decreasing applied load in the strain softening regime. The increasing COD at the relatively blunt notch tip with increasing applied displacement but with decreasing load level did not reach a critical value and demonstrates that in these small specimens, the remaining ligament of 20.3 mm in length was insufficient to permit the fracture process zone to fully develop.

The increasing COD at the blunt notch tip with extension of the micro-crack length for a partially developed fracture process zone in concrete corresponds to the crack tip blunting behavior that occurs prior to stable crack growth in a ductile metal. The data points in Figure 9 shows the variations in COD values at the blunt notch tip with micro-crack tip extensions for four of the concrete bend specimens. The data points for the four specimens all fall on the same continuous curve, as is to be expected if there is a unique R-curve (COD-resistance curve) for this specimen's configuration and material.

NUMERICAL PROCEDURE

Static finite element analysis were made of the fracture process for these three point concrete bend specimens. Figure 10 shows the finite element model used for one half of the specimen shown in Figure 2. The three-segment fracture process zone, shown in Figure 1, was also incorporated into that model as a constitutive relationship for the fracture process zone above the notch. Although characteristic properties for the concrete were not measured directly, realistic values for the tensile strength, and elastic modulus for use in the finite element model are shown in Table 1 and were taken from the results for the CLWL-DCB specimens from which these test specimens were cut. Tensile strength values are needed because micro-crack extension is assumed to initiate when that value is exceeded at the element beyond the notch tip. The crack closure stress distribution for the three segments of the process zone was determined through a trial-and-error procedure that required satisfaction of the variations in values along the crack length determined by moire interferometry. That iteration procedure was similar to the incremental iterative procedure used in numerical elastic-plastic stress analysis.

## NUMERICAL RESULTS

The foregoing finite element model was used in its application mode [3] to determine the crack closure stresses for the fracture process zones of the various specimens. The resultant crack closure stress versus COD relation shown in Figure 11 provided the COD variations along the fracture process zone that are indicated by the continuous curves in Figures 7 and 8. A comparison of Figure 11 with the corresponding constitutive relation of Ref. [4] show that the two are almost identical in shape, except for the adjustment in maximum tensile strength, despite differences in the processing of the two concrete specimens. The COD's at the machined notch tip and hence the extent of the fracture process zone developments for Specimen Nos. 1 through 4 are also identified in Figure 11. These COD's average about 50 microns and are therefore much smaller than the macro-crack tip COD of 250 microns. That result indicates clearly only partial development of the fracture process zones in these subsize specimens.

The continuous curve in Figure 9 also labelled computed, shows the variations in the computed COD at the notch tip with micro-crack extension. Again, the agreement between the measured and computed COD's is excellent. Figure 12 shows the typical applied load versus crack mouth opening displacement relation obtained using the finite element model for Specimen No. 4. There is a prominent strain softening effect, typical for a concrete specimen loaded in a rigid test fixture.

## CONCLUSIONS

1. White light moire interferometry is a simple procedure for determining the micro-crack tip location and the variation in the COD relationship along the fracture process zone of concrete fracture specimens.

2. The finite element method in its application mode together with COD data collected using white light moire interferometry can be used effectively to determine the crack closure stresses along the fracture process zone.

3. The three-segment crack closure stress versus COD relationship for the fracture process zone established previously by hybrid experimental- numerical analysis of CLWL-DCB specimens was confirmed by this alternate procedure involving tests on three point concrete bend specimens.

## ACKNOWLEDGEMENTS

The results reported in this paper were obtained under NSF Grant MSM 831053 and AFOSR Grant No. 86-0204.

## REFERENCES

1. Barker, D.B., Hawkins, N.M., Jeang, F.-L., Cho, K.-Z. and Kobayashi, A.S. "Concrete Fracture in CLWL Specimen," ASCE Journal of Engineering Fracture, Vol. 111, No. 5, pp. 623-638, May 1985.

2. Cho, K.-Z., Kobayashi, A.S., Hawkins, N.M., Barker, D.B. and Jeang, F.-L. "Fracture Process Zone of Concrete Cracks," ASCE Journal of Engineering Fracture, Vol. 110, No. 8, pp. 1174-1184, August 1984.

3. Kobayashi, A.S., "Dynamic Fracture Analysis by Dynamic Finite Element Method--Generation and Propagation Analyses," Nonlinear and Dynamic Fracture Mechanics, ed. by N. Perrone and S.N. Atluri, AMD-Vol. 35, ASME, 1979.

4. Jeang, F.-L. and Hawkins, N.M., "Non-linear Analysis of Concrete Fracture," Report SM 85-2, Department of Civil Engineering, University of Washington, July 1985.

5. Hillerborg, A., Modeer, M. and Petersson, R.E., "Analysis of Crack Formation and Growth in Concrete by Means of Fracture Mechanics and Finite Elements," Cement and Concrete Research," Vol. 6, No. 6, pp. 773-782, Nov. 1976.

6. Modeer, M., "A Fracture Mechanics Approach to Failure Analysis of Concrete Materials," Report TVBM-1001, University of Lund, Sweden, 1979.

7. Petersson, P.E., "Crack Growth and Development of Fracture Zones in Plain Concrete and Similar Materials," Report TVBM-1006, University of Lund, Sweden, 1981.

8. Wecharatana, M. and Shah, S.P., "Prediction of Non-linear Fracture Process Zone in Concrete," ASCE Journal of Engineering Mechanics, Vol. 109, No. 5, pp. 1231-1246, Oct. 1983.

9. Mindess, S. and Bentur, A., "A Preliminary Study of the Fracture of Concrete Beams Under Impact Loading Using High Speed Photography," Cement and Concrete Research, Vol. (??), pp. 474-484, 1985.

10. Kobayashi, A.S., Hawkins, N.M., and Du, J.J., "An Impact Damage Model of Concrete," Material Research Society Symposia Proceedings, Vol. 64, pp. 203-215, Cement-Based Composites: Strain Rate Effects on Fracture, ed. S. Mindess and S.P. Shala.

11. Kobayashi, A.S., Du, J.J., Hawkins, N.M. and Bradt, R.C., "Nonlinear Fracture of Concrete and Ceramics," to be published in the Proc. of 3rd National Symposium on Nonlinear Fracture Mechanics, ASTM STP.

12. Cedolin, S., Poli, S.D. and Iori, I., "Experimental Determination of the Fracture Process Zone in Concrete," Cement and Concrete Research, Vol. 13, pp. 557-567, 1983.

13. Ingraffea, Anthony R., "Non-linear Fracture Models for Discrete Crack Propagation," Martinus Nighoff Publishers, pp. 247-285, 1985.

14. Post, D., "Moire Interferometry with White Light," Applied Optics, Vol. 18, No. 24, pp. 4136-4167, 1979.

15. Visalvanich, K. and Naaman, A.E., "Fracture Model for Fiber Reinforced Concrete," American Concrete Institute Journal, pp. 128-138, Mar./Apr. 1983.

5,15

TABLE 1. MATERIAL PROPERTIES

| | |
|---|---|
| Elastic Modulus | 2.42 GPa ($3.5\times10^6$ psi) |
| Poisson Ratio | 0.15 |
| Tensile Strength | 3.80 MPa ($5.5\times10^2$ psi) |

TABLE 2. MAXIMUM LOAD

| | | |
|---|---|---|
| # 1 | 0.669 kNs | (150 lbs) |
| #2 | 0.655 kNs | (147 lbs) |
| #3 | 0.678 kNs | (152 lbs) |
| #4 | 0.664 kNs | (149 lbs) |
| Average | 0.666 kNs | (149.5 lbs) |

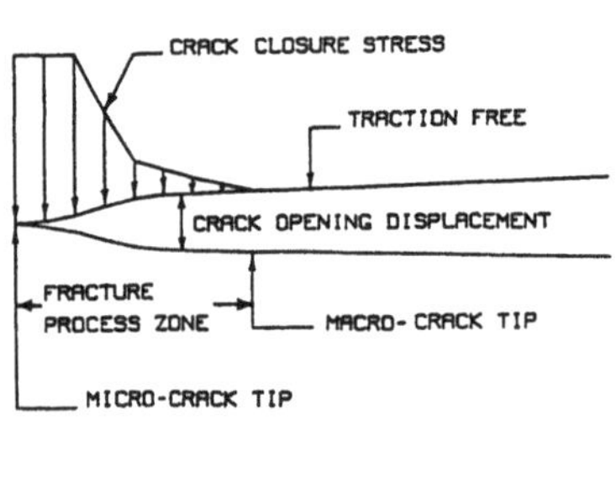


Figure 1. Fracture Process Zone of Concrete

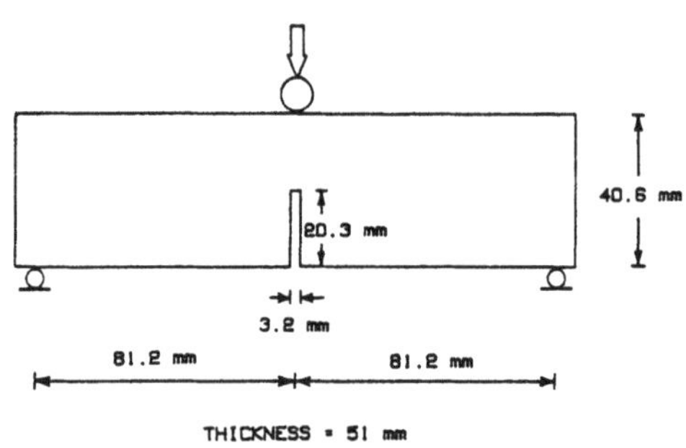


Figure 2. Concrete Three-Point Bend Specimen

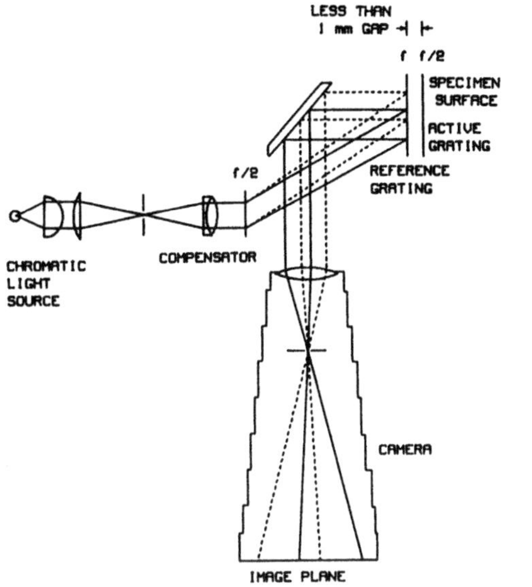


Figure 3. Optical Set-up for White Light Moire Interferometry (f = 1200 lines/mm)

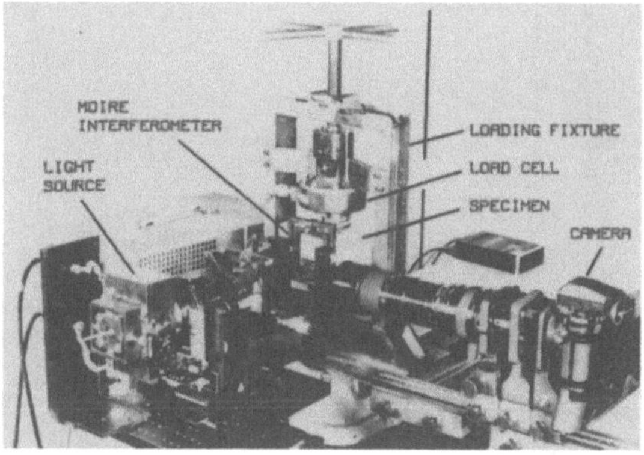


Figure 4. Experimental Set-up

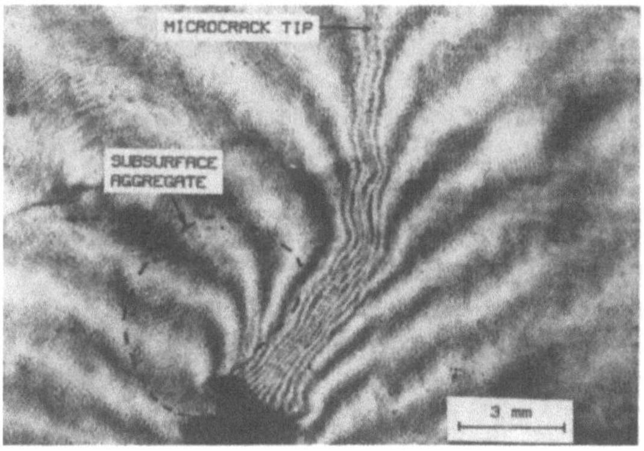


Figure 5. Moire Interferometric Pattern at Crack Tip of Concrete Three-Point Bend Specimen No. 1 at Load 624 N

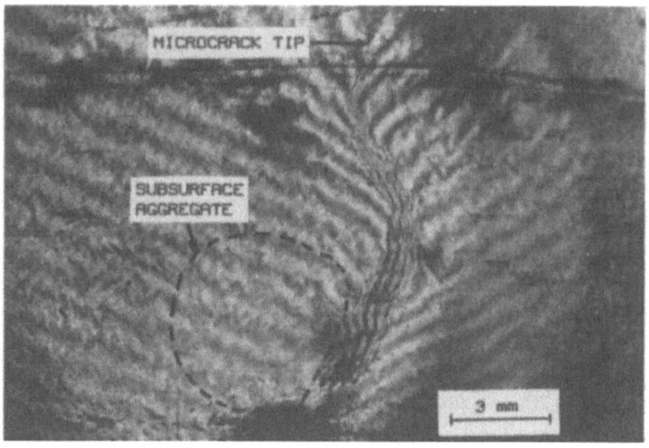


Figure 6. Moire Interferometric Pattern at Crack Tip of Concrete Three-Point Bend Specimen No. 4 at Load 557 N.

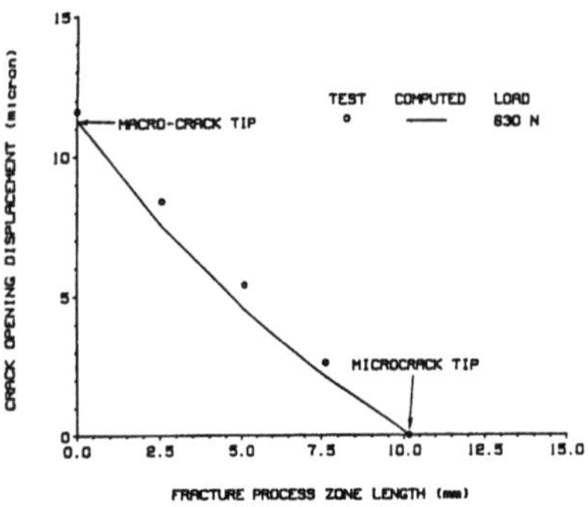


Figure 7. COD Variations Along Fracture Process Zone Specimen No. 1

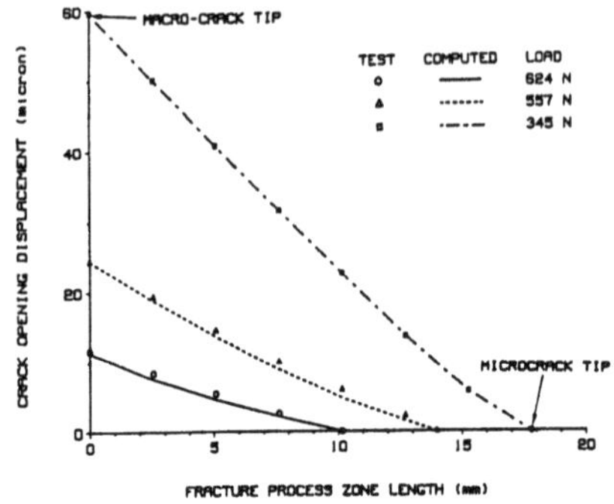


Figure 8. COD Variations Along Fracture Process Zone Specimen No. 4.

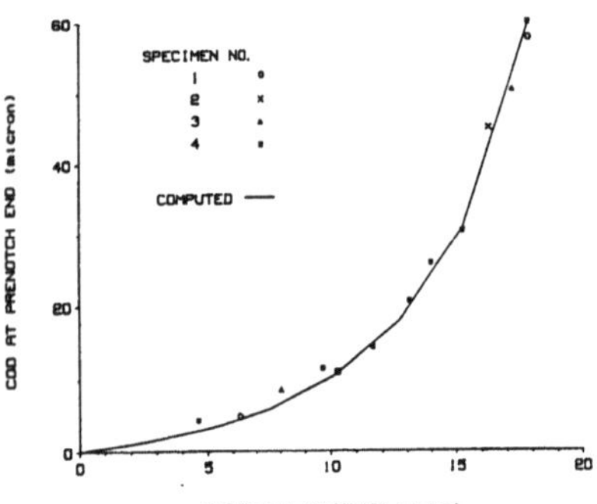


Figure 9. COD at Prenotch End Versus Microcrack Extension Concrete Three-Point Bend Specimen

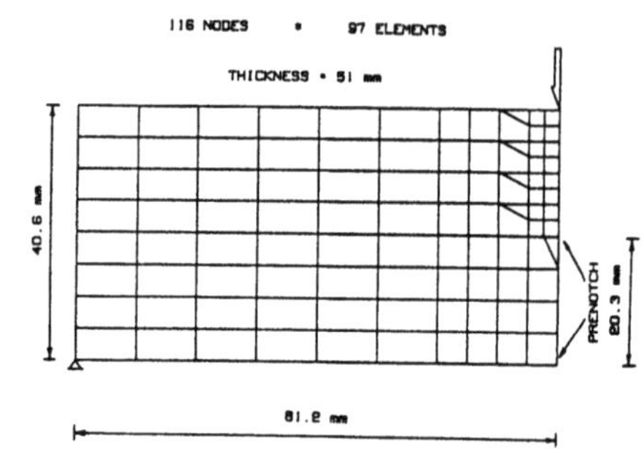


Figure 10. Finite Element Model of Concrete Three-Point Bend Specimen

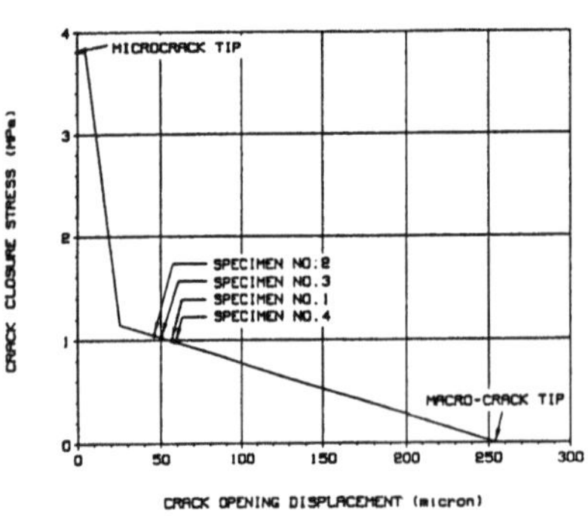


Figure 11. Crack Closure Stress Versus COD Relation in Concrete Fracture Process Zone

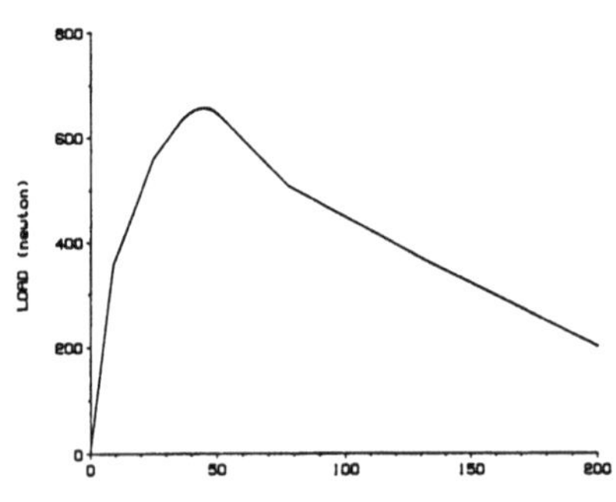


Figure 12. Load Versus Crack Mouth Opening Displacement in Concrete Three-Point Bend Specimen

SEM/RILEM International Conference
on Fracture of Concrete and Rock
Houston, Texas, June 1987
Editors: S.P. Shah, S.E. Swartz

# FRACTURE PROCESS AND BRIDGING ZONE MODEL AND INFLUENCING FACTORS IN FRACTURE OF CONCRETE

H. Horii, A. Hasegawa and F. Nishino

ABSTRACT

In this paper simple models of fracture process zone are proposed. Firstly, a model of the strain-softening is proposed based on the growth and the interaction of micro-cracks. Two different mechanisms, micro-cracking and bridging, play important roles in the strain-softening behavior. Secondly, Dugdale-type models for fracture process zone are investigated. In the usual model, different influencing factors are involved. A simplified model is proposed to study the effect of each influencing factor. It is seen that each influencing factor such as the size and shape of the specimen and the loading condition appears in the formulation differently. Among those influencing factors the distribution of the stress ahead of the crack tip is shown to be one of the dominant factors. The present model shows the possibility of obtaining approximate solutions to general process zone problems.

## 1 INTRODUCTION

Recently, great efforts have been devoted to apply fracture mechanics to fracture phenomena of concrete. Fracture toughness of concrete has been measured for the application of the linear fracture mechanics. It is reported that the measured fracture toughness depends on the experimental methods, the size of the specimen and other factors; see e.g. (1).

This dependence is considered to be caused by the existence of a fracture process and bridging zone (2). There seems to exist a confusion on the definition of the fracture process zone in the literature. A "crack" itself is ambiguous in concrete since macroscopic (visible) crack may not be a stress free crack. In the present paper, the "fracture process" zone refers to a zone in which the initiation of micro-cracks and their growth are dominant. In the "bridging" zone, the bridging by the aggregate interlocking and the nonlinear behavior of the ligament is the main mechanism. They are called the fracture process and bridging zone, altogether. A crack (or stress free crack) refers to a crack whose surfaces are completely free of stress.

The size of the fracture process and bridging zone ranges from a few centimeters to more than 10 centimeters. If the size of the specimen is not large enough compared with that of the fracture process and bridging zone, the direct application of the linear fracture mechanics may not be possible and the effect of the fracture process and bridging zone must be evaluated.

With increasing loading the fracture process and bridging zone is formed and grows at the tip of the macro-crack. At the critical load the macro-crack starts to grow. The toughness to fracture increases with the growth of the macro-crack. The main task in the application of fracture mechanics to concrete is to construct an analytical model of the fracture process and bridging zone which can predict those fracture phenomena. The material parameters involved in the model must be determined from material tests independent of the size of the specimen and other factors.

Numerous models for the fracture process and bridging zone have been proposed. One of the most popular types of models is the Dugdale-type model. It has been successfully used in fracture mechanics to represent the crack tip nonlinear zone. The nonlinear zone is modeled by the extended part of the crack along which stresses are transmitted. The behavior of the material in the nonlinear zone is characterized by the relation between the transmitted stress (sometimes called cohesive force or crack closing pressure) and the crack opening displacement.

---

H. Horii and A. Hasegawa are Associate Professors, and F. Nishino is a Professor, Department of Civil Engineering, University of Tokyo, Bunkyo-ku, Tokyo, JAPAN.

It is quite natural to apply this model to the fracture process and bridging zone in concrete. In fact numbers of studies are reported based on this type of model; see for example (2-5). For this model the identification of the material properties in the fracture process and bridging zone is the important point. In the fracture process zone micro-cracks are initiated and grow interacting with each other. After the growth and the connection of micro-cracks, forces are transmitted across the fracture plane by the aggregate interlocking and the nonlinear behavior of the ligament. This region is called the bridging zone. The state inside the fracture process and bridging zone is considered to be strain-softening. The strain-softening curve obtained from the uniaxial tensile fracture test is sometimes used as the property of the fracture process and bridging zone.

The focus of the studies with this model is placed on the following points. What kind of relationship between the transmitted stress and the crack opening displacement represents the behavior of the fracture process and bridging zone? How will the predicted results for different loading conditions and different sizes and shapes of specimens depend on the strain-softening property? Is the relationship determined uniquely from material experiments with different specimen sizes?

Dugdale-type models are simple and are considered to predict the growth of the fracture process and bridging zone and the condition for initiation of the stress free crack growth. However, the increase in resistance to fracture after the main crack starts to grow can not be predicted by the Dugdale-type models, because the toughening occurs due to the thickness effect of the process zone, that is, the micro-cracks left behind the crack tip decrease the stress intensity factor at the crack tip.

The shape and the size of the fracture process zone and the change in the stress intensity factor at the macro-crack tip may be predicted by extending and applying the study of the interaction between micro-cracks and the macro-crack. This macro-micro interaction model has been studied by Kachanov and co-workers(6-7), Tauzs and Bomalis(8) and Rose (9). Applications of models of this type have been tried in fracture of ceramics (10).

To investigate the extent of the process zone, a damaged material model has been studied. The stiffness of cracked material is known to be decreased. The overall properties of cracked solids have been studied by many researchers(11-14). The process zone is modeled as a continuous material with varying stiffness. Along this line Charalambides and McMeeking(15) simulated the growth of the process zone by the finite element method.

Other than those mentioned above, many models of fracture process zone have been proposed; see, for example, (16-17).

In this paper the direct consequence of simple models of the fracture process and bridging zone is investigated. Firstly, a model of the strain-softening is proposed based on the growth and the interaction of micro-cracks. Secondly, a Dugdale-type model for fracture process and bridging zone is investigated. A simplified model is proposed to study the effect of each of the influencing factors which are mixed up in the usual model.

## 2 STRAIN-SOFTENING MODEL

Fig. 1 shows a typical stress displacement curve for a uniaxial tensile fracture test (18). In the strain-softening part after the peak point, the deformation is localized near the fracture plane where initiation, growth and interaction of micro-cracks are first dominating. After micro-cracks grow up to certain extent, the major mechanism is replaced by the bridging due to the aggregate interlocking, nonlinear behavior of the ligament and others (or fiber reinforcement). Many studies to observe the strain-softening curve have been reported; see, for example, (18-20).

We consider a problem of an infinite row of collinear cracks; see Fig. 2. (The present model is also investigated independently by Ortiz (21) to estimate the effect of micro-cracking on crack growth initiation in brittle solids. A similar model in three dimension is considered by Bazant (22).) Along the failure plane, initial cracks of size $2a_0$ with equal distance d are considered. The solid is under uniaxial tension $\sigma$ at infinity. The stress intensity factor $K_I$ at the crack tip and the crack opening displacement $\langle v\rangle = v^+ - v^-$ are given by

$$K_I = \sigma\sqrt{d\tan\frac{\pi a}{d}}\,, \qquad \langle v\rangle = \frac{4\sigma d}{\pi E'}\log\left(\left(\cos\frac{\pi x}{d} + \sqrt{\cos^2\frac{\pi x}{d} - \cos^2\frac{\pi a}{d}}\right)\Big/\cos\frac{\pi a}{d}\right)\,, \tag{1}$$

where 2a is the crack length, $E' = E$ for plane stress and $E' = E/(1-\nu^2)$ for plane strain with Young's modulus E and Poisson's ratio $\nu$.

We assume that the linear fracture mechanics is applicable at the microscopic level, that is, micro-cracks grow at $K_I = K_c$; $K_c$ represents microscopic fracture toughness. Then the relation between stress and crack size after the peak is given by

$$\frac{\sigma}{\sigma_c} = \sqrt{\tan\frac{\pi a_0}{d}/\tan\frac{\pi a}{d}} , \qquad \text{with} \qquad \sigma_c = K_c/\sqrt{d\tan\frac{\pi a_0}{d}} , \tag{2}$$

$\sigma_c$ being the ultimate strength, at which the micro-cracks start to grow. The stiffness of the cracked solids decreases because of the opening of micro-cracks. In other words, the average strain and displacement are larger than those for the uncracked solids. They are calculated by integrating the crack opening displacement (12). The inelastic displacement $\delta_t$ due to the opening displacement along micro-cracks is obtained from (1) and (2) as (12)

$$\delta_t = \frac{1}{d}\int_{-a}^{a} (v)\ dx = \frac{4\sigma d}{\pi E'}\log(\sec\frac{\pi a}{d}) . \tag{3}$$

The inelastic displacement $\delta = \delta_t - \delta_0$ after the peak point with $\delta_0$ at the peak stress is given by

$$\frac{\delta}{\delta_0} = \frac{\sigma}{\sigma_c}\frac{\log(\sec\frac{\pi a}{d})}{\log(\sec\frac{\pi a_0}{d})} - 1 , \qquad \text{with} \qquad \frac{\delta_0}{a_0} = \frac{4\sigma_c d}{\pi E' a_0}\log(\sec\frac{\pi a_0}{d}) . \tag{4}$$

Equations (4) and (2) provide the strain-softening relation between the stress $\sigma$ and the inelastic displacement $\delta$. In this strain-softening model the involved material parameters are only the normalized tensile strength $\sigma_c/E'$, initial crack density $a_0/d$, and initial crack size $2a_0$. The tensile strength is related to the microscopic fracture toughness $K_c$ through (2).

The strain-softening curves predicted by the present model (4) and (2) are plotted for indicated values of the crack density $a_0/d$ in Fig. 3. With decreasing stress the inelastic displacement increases and the steep slope of the curve gradually becomes gentle. At low stress the inelastic displacement attains its maximum value $\delta_c$. At stress lower than that point, displacement must be decreased to keep the condition $K_I = K_c$ (the dashed lines). This implies that unstable crack growth and complete separation occur at the maximum inelastic displacement $\delta_c$. At this critical state the size of the micro-crack is calculated from (2) and (4) as $a_c/d = 0.455$. With this value the maximum inelastic displacement $\delta_c$ is obtained from (2) and (4). With decreasing crack density $a_0/d$ the value of $\delta_c/a_0$ increases. Values of $\delta_c$ are plotted as a function of the crack density in Fig. 4.

With (2) and (4) the energy required to separate the material per unit area, $G_f$, is obtained as

$$G_f = \int_0^{\sigma_c} (\ \delta + \delta_0(1-\sigma/\sigma_c)\ )\ d\sigma = G_c(1-2a_0/d) , \qquad \text{with} \qquad G_c = K_c^2/E' . \tag{5}$$

If the interaction between micro-cracks is ignored, the expressions from (1) to (4) are simplified as

$$K_I = \sigma\sqrt{\pi a} , \quad (v) = \frac{4\sigma}{E'}\sqrt{a^2-x^2} , \quad \frac{\sigma}{\sigma_c} = \sqrt{\frac{a_0}{a}} , \quad \sigma_c = \frac{K_c}{\sqrt{\pi a_0}} , \quad \frac{\delta}{\delta_0} = \frac{a^2\sigma}{a_0^2\sigma_c} - 1 , \quad \frac{\delta_0}{a_0} = \frac{2\pi a_0\sigma_c}{E' d} . \tag{6}$$

These simplified expressions coincide with the leading terms in the Taylor expansion of expressions from (1) to (4) with respect to a/d for a/d << 1.

It is seen from the above formulation that, for the description of behavior with localized deformation, the elastic part and inelastic part must be separated and, to be independent of the specimen size, the elastic part must be described by strain and the inelastic part by displacement. The relation between the stress and the total displacement (or strain) like the one shown in Fig. 1 is dependent on the specimen size (or the gage length) since the elastic and inelastic components are mixed up.

It is shown that strain-softening curve can be predicted from a simple analytical model based on the micro-crack growth. Obtained results are direct consequences of the modelling and associated assumptions. The obtained strain-softening curve has a steep slope near the peak stress, which seems to agree with experimental observations (18-20). In this region, the micro-cracking is considered to be the major mechanism. The present model predicts unstable crack growth and complete separation at the critical opening displacement, which may not agree with experimental observations. It has been reported that the observed strain-softening curves have long tails; i.e. the stresses are transmitted at larger opening displacement. This is explained by the different mechanism, the bridging by the aggregate interlocking and nonlinear behaviors of the ligament, which is not covered by the present model.

In the strain-softening behaviors of concrete, two mechanisms, micro-cracking and bridging, are important. They are schematically shown in Fig. 5(a). At the tip of a stress free crack in a concrete specimen, there exists a zone which is in the strain-softening state; Fig. 5(b). Each point in the zone corresponds to certain state in the strain-softening curve. For example the tip of the fracture process zone, A, corresponds to the peak point of the strain-softening curve. The tip of the stress free crack (the end of the bridging zone), C, corresponds to the end of the strain softening curve. The boundary of the fracture process zone and the bridging zone, B, (although it is difficult to identify such point) may correspond to a certain point in the strain-softening curve separating the regions in which the micro-cracking and the bridging are major mechanisms. Those speculations must be verified by experimental studies.

As is seen in the next section, it is not necessary to distinguish the fracture process zone and the bridging zone to model those zones if the strain-softening curve is used to represent the behavior of materials in those zones.

## 3 DUGDALE-TYPE MODEL

A number of studies have been reported which are based on Dugdale-type models. The finite element method is often employed to solve problems of specimen of finite size containing a stress free crack and a fracture process and bridging zone. Their main object is to study effects of specimen shape, type of loading, specimen size, and strain-softening curve (relation between the transmitted stress and the crack opening displacement along the fracture process and bridging zone); see, for example, (2-5).

If we consider a fracture process and bridging zone at a tip of a stress free crack in a specimen of finite size, many interacting and influencing factors are involved in the problem. For example, obtained results may include interactions between the process zone and external boundary of the specimen or the other end of a finite crack, the variation of the stresses ahead of the crack, the dependence of the applied stress intensity factor (as is defined later, it is the stress intensity factor when the process zone does not exist) on the crack length, and so on.

It seems necessary to start analyses and discussions from very simplified problems in which influencing factors are excluded. Then each influencing factor is investigated one by one. In the following we construct a simple model and derive the consequences of the used assumptions. Then some influencing factors are investigated.

In general problems of a finite body with a stress free crack and a fracture process and bridging zone, several complicating factors exist. They are, for example,

1) the dependence of the applied stress intensity factor on the crack size,
2) the distribution of the stress ahead of the crack tip,
3) the interaction from the other end of the crack,
4) the interaction between the outer boundary and the fracture process and bridging zone,
5) the strain-softening property.

It is not easy to separate and classify each influencing factors because they are correlated with each other. The effect of the loading condition, for example, is related to several of the factors listed above.

If we consider a crack in an infinite plane, the factor 4) is excluded. If we take the applied stress intensity factor $K_{IA}$ as the loading parameter (instead of the applied stress or forces), the factor 1) is not significant. In addition, Dugdale-type model is valid only before the main crack starts to grow. Hence we do not pay special attention to the factor 1). To introduce the factor 2), consider a single crack in an infinite plane under two different loading conditions shown in Fig. 6 (a) and (b). The stress distributions ahead of the crack tip are given by

$$\sigma_y = \sigma\frac{x}{\sqrt{x^2-a^2}} \approx \frac{K_{IA}}{\sqrt{2\pi r}}(1 + \frac{3r}{4a}) \qquad \text{for } r/a \ll 1 \text{ in problem (a), and}$$

$$\sigma_y = \frac{aP}{x\pi\sqrt{x^2-a^2}} \approx \frac{K_{IA}}{\sqrt{2\pi r}}(1 - \frac{5r}{4a}) \qquad \text{for } r/a \ll 1 \text{ in problem (b).} \tag{7}$$

These stress distributions are expanded with respect to r/a and higher order terms are neglected. First terms in the expressions correspond to the near tip stress distribution. The size effect (in the sense that the size of the fracture process and bridging zone changes the results) appears in the second term. The factors 3) and 4) also lead to the size effect. However, as is clearly demonstrated, the factors 2) and 3),4) appear differently in the formulation.

In the subsection 3.1, a simple model is investigated with attention only to the influencing factor 2). In the subsection 3.2, influencing factors 3) and 4) are discussed and the fracture process and bridging zone at the tip of a finite crack is investigated. It is seen that the difference in the stress distribution ahead of the stress free crack is one of the major factors. Although the factor 5) is very important, it is not included in the present paper.

### 3.1 SIMPLE MODEL

To exclude (for the time being) the factors 3) and 4), we consider a semi-infinite crack in an infinite plane as shown in Fig. 7. The fracture process and bridging zone is modeled by the extended part of the main crack where stresses are transmitted. In the fracture process and bridging zone strain-softening relation, i.e. the relation between the transmitted stress $\sigma_t$ and the crack opening displacement $\delta$,

$$\delta = f(\sigma_t) , \quad \text{or} \quad \sigma_t = f^{-1}(\delta) \qquad \text{with} \quad \delta_c = f(0) \text{ and } \sigma_c = f^{-1}(0) , \tag{8}$$

is satisfied. $\sigma_c$ and $\delta_c$ stand for the tensile strength and the critical opening displacement at which complete separation occurs, respectively. The stress at the end of the process zone, O', must be bounded, that is,

$$K_I = 0 \qquad \text{at O'}. \tag{9}$$

As a loading parameter we consider the applied stress intensity factor $K_{IA}$ which is the stress intensity factor at the tip of the stress free crack O when the fracture process and bridging zone is not present (i.e. when the material is elastic without micro-cracking). In addition to $K_{IA}$ we introduce another parameter $\beta$ (of the length dimension) to represent the variation in the stress distribution for different problems as

$$\sigma_y = \sigma_0(r) = \frac{K_{IA}}{\sqrt{2\pi r}}(1 + r/\beta) , \qquad \text{along } y = 0 \text{ (without process zone).} \tag{10}$$

To find the solution, i.e. the distribution of the transmitted stress $\sigma_t$, which satisfies the conditions (8) and (9) under the loading defined by $K_{IA}$ and $\beta$ through (10), we consider a semi-infinite crack with unit concentrated forces as shown in Fig. 8. The opening displacement g(s,t) at t and the stress intensity factor k(s) at the crack tip due to the concentrated forces at s are easily derived as

$$g(s,t) = \frac{4}{\pi E'}\log\left|\frac{\sqrt{s} - \sqrt{t}}{\sqrt{s} + \sqrt{t}}\right| , \qquad k(s) = -\sqrt{2/\pi s} . \tag{11}$$

It should be noticed that expressions (11) are very simple since we are dealing with a semi-infinite crack. If we consider a stress free crack of finite size or a body with external boundaries, the expressions corresponding to (11) depend on the size of the crack or the specimen and other parameters. This type of influencing factors is distinguished from that treated by (10). They are considered in the next section.

To derive equations for the transmitted stress $\sigma_t$, we consider the decomposition of the problem as shown in Fig. 9. The first sub-problem is the elastic one without the process zone; the stress intensity factor at the tip of the stress free crack O is equal to $K_{IA}$ and the stress distribution ahead of the crack tip is $\sigma_0(r)$ of (10). In the second sub-problem, the stress and opening displacement are $\sigma_t(s)-\sigma_0(\ell_p-s)$ and $\delta(s)$, respectively. The conditions (8) and (9) lead to the integral equations for the transmitted stress $\sigma_t$,

$$\int_0^{\ell_p} g(s,t)\,(\sigma_t(s)-\sigma_0(\ell_p-s))\,ds = f(\sigma_t(t)) , \qquad \int_0^{\ell_p} k(s)\,(\sigma_t(s)-\sigma_0(\ell_p-s))\,ds = 0 . \tag{12}$$

The left hand sides of (12) represent the crack opening displacement and the stress intensity factor at O' due to the stress $\sigma_t(s)-\sigma_0(\ell_p-s)$, respectively. Equations (12) are nondimensionalized with (10) and (11) leading to,

$$\hat{\ell}\int_0^1 \log\left|\frac{\sqrt{\hat{s}}-\sqrt{\hat{t}}}{\sqrt{\hat{s}}+\sqrt{\hat{t}}}\right|\hat{\sigma}(\hat{s})\,d\hat{s}+\hat{K}\sqrt{\hat{\ell}\hat{t}}\left(1+\frac{3-2\hat{t}}{6\hat{\beta}}\right) = \hat{f}(\hat{\sigma}(\hat{t})), \qquad \hat{K}\left(1+\frac{1}{2\hat{\beta}}\right)-2\sqrt{\hat{\ell}}\int_0^1 \frac{\hat{\sigma}(\hat{s})}{\sqrt{\hat{s}}}d\hat{s} = 0 , \tag{13}$$

with

$$\hat{\ell} = \frac{4\ell_p\sigma_c}{\pi E'\delta_c} , \quad \hat{K} = \frac{K_{IA}}{\sqrt{E'\sigma_c\delta_c/8}} , \quad \hat{f} = \frac{f}{\delta_c} , \quad \hat{\sigma} = \frac{\sigma_t}{\sigma_c} , \quad \hat{\beta} = \frac{\beta}{\ell_p} , \quad \hat{s} = \frac{s}{\ell_p} , \text{ and } \quad \hat{t} = \frac{t}{\ell_p} . \tag{14}$$

The integral equations (13) are solved numerically for the given value of $\hat{K}$ (or $\hat{\ell}$) to obtain $\hat{\sigma}(\hat{s})$ and $\hat{\ell}$ (or $\hat{K}$). Note that there is only one independent variable; the normalized applied stress intensity factor $\hat{K}$ or the normalized process zone length $\hat{\ell}$ defined in (14). As is shown later, with increasing load the length of the fracture process and bridging zone increases. At the critical load the crack tip opening displacement (CTOD) $\delta(\ell_p)$ attains the critical opening $\delta_c$ and the stress free crack starts to grow.

In addition the energy spent to form the fracture process and bridging zone $W_d$ and the J-integral J are given by

$$W_d = \int_0^{\ell_p}\int_0^{\delta(s)} f^{-1}(\delta)\,d\delta\,ds , \qquad J = \int_0^{\delta(\ell_p)} f^{-1}(\delta)\,d\delta , \tag{15}$$

where $W_d$ is calculated from the solution of (12) and the expression for J is the well-known results for Dugdale-type models. It is known that the critical value of J-integral $J_c$, at which the stress free crack starts to grow, equals the energy required for the unit area separation, $G_f$, which is the area under the strain-softening curve. (This is always true as far as Dugdale-type models are employed, whereas critical values of the applied stress intensity factor and the associated energy release rate are variable as is seen later.)

A linear strain-softening relation shown in Fig. 10 is the simplest assumption and is often used for the fracture process zone. It is important to investigate the effect of the strain-softening property. In the present study, however, the considerations are restricted to the influencing factor 2), i.e. the effect of stress distribution ahead of the crack tip. For this curve (8) becomes

$$\delta = f(\sigma_t) = \delta_c(1-\sigma_t/\sigma_c) , \qquad \sigma_t = f^{-1}(\delta) = \sigma_c(1-\delta/\delta_c). \tag{16}$$

With (16) integral equations (13) are solved numerically. At first the parameter $\beta$ is set to be infinitely large, that is, the second term in (10) is ignored. Obtained results are shown in Figs. 11 and 12. Fig. 11 shows the distribution of the transmitted stress for indicated values of the fracture process and bridging zone length. Fig. 12 shows that the size of the zone increases with increasing load. At the critical load the crack tip opening displacement $\delta(\ell_p)$ reaches its critical value $\delta_c$ and the stress free crack starts to grow. At this critical state, we have

$$\hat{\ell} = \frac{4\ell_{pc0}\sigma_c}{\pi E'\delta_c} = 0.466\ , \quad \hat{K} = \frac{K_{IAc0}}{\sqrt{E'\sigma_c\delta_c/8}} = 2\ , \quad \frac{W_d}{\sigma_c\delta_c\ell_{pc0}} = 0.287\ , \quad J_c = \frac{1}{2}\sigma_c\delta_c\ , \tag{17}$$

where $K_{IAc0}$ and $\ell_{pc0}$ are critical values of $K_{IA}$ and $\ell_p$ (subscript c is for the critical value at the stress free crack growth and subscript 0 is for a semi-infinite crack with $\beta >> \ell_p$). Note that the energy release rate calculated from the critical value of the applied stress intensity factor $K_{IA}$ given by (17) is equal to the energy required for the material separation, $G_f$, which is true only for this case (semi-infinite crack with $\beta >> \ell_p$). It is interesting to notice that results (17) hold for different materials with different parameters $E'$, $\sigma_c$, $\delta_c$. The critical zone size $\ell_{pc0}$ is used to represent the characteristic length of the material. The nondimensional parameter $\lambda = \ell_{pc0}/\beta$ is introduced to represent the difference in the crack tip stress distribution.

Relationships between the applied stress intensity factor $K_{IA}$ and the fracture process and bridging zone length $\ell_p$ for $\lambda \neq 0$ are shown in Fig. 13. The end of each line corresponds to the critical state at which the crack tip opening displacement $\delta(\ell_p)$ reaches its critical value $\delta_c$ and the stress free crack starts to grow. When the stress distribution ahead of the crack tip is different from the near tip stress field, that is, when $\lambda$ is not zero, the critical value of the applied stress intensity factor $K_{IAc}$ and the critical zone size $\ell_{pc}$ at the initiation of the stress free crack growth are different from $K_{IAc0}$ and $\ell_{pc0}$ for $\lambda = 0$ given in (17).

On the other hand, as mentioned before, the critical value of J-integral at the stress free crack growth, $J_c$, is identical for different $\lambda$. It implies that $J_c$ is uniquely determined, but the critical load can not be predicted by $J_c$ and elasticity solutions. It should be also noticed that the applied stress intensity factor $K_{IA}$ attains the peak value before the stress free crack starts to grow. As is shown in the next section, this is important for the response of finite sized specimens. It is interesting to note that the length of the fracture process and bridging zone at the peak load happens to be the same as $\ell_{pc0}$.

The critical values of $K_{IA}$ and $\ell_p$ at which the stress free crack starts to grow are shown in Fig. 14 and 15. It is seen that with increasing value of $\lambda$ the critical applied stress intensity factor decreases while the critical length of the fracture process and bridging zone increases. These variations in the solution are just due to the difference in the stress distribution ahead of the crack tip.

### 3.2 FRACTURE PROCESS AND BRIDGING ZONE AT THE TIP OF A FINITE CRACK

If we consider the fracture process and bridging zone at the tip of a finite crack shown in Fig. 6 (a) and (b), the influencing factor 3) is involved and the formulation given in the previous section must be modified. For those problems equations (12) are solved with

$$g(s,t;c) = \frac{4}{\pi E'}\log\frac{(ct+cs-st-2c^2-\sqrt{st(2c-s)(2c-t)})\,(s-t)}{(st-ct-cs-\sqrt{st(2c-s)(2c-t)})\,(2c-s-t)}\ , \quad k(s;c) = -\sqrt{\frac{2}{\pi s(1-s/2c)}}\ ,$$

and

(18)

$$\sigma_0(r;a) = \frac{K_{IA}}{\sqrt{2\pi r}}\frac{1+r/a}{\sqrt{1+r/2a}}, \qquad \text{for the problem (a)}$$

$$\sigma_0(r;a) = \frac{K_{IA}}{\sqrt{2\pi r}}\frac{1}{(1+r/a)\sqrt{1+r/2a}}, \qquad \text{for the problem (b).}$$

Note that the size a of the finite crack comes in the equation in different places. The effect in $\sigma_0(r;a)$ is already included (approximately) in the previous model in (10) with $\beta = 4a/3$, $\lambda = 3\ell_{pc0}/4a$ for the problem (a) and $\beta = -4a/5$, $\lambda = -5\ell_{pc0}/4a$ for the problem (b); see equations (7). With (18) equations (12) are solved numerically and results are shown in Figs. 14 and 15 with dashed lines. Note that for small $\lambda$ the results of the previous model are seen to be good approximate solutions. Since the difference between problems (a) and (b) appears only in $\sigma_0(r;a)$, it is concluded that the difference in the crack tip stress distribution is one of the most significant controlling factors.

In general problems of finite specimen with process zone, functions $g(s,t)$, $k(s)$ and $\sigma_0(r)$ in (12) depend on the size and the geometry of the specimen and the crack, the loading condition, and others in a complex manner. Since it is not easy to find functions $\sigma_0(s,t)$ and $k(s)$ for each problems, it is not practical to solve equations (12) for each problem. The present study revealed that the variation in the stress distribution is one of the major controlling factors. It is possible, under certain conditions, to obtain good approximate solutions by determining one parameter $\lambda$ (or $\beta$) for each problem and applying the results of the present simple model. Since many crack problems (without process zones) have been solved, it is not difficult to determine the parameter $\lambda$. If the stress intensity factor is provided as a function of the crack length, $\lambda$ is approximately obtained through the relation

$$K_I(r) = -\int_0^r k(s)\sigma_0(r-s)\,ds = K_{IA}\left(1+\frac{r}{2\beta}\right). \tag{19}$$

This is the variation of the stress intensity factor with the crack extension length, r, when the stress distribution $\sigma_0$ ahead of the original crack tip is given by (10).

In the next section, the simple model proposed in the previous section is used to investigate the growth of the fracture process and bridging zone in a specimen of finite size. Results are compared with experimental data.

### 3.3 FRACTURE PROCESS AND BRIDGING ZONE IN A SPECIMEN OF FINITE SIZE

A Dugdale-type model for the fracture process and bridging zone in a finite sized specimen requires explicit form of $g(s,t)$ and $k(s)$ in integral equations (12). They are elasticity solutions of a finite body with concentrated forces on the crack surface and are difficult to obtain. Finite element methods have been used instead in the literature. However, careful attention must be paid to the treatment of stress singularity and accurate satisfaction of conditions (8) and (9).

To have approximate solutions of those problems, the simple model proposed in the subsection 3.1 can be used. For the application of the simple model, only the identification of the parameter $\lambda$ is necessary. In general, only the stress intensity factor is provided for a finite sized specimen such as the three point bending specimen and the compact tension specimen. It is sufficient information since the parameter $\lambda$ is approximately determined through (19).

As an example, a crack line wedge loaded (CLWL) specimen shown in Fig. 16 is considered. The stress intensity factor is given by (23),

$$K_{IA} = \frac{P}{t\sqrt{W}}F(a/W), \qquad F(\alpha) = (2+\alpha)(0.886+4.64\alpha-13.32\alpha^2+14.72\alpha^3-5.6\alpha^4)(1-\alpha)^{-3/2}. \tag{20}$$

Then the parameter $\lambda$ is obtained through (19) as follows;

$$K_I(r)/K_{IA} = F\left(\frac{a+r}{W}\right)/F\left(\frac{a}{W}\right) \approx 1 + \frac{F'(a/W)}{F(a/W)}\frac{r}{W} = 1 + \frac{r}{2\beta} = 1 + \lambda\frac{r}{2\ell_{pc0}}, \tag{21}$$

which leads to

$$\lambda = \frac{F'(a/W)}{F(a/W)} 2\ell_{pc0}/W \ . \qquad (22)$$

Note that $\lambda$ is positive for the CLWL specimen.

For given properties of the material, the characteristic length $\ell_{pc0}$ (the critical length of the fracture process and bridging zone at the tip of a semi-infinite crack with $\lambda = 0$) is calculated through (17). For simplicity, we assume that the tensile strength $\sigma_c$, the Young's modulus E and the Poisson's ratio $\nu$ of concrete are give by $\sigma_c = 0.1f'_c$ (kgf/cm²), $E = 15000\sqrt{f'_c}$ (kgf/cm²), $\nu = 0.2$ where $f'_c$ stands for the compressive strength of the concrete. The characteristic length $\ell_{pc0}$ is shown in Fig. 17 as a function of the tensile strength $\sigma_c$ for indicated values of the critical opening displacement $\delta_c$.

Then, for given values W and a, the parameter $\lambda$ is obtained through (22). From (20) the applied stress intensity factor $K_{IA}$ is obtained for the applied load P. Then the normalized stress intensity factor $\hat{K}$ is calculated through (17). Solving integral equations (13) with those $\hat{K}$ and $\lambda$, we obtain the normalized length of the fracture process and bridging zone $\hat{\ell}$ and accordingly $\ell_p$ through (17). Typical results are shown in Fig. 18. With increasing load P, the length of the fracture process and bridging zone $\ell_p$ increases. The load attains its peak value before the stress free crack starts to grow. This is an important fact because the peak point does not satisfy the critical condition $\delta(\ell_p) = \delta_c$ and $J = J_c = \sigma_c\delta_c/2$.

After the stress free crack starts to grow, the total process zone size $(l_p+a-a_0)$ increases keeping the condition $\delta(\ell_p) = \delta_c$ (dashed lines). Note that the stress free crack length, a, itself increases from its initial value $a_0$ in this state and the value of $\lambda$ changes accordingly. After the stress free crack extension, the load decreases drastically. In the present model, a simple strain-softening curve, which is straight and does not have a long tail, is assumed. If we introduce a strain-softening curve which has a long tail, the bridging zone becomes longer and the growth of the stress free crack is delayed. Then the solid curve in Fig. 18 is extended longer. In the actual situation, toughening may also be present after the stress free crack growth. This is because of the residual strain (cracking) left behind the crack tip which can not be presented by the Dugdale-type model.

In Fig. 19, the peak load and the critical load for the stress free crack extension are plotted as a function of $\ell_{pc0}/W$ with solid and dashed lines, respectively. $P_0$ is the load when $\ell_{pc0}/W = 0$. This shows the size effect due to the difference in the stress distribution ahead of the crack tip which is represented by the parameter $\lambda$.

In Figs. 20-23, results are compared with experimental results by Kobayashi et al. (2) for the CLWL specimens of two sizes. The total fracture process and bridging zone size is observed with the replicating technique (2). Open symbols are for the fracture process and bridging zone growth and closed symbols for the stress free crack extension; they are distinguished by the results obtained from the double displacement compliance plots (2). The experimental data are best fitted with $\sigma_c = 25$ kgf/cm², $\delta_c = 0.0025$ cm for large specimens and $\sigma_c = 45$ kgf/cm², $\delta_c = 0.0025$ cm for small specimens. The reported average compressive strengths are $f'_c = 240$ kgf/cm² and 640 kgf/cm² for large and small specimens, respectively. The tensile strength estimated with reported relation $\sigma_c = 1.46\sqrt{f'_c}$ (kgf/cm²) (2) is 23 kgf/cm² for large specimens and 37 kgf/cm² for small specimens.

It has been shown that the simple model which includes the parameter $\lambda$ is used to predict fracture phenomena with the fracture process and bridging zone in a finite sized specimen. As an example, the CLWL specimen is considered. The procedure is the same for other types of specimen such as the three point bending specimen and the compact tension specimen. The size effect is reproduced by the present model. The validity of the present model and associated conclusions must be verified by further comparison with experimental results.

In this study the influencing factors 4) and 5), i.e., the interaction between outer boundary of the specimen and the process zone and the effect of the strain-softening curve are not included. Those are under investigation and results will be presented later.

ACKNOWLEDGMENT

This study is supported in part by the Grant-in Aid for Scientific Research from the Japanese Ministry of Education, Science and Culture and by the Mitsubishi Funds.

REFERENCES

(1) Nallathambi, P., B.L. Karihaloo, and B.S. Heaton, "Effect of specimen and crack sizes, water/cement ratio and coarse aggregate texture upon fracture toughness of concrete" Magazine of Concrete Research, 36(129) 227-236 (1984).
(2) Kobayashi, A.S., N.M. Hawkins, D.B. Barker, and B.M. Liaw, "Fracture process zone of concrete" Application of Fracture Mechanics to Cementitious Composites, Proceedings of Nato Advanced Research Workshop September 4-7 (1984), 25-48 (1984).
(3) Hillerborg, A., "Analysis of one single crack" Fracture Mechanics of Concrete, edited by F.H. Wittmann, Elsevier Sci. Pub., 223-249 (1983).
(4) Ballarini, R., S.P. Shah, and L.M. Keer, "Crack growth in cement-based composites" Engineering Fracture Mechanics, 20(3) 433-445 (1984).
(5) Li, V.C., and E. Liang, "Fracture processes in concrete and fiber reinforced cementitious composites" J. Engineering Mechanics, 112(6) 566-586 (1986).
(6) Kachanov, M., "On crack - microcrack interactions" Int. J. Fracture, 30, R65-R72 (1986).
(7) Chudnovsky, A., A. Dolgopolsky and M. Kachanov, "Elastic interaction of a crack with microcracks" Proceedings Sixth Int. Conference on Fracture, New Delhi, India, 825-832 (1984).
(8) Tamuzs, V.P., and N.B. Romalis, "Interaction of a macrocrack with microdamages" Proceedings Sixth Conference on fracture, New Delhi, India, 833-840 (1984).
(9) Rose, L.R.F., "Microcrack interaction with a main crack" Int. J. of Fracture, 31, 233-242 (1986).
(10) Hoagland, R.G. and J.D. Embury, "A treatment of inelastic deformation around a crack tip due to microcracking" J. of the American Ceramic Society, 63, 404-410 (1980).
(11) Budiansky, B. and R.J. O'Connel, "Elastic moduli of a cracked solid" Int. J. of Solids and Structures, 12, 81-97 (1976).
(12) Horii, H., and S. Nemat-Nasser, "Overall moduli of solids with microcracks: load-induced anisotropy" J. of Mechanics and Physics of Solids, 31(2) 155-171 (1983).
(13) Sumarac, D., and D. Krajcinovic, "A Self-consistent model for microcrack weakened solids" Mechanics of Materials; to appear.
(14) Sahasakmontri, K., H. Horii, A. Hasegawa and F. Nishino, "Mechanical properties of solids containing a doubly periodic rectangular array of cracks" Structural Eng./Earthquake Eng., JSCE, (1987); to appear.
(15) Charalambides, P., and R.M. McMeeking, "Finite element method simulation of crack propagation in a brittle microcracking solid" Mechanics of Materials; to appear.
(16) Bazant, Z.P., and B.H. Oh, "Crack band theory for fracture of concrete" Materiaux et Constructions, RILEM, 16(93) 155-177 (1983).
(17) Yeoushang Jenq, S.M., and S.P. Shah, "Two parameter fracture model for concrete" J. of Engineering Mechanics, ASCE, 111(10) 1227-1241 (1985).
(18) Willam, K., B.H. Synopsis and S. Sture, "Experimental, constitutive and comutational aspects of concrete failure" Seminar on Finite Element Analysis of Reinforced Concrete Structures, May 1985, Tokyo, 1, 149-171 (1985).
(19) Gopalaratnam, V.S., and S.P. Shah, "Softening response of plain concrete in direct tension" ACI Journal, 82(3) 310-323 (1985).
(20) Li, V.C., C.M. Chan and C. Leung, "Experimantal determination of the tension-softening curve in cementitious composites" Cement and Concrete Research, 17, 441-452 (1987).
(21) Ortiz, M., "Microcrack coalescence and macroscopic crack growth initiation in brittle solids" Int. J. Solids and Struct.; to appear.
(22) Bazant, Z.P., "Snapback instability at crack ligament tearing and its implication for fracture micromechanics" Report No. 87-6/498s, Center for Concrete and Geomaterials, Northwestern University, June (1987).
(23) Stress Intensity Factors Handbook, The Society of Materials Science, Japan, Editor-in-chief Y. Murakami, Pergamon Press, 1, (1986).

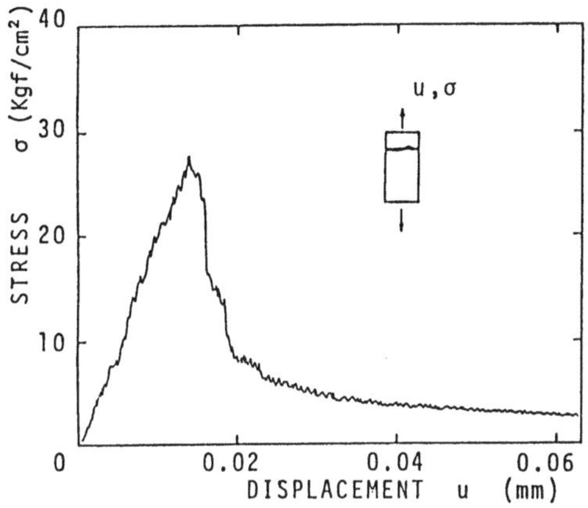


Fig. 1. Stress-displacement relation in uniaxial tension test

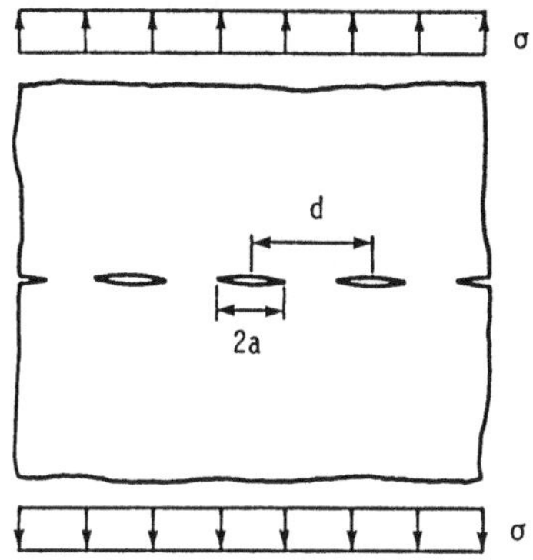


Fig. 2. Strain-softening model

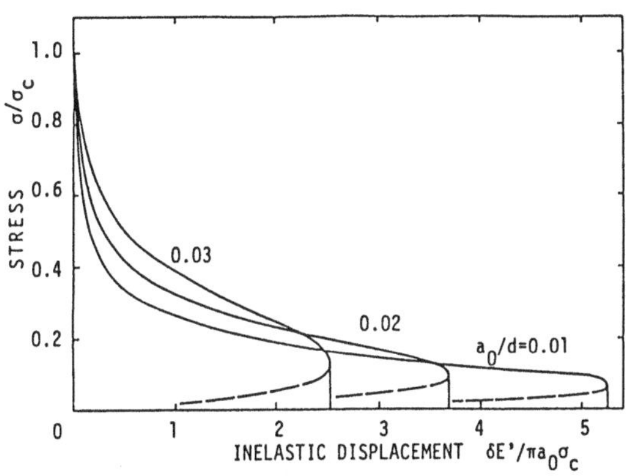


Fig. 3. Strain-softening curves predicted by the present model

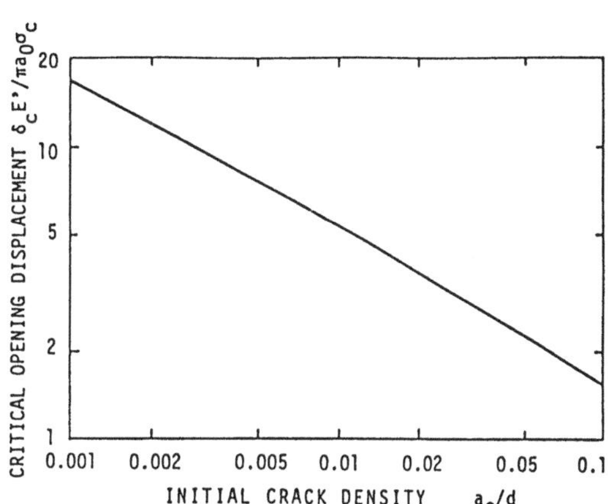


Fig. 4. Critical opening displacement as a function of the initial crack density

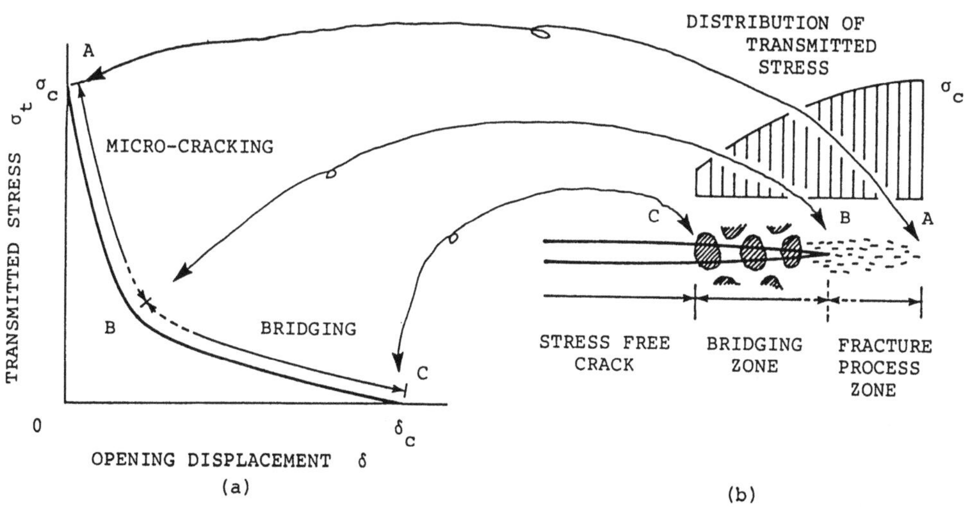


Fig. 5. Relationship between (a) the strain-softening curve and (b) the fracture process and bridging zone at the tip of a stress free crack

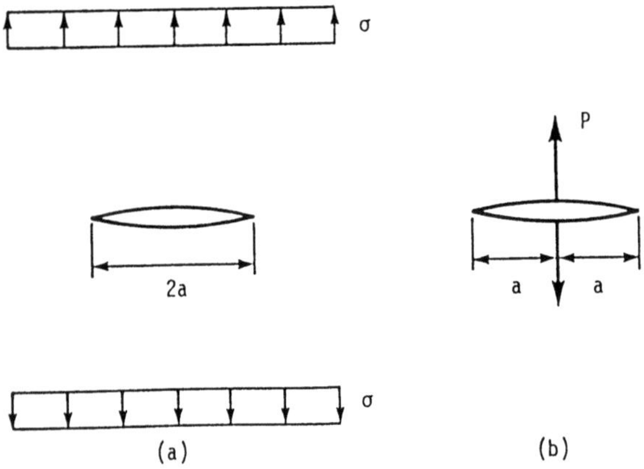


Fig. 6. Finite cracks with (a) uniform stress at infinity and (b) concentrated forces at the center of the crack

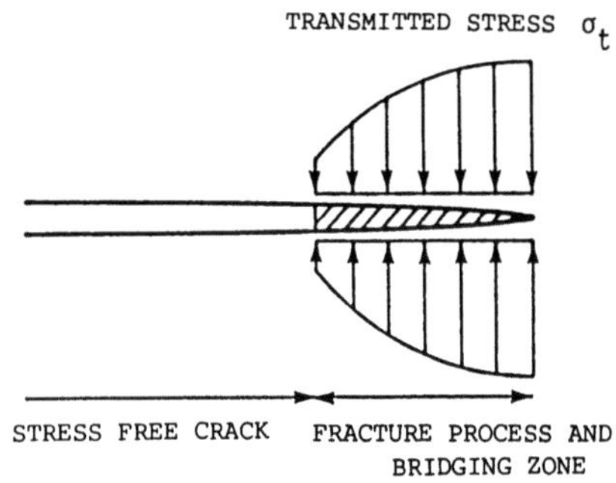


Fig. 7. Fracture process and bridging zone at the tip of a semi-infinite crack

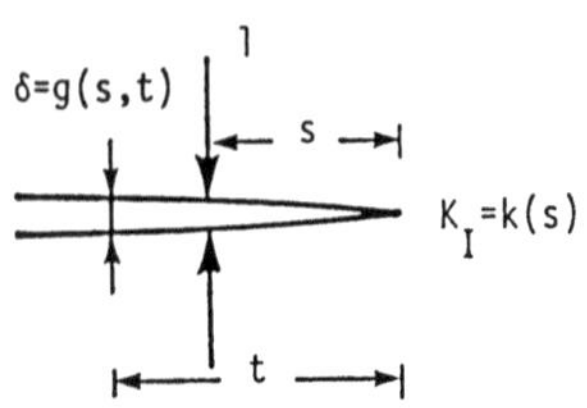


Fig. 8. A semi-infinite crack with concentrated forces

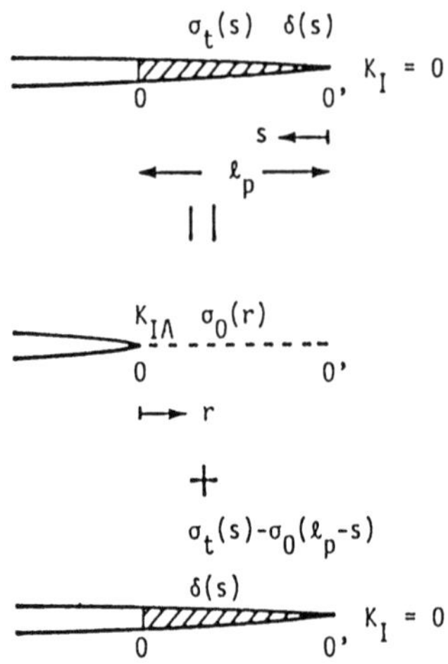


Fig. 9. Decomposition of the original problem

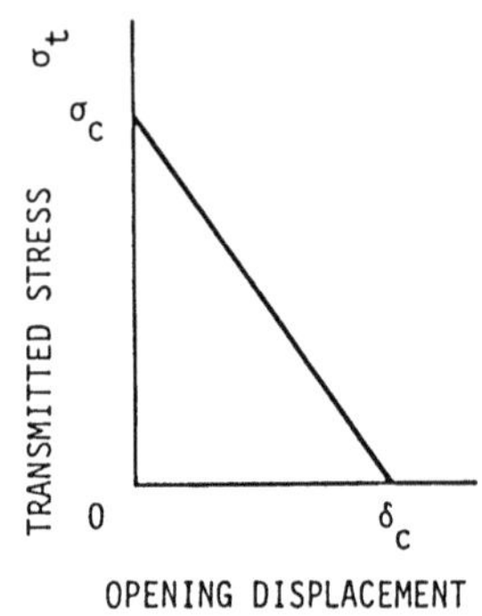


Fig.10. Linear strain-softening relation

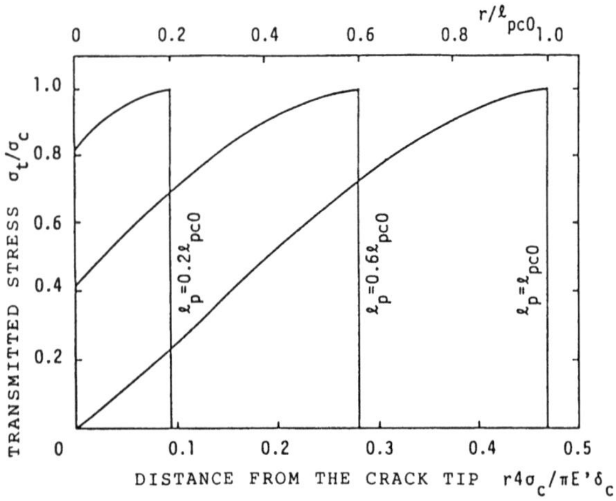


Fig. 11. Distribution of the transmitted stress

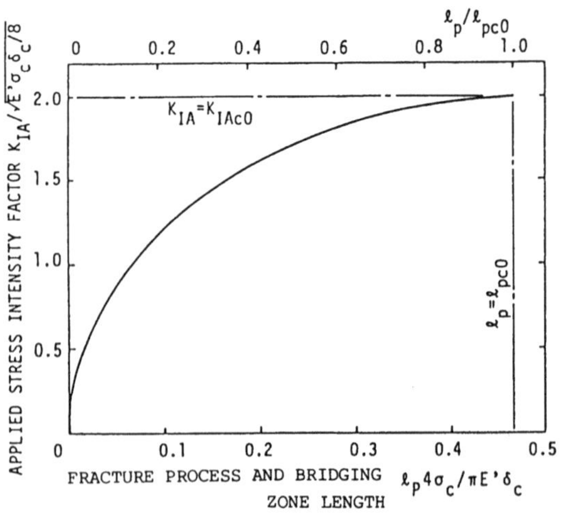


Fig. 12. Applied stress intensity factor as a function of the fracture process and bridging zone length

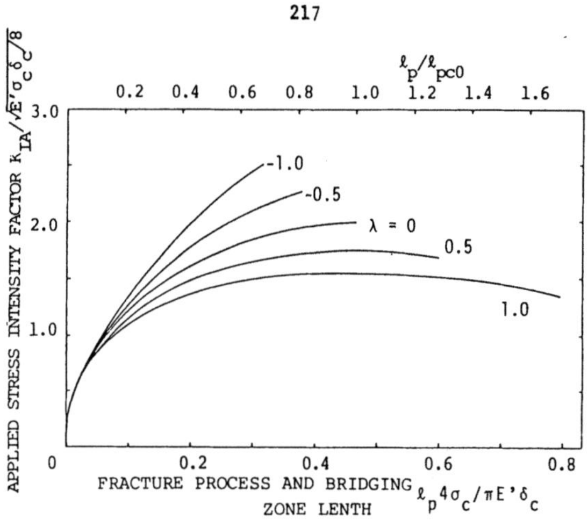


Fig. 13. Applied stress intensity factor as a function of fracture process and bridging zone length

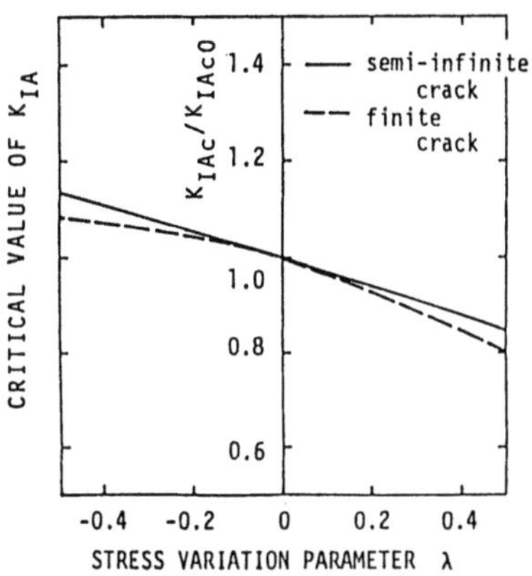


Fig. 14. Critical value of the applied stress intensity factor

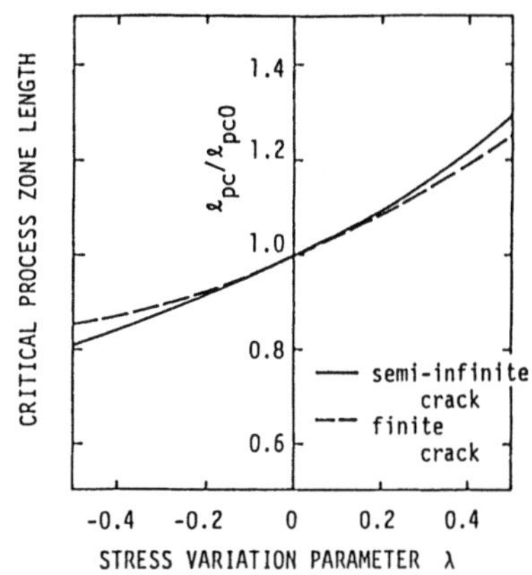


Fig. 15. Critical length of the fracture process and bridging zone

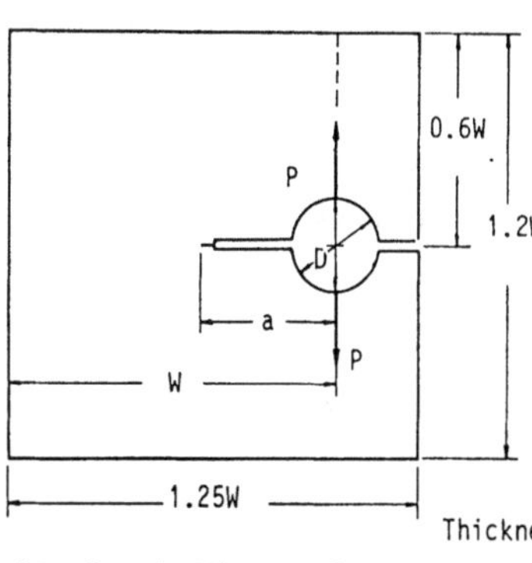


Fig. 16. Crack Line Wedge Loaded Specimen

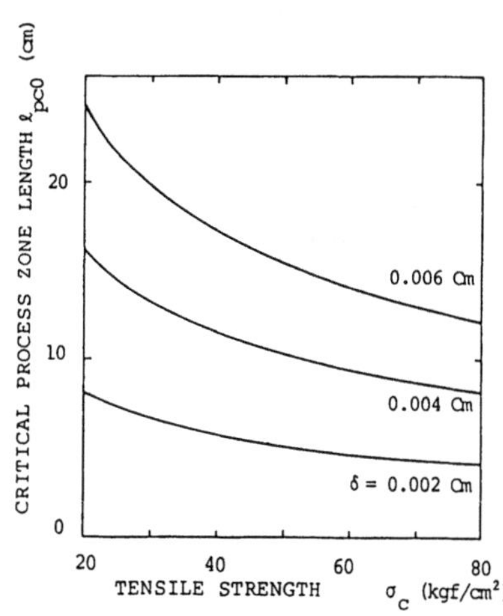


Fig. 17. Critical length of the fracture process and bridging zone for a semi-infinite crack with λ = 0

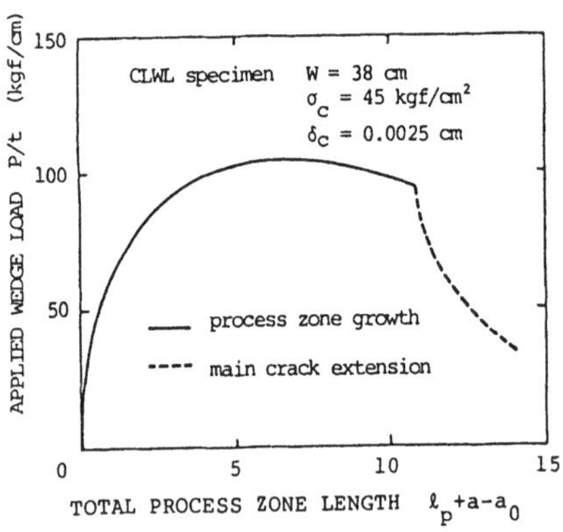


Fig. 18. Applied wedge load vs total fracture process and bridging zone length for the CLWL specimen

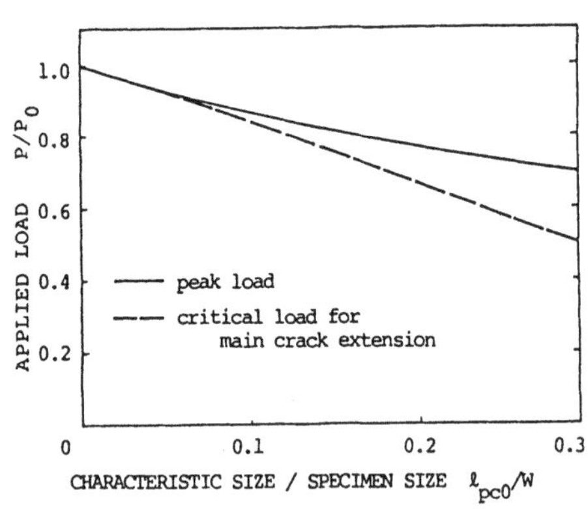


Fig. 19. Size effect in the critical load for the CLWL specimen

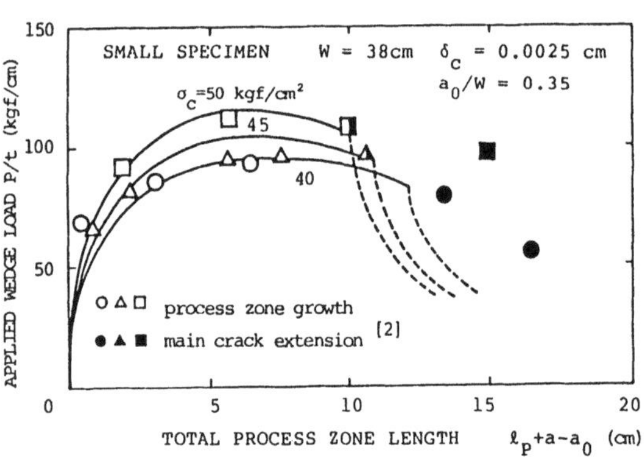


Fig. 20. Applied wedge load vs total process zone length for the CLWL specimen

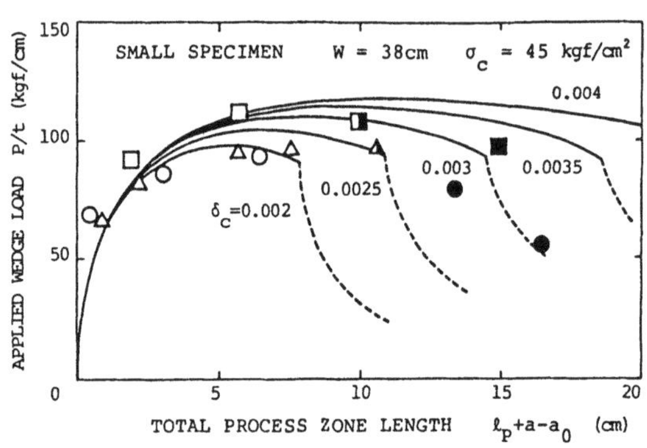


Fig. 21. Applied wedge load vs total process zone length for the CLWL specimen

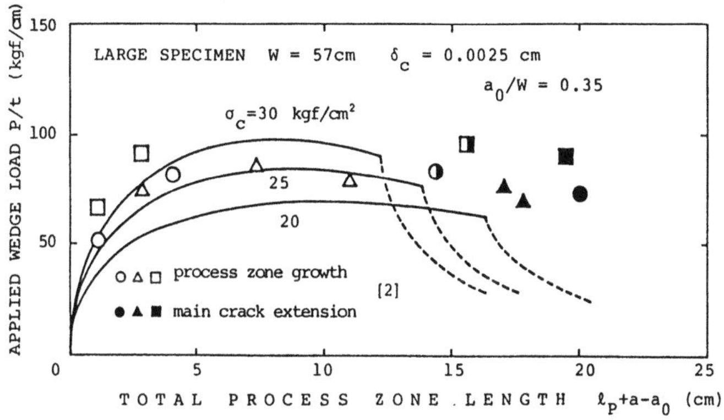


Fig. 22. Applied wedge load vs total process zone length for the CLWL specimen

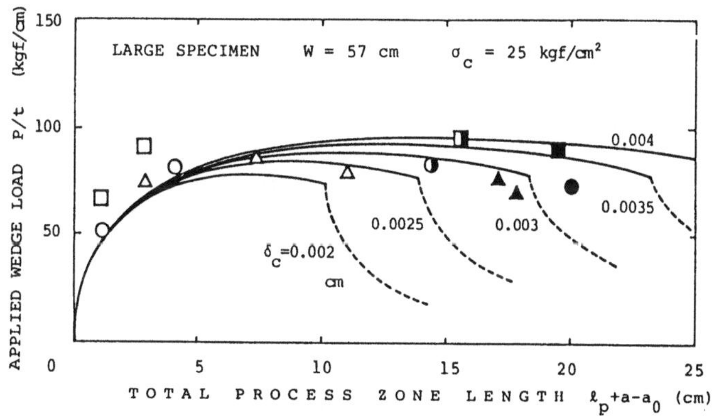


Fig. 23. Applied wedge load vs total process zone length for the CLWL specimen

SEM/RILEM International Conference
on Fracture of Concrete and Rock
Houston, Texas, June 1987
Editors: S.P. Shah, S.E. Swartz

# 3-D-MODELING OF PROCESS ZONE IN CONCRETE BY NUMERICAL SIMULATION

H. Schorn, U. Rode

## ABSTRACT

A new simulation method for the calculation of crack formation and crack growth in concrete has been developed at University of Bochum. The method uses a large number of struts in an equidistant lattic point as a mechanical model. In the process of numerical simulation microcrack opening and microcrack growing can be studied up to the totally collapsed system.

## 1. INTRODUCTION

As a useful characteristic value in fracture mechanics of concrete the absorbed energy ($G_F$) can be evaluated from tests on bending or in tensile loaded specimens. The term $G_F$ is based on Hillerborg's theory /1/ which presumes an existing process zone consisting of microcracks, see fig. 1, left. The abstraction of that process zone is the idea of a macrocrack, see fig. 1 right, with a load carrying characteristic. Thus, the fracture energy of a great number of microcracks has been reduced to only one hypothetic crack; microcracks outside of the process zone are not taken in account.

Evaluating the stress strain curve including the unloading branch of tensile loaded specimens the $G_F$-value results from the $\sigma$-$\omega$-diagram, see fig. 2, as the integral of the curve.

In an experimental study Budnik /2/ has pointed out the dependence of $G_F$ from the type of material structure of different concrete mixes and proposed the use of $G_F$ in relation to the strength, called length of ductility ($l_D$).

Hillerborgs characteristic value $G_F$ as well as the $l_D$-value by Budnik are not independent from the geometry of the used specimens. The authors have studied this effect by test results of the often used notched type of specimens, compared with the - due to their difficult conditions in measuring - seldom used unnotched type of specimens of the same material structure.

The results of experimental works make uncertain $G_F$ and $l_D$ as a real specific characteristic value in fracture mechanics. Further investigations for description of the process zone and its alteration due to loading must be carried out. But doing this, more knowledge of microcrack opening in the process zone inside the body is necessary. That cannot be obtained by experiments only. As a supplement, a theoretical method should be used which is able to show occuring and growing microcracks due to parameters of material structure. Such a method has been developed at the University of Bochum. It is a numerical simulation method using a mechanical model for delineating material structure of concrete and showing stress strain behaviour as well as crack states on the surface and inside the body.

## 2. NUMERICAL METHOD

### 2.1 MECHANICAL MODEL

The basis of each simulation method is the existence of a suitable mechanical model. Often the use and the effectiveness of such models are seen under the view of mathematical aspects primarily. In cases of problems of material research a mechanical model can only work satisfactorily if the real material structure is represented in a sufficient approximation. Those structure based-models we call "synthetic" models.

Some authors, e.g. /3/, use a mesh for FEM which delineates all aggregate particles in concrete in a geometrically exact state. Calculating stress strain behaviour and crack states on such a model as a three dimensional one, the limits of capacity of today's computers will be reached soon, or it needs tremendous time for operating, respectively. At Bochum another way has been chosen.The typical material structure is abstracted on forces in space, forces between neighboured points. The idea is: modeling stress trajectories between equidistant lattice points. Thus, the model is characterized as a three dimensional point lattice, see fig. 3. Each point is connected by struts with all neighboured ones. If a strut pervades the border of a gravel particle, its behaviour is determined by both, gravel characteristics as well as binder characteristics. Thus, the material structure is established in the model without any necessity for delineating the geometry of particles exactly. The circles shown in fig. 3 only visualize the position and the shape of the aggregate particles; they are unnecessary for the model itself.

This framework has its roots in the method of Hrennikoff, published already in 1941. But there are important differences. The Bochum model does not base on continuum mechanics. In the model, see fig. 3, the strut behaviour is determined by different strut characteristics due to inhomogeneous material structure and allows the simulation of alteration in material structure due to crack opening. Even if all struts are described by a linear material law, the result of the calculation shows a nonlinear stress-strain relationship.

### 2.2 MODELING CONCRETE

As partly mentioned before, the 3D-simulation framework consists of a large number of struts arranged according to a regular cubic pattern with equidistant nodal points /4/. The basic cell is a cube with 12 edge struts, 12 surface diagonal struts and 4 space diagonal struts. In a concrete body the forces flow through the components. Thus, a sample strut represents the stress flow through an aggregate particle, another one only through the binder and a third one represents the bond between aggregate and binder. Since exact geometrical modeling of the material structure is not required the strut parameters are endowed with values by a given distribution. The struts itself behave linear elastic up to given strain rates. On exceeding the maximum tensile or compressive strain during a loading process the affected struts are irreversibly removed from the system representing cracks, and new equilibrium states of the degenerated framework yield in the following iteration steps.

The input parameters for the program system define the stiffness, tensile strength, compressive strength and the viscosity for any strut element. Fig. 4 shows the stiffness distribution for all struts and the input parameters tensile strength and compressive strength as a function of element stiffness. The strut parameters are stochastically mixed, modeling an inhomogenous concrete. Viscosity has not been taken in account.

### 2.3 CALCULATIONS

The modeling of the crack opening and the failure process requires the exact consideration of the geometrical and physical nonlinear behaviour of the strut elements. Large nodal displacements may hardly occur, but strut series may buckle or snap through in heavily degenerated states of the simulation framework. Thus, numerical stability is the fundamental aspect of the realisation of the simulation framework analysis.

On the base of the finite element displacement formulation MESY /5/, Diekkämper /6/ developed the program package VMESY to realize the simulation framework analysis on a vector computer CDC Cyber 205. This machine provides the computing power required for frameworks with large system sizes. However, the vector computer does not supply interactive services. Thereby interactive control of the system analysis may not be applied.

Due to the crack growth during the loading process parts of the system may become isolated forming mechanisms or instable strut series. Fig. 5 shows an example for the creation of a mechanism consisting of the strut 1 due to the elimination of the struts 2 and 3. For that reason special algorithms have been implemented in the program package to identify indefinities, pertubate instabilities or eliminate mechanisms. This guaranties the fully automatized analysis of the system up to the total collapse.

The fundamental FE-algorithms may be written as follows:

a) Computing of element stiffness matrices and element load vectors

b) Assembly of the system equations

c) Solution of the linearized system equations.

A Newton-Raphson iteration scheme is employed here to solve the nonlinear system equations.

The major part of computing cost per NR-cycle has to be paid for solving the linearized system equations. Diekkämper /7/ pointed out the applicability of the conjugate gradient algorithm (CG) for solving the system equations in structural mechanics iteratively on a vector computer. This method only performs matrix-vector multiplications and so it makes better use of the vector processor's computing power than an elimination algorithm. Convergence to the correct solution is guaranteed for symmetric and positiv definit matrices.

CG-methods keeps the system stiffness matrix unchanged, so it may be completely kept in main memory. Due to the banded matrix structure and only 49 non-zero diagonals for the 3-D-framework a special diagonal vector matrix storage scheme (DVS) is used here. Table 1 shows the number of memory words for the system stiffness matrix for a simulation framework with n·n·n nodes, using different algorithms for solving the system equations. This table illustrates the superiority of CG-method with diagonal vector matrix storage scheme compared with elimination methods for regular systems used here.

## 2.4 EXAMPLE

The efficiency of the implemented algorithms will be demonstrated by the analysis of a square plain framework with 21·21 nodes. It consists of 400 basic squares with 441 nodes and 1640 struts. The strut parameters are distributed as shown in Fig. 7 and then stochastically mixed, thus yielding a very inhomogeneous data set. To calculate also the unloading branch a displacement control is used here. The displacements of the upper edge are prescribed in a way, that all nodes on this edge displace equally in vertical and freely in horizontal direction. The loads are computed as a sum of system reaction forces.

Fig. 9 shows the graphical interpretation of the uniaxial tension test simulation: load-displacement curve, system deformation states (a) and crack plots (b) for four selected load levels. Deformation states are equally drawn to a large scale to visualize better the deformations.

At load level 1 the cracks appear at the upper left corner. At load level 2 a second crack band at the right edge can be identified. Load levels 3 and 4 show the upper crack running in horizontal direction near the upper system edge and marking the typical tension failure mechanism, while the second crack band closes just before the total collapse (load level 4).

## 3. SIMULATION STUDIES

Only a three-dimensional mechanical model can give realistic information about the crack opening and crack growth inside the specimens. Fig. 7 shows a cube, which has been cutted into four bodies. Each of them consists of 9 · 9 · 3 nodes with 8 · 8 · 2 basic cubes and 3584 struts. The input parameters have been choosen as shown in Fig.4. This distribution is fixed for all specimens but each data set has been stochastically mixed by a computerized random number process modeling an inhomogenous concrete.

The loading acts on the upper surface of the simulation framework according to the example in chapter 2.4 (rigid load induction body without friction effects).

Fig 7. shows the graphical interpretation of the results of the four uniaxial tension test simulations: load-displacement curves and the crack formations in three sections of each specimen for two different load levels on the unloading branch (level 1 ≈ 0,5 x maximum load, level 2 ≈ 0,1 x maximum load).

As exspected, the load-displacement relationships and the occuring fracture zones in several specimens are different from each other. According to the strut parameters the first cracks appear randomly. At the unloading branch they soon form continuous crack bands and finally mark the failure mechanism. Parts outside of this zones remain relatively inaffected and cracks inside this zones close at the total collapse.

## REFERENCES

/1/ Hillerborg, A.: Analysis of One Single Crack, Fracture Mechanics of Concrete, pp. 223-249, edited by F.H. Wittmann, Elsevier Science Publishers B.V., Amsterdam 1983.

/2/ Budnik, H.-J.: Bruch- und Verformungsverhalten harzmodifizierter und faserverstärkter Betone bei einachsiger Zugbeanspruchung, Dissertation, Ruhr-Universität Bochum, 1985.

/3/ Wittmann, F.H.; Roelfstra, P.E.; Sadouki, H.: Simulation and Analysis of Composite Structures, Summary Report on Research Activities, Swiss. Federal Institute of Technology, Lausanne 1986.

/4/ Schorn, H.: Zur Einführung numerischer Berechnungsverfahren in der Ermittlung strukturorientierter Stoffgesetze, Strukturmechanik und Numerische Verfahren, Hrsg. R. Diekkämper, H.-J. Niemann, Verlag R. Müller, Köln 1982.

/5/ Schrader, K.-H.: Mesy-Einführung in das Konzept und Benutzeranleitung für das Programm MESY-MINI, Techn. Wiss. Mitteilungen aus dem Institut für KIB, Nr. 78-11, Ruhr-Universität Bochum 1978

/6/ Diekkämper, R.: Ein Verfahren zur numerischen Simulation des Bruch- und Verformungsverhaltens spröder Werkstoffe, Dissertation, Techn. Wiss. Mitteilungen aus dem Institut für KIB, Nr. 84-7, Ruhr-Universität Bochum 1984.

/7/ Diekkämper, R.: Vektorrechnerorientierte Finite-Element-Analyse bei nichtlinearen Problemen, Strukturmechanik und Numerische Verfahren, Hrsg. R. Diekkämper, H.-J. Niemann, Verlag R. Müller, Köln 1982.

| nodes per edge | total nodes | total displace-ments | ELIMINATION PROCESS memory words | | CG-ALGORITHM memory words | |
|---|---|---|---|---|---|---|
| | | | banded matrix | CCS | max. non-zero words | DVS |
| 4 | 64 | 192 | 12.672 | 9.456 | 4.596 | 7.869 |
| 5 | 125 | 375 | 36.000 | 28.650 | 10.074 | 16.095 |
| 6 | 216 | 648 | 85.536 | 70.956 | 18.756 | 28.623 |
| 7 | 343 | 1.029 | 179.046 | 152.880 | 31.380 | 46.353 |
| 8 | 512 | 1.536 | 340.992 | 297.408 | 48.684 | 70.185 |
| 9 | 729 | 2.187 | 603.612 | 570.078 | 71.406 | 101.019 |
| 10 | 1.000 | 3.000 | 1.008.000 | 959.100 | 100.284 | 139.755 |

Table 1: Comparison between different algorithms for solving the system equations with regard to memory words

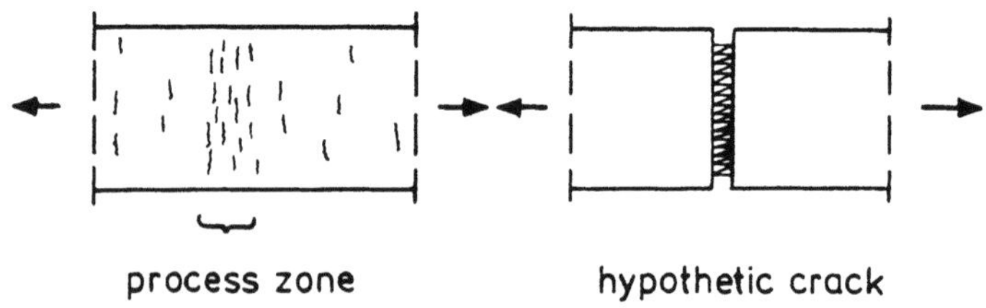


Fig. 1: Process zone (left) and idealized macrocrack (right) in a tensile loaded concrete specimen due to Hillerborg's theory

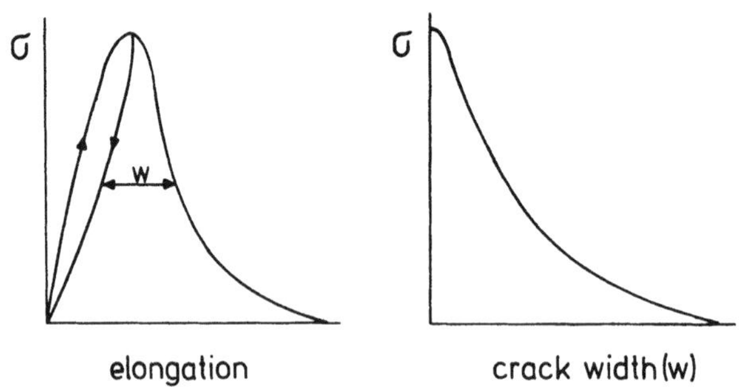


Fig. 2: Stress-strain (left) and stress-crack width of the hypothetic crack (right) due to Hillerborg's theory

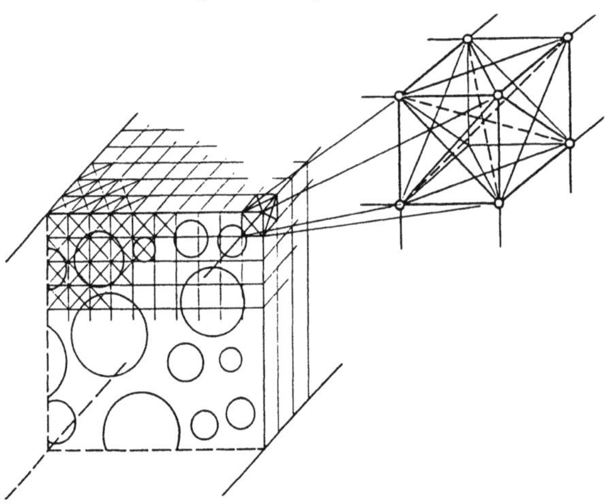

Fig. 3: 3D-model for numerical simulation of concrete

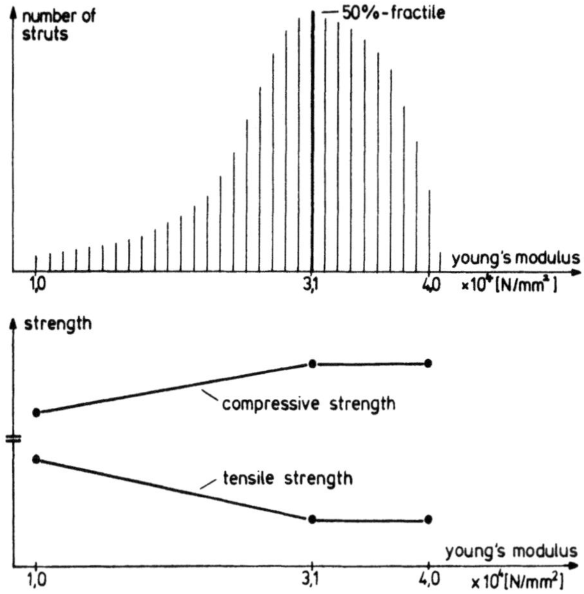


Fig. 4: Input parameters for the simulation framework analysis

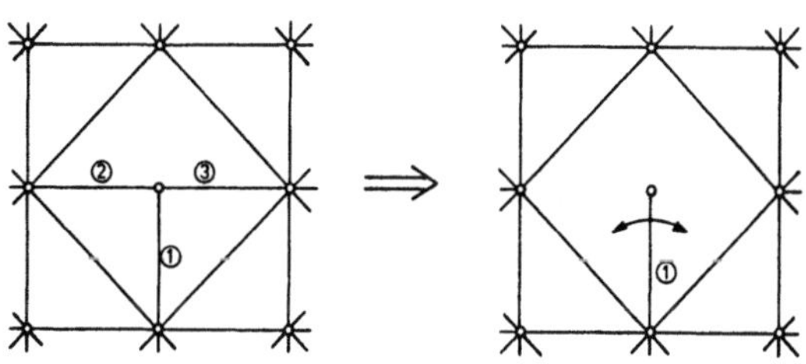


Fig. 5: Example for mechanism forming

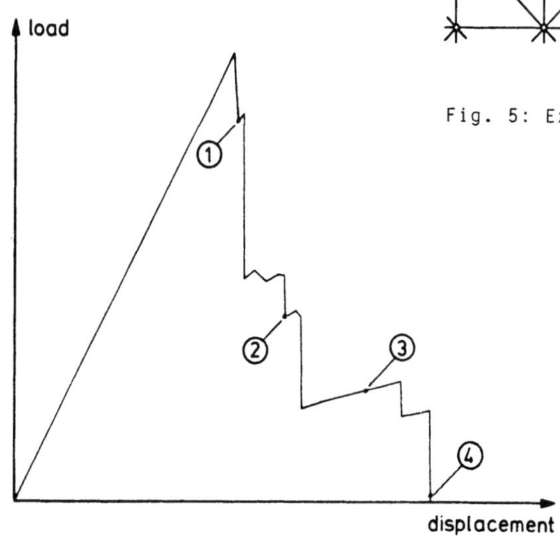


Fig. 6: Simulation of an uniaxial tension test at a plane framework

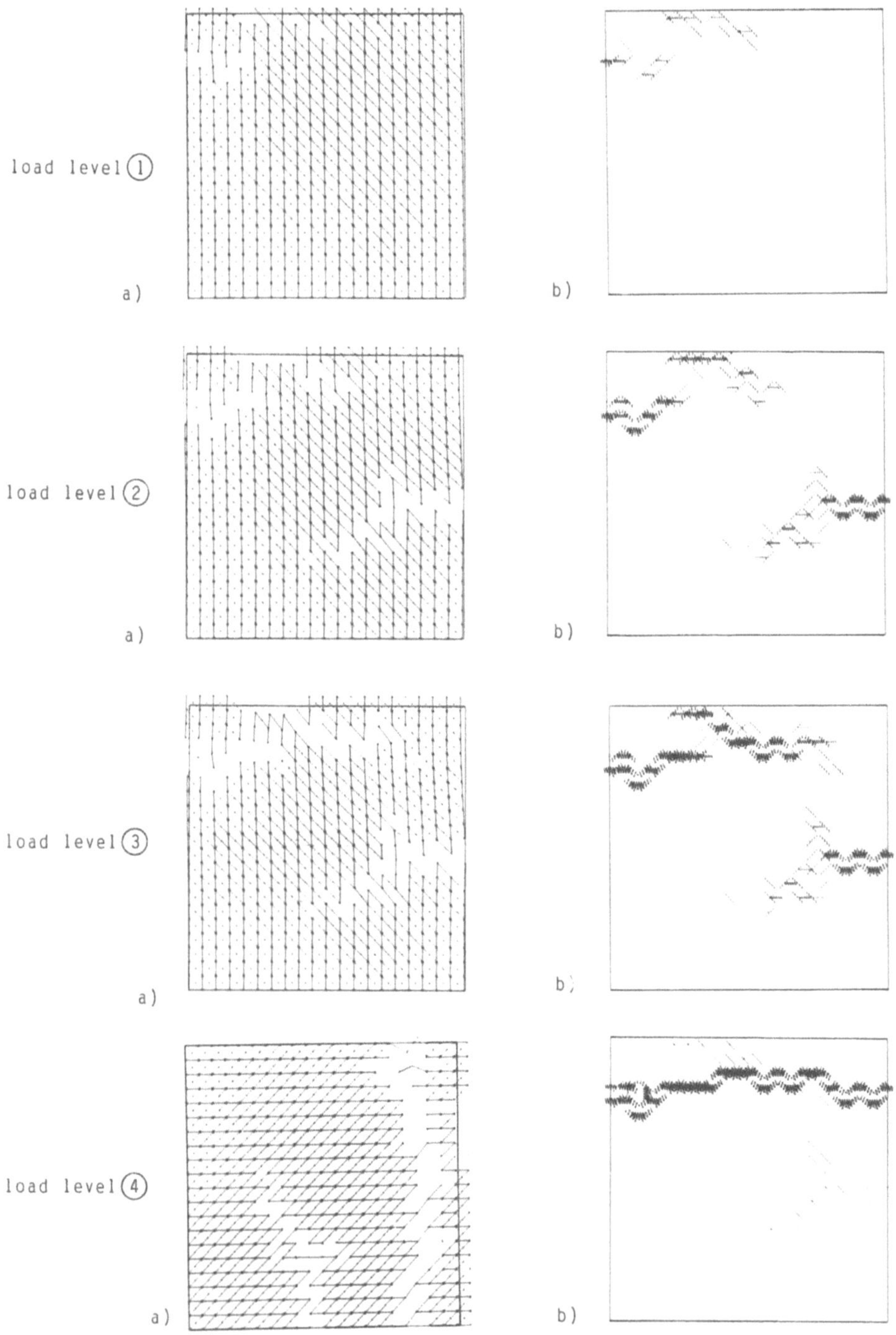


Fig. 6: continued

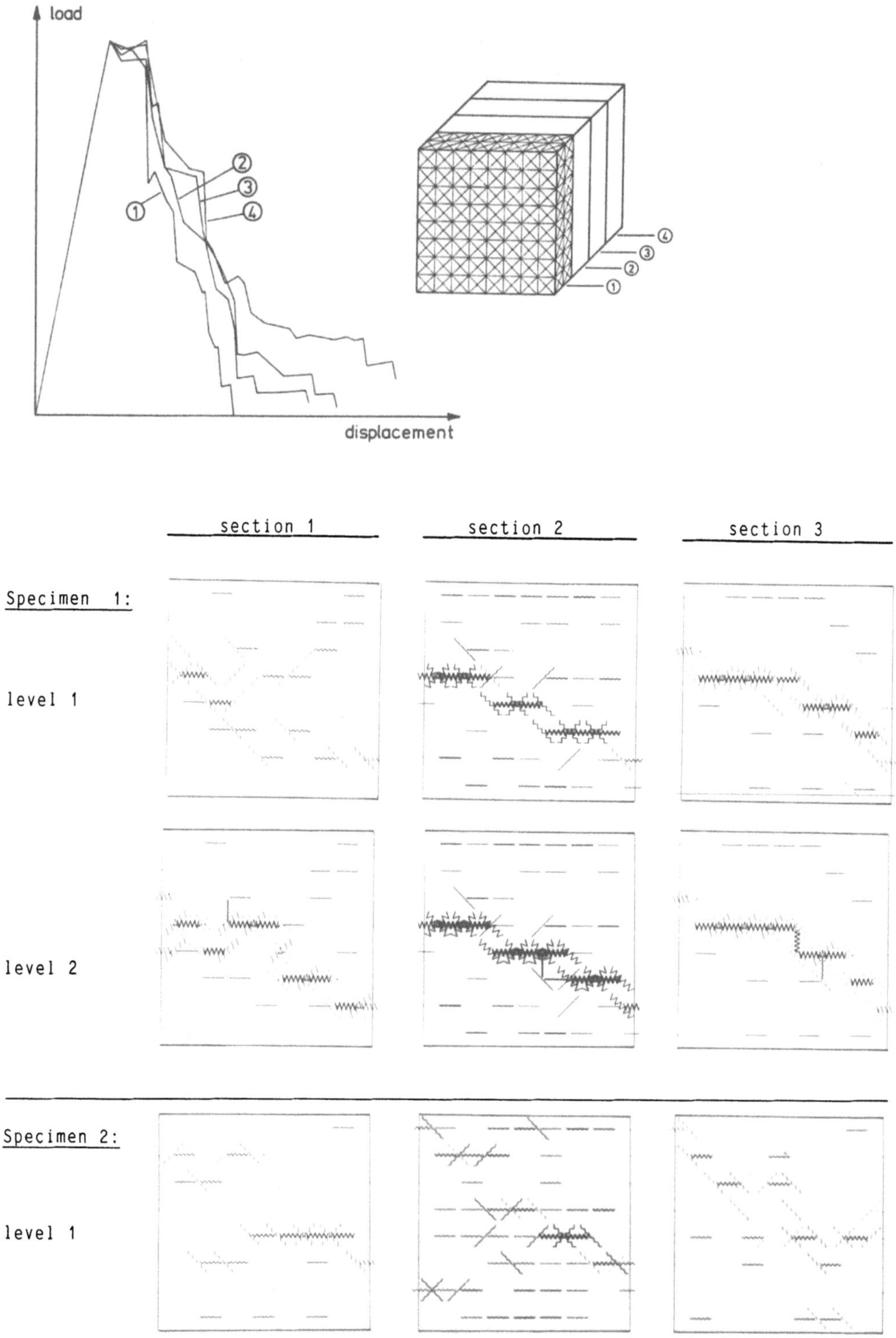


Fig. 7: 3D-simulation of uniaxial tension tests

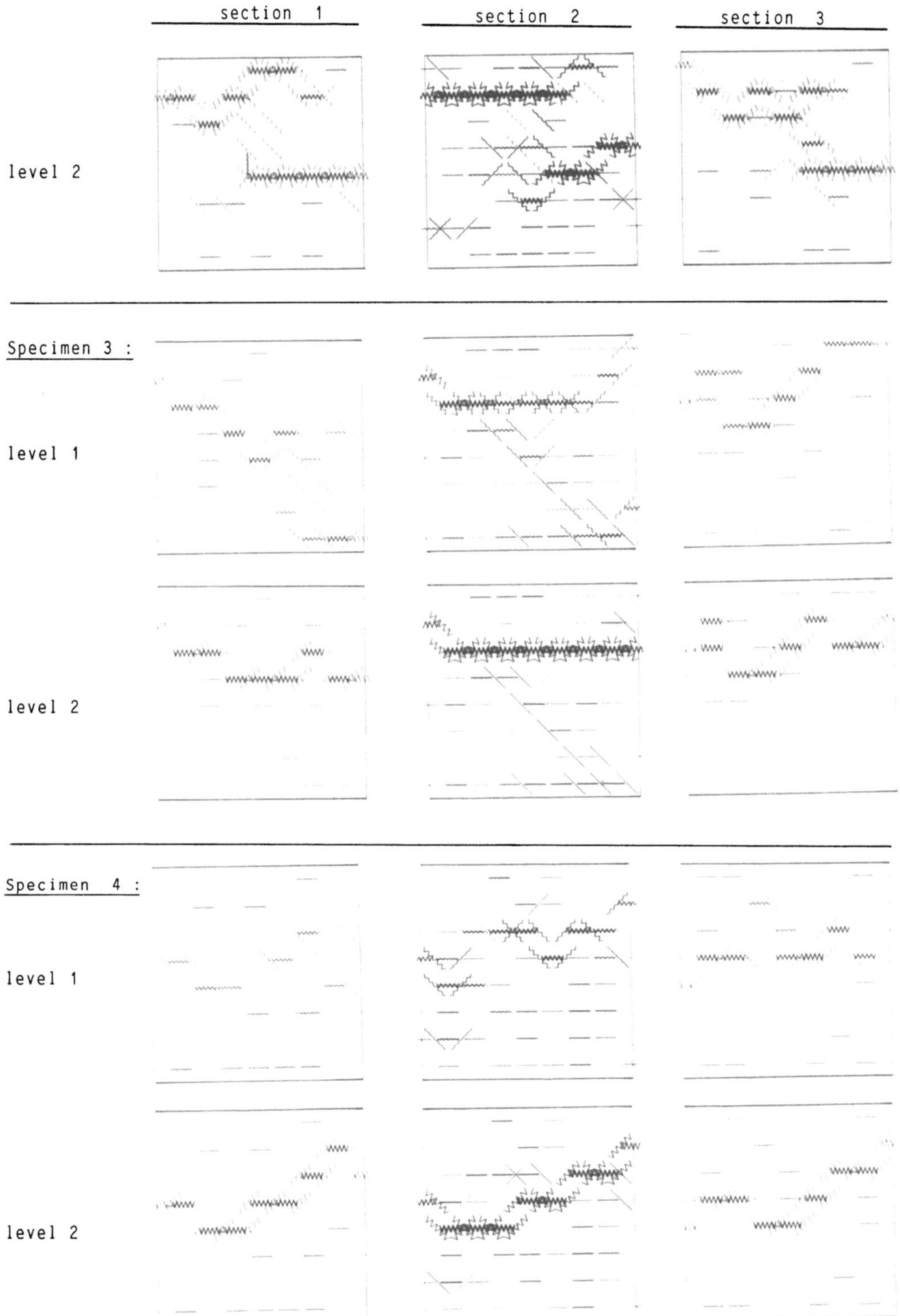


Fig. 7: continued

SEM/RILEM International Conference
on Fracture of Concrete and Rock
Houston, Texas, June 1987
Editors: S.P. Shah, S.E. Swartz



# FRACTURE ENERGY OF HETEROGENEOUS MATERIALS AND SIMILITUDE

Zdeněk P. Bažant*

## ABSTRACT

The paper reviews the size effect accompanying blunt fracture in brittle heterogeneous materials such as concrete and its applications to structural analysis and design. Various refinements of Bažant's size effect law are summarized, and recent results of fracture tests of Bazant and Pfeiffer which establish measurements of concrete fracture energy on the basis of size effect are given. These test results confirm that fracture energy may be uniquely defined by size extrapolation to infinity. Furthermore, changes in the size effect caused by temperature and humidity are considered, a new formula for the effect of temperature on fracture energy is derived on the basis of the activation energy concept, and some of the test results of Prat and Bazant which determine the temperature and humidity effect on fracture energy are given. Finally, similitude of brittle failures due to blunt fracture is analyzed in three dimensions and it is shown that the same form of the size effect law as for two dimensions is applicable.

## INTRODUCTION

The size effect, which consists in the fact that the nominal stress at failure of geometrically similar structures of different sizes is not constant but decreases with the size, is no doubt the most important aspect of fracture mechanics, setting it apart from stress-based failure theories such as plasticity. The present paper will review various recent advances in this subject and focus particular attention on the use of the size effect for determining the fracture energy, as well as the changes in the size effect brought about by changes of temperature, and their consequence for the dependence of fracture energy on temperature. A generalization of the size effect law to three dimensions will be also studied.

## SIZE EFFECT LAW AND ITS REFINEMENTS

The size effect may be isolated from other influences by considering geometrically similar structures of different sizes. For fractures blunted by distributed cracking, the simplest description of the size effect is provided by the approximate size effect law [1] (Fig. 1):

$$\sigma_N = Bf_t'\left(1 + \frac{d}{\lambda_0 d_a}\right)^{-1/2} \quad (1)$$

in which $\sigma_N$ is the nominal stress at failure (the failure load divided by the characteristic dimension and structure thickness), d is the characteristic dimension of the structure, $d_a$ is the maximum size of material inhomogeneities, e.g., the aggregate size in concrete; $f_t'$ = tensile stength (from direct tensile tests), and B, $\lambda_0$ = empirical constants. The size effect law has been shown to follow by dimensional analysis and similitude arguments from the following two simplifying hypotheses:

1. The energy release W due to failure is a function of the length a of the fracture.

2. At the same time, W is a function of the volume of the zone of cracking or, alternatively, of the size of the fracture process zone at fracture front.

The second hypothesis alone leads to plastic limit analysis, which exhibits no size effect, and the first hypothesis alone leads to classical linear elastic fracture mechanics, which exhibits the strongest possible size effect. We deal with the type of failure which requires a combination of both.

Eq. (1) represents the simplest possible formula for the transition between failures dominated by the strength limit and failures dominated by energy dissipation at fracture front and characterized by fracture

*Professor of Civil Engineering, Northwestern University, Evanston, Illinois 60201.

energy. It has been shown that Eq. (1) correctly describes the results of fracture tests of concrete within the scatter range of measurements [5], and that it can also be applied to the failure of various concrete structures which fail in a brittle manner. These include the diagonal shear failure of prestressed and unprestressed reinforced concrete beams without stirrups, the torsional failures of concrete beams, the beam and ring failures of unreinforced pipes, the punching shear failure of reinforced concrete slabs, etc. In making these applications, two further hypotheses are implied in the use of the size effect law in Eq. (1):

1. The shapes of the final fractures at failure in specimens of different sizes are geometrically similar; and

2. The failure does not occur at crack initiation.

The latter hypothesis is practically always satisfied since it is prohibited by codes to design concrete structures which would fail at the first crack initiation. The approximate applicability of the first hypothesis appears to be verified by the existing test data.

The scatter of existing measurements is not sufficiently small to make it possible to detect significant deviations from the size effect law in Eq. (1). However, finite element computations can be carried out to make comparisons with the size effect law [3,4]. One such calculation has recently been carried out by Hillerborg [4] using his fictitious crack model for a three-point bent fracture test. Considering specimen size range of the ratio 1:250 (which is much broader than one can realize in practice, due to cost limitations), he detected appreciable deviations from Eq. 1; see the results in Table 1.

TABLE 1. Comparison of size effect law (with r=0.44, B=306, $\lambda_0$=0.608) and Hillerborg's finite element results with fictitious crack model

| $df_t'^2/EG_f$ | $\sigma_N/f_t'$ | |
|---|---|---|
| | Finite Elements (Hillerborg) | Size Effect Law |
| 0.02 | 2.43 | 2.44 |
| 0.05 | 2.22 | 2.21 |
| 0.1 | 2.01 | 2.00 |
| 0.2 | 1.77 | 1.78 |
| 0.5 | 1.46 | 1.46 |
| 1 | 1.222 | 1.222 |
| 2 | 0.992 | 0.995 |
| 5 | 0.725 | 0.739 |

It has been shown, however, [3], that a refinement of the size effect law is possible such that it closely agrees with Hillerborg's finite element results; see the last column in Table 1 which is so close to Hillerborg's results that a graphical distinction is hardly possible. These results are based on the following generalized size effect law [2,20]:

$$\sigma_N = Bf_t^* \left[1 + \left(\frac{d}{\lambda_o d_a}\right)^r\right]^{-1/2r} \tag{2}$$

which itself is a special case of the general asymptotic series expansion [2]:

$$\sigma_N = Bf_t^* (c_0\xi^{-1} + 1 + c_1\xi + c_2\xi^2 + c_3\xi^3 + \dots)^{-1/2r}, \quad \xi = \left(\frac{d_a}{d}\right)^r \tag{3}$$

in which $f_t^* = f_t'$ if the aggregate size $d_a$ is the same for all specimens, and B, $\lambda_0$, r are empirical parameters, and so are the coefficients $c_0, c_1, c_2, \dots$ If the aggregate size $d_a$ is varied, then theoretical analysis leads to the formula [5]:

$$f_t^* = f_t' \left[1 + \left(\frac{c_0}{d_a}\right)^{1/2}\right] \tag{4}$$

which is similar to the Petch formula [6,7,8] for the effect of grain size on the yield limit of polycrystalline metals.

It has been established [8] that there is a one-two-one relationship between the size effect law and the shape of the softening portion of the stress-displacement diagram used in Hillerborg's type models. When one of these relations is known, the other one can be determined. The same is true for the crack band model, in which the front of the strain-softening damage is assumed to have a certain constant width which is a material property; to each shape of the strain-softening stress-strain diagram there corresponds a certain size effect law and vice versa. No doubt this is also true of the damage laws. Furthermore, if the front of the band of strain-softening damage is variable, this has a direct effect on the size effect law, and from size effect observations it is possible to make inferences on the size of the strain-softening zone.

A similar one-to-one relationship was previously established between the size effect law and the R-curves from blunt fracture tests [8].

The size effect is of interest not only with regard to fracture testing and design of structures. The size effect is equaly valuable for checking the soundness of finite element models. At present, models of cracking which are formulated strictly on the basis of stress-strain relations and pay no attention to strain-localization instabilities and energy aspects of failure are still in prevalent use. All these finite element codes predict the nominal stress at failure for structures of different sizes which are geometrically similar and are analyzed with geometrically similar meshes to be the same. Experimental evidence clearly indicates that for brittle failures such predictions are incorrect. This may be one reason that despite two decades of effort, the existing finite element codes still cannot reliably predict brittle failures of concrete structures, except perhaps when the parameters of the model are calibrated for one structure size and predictions are made for roughly the same size.

A check for the size effect, and comparisons with the theoretically derived size effect law or experimental evidence, if available, should be an integral part of evaluation of the applicability of every finite element code to brittle failures due to cracking of concrete as well as rock.

## DEFINITION AND DETERMINATION OF FRACTURE ENERGY ON THE BASIS OF SIZE EFFECT

The size effect observed on geometrically similar specimens appears as the best means for identifying the material properties which govern fracture. The most important among these properties is the fracture energy. The fracture energy of materials such as concrete has proven difficult to determine as well as define. Various testing methods currently in use yield results which may differ by several hundred percent, and aside from that, none of the existing definitions of the fracture energy appears to yield unique results.

Based on the size effect, it now appears that a unique definition of fracture energy can be provided as follows [2,9]:

The fracture energy $G_f$ of a heterogeneous brittle material is the specific energy required for fracture propagation in a geometrically similar specimen of infinite size.

It has been shown [2,8] that this definition leads to the formula:

$$G_f = \frac{g_F}{A\,E_c} f_t'^2\, d_a = \frac{g_F}{E_c} B^2 \lambda_0 f_t'^2\, d_a \tag{5}$$

in which $B$, $\lambda_0$, $d_a$ are the parameters of the size effect law, $E_c$ is the elastic modulus of the material, $A$ = slope of the regression line in Fig. 1, and $g_F$ is the nondimensional energy release rate for a sharp fracture, calculated according to linear elastic fracture mechanics. It can be theoretically shown that the fracture energy for specimen size extrapolated to infinity must be the same for all specimen shapes [9]. Indeed, for an infinitely large specimen, the relative size of the strain-softening damage zone is infinitely small, and the zone is surrounded by the asymptotic elastic field known from linear elastic fracture mechanics, which is the same regardless of the structure geometry. Therefore, at extrapolation to infinity, the detailed picture of the fracture process zone must be the same for all structure geometries.

This theoretical conclusion has been verified experimentally [9]. Specimens made of the same concrete were cast in different sizes, and different types of notches as well as different types of loading were used. The test series included three point bent specimens, centrically tensioned edge-notched specimens, and eccentrically compressed specimens (Fig. 1). These shapes include just about the extreme of the range of conditions to which the ligament cross section may be exposed: bending moment over the ligament, tensile force over the ligament, and a combination of bending moment and force over the ligament.

The results of these tests conducted at Northwestern University are exhibited in Fig. 1 in terms of the plots of $\sigma_N^{-2}$ versus $d/d_a$. According to the size effect law, these plots should ideally be straight lines, which makes it possible to use linear regression for the determination of the parameters of the size effect law. The slope of the regression line is proportional to the inverse of the fracture energy, in view of Eq. (5). Based on slopes of the regression lines of the test results in Fig. 1, it is found that the fracture energy for the three types of specimens are about the same, and do not deviate from each other more than is inevitable for a heterogeneous material such as concrete (the deviations from the mean are within ± 3%); see Table 2.

TABLE 2. Fracture Energy Values Obtained from Measurements Evaluated by Size Effect Law (lb./in.)

| Specimen Type | Concrete | Mortar |
|---|---|---|
| 1) Three-Point Bent | 0.229 | 0.129 |
| 2) Edge-Notched Tension | 0.210 | 0.118 |
| 3) Eccentric Compression | 0.233 | 0.132 |

Noting at the same time that, by definition, this method of determining fracture energy is independent of the specimen size, it appears that the determination of fracture energy on the basis of the size effect indeed yields unique results. This cannot be said of other existing methods.

The size effect law in Eq. (1) is nevertheless an approximation, and deviations occur when a broad range of sizes is considered or accurate results are desired. Important work in this direction has been recently carried out by Planas and Elices at Technical University in Madrid (private communication, 1986). They considered various strain-softening formulations and calculated the corresponding size effect curves for certain types of specimens using the Green's function approach. By matching such curves to test results it should be possible, in principle, to gain further information on the material parameters that govern strain-softening.

SIZE EFFECT IN THREE DIMENSIONS

The size effect due to blunt fracture has so far been treated in a two-dimensional context, assuming the state of plane stress or plane strain and uniform conditions throughout the thickness. In two experimental studies [9,10], however, the size effect law (Eq. 1) was used for brittle three-dimensional failures -- the punching shear failure of circular reinforced concrete slabs and the torsional failure of unreinforced or longitudinally reinforced concrete beams. In the course of these studies it was experimentally verified that Eq. 1 can be extended to three-dimensions, and the theoretical proof will be presented now.

The two fundamental similifying hypotheses listed under Eq. 1 remain applicable in three-dimensions. Similar to the previous derivation [1,2], these fundamental hypotheses mean that the failure is describable in terms of two parameters, a and $k_0 a^2 n d_a$, the latter one representing the volume of the cracking zone (Fig. 2). Coefficient $k_0$ characterizes the geometry of the failure zone and is constant if geometrically similar bodies with geometrically similar failure modes are considered; $d_a$ = maximum aggregate size (or more generally inhomogeneity size), and n = empirical coefficient, for concrete typically about $n \approx 3$, such that $nd_a$ represents the effective width of the front of cracking or alternatively the length of the fracture process zone (these two meanings were shown to lead to the same result; see Ref. 21). The foregoing two parameters, however, are not nondimensional. It must, of course, be possible to describe the failure in terms of nondimensional parameters, and for this purpose length a must be divided by a quantity of the dimension of length, and volume $k_0 a\, nd_a$ by a quantity of the dimension of volume. We choose to divide these two parameters by d and $k_0 a^2 d$, respectively ( other choices are equally possible and yield the same result, but the derivation is the simplest for this choice). Thus, the nondimensional parameters for our problems are:

$$\theta_1 = \frac{a}{d}, \quad \theta_2 = \frac{k_0\, a^2\, nd_a}{k_0\, a^2\, d} = \frac{nd_a}{d} \tag{6}$$

Parameter $\theta_1$ represents the relative size of the fracture, and parameter $\theta_2$ the relative volume of the cracking zone. Fig. 2a shows an example of three-dimensional fracture which is axisymmetric, but the foregoing description holds for arbitrary three-dimensional geometries provided there is geometrical similarity between various sizes; see Fig. 2b.

We now consider geometrically similar specimens or structures of different sizes d (Fig. 2b), characterized by a constant ratio a/d. Thus, parameter $\theta_1$ is size-independent, while parameter $\theta_2$ characterizes the size. As in the previous studies [1,2], the total release of potential energy W from the structure may always be given in the form:

$$W = \left[\frac{1}{2E_c}\left(\frac{P}{d^2}\right)^2\right] d^3\, F(\theta_1, \theta_2) = \frac{P^2}{2E_c}\,\frac{F(\theta_1, \theta_2)}{d} \tag{7}$$

in which $E_c$ is the Young's elastic modulus of concrete, P is the maximum load (i.e., the failure load under load-controlled conditions), and F is a certain function of the nondimensional parameters, characteristic of the given geometry of the structure and the cracking zone. The essential point is that function F must be the same for different sizes if the structures or specimens are geometrically similar. Energy balance during failure requires that:

$$\left[\frac{\partial W}{\partial a}\right]_{d\, =\, \text{const.}} = kd\, G_f \tag{8}$$

in which kd represents the length of the perimeter of the fracture front, k being an empirical coefficient independent of size d, and $G_f$ is the fracture energy of the material. Differentiating Eq. 7 with respect to a and substituting into Eq. 8, we have:

$$\frac{P^2}{2E_c d}\,\frac{\partial F}{\partial \theta_1}\,\frac{1}{d} = k\, d\, G_f \tag{9}$$

The nominal stress at failure may be defined in three-dimensions as $\sigma_N = P/d^2$, and so we get:

$$\sigma_N = \frac{P}{d^2} = \left(\frac{2k\ G_b\ E_c}{(\partial F/\partial \theta_1)\ d}\right)^{1/2} \tag{10}$$

The state $\theta_2 = 0$, which corresponds to an infinitely large structure ($d/d_a \to \infty$) may now be chosen as the reference state, and $\partial F/\partial \theta_1$ may be expanded as a function of $\theta_2$ in Taylor series:

$$\frac{\partial F(\theta_1, \theta_2)}{\partial \theta_1} = f_0(\theta_1) + f_1(\theta_1)\ \theta_2 + f_2(\theta_1)\ \theta_2^2 + f_3(\theta_1)\ \theta_2^3 + \ldots \tag{11}$$

in which we introduce the notations:

$$f_1(\theta_1) = \frac{1}{1!}\left[\frac{\partial F}{\partial \theta_2}\right]_{\theta_2 = 0}, \quad f_2(\theta_1) = \frac{1}{2!}\left[\frac{\partial^2 F}{\partial \theta_2^2}\right]_{\theta_2 = 0}, \quad f_3(\theta_1) = \frac{1}{3!}\left[\frac{\partial^3 F}{\partial \theta_2^3}\right]_{\theta_2 = 0}, \tag{12}$$

Substitution of Eq. 11 into Eq. 10 then yields for the size effect the asymptotic expansion:

$$\sigma_N = Bf_t'\left[1 + \frac{d}{\lambda_0 d_a} + \lambda_1 \frac{d_a}{d} + \lambda_2 \left(\frac{d_a}{d}\right)^2 + \lambda_3 \left(\frac{d_a}{d}\right)^3 + \ldots\right]^{-1/2} \tag{13}$$

in which we introduced the following constants:

$$B = \frac{2k}{f_1\ nd_a} \frac{(G_f\ E_c)^{1/2}}{f_t'}, \quad \lambda_0 = \frac{f_1}{f_2}, \quad \lambda_1 = \frac{f_2}{f_1}, \quad \lambda_3 = \frac{f_3}{f_1}, \ldots \tag{14}$$

and $f_t'$ denotes the direct tensile strength of the material. The constants in Eq. 14 depend on the geometry of the structure and of the cracking zone.

If we adopt the viewpoint of a cracking band of effective width $nd_a$ at the cracking front, we may express the fracture energy as:

$$G_f = nd_a \frac{f_t'^2}{2} \left(\frac{1}{E_c} - \frac{1}{E_t}\right) \tag{15}$$

in which $E_t$ denotes the mean softening modulus, i.e., the mean slope of the descending branch of the stress-strain diagram. Then the expression for B becomes

$$B = \left[\frac{k}{f_1}\left(1 - \frac{E_c}{E_t}\right)\right]^{1/2} \tag{15a}$$

Eq. 13, which is of the same form as derived before for two dimensions (Ref. 2, Eq. 39), describes the most general possible size effect. The first-degree approximation is obtained by deleting the terms with coefficients $\lambda_1, \lambda_2, \ldots$, and this yields Eq. 1 stated at the outset and originally derived for two-dimensions.

It may be noted that the function which is expanded into the asymptotic series in Eq. 11 could be chosen as $(\partial F/\partial \theta_1)^r$ instead of $\partial F/\partial \theta_1$. In that case, one obtains in the same manner a still more general expression for the size effect:

$$\delta_N = Bf_t'(c_0 \xi^{-1} + 1 + c_2 \xi + c_3 \xi^2 + c_4 \xi^3 + \ldots)^{-1/2r}, \quad \xi = (d/d_a)^r \tag{16}$$

in which $c_0, c_2, c_3 \ldots$ are certain constants. This expansion makes it possible to describe the size effect over a broader range with fewer terms, as illustrated in the example of Table 1. However, it is not clear whether Eq. 16 for $r = 1$ yields the correct initial asymptotic behavior for a very small size ($d \to 0$).

The idea that the approximate size effect law should be of general applicability was initially conceived [1] upon noting that calculations for various specific geometries yield the same result. Let us now carry out a similar calculation, choosing as an example the axisymmetric specimen in Fig. 2a with a cylindrical cracking zone 123456 of diameter a (Fig. 2a). Before cracking, the specimen is under uniform uniaxial stress $\sigma_N$, with resultant $P = \sigma_N d^2/4$ where d is chosen to represent the diameter of the specimen. The energy release may be considered to approximately equal the strain energy initially contained within the cracking layer (cylinder 123456) and in cones 1237 and 4568 (Fig. 2a) adjacent to this layer. The height of these cones is considered as $k_1 a$, $k_1$ being an empirical constant, imagined to represent the slope of the "stress diffusion" lines, represented here by the slope of the cone. The stress diffusion concept is of course approximate, however, it yields dimensionally correct results, and this is all that matters for our purpose. The total energy release may thus be expressed as:

$$W = \left(2 \frac{k_1 a}{3} \frac{\pi a^2}{4} + n d_a \frac{\pi a^2}{4}\right) \frac{\sigma_N^2}{2E_c} \tag{17}$$

Eq. 8 for energy balance now takes the form:

$$\partial W/\partial a = \pi \, a \, G_f \tag{18}$$

Substituting for $G_f$ Eq. 15, evaluating the derivative of Eq. 17 with respect to a, and solving for $\sigma_N$, we now get again Eq. 1 in which:

$$B = \sqrt{2}\left(1 - \frac{E}{E_t}\right)^{1/2}, \quad \lambda_0 = \frac{n \, d_a}{k_1} \frac{d}{a} \tag{19}$$

So the result agrees with the first order approximation according to Eq. 13. A similar agreement can be demonstrated for various other simple geometries such as edge-cracked cylinder, or square prism with a square crack layer, etc.

## SIMILITUDE OF FRACTURE RATES AND DEPENDENCE OF FRACTURE ENERGY ON TEMPERATURE

It is generally assumed that the fracture energy of concrete depends on temperature, but no formula is apparently available. While heating causes in metals abrupt increases of fracture toughness, due to brittle-ductile transitions which change the fracture mechanism, the fracture toughness of purely brittle materials, such as glass, ceramics, graphite and rocks is known to smoothly decrease with increasing temperature [11,12,13]. We will now show the derivation of a simple formula, exploiting the similitude of fracture rates at various temperatures. This is, of course, a completely different type of similitude than that which governs the size effect.

It is generally accepted that fracture is a thermally activated rate process. This means that the atomic bond ruptures which constitute the mechanism of fracture are provoked by the energies of thermal vibrations [14]. These energies are statistically distributed, which is known to be described by the Maxwell distribution, and a rise of temperature causes an increase of the probability (or frequency) that the atom's energy would exceed the activation energy barrier of the bond. Therefore, a rise of temperature produces an increase in the growth rate of fracture, which generally follows a formula of the type [15]:

$$\dot{a} = \phi(K) \, e^{-U/RT} \tag{20}$$

in which $\dot{a}$ = rate of growth of fracture length, U = activation energy of bond rupture, R = universal gas constant, T = absolute temperature, K = stress intensity factor, and $\phi(K)$ = an empirical function, increasing monotonically.

Recently, Evans [16,17] and Thouless et al. [18] proposed and verified for ceramics a special form of Eq. 20:

$$\dot{a} = v_c \left(\frac{K}{K_c}\right)^n e^{-U/RT} \tag{21}$$

where $K_c$ = critical value of fracture toughness, and $v_c$, n = empirical constants characterizing the given material.

A formula of this type may be expected to apply also for concrete. Eq. 21 is not exact but only approximate, for two reasons: 1) the proportionality of a to $K^n$ is only empirical, and 2) more than one mechanism of atomic bond rupture, with different activation energies, might be involved, and the type of fracture mechanism might change with temperature.

Because of the well-known relations $K = (GE_c)^{1/2}$ and $K_c = (G_f^0 E_c)^{1/2}$, where G = rate of energy release from the structure into the fracture process zone [19,20], Eq. 21 may be rewritten as:

$$\dot{a} = v_c\left(\frac{G}{G_f^0}\right)^{n/2} \exp - \frac{U}{R}\left(\frac{1}{T} - \frac{1}{T_0}\right) \tag{22}$$

Eq. 22 may serve as the basic relation for determining the crack growth in time. Although the time-dependent fracture description in terms of the crack growth rate is no doubt physically more fundamental, the time-independent fracture description prevails in applications. In fact, what is known as fracture mechanics is a time-independent theory. So we need to deduce the consequences of Eq. 23 for time-independent fracture description, using a comparison of the fracture growth rate as the basis for similitude at various temperatures.

The choice of the reference temperature in Eq. 22 is of course arbitrary. If we choose temperature T as the reference temperature, then according to Eq. 22, the crack growth rate at temperature T is simply expressed as:

$$\dot{a} = v_c(G/G_f)^{n/2} \tag{23}$$

because $1/T_0 - 1/T = 0$ in this case.

As a basic condition of similitude of fracture at different temperatues, we must now require that the crack growth rate at temperature T must be the same whether expressed on the basis of reference temperature $T_0$, or reference temperature T. Thus, equating the expressions in Eqs. 23 and 22, we obtain the following approximate formula:

$$G_f = G_f^0 \exp\left(\frac{\gamma}{T} - \frac{\gamma}{T_0}\right) \tag{24}$$

in which

$$\gamma = 2U/nR \tag{25}$$

$\gamma$ is a constant characterizing the given material.

Eq. 24 may be simply transformed to a linear plot of $\ell n\, G_f$ versus $T^{-1}$, and linear regression in this plot then yields $G_f^0$ as well as $\gamma$. However, the activation energy values U cannot be determined unless constant n is obtained separately.

Extensive experiments aimed at determining the effect of temperature as well as humidity on the fracture energy of concrete have recently been conducted at Northwestern University by Prat [19]. Fracture energies were obtained for different temperatures according to the size effect (Eq. 5), and it was checked that different types of specimens yield at various temperatures about the same fracture energy values. The dependence of the fracture energy on temperature was found to closely agree with Eq. 24 for both dried concrete and concrete with the saturation water content. The test results obtained by Prat [19] are exhibited in Fig. 3; for a detailed description of the tests and their analysis, see Ref. 19.

## CONCLUSION

The size effect, the salient property of fracture mechanics, represents for brittle heterogeneous material such as concrete a smooth transition between failures governed by the strength or yield limit and failures governed purely by fracture energy. This transition is describable by a relatively simple size effect law, which has proven useful as an improvement for various design formulas for concrete structures (diagonal shear in beams, punching shear in slabs, torsional failure of beams, ring and beam failures of pipes, etc.) as well as for the determination of nonlinear fracture parameters of the material, especially the fracture energy. The simple formula represented Bazant's size effect law is also applicable in three dimensions provided that structures that are geometrically similar in three dimensions are considered. A different type of similitude, based on the notion of equivalent crack growth rates, may be exploited to determine the dependence of fracture energy on temperature in the context of time-independent fracture mechanics theory.

ACKNOWLEDGMENT. - Partial financial support under U. S. Air Force Office of Scientific Research Contract No. F49620-87-C-0030DEF with Northwestern University, monitored by Dr. Spencer T. Wu, is gratefully acknowledged. Furthermore, thanks for valuable assistance in the study of the temperature effect are due to graduate research assistsant P. C. Prat, and for stimulating discussions on the three-dimensional size effect to doctoral student M. Kazemi.

## REFERENCES


1. Bažant, Z. P., "Size Effect in Blunt Fracture: Concrete, Rock, Metal," J. Engrg. Mech., ASCE Vol. 110, 1984, 518-535.

2. Bažant, Z. P., "Fracture Mechanics and Strain-Softening of Concrete," in Proceedings, U. S. - Japan Seminar on Finite Element Analysis of Reinforced Concrete Structure, held in Tokyo in May 1985, publ. by ASCE, New York, 1986, pp. 121-150.

3. Bažant, Z. P., "Comment on Hillerborg's Comparison of Size Effect Law with Fictitions Crack Model," in Dei Poli Anniversary Volume, L. Cedolin, Ed., Politecnico di Milano, Milano, 1985, pp. 335-338.

4. Hillerborg, A., "A Comparison Between the Size Effect Law and the Fictitious Crack Model", ibid., pp. 331-334.

5. Bažant, Z. P., and Kim, J. K., Discussion Closure to "Size Effect in Shear Failure of Longitudinally Reinforced Beams," Am. Concrete Institute Journal, Vol. 81, July-August 1985, pp. 580-538.

6. Petch, N. J., "The Cleavage Strength of Polycrystals," Journal, Iron and Steel Institute, Vol. 173, 1953, pp. 25-28.

7. Tetelman, A. S., and McEvily, A. J., "Fracture of Structural Materials," J. Wiley, New York, 1967, p. 186.

8. Cottrell, A. H. "The Mechanical Properties of Matter," J. Wiley, New York, 1964, p. 282.

9. Bažant, Z. P., and Cao, Z., "Size Effect in Punching Shear Failure of Slabs," ACI Materials Journal, Vol. 84, No. 1, Jan.-Feb. 1987.

10. Bažant, Z. P., Sener, S., and Prat, P. C., "Size Effect Tests of Torsional Failure of Concrete Beams," Report No. 86-12/428s, Center for Concrete and Geomaterials, Northwestern University, Evanston, Illinois, Dec. 1986.

11. Tetelman, A. S., and McEvily, A. J., "Fracture of Structural Materials," John Wiley & Sons, New York, 1967, pp. 274-276.

12. Kanninen, M. F., and Popelar, C. H., "Advanced Fracture Mechanics," Oxford University Press, New York, 1985, pp. 17-18, 26-30.

13. Cherepanov, G. P., "Mechanics of Brittle Fracture," MacGraw Hill, 1979, pp. 174, 455 (translated from Russian original published by Nauka, Moscow, 1974).

14. Cottrell, A. H., "The Mechanical Properties of Matter," John Wiley & Sons, New York, 1964.

15. Evans, A. G., and Fu, Y., "The Mechanical Behavior of Alumina," in Fracture in Engineering Materials, ed. by A. G. Evans, Noyes Publications, Park Ridge, N. J., 1984, p. 71.

16. Thouless, M. D., Hsueh, C. H., and Evans, A. G., "A Damage Model of Creep Crack Growth in Polycrystals," Acta Metalurgica, Vol. 31, No. 10, 1983, pp. 1675-1687.

17. Knott, J. F., Fundamentals of Fracture Mechanics, Butterworths, London, England, 1973.

18. Broek, D., "Elementary Engineering Fracture Mechanics," Noordoff International Publishing, Leyden, Netherlands, 1974.

19. Bažant, Z. P., and Prat, P. C., "Effects of Temperature and Humidity on Fracture Energy of Concrete," Report No. 86-12/692e, Center for Concrete and Geomaterials, Northwestern University, Evanston, Illinois, Dec. 1986.

20. Bažant, Z. P., "Mechanics of Distributed Cracking," ASME Applied Mechanics Reviews, Vol. 39, No. 5, May 1986, pp. 675-705.

21. Bažant, Z. P., "Crack Band Model for Fracture of Geomaterials," Proc. 4th Intern. Conf. on Numer. Meth. in Geomechanics, held in Edmonton, Alberta, ed. by Z. Eisenstein, University of Alberta, 1982, Vol. 3, pp. 1137-1152.

22. Bazant, Pfeiffer (1987), "Determination of Fracture Energy from Size Effect and Brittleness Number" in A.C.I. Materials Journal, 86, pp. 463-480.

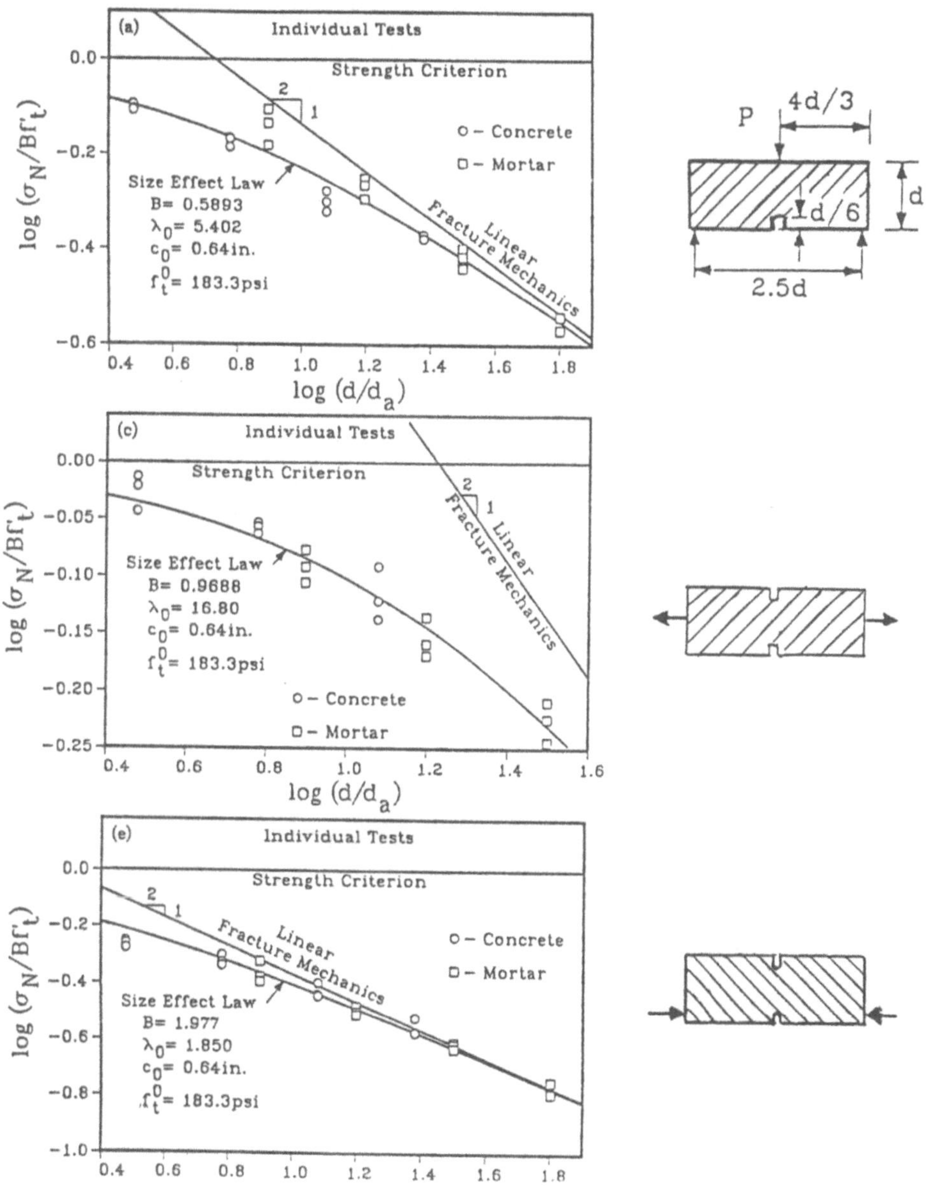


Fig. 1. Nominal stresses at failure vs. specimen size as measured by Bazant and Pfeiffer (1986), and fits by size effect law.

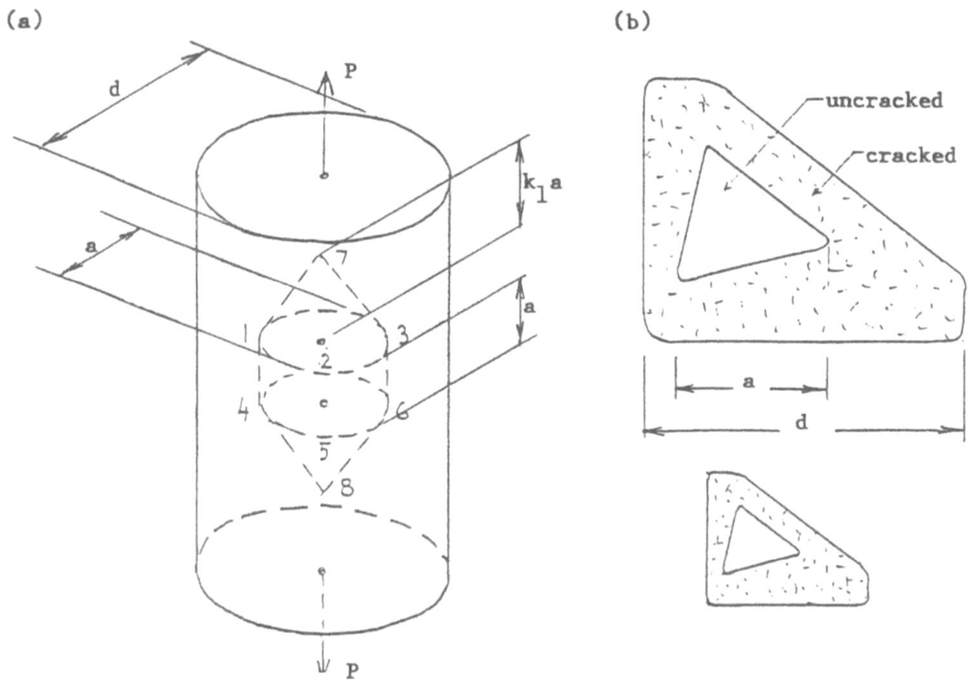


Fig. 2. Three-Dimensional Fracture and its Geometrical Similarity

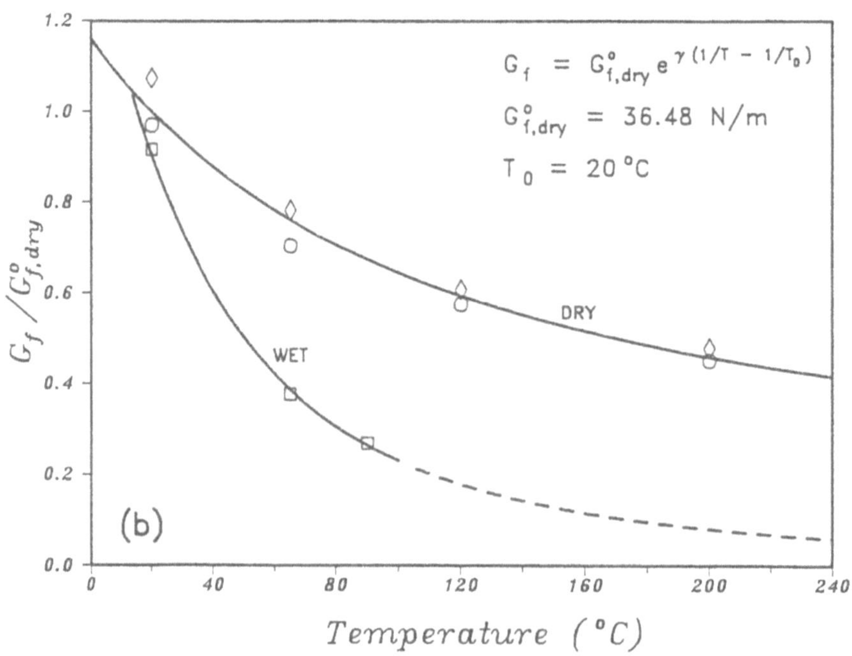


Fig. 3. Test Results of Prat and Bazant (1986) on Temperature Dependence of Fracture Energy of Dry and Wet Concrete

## Appendix I. - Deviations from Size Effect Law and Other Size Effects

### Transition to Yielding

The similitude of brittle fracture as described by the size effect law may be either limited or entirely overridden by plastic energy dissipation which can occur in plain concrete if there is high confining pressure. E.g., consider the split-cylinder brazilian test on a cylinder of diameter d and thickness (length) b. The formation of the splitting fracture is a brittle event, and so the corresponding load $P_f$ and nominal stress $\sigma_N = \sigma_N^f = P_f/bd$ approximately follows the size effect law. The splitting fracture, however, does not represent a complete failure because a wedge-shaped zone under the load (Fig. 4) must subsequently also slip in order to obtain complete failure. Let the limit load for shearing the wedge off be $P_y$. If the cylinder is not thin but long, the slip failure of the wedge is essentially plastic since it occurs at high confining pressure (this would not be true for a thin cylinder in which the normal stress in the direction of cylinder axis is small). So the normal stress $\sigma_N^y = P_y/bd$ must be essentially size-independent.

For small enough sizes d we have $P_b > P_y$. Then the maximum load is governed by the size effect law and the plastic wedge shear-off occurs later under a smaller load. However, for a certain sufficiently large size $d_{max}$, $\sigma_N$ becomes smaller than $\sigma_y$, and then the load must further increase after the formation of the splitting fracture until it reaches maximum $P_y$ when the wedge shears off plastically. For this type of behavior the size effect plot has the shape shown in Fig. 4b, given as

$$\sigma_N = \sigma_N^f = Bf_t' \left(1 \quad \frac{d}{\lambda_0 d_a}\right)^{-1/2} \text{ if } \sigma_N > \sigma_N^y \tag{26}$$

Otherwise $\sigma_N = \sigma_N^y$

The size effect in the brasilian test was recently measured by T. Shioya and T. Hasegawa at Shimizu Institute of Technology, Tokyo (private communication in May 1985 and Jan. 1987). Their results appear to agree with the behavior shown in Fig. 4b, although there is inevitably considerable random scatter in this kind of test. It may be noted that if the specimens for all the sizes were thin and had the same thickness, sufficient confining pressure could not develop to permit plastic failure of the wedge, and then $\sigma_N^y$ would be negligibly small. It should also be noted that when the large specimens are much thicker than the small ones, they can heat up significantly due to hydration, and this effect as well as the effect of heating on moisture loss can then superimpose a different size effect.

A different type of yielding effect may apparently be obtained when the yielding occurs simultaneously with the propagation of the cracking zone (fracture) at maximum load. This is apparently the case for the diagonal shear ailure of beams with stirrups. As shown in Ref. 1, the size effect for this type of yielding effect should follow the equation:

$$\sigma_N = Bf_t' \left(1 + \frac{d}{\lambda_0 d_a}\right)^{1/2} + \sigma_y \tag{27}$$

as one can show by analyzing the energy balance at failure.

### Other Types of Size Effect

Three types of size effects can be distinguished in concrete structures:

1. Fracture mechanics size effect, due to fracture propagation and energy release.

2. Diffusion-type size effect due to heat conduction and water diffusion.

3. Statistical size effect due to randomness of material properties.

The size effects due to heat condition and water diffusion play a major role in creep and shrinkage but they can also interfere with the size effect law for blunt fracture. Particularly when specimens that are geometrically similar in three rather than two dimensions are compared, the large specimens may heat up significantly due to hydration, which accelerates the hardening of concrete. This then leads to higher $f_t'$ as well as $G_f$, thus altering the parameters of Eq. 1. At the same time, the heating may produce thermal stresses which cause cracking or at least tensile strain-softening. This may then reduce the effective strength and fracture energy for externally applied load. Drying arrests hydration, and, therefore, a smaller specimen has a smaller gain in strength and in fracture energy than a larger specimen over the same period of time. But in a larger specimen, nonuniform drying may cause more cracking or tensile strain-softening, which could cause an opposite tendency. The diffusion type size effects may be described by simple formulas only in some simple situations (e.g., in the BP Model for shrinkage and drying creep). Generally they defy description by a simple law and call for complete stress and cracking analysis.

## Statistical Size Effect

The statistical size effect in failure is described by Weibull theory. This theory is applicable to the failure of a chain of brittle elements. It predicts the strength of the chain to decrease as the length of the chain is increased, because the chance of encountering a lower strength in a longer chain increases as the length of the chain increases. Thus, e.g., Weibull theory no doubt applies to the length effect in tensile failure of a long and thin uniaxially stressed concrete bar without notches.

However, for failures that occur after a large crack has already developed, which is the majority of failures in concrete structures, the Weibull-type statistical effect plays probably an insignificant role. The reason is that the zone in which the fracture front at failure can be located is, due to the mechanics of failure, rather small compared to the volume of the structure, even if the structure is rather large.

The Weibull distribution of strength R is defined as $F(R) = 1 - \exp[-k_1 V(R-R_0)^{\alpha}]$ where V = volume of the structure, and $R_0$, $k_1$, $\alpha$ = constants, $R_0$ = minimum strength. The mean of R is the tensile strength $f'_t$. It can be shown that the statistical aspect of fracture would cause in the size effect law (Eq. 1) a modification of $f'_t$. With the inclusion of the effect of $d_a$ (Eq. 4), this modification may be written as follows:

$$\sigma_N = \frac{B}{\sqrt{1 + \frac{d}{\lambda_0 d_a}}} \left[R_0 + c\left(\frac{V_{fr}}{V}\right)^{1/\alpha} \left(\frac{d}{d_a}\right)^{-2/\alpha}\right] \qquad (28)$$

$V_{fr}$ is the volume of the zone in which the mechanics of failure permits the crack front a the moment of failure to be possibly located (Fig. 5), and V = total volume of the specimen.

Now, an important point is that, for brittle structural failures with fairly reproducible failure modes, such as the diagonal shear failure of beams, the volume $V_{fr}$ is very small compared to V. This means that the effect of statistical variation of strength must be much smaller than it is for a long bar in tension. If the material parameters $R_0$, k, and $\alpha$ are calibrated to give a reasonable statistical size effect for a long thin concrete bar in tension, the use of the same material parameters for structural failures such as the diagonal shear failure given in Eq. 27 yields only a very small correction, because the ratio $V_{fr}/V$ is rather small. It is for this reason that the statistical size effect in diagonal shear failure appears to be unimportant.

## Derivation of Eq. 5

For $d \to \infty$, the size effect law (Eq. 1) yields:

$$\sigma_N = B_t (\lambda_0 \; _a)^{2} \qquad (29)$$

This must be equivalent to linear elastic fracture mechanics, which yields for all structure geometries:

$$\sigma_N = [G_f E/g(\alpha_0)]^{1/2} d^{-1/2} \qquad (30)$$

Eqs. 29 and 30 must be equivalent, and equating them one gets the last expression in Eq. 5. A slightly different derivation was originally given on p. 293 in Bažant, Kim and Pfeiffer (1986, J. of Str. Eng. ASCE 112).

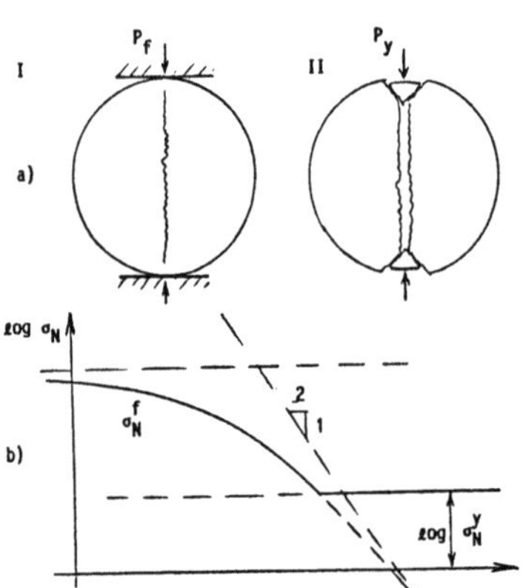


Fig. 4 Brittle Failure with Yielding (a), and the corresponding size effect (b).

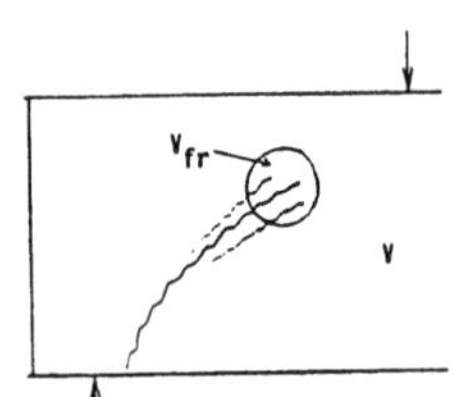


Fig. 5 Volume $V_{fr}$ in which crack front can be located (due to random scatter of material properties).

## Appendix II. - Brittleness Number

The fact that the size effect law yields approximately the same fracture energy values regardless of specimen size and shape makes it possible to base on the size effect law a nondimensional characteristic for the type of fracture behavior, which we may call the brittleness number. It may be defined as

$$\beta = \frac{d}{\lambda_0 d_a} \tag{31}$$

and can be calculated after $\lambda_0$ has been determined either experimentally or by finite element analysis. The value $\beta = 1$ indicates the value of $d/d_a$ at the point where the horizontal asymptote for the strength criterion intersects the inclined straight-line asymptote for linear elastic fracture mechanics (Fig. 1). So, $\beta$ represents the center of the transition between these two extreme types of fracture behavior. For $\beta > 1$, the behavior is closer to plastic limit analysis, and for $\beta < 1$ it is closer to linear elastic fracture mechanics. For $\beta < 0.1$, the plastic limit analysis may be used as an approximation, and for $\beta > 10$, linear elastic fracture mechanics may be used as an approximation. For $0.1 < \beta < 10$, nonlinear fracture analysis must be used. The brittleness number can serve as a basic qualitative indicator of the type of fracture response, and in this sense it is in fact analogous to the nondimensional characteristics used, e.g., in fluid mechanics, such as the Reynolds number.

A. Carpinteri (Engng. Fract. Mech. 16, 1982, p. 467-481) previously proposed to characterize the efect of the structure size on its brittleness by the ratio $s = G_f/b\, f_t'$. A. Hillerborg (Mat. Str., RILEM, 18, 1985, pp. 25-30) characterized the effect of size on structure brittleness by the ratio of the structure size to the characteristic length defined as $\ell_{ch} = E_c\, G_f/f_t'^2$. The use of these characteristics is, however, limited, since they can correlate only specimens or structures of the same geometry. They have the disadvantage that for the same value of this brittleness characteristic, a specimen of one shape may be quite brittle (i.e., close to linear elastic fracture mechanics), while a specimen of another shape may be quite ductile (i.e., close to plastic limit analysis). The brittleness number defined by Eq. 31 is free of this limitation, making it possible to compare in brittleness a small structure of one shape with a large structure of another shape. The greater generality of $\beta$ is due to the fact that it is related not only to $G_f$, $E_c$ and $f_t'$ (through Eq. 1) but also to the size (width or length) of the fracture process zone (see Ref. 7) which is independent of the aforementioned characteristic length $\ell_{ch}$, and also to geometry of the structure.

The change of load-deflection response of a structure of a certain geometry as a function of $\beta$ is illustrated in Fig. 6. An increase in structure size causes that a further elastic deformation (Fig. 6b) is superimposed on the original deformation for small size at the same load P (Fig. 6a). This means that the displacements A and B or A and D in Fig. 6a and b or 6a and d (lines I and II) must be added. The resulting response diagrams are shown in Fig. 6c for the case that the additional elastic deformation is of medium value, or large. It is clear that for a sufficiently large structure the resulting load-deflection diagram must exhibit snap-back, and when this structure is loaded in a displacement-controlled manner this must then lead to snap-down instability represented by the dynamic passage from point 2 to point 4 in Fig. 6e, in which the structure acquires the kinetic energy indicated by the cross-hatched area which is ultimately dissipated as heat. This graphical construction clearly illustrates that a sufficient increase of size must lead to purely brittle, i.e., dynamic, explosive failure.

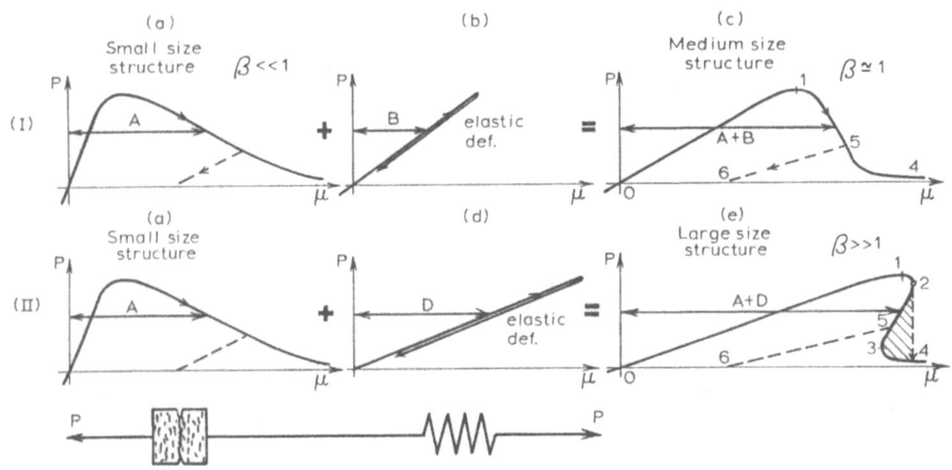


Fig. 6 - Explanation of the Effect of the Size on Brittleness

SEM/RILEM International Conference
on Fracture of Concrete and Rock
Houston, Texas, June 1987
Editors: S.P. Shah, S.E. Swartz

# INFLUENCE OF SIZE EFFECTS ON OPENING MODE FRACTURE PARAMETERS FOR PRECRACKED CONCRETE BEAMS IN BENDING

S. E. Swartz and T. M. E. Refai

## ABSTRACT

Results of three-point bending tests on three sizes of single-edge-notch (SEN) concrete beams are presented. The beams all had the same width (76.2 mm) and span/depth ratio (3.75) with depths of 102 mm, 203 mm and 305 mm. The fracture energy results show wide scatter but apparent independence with beam depth although not with crack depth. The energy release rate results based on LEFM show strong invariance with crack depth and also are invariant between the two larger beam sizes. Good agreement between results of different methods of data evaluation was obtained.

## NOTATION

| | |
|---|---|
| $a_o$ | notch depth |
| $a, a_i, a_e$ | crack length, initial crack length, extended crack length |
| B,W,S | beam width, depth, span length |
| CMOD,LPD | crack-mouth-opening displacement, load-point-displacement |
| $E_c$ | modulus of elasticity of concrete |
| $f'_c, f'_t$ | uniaxial compressive strength, split cylinder tensile strength of concrete |
| $G_F, G_f$ | fracture energy following RILEM and size effect method |
| $\bar{G}_{IC}, G_{IC}$ | energy release rate by direct energy and fracture toughness methods |
| $J_{IC}$ | energy release rate by J-Integral method |
| $K_{IC}, K^S_{IC}, K^G_{IC}$ | fracture toughness and by Jenq/Shah, Go methods |
| mg | beam weight between supports |
| $P, P_m, P^o$ | applied load, maximum load, net load by size effect method |
| $W_o, U$ | total energy absorbed to beam separation and energy absorbed to point of instability |
| $\delta_o, \bar{\delta}_o$ | load-point-displacement at beam separation and point of instability |

## INTRODUCTION

The influence of size effects in the application of fracture mechanics methods to predicting crack growth and fracture in concrete has occupied the attention of many investigators. The generally-accepted concept is that linear elastic fracture mechanics (LEFM) is valid only for very large structures--or structural components--or test specimens. Unfortunately, there is virtually no agreement as to what is meant by the term "large".

These views have evolved from contradictory and size-dependent results for fracture toughness, $K_{IC}$; critical energy release rate $G_{IC}$ or $J_{IC}$; fracture energy, $G_F$. The first author has contended the LEFM methods should be satisfactory even for specimens with depth as small as four to six times the maximum aggregate size if the test specimens are precracked or if slow crack growth is considered (1). This view was based primarily on the good agreement in results obtained between different methods and relative invariance with crack length for tests performed on 4 in. (102 mm) deep beams (2) which had a maximum aggregate size of 12 mm.

S. E. Swartz is Professor of Civil Engineering, Kansas State University, Manhattan, KS 66506, U.S.A.
T. M. E. Refai is a Graduate Research Assistant and Ph.D. candidate, Department of Civil Engineering, Kansas State University, Manhattan, KS 66506, U.S.A.

However, in order to justify this argument it is necessary to show invariance of results with beam size as well as with crack length. Thus, a testing program using single-edge-notch (SEN) beams in three-point bending was conducted with three beam sizes.

## TEST SPECIMENS

All test specimens were beams conforming to the geometry given in Fig. 1 and all had $\frac{S}{W}$ = 3.75 with beam width B = 76.2 mm. The beams had the following dimensions:

Series A: W = 102 mm (4 in.), S = 0.381 m (15 in.)
Series B: W = 203 mm (8 in.), S = 0.762 m (30 in.)
Series C: W = 305 mm (12 in.), S = 1.143 m (45 in.)

The beams in series A were tested by Rood (3) and were evaluated following the methods given in Ref. 2. Additional results for these beams are presented here.

All the beams had the same mix design which is given in Table 1. It is seen that the maximum aggregate size was 19 mm. The coarse and fine aggregates were well-graded. The coarse aggregate was crushed limestone.

The beams were tested to failure using a closed-loop system (MTS) operating in load control (for most of the beams). All beams were tested in a manner that caused upward deflection. The suitability of the use of load control instead of strain control was investigated (4). It was found that for the load rates used--0 to $P_m$ in greater than 30 sec.--that the load versus crack-mouth-opening displacement (CMOD) traces were virtually the same for the two types of control--even for the descending branch of the trace when the crack propagation was clearly unstable (although measurable). The same was true for the load versus load-point-displacement (LPD) traces.

It should be noted however that the initial precracking was done using strain control in order to grow a crack to a desirable size. After that, the beam was then tested to failure using load control to obtain the P-CMOD and P-LPD traces. The necessity for pre-cracking was first justified in Ref. 5 in which a description of the precracking procedure may be found as well.

### *Crack Length Measurement*

It is necessary to have an accurate method of measuring the average crack length corresponding to the initial part of the load-deformation diagram. A modified compliance calibration procedure was proposed by Swartz and Go (6) which used dye to show the crack front after the beam was broken. For the 4 in. (102 mm) deep beams they tested, and also for the ones tested by Rood (3), the crack front appeared to be planar through about the middle 80% of the specimen width with upward "lips" at the sides as shown in Fig. 2.

However, the beams in series B and C with W = 203 mm and 305 mm respectively showed a different pattern for $\frac{a_i}{W}$ less than about 0.65. This pattern is also shown in Fig. 2. For longer crack lengths, the pattern changed and began to look the same as for the series A beams with W = 102 mm. This must be due to the fact that the process zone is not able to develop to its fullest extent as the crack front approaches the free- and compressed-surface of the beam. If the process zone is estimated to be the distance from peak to valley of the crack front then the average value of this is 2.83 in. (72 mm) for series B with $\frac{a_i}{W}$ < 0.75 and 3.84 in. (98 mm) for series C with $\frac{a_i}{W}$ < 0.62. This is 4 to 5 times the maximum aggregate size. For series A this distance is about 25 mm or 1.3 times the maximum aggregate size. Thus, there is clearly a size effect on the shape of the crack front.

The value of $a_i$ was determined in each case by carefully drawing a sketch of the cracked beam cross-section on graph paper, measuring the area with a planimeter and dividing by the beam width. The beam compliance $C_i$ was also measured and curves of $C_i$ versus $a_i$ were plotted.

It is desirable to use the compliance curve to estimate an extended crack length, $a_e$ at some other point on the load versus CMOD plot. Such an approach is illustrated in Fig. 3 where it is desired to determine the crack length at point C. If the actual unload trace is available, $C_u$ can be readily measured. If the unload trace is not available then $C_u$ must be estimated. In the work presented by Go and Swartz (2), this was done by using the inverse slope of line OC. Clearly this leads to an overestimate of $C_u$ and hence $a_e$.

In the present work an alternative approach is proposed by the second author. For each measured $a_i$, the maximum load $P_m$ is noted and the relationship plotted as in Figs. 4-6. The error in using these curves to find $a_i$ or $a_e$ subsequently is also illustrated in Fig. 3 in which the actual compliance, the inverse slope of line OB is seen to be higher than the initial compliance $C_i$. However, the error is not great. In Table 2 are given comparisons between the compliance method and peak load method to find $a_i$. The values from the peak load method are obtained from the curve fit formulas presented in Figs. 4-6. It is seen that the correspondence between the two methods is excellent.

The advantage of using the peak load method is that the extended crack length can be obtained from any point on the load-CMOD diagram simply by considering the load to be $P_m$ for a cracked beam with initial crack length $a_i = a_e$. This method has been used throughout to determine $a_e$ for beam series B and C (all evaluation methods) and beam series A for the Go fracture toughness method.

However, in the succeeding tables the values of $a_i$ are based on the actual dye measurements when available. Otherwise, they are based on the fitted $P_m$-$a_i$ curves.

## DATA EVALUATION METHODS

Of the various methods proposed by different investigators and groups (2, 7, 8, 9, 10) the following are used here: RILEM (7), Direct Energy (2), J-Integral (2), Jenq/Shah (8), Go-$K_{IC}$ (2) and the Size Effect Method (9). It should be noted that many of these methods were proposed to be used with notched-not pre-cracked-beams thus it might be more accurate to state that the data have been evaluated in the spirit or concept of each method.

### RILEM Method

The principal feature of this method is that the fracture energy $G_F$ is the total energy required to fracture a notched beam specimen divided by the net cross-sectional area at the notch with self-weight effects considered. The formula is (7)

$$G_F = \frac{W_0 + mg\ \delta_0}{A_{lig}} \tag{1}$$

in which $W_0$ = area under the load-LPD curve from P=0 to P=0 again; $\delta_0$ = the measured displacement at P=0 of the unload diagram; $A_{lig}$ = area of ligament at notch, mg = weight of beam between supports plus twice the weight of fixtures supported by the beam.

The beams should be loaded in deformation or strain control, the notch depth should equal one half the beam depth and the span/depth ratio should be about 5.6 to one corresponding to W = 200 mm and maximum aggregate size between 16.1 and 32 mm.

Here, the beams are precracked, S/W = 3.75 and $a_i$/W varies. Furthermore, as shown in Ref. 11, the RILEM formula must be modified for the upward-loaded beams tested. The modified formula which was used is

$$G_F = \frac{W_0 - \frac{1}{2} mg\ \delta_0}{BW\ (1 - \frac{a_i}{W})} \tag{2}$$

In this $a_i$ is the initial crack length, mg is the weight of the beam only (between supports) and $\delta_0$ is the displacement corresponding to $P = \frac{1}{2}$ mg on the unload diagram. Also, as mentioned earlier the beams were tested using load control.

The results obtained for beam series B and C are presented in Tables 3 and 4. In these tables the type of testing control is indicated. All but one of the beams were tested in load control after the initial precracking was completed. The results are plotted in Fig. 7 along with those obtained earlier (11) for the 102 mm deep beams. In this figure it is seen that there is a wide scatter of results, that a size effect is not clearly defined and that the values vary with $\frac{a_i}{W}$. The same variation with $a_0$/W is shown for notched beams in Fig. 8. Average values at $\frac{a_i}{W}$ = 0.5 for precracked beams and $\frac{a_0}{W}$ = 0.5 for notched beams are given in Fig. 9. The results between the precracked and notched beams agree only for the largest depth tested, W = 305 mm.

### Direct Energy Method

This method was proposed by Go (2) and estimates the energy release rate $\bar{G}_{IC}$ as the energy per unit area of uncracked ligiment necessary to trigger unstable crack growth. The point of unstable crack growth is taken here to be when the load has dropped 5% from its peak value. The energy release rate is determined from

$$\bar{G}_{IC} = \frac{U - \frac{1}{2} mg\ \bar{\delta}_0}{1.15\ BW(1 - \frac{a_e}{W})} \tag{3}$$

In this, U is the energy obtained from the area under the P-LPD curve to $\bar{\delta}_o$; $\bar{\delta}_o$ is the load-point-displacement at 0.95 $P_m$ on the unload line; $a_e$ is the crack length at 0.95 $P_m$ and the factor 1.15 represents the effect of surface roughness on crack length.

The results obtained from this are also given in Tables 3 and 4 and are plotted in Fig. 10 for series B and C precracked beams and Fig. 11 for all the notched beams. The trends are the same as those obtained from the RILEM method.

J-Integral Method

This method--also proposed by Go (2)--is based on the idea that the energy release rate can be determined directly by change in energy versus change in crack extension. To do this in an average sense the energy value U is plotted versus $\frac{a}{W}$ where a may be $a_i$ or $a_e$. The slope of the plot gives the energy release rate as

$$J_{IC} = \frac{\left|\frac{dU}{d(\frac{a}{W})}\right|}{1.15\ BW} \tag{4}$$

A typical plot--for precracked beams series B--is shown in Fig. 12. It is seen that the data points are fairly consistent and that the same results are obtained if $a_e$ is used instead of $a_i$. The correlation coefficient for the straight line is .94. The average value of $J_{IC}$ is 100.1 N/m. For beam series C it is 95.2 with a correlation coefficient of .96. However, for beam series A, $J_{IC}$ = 54.1 N/m. For notched beams--using $a_e$--the values for $J_{IC}$ are: 103.6 N/m, W = 102 mm; 87.7 N/m, W = 203 mm; and 164.7 N/m, W = 305 mm.

Fracture Toughness Methods

In these methods, the fracture toughness, $K_{IC}$ is determined by LEFM using the appropriate load and crack length. The energy release rate then is

$$G_{IC} = \frac{K_{IC}^2}{E_c} \tag{5}$$

in which the effect of Poisson's Ratio is neglected.

The only difference between the two methods presented here is in the determination of $a_e$. Once this is found, $K_{IC}$ is obtained from the following (12):

$$K_{IC} = \frac{P_m\ S}{4\ BW^{1.5}} \left(-0.065\ z^2 - 3.483\ z - 0.120 + 5.706\ z^{-1} + 0.166\ z^{-2}\right) \tag{6}$$

and

$$Z = 1 - \frac{a_e}{W}. \tag{7}$$

This formula is derived for beams in three-point bending with $\frac{S}{W}$ = 3.75. Similar formulas for beams with other span to depth ratios are given in Ref. 12.

*Jenq/Shah Method*

In this method, which is presented in Ref. 8, it is necessary to unload the specimen at or near the peak load and measure the initial compliance $C_i$ and unload compliance $C_u$. In the tests reported here, since load control was used, unloading took place just at the onset of instability. The beams that were tested this way were only those in series B and C and are denoted LC in the various tables.

Knowing $C_i$ and $C_u$, the effective crack length $a_e$ is calculated from LEFM by (8)

$$\frac{a_e}{W} = \left(\frac{a_i}{W} + \frac{HO}{W}\right) \frac{C_u}{C_i} \frac{V_1(a_i/W)}{V_1\ (a_e/W)} - \frac{HO}{W} \tag{8}$$

and

$$V_1(a/W) = 0.76 - 2.28\ (a/W) + 3.87\ (a/W)^2 - 2.04\ (a/W)^3 + \frac{0.66}{(1 - a/W)^2} \tag{9}$$

In this HO is the distance from the edge of the beam to the knife edge to which the displacement transducer is attached. This is HO = 0.25 in. (6.4 mm) for the beams tested in this program. Eq. 8 is solved by trial and error to get $a_e/W$ and this is substituted into Eqs. 6 and 7 to obtain $K_{IC} = K_{IC}^S$. Values of $K_{IC}^S$ are tabulated in Tables 6 and 7 for series B and C precracked beams. Values of $G_{IC}$ obtained from Eq. 5 are also presented in these tables. It is seen that these results are essentially invariant with crack length for $a_e/W$ less than 0.6 to 0.7 and are also invariant with respect to beam size.

*Go Method*

In this method, originally presented in Ref. 2, $a_e$ is determined by use of the maximum load-crack relationship described earlier. The determination was made at P = 0.95 Pm . Then $K_{IC} = K_{IC}^G$ and $G_{IC}$ are calculated from Eqs. 5,6,7. The results obtained from all series of precracked beams are presented in Tables 5-7. It may be noted that in those cases where Jenq/Shah data are available that agreement in results is excellent. For the series B beams the average ratio of $a_e/W$ by Jenq/Shah and $a_e/W$ by Go is 1.00 with a standard deviation of 0.038 and a coefficient of variation of 3.8%. For the series C beams the average is 1.03 with a coefficient of variation of 5.6%. This agreement is remarkable when one considers that these are independent methods based on different concepts. A corresponding agreement exists between values of $K_{IC}$ and $G_{IC}$.

The values of $G_{IC}$ obtained using Go's method are plotted in Fig. 13 for precracked beams and Fig. 14 for notched beams. Using a cutoff of $a_e/W$ = 0.6 to 0.7 it is seen that the values of $G_{IC}$ are essentially constant--with some scatter--with respect to crack length. The average values agree well between series B and series C beams (34.9 N/m and 33.7 N/m respectively) but the average value of 24.0 N/m for series A is considerably lower. For the notched beams the results seem to be invarient for $a_e/W < .7$ but far fewer beams were tested. The average $G_{IC}$ values are 24.6 N/m, W = 102 mm; 28.3 N/m, W = 203 mm; and 32.8 N/m, W = 305 mm.

Size Effect Method

This method, proposed by Bazart and Pfeiffer (9) is intended to be used with notched beams and requires at least three sizes of beams with proportional geometry except that the width should be the same in all cases. These requirements were satisfied in these tests with the exception that the crack length was used instead of the notch depth. In this method average values of $(\frac{BW}{P^o})^2$ are plotted (ordinate) versus W (abscissa) and the best straight line fitted through the three points. In this $P^o$ is the maximum beam load plus or minus the effect of carried weight. For the upward loaded beams

$$P^o = P_m - \frac{1}{2} mg \tag{10}$$

The slope of the line, denoted by A, is used to determine the fracture energy $G_f$ by

$$G_f = \frac{g(\alpha_o)}{E_c A} \tag{11}$$

where

$$g(\alpha_o) = (\frac{S}{W})^2 \; \pi\alpha_o[1.5 \; F(\alpha_o)]^2 \tag{12}$$

Here, $\alpha_o = \frac{a_o}{W}$ for notched beams and $\frac{a_i}{W}$ for precracked beams and

$$F(\alpha_o) = 1.089 - 1.746 \; \alpha_o + 8.231 \; \alpha_o^2 - 14.22 \; \alpha_o^3 + 14.59 \; \alpha_o^4 \tag{13}$$

for beams with $\frac{S}{W} = 3.75$.

A typical plot for $\alpha_o = a_i/W = 0.52$ and precracked beams is displayed in Fig. 15. The correlation coefficient is 0.94 and the resulting $G_f$ is 42.4 N/m. A similar plot for $\alpha_o$ = 0.403 gave a correlation of 0.99 and $G_f$ of 44.7 N/m. Another approach to this method is to obtain $\alpha_o$ data from the equations relating load to $a_i$ for the three beam sizes (given in Figs. 4-6) and then determine $G_f$ for each value of $\alpha_o$. This was done for $\alpha_o$ ranging from 0.2 to 0.8 in 0.1 increments. The correlation coefficients on the line fits varied as 0.86, 0.92, 0.98, 1.00, 0.88, 0.75, 0.58. Using the best two--0.98, 1.00 at $\alpha_o$ = 0.4 and 0.5 respectively gave $G_f$ values of 35.0 N/m and 35.7 N/m.

SUMMARY AND CONCLUSIONS

Average values corresponding to the six methods used are presented in Table 8. From these results the following can be concluded.

1. A size effect is apparent when notched beams are used in conjunction with energy methods (RILEM, Direct Energy, J-Integral). However, this effect--while still present--is not as great when the Jenq/Shah and Go methods are employed. The size effect method of Bazant/Pfeiffer gives results in reasonable agreement with these two methods.

2. When precracked beams are used the RILEM and Direct Energy Methods still indicate a size effect. The J-Integral Method gives comparable results for the 203 mm and 305 mm beams. The Jenq/Shah and Go methods also indicate the present of a size effect but give very consistent results between each other and also between the two larger beam sizes. These results also agree very well with those obtained from the size effect method.

3. The energy results--whether fracture energy or energy release rate--are all higher by two to four times the results obtained from Jenq/Shah, Go, or the size effect methods. The reason for this is not presently apparent since both groups of results are internally consistent.

4. Both the calculated results and the physical appearance of the failed beam surface indicate that the 102 mm beams are too small. The excellent agreement in results from the 203 mm and 305 mm deep beams indicates that a minimum beam size of 203 mm--or a depth of about eleven times the maximum aggregate size--is suitable for testing in three point bending to obtain valid fracture parameters.

   For this size of beam and larger the methods of LEFM appear to give acceptable results provided the ratio $\frac{a_e}{W}$ is less than 0.6 even though the size of the process zone is 4 to 5 times the size of the maximum aggregate.

ACKNOWLEDGEMENTS

The work reported herein has been supported by the National Science Foundation through Grants CEE-8317136 and MSM-8317136. This support is gratefully acknowledged.

The work of Ph.D. candidate Sze-Ting Yap in helping with the data reduction is appreciated.

REFERENCES

1. Swartz, S. E., "Fracture Toughness Testing of Concrete at Kansas State University: Is LEFM Acceptable?," Proceedings, International Conference on Fracture Mechanics of Concrete, Lausanne, Switzerland, October 1-3, 1985.

2. Go, C. G. and Swartz, S. E., "Energy Methods for Fracture-Toughness Determination in Concrete," Experimental Mechanics, Vol. 26, No. 3, Sept. 1986.

3. Rood, S. M., "Fracture Toughness Testing of Small Concrete Beams," M.S. Thesis, Kansas State University, Manhattan, KS, 1984.

4. Swartz, Stuart E. and Siew, Hoi-Choong, "Time Effects in the Static Testing of Concrete to Determine Fracture Energy," Report No. 182, Engineering Experiment Station, Kansas State University, June 1986.

5. Swartz, S. E., Hu, K. K., Fartash, M. and Huang, C.-M.J., "Stress-Intensity Factor for Plain Concrete in Bending-Prenotched Versus Precracked Beams," Experimental Mechanics, Vol. 22, No. 11, Nov. 1982.

6. Swartz, S. E. and Go, C. G., "Validity of Compliance Calibration to Cracked Concrete Beams in Bending," Experimental Mechanics, Vol. 24, No. 2, June 1984.

7. RILEM Technical Committee 50-FMC, "Determination of the Fracture Energy of Mortar and Concrete by Means of Three-Point Bend Tests on Notched Beams," Materials and Structures, Vol. 18, No. 106, July-August 1985.

8. Jenq, Y. S. and Shah, S. P., "Two Parameter Fracture Model for Concrete," Journal of Engineering Mechanics, ASCE, Vol. 111, No. 10, Oct. 1985.

9. Bazant, Z. P. and Pfeiffer, P. A., "Determination of Fracture Energy From Size Effect," Report No. 86-8/428d, Center for Concrete and Geomaterials, Northwestern University, Sept. 1986.

10. Nallathambi, P. and Karihaloo, B. L., "Determination of Specimen-Size Independent Fracture Toughness of Concrete," accepted for publication in Magazine of Concrete Research, 1986.

11. Swartz, Stuart E. and Yap, Sze-Ting, "Evaluation of Proposed Methods to Determine Fracture Parameters for Concrete in Bending," Report No. 181, Engineering Experiment Station, Kansas State University, June 1986.

12. Go, Cheer-Germ, Swartz, Stuart E., and Hu, Kuo-Kuang, "Stress Intensity Factors for Single-Edge-Notch Beam," Technical Note, Journal of Engineering Mechanics, ASCE, Vol. 110, No. 4, April 1984.

Table 1. Mix Design

| | |
|---|---|
| Water/Cement | 0.5 |
| Cement Type | 1 |
| S.G. Coarse | 2.56 |
| S.G. Fine | 2.65 |
| S.G. Cement | 3.15 |
| Sand Fineness Modulus | 2.91 |
| Max. Aggregate Size | 19 mm |
| % Coarse by wt. | 47.46 |
| % Fine by wt. | 32.68 |
| % Cement by wt. | 13.24 |
| % Water by wt. | 6.62 |

| Series: | A | B | C |
|---|---|---|---|
| Compressive Strength, $f'_c$ (MPa) | 55.8 | 53.1 | 54.4 |
| Splitting Strength, $f'_t$ (MPa) | -- | 4.0 | 3.8 |
| Modulus of Elasticity, $E_c$ (GPa) | 36.8 | 38.4 | 39.3 |

Cylinder size 3 in. x 6 in. (76x152 mm)

Table 2. Crack Length Estimates by Two Methods
1. Compliance Calibration by fitted curve
2. Peak Load Calibration by fitted equation

| Series B, W = 8 in. = 203 mm | | | | | | Series C, W = 12 in. = 305 mm | | | | | |
|---|---|---|---|---|---|---|---|---|---|---|---|
| No. | $P_m$ kN | (1.) $\frac{a_i}{W}$ | (2.) $\frac{a_i}{W}$ | Dye $\frac{a_i}{W}$ | $\frac{(2.)}{(1.)}$ | No. | $P_m$ kN | (1.) $\frac{a_i}{W}$ | (2.) $\frac{a_i}{W}$ | Dye $\frac{a_i}{W}$ | $\frac{(2.)}{(1.)}$ |
| B16 | 5.78 | .305 | .30 | | .984 | C23 | 6.61 | .404 | .42 | .375 | 1.040 |
| B4 | 5.61 | .305 | .31 | | 1.016 | C22 | 7.65 | .417 | .33 | .381 | .791 |
| B36 | 4.41 | .424 | .43 | .375 | 1.014 | C1 | 6.54 | .494 | .39 | | .789 |
| B31 | 4.85 | .468 | .38 | | .812 | C21 | 5.74 | .435 | .43 | .418 | .989 |
| B3 | 4.36 | .400 | .43 | | 1.075 | C2 | 6.05 | .521 | .42 | | .806 |
| B37 | 4.72 | .435 | .40 | .436 | .920 | C3 | 5.87 | .500 | .43 | | .860 |
| B18 | 4.05 | .445 | .47 | | 1.056 | C24 | 6.12 | .425 | .42 | .433 | .988 |
| B40 | 3.83 | .500 | .49 | .504 | .980 | C4 | 5.52 | .528 | .45 | | .852 |
| B45 | 3.56 | .528 | .52 | .528 | .985 | C20 | 4.67 | .552 | .51 | .472 | .924 |
| B5 | 3.20 | .570 | .56 | | .982 | C15 | 4.90 | .569 | .50 | | .879 |
| B44 | 3.03 | .575 | .58 | .575 | 1.009 | C16 | 5.07 | .577 | .48 | | .832 |
| B25 | 2.80 | .646 | .61 | | .944 | C19 | 4.50 | .567 | .53 | .520 | .935 |
| B38 | 2.54 | .634 | .64 | .619 | 1.009 | C6 | 6.05 | .552 | .52 | | .942 |
| B21 | 2.45 | .650 | .65 | | 1.00 | C5 | 4.54 | .569 | .52 | | .914 |
| B22 | 2.45 | .670 | .65 | | .970 | C17 | 4.27 | .585 | .54 | .560 | .923 |
| B39 | 2.54 | .624 | .64 | .656 | 1.026 | C26 | 4.23 | .538 | .54 | .568 | 1.004 |
| B8 | 2.22 | .600 | .68 | | 1.133 | C7 | 3.47 | .631 | .60 | | .951 |
| B7 | 2.25 | .658 | .68 | | 1.033 | C27 | 2.80 | .617 | .65 | .652 | 1.053 |
| B10 | 2.00 | .610 | .71 | | 1.164 | C8 | 3.20 | .667 | .62 | | .930 |
| B24 | 1.98 | .680 | .71 | | 1.044 | C28 | 2.58 | .643 | .67 | .638 | 1.042 |
| B28 | 1.40 | .800 | .79 | | .988 | C30 | 3.12 | .612 | .63 | .647 | 1.029 |
| B26 | 1.56 | .805 | .77 | | .957 | C29 | 2.98 | .623 | .64 | .669 | 1.027 |
| B9 | 1.56 | .705 | .77 | | 1.092 | C10 | 2.49 | .689 | .68 | | .987 |
| B27 | 1.20 | .790 | .82 | | 1.038 | C9 | 2.45 | .704 | .68 | | .966 |
| B30 | 1.29 | .812 | .81 | | .998 | C11 | 1.51 | .705 | .76 | | 1.078 |
| B29 | 1.07 | .830 | .84 | | 1.012 | C31 | 1.20 | .792 | .79 | .761 | .997 |
| B11 | .98 | .800 | .85 | | 1.063 | C12 | 1.11 | .817 | .80 | | .979 |
| B12 | .89 | .808 | .87 | | 1.077 | C14 | .87 | .867 | .82 | | .946 |
| B43 | .58 | .882 | .91 | .891 | 1.032 | C13 | .85 | .867 | .82 | | .946 |
| B13 | .58 | .820 | .91 | | 1.110 | C25 | .67 | .867 | .84 | .884 | .969 |
| B41 | 1.51 | .798 | .78 | .798 | .977 | | | | | | |
| B42 | 1.51 | .775 | .78 | .794 | 1.006 | | | | | | |
| | | | | Mean | 1.0159 | | | | | | 0.9457 |
| | | | | Std. Dev. | 0.070 | | | | | | 0.080 |
| | | | | C.V. | 6.9% | | | | | | 8.5% |

Table 3. Energy Data for Series B Precracked Beams
W = 8 in. = 203 mm.

| No. | $\frac{a_i}{W}$ | $P_m$ kN | $\delta_o$ mm | $W_o$ N-m | $G_F$ N/m | $\bar{\delta}_o$ mm | U N-m | $\frac{a_e}{W}$* | $\bar{G}_{IC}$ N/m | Control Type |
|---|---|---|---|---|---|---|---|---|---|---|
| B16 | .295 | 5.78 | .621 | 1.73 | 150 | .336 | 1.08 | .32 | 85 | L |
| B4 | .310 | 5.61 | .610 | 1.74 | 155 | .339 | 1.05 | .34 | 85 | L |
| B36 | .375 | 4.41 | .660 | 1.31 | 126 | .457 | 1.12 | -- | --- | LC |
| B31 | .380 | 4.85 | .917 | 1.55 | 148 | .369 | .833 | .41 | 74 | LC |
| B3 | .424 | 4.36 | .701 | 1.35 | 141 | .417 | .924 | .46 | 90 | L |
| B37 | .436 | 4.72 | .508 | 1.28 | 138 | .259 | .751 | .43 | 70 | L |
| B18 | .459 | 4.05 | .531 | 1.12 | 125 | .323 | .746 | .49 | 77 | L |
| B40 | .504 | 3.83 | .665 | 1.28 | 154 | .462 | 1.00 | .51 | 107 | L |
| B45 | .528 | 3.56 | .691 | 1.37 | 174 | .389 | 1.01 | .54 | 117 | L |
| B5 | .560 | 3.20 | .593 | 1.13 | 153 | .496 | .814 | .58 | 100 | L |
| B44 | .571 | 3.03 | .568 | .892 | 123 | .348 | .642 | .60 | 83 | L |
| B25 | .600 | 2.80 | .847 | .933 | 132 | .525 | .657 | .63 | 89 | SC |
| B38 | .619 | 2.54 | .671 | .881 | 133 | .330 | .625 | .66 | 96 | L |
| B21 | .646 | 2.45 | .714 | .805 | 129 | .426 | .592 | .67 | 91 | L |
| B22 | .646 | 2.45 | .629 | .749 | 121 | .299 | .444 | .67 | 68 | L |
| B39 | .656 | 2.54 | .548 | .628 | 103 | .199 | .292 | .66 | 44 | L |
| B8 | .670 | 2.22 | .540 | .582 | 99 | .285 | .351 | .70 | 58 | L |
| B7 | .675 | 2.25 | .654 | .712 | 123 | .376 | .425 | .69 | 67 | L |
| B10 | .708 | 2.00 | .654 | .573 | 106 | .358 | .351 | .72 | 60 | L |
| B24 | .711 | 1.98 | .531 | .509 | 97 | .254 | .277 | .73 | 50 | LC |
| B28 | .760 | 1.40 | .471 | .345 | 75 | .240 | .203 | .80 | 48 | LC |
| B26 | .770 | 1.56 | .598 | .499 | 117 | .279 | .277 | .78 | 61 | LC |
| B9 | .772 | 1.56 | .626 | .462 | 106 | .322 | .259 | .78 | 55 | L |
| B27 | .788 | 1.20 | .529 | .342 | 82 | .175 | .148 | .83 | 41 | LC |
| B30 | .800 | 1.29 | .457 | .301 | 77 | .217 | .177 | .82 | 46 | L |
| B29 | .835 | 1.07 | .494 | .366 | 116 | .240 | .203 | .85 | 63 | L |
| B11 | .854 | 0.98 | .711 | .370 | 120 | .497 | .259 | .86 | 76 | L |
| B12 | .865 | 0.89 | .680 | .351 | 122 | .310 | .175 | .87 | 57 | L |
| B43 | .891 | 0.58 | .340 | .131 | 49 | .130 | [illegible] | .92 | 31 | LC |
| B13 | .908 | 0.58 | .542 | .166 | 63 | .339 | .111 | -- | --- | LC |

*Values at 0.95 $P_m$

$E_c$ = 5.57 x $10^6$ psi = 38.4 GPa

mg = 62.5 lb = 0.28 kN

L = load control, LC = load control-cycled

SC = strain control-cycled

Table 4. Energy Data for Series C Precracked Beams
W = 12 in. = 305 mm.

| No. | $\frac{a_i}{W}$ | $P_m$ kN | $\delta_o$ mm | $W_o$ N-m | $G_F$ N/m | $\bar{\delta}_o$ mm | U N-m | $\frac{a_e}{W}$* | $\bar{G}_{IC}$ N/m | Control Type |
|---|---|---|---|---|---|---|---|---|---|---|
| C23 | .375 | 6.61 | .610 | 2.14 | 134 | .310 | 1.02 | .44 | 62 | LC |
| C22 | .381 | 7.65 | .536 | 1.99 | 127 | .356 | 1.48 | .35 | 79 | LC |
| C1 | .396 | 6.54 | .970 | 2.70 | 171 | .356 | 1.28 | .41 | 74 | LC |
| C21 | .418 | 5.74 | .582 | 1.75 | 116 | .345 | 1.18 | .45 | 73 | L |
| C2 | .423 | 6.05 | .678 | 1.85 | 123 | .310 | 1.04 | .44 | 63 | LC |
| C3 | .432 | 5.87 | .886 | 2.36 | 158 | .361 | 1.06 | .45 | 64 | L |
| C4 | .458 | 5.52 | .752 | 2.06 | 144 | .305 | .999 | .47 | 64 | L |
| C24 | .433 | 6.12 | .960 | 2.03 | 104 | .384 | 1.28 | .44 | 77 | LC |
| C20 | .472 | 4.67 | .564 | 1.41 | 101 | .300 | .869 | .53 | 62 | LC |
| C15 | .479 | 4.90 | .620 | 1.63 | 119 | .361 | .955 | .51 | 64 | LC |
| C16 | .482 | 5.07 | .589 | 1.57 | 115 | .338 | .991 | .50 | 66 | L |
| C19 | .520 | 4.50 | .564 | 1.36 | 106 | .335 | .989 | .54 | 72 | L |
| C6 | .527 | 6.05 | .643 | 1.56 | 124 | .305 | .814 | .54 | 58 | L |
| C5 | .527 | 4.54 | .955 | 1.92 | 148 | .305 | .712 | .54 | 50 | LC |
| C17 | .560 | 4.27 | .488 | 1.15 | 98 | .277 | .689 | .56 | 51 | L |
| C26 | .568 | 4.23 | .632 | 1.40 | 120 | .401 | .955 | .56 | 71 | LC |
| C7 | .598 | 3.47 | .947 | 1.25 | 103 | .259 | .548 | .62 | 46 | LC |
| C27 | .608 | 2.80 | .696 | 1.1 | 87 | .307 | .566 | .66 | 52 | LC |
| C8 | .618 | 3.20 | .516 | 1.06 | 101 | .503 | .712 | .63 | 56 | L |
| C28 | .638 | 2.58 | .632 | .954 | 86 | .373 | .636 | .68 | 61 | L |
| C30 | .647 | 3.12 | .582 | .938 | 92 | .300 | .601 | .64 | 53 | LC |
| C29 | .669 | 2.98 | .655 | 1.28 | 139 | .305 | .654 | .65 | 60 | LC |
| C10 | .673 | 2.49 | .493 | .742 | 78 | .251 | .407 | .69 | 40 | LC |
| C9 | .678 | 2.45 | .607 | .920 | 98 | .257 | .407 | .69 | 39 | LC |
| C11 | .758 | 1.51 | .437 | .425 | 51 | .246 | .244 | .77 | 27 | L |
| C31 | .761 | 1.20 | .589 | .451 | 48 | .366 | .270 | .79 | 28 | LC |
| C12 | .783 | 1.11 | .403 | .314 | 37 | .251 | .175 | .80 | 18 | LC |
| C14 | .823 | .87 | .356 | .203 | 22 | .220 | .115 | --- | --- | LC |
| C13 | .825 | .85 | .396 | .247 | 30 | .214 | .132 | .82 | 13 | LC |
| C25 | .883 | .67 | .503 | .203 | 17 | .315 | .147 | .84 | 11 | LC |

*Values at 0.95 $P_m$

$E_c$ = 5.71 x $10^6$ psi = 39.3 GPa

mg = 140.6 lb = 0.63 kN

L = load control, LC = load control-cycled

Table 5. Fracture Toughness Data for Series A Precracked Beams, W = 4 in. = 102 mm

| No. | $\frac{a_i}{W}$ | $P_m$ kN | Go (2) $\frac{a_e^*}{W}$ | $K_{IC}^G$ $kN\text{-}m^{-3/2}$ | $G_{IC}$ N/m |
|---|---|---|---|---|---|
| B9 | .28 | 4.03 | .30 | 918 | 22.9 |
| B1 | .30 | 3.87 | .32 | 931 | 23.6 |
| C7 | .30 | 3.85 | .33 | 952 | 24.6 |
| B11 | .31 | 4.27 | .28 | 921 | 23.1 |
| C1 | .32 | 3.96 | .31 | 927 | 23.4 |
| B10 | .33 | 4.05 | .30 | 923 | 23.2 |
| C2 | .33 | 3.47 | .37 | 956 | 24.8 |
| B2 | .36 | 3.47 | .37 | 956 | 24.8 |
| C8 | .40 | 2.58 | .48 | 960 | 25.0 |
| B14 | .50 | 2.14 | .54 | 946 | 24.3 |
| B16 | .52 | 2.31 | .52 | 963 | 25.2 |
| B17 | .52 | 1.69 | .61 | 929 | 23.5 |
| C4 | .52 | 1.89 | .58 | 944 | 24.2 |
| C9 | .59 | 1.78 | .59 | 917 | 22.9 |
| C10 | .59 | 1.56 | .63 | 916 | 22.8 |
| B4 | .60 | 2.18 | .54 | 964 | 25.2 |
| B20 | .67 | .979 | .71 | 775 | 16.3 |
| C6 | .67 | .846 | .73 | 730 | 14.5 |
| B18 | .68 | .846 | .73 | 730 | 14.5 |
| B19 | .69 | .979 | .71 | 775 | 16.3 |
| C11 | .75 | .712 | .75 | 673 | 12.3 |
| B8 | .79 | .490 | .79 | 569 | 8.8 |
| B7 | .81 | .445 | .79 | 517 | 7.3 |
| C12 | .81 | .445 | .79 | 517 | 7.3 |

*Values at 0.95 $P_m$; $E_c = 5.34 \times 10^6$ psi = 36.8 GPa;

All beams tested in load control.

Table 6. Fracture Toughness Data for Series B Precracked Beams, W = 8 in. = 203 mm

| No. | $\frac{a_i}{W}$ | $P_m$ kN | Jenq/Shah (8) $\frac{a_e}{W}$ | $K_{IC}^S$ $kN\text{-}m^{-3/2}$ | $G_{IC}$ N/m | Go (2) $\frac{a_e^*}{W}$ | $K_{IC}^G$ $kN\text{-}m^{-3/2}$ | $G_{IC}$ N/m | Control Type |
|---|---|---|---|---|---|---|---|---|---|
| B16 | .295 | 5.78 | | | | .32 | 983 | 25.2 | L |
| B4 | .310 | 5.61 | | | | .34 | 1007 | 26.4 | L |
| B36 | .375 | 4.41 | .48 | 1160 | 35.0 | | | | LC |
| B31 | .380 | 4.85 | .44 | 1142 | 34.0 | .41 | 1052 | 28.8 | LC |
| B3 | .424 | 4.36 | | | | .46 | 1085 | 30.7 | L |
| B37 | .436 | 4.72 | | | | .43 | 1081 | 30.4 | L |
| B18 | .459 | 4.05 | | | | .49 | 1096 | 31.3 | L |
| B40 | .504 | 3.83 | | | | .51 | 1097 | 31.3 | L |
| B45 | .528 | 3.56 | | | | .54 | 1113 | 32.3 | L |
| B5 | .560 | 3.20 | | | | .58 | 1129 | 33.2 | L |
| B44 | .571 | 3.03 | | | | .60 | 1139 | 33.8 | L |
| B25 | .600 | 2.80 | .62 | 1124 | 32.9 | .63 | 1163 | 35.2 | SC |
| B38 | .619 | 2.54 | | | | .66 | 1172 | 35.8 | L |
| B21 | .646 | 2.45 | | | | .67 | 1173 | 35.8 | L |
| B22 | .646 | 2.45 | | | | .67 | 1173 | 35.8 | L |
| B39 | .656 | 2.54 | | | | .66 | 1172 | 35.8 | L |
| B8 | .670 | 2.22 | | | | .70 | 1193 | 37.1 | L |
| B7 | .675 | 2.25 | | | | .69 | 1162 | 35.2 | L |
| B10 | .708 | 2.00 | | | | .72 | 1168 | 35.5 | L |
| B24 | .711 | 1.98 | .72 | 1156 | 34.8 | .73 | 1208 | 38.0 | LC |
| B28 | .760 | 1.40 | .77 | 1033 | 27.8 | .80 | 1217 | 38.6 | LC |
| B26 | .770 | 1.56 | .78 | 1212 | 38.3 | .78 | 1212 | 38.3 | LC |
| B9 | .772 | 1.56 | | | | .78 | 1212 | 38.3 | L |
| B27 | .788 | 1.20 | .83 | 1264 | 41.6 | .82 | 1181 | 36.3 | LC |
| B30 | .800 | 1.29 | | | | .83 | 1359 | 48.1 | L |
| B29 | .835 | 1.07 | | | | .85 | 1307 | 44.5 | L |
| B11 | .854 | .98 | | | | .86 | 1300 | 44.0 | L |
| B12 | .865 | .89 | | | | .87 | 1291 | 43.4 | L |
| B43 | .891 | .58 | | | | .92 | 1533 | 61.2 | LC |
| B13 | .908 | .58 | .91 | 1321 | 45.4 | | | | LC |

*Values at 0.95 $P_m$; $E_c = 5.57 \times 10^6$ psi = 38.4 GPa

Table 7. Fracture Toughness Data for Series C Precracked Beams
W = 12 in. = 305 mm

| | | | Jenq/Shah (8) | | | Go (2) | | | |
|---|---|---|---|---|---|---|---|---|---|
| No. | $\frac{a_i}{W}$ | $P_m$ kN | $\frac{a_e}{W}$ | $K^S_{IC}$ $kN\text{-}m^{-3/2}$ | $G_{IC}$ N/m | $\frac{a_e}{W}$* | $K^G_{IC}$ $kN\text{-}m^{-3/2}$ | $G_{IC}$ N/m | Control Type |
| C23 | .375 | 6.61 | .42 | 1203 | 36.8 | .44 | 1271 | 41.1 | LC |
| C22 | .381 | 7.65 | .40 | 1319 | 44.3 | .35 | 1152 | 33.8 | LC |
| C1 | .396 | 6.54 | .43 | 1224 | 38.1 | .41 | 1159 | 34.2 | LC |
| C21 | .418 | 5.74 | | | | .45 | 1134 | 32.7 | L |
| C2 | .423 | 6.05 | .49 | 1337 | 45.5 | .44 | 1163 | 34.4 | LC |
| C3 | .432 | 5.87 | | | | .45 | 1160 | 34.2 | L |
| C24 | .433 | 6.12 | .46 | 1243 | 39.3 | .44 | 1177 | 35.2 | LC |
| C4 | .458 | 5.52 | | | | .47 | 1153 | 33.8 | L |
| C20 | .472 | 4.67 | .55 | 1228 | 38.4 | .53 | 1157 | 34.1 | LC |
| C15 | .479 | 4.90 | .54 | 1251 | 39.8 | .51 | 1146 | 33.4 | LC |
| C16 | .482 | 5.07 | | | | .50 | 1152 | 33.8 | L |
| C19 | .520 | 4.50 | | | | .54 | 1149 | 33.6 | L |
| C6 | .527 | 4.54 | | | | .54 | 1159 | 34.2 | L |
| C5 | .527 | 4.54 | .49 | 1003 | 25.6 | .54 | 1159 | 34.2 | LC |
| C17 | .560 | 4.27 | | | | .56 | 1157 | 34.1 | L |
| C26 | .568 | 4.23 | .60 | 1299 | 42.9 | .56 | 1146 | 33.4 | LC |
| C7 | .598 | 3.47 | .62 | 1138 | 33.0 | .62 | 1138 | 32.9 | LC |
| C27 | .608 | 2.80 | .65 | 1018 | 26.4 | .66 | 1055 | 28.3 | LC |
| C8 | .618 | 3.20 | | | | .63 | 1085 | 30.0 | L |
| C28 | .638 | 2.58 | | | | .68 | 1047 | 27.9 | L |
| C30 | .647 | 3.12 | .70 | 1369 | 47.7 | .64 | 1095 | 30.5 | LC |
| C29 | .669 | 2.98 | .69 | 1257 | 40.2 | .65 | 1083 | 29.9 | LC |
| C10 | .673 | 2.49 | .71 | 1138 | 33.0 | .69 | 1050 | 28.1 | LC |
| C9 | .678 | 2.45 | .69 | 1033 | 27.2 | .69 | 1033 | 27.2 | LC |
| C11 | .758 | 1.51 | | | | .77 | 909 | 21.0 | L |
| C31 | .761 | 1.20 | .77 | 723 | 13.3 | .79 | 804 | 16.5 | LC |
| C12 | .783 | 1.11 | .80 | 788 | 15.8 | .80 | 788 | 15.8 | LC |
| C14 | .823 | .87 | .83 | 748 | 14.2 | --- | ---- | ---- | LC |
| C13 | .825 | .85 | .84 | 785 | 15.7 | .82 | 683 | 11.9 | LC |
| C25 | .883 | .67 | | | | .84 | 619 | 9.8 | LC |

*Values at 0.95 $P_m$

$E_c = 5.71 \times 10^6$ psi = 39.3 GPa

Table 8. Summary of Averaged Values

| | W mm | RILEM[1] Method $G_F$ N/m | Direct[2] Energy Method $\bar{G}_{IC}$ N/m | Jenq/ Shah[2] Method $G_{IC}$ N/m | Go[2] Method $G_{IC}$ N/m | Bazant[1] Method $G_f$ N/m | J-Integral Method $J_{IC}$ N/m |
|---|---|---|---|---|---|---|---|
| Notched | 102 | 59.0 | 58.5 | --- | 24.6 | | 103.6 |
| | 203 | 68.0 | 67.8 | 35.8 | 28.3 | 28.2 or 20.6 | 87.7 |
| | 305 | 112.0 | 94.3 | 54.7 | 32.8 | | 164.7 |
| Precracked | 102 | 93.0 | 52.5 | --- | 24.0 | | 54.1 |
| | 203 | 151.0 | 79.5 | 34.2 | 34.9 | 35.7 or 42.4 | 100.1 |
| | 305 | 119.0 | 63.4 | 37.3 | 33.7 | | 95.2 |

1. Values averaged for $\frac{a}{W} = 0.5$

2. Values averaged for $\frac{a}{W} \leq .65$

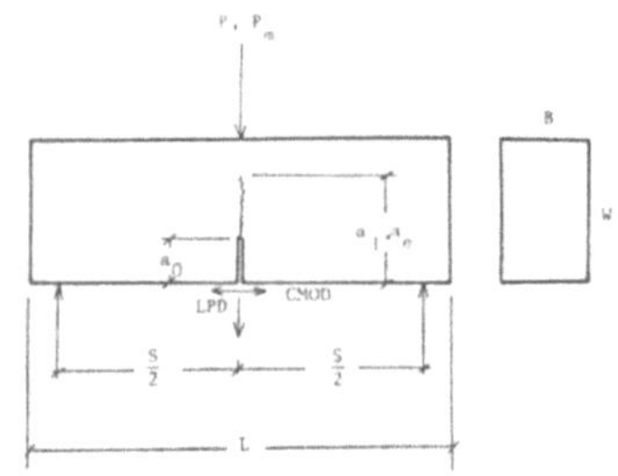


Fig. 1  Test Beam Geometry

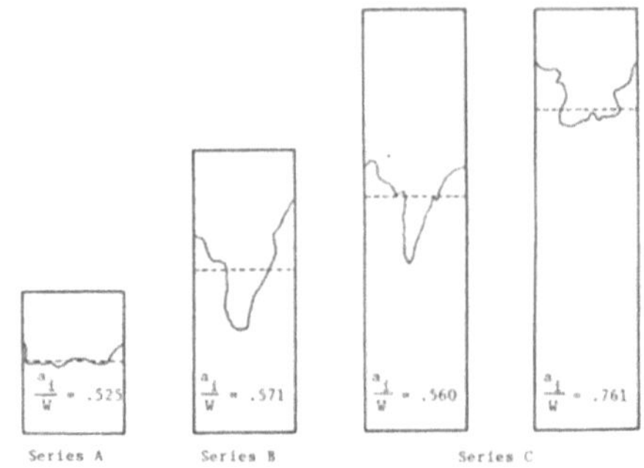


Fig. 2  Appearance of Precracked Surface for Different Beam Sizes

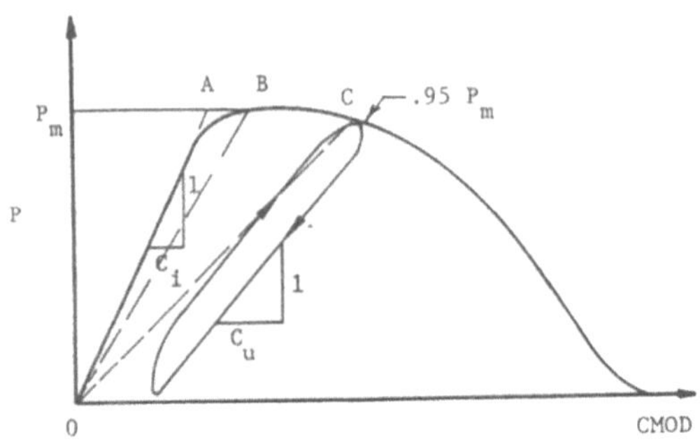


Fig. 3  Load Versus Crack-Mouth-Opening-Displacement

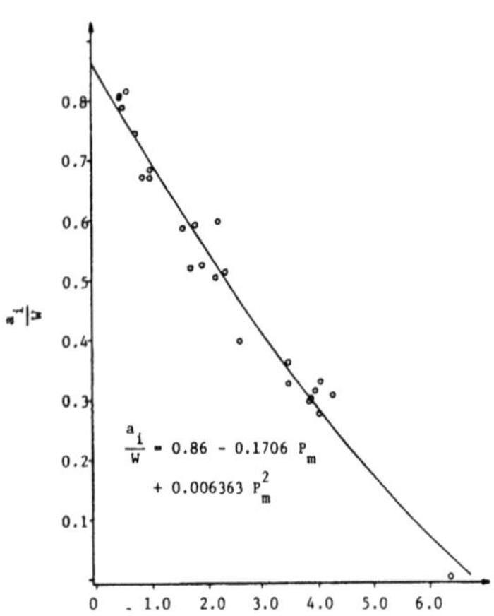


Fig. 4  $\frac{a_i}{W}$ V $P_m$, W = 102 mm

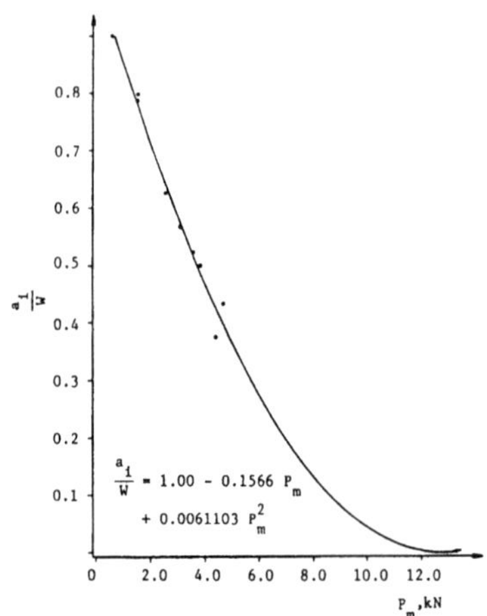


Fig. 5  $\frac{a_i}{W}$ V $P_m$, W = 203 mm

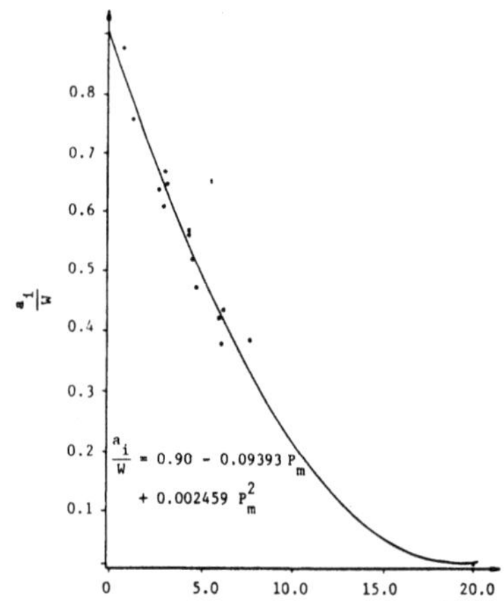


Fig. 6  $\frac{a_i}{W}$ V $P_m$, W = 305 mm

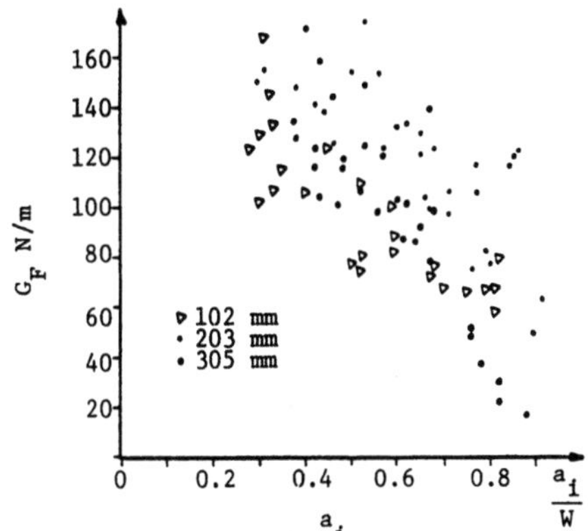


Fig. 7 $G_F \vee \frac{a_i}{W}$ Precracked Beams

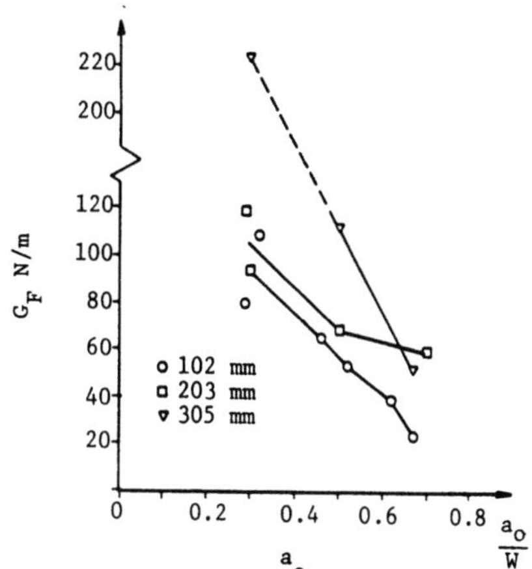


Fig. 8 $G_F \vee \frac{a_o}{W}$, Notched Beams

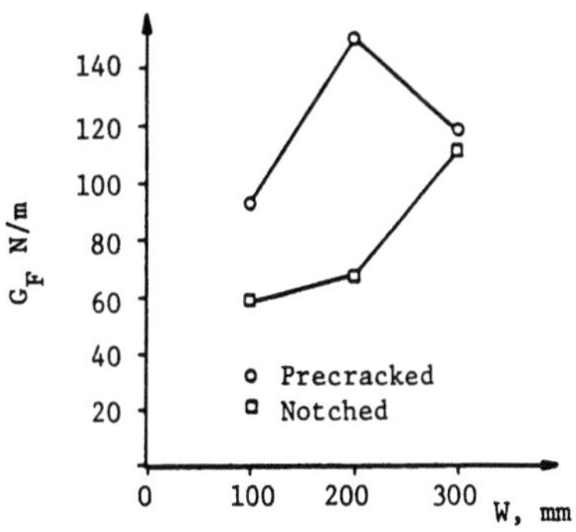


Fig. 9 Average Values of $G_F$ Versus Beam Size For a/W = 0.5

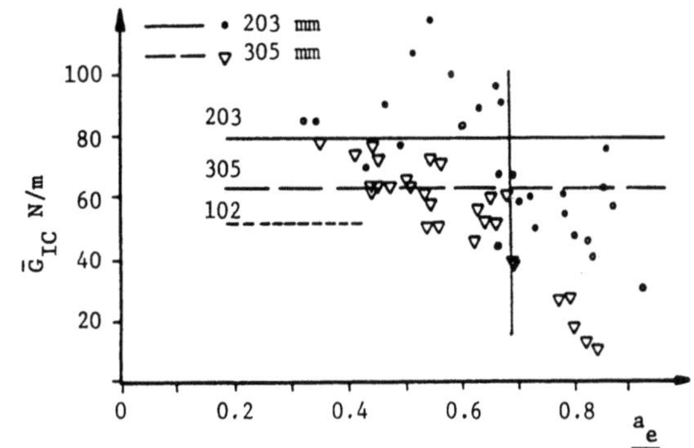


Fig. 10 $\bar{G}_{IC}$ Versus Extended Crack Length - Precracked Beams

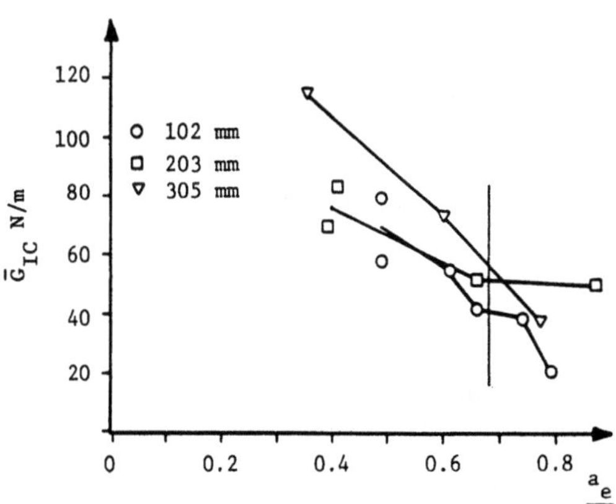


Fig. 11 $\bar{G}_{IC}$ Versus Extended Crack Length - Notched Beams

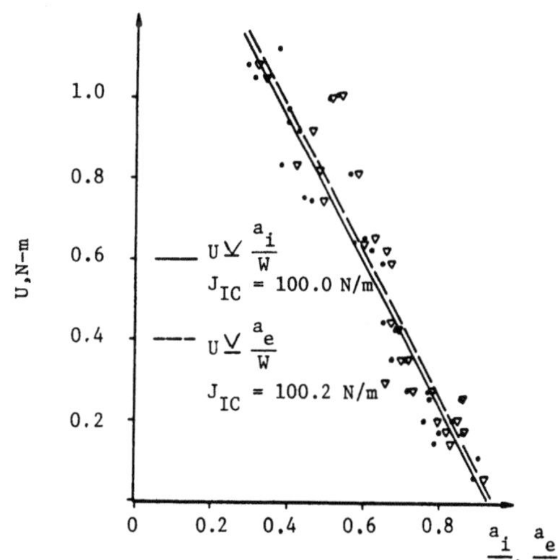


Fig. 12 $U \vee a/W$, Beam Series B W = 8 in. = 203 mm

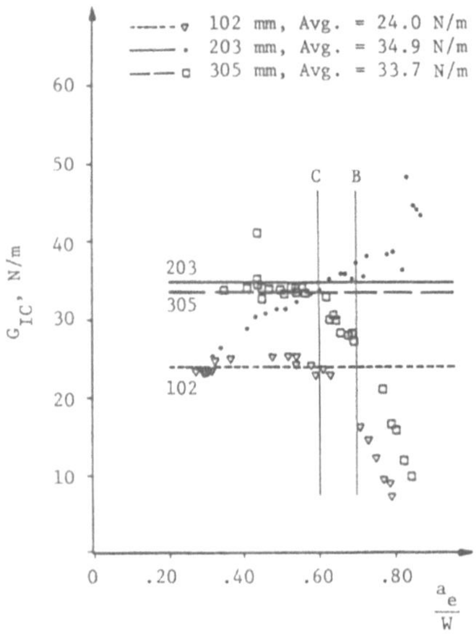


Fig. 13 $G_{IC}$ Based On LEFM And Extended Crack Length, Go Method (2) - Precracked Beams

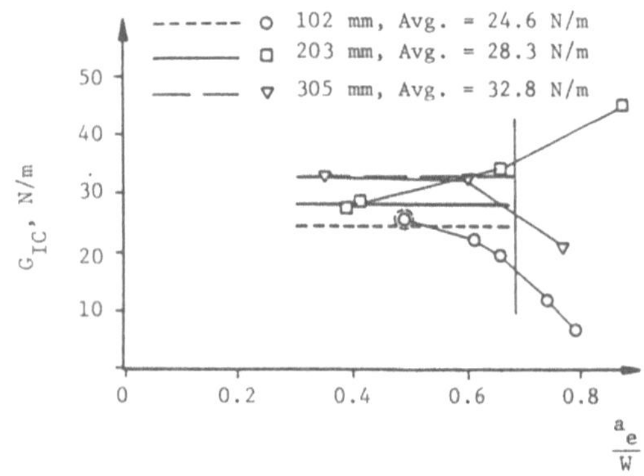


Fig. 14 $G_{IC}$ Based on LEFM And Extended Length, Go Method (2) - Notched Beams

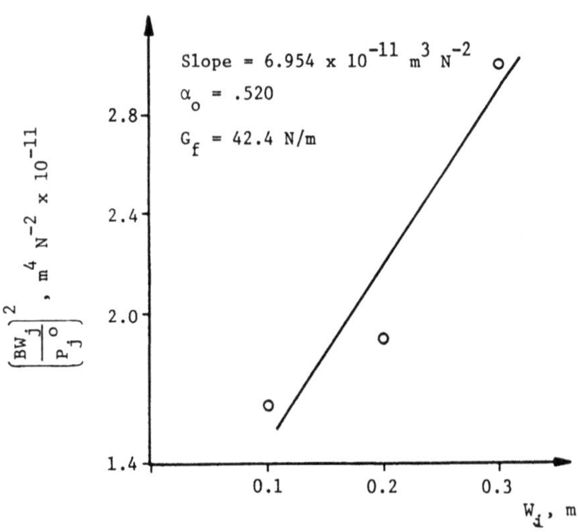


Fig. 15 Size Method (9), Precracked Beams, $a_i/W = 0.52$

SEM/RILEM International Conference
on Fracture of Concrete and Rock
Houston, Texas, June 1987
Editors: S.P. Shah, S.E. Swartz

# TENSION SOFTENING AND SIZE EFFECTS ON THE FRACTURE DETERMINATION OF GEOMATERIALS

V. C. Li, K. P. Chong, and H. H. Einstein

## ABSTRACT

While the linear elastic fracture mechanics (LEFM) based $K_{Ic}$ fracture toughness test has been well established for metals, its direct application to toughness testing of most quasi-brittle geomaterials has not been consistently successful.

In this paper we briefly review experimental and field observations of $K_{Ic}$ size dependence in concrete and rock, and suggest that a valid $K_{Ic}$ test for some such materials may require impractically large specimen sizes. An alternative material tensile property - the tension softening curve - is given prominence in its ability to characterize crack formation and growth in geomaterials, and could be determined in the laboratory with reasonable specimen sizes. An experimental technique, originally developed for concrete, is suggested to be of value in determining the tension-softening curves for rocks.

## INTRODUCTION

Mode I fracture is important for the understanding, design, and optimization of rock fracture processes. For rocks with small grains, such as oil shale the elastic fracture toughness, $K_{Ic}$, has been found to be a useful parameter for the characterization of intact rock with respect to its resistance to crack propagation and as an index of fragmentation processes in the analysis of fracturing [1-4].

Geomaterials, including concrete, rocks and possibly ice, generally exhibit tension-softening behavior during fracturing. While $K_{Ic}$ fracture toughness tests may be meaningful for some metals and other materials with small-grains, they are generally inadequate for toughness testing of geomaterials which exhibit quasi-brittleness, resulting in large process zones [5,6] at crack tips. Existing experimental data indicate that $K_{Ic}$ increases with respect to crack lengths for

V. C. Li and H. H. Einstein are Associate Professor and Professor, respectively, of the Department of Civil Engineering, Massachusetts Institute of Technology, Cambridge, MA 02139.

K. P. Chong (SEM Member) is a Professor, Department of Civil Engineering, University of Wyoming, Laramie, WY 82071.

limestone, concrete, and other geomaterials. A valid $K_{Ic}$ test generally requires impractically large specimen sizes.

Concrete and rock fracture resistance has been mostly characterized in terms of toughness. In this paper, we review the observations of specimen size dependence in fracture toughness determined by LEFM theory in the laboratory, and (for rocks)the discrepancies between toughness estimation in the laboratory and in the field. Such discrepancies are observed both in Mode I tension and in mode II shear. Next, we present a tension-softening constitutive model which is consistent with several macro-mechanical and micro-mechanical observations of careful experimental studies. This model is shown to predict the size dependence mentioned above. A non-linear J-integral based technique is then proposed for accurate determination of this tension-softening curve and fracture toughness using laboratory size rock specimens. Experience in the use of this technique in testing mortar and fiber-reinforced mortar is briefly described.

## FIELD AND LABORATORY OBSERVATIONS

It has been observed that the propped fracture length estimated using pressure build-up tests and/or production data is often much less than predicted by hydraulic fracture simulators [7, 8, 9]. In-situ toughness determined from field pressure data by Shlyapobersky [10] suggests that the in-situ values can often be as much as one to two orders of magnitude higher than LEFM based laboratory results.

Underestimation of fracture toughness based on linear elastic fracture mechanics applied to laboratory size rock specimens have been known for some time [11]. Fig. 1 shows the increasing apparent toughness of rock with a planar dimension of the specimens, in this case the crack length. This suggests that the steady state value or the true $K_{Ic}$ could be achieved only when the specimen size becomes impractically large.

In mode II, Li [12] reported four orders of magnitude difference between fracture toughness data 'measured in-situ' from observations of seismic events (~ $10^6$ $Jm^{-2}$) and that measured in the laboratory based on LEFM (~ $10^1$-$10^2$ $Jm^{-2}$, [13]). However in this case part of the discrepancy may be due to the very different dimensions of heterogenities associated with the linking up of diagonal tensile cracks in the shear rupture process sometimes running up to hundreds of km size scale in comparison to the much smoother rupture in laboratory experiments.

Indeed, the size dependence of apparent fracture toughness is common in any quasi-brittle material (sometimes known as non-yielding material). For example, Francois [14] pointed out that the apparent toughness of concrete increases with the dimension of the specimen used in the test, as shown in Fig. 2.

The above observation of size dependence of apparent fracture toughness in laboratory rocks is consistent with the findings of Shlyapobersky [10], in that the in-situ field measurements could be interpreted as experimental results from 'specimens' many times larger than typical laboratory size specimens, and hence also result in a toughness value many times larger than that obtained from a laboratory test based on LEFM.

The question remains as to how laboratory tests could produce a toughness value closer to the in-situ true fracture toughness. We can either build a huge laboratory and test huge specimens; or we can abandon the concept of linear elastic fracture mechanics (LEFM). We propose to develop an experimental technique based on non-linear elastic fracture theory which extracts not only the fracture toughness of rocks accurately but also provides additional information on the tensile softening behavior. Such information are often extremely important in analyzing the failure stability of a rock mass.

## TENSION-SOFTENING BEHAVIOR IN ROCK

By tension-softening we mean the gradual decrease in tensile load bearing capacity as material separate across an eventual failure surface (it is easier to think of this for a uniaxial tension specimen). The micromechanisms responsible for tension-softening could be grain sliding, microcrack linking and possibly frictional interlocking. This tension-softening behavior is responsible for the development of the crack tip fracture process zone.

Measurement of tension-softening curve for rocks has not been attempted. However, stable unaxial tests of Barre granite, Berea sandstone, Valdor limestone and Tennessee marble has been carried out by Peng [15]. A set of this result for Barre granite is shown in Fig.3a. The corresponding tension-softening curve can be deduced by assuming elastic unloading from the softening regime, and is shown (for the $2.2 \times 10^{-4}$ $sec^{-1}$ strain rate data) in Fig.3b. The limited data set from [15] suggests that these diverse rock type exhibits quite different tension-softening behavior, depending on the material microstructure (and loading rate).

## THEORETICAL WORK ON COHESION FRACTURE MODELS AND PROCESS ZONE DEVELOPMENT

The link between tension-softening and fracture behavior is provided (amongst others), by Li and Liang [16] who analyzed the process zone development of a simple center cracked panel made of a quasi-brittle material. They demonstrate that the development of a long process zone is responsible for the inapplicability of LEFM, and that for such a material, an indiscrete use of LEFM will lead to an underestimate of the true fracture toughness. Li [12] used this model to predict the increase of rock fracture toughness as a function of crack length, shown as the solid curves in Fig. 1.

Presence of a long process zone at a crack tip has been repeatedly observed in various quasi-brittle materials. For example, using acoustic emission sensors, Labuz et al [17] tracked development of the process zone in double cantilever beam specimens of Charcoal and Rockville granite (Fig. 4). They concluded that the length of the ligament process zone could form a substantial portion of the effective crack. Many other investigators [6] have found similar results.

The tension softening curve is a very useful property of quasi-brittle materials. Apart from providing size-independent fracture parameters, it may also be used in numerical fracture simulations where LEFM is not applicable. Such situations arise, e.g., in fracture running into geometric or material boundaries in a rock mass, when crack and/or joints interact, and when cracks are forming at a stress concentrator. In those situations, the small scale yielding condition required by LEFM may be violated.

## TENSION SOFTENING CURVE AS A NON-LINEAR FRACTURE PARAMETER

Using the J-integral with a contour around the process zone in which tension-softening occurs, Rice [18] showed that the area under the tension-softening curve can be related to the critical energy release rate for imminent fracture. Thus if the tension softening curve can be determined experimentally, then it is possible to deduce the critical energy release rate and the fracture toughness. Note that in this scheme, the experimental portion does not involve the use of LEFM theory and hence there is usually no specimen size requirements as in the case of a regular fracture toughness test.

## EXPERIMENTAL WORK IN TENSION-SOFTENING CURVE MEASUREMENTS

Direct tension test for quasi-brittle materials are usually problematic when it is necessary to carry the test stably in the post-peak regime. In addition, the problem of instrumentation (location of displacement transducers or strain gages across potential crack planes) and the sensitivity of slight loading eccentricity on the test result make the direct tension test difficult to carry out.

In the last several years a method has been developed at the Center of Advanced Construction Materials at MIT [19, 20] for the indirect determination of tension-softening curves for quasi-brittle materials. The theoretical basis employs the J-integral and the concept of tension-softening, and has been described in detail in Li et al [19]. (To provide perspective, the J-integral test method for metal has now been well defined and accepted, and has had a strong impact on the toughness testing of ductile metals. Our work aims at developing along parallel lines the theoretical base and experimental technique for toughness evaluation in quasi-brittle materials which do not plastically yield, but nevertheless has a process zone large enough to invalidate

LEFM.) The test technique has been applied to mortar and steel fiber reinforced concrete [19,20] using a compact tension specimen and a 4-point bend specimen. The results have been compared to those from stable direct tension tests in similar materials, and has been shown to be relatively consistent. It may be necessary, however, to carry out an independent direct tension test just to obtain the tensile strength, as the indirect test often produces some scatter of this value from test to test [20]. The test technique can also be adapted for the semi-circular bend specimen [21, 22] which may be more suitable for rock testing.

## VERIFICATION OF TESTING PROCEDURE

An independent verification for our testing procedure was provided by A. Hillerborg (private communications, 1985). He employed his fictitious crack model in a finite element scheme to simulate the load-deformation curves and load-crack tip opening curves of a pair of three point bend specimens of slightly different notch lengths. He used an artificial bi-linear curve as input for the tension-softening behavior in the material ahead of the notches. The objective of the exercise is to extract this same curve using our procedure [19] and his numerically derived 'test' results. If the J-integral based test technique is theoretically sound, then he should get a predicted tension-softening curve overlapping the bi-linear curve he input into his finite-element program. The result is most encouraging. As shown in Fig. 5, the two curves essentially overlap one another.

## CONCLUSION

The theoretical work by Li and Liang [16] provides the physical basis for understanding the expected underestimation of fracture toughness of quasi-brittle materials using LEFM theory. We have now a methodology in hand to obtain fracture test results which are size independent. We suggest that the application and refinement of this test methodology to rocks can eliminate the discrepancies between model predictions and field observations in rock masses and other applications, such as overpressures in hydraulic fracturing processes. Once this method is established, an updated data base of rock fracture toughness data can then be obtained for use by engineers.

## ACKNOWLEDGEMENT

VCL would like to thank A. Hillerborg for performing the FEM numerical verification of the test technique, which results in Fig. 5. Fig. 3b is from a term project report by R. Ty. This work has been supported by a grant (MSM-8516893) from the Solids and Geomechanics Program of the National Science Foundation and an award (DACA88-86-D-0013 #07) from the Air Force Engineering and Systems Laboratory to the Massachusetts Institute of Technology.

# REFERENCES


[1] Chong, K.P. and J.W. Smith, eds, Mechanics of Oil Shale, Elsevier (1984).

[2] Simonson, E.R., A.S. Abou-Sayed and R.J. Clifton, "Constrainment of Massive Hydraulic Fracture", SPE paper No. 6089, 51st Annual Fall Meeting of SPE of AIME (1976).

[3] Schmidt, R.A., "Fracture Toughness of Limestone", Exper. Mech. 16, 161-167 (1976).

[4] Kusmaul, J.S., "Numerical Modeling of Oil Shale Fragmentation Experiments", Proceedings 11th Annual Society of Explosives Engineers Conf., C.S. Konya, ed., San Diego, California (1985).

[5] Wecharatana M. and S.P. Shah, "Prediction of Nonlinear Fracture Process Zone in Concrete", Journal of EMD of ASCE 109, 1231-1246 (1983).

[6] Goldsmith, W, ed. "Assessment of Experimental Techniques Applicable to Rock Fracture", a report on an NSF workshop, Utah, (1986).

[7] Medlin W.L. and Fitch J.L., "Abnormal Treating Pressures in MHF Treatments". SPE Paper 12108. 58th Annual Conference. San Francisco, CA, Oct., (1983).

[8] Warpinski, N.R., "MWX Paludal In Situ Stress Measurements and Hydraulic Fracture Behavior". Memorandum. Sandia National Labs. Oct, (1984).

[9] Warpinski, N.R., "Summary of Results of MWX Paludal Zone Phase I Stimulation". Memorandum of Record. Sandia National Labs., Jan, (1984).

[10] Shlyapobersky, J., "Energy Analysis of Hydraulic Fracturing", in the 26th US Symposium on Rock Mechanics, Rapid City, SD, pp. 539-546, June, (1985).

[11] Ingraffea, A.R., Gaunsallus, K.L., Beech, J.F. and Nelson, P.P., "A Short Rod Based System for Fracture Toughness Testing of Rock", ASTM STP 855 Chevron-Notched Specimens: Testing and Stress Analysis, Louisville, p.152-166, (1983).

[12] Li, V.C., "Mechanics of Shear Rupture Applied to Earthquake Zones", book chapter in Rock Fracture Mechanics, ed. B. Atkinson, Academic Press, (1986).

[13] Atkinson, B., "Experimental Methods in Rock Fracture Testing", in Rock Fracture Mechanics, edited by B. Atkinson, Academic Press, (1986).

[14] Francois, D., "Fracture and Damage Mechanics of Concrete", in Application of Fracture Mechanics to Cementitious Composites, edited by S.P.Shah, Martinus Nijhoff Publishers, p141-156, 1985.

[15] Peng, Syd S., "A Note on the Fracture Propagation and Time-Dependent Behavior of Rocks in Uniaxial Tension", Int. J. Rock Mech. Min. Sci. & Geomech. Abstr., Vol 12, 125-127, Pergamon Press, (1975).

[16] Li, V.C. and E. Liang, "Fracture Processes in Concrete and Fibre-Reinforced Cementitious Composites", ASCE. J. Engineering Mechanics, Vol. 122 No. 6, pp. 566-586, (1986).

[17] Labuz, J.F., Shah, S.P., and C.H. Dowding, "Measurement and Description of the Tensile Fracture Process in Rock", submitted to J. Geophys. Res, (1986).

[18] Rice, J.R, "Mathematical Analysis in the Mechanics of Fracture", in Fracture: An Advanced Treatise, Vol. 2, Academic Press, pp. 191-311, (1968).

[19] Li, V.C. and C.M. Chan and C. Leung, "Experimental Determination of the Tension-Softening Curve in Cementitious Composites", accepted for publication in Cement and Concrete Research, (1986).

[20] Leung, C.K.Y. and V. C. Li, "Determination of Fracture Toughness Parameter of Quasi-Brittle Materials with Laboratory-Size Specimens", Proc. Int'l Conf. on Fracture Concrete and Rock, SEM, Houston, TX, (1987).

[21] Chong K.P. and M.D. Kuruppu, "New Specimen for Fracture Toughness Determination of Rock and Other Materials", Int. J. Fract., 26, R 59 - R 62 (1984).

[22] Chong, K.P., M.D. Kuruppu and J.S. Kuszmaul, "Fracture Toughness Determination of Rocks with Core-Based Specimens", keynote paper, Proc. Int'l Conf. on Fracture of Concrete and Rock, SEM, Houston, TX (1987).

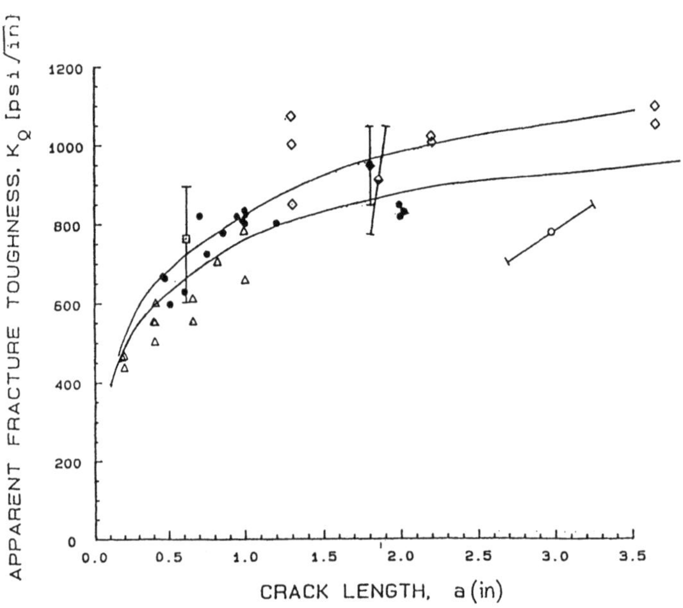


Figure 1 : Increasing apparent fracture toughness $K_Q$ with crack length of Indiana Limestone specimens (from Ingraffea et al [117]). Solid curves are size-dependent $K_Q$-predictions based on cohesion model of Li and Liang [16], reported in Li [12].

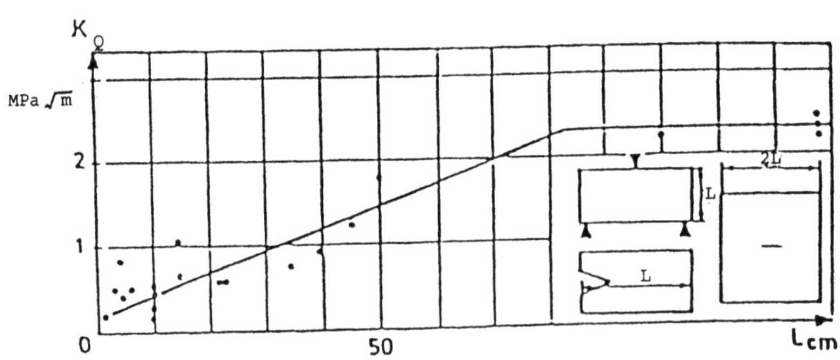


Figure 2 : Various results of the fracture toughness $K_Q$ taken from the literature as a function of the size of the specimens (from Francois [14]).

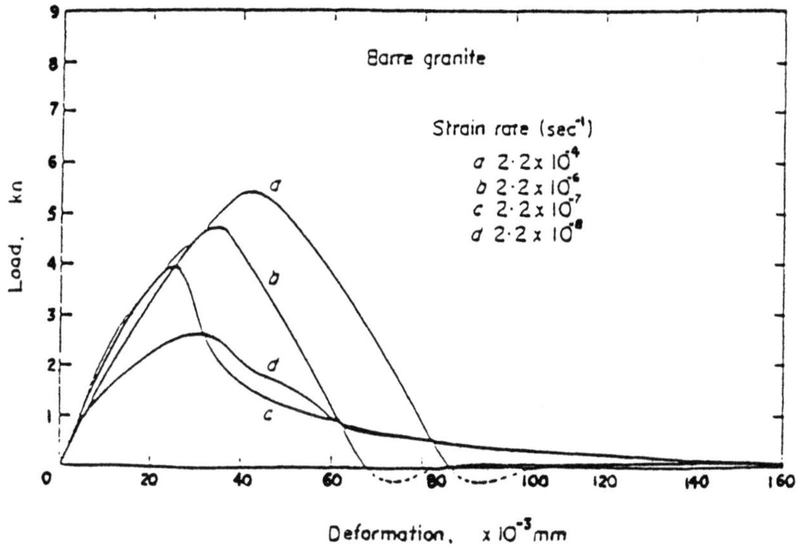


Figure 3a : Complete load-deformation curves for Barre granite tested under various strain rates in uniaxial tension (from Peng [15]).

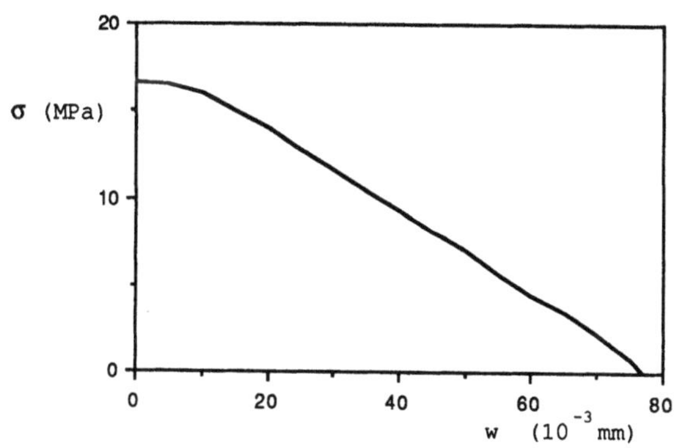


Figure 3b : Tension-softening curve for Barre granite deduced from load-deformation curve of uniaxial tension test shown in Figure 3a (strain rate = 2.2 x $10^{-4}$ $sec^{-1}$).

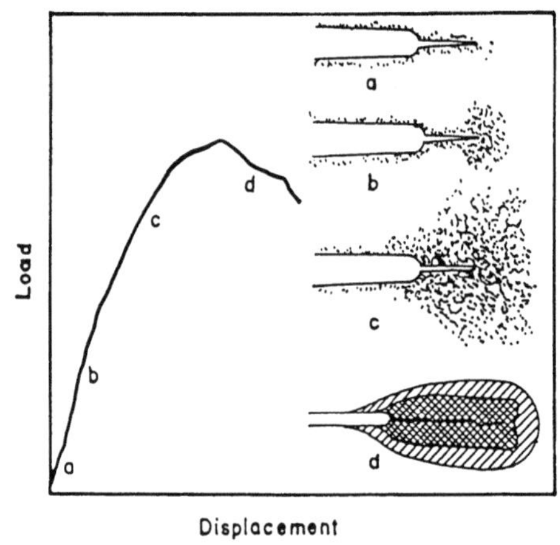


Figure 4 : Micro-cracking identified by acoustic emission, and process zone development (from Labuz et al [17]).

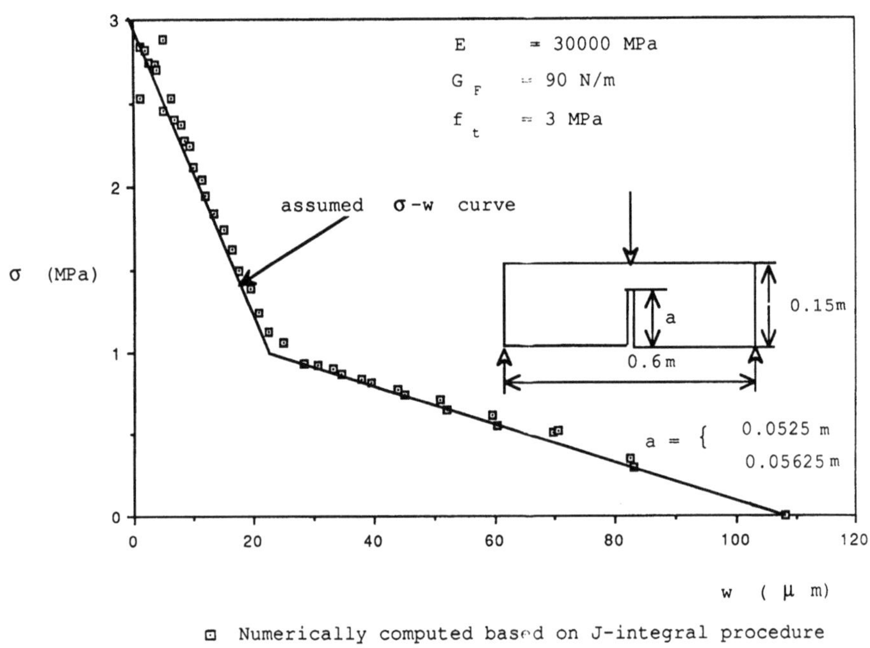


Figure 5 : Assumed bi-linear and computed tension-softening curve (Hillerborg, personal communications, 1985). The good agreement supports the J-integral procedure proposed by Li et al [19] for experimentally deducing the tension-softening curve.

SEM/RILEM International Conference
on Fracture of Concrete and Rock
Houston, Texas, June 1987
Editors: S.P. Shah, S.E. Swartz

# INFLUENCE OF FRACTURE PROCESS ZONE ON INELASTIC BEHAVIOUR OF CONCRETE

Lucio NOBILE

## ABSTRACT

In the framework of the constructed model of behaviour of a compressed concrete containing distributed flat microcracks,slow microcrack growth is included in deriving stress-strain relations.

## INTRODUCTION

Recently, a model of behaviour of a compressed concrete containing randomly distributed flat microcracks of the same size has been proposed to explain the non linearity of the stress-strain relations, the dilatancy and the stress induced anisotropy(1).

In particular the overall mechanical response of concrete in compression strongly depends on the distribution of the microcracks that undergo relative frictional sliding ( First stage of inelasticity ) and those that propagate ( Second stage of inelasticity ). Therefore the average strain can be decomposed into the sum of the average elastic strain, related to the elastic matrix, and the average additional strain,related to the microcracks.

Based on the model proposed by Wecharatana & Shah (2) and Jenq & Shah (3) to include non linear slow crack growth, in this paper an attempt is made to modify the average additional strain related to microcrack propagation during the second stage of inelasticity.

## STRAIN ENERGY RELEASE RATE

In order to include the non linear slow crack growth, R-curve analysis has been proposed (2), equivalent to the analysis of instability based on the elastic-plastic extending crack solutions (4). Thus R-curves, obtained by plotting the strain energy release rate at each crack initiation ( $G_R$) against the actual crack extension, can be considered a material property.

In the framework of the proposed model of concrete behaviour (1), the second stage of inelasticity is determined by penny-shaped microcrack growth.

Let us consider a penny-shaped microcrack with any orientation lying within an isotropic elastic body. If P is the total energy released by self-similar growth of the microcrack,then (5)

$$\frac{\delta P}{\delta r} = \frac{1-\upsilon^2}{E} \oint_C \left[ K_I^2 + K_{II}^2 + K_{III}^2/(1-\upsilon) \right] ds \qquad (1.1)$$

because the path indipendence applies strictly to surfaces that enclose the microcrack outside the process zone. In (1.1)r is microcrack radius, $\delta/\delta r$ denotes partial differentiation with respect to r,E and $\upsilon$ are Young's modulus and Poisson's ratio, respectively, and $K_I$, $K_{II}$ and $K_{III}$ are the stress-intensity factors corresponding to the Opening, sliding and tearing modes of fracture, respectively.

---

L.NOBILE, research engineer,Istituto di Scienza e Tecnica delle Costruzioni, Facoltà di Ingegneria,Università di Bari (Italy).

In compression the symmetric singularities disappear and then (1)

$$\frac{\delta P}{\delta r} = \frac{16(1-\upsilon^2)}{E(2-\upsilon)} r^2 \tau_n^2 \tag{1.2}$$

with $\tau_n$ the shear traction reduced by the amount of Coulomb friction, $\mu|\sigma_n|$, with $\sigma_n$ the normal traction.

The strain energy release rates for increasing applied loads are shown in Fig.1, as dashed lines originating at the origin. The instability load corresponds to the member of these lines which tangentially touches the resistance curve, and the amount of stable microcrack growth is determined by the abscissa of the point of tangency, beyond which a load drop is required to maintain quasistatic microcrack growth.

Since the R-curve would be a universal curve for a given material and condition, Jenq & Shah(3) proposed a two parameter fracture model in order to obtain a size-independent resistance curve. This curve is shown in Fig.2 and is related to concrete of given mix-proportions and maximum aggegate size.

Let us consider a typical size-independent R-curve, as in Fig.3. If $G_c=(\delta P/\delta r)/(2\pi r)$ denotes the applied energy release rate, slow crack growth begins approximately when $G_c=G^s_{IC}/2$, where $G^s_{IC}$ is the steady-state value at which the strain energy release rate becomes independent of the crack length.

The value $\tau_{no}$ of the shear traction, corresponding to the beginning of slow microcrack growth, can be derived from

$$\tau^2_{no} = \frac{\pi E(2-\upsilon)}{16(1-\upsilon^2)} \frac{G^s_{IC}}{r_o} \tag{1.3}$$

where $r_o$ is the initial microcrack radius.

The value $\tau_{ni}$ of the shear traction, corresponding to instability, is determined by the line which tangentially touches the resistance curve.

The current value r of the microcrack radius is determined by the family of dashed lines from $\tau_{no}$ to $\tau_{ni}$, as shown in Fig.3.

MACROSCOPIC STRESS-STRAIN RELATIONS

According to the proposed model(1), the average strain can be decomposed into the sum of the average elastic strain, related to the linear elastic matrix, and the additional average strain, related to the microcracks.

In the first stage of inelasticity - frictional sliding on preexisting microcracks -, the additional average strain is (1)

$$\bar{\varepsilon}^M_{ij} = \frac{8}{3\pi} \frac{1-\upsilon^2}{2-\upsilon} \frac{\rho_o}{E} \int_{\Phi_1}^{\Phi_2} \int_{\alpha_1}^{\alpha_2} (\tau_{ni}n_j + \tau_{nj}n_i)\, d\alpha \cos\Phi\, d\Phi \tag{2.1}$$

where $n_i$ and $n_j$ are the components of the exterior unit normal to the microcrack faces, $\rho_o = N r^3/V$ ( N is the number of microcracks in the solid of total volume V ) is the initial microcrack concentration parameter and the limits of intagration determine the set of activated microcracks, for which $\tau_n > 0$.

In the second stage of inelasticity- microcrack growth - the additional average strain consists of two parts:

- the additional average strain caused by frictional sliding, given by (2.1) but replacing $\rho_o$ with $\rho = Nr^3/V$ - current value of microcrack concentration parameter related to the current value of microcrack radius;
- the additional average strain (1)

$$\bar{\bar{\varepsilon}}^M_{ij} = \frac{\delta}{\delta\sigma^M_{ij}} \frac{16(1-\upsilon^2)}{3(2-\upsilon)} \frac{\rho}{E} \int_{\Phi'_1}^{\Phi'_2} \int_{\alpha_1}^{\alpha_2} \tau^2_n \, d\alpha \cos\Phi\, d\Phi \tag{2.2}$$

where $\delta/\delta\sigma^M_{ij}$ denotes partial differentiation with respect to the average stress component and the limits of integration determine the set of activated microcracks for which $\tau_n > \tau_{no}$.

This stage ends when $\tau_n = \tau_{ni}$.

In the case of axisymmetric loading $\alpha_1=0$ and $\alpha_2 = 2\pi$.

In particular, for the triaxial compression test with confining pressure q kept constant and axial pressure p ( in the $x_3$-direction) increasing, the limits of integration $\Phi_1, \Phi_2$ in (2.1) are to be determined from the condition

$$(p-q)\sin\Phi\cos\Phi - \mu(p \sin^2\Phi + q \cos^2\Phi) > 0 \tag{2.3}$$

while the limits of integration $\Phi_1', \Phi_2'$ in (2.2) are to be determined from the condition

$$(p-q)\sin\Phi\cos\Phi - \mu(p \sin^2\Phi + q \cos^2\phi) > \tau_{no} \tag{2.4}$$

Thus the only non-zero components are

$$\bar{\varepsilon}^M_{11} = \bar{\varepsilon}^M_{22} = \frac{16(1-\upsilon^2)}{3(2-\upsilon)} \frac{\rho}{E} q \left[ \mu \frac{p}{q} \frac{\cos^3\Phi}{3} + \left(\frac{p}{q} - 1\right)\left(\frac{\sin^3\Phi}{3} - \frac{\sin^5\Phi}{5}\right) - \mu \frac{\cos^5\Phi}{5} \right]_{\Phi_1}^{\Phi_2} \tag{2.4}$$

$$\bar{\varepsilon}^M_{33} = -2\bar{\varepsilon}^M_{11} \tag{2.5}$$

$$\bar{\bar{\varepsilon}}^M_{11} = \bar{\bar{\varepsilon}}^M_{22} = \frac{16(1-\upsilon^2)}{3(2-\upsilon)} \frac{\rho}{E} 2\pi q \left[ 2\left(1 - \frac{p}{q} + 2\mu^2 \frac{p}{q}\right)\left(\frac{\sin^3\Phi}{3} - \frac{\sin^5\Phi}{5}\right) - 4\mu \frac{\cos^5\Phi}{5}\left(1 - \frac{p}{q}\right) + -2\mu \frac{p}{q} \frac{\cos^3\Phi}{3} + 2\mu^2\left(\frac{\sin\Phi\cos^4\Phi}{5} + \frac{4}{5}\sin\Phi - \frac{4}{15}\sin^3\Phi\right) \right]_{\Phi_1'}^{\Phi_2'} \tag{2.6}$$

$$\bar{\bar{\varepsilon}}^M_{33} = \frac{16(1-\upsilon^2)}{3(2-\upsilon)} \frac{\rho}{E} 2\pi q \left[ 2\left(\frac{p}{q} + \mu^2 - 1\right)\left(\frac{\sin^3\Phi}{3} - \frac{\sin^5\Phi}{5}\right) + 2\frac{p}{q}\frac{\sin^5\Phi}{5} + 2\mu\frac{\cos^5\Phi}{5} + 2\mu\left(\frac{p}{q} - 1\right)\left(\frac{\cos^3\Phi}{3} - \frac{\cos^5\Phi}{5}\right) \right]_{\Phi_1'}^{\Phi_2'} \tag{2.7}$$

The condition determining fracture instability is given by

$$(p-q)\sin\Phi\cos\Phi - \mu(p \sin^2\Phi + q \cos^2\Phi) = \tau_{ni} \tag{2.8}$$

REFERENCES

(1)L.Nobile : A microcrack model of concrete's behaviour under applied loads.Fracture Toughness and Fracture Energy of Concrete,edited by F.H.Wittmann.Elsevier Science Pub.(1986)

(2)M.Wecharatana & S.P.Shah: Slow crack growth in cement composites.Proceedings,Journal of the Structural Division,ASCE,Vol.108,June 1982,pp.1400-1413.

(3)Y.Jenq & S.P.Shah: Two Parameter Fracture Model for Concrete. Journ. of Engin.Mechanics. Vol.111,No.10,Oct.85,pp.1227-1241.

(4)J.R.Rice:Mathematical Analysis in the Mechanics of Fracture.Fracture,edited by H.Liebowitz. Academic Press(1968)

(5)B.Budiansky & J.R.Rice: Conservation laws and Energy release rates.Journ.Appl.Mechanics. March 1973.

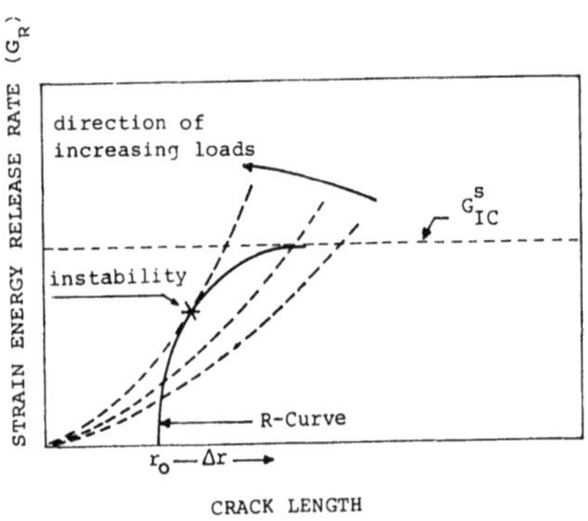


FIG.1 - RESISTANCE, R-CURVE.

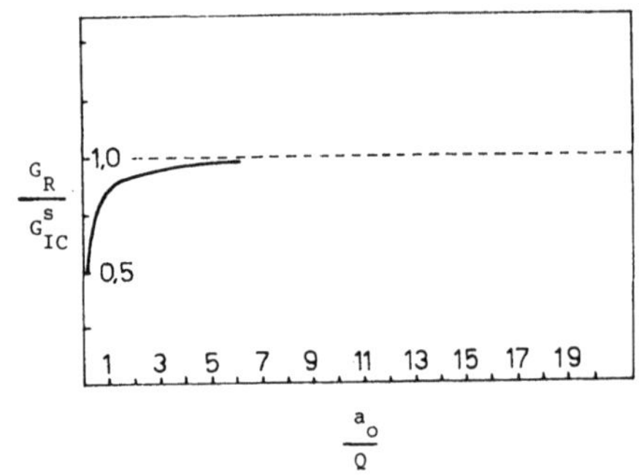


$C_R = \bar{K}_{IC}^2/E$ $\qquad Q = \left(\frac{E\ CTOD}{K^s_{IC}} c\right)^2$

FIG.2 - R-CURVE BY JENQ & SHAH (3).

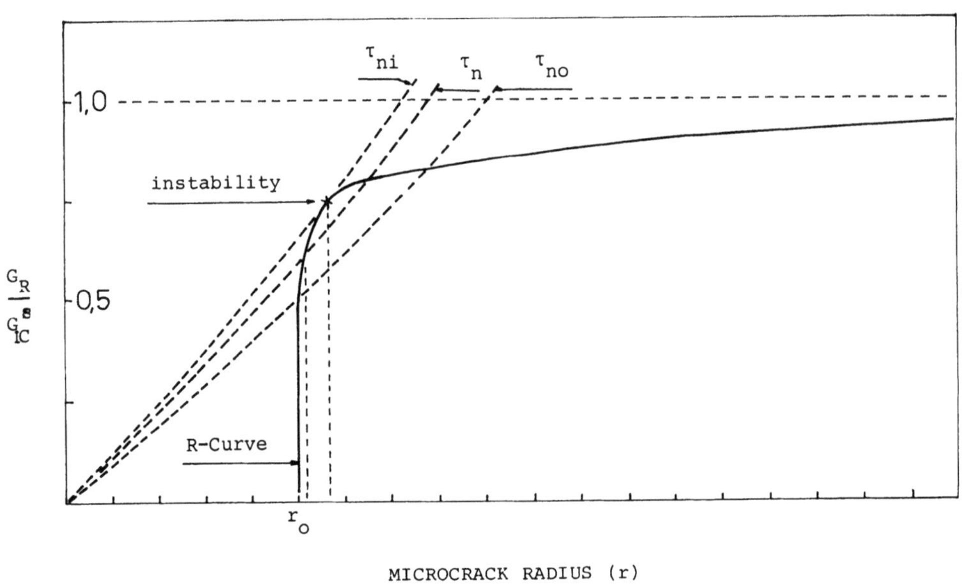


FIG.3 - GRAPHICAL REPRESENTATION OF CONDITION DETERMINING MICROCRACK GROWTH AND FRACTURE INSTABILITY.

SEM/RILEM International Conference
on Fracture of Concrete and Rock
Houston, Texas, June 1987
Editors: S.P. Shah, S.E. Swartz



# ANCHOR BOLTS ANALYSED WITH FRACTURE MECHANICS

Lennart Elfgren, Ulf Ohlsson and Kent Gylltoft

## ABSTRACT

A fracture mechanics analysis is presented of anchor bolts embedded in concrete or grouted in concrete or rock. A discrete crack approach is used with a bi-linear softening material model.

## INTRODUCTION

Anchor bolts, see Fig. 1, embedded in concrete or grouted in concrete or rock are very important structural details. Various design procedures have been developed over the years based on simplified assumptions and empirical test results, see e.g. Klinger and Mendonca 1982 [1].

With the help of fracture mechanics it is possible to analyse the interaction between an anchor and the surrounding concrete or grout in a more efficient way than was possible before. This may in the future form a basis for improved design methods.

Early non-linear analysis of the anchor bolt problem was performed by Ottosen in 1981 [2] and Peier in 1983 [3]. A recent investigation was presented by de Borst in 1986 [4].

Fracture mechanics analysis with a discrete crack approach was used by Elfgren et al in 1982 and 1986 [5], [6], by Ballarini et al in 1986 [7] and by Hellier et al in 1987 [8].

## ELASTIC ANALYSIS

In order to get a general idea of the stresses around an anchor bolt, before cracking of the concrete, an elastic analysis can be performed. In Fig. 2 plots of isostresses for a typical case is given [9]. The inclinations of the principle tensile and compressive stresses is also indicated in the figure. In the neighbourhood of the top corner of the disk the inclination of the principle tensile stresses varies around $20^o$ to $40^o$ to the vertical axis. The stress condition can also be illustrated with trajectories of principle stresses, see Fig. 3 [10].

It can be seen from Figs. 2 and 3 that high tensile stresses occur at the top of the disk and that tensile cracking is likely to appear here. Results from pull-out tests by Stone and Carino [11], [12] and by Krenchel and Shah [13] support this.

## FRACTURE MECHANICS MODEL

A discrete crack approach as outlined by Hillerborg et al in [14], [15] will be used. The crack will be modelled by a non-linear two-dimensional linkage element developed by Gylltoft [16] using the finite element program FEMP [17]. The linkage element consists of two non-linear springs, see Fig. 4 with material properties according to Fig. 5. In tension a normal bi-linear softening diagram has been chosen with a tensile strength of 3 MPa and a fracture energy $G_F$ of 100 $Nm/m^2$. In shear a bi-linear diagram has also been chosen. The same value was chosen for characteristic shear deformations as for characteristic tensile deformations. As frictional forces play a dominant part it is assumed that the shear diagram has an almost constant part for small deformations. As the crack widens, the shear capacity is gradually decreased.

---

Lennart Elfgren and Ulf Ohlsson are Professor and Research Assistant, respectively, Division of Structural Engineering, Luleå University of Technology, S-951 87 Luleå, Sweden; Kent Gylltoft is Head of the Section for Building Structures, The Swedish National Testing Institute, Box 857, S-501 15 Borås, Sweden

When fracture occurs in one direction, this must also influence the behaviour in the other direction. The principle used here is that each fracture stage should be reached in the two directions at the same time. This means that if fracture is first reached in tension, point A in Fig. 6a, line AB in Fig. 6b will be lowered to position A'B' until fracture is also reached in shear. Further, when points B and C are reached in tension, Fig. 6a, point B and C will be changed to B' and C' in shear, Fig. 6b.

MODELS WITH DIFFERENT CRACK PATHS

For the geometry shown in Fig. 7 calculations were performed with three different crack paths:

(a) a straight crack with an inclination of $45^0$ to the vertical axis. This is in conformity with many approximate design procedures

(b) a straight crack with an inclination of $67^0$ to the vertical axis. This is in conformity with the direction of the maximum tensile stresses in the concrete close to the top of the disk

(c) a curved crack with an initial angle of $73^0$ to the vertical axis. At the point when the crack was to propagate (that is the tensile strength was reached in the region at the top of the crack) the direction of the maximum tensile stress was calculated and the element net was changed so that the next crack element got this direction. This procedure was performed manually after each calculation step.

Deformed configurations of the anchor bolt are shown in Fig. 8 for two load levels: one corresponding to a load close to 200 kN and one close to failure. Load-deformation diagrams are shown in Fig. 9 and tensile and shear stresses along the cracks are shown in Fig. 10. Numerical inaccuracies may occur in some of the solution algoritms for high load levels. Accordingly, the interpretation of the results should be done with due caution.

For the crack path with a straight $45^0$ crack, the maximum stress (3 MPa) was first reached in the shear direction. This is not likely to happen in a real structure. According to the failure laws of Mohr-Coulomb, the shear capacity is usually higher than the tensile capacity. This can be taken into consideration by modifying the material properties of the linkage elements. However, for a case where the maximum stress is first reached in the tensile direction, it is not likely that the shear contribution ought to be higher than what has been assumed here. Better knowledge of the interaction between the tensile and shear capacity in a crack zone may clarify this.

For the crack paths with an initial inclination of $67^0$ to the vertical axis, the maximum stress is first reached in tension. This is as it should be as the initial inclination was choosen to be perpendicular to the principle tensile stress. The differences between the straight and the curved crack paths are here small.

The failure load for the three studied crack paths amounts to 283 kN ($45^0$), 342 kN ($67^0$), and 340 kN (curved crack). For a bolt with similar geometrical conditions but located only 300 mm from an edge, tests gave a failure load of 206 kN. The concrete in the test had a compressive strength of $f_c$ = 58 MPa (150 mm cubes) and a tensile strength of $f_t$ = 2.9 MPa (cube splitting test) [9]. Taking the weakening effect of the edge into consideration the test result is of near the same magnitude as the calculated value for a $45^0$-crack.

DISCUSSION AND CONCLUSIONS

The straight crack path with an inclination of $45^0$ gave the lowest load (283 kN). This is probably due to the fact that this crack path is the shortest one. Thus the fracture energy for this path is lower than the fracture energy for the other, longer crack paths.

As an alternative a criterion for the crack direction based on minimum of fracture could be used. There ought to be a balance point where the positive influence of a steep crack (small crack area) is counteracted by high shear stresses as well as where the negative influence of a slightly inclined crack (big crack area) is compensated by low shear stresses.

Pull-out tests by Krenchel and Shah [13] indicate that primary cracking can occur at an angle of 70 to $80^0$ to the vertical axis. At higher load levels they observed secondary cracks starting from the upper edge of the disk and going to the support. In their case the support was located above the edge of the disk at an angle of $31^0$ to the vertical axis. A recent fracture mechanics analysis of this problem using two discrete cracks have been performed by Hellier et al [8]. In our case secondary cracks may appear as in Fig. 11.

Further studies should be directed at:

o the interaction between shear and tension in crack zones with strain-softening

o finding critical crack paths for different geometries and varying material properties.

o establish rational design procedures based on the real failure behaviour.

REFERENCES

[1] R.E. Klinger and J.A, Mendonca: Tensile capacity of short anchor bolts and welded studs: a literature review. Technical paper No. 79-27, ACI Journal, Vol. 79, July-August 1982, pp 270-279

[2] Niels Saeby Ottosen: Nonlinear finite element analysis of pullout test. American Society of Civil Engineering, ASCE, Journal of the Structural Division, Vol. 107, No. ST4, April 1981, Paper No. 16197, pp 591-603

[3] Walter H. Peier: Model for pull-out strength of anchors in concrete. ASCE, Journal of Structural Engineering, Vol. 109, No. 5, May 1983, Paper No. 17949, pp 1155-1173

[4] René de Borst: Non-linear analysis of frictional materials. Proefschrift ter verkrijging van de graad van doctor in de technische wetenschappen aan de Technische Hogeschool Delft... op dinsdag 22 april 1986 te 14.00 uhr. Delft 1986, 140 pp

[5] Lennart Elfgren, Krister Cederwall, Kent Gylltoft and Carl Erik Broms: Fatigue of anchor bolts in reinforced concrete foundations. IABSE Colloquium "Fatigue of steel and concrete structures", Lausanne March 1982, Proceedings, IABSE Report Vol. 37, Zürich 1982, pp 463-470

[6] Lennart Elfgren and Ulf Ohlsson: Modelling of hook anchors. Paper in "Fracture mechanics of concrete. Nordic seminar held at Division of Building Materials, November 6, 1986", Lund Institute of Technology, Division of Building Materials, Report TVBM-3025, Lund 1986, pp 105-117

[7] R. Ballarini, S.P. Shah and L.M. Keer: Failure characteristics of short anchor bolts embedded in a brittle material. Proceedings of the Royal Society of London, Vol. A404, 1986, pp 35-54

[8] A.K. Hellier, M. Sansalone, N.J. Carino, W.C. Stone and A.R. Ingraffea: "Finite-element analysis of the pullout test using a nonlinear discrete cracking approach". Cement, Concrete and Aggregates, CCAGDP, Vol. 9, No. 1, Summer 1987, pp 20-29

[9] Lennart Elfgren, Carl Erik Broms, Håkan Johansson and Arne Rehnström: Anchor bolts in reinforced concrete foundations. Short time tests. Luleå University of Technology, Research Report TULEA 1980:36, Luleå 1980, 117 pp

[10] G. Utescher: Anchorage in walls (Beurteilungsgrundlagen für Fassadenverankerungen. In German). Ernst & Sohn, Berlin 1978, 99 pp

[11] Willian C. Stone: Internal strain, deformation, and failure of large scale pullout tests in concrete. Report NBSIR 82-2484, National Bureau of Standards, National Engineering Laboratory, Center for Building Technology, Washington D.C., March 1982, 122 + 44 pp

[12] W.C. Stone and N.J. Carino: Deformation and failure in large-scale pullout tests. Technical paper No. 80-46, ACI Journal, Vol. 80, Nov-Dec 1983, pp 501-513

[13] Herbert Krenchel and Surendra P. Shah: Fracture analysis of the pullout test. Materials and Structures, Vol. 18, No. 108, Nov-Dec 1985, pp 439-446

[14] Arne Hillerborg, Mats Modéer and Per-Erik Petersson: Analysis of crack formation and crack growth in concrete by means of fracture mechanics and finite elements. Cement and Concrete Research, Vol. 6, 1976, pp 773-782

[15] Arne Hillerborg: Analysis of one single crack. Chapter 4.1 in Folker Wittman (Ed.): "Fracture mechanics of concrete", Elsevier, Amsterdam 1983, pp 223-249

[16] Kent Gylltoft: Fracture mechanics models for fatigue in concrete structures. Division of Structural Engineering, Luleå University of Technology, Doctoral Thesis 1983:25D, Luleå 1983, 210 pp

[17] Larsgunnar Nilsson and Mats Oldenburg: FEMP. An interactive graphic finite element program for small and large computer systems. Division of Structural Engineering, Luleå University of Technology, Technical Report 1983:07T, Luleå 1983, 116 pp

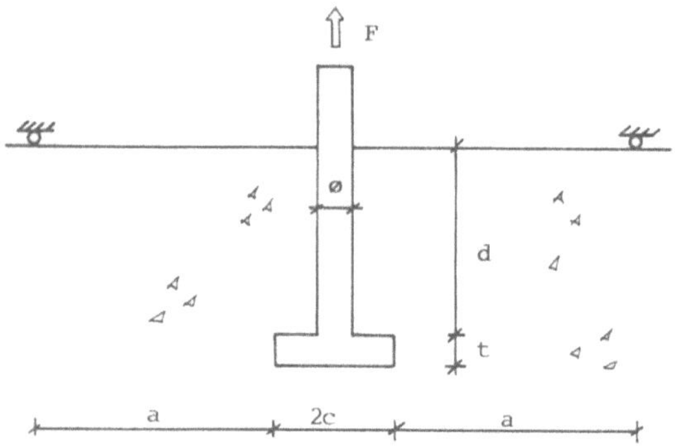


Fig. 1. Anchor bolt in concrete. Notations. The indicated roller supports can be ordinary external supports. They may also represent reinforcement bars embedded in the concrete.

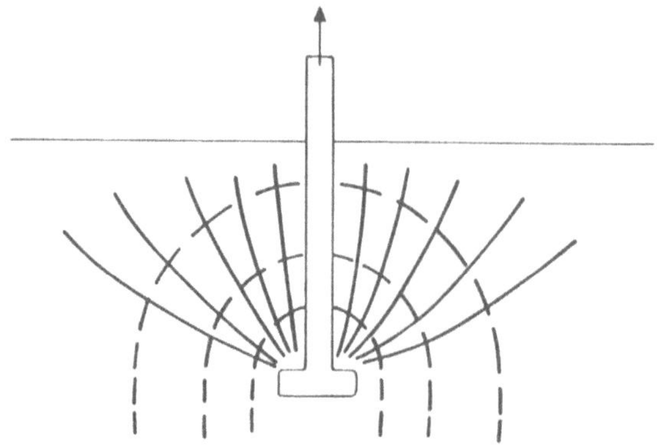

Fig. 3. Trajectories of principal stresses around an anchor bolt [10]. Full lines indicate the direction of the principal compressive stresses and dotted lines indicate the direction of the principal tensile stresses.

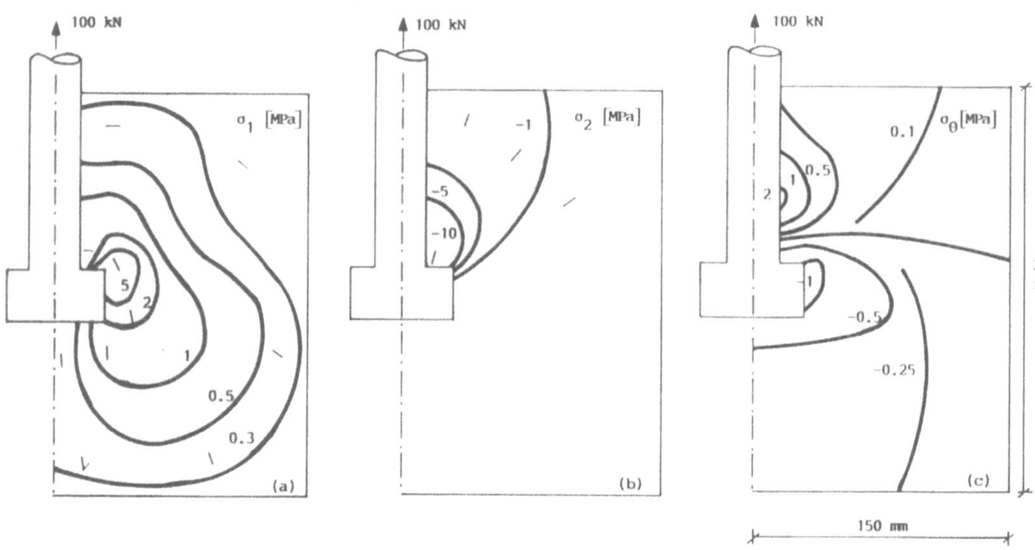


Fig. 2. Plots of isostress curves for the principal elastic stresses of an anchor bolt [9]. Short lines indicate directions of the principal stresses. The stresses were calculated with a finite element model with the following dimensions (see Fig. 1 for notations): d = 400 mm, t = 30 mm, ø = 30 mm, 2c = 60 mm, a = 500 mm. The load is applied directly to the concrete in the contact zone at the top of the disk. Modulus of elasticity for concrete $E_c$ = 25 GPa, Poisson's ratio $\nu$ = 0.

(a) First principal stresses $\sigma_1$ (maximum tensile stress)

(b) Second principal stress $\sigma_2$ (maximum compressive stress)

(c) Tangential stress $\sigma_\theta$

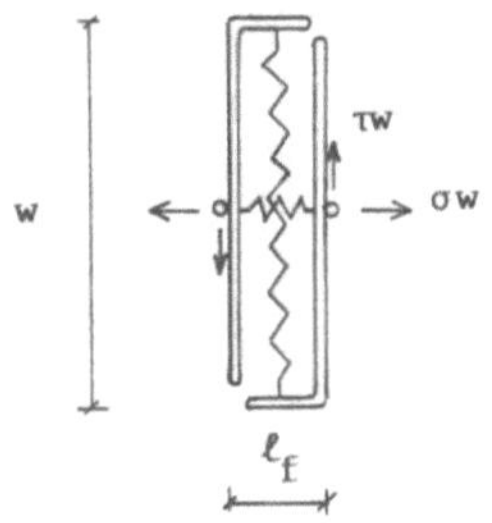


Fig. 4. Two-dimensional linkage element with unit thickness [16].

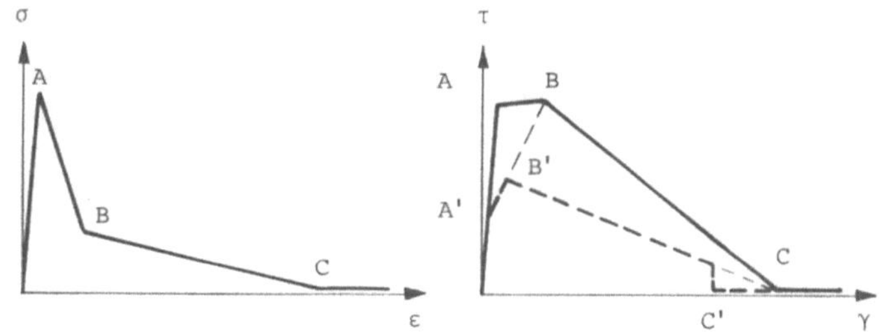


Fig. 6. Change of shear properties when tensile stresses reach points A, B and C

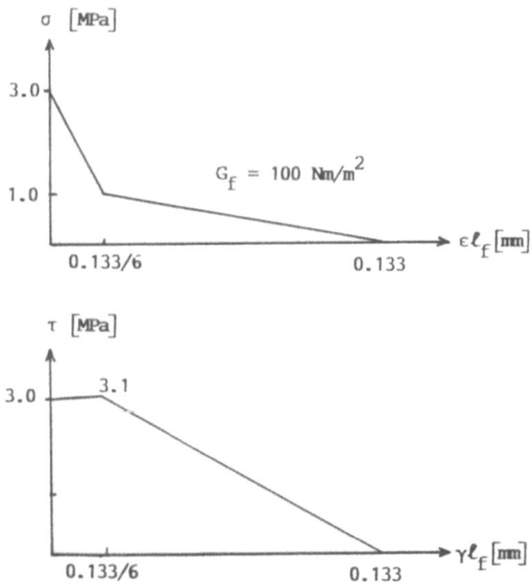


Fig. 5. Material properties used for linkage elements in finite element model of anchor bolts. Before reaching maximum values elastic conditions are assumed ($\sigma = E\varepsilon$, $\tau = G\gamma$, $E = 30$ GPa, $G = 12.5$ GPa). The length of the crack element $\ell_f$ is 0.01 mm.

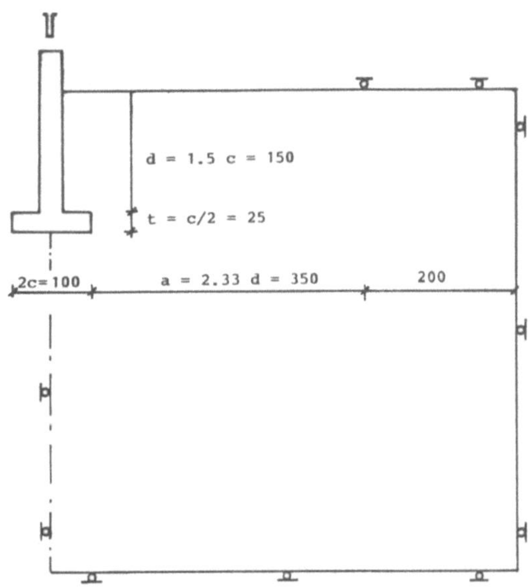


Fig. 7. Dimensions and boundary conditions for finite element model. The overall diameter of the axisymmetrical model is 1200 mm and its height is 600 mm.

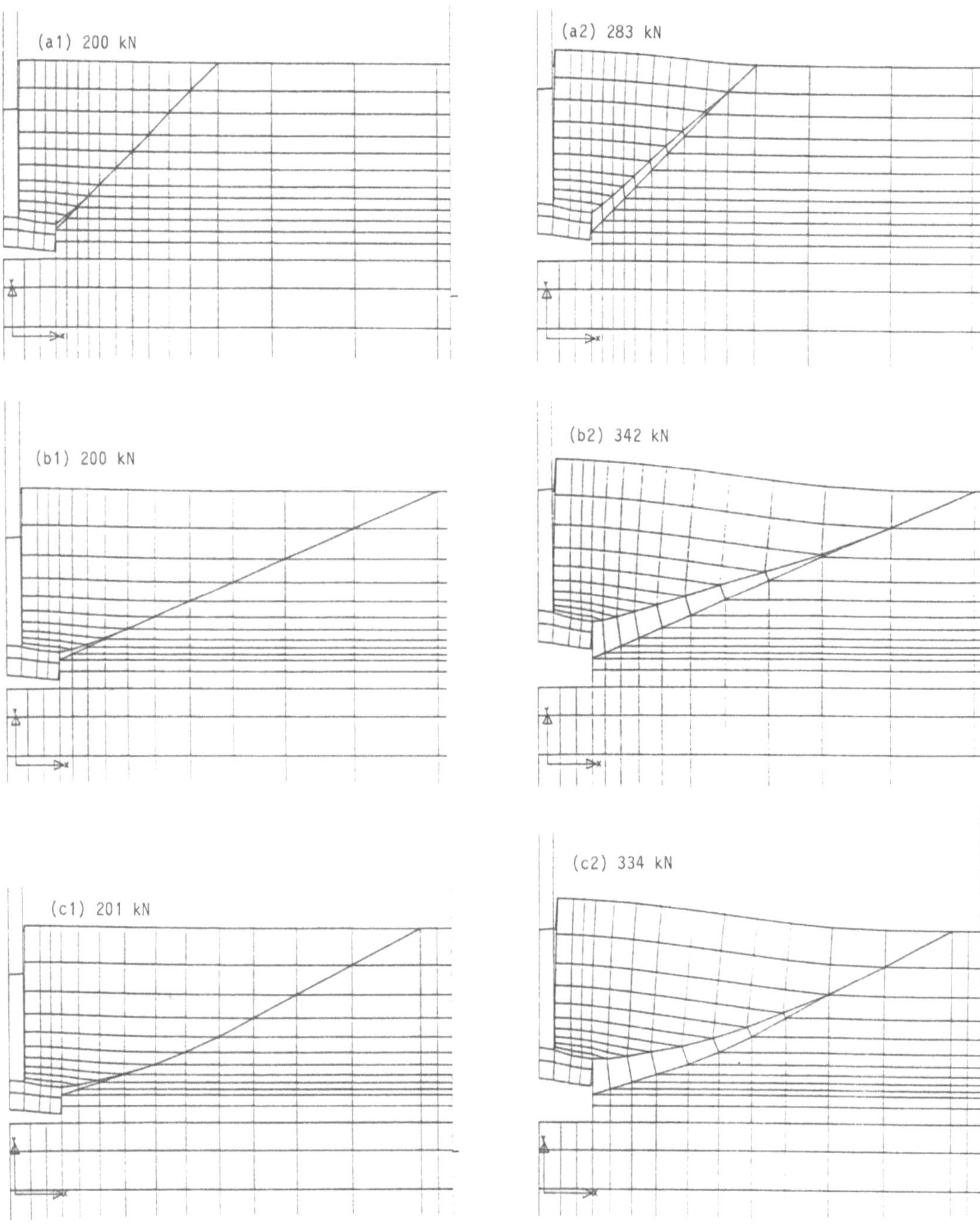


Fig. 8. Deformation plots

(a) $45^0$ straight crack

(b) $67^0$ straight crack

(c) curved crack

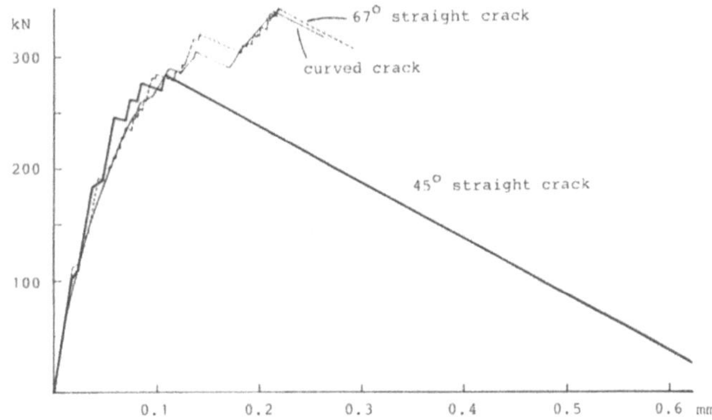


Fig. 9. Load-deformation curves. The displacement is measured at the top edge of the disk.

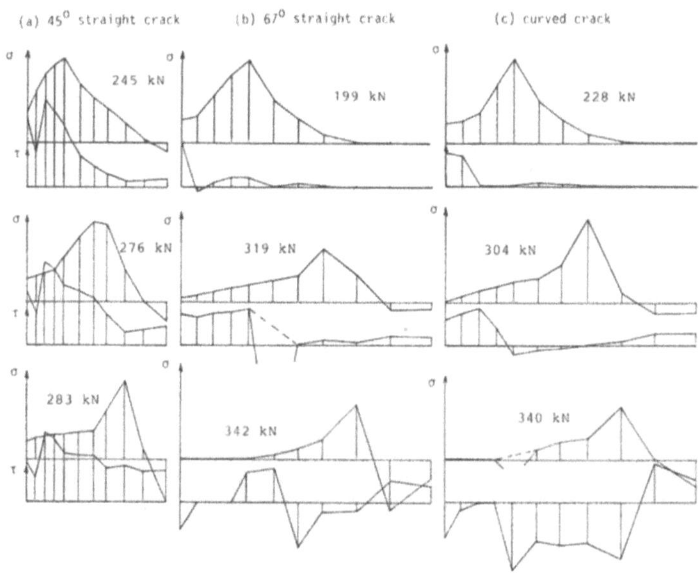


Fig. 10. Tensile and shear stresses along crack path.

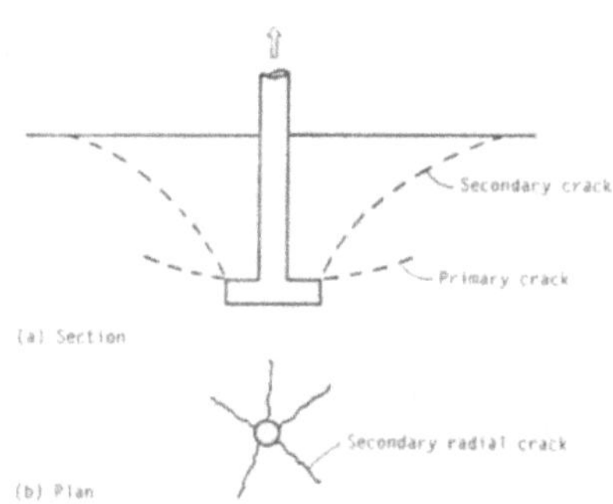


Fig. 11. Possible primary and secondary cracks.

SEM/RILEM International Conference
on Fracture of Concrete and Rock
Houston, Texas, June 1987
Editors: S.P. Shah, S.E. Swartz

# MODELING OF REINFORCED CONCRETE USING THE DISTINCT ELEMENT METHOD

L. J. Lorig and P. A. Cundall

## ABSTRACT

In order to study the mechanisms that operate under high intensity dynamic loading, a numerical method is presented in which a concrete member is modeled as an assemblage of elastic blocks bonded together at common boundaries. Cracking and failure is simulated by the progressive breakage of bonds. Examples are given of static and dynamic tests done numerically on beams with and without reinforcement.

## INTRODUCTION

Traditional procedures for analyzing reinforced concrete members (i.e., beams or columns) are often based on independent determinations of axial loads in concrete and steel using assumed strain profiles and stress-strain relations for both steel and concrete. Such procedures yield reasonable results for members loaded quasi-statically at or below ultimate capacity. However, such procedures may not adequately predict behavior of members subject to high intensity dynamic loading involving inertial effects, large amounts of energy, large displacements, and post-peak behavior.

In an attempt to understand the behavior of reinforced concrete under these conditions, the distinct element method [1] is used to model the member as an assemblage of intact blocks that are bonded along their common boundaries. Progressive failure of the member takes place as the bonds break and slip occurs between blocks. The phenomena of dynamic crack propagation and spalling are represented numerically by the breaking of pre-existing bonds between elements which remain intact. This is thought to be a realistic analog for concrete, in which fractures often pass along bonds between different components such as cement and aggregate. No attempt is made to model the actual size, shape or number of particles in real concrete. The assemblages of blocks are generated by creating Voronoi polygons [2, 3] around a quasi-random array of points within each concrete body. Each block is further discretized into triangular finite difference zones of constant strain.

This paper presents a brief description of the distinct element formulation and its application to the study of the static and dynamic behavior of reinforced concrete. All of the results shown here were obtained using MUDEC, which is the microcomputer version of UDEC (the Universal Distinct Element Code) [4, 5].

## THE DISTINCT ELEMENT METHOD

The essential feature of the distinct element method is its ability to model arbitrary motion of each block with respect to any other. Large translations and rotations of blocks are allowed. The interaction forces between blocks are obtained by applying constitutive relations at interfaces, normally expressed in incremental form. Interface displacement is defined as the relative motion between the edges of adjacent blocks; the interface constitutive relation provides the corresponding contact stresses and forces.

---

L. J. Lorig is a Research Engineer and P. A. Cundall is an Associate Consultant for Itasca Consulting Group, Inc., Minneapolis, Minnesota 55414; Dr. Cundall is an Associate Professor, Department of Civil Engineering, University of Minnesota.

The original distinct element codes assumed the blocks to be rigid. Later, simply-deformable (constant stress) blocks were formulated. The examples reported in this paper make use of a more general representation of block deformability, in which each "fully-deformable" block is discretized into a mesh of triangular (2-D) finite-difference zones. An automatic mesh generator provides any degree of refinement, depending on the desired accuracy of the internal stress distribution. Because each zone is assumed to be in a state of constant strain, block boundaries remain defined by piecewise straight lines, allowing a simple determination of relative displacements between adjacent blocks.

The method employs an explicit time-marching algorithm using a central difference scheme. For each timestep, two sets of calculations are executed: first, the application of the equations of motion allow the new kinetic quantities (accelerations, velocities, displacements) to be determined; then, the block and interface constitutive relations provide the new internal stresses and interaction forces, respectively. The computational steps within each time increment are as follows.

Gridpoint accelerations, $\ddot{u}_i$, are obtained from the equations of motion:

$$\ddot{u}_i = \frac{\int_S \sigma_{ij} n_j ds + F_i}{m} + g_i \tag{1}$$

where S is a path that surrounds each gridpoint, and m is the mass lumped at each gridpoint. The forces $F_i$ include applied external loads and, for gridpoints on a block boundary, the contact forces.

Integration of the above accelerations provides the velocity and displacement increments. Hence, new zone strain increments can be determined. The application of the constitutive relation for the block material gives the new internal stresses. In the following examples, only linear elastic relations are used for intact material.

The updated location of the block boundaries defines the new interface displacements; the new interface stresses (and forces) follow from the interface constitutive relations. Once all new stresses and forces are known, a new calculation cycle begins.

Viscous damping is included in the equations of motion. By making use of artificially high values for this damping, the distinct element method can to obtain solutions for quasi-static problems. In dynamic analyses, both mass-proportional and stiffness-proportional damping are utilized, as in the Rayleigh damping scheme.

## INTERFACE CONSTITUTIVE RELATIONS

The deformability and strength properties of interfaces are represented in the numerical model by spring-slider elements located at contact points between a block corner and an adjacent block edge. A simple force-displacement law relates normal forces directly to the amount of notional penetration—i.e.,

$$F_n = k_n u_n \tag{2}$$

where $F_n$ = normal force at the contact,

$k_n$ = normal stiffness at a point, and

$u_n$ = total normal penetration.

For work described here, the amount of tension that can occur between blocks is limited. If the tension limit is exceeded, failure of the interface is assumed, and shear and normal forces, as well as the tension and cohesion strength parameters, are set to zero. However, the interface may subsequently take compressive load and may slide.

Shear forces are considered to depend on load-path. Incremental shear forces develop in proportion to incremental changes in relative shear displacement—i.e.,

$$\Delta F_s = k_s \, \Delta u_s \tag{3}$$

where $\Delta F_s$ = change in shear force,

$k_s$ = shear stiffness at a point, and

$\Delta u_s$ = incremental shear displacement.

The maximum shear force is limited according to the Mohr-Coulomb criterion:

$$|F_s| \le c + F_n \tan \phi \tag{4}$$

where c = point cohesion, and

$\phi$ = basic joint friction angle.

Shear failure occurs when the shear force reaches the maximum value. Again, for the work described here, the tension and cohesion parameters are set to zero following shear failure.

In addition to point contacts specified by force-displacement relations, edge contacts are important physically because they correspond to the case of an interface closed along its entire length. For such cases, the previous expressions are written in terms of stress rather than force, and representative interface lengths are taken into account.

## MODELING OF UNCONFINED COMPRESSION TESTS

The distinct element method using Voronoi discretization introduces a new method for studying the behavior of concrete. The concept of modeling the microstructure of solid brittle materials using a bonded assembly of elements was introduced by Plesha and Aifantis [6]. These authors used a rigid block formulation to produce a non-linear global stress-strain relation for a hypothetical rock. They found that important quantitative information concerning interrelations between crack morphology, crack orientation, degree of fracturing, applied stress and effect elastic moduli could be extracted.

Part of the procedure for modeling concrete involves selection of values for interface strength that enables simulated test samples to exhibit reasonable agreement with the observed stress-strain relations of real samples. The interface strength parameters to be determined are the cohesion, friction and tensile strength. Results for three numerical simulations of a uniaxial laboratory test are presented here. The geometry and discretization shown in Fig. 1 was used for all simulations. The specimen in Fig. 1 has a height of 300mm and a width of 150mm and was loaded by applying a constant vertical velocity to each end, thereby simulating a displacement controlled test with rough plattens. Joint normal and shear stiffnesses were 2700 GPa/m for all trials.

The stress-strain relations obtained numerically are shown in Fig. 2, together with a typical laboratory result. Figure 3 shows the displaced positions of the polygonal blocks and the principal stresses at the center of each constant strain finite difference triangle for Trial A at an axial strain of 0.0027. The de-stressed areas on each side are typical of those observed in laboratory tests, which show "hour-glass" failure patterns.

## QUASI-STATIC MIDPOINT LOADING OF UNREINFORCED CONCRETE BEAM

The next step in applying the distinct element method to concrete beams involved demonstration that the model produced displacements of an elastic beam (at pre-failure load levels) and failure load in agreement with elementary flexural stress analysis. For the demonstration problem, the model beam was loaded at its midpoint by applying a constant vertical velocity at that point. The beam dimensions were: length = 3.66m; depth = 0.56m. The discretization and boundary conditions are shown in Fig. 4. There are 244 blocks, 876 finite difference triangles, 1364 gridpoints, and 1247 contacts. For this problem, a 25mm problem width is assumed. It should be noted that the end supports (i.e., zero y-velocity) span approximately 200mm, in contrast to the single support point assumed by elastic beam solutions.

The interface parameters were:

| | |
|---|---|
| friction coefficient | 0.3 |
| cohesion | 8.3 MPa |
| tension | 4.1 MPa |
| shear and normal stiffness | 2700 GPa/m |

The predicted load at failure, $P_f$, assuming pinned ends, is given by

$$P_f = \frac{4\sigma_t I}{Lc} = 6.7 \text{ kN} \tag{5}$$

where L = 3.3m, $\sigma_t$ = 4.1 MPa, and c = 0.28m.

The predicted midspan deflection, $\Delta_{CL}$, at failure is given by

$$\Delta_{CL} = \frac{P_f L^3}{48EI} = 0.6\text{mm} \tag{6}$$

where E = 20.7 GPa.

The results summarized in Table I indicate that the numerical model yielded results in reasonable agreement with the expected results, although the centerline displacement at failure (about 0.53mm) was less than predicted. This may be the result of

(a) too high a load rate (i.e., quasi-static beam conditions not fully achieved) [Plots of principal stresses indicated vertical stresses greater than horizontal directly beneath the midspan load.], or

(b) support conditions inhibiting rotation at beam ends.

The crack pattern at failure shown in Fig. 5 exhibits the expected near vertical centerline crack resulting from tensile failure of interfaces at the bottom of the beam. However, Fig. 5 also shows near horizontal cracks at each end and at the center. The cracks at each end likely result from the specified end support. The midspan horizontal crack may result from the previously referenced load rate.

Table I

RESULTS OF UNREINFORCED BEAM DEMONSTRATION PROBLEM

| MIDSPAN DISPLACEMENT (mm) | SUPPORT LOAD (kN) | NUMBER OF FAILED INTERFACES |
|---|---|---|
| 0.41 | 5.5 | 0 |
| 0.48 | 6.2 | 0 |
| 0.56 | 3.2 | 21 |
| 0.61 | 2.2 | 30 |

## QUASI-STATIC MIDPOINT LOADING OF REINFORCED CONCRETE BEAM

In order to model reinforced concrete, it was necessary to introduce reinforcement logic into the distinct element formulation. In the work described here, reinforcement was modeled as a special contact between two blocks using force-displacement relations similar to those outlined for interfaces to describe both shear and axial behavior of a reinforcement element. For each element, the forces are generated in the reinforcement by displacements across the interface through which the element passes. Forces generated in reinforcing elements are applied to the appropriate neighboring blocks. Large shear displacements are assumed to be accompanied by simple geometric change in an element where it crosses an interface. The formulation, described in detail by Lorig [7], was developed for modeling fully-bonded (i.e., grouted) reinforcement in jointed rock masses.

Quasi-static behavior involving reinforcement was evaluated by comparing numerical results with the results of a well documented physical test (Beam 0A1). Beam OA1 had two layers of tension reinforcing and no stirrups. In the numerical simulation, the beam was loaded to failure by applying a gradually increasing midspan displacement. Figure 6 shows the boundary conditions, discretization, and reinforcing location for the numerical simulations. Again, a 25mm beam width was assumed in the numerical simulation. Reinforcing properties, as well as resultant loads were accordingly modified to account for the actual 310mm beam width. The problem parameters used for the reinforcement were:

| | |
|---|---|
| initial axial stiffness | 2.8 GN/m |
| initial shear stiffness | 2.8 MN/m |
| ultimate axial capacity | 6.7 kN |
| ultimate shear capacity | 4.5 kN |
| "active" length | 8 mm |

It should be noted that there is no <u>a priori</u> method for selecting the reinforcement parameters because the original formulation was derived for reinforcement grouted into drilled holes. The strength parameters for the concrete were the same as those listed in the previous section.

The midspan load deflection curves for the physical test and numerical simulation are shown in Fig. 7. This figure shows that the ultimate load for the numerical simulation agrees reasonably well with the physical result but, again, the centerline displacement is less than physical results. In this case, the difference may be attributed to the reinforcement axial stiffness being too great.

Figures 8(a-f) show the development of failed interfaces with centerline displacement. The apparent irregularities or notches in the boundary of the beam are the result of removing some of the very small blocks at the boundary of the beam. These small blocks do normally not affect the results of the simulation; they are removed simply to improve the computational efficiency.

Figure 8 shows that, as a result of reinforcing, the cracked region is more distributed than in the unreinforced beam; it also shows that the beam failed by a diagonal tension crack near the center of the beam. The physical beam also experienced a diagonal tension failure which is typical of reinforced beams without stirrups where the shear span is more than about 3d (d = vertical distance from the upper surface to the reinforcing centroid). Normally, however, the diagonal tension crack begins closer to the end support than was evident in the numerical simulation.

## DYNAMIC ANALYSIS OF A REINFORCED CONCRETE BEAM

Finally, the distinct element method was used to simulate the behavior of a laboratory-scale dynamic test (beam OE2). This beam had both top and bottom longitudinal reinforcement as well as two sizes of vertical (stirrup) reinforcement. A unique feature of the test is that stirrups were not present over a section of the left end of the beam. For the numerical simulation shown here, a time-varying, uniformly distributed vertical stress was applied to the top surface of the beam. The vertical stress, p(t), on the top surface of the beam, was applied in the following manner:

$$p(t) = 0.4 * \sin 392t \qquad 0<t<0.004 \text{ sec}$$

$$p(t) = 0.4 \text{ MPa} \qquad t>0.004 \text{ sec}$$

Both mass- and stiffness-proportional damping were specified (approximately 1.4% of critical damping at 40 Hz). The discretization and support points are shown in Fig. 9.

Figure 10 shows gridpoint velocity vectors for the right half of the beam, and Fig. 11 shows the spalling at 6 milliseconds along the bottom right-hand side of the beam in the region of highest velocity. Figure 12 compares the early time histories of centerline displacement of the physical test and numerical simulation.

The noticeable difference between numerical results and physical results in Fig. 12 is that the numerical results show lower centerline displacement when compared to the physical results. Two factors may have contributed to these differences:

(1) The vertical stress applied to the numerical model was applied at a slower rate when compared to physical tests. Ideally, the same pressure history should have been applied to the numerical model as measured in the physical test; and/or

(2) Damping parameters were too great.

## DISCUSSION

The study has shown that the distinct element method can provide a realistic analog of concrete behavior by representing the material as a bonded assemblage of grains or blocks. The method has the following characteristics which are useful in analyzing the dynamic behavior of reinforced concrete members:

(1) the ability to visualize deformed shape and crack development with displacement and time (The importance of visualizing the deformed structure has long been recognized [8].);

(2) the ability to model large displacements and post-peak behavior;

(3) the ability to model stress wave propagation and resultant tensile spalling; and

(4) the ability to represent energy losses through crack propagation, slip and spalling.

The most appropriate use of the distinct element method in the analysis of reinforced concrete is likely to be in (1) the development and validation of simplified numerical procedures such as those proposed by Shanaa and Krauthammer [9], and (2) the study of various design options involving structure geometry, concrete and reinforcing properties, and reinforcement location.

## ACKNOWLEDGEMENT

A portion of this work was funded through a research study sponsored by the Engineering and Services Laboatory at Tyndall Air Force Base, Florida.

## REFERENCES

1. Cundall, P. A., and O.D.L. Strack. "A Discrete Numerical Model for Granular Assemblies," Geotechnique, 29, 47-65 (1979).

2. Brostow, W., and J. P. Dussault. "Construction of Voronoi Polyhedra," J. Comp. Phys, 29, 81-92 (1978).

3. Finney, J. L. "A Procedure for the Construction of Voronoi Polyhedra," J. Comp. Phys, 32, 137-143 (1979).

4. Cundall, P. A. "UDEC - A Generalized Distinct Element Program for Modelling Jointed Rock," Report PCAR-1-80, Peter Cundall Associates; Contract DAJA37-79-C-0548, European Research Office, U.S. Army, March 1980.

5. Cundall, Peter A., and Roger D. Hart. "Development of Generalized 2-D and 3-D Distinct Element Programs for Modeling Jointed Rock." Itasca Consulting Group; Misc. Paper SL-85-1, U.S. Army Corps of Engineers, 1985.

6. Plesha, M. E., and Aifantis, E. C. "On the Modeling of Rocks with Microstructure," Proceedings of the 20th U.S. Symposium on Rock Mechanics (Austin, Texas, June 1979). New York: American Society of Civil Engineers, 1979.

7. Lorig. L. J. "A Simple Numerical Representation of Fully Bounded Passive Rock Reinforcement for Hard Rocks." Computers and Geotechnics, 1, 79-97 (1985).

8. Cross, H., and N. D. Morgan. Continuous Frames of Reinforced Concrete. New York: John Wiley & Sons, Inc., 1932.

9. Shanaa, Hussain M., and Theodor Krauthammer. "A Numerical Procedure for Analysing Reinforced Concrete Beams," University of Minnesota, Department of Civil and Mineral Engineering, Structural Engineering Report ST-86-04, 1986.


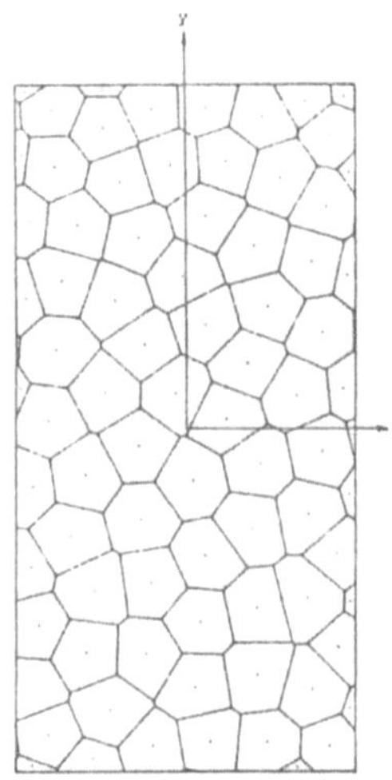


Fig. 1 Voronoi Discretization for Unconfined Compression Numerical Simulations with Ends Laterally Confined (x-velocity = 0 and y-velocity = constant at each end)

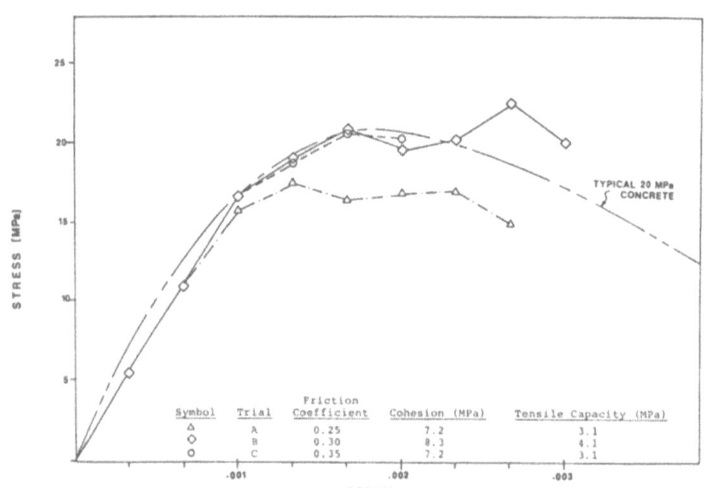


Fig. 2 Stress-Strain Curves for Problem Geometry Shown in Fig. 1 for Three Interface Strength Assumptions

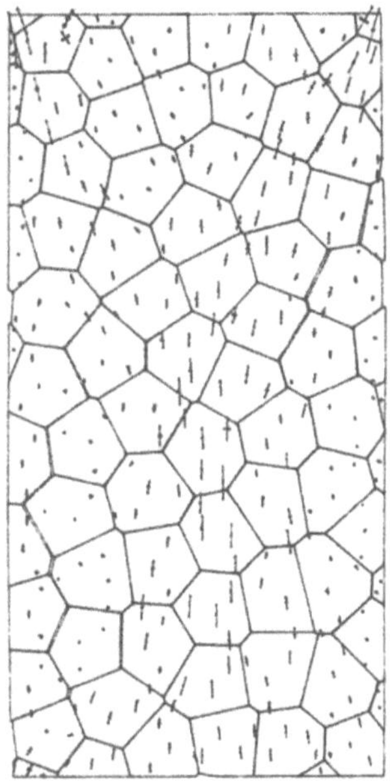

Fig. 3 Voronoi Polygons and Principal Stresses for Trial A at an Axial Strain of 0.0027

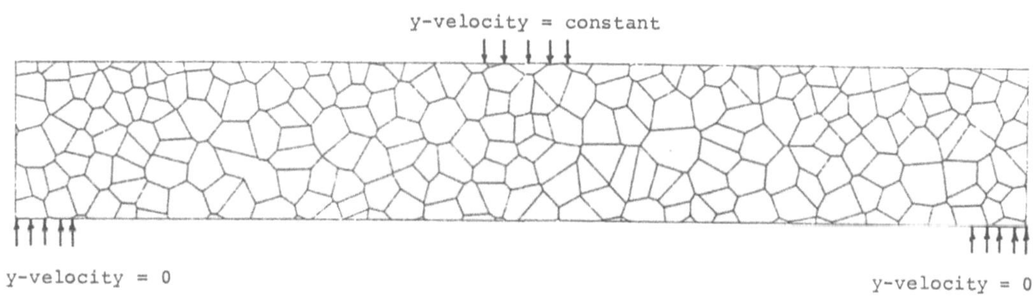


Fig. 4 Discretization and Boundary Conditions for Simulation of Unreinforced Beam Subject to Midspan Loading

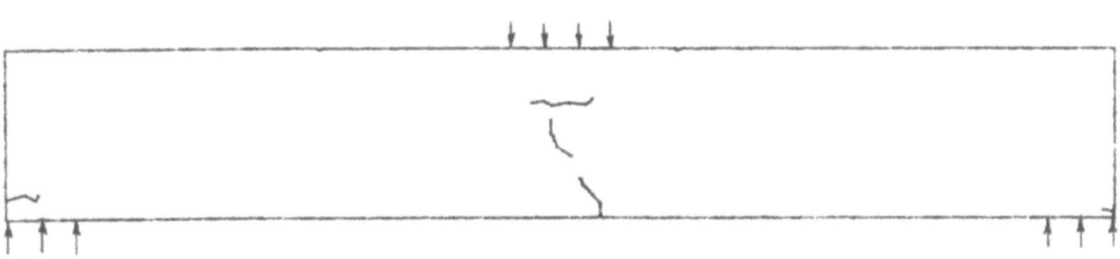

Fig. 5 Failed Interfaces for Problem Shown in Fig. 4 (Midspan Deflection = 0.7 mm)

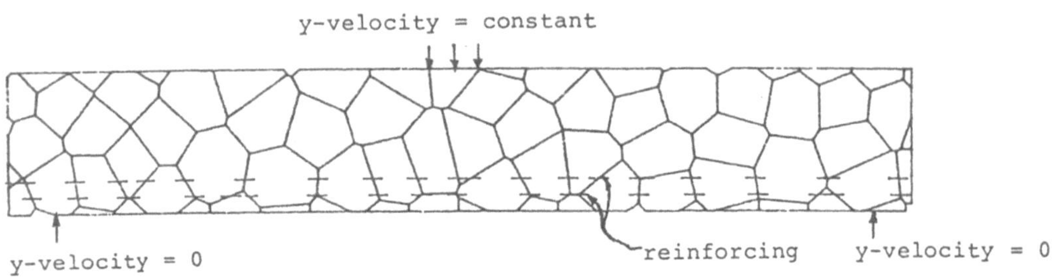


Fig. 6 Problem Geometry, Reinforcement Location, and Boundary Conditions for Reinforced Concrete Beam Simulation

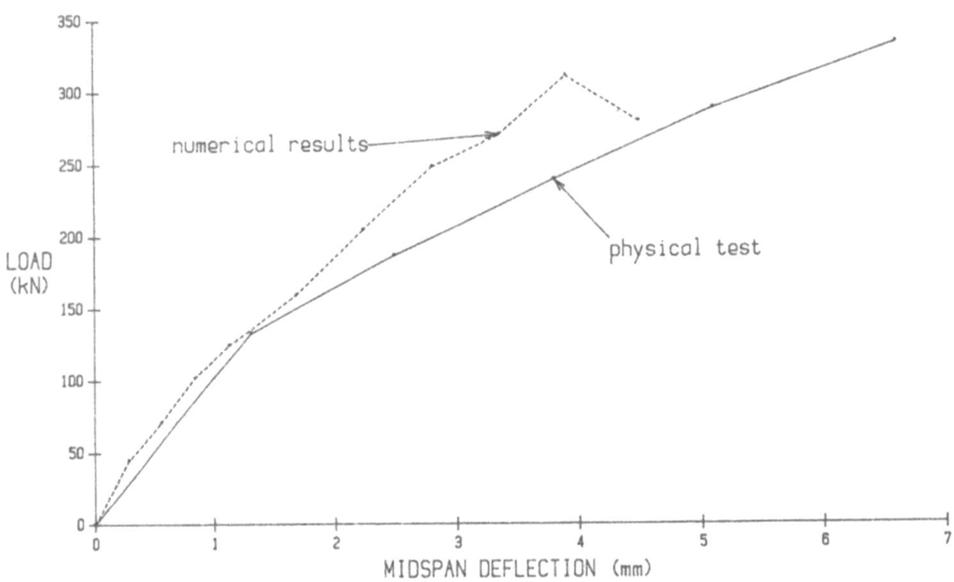


Fig. 7 Plot of Midspan Deflection vs Support Load for the Problem in Fig. 6

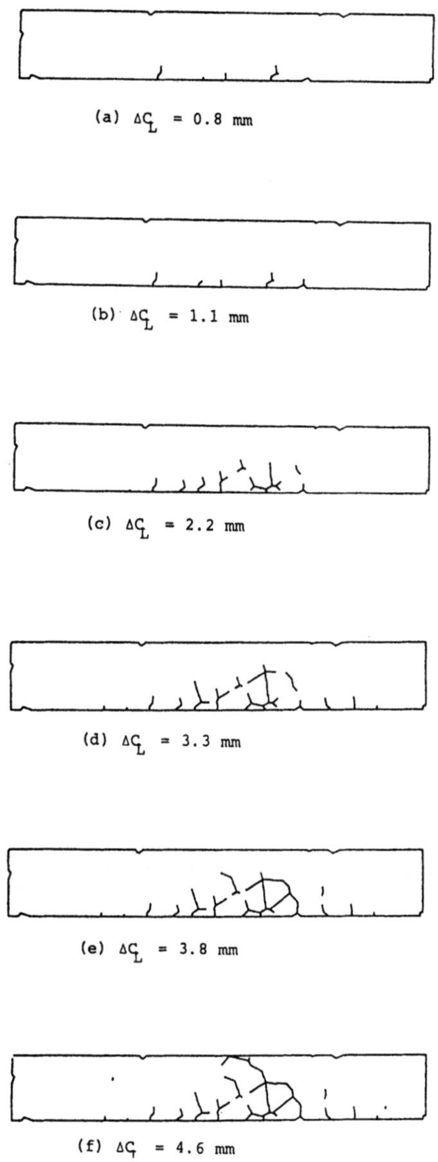


Fig. 8 "Failed" Interfaces in Beam in Fig. 6

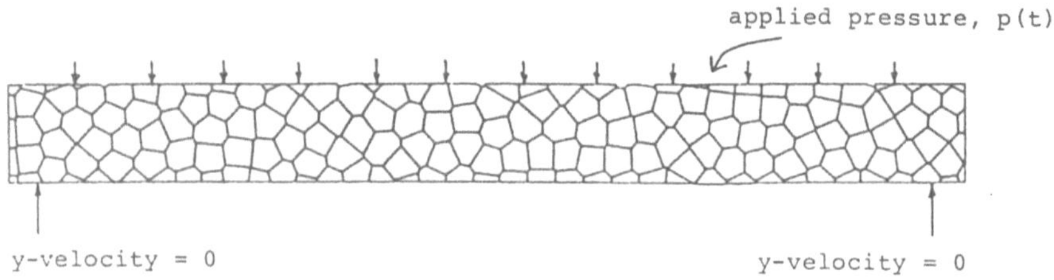


Fig. 9 Discretization Used for Simulation of Beam Test 0E2

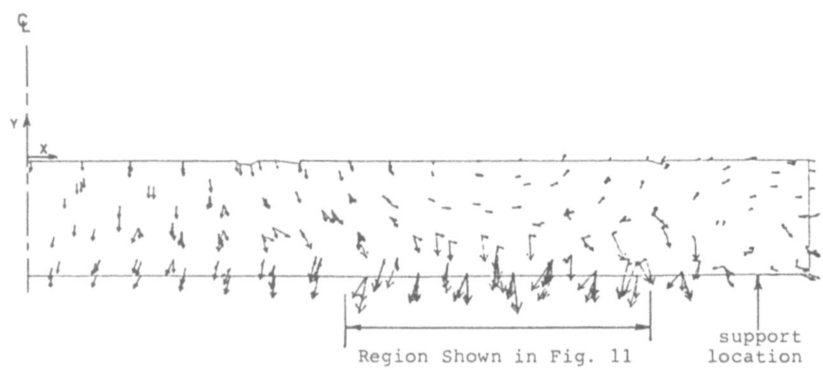


Fig. 10 Right-Hand Side of Numerical Simulation of Beam 0E2 at 6 Milliseconds (Gridpoint velocities, shown by arrows, are scaled to the maximum value of 8.2m/sec.)

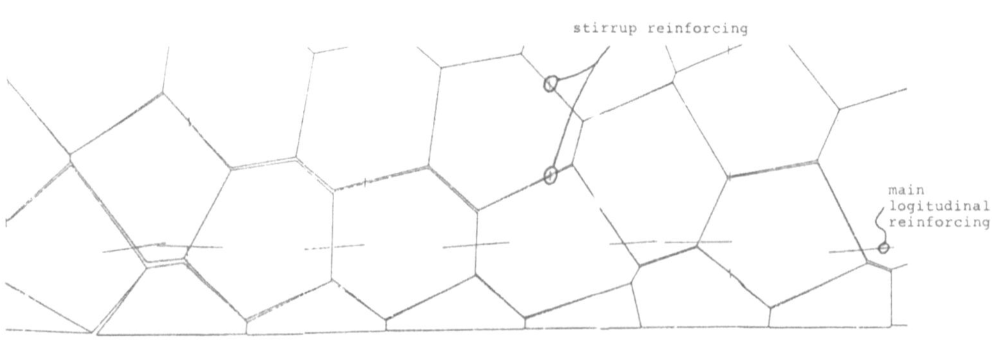


Fig. 11 Displaced Position of Elements and Reinforcing at 6 Milliseconds at Bottom of Beam (see Fig. 10)

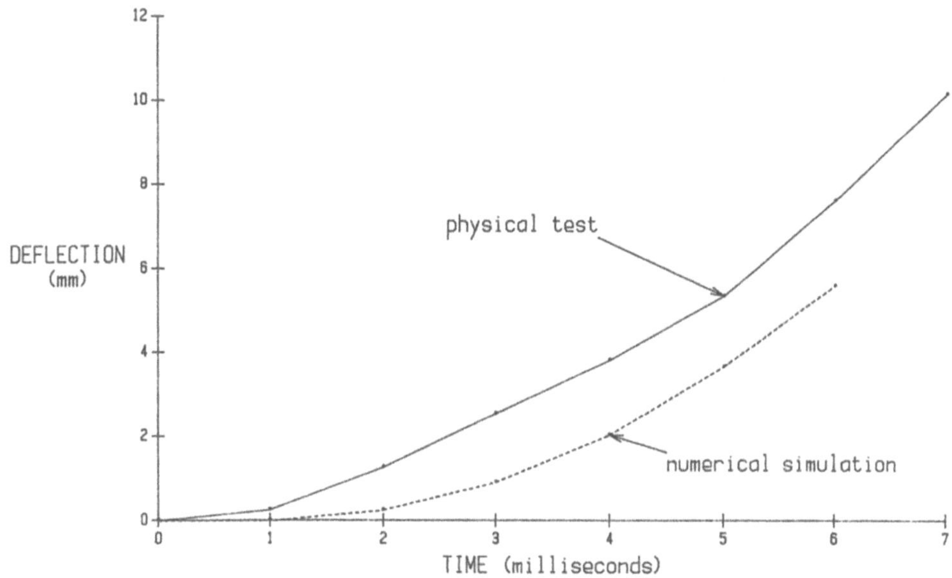


Fig. 12 Comparison of Numerical and Physical Results of Centerline Displacement versus Time for Simply-Supported Beam Subject to a Uniformly-Distributed Dynamic Load

SEM/RILEM International Conference
on Fracture of Concrete and Rock
Houston, Texas, June 1987
Editors: S.P. Shah, S.E. Swartz

# TOUGHNESS EVALUATION AND ITS APPLICATION TO CONCRETE STRUCTURAL DESIGN

Wataru KOYANAGI, Keitetsu ROKUGO, Hiroyuki IWASE

ABSTRACT

The evaluation, improvement and application of concrete toughness are discussed. It is shown that the differential signal between factored load and displacement is effective to control unstable failure of concrete specimens at loading tests. The relations among impact load, displacement, kinetic energies and duration of concrete beams subjected to impact are presented. Concrete having both high toughness and high strength can be produced by adding steel fibers at high percentage into high strength cement concrete or polyester resin concrete. The analytical relation between the compressive toughness of concrete and the toughess of RC (reinforced concrete) beams in bending is derived. The ductility index is proposed and its validity is proved by experiments of PC (prestressed concrete) beams.

## 1. INTRODUCTION

Generally, concrete structures have been designed mainly based on the strength of both materials and structural members, because the strength has been the most important index when describing the failure behavior of concrete and concrete structural members. Recently, however, in addition to the strength, the whole shape of the load-displacement curves (or stress-strain curves) and the toughness, which is expressed by the area under the curves, have become important in the field of structural designs and investigations for the following reasons.

i) Materials: New kinds of concretes, for example, fiber reinforced concrete (FRC) and high strength concrete, have appeared and been used. It has been well recognized that their failure behavior is apparently different from that of conventional concrete.

ii) Structures: Detailed information on the fracture behavior has been needed for huge, important and special concrete structures such as reactor containment vessels, LNG (liquid natural gas) tanks, large dams, etc.

iii) Measurements: The development of testing machines has made it possible and easier to measure complete load-displacement curves including falling branch. For example, closed-loop servo controlled testing machines are widely used.

iv) Simulations: The simulation of complete load-displacement curves became widely accepted because of the development of numerical simulation methods. Especially, the concept of Fracture Mechanics has been adopted.

The purpose of this paper is to discuss the toughness of concrete based on the complete load-displacement curves. Our recent studies on toughness are gathered and presented in this paper. The contents of standards and recommendations in Japan having relation to toughness are also introduced. This paper consists of three parts; evaluation, improvement and application of toughness of concrete.

---

W. Koyanaagi, K. Rokugo and H. Iwase are Professor, Associate Professor and Research Associate, respectively, Department of Civil Engineering, Gifu University, Yanagido, Gifu, Japan

Firstly, following the discussions on toughness and toughness parameters, some topics in the measuring techniques of toughness of concrete at static and impact loading tests are presented. Secondly, several attempts to improve toughness of concrete are introduced. Test results are presented for high toughness concretes, which are produced not only by steel fiber additon of a large amount but also by the replacement of the matrix with high strength concrete or polyester resin concrete. In the third part, the application of toughness to structual design of concrete is analytically discussed. The toughness of reinforced concrete beams in bending is related to the toughness of concrete in compression by equations. Ductility parameters are derived from the equations. Test results of prestressed concrete (PC) beam are given to demonstrate the validity of the equations and the parameters. It is stressed that the failure behavior of both concrete (material) and reinforced concrete beams (members) can be connected with each other by means of the energy value, that is, toughness. This would be also one of the aims of fracture mechanics approach to concrete in a wide sense.

## 2. TOUGHNESS EVALUATION AND LOAD-DISPLACEMENT CURVES

### 2.1 TOUGHNESS AND TOUGHNESS PARAMETERS

In general the term "toughness" is used in a broad sense including the concept of ductility and deformability of materials, members or structures. But in a narrow sense, it means the ability to absorb energy. Therefore, toughness can be evaluated by the area under the load-displacement curve up to a certain limiting point. Many points can be taken as the limit. In the case of the load-displacement curve of a steel fiber reinforced concrete (SFRC) beam, for example, cracking point, maximum load point, breaking point, etc. can be adopted.

In the "Standards for testing methods of fiber reinforced concrete" by Japan Concrete Institute (JCI) [1], the limit for toughness measurement is selected at the point where the specimen displacement is 1/150 of the span (2mm for 10x10x40cm beams with 30cm span) for bending tests and 3/400 of the gage length (0.75mm for /10x20cm cylinders with 10cm gage length) for compression tests.

Since toughness is expressed by the area under the load displacement curve of the specimen, it directly depends on the specimen size. Toughness parameters or indexes, which are relatively insensitive to the specimen size, are often used. For example, the toughness index for SFRC proposed by ACI Committee 544 [2] is given by the ratio of flexural toughness up to the point of 1.9 mm in deflection to that up to the first cracking point. In JCI Standards [1], the equivalent flexural strength $\sigma_b$ is proposed and calculated by the following equation,

$$\sigma_b = T_b \cdot \ell / (\delta_{tb} \cdot b \cdot h^2) \qquad (2\text{-}1)$$

where $T_b$ is the flexural toughness expressed by the area under the flexural load-displacement curves up to the limit $\delta_{tb}$ (2mm when the span $\ell$ is 30 cm, as mentioned above) as shown in Fig. 2.1, b is the width and h is the height of the beam specimen. $\sigma_b$ is determined from the average load ($P_a = T/\delta_t$) up to the limit.

The appearance of SFRC, which is able to be used as a tension member without steel reinforcing bars, stimulated the early establishment of the evaluation methods of the toughness in order to control the quality of SFRC, to improve the toughness further and also to use the toughness in the structural design.

In the case of concrete beam specimens, the material toughness for a unit fracture area is given by dividing the flexural toughness up to the rupture (separation) point of the specimen by the area of the fractured cross-section. This corresponds to the fracture energy $G_F$ in Fracture Mechanics of Concrete.

In the case of concrete cylinders, the compressive load-displacement curve can be immediately converted into the compressive stress-strain curve and then the area under the stress-strain curve gives the material

In the case of concrete cylinders, the compressive load-displacement curve can be immediately converted into the compressive stress-strain curve and then the area under the stress-strain curve gives the material toughness for a unit fracture volume. This material toughness in compression is used in chapter 4.

The material toughness in compression or in flexure, if necessary, can be distinguished from the specimen (or member) toughness, which is given by the area under load-displacement curves. However, the toughness in its general meaning can include both material toughness and specimen toughness, if there is no confusion. This relation is similar to that of the word "strength" which covers not only material strength having a stress unit but also member strength having a force unit. Therefore, in this paper both specimen (or member) toughness and material toughness are called just "toughness".

## 2.2 COMPLETE LOAD-DISPLACEMENT CURVES AT STATIC TESTS

In order to evaluate toughness of concrete by the area under the load-displacement curves, complete load-displacement curves of concrete specimens including the so-called falling branch after the maximum load point must be measured in all cases. There are two types of load-displacement curves as illustrated in Fig. 2.2. Type-A curve is normal. Type-B curve has an indented portion after the maximum load point. Type-B curve is rare but it exists[3,4,5].

Stiff testing machines are generally used to measure Type-A curve by a displacement control. In the case of Type-B curve, however, the use of stiff testing machines alone is not sufficient. A monotonically increasing parameter in the course of the fracture process should be adopted. Otherwise the unstable failure of the concrete specimen can not be prevented. For example, crack opening displacement is used in flexural tests and lateral deformation in compression tests[4].

The authors have proposed a new technique to control the failure and then to measure the load-displacement curve[3]. As shown in Fig. 2.3, a new control variable X can be produced by rotating the old co-ordinate system x-y about the origin at $\Theta$. This new control variable X consists of only load x and displacment y, that is,

$$X = x \cdot \cos\Theta - y \cdot \sin\Theta \qquad (2\text{-}2)$$

The differential signal between factored load signal and factored displacement signal can be adopted as a feedback signal to a closed-loop servo-controlled testing machine in order to automatically control the specimen failure.

Fig. 2.4 shows load-displacement curves of high strength light-weight concrete specimens (size: /10x20cm) in compression. These curves were measured by the new technique. These curves have indented portions but were able to be measured without causing any violent failure in the specimens and hence in perfect safety.

## 2.3 LOAD-DISPLACEMENT CURVES AT IMPACT TESTS

Examples of measured relations among time, load, displacement and energies for concrete beam specimens subjected to impact load are presented in this section.

The impact fracture process of concrete can be treated as the energy transformation process as shown in Fig. 2.5. A part or most parts of the potential energy $W_w$ of the striker is transferred to the specimen as the applied energy $W_e$. The applied energy $W_e$ consists of the absorbed energy $W_a$ and the kinetic energy $W_k$. The absorbed energy $W_a$ consists of the dissipated energy $W_i$ and the static energy $W_s$. The irrecoverable dissipated energy $W_i$ is the energy consumed in the crack formation and viscous friction. When the specimen vibrates, the strain energy $W_s$ is transformed into the kinetic energy $W_k$ along the dotted arrow in the figure. The applied energy here can be called the toughness in impact or the impact toughness.

Impact loading tests were carried out on concrete beam specimens (size: 7.5x7.5x66cm, span: 60cm) by dropping a steel striker (mass: 14.1kg) onto the mid-span, as illustrated in Fig. 2.6[6]. Each specimen was subjected to a single impact. The striker was dropped from different heights, including from the minimum height required just to cause rupture to the specimen. The accelerations of the striker and the specimen at the midspan were measured. The strain of the specimen was also measured. The data were recorded in a digital wave memory and then sent to a micro computer for the subsequent analysis.

Examples of the measured accelarations and strain are shown in Fig. 2.7. The relations among load, displacement, strain, energies, and time are shown in Fig. 2.8. Figs. 2.7 and 2.8 show the results of a concrete specimen just ruptured due to a single impact. The strain energy of the whole specimen was calculated from the strain in the specimen. The kinetic energy was calculated from the velocity which was given by single integration of the specimen acceleration. The displacement at the mid-span of the specimen (Fig. 2.8(d)) was calculated by double integration of the acceleration of the specimen. The accuracy was confirmed by another measuring method. The impact load applied to the specimen (= the reaction from the specimen to the striker) was obtained as the product of the mass of the striker and the negative acceleration of the striker after contact. The impact load-displacement curve (Fig. 2.8(f)) was obtained from the displacement-time curve (d) and the impact load-time curve (e). The applied energy (c) was calculated from the area under the impact load-displacement curve (f).

As shown in Fig. 2.8, there was little kinetic and strain energies remaining after the moment of rupture, because the minimum impact energy was applied to the specimen. Fig. 2.9 shows the kinetic and strain energies of a concrete beam specimen, which was subjected to a larger impact energy than the required minimum impact energy to just rupture it and hence ruptured violently. In Fig. 2-9 the kinetic energy remained after the strain energy was released due to the complete separation.

The dissipated energy of the specimen due to impact can be obtained by subtracting the kinetic energy (and also the strain energy, if necessary) from the applied energy. The dissipated energies, both of the exactly ruptured specimens and of the violently ruptured specimens were almost equal each other. The dissipated energy of the specimens ruptured due to impact was about 1.5 times as large as the corresponding dissipated energy obtained from static tests in the range of this experiment.

## 3. IMPROVEMENT OF TOUGHNESS

Several attempts to obtain high toughness concrete are presented in this sections. Generally, steel fibers are added to concrete to increase not only the flexural (or tensile) strength but also the toughness. The toughness of steel fiber reinforced concrete (SFRC) can be improved by,

i) increasing fiber content,

ii) increasing bond strength between steel fibers and matrix,

iii) increasing tensile strength of steel fibers, if fibers tend to break at fracture of the concrete specimen.

In order to obtain high toughness concrete, high strength concrete or resin concrete were used as a matrix of SFRC [7,8].

Steel fiber reinforced high strength concrete (SFRHC) was produced by adding indented steel fibers (size: ∅1.0mm x 40mm) with low aspect ratio (40) at high percentage (up to 6% by volume) into a matrix of high strength concrete. Superplasticizer was used and the water-cement ratio was 0.28. Since the aspect ratio of steel fibers was low, no fiber balls were found at mixing even at high content of fibers.

Fig. 3.1 shows flexural load-displacement curves of five kinds of SFRHC beam specimens (size: 10x10x40cm, span: 30cm) differing in fiber contents from 0% to 6%. The displacement (deflection) was measured at loading points, that is, at third points. Five specimens were used for each test conditions and the results are expressed by the average. As shown in Fig. 3.1, both the ultimate flexural load and the flexural toughness (area under the curves) increased remarkably with an increase in fiber content. SFRHC with steel fibers at 6% had the flexural strength of more than 20MPa and also the high flexural toughness.

Fig. 3.2 shows compressive load-displacement curves of SFRHC cylindrical specimens (size: ø10x20cm, gage length: 18cm). The fiber addition was not so effective to the increase in the compressive strength but quite effective to the increase in the compressive toughness.

Since the high strength matrix has high bond strength between steel fibers and matrix, the aspect ratio can be reduced to 40 from the normal value of 50 to 60 without losing the anchor effect. When the aspect ratio is low, steel fibers are able to be mixed at more that 4% by volume without fiber balls. As a result, SFRC with high strength and high toughness can be produced.

SFRHC can be interpreted in two ways; (1) the properties of high strength concrete are improved by steel fiber addition, or (2) the properties of SFRC are improved by strengthening the matrix. Test results of PC beams made with SFRHC are described in section 4.2(g).

Steel fiber reinforced resin concrete (SFRREC) was produced by adding steel fibers (size: ø1.0mmx40mm) into polyester resin concrete to increase its flexural toughness. Unsaturated polyester resin was used as a binder (instead of cement paste). Cobalt naphtenate and methyl ethyl keton peroxide were used as an accelerator and a catalyst, respectively. Generally, polyester resin concrete has been mainly used in flexural members, because the flexural strength is three to four times larger than that of high strength cement concrete. It is well known, however, that resin concrete fails in a brittle manner.

SFRREC beam specimens (size: 10x10x40cm) were tested under third point loading. Fig. 3.3 shows flexural load-displacement curves for four kinds of SFRREC varied in fiber contents from 0% to 3.8%. The sudden decrease in load after cracking was prevented by steel fiber addition of more than 2%. The flexural toughness increased as the fiber content increased. The curve of resin concrete with no steel fibers became Type-B curve having an indented portion (see Fig. 2.2 in section 2.2).

## 4. FAILURE BEHAVIOR OF RC MEMBERS AND APPLICATION OF TOUGHNESS TO STRUCTURAL DESIGN

### 4.1 TOUGHNESS IN ALLOWABLE STRESS DESIGN

The methods to use the toughness of SFRC in structural designs have been researched. For example, it is proposed in "Recommendation of design and construction criteria for steel fiber reinforced concrete" by Japan Society of Civil Engineers (JSCE) [9] that the allowable flexural stress $\sigma_{ba}$ shall be increased with the flexural toughness of SFRC, as expressed in the following equation,

$$\sigma_{ba} \leqq \sigma_{bk}/4 \qquad (4\text{-}1)$$
$$\leqq \sigma_b/2$$

where, $\sigma_{bk}$ is the flexural strength and $\sigma_b$ is the equivalent flexural strength, which is defined in equation (2-1) in terms of toughness. Generally, the allowable flexural stress for plain concrete is limited to 0.3 MPa in Japan. According to equation (4-1), it is possible to take about ten times higher value for $\sigma_{ba}$ of SFRC as compared with that of conventional concrete.

4.2 APPLICATION OF TOUGHNESS TO DUCTILITY DESIGN OF RC AND PC BEAMS

(a) Failure behavior of RC beams and yield ending point

Not only the strength but also the toughness of high toughness concrete such as SFRC and SFR high-strength concrete should be taken into account when designing reinforced concrete structures. First of all, the failure behavior of RC beams is outlined.

As shown in Fig. 4.1, moment-curvature (M-φ) or load-displacement (P-δ) relationships of RC beams in bending can be generally characterized by cracking point, yield starting point of tensile reinforcing bars, strength failure (maximum load) point, yield ending point, rupture point, etc. At the yield ending point of tensile reinforcing bars, the yielding stops and then the stress and strain in the reinforcing bars begin to decrease. After the yield ending point the load of RC beams decrease suddenly. Since most of the absorbed energy of RC beams in bending is dissipated in tensile reinforcing bars by the yielding [11], the yield ending point as well as the yield starting point are considered to be the most suitable points for the ductility evaluation of RC beams.

(b) Experimental results of load-displacement curves of RC beams

Examples of measured load-displacement curves of various kinds of RC beams are shown in Figs. 4.2 and 4.3. The dimension of the tested beams is given in Fig. 4.4. Three types of concrete (plain, SFR, light-weight) and three kinds of tension reinforcement ratio (p=0.92%, 2.61%, 6.88%) were adopted in the tests of singly reinforced concrete beams in Fig. 4.2. Three different compression reinforcement ratios (p'=0%, 0.93%, 1.64%) were chosen at constant tension reinforcement ratio (p=1.64%) for doubly reinforced concrete beams in Fig. 4.3. The steel fiber content of SFRC was 2% by volume. The test results can be summarized as follows;

i) The ductility (displacement at high load level after yielding) of RC beams increased in the order of light-weight concrete, plain concrete, and SFRC (Fig. 4.2).

ii) The yield load increased but the ductility decreased with increasing tension reinforcement ratio (Fig. 4.2).

iii) With an increase in compression reinforcement ratio, the ductility increased and the maximum load slightly increased (Fig. 4.3).

iv) The load at yield starting point for RC beams made of SFRC was higher than that for RC beams made of plain concrete (Fig. 4.2 (a),(b)).

v) The yield ending point was observed. After this point, the load-carrying capacity decreased suddenly (Fig. 4.3(a)).

vi) In the case of RC beams made of SFRC, the tensile steel bars ruptured after large displacement (Fig. 4.3(b)).

The failure behaviors i) to iii) are well known. However, there are few reports on the yield ending points and the rupture point after large displacement. In addition to the above, of course, there are many factors affecting on the failure behavior of RC beams: axial force, bond properties, specimen size, confinement, etc.

(c) SIMULATION OF LOAD-DISPLACEMENT CURVES OF RC BEAMS

Fig. 4.5 shows examples of simulated load-displacement curves of RC beams with different tension reinforcement ratios, compression reinforcement ratios and axial forces. The moment-curvature curves of the cross-section were calculated from the force balance by using stress-strain curves of both concrete and steel bars as given in Fig. 4.6, and then the load-displacement curves for RC specimens with dimensions shown in Fig. 4.4 were calculated from the moment-curvature curves. Conditions for the simulations are given in the figure.

The simulation results exhibit well the features of the failure process of RC beams: namely, the increase in maximum load and the decrease in ductility due to an increase in tension reinforcement ratio, the decrease in ductility with an increasing axial force, and the increase in both maximum load and ductility due to an increase in compression reinforcement ratio.

The relationships between points on the stress-strain curves and points on the simulated load-displacement curves are indicated with several special points (the yield ending point, etc.) on the curves in Fig. 4.5. For example, it can be seen from these relationships that the increase of the maximum load with an increase of compression reinforcement ratio is mainly caused by the work-hardening of tensile steel bars.

In the calculation of load-displacement curves including falling branch after the yield ending point shown in Fig. 4.5, the strain in tensile steel bars is not able to be used as a control parameter, because it begins to decrease at the yield ending point, that is, it does not increase monotonously in the failure process of RC beams. The nominal strain in the extreme fiber of concrete in compression is used in simulations in Fig. 4.5 to obtain stable simulation results. The dependence of stability in simulations on the control parameter is similar to that in experiments as discussed in section 2.2.

(d) Stress and strain at yield-ending point

Prior to the discussions on the relation between the toughness of concrete in compression and the toughness of RC beams, the stress and strain in concrete and steel bars at the yield ending point of RC beams are presented below.

Consider a RC beam with rectangular cross section under bending. The stress $\sigma_{sr}$ and the strain $\varepsilon_{sr}$ in tensile reinforcing bars, as well as the stress $\sigma_{cr}$ and the strain $\varepsilon_{cr}$ in the extreme fiber of concrete in compression at the yield ending point can be expressed as follows;

$$\sigma_{sr} = \sigma_{sy} \qquad (4\text{-}2)$$

$$\varepsilon_{sr} = S(\varepsilon_{cr})/[p \cdot \sigma_{sy} - p' \cdot \sigma_{sy}' + N'/(bd)] - \varepsilon_{cr} \qquad (4\text{-}3)$$

$$\sigma_{cr} = p \cdot \sigma_{sy} - p' \cdot \sigma_{sy}' + N'/(bd) \qquad (4\text{-}4)$$

where, p and p' are tension and compression reinforcement ratios, $\sigma_{sy}$ and $\sigma_{sy}'$ are yield strengths of tensile and compressive reinforcing bars, b and d are the width and effective depth of a RC rectangular cross section and N' is the axial force normal to the cross section. $\varepsilon_{cr}$ is the strain corresponds to the point where the stress becomes $\sigma_{cr}$ in the falling branch of the stress-strain curve of concrete (see Fig.4.7(a)). $S(\varepsilon_{cr})$ is the toughness of concrete in compression up to the strain of $\varepsilon_{cr}$ and is given by the area below the stress-strain curve up to $\varepsilon_{cr}$, as shown in Fig. 4.7(a).

The stress-strain curve of reinforcing steel bars is idealized as two straight lines, as shown in Fig. 4.7(b), by ignoring the increase in stress due to the strain-hardening. It is assumed that the compressive reinforcing bars have already yielded at the yield ending point of the tensile reinforcing bars. The process of obtaining these equations is given in ref.[11].

The effects of the steel fiber additions and the confinement by using hoop reinforcement are included in the term of toughness $S(\varepsilon_{cr})$. It would be practical to consider $S(\varepsilon_{cr})$ to be a material constant by fixing $\varepsilon_{cr}$ (for example $\varepsilon_{cr} = 0.005$). From equation (4-3), the steel strain at the yield ending point $\varepsilon_{sr}$ does not depend on the compressive strength of concrete. It depends only on the area under the stress-strain curve of concrete.

There exist two limits of reinforcement ratio. One is yield occurring (upper) limit $p_y$ and the other is rupture (lower) limit $p_f$. These two limits for RC beams are derived from equations (4-2) to (4-4) and discussed with examples in ref.[12].

$$p_y = [S(\varepsilon_{cr})/(\varepsilon_{sy} + \varepsilon_{cr}) + p'\cdot\sigma_{sy}' - N'/(bd)]/\sigma_{sy} \quad (4\text{-}5)$$

$$p_f = [S(\varepsilon_{cr})/(\varepsilon_{sf} + \varepsilon_{cr}) + p'\cdot\sigma_{sy}' - N'/(bd)]/\sigma_{sy} \quad (4\text{-}6)$$

where, $\varepsilon_{sf}$ is the strain in tensile reinforcing bars at rupture.

(e) Toughness of cross-section of RC beams

Following the experimental results and the simulation results in the previous sections, the analytical results on the toughness of RC beams are presented here.

The toughness of RC beam cross-section $W_{it}$ can be represented by the dissipated energy between the yield starting and yield ending points in tensile reinforcing steel bars (ref.[11]). Hence, $W_{it}$ can be written:

$$W_{it} = A_s\cdot\sigma_{sy}(\varepsilon_{sr} - \varepsilon_{sy}) \quad (4\text{-}7)$$
$$= p\cdot b\cdot d\cdot\sigma_{sy}(\varepsilon_{sr} - \varepsilon_{sy})$$

where, $A_s$ is the area of tension reinforcement. Since the steel strain at the yield ending point $\varepsilon_{sr}$ is given by equation (4-3), $W_{it}$ can be written:

$$W_{it} = p\cdot b\cdot d\cdot\sigma_{sy}[S(\varepsilon_{cr})/\{p\cdot\sigma_{sy} - p'\cdot\sigma_{sy}' + N'/(bd)\} - (\varepsilon_{cr} + \varepsilon_{sy})] \quad (4\text{-}8)$$

The above equation indicates that the toughness of RC beam cross-section $W_{it}$ increases as both the toughness of concrete $S(\varepsilon_{cr})$ and the compression reinforcement ratio p' increase, and also as the axial force N' decreases. These coincide with the facts observed in the experiments and the simulations. When p'=0 and N'=0, it is simplified as:

$$W_{it} = \{S(\varepsilon_{cr}) - (\varepsilon_{cr} + \varepsilon_{sy})\cdot p\cdot\sigma_{sy}\}\cdot b\cdot d \quad (4\text{-}9)$$

When singly reinforced concrete beams have a relatively small value of $p\cdot\sigma_{sy}$, the second term in the parenthesis is sufficiently smaller than the first term. Therefore, if the tension steel does not rupture, $W_{it}$ can be expressed as:

$$W_{it} = S(\varepsilon_{sr})\cdot b\cdot d \quad (4\text{-}10)$$

This equation indicates that the toughness $W_{it}$ for singly reinforced concrete beams with a moderate value of $p\cdot\sigma_{sy}$ is proportional only to the toughness of concrete in compression $S(\varepsilon_{cr})$ and the size of the cross-section b·d. In other words, since the value of $\varepsilon_{sr}$ decreases with the increasing value of $p\cdot\sigma_{sy}$, the toughness $W_{it}$ becomes almost constant (see equation (4-7)).

(f) Ductility index $\gamma$

The reinforcement index $\eta$ expressed by equation (4-11) has been used to indicate roughly the properties of RC members.

$$\eta = f_c'/(p\cdot\sigma_{sy} - p'\cdot\sigma_{sy}') \quad (4\text{-}11)$$

where $f_c'$ is the compressive strength of concrete. A new reinforcement index $\gamma$ can be defined by using a part of equation (4-3), and is named ductility index here,

$$\gamma = S(\varepsilon_{cr})/\{p\cdot\sigma_{sy} - p'\cdot\sigma_{sy}' + N'/(bd)\} \quad (4\text{-}12)$$

The compressive toughness of concrete $S(\varepsilon_{cr})$ is contained in the index $\gamma$. $\gamma$ indicates the final plastic strain in tensile reinforcing bars and hence can be used to evaluate the ductile failure behavior of RC and PC beams.

(g) Failure behavior of PC beams

Generally, new design methods are desired when new types of structures or new kinds of materials appear. In this section, the failure behavior of PC beams made of SFRHC is discussed to demonstrate the validity of the proposed ductility index $\gamma$.

Two PC beams each were tested for the nine kinds of testing conditions differing in concrete types, reinforcment ratios, and prestressing force as tabulated in Table 4.1[7]. Three kinds of concrete were used, namely, normal strength concrete N, high strength concrete H, and steel fiber reinforced high strength concrete (SFRHC) F, where steel fibers (size: ø 1x40mm) were added at 4% by volume. Fig. 4.8 shows the compressive load-displacement curves of cylindrical specimens (size: ø10x20cm) of the concretes and also the simplified stress-strain curves. SFRHC has high compressive toughness as compared with the other types of concrete. The dimensions of PC beams and the loading manner are illustrated in Fig. 4.9. Two kinds of prestressing steel bars were adopted; the tension reinforcement ratio was 2.4% for the large bar ($\sigma_{py}$=1.19GPa) and 1.1% for the small bar ($\sigma_{py}$=1.39GPa).

Test results of PC beams are given in Fig. 4.10 and Table 4.2. The ductility (deformability) and the toughness (area under the curve) of PC beams increased with an increase in the compressive toughness of concrete; for example, it increased in the order of N-4, H-2 and F-3, while they have the same reinforcement conditions. In the case of H-2, the ultimate load increased as compared with N-4 having the same reinforcement, but the ductility did not increase and failed in a brittle manner after the peak load.

The values of the proposed ductility index $\gamma$ are given in Table 4.2. The toughness of concrete in compression $S_c$ in Table 4.2 is given by the total area under stress-strain curve models in Fig. 4.8(b). If $\gamma$ was too large as in the case of F-3, the tension (prestressing) bar broke finally. On the other hand, if $\gamma$ was small as in the cases of N-1, N-2 and H-1, PC beams failed in a brittle manner. PC beams of F-1 and H-2 were deformed enough at high load level without bar-breaking. The combination of concrete and reinforcement for F-1 and H-2 having $\gamma$ of about 0.02 seems to be the most appropriate within the range of these experiments. These test results indicate that the proposed index $\gamma$ is useful to evalulate the ductile behavior of PC beams differing in combination of concrete and reinforcement.

It is indicated by equation (4-12) that when the compressive toughness of concrete $S_c$ is high, the value of $p \cdot \sigma_{sy}$ should be also high to keep $\gamma$ constant, that is, to keep ductility. When PC beams and RC beams have high values of $S_c$ and $p \cdot \sigma_{sy}$ and a moderate value of $\gamma$, they are superior to others in the ultimate strength as well as in the ductility.

## CONCLUSIONS

Our recent investigations on evaluation, improvement and application of concrete toughness were compiled in this paper. Several topics in Standards and Recommendation in Japan in relation to toughness were also included.

In the paper, toughness was considered to be the ability to absorb energy and therefore can be evaluated by the area below the load displacement curve up to a limit point. It was shown that the differential signal between factored load signal and factored displacement signal was effective to control unstable failure of concrete specimens and to measure complete load-displacement curves even though it had an indented portions after the maximim load point. The relations among load, displacement, energies and time during impact tests were obtained from the data of acceleretions and strain.

High toughness concrete having not only high strength but also high toughness was produced by introducing steel fibers at high percentage into matrix of high strength cement concrete or polyester resin concrete.

The toughness of reinforced concrete beams in bending was related to the toughness of concrete in compression by equations. The ductility index $\gamma$ was proposed to evaluate ductile behavior of RC and PC beams. The validity of the index $\gamma$ was confirmed by experiments of PC beams differing in combination of concrete and reinforcement.

REFERENCES

[1] JCI Standards for Test Methods of Fiber Reinforced Concrete, Japan Concrete Institute, 1984.

[2] ACI Committee 544; Measurement of Properties of Fiber Reinforced Concrete, Jour. American Concrete Institute, Vol.75, No.7, pp.283-298, 1987.

[3] K.Rokugo, S.Ohno and W.Koyanagi; Automatical Measuring System of Load-displacement Curves Including Post-failure Region of Concrete Specimens, Int. Conf. on Fracture Mechanics of Concrete, Lausanne, pp.291-299, 1985.

[4] J.A.Hudson, S.L.Crouch and C.Fairhurst; Soft, Stiff and Servo Controlled Testing Machines, - A Review with Reference to Rock Failure, Dept. of Civil and Mineral Eng., Univ. of Minnesota, 1971.

[5] J.A.Hudson, M.P.Harly and C.Fairhurst; The Failure of Rock Beams Part 1 and Part 2, Int. Jour. Rock Mech. Min. Sci, Vol. 10, pp.53-82, 1973.

[6] W.Koyanagi, K.Rokugo, Y.Uchida and H.Iwase; Energy Approach to Deformation and Fracture of Concrete under Impact Load, Trans. of Japan Concrete Institute, pp.161-168, 1983.

[7] K.Rokugo, H.Iwase and W.Koyanagi; High Quality Fiber Reinfored Concrete and Its Application to Prestressed Concrete Beams, Symp. on Developments in Fiber Reinforced Cement and Concrete, Sheffield, No.8-16, 1986.

[8] W.Koyanagi, K.Rokugo, N.Murai and F.Hayashi; Evaluation of Toughness of Resin Concrete and Its Improvement by Steel Fibers, 4th Int. Conf. on Polymers in Concrete, Darmstadt, pp.93-98, 1984.

[9] JSCE SFRC Reserch Subcommittee; Recommendation for Design and Construction of Steel Fiber Reinforced Concrete, No.50, 1983.

[10] W.Koyanagi, K.Rokugo, Y.Uchida and H.Iwase; Evaluation of Toughness of Doubly Reinforced Concrete Beams, Trans. of Japan Concrete Institute, Vol.5, pp.425-432, 1983.

[11] K.Rokugo, H.Iwase and W.Koyanagi; Application of Fracture Energy of Concrete in Compression to Section Design of Reinforced Concrete Members, Int. Conf. on Fracture Mechanics of Concrete, Lausanne, pp.441-450, 1985.

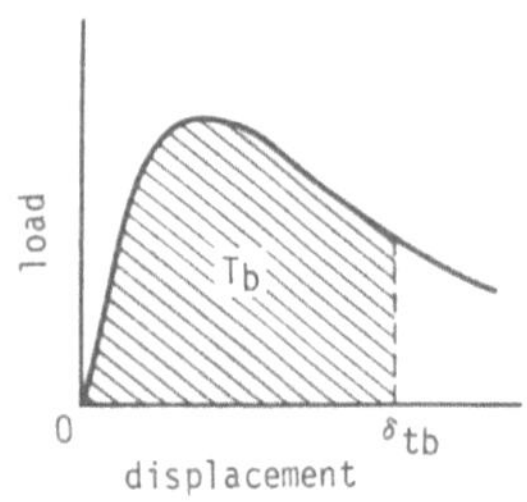


Fig. 2.1 Flexural toughness $T_b$ by JCI.

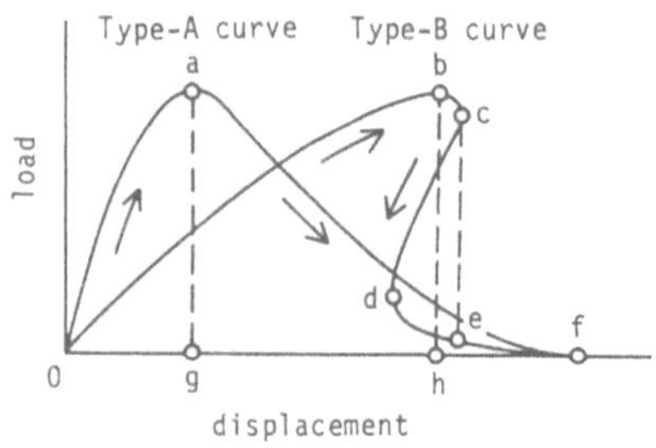


Fig. 2.2 Two types of load-displacement curves.

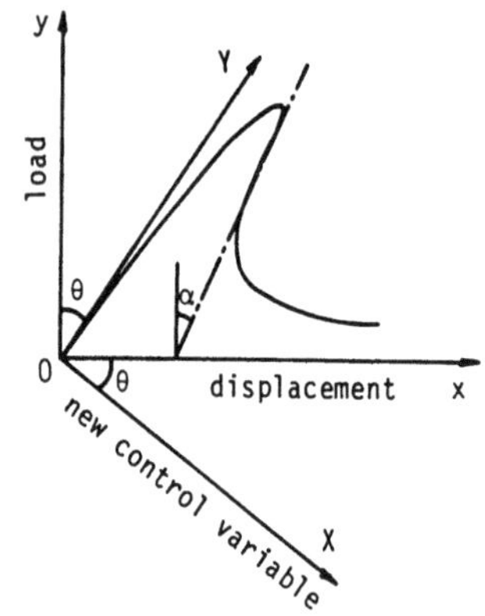


Fig. 2.3 New control variable.

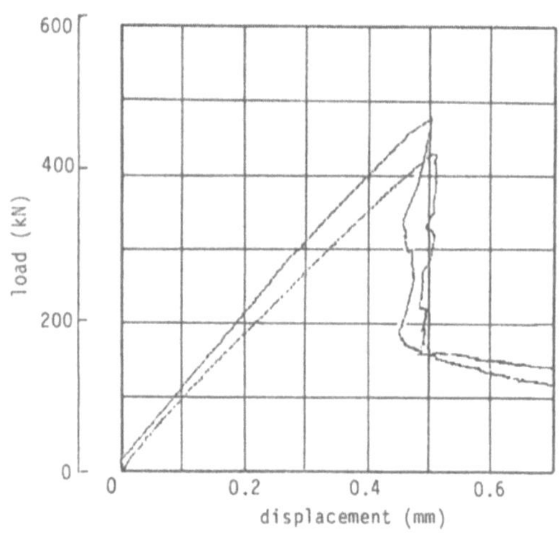


Fig. 2.4 Compressive load-displacement curves of high strength light-weight concrete.

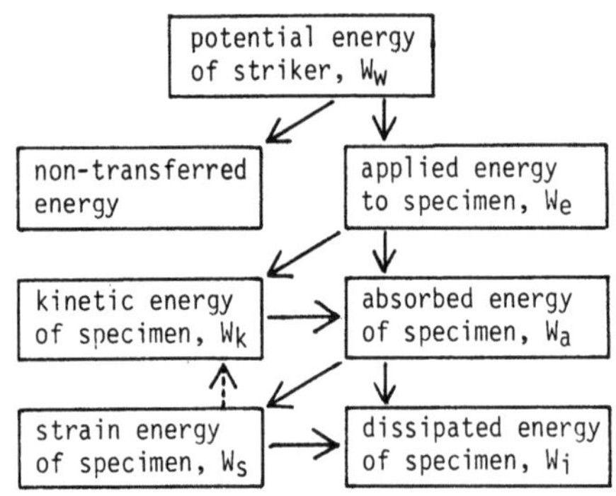


Fig. 2.5 Energy transformation process of specimens under impact.

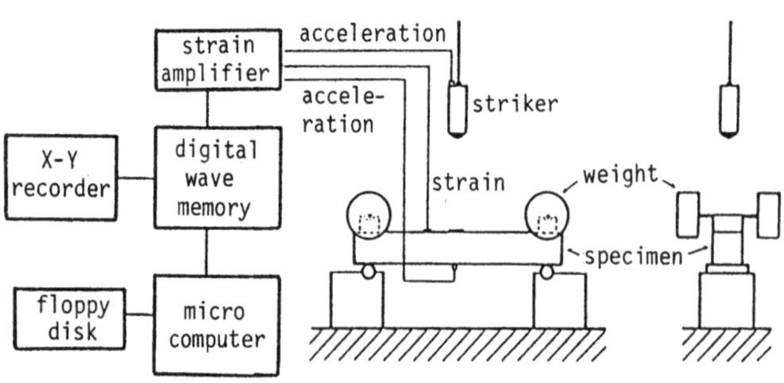


Fig. 2.6 Setup of impact tests.

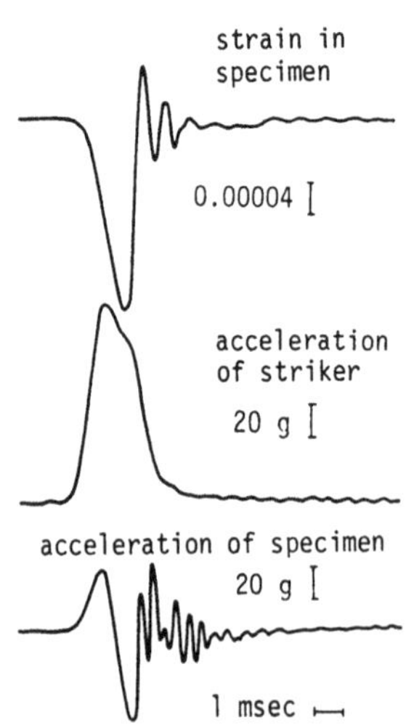


Fig. 2.7 Measures strain and accelerations.

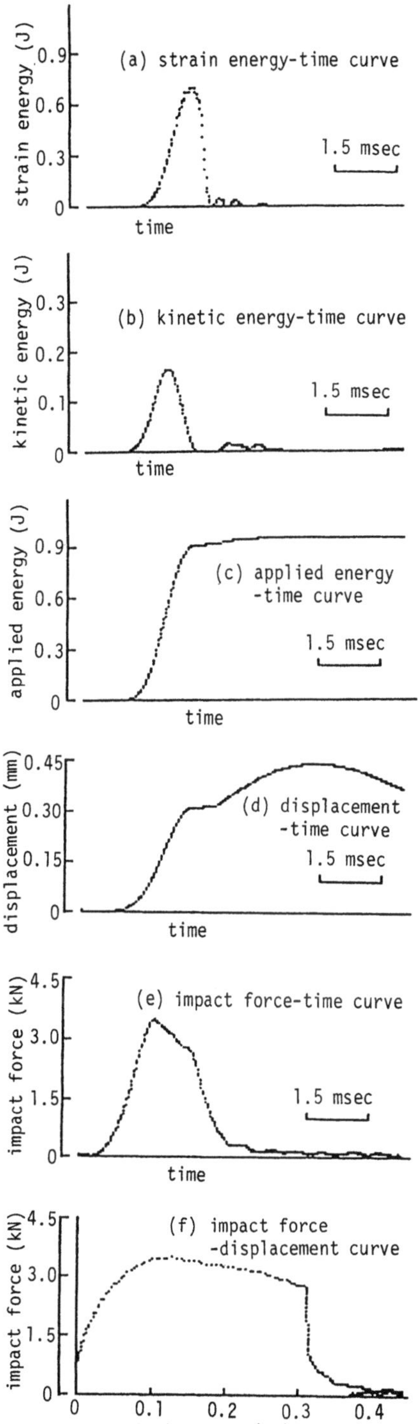


Fig. 2.8 Relations among impact load, displacement, energies and time at impact test.

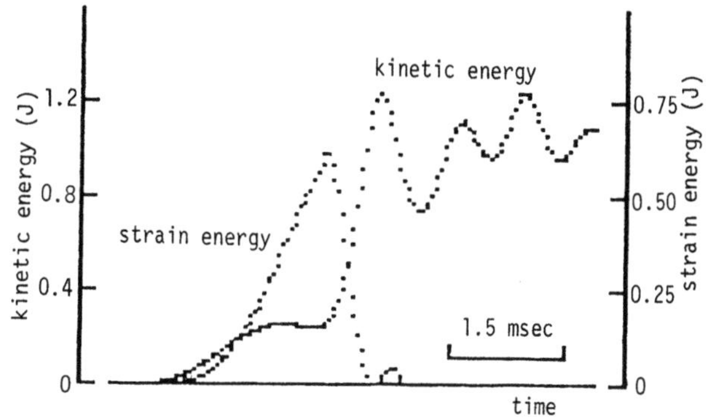


Fig. 2.9 Strain energy and kinetic energy of violently ruptured specimen.

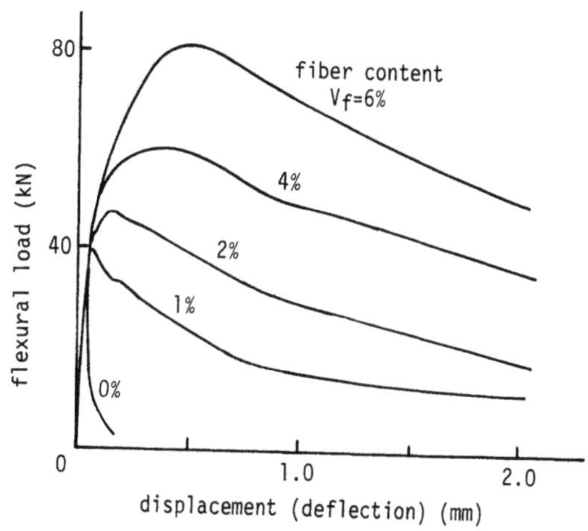


Fig. 3.1 Flexural load-displacement curves of SFRHC beam specimens.

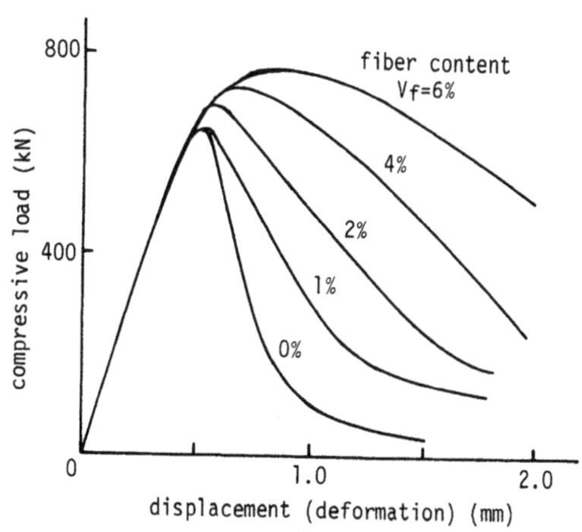


Fig. 3.2 Compressive load-displacement curves of SFRHC cylindrical specimens.

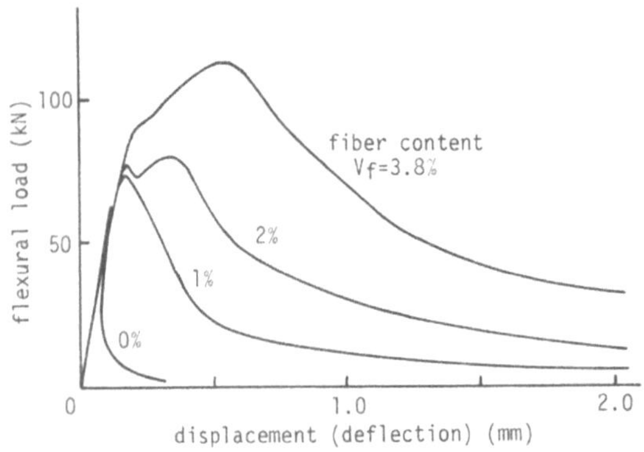


Fig. 3.3 Flexural load-displacement curves of SFRREC beam specimens.

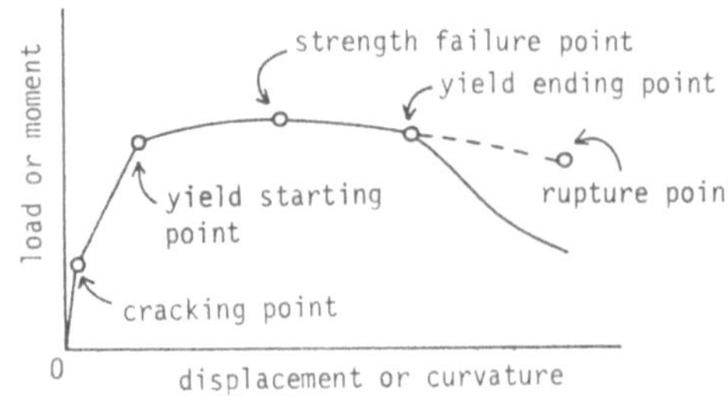


Fig. 4.1 Failure process of RC beams.

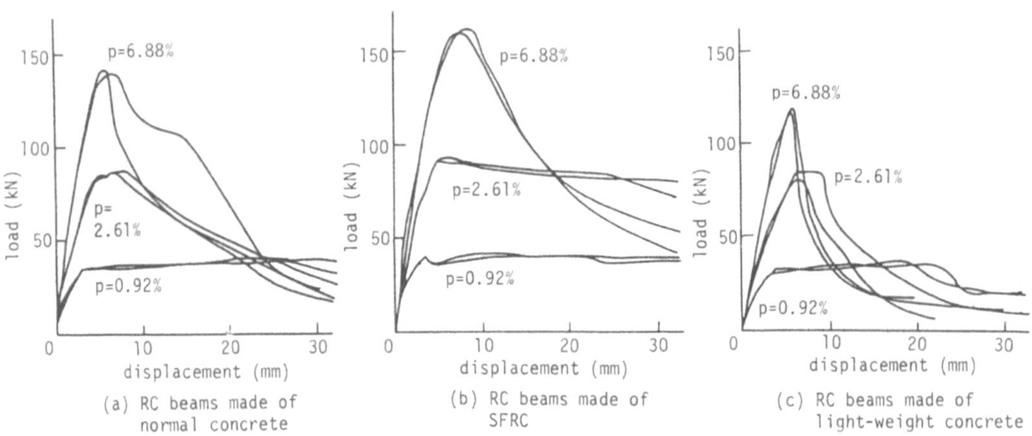


Fig. 4.2 Measured load-displacement curves of singly reinforced concrete beams.

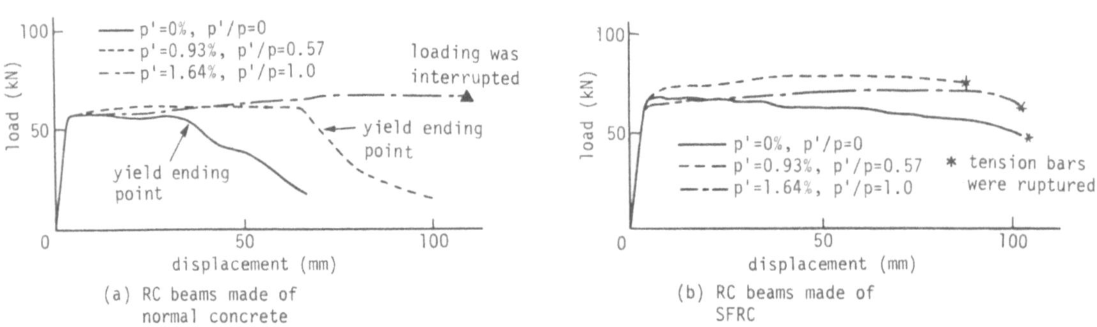


Fig. 4.3 Measured load-displacement curces of doubly reinforced concrete beams.

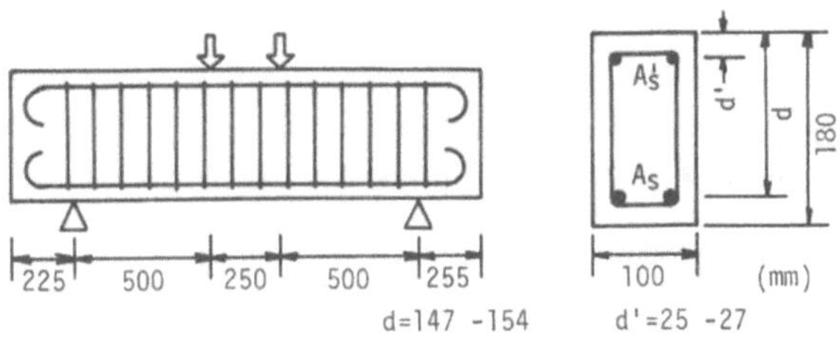


Fig. 4.4 Dimensions of RC beam specimen.

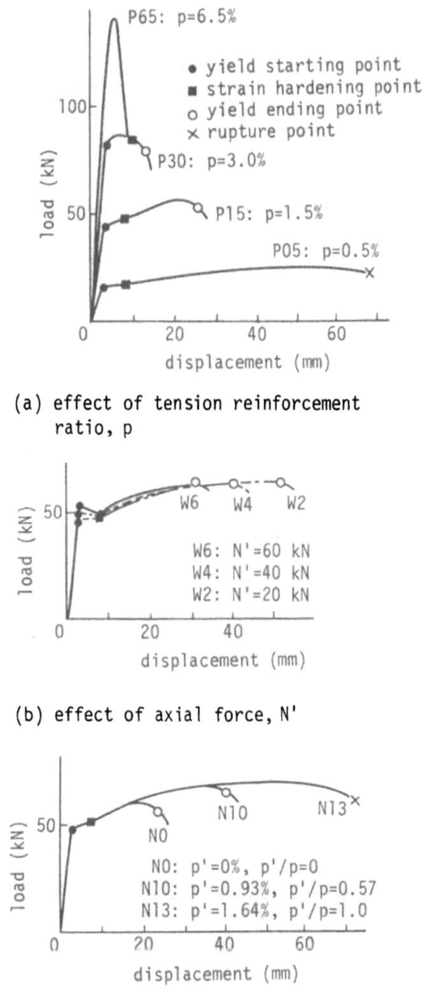


(a) effect of tension reinforcement ratio, p

(b) effect of axial force, N'

(c) effect of compression reinforcement ratio, p'

Fig. 4.5 Simulations of load-displacement curves of RC beams.

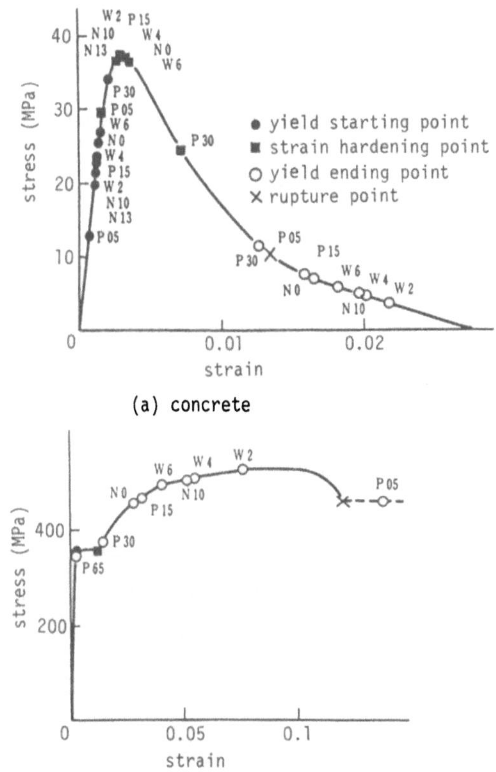


(a) concrete

(b) steel

Fig. 4.6 Stress-strain curves for simulation.

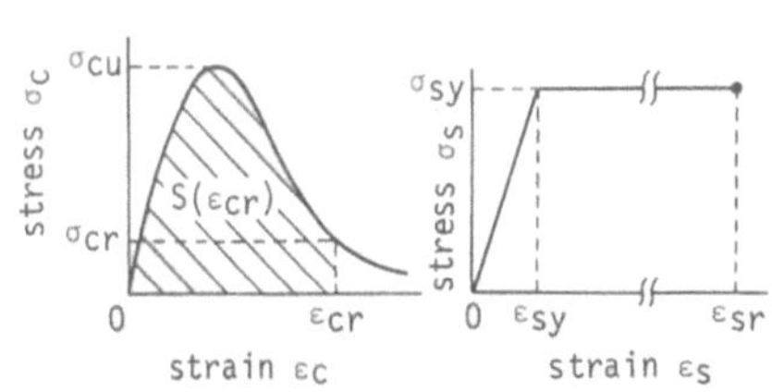


(a) concrete (b) steel

Fig. 4.7 Stress-strain curves for analysis.

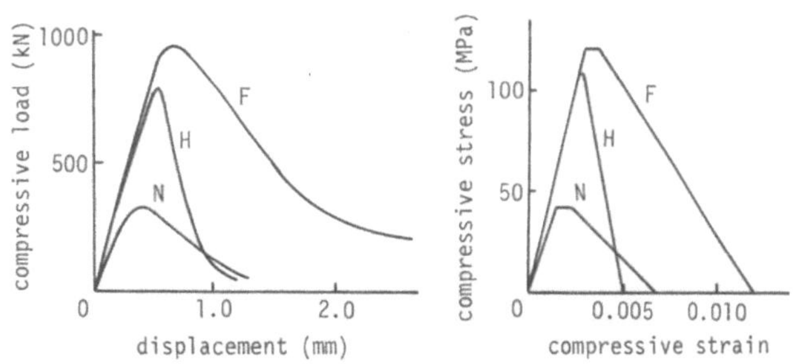


(a) measured curves (b) models

Fig. 4.8 Load-displacement curves of concrete cylinders ($\phi$10x20 cm) and stress-strain curve models.

Table 4.1 Kinds of PC beams.

| kind of beam | concrete | area of prestressed reinforcement ($mm^2$) | pre-stressing force (kN) | area of compression reinforcement ($mm^2$) |
|---|---|---|---|---|
| N-1 | N | 284 | 88 | 0 |
| N-2 | N | 284 | 88 | 143 |
| N-3 | N | 284 | 88 | 397 |
| N-4 | N | 133 | 88 | 0 |
| H-1 | H | 284 | 176 | 0 |
| H-2 | H | 133 | 88 | 0 |
| F-1 | F | 284 | 176 | 0 |
| F-2 | F | 284 | 176 | 397 |
| F-3 | F | 133 | 88 | 0 |

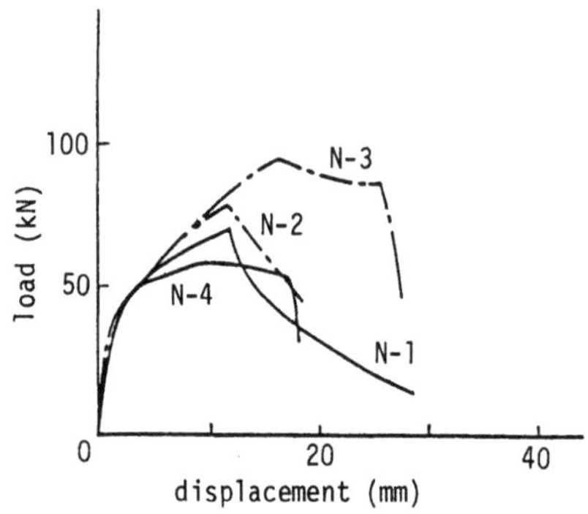


(a) PC beams made of normal concrete

Table 4.2 Test results of PC beams.

| kind of beam | $P_{cr}$ (kN) | $P_{max}$ (kN) | $S_c$ (MPa) | $p\cdot\sigma_{sy}-p'\cdot\sigma_{sy}'$ (MPa) | $\gamma$ |
|---|---|---|---|---|---|
| N-1 | 34 | 73 | 0.161 | 28.0 | 0.006 |
| N-2 | 34 | 77 | 0.129* | 23.7 | 0.005 |
| N-3 | 34 | 96 | 0.129* | 14.5 | 0.009 |
| N-4 | 37 | 62 | 0.161 | 15.4 | 0.010 |
| H-1 | 52 | 120 | 0.263 | 28.0 | 0.009 |
| H-2 | 39 | 76 | 0.263 | 15.4 | 0.017 |
| F-1 | 57 | 137 | 0.748 | 28.0 | 0.027 |
| F-2 | 67 | 138 | 0.748 | 14.5 | 0.052 |
| F-3 | 47 | 97 | 0.748 | 15.4 | 0.049 |

* area for N-2 and N-3 was somewhat different from that for N-1 and N-4

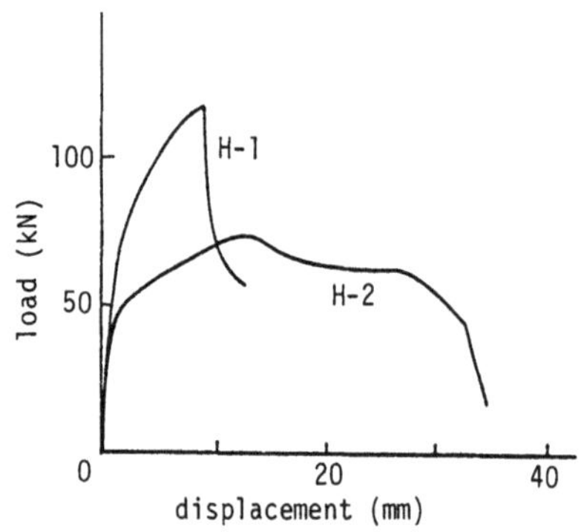


(b) PC beams made of high strength concrete

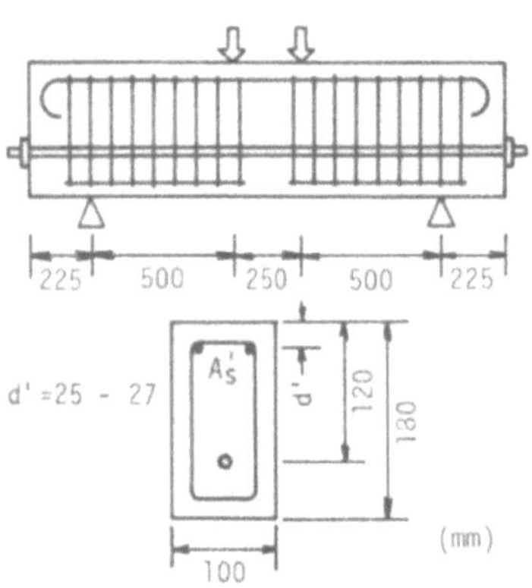


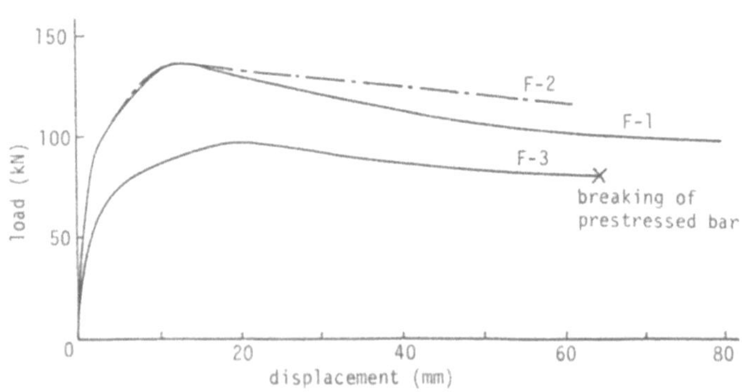


(c) PC beams made of SFRHC

Fig. 4.9 Dimensions of PC beams.

Fig. 4.10 Load-displacement curves of PC beams.

SEM/RILEM International Conference
on Fracture of Concrete and Rock
Houston, Texas, June 1987
Editors: S.P. Shah, S.E. Swartz

# MODELING OF THE FRACTURE PROCESS IN PRENOTCHED CONCRETE BEAMS TO EXAMINE SIZE EFFECTS

T. M. E. Refai, D. V. Swenson

## ABSTRACT

Concrete is a strain-softening material with a relatively low tensile strength. For laboratory specimens the resulting fracture process zone is large relative to the size of the K-dominated region calculated using linear elastic fracture mechanics. This violates a fundamental assumption of linear elastic fracture mechanics. In this paper, we analyze three-point beam specimens using a direct model of the process zone, and compare the results to experimental data. The results show the necessity of using such a model for laboratory specimens. Some difficulties with interpreting the experimental data are also noted. This is a report of work in progress.

## 1.0 INTRODUCTION

Many investigations have been undertaken to determine the validity of linear elastic fracture mechanics (LEFM) as applied to mortars and concrete [1,2]. Since Kaplan [3] in 1961, many specimen geometries, crack lengths, aggregate sizes, compressive and tensile strengths and other characteristics have been tested in mode I to examine the validity of LEFM. Few researchers have considered the influence of specimen size when applying LEFM to concrete. However, a significant size effect in the determination of $K_{IC}$ has been observed using conventional LEFM approaches. In fact, in most concrete laboratory specimens, fundamental assumptions of LEFM are related.

This has been argued by Peterson [4] and Gerstle, et. al., [5] for concrete, and by Boone, et. al., [6] for granite. Concrete behaves as a strain-softening material when loaded in tension. Two features of this response are significant: first, the maximum tensile stress is relatively low; and, second, after the peak stress is reached, the material does not fail abruptly, but softens with increasing strain. If this behavior is accepted, it follows that for the experiments discussed in this report, linear elastic fracture mechanics concepts are not applicable.

The reason is that, the resulting process zone is large relative to the region dominated by the singular K-field calculated using LEFM. (As an approximation, the K-field region is 1/5 the crack length, and the process zone should be 1/5 the K-field for linear fracture mechanics to apply.) Thus, the stresses in the vicinity of the crack tip do not correspond to those predicted by linear elastic fracture mechanics. Instead, the stresses are conceptually those shown in Figure 1. Ahead of the crack lies an elastic region, where the material has not yet reached the peak failure stress. Between the elastic region and the stress-free crack face lies a process zone in which the material is strain softening. Finally, there is a region where no load is being carried on the crack face and there exists the traditional stress-free crack. Clearly, even the definition of the crack tip now becomes uncertain when evaluating experimental results. For example, if dye is used to locate that crack tip, it is not clear how far into the process zone the dye will penetrate.

An alternative approach to using linear fracture mechanics is a direct model of the process zone, where the material is assumed to follow a specified stress-crack opening displacement (COD) relation. The process zone is then allowed to grow and propagate within the constraints of the material response and loading. This is the approach used in the present paper. Analytical results are compared with experimental results for prenotched beams loaded in three-point bending.

T.M.E. Refai is a Research Assistant, Civil Engineering Department, D. V. Swenson is an Assistant Professor, Mechanical Engineering Department at Kansas State University, Manhattan, KS 66506

## 2.0 DESCRIPTION OF EXPERIMENTS

### 2.1 Testing of Beams in Bending

A comprehensive program of testing plain concrete beams in three-point bending (Fig. 2) was conducted and is reported in 7. Three beam sizes were used in three series of tests.

Series 1: 76 mm (3 in.) wide, 102 mm (4 in.) deep, 381 mm (15 in.) span;
Series 2: 76 mm (3 in.) wide, 203 mm (8 in.) deep, 762 mm (30 in.) span;
Series 3: 76 mm (3 in.) wide, 305 mm (12 in.) deep, 1143 mm (45 in.) span;

The mix design, strength, testing condition and technique were the same for each series. The average compressive strength was about 55.2 MPa (8000 psi) and the corresponding value $E_c$ was 38.1 GPa (5.53 x $10^6$ psi).

All tests were performed using an electro-hydraulic materials testing system (MTS). Crack-month-opening displacement (CMOD) and load-point displacement (LPD) were monitored simultaneously through the use of commercially available displacement transducers (MTS 632.05 B-60).

The majority of the beams in the three series were precracked before testing, and were used for a different investigation. However, some beams from each series were prenotched (saw cut to simulate cracks) to $a_o$/w ratios of 0.3, 0.5 and 0.7, and were tested in load control to failure. Plots of load vs. crack-month-opening displacement (CMOD) and load vs. load-point displacement (LPD) were obtained simultaneously. In this paper, we compare analytical results to experimental results for the $a_o$/w ratio of 0.3 (beams of series 2 and 3) and 0.5 (beams of series 1).

### 2.2 Testing of Bars in Direct Tension

An additional experiment was carried out to obtain a stress vs. crack-opening displacement (COD) relation, which was used as input to the analytical model. Eight concrete specimens of dimensions 76 x 102 x 406 mm (3 x 4 x 16 in.) were tested in direct tension as shown in Fig. 3. The tension specimens had a mid-height notch on each side. Four specimens each for notch depth ratio of 0.5 were tested in load and strain control. Two simultaneous plots of load vs. CMOD were obtained.

Figures 4 and 5 show the results of three tests, two performed under strain control and one under stroke control. The stroke control and one of the strain control tests correlate well (Fig. 5). However, strain control test #2 did not strain soften rapidly, but sustained load much larger than strain control test #1. The reason for the difference in the measured response of these tests is not known. Clearly, this curve absorbs much more energy during the softening than the other two curves. Although we are verifying the validity of these tests, we believe that the tests represent the true material response and are not faulty tests. If the data are correct this implies considerable spread in the results.

The stroke control load vs. COD curve was used to obtain a stress vs. COD material response. To do this, the load was divided by the area of the beams (excluding the notch) and the measured COD was assumed to give the true COD (this neglects any straining away from the crack face). The unnotched area was 0.0039 $m^2$ (6 $in^2$). The resulting stress vs. COD relation used in the analysis is shown in Fig. 6. The peak load is about 70% of that given by Petersson [4], while the measured COD at failure is about the same.

## 3.0 FINITE ELEMENT MODEL

The finite element model consists of two parts, the model of the continuum of the beam and the model of the process zone. The analysis was performed using the finite element code CRACKER [8], which was developed to perform analyses of dynamic crack propagation. It uses explicit time integration to integrate nodal accelerations through time. Its advantage for dynamic crack propagation analyses is that no global matrices are assembled, instead the solution is obtained at the element level. This makes it possible to have many small time steps to follow stress waves propagating through the structure and it allows remeshing to be performed without regard to the order of node or element numbering. Thus automatic remeshing is simplified. The approach is not intended for static solutions, but dynamic relaxation can be used to obtain static solutions. As presently implemented, the user must specify the damping, no optimum damping parameters are calculated.

In the present analysis, a static loading was approximated by applying the load over a time interval that was long with respect to the natural frequencies of the beam and by including a small amount of damping in the solution. This ensures that at any time in the analysis, the solution is approximately in static equilibrium. One advantage of using a dynamic analysis is that no difficulties are encountered obtaining converged solutions as the beam softens during fracture. The disadvantage is the relatively long solution times required to approximate a static solution.

The body of the beam was modeled using standard size-noded triangles with quadratic shape functions. Experience with these elements shows that they perform well in elastic analyses. The process zone was modeled directly using a six-noded "interface" or "surface traction" element. In this element, a stress-displacement relationship may be specified between the two surfaces. In the present analysis, we use a Dugdale type approach to modeling the process zone, where tractions on the crack face are a function of crack opening. Although the implementation is usually described as an "interface" element, it is more properly thought of as a special case of surface tractions, where the surface tractions are prescribed as a function of nodal displacements. The stress vs. COD relation is that shown in Fig. 6.

Figure 7 shows the finite element model of the 0.203 m (8 inch) deep beam. The beam is supported in the Y direction at the ends, and in the center, the loading platen is approximated by element which contact the bottom of the beam. Interface elements transmit normal forces from the platen to the beam, but no sliding forces are transmitted. The velocity of the platen was specified to be constant during the analysis (a value of 0.0254 m/s (1.0 in/sec)). Plane stress assumptions were used for the beam.

## 4.0 RESULTS

Figures 8 through 10 compare experimental and analytical results for load on the beams vs. CMOD measured at the top of the beams. Figure 11 shows a similar plot for load versus load point displacement. The comparison between experiment and analysis is moderately acceptable for load versus CMOD, but the comparison is poor for load versus LPD. In this case, the peak analytical load is reached before the experimental data and the load then decreases much more rapidly. The reasons for the discrepancy are not known.

One possible explanation is that the stress-COD relation used in the analysis should be replaced with the measured response labeled as strain 2 in Figure 4. In this case, the peak load would remain about the same, but the post-peak behavior would be elongated to more closely match the experimental data.

However, the situation is not that clear. At low loads (near the end of the test), we would expect a linear relation between CMOD and LPD. This follows from assuming a hinge point at the bottom surface of the beam (Fig. 7). For the 0.203 m (8-inch) beam, the expected relation would be LPD = (0.94) CMOD. The finite element model does follow this relation, however, the experimental data shows a much larger LPD than predicted. The experimental results are consistent with some inelastic deformation (crushing) at the hinge point, with the result that the CMOD is less than expected. If this is true, the model of the process zone must also include non-linear behavior in compression as well as tension. These effects are being investigated.

Figure 12 and 13 show displacement plots for the 0.203 m (8-inch) beam at the peak load and near failure. Figures 14 and 15 show corresponding maximum principal stress contours. The normal stress ($\sigma_x$) on lines from the top of the beam to the bottom are shown in Figures 16 and 17. These figures indicate the propagation of the process zone through the beam as failure occurs. The typical length of the zone is about 50 mm (2 inches). This size is similar to that of the 0.102 and 0.305 m (12 and 4 inch) beams. Clearly, the stresses do not correspond to LEFM predictions.

## 5.0 CONCLUSIONS

These results demonstrate that a direct model of the process zone is necessary to evaluate fracture in laboratory concrete specimens. Because the process zone is relatively large (approximately 5 cm (2 inches)) relative to the specimen dimensions, the stresses near the crack tip do not correspond to those predicted using LEFM. Thus, a basic assumption of LEFM is violated in these experiments and an alternate model is necessary. Nevertheless, the size of the process zone was about the same for the three different specimen sizes.

Questions still remain with respect to the large variation in measured tensile data for the concrete. In addition the experimental data suggests that non-linear compressive behavior occurred at the hinge point of the three-point beams. If so, the model of the process zone must include this effect.

## REFERENCES

1. V. E. Saouma, A. R. Ingraffea, and D. M. Catalano, "Fracture Toughness of concrete - $K_{IC}$ Revisited," J. Eng. Mech. Div., ASCE, 108, EM6, 1152-1166, 1982.

2. S. E. Swartz, "Fracture Toughness Testing of Concrete at Kansas State University: Is LEFM Acceptable," Proceedings, International Conference on Fracture Mechanics of Concrete, Lousanne, Switzerland, Oct. 1-3, 1985.

3. E. Kaplan, "Crack Propagation and the Fracture of Concrete," ACI Journal, Proceedings, Vol. 58, No. 5, pp. 591-610, 1961.

4. P. Petersson, "Crack Growth and Development of Fracture Times in Plain Concrete and Similar Materials," Report TVBM-1006, Division of Building Materials, Load Institute of Technology, Sweden, 1981.

5. W. H. Gerstle, A. R. Ingraffea and P. Gergely, "The Fracture Mechanics of Bond in Reinforced Concrete," Civil Engineering Report 8-7, Cornell University, Ithaca, NY, June 1982.

6. T. J. Boone, P. A. Wawrzynek, A. R. Ingraffea, "Simulation of Fracture Process in Rock with Application to Hydrofracturing," Int. J. Rock Mech. Min. Sci. Geomech., Abstr., Vol. 23, No. 3, pp. 255-265, 1986.

7. S. E. Swartz and T. M. E. Refai, "Influence of Size Effects on Opening Mode Fracture Parameters for Precracked Concrete Beams in Bending," International Conference on Fracture of Concrete and Rock, Houston, TX, June 17-19, 1987.

8. D. V. Swenson, A. R. Ingrafea, "Modeling Mixed-Mode Dynamic Crack Propagation Using Finite Elements: Theory and Applications," to appear in Computational Mechanics.


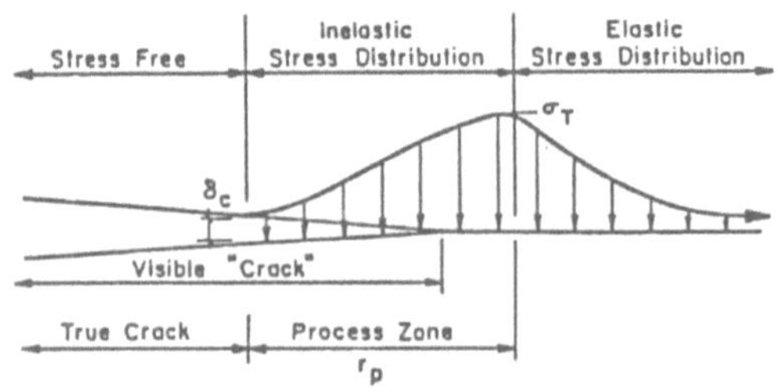


Fig. 1 Schematic of Hypothesized Process Zone in Concrete (After Boone, et al. (6) ).

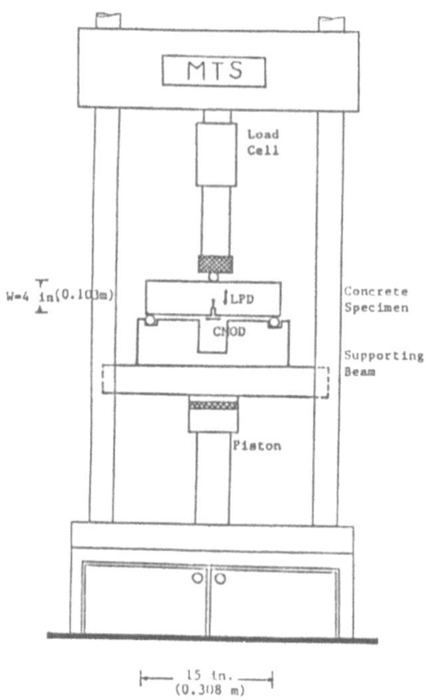


Fig. 2 The Set-Up for 0.102 m(4 in.) Deep Beam in 3-Point Bending.

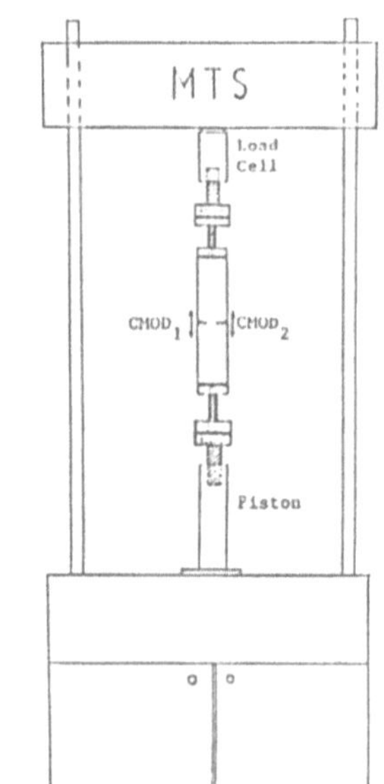


Fig. 3 The Set-Up for Specimens in Tension.

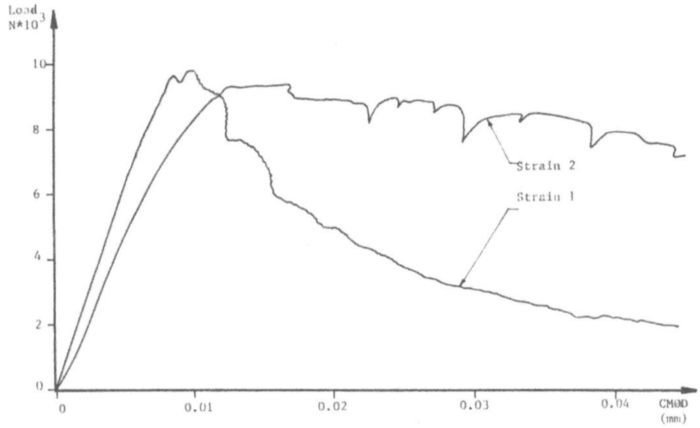


Fig. 4 Load vs CMOD for Strain 1 & Strain 2.

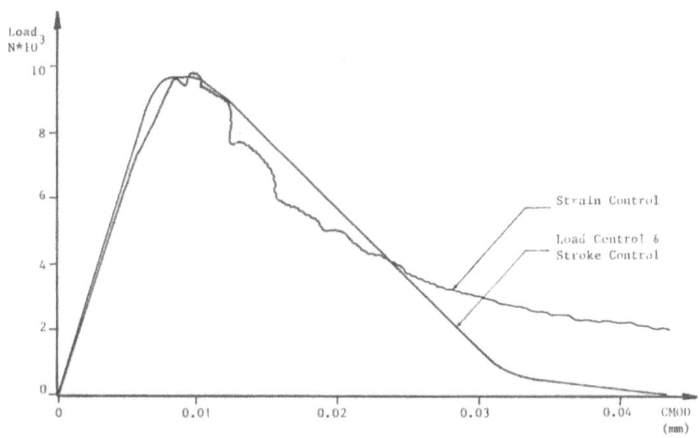


Fig. 5 Load vs CMOD for Different Types of Control.

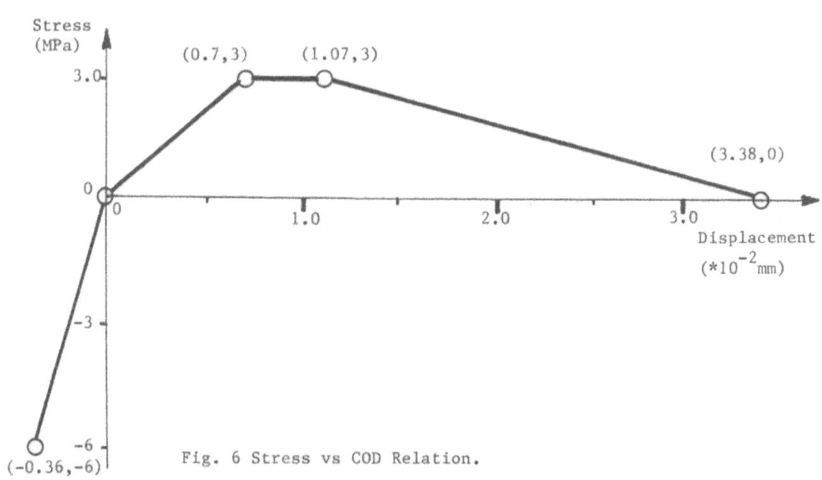


Fig. 6 Stress vs COD Relation.

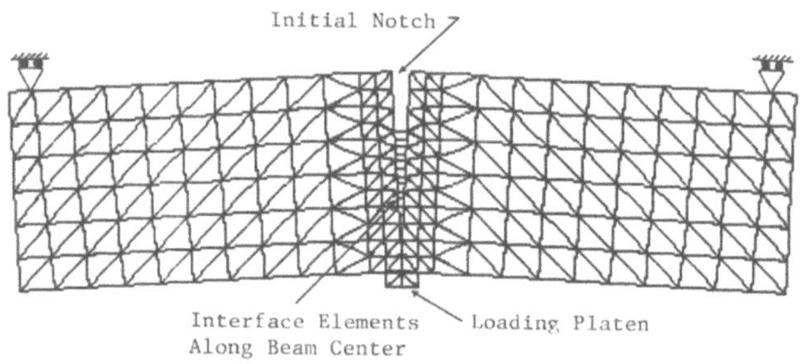


Fig. 7 Finite Element Model of 0.203 m Deep Beams. (The beam is shown at the end of the analysis with displacement magnified 100 times.)

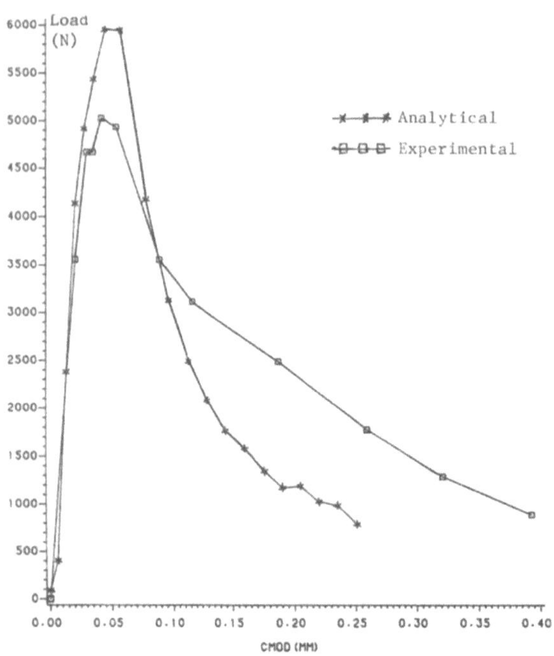


Fig. 9 Load vs CMOD for 0.203 m Deep Beams.

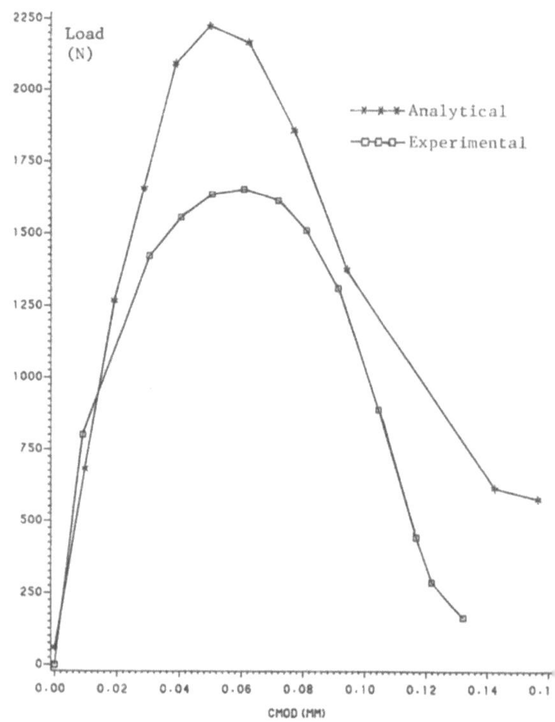


Fig. 8 Load vs CMOD for 0.102 m Deep Beams.

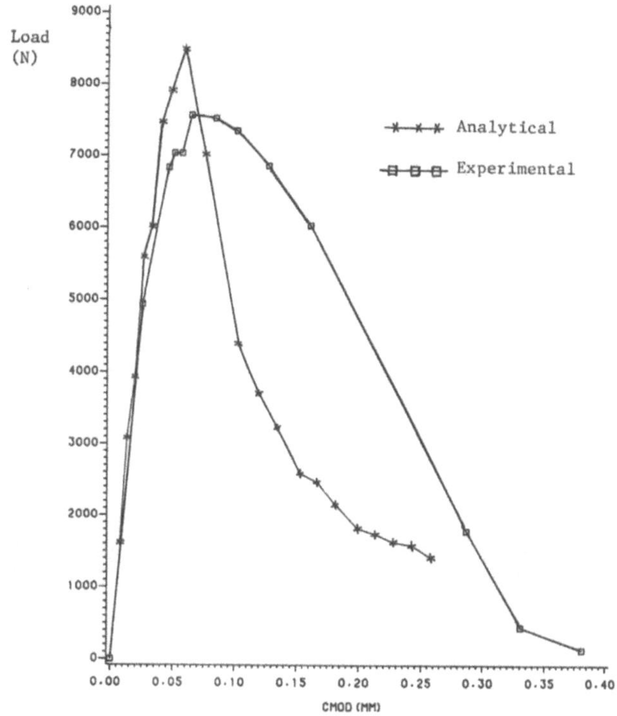


Fig. 10 Load vs CMOD for 0.305 m Deep Beams.

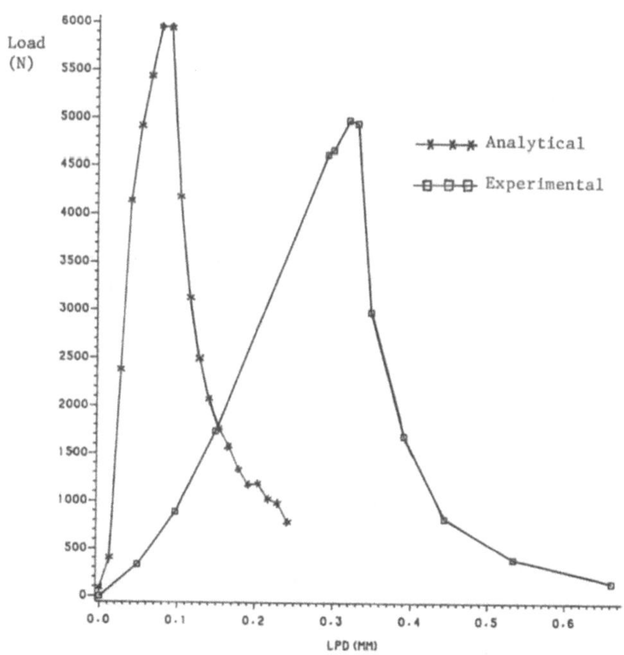


Fig. 11 Load vs LPD for 0.203 m Deep Beams.

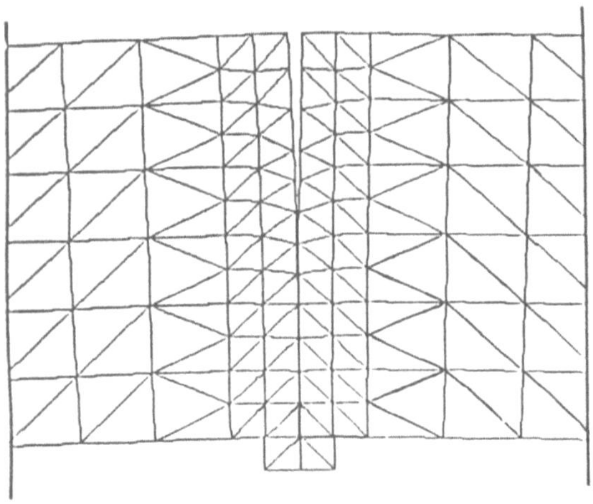

Fig. 12 Displacement Plot for 0.203 m Deep Beams Around The Crack (At Peak Load).

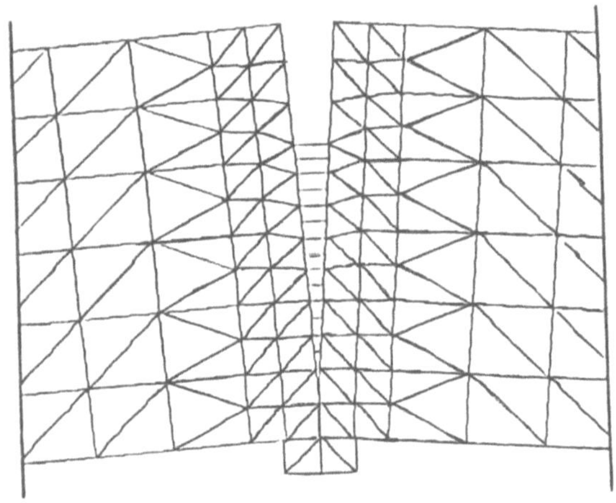

Fig. 13 Displacement Plot for 0.203 m Deep Beams Around The Crack (Near Failure).

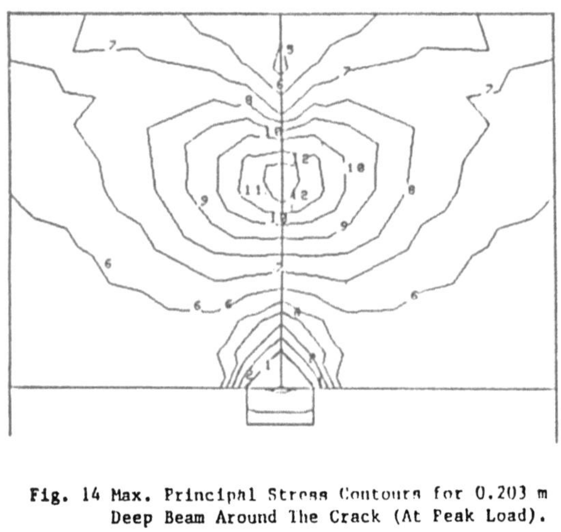

Fig. 14 Max. Principal Stress Contours for 0.203 m Deep Beam Around The Crack (At Peak Load).

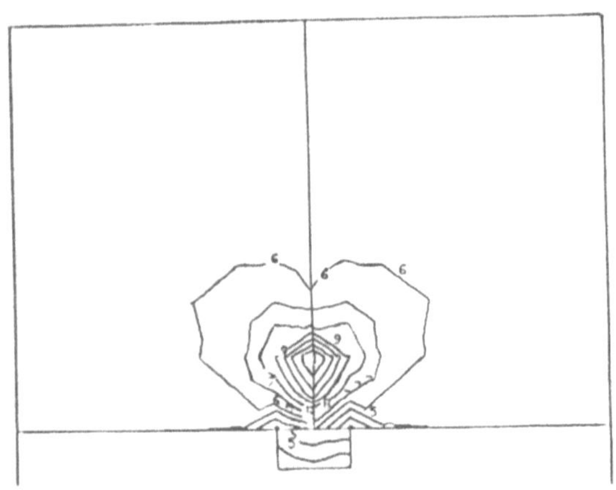

Fig. 15 Max. Principal Stress Contours for 0.203 m Deep Beam Around The Crack (Near Failure).

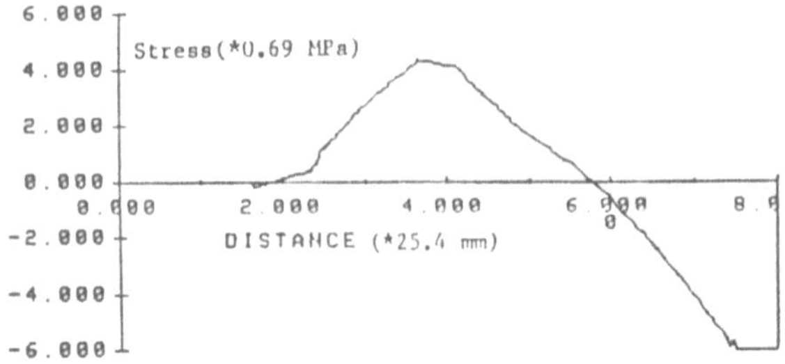


Fig. 16 Normal Stress Distribution Along The Crack Cross Section (At Peak Load).

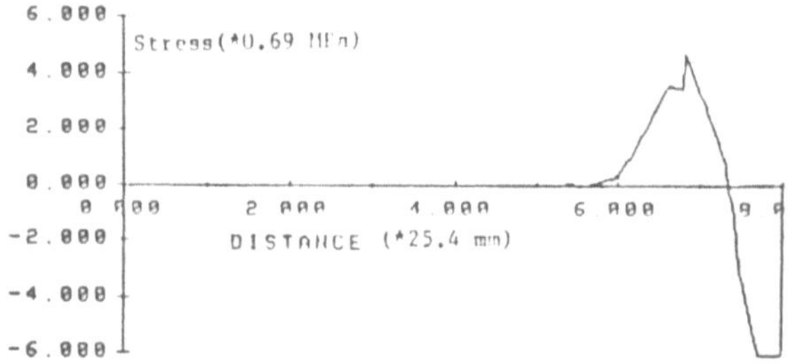


Fig. 17 Normal Stress Distribution Along The Crack Cross Section (Near Failure).

SEM/RILEM International Conference
on Fracture of Concrete and Rock
Houston, Texas, June 1987
Editors: S.P. Shah, S.E. Swartz



# FRACTURE MECHANICS OF CONCRETE GRAVITY DAMS

by:
Victor E. Saouma[1]
Mohamed L. Ayari[2]
Howard Boggs[3]

April 6, 1987

**Abstract**

**A comprehensive investigation on the behavior of cracked concrete gravity dams using fracture mechanics concepts through a series of parametric studies is presented. The effect of various forms of loading, concrete age and anisotropy on the stress intensity factors, direction of crack profiles, crack lengths and stress redistribution is assessed.**

## 1 INTRODUCTION

It is ironic that while fracture mechanics foundations were laid by Griffith in the early thirties for brittle materials [13], it has been mostly applied to metallic materials. While there has been some pioneering efforts to apply fracture mechanics to concrete [18], it was not until the mid-seventies that a number of researchers from the academic community focused their attention on various aspects of this application.

In applying fracture mechanics to concrete, much was borrowed from the wealth of information and research previously undertaken in conjunction with metals by metallurgists or mechanical engineers. However it quickly became evident that by its very heterogeneous nature concrete has some unique fracture characteristics which required the alteration of previous models.

By now fracture mechanics is universally acknowledged as the tool of analysis for investigation of concrete cracking. And after many years of development on numerous constitutive models for uniaxial, biaxial, and triaxial models of concrete under (primarily) compressive stresses, and their various refinements, researchers have finally turned their attention to the other side of the stress

[1] Assistant Professor, Department of Civil Engineering, University of Colorado, CO.
[2] Graduate Research Assistant, Department of Civil Engineering, University of Colorado, CO.
[3] Head, Analytical Design Section, Concrete Dams Branch, Bureau of Reclamation, Denver, CO.

strain curve and are giving more attention to the effect of cracking (micro and macro) on the nonlinear response of concrete structures.

## 1.1 Fracture of Concrete Dams

Dams, like all concrete structures, do crack. In many cases cracking is caused by non-structural causes such as shrinkage, and in those instances the safety of the dam is rarely jeopardized by cracking.

On the other hand there may be numerous structural causes for cracking, such as: a) Internal release of heat due to the hydration process, and resulting internal thermal gradient; b) Uplift water pressure along internal cracks; c) Upstream water pressure; d) Differential settlement.

As such it should not come as a surprise that numerous dams in the United States and in Europe have shown disturbing signs of extensive cracking. Because of the human and economical consequences of dam failure, dams with signs of cracks are thoroughly investigated and rarely publicized.

It is interesting to note that the first reported Civil Engineering application of finite element was made by Clough [32] in his analysis of the Norfork dam for the US Corps of Engineers, Little Rock (Arkansas) district. Again, this analysis was undertaken after major (vertical) crack was found inside the dam, and its safety seemed jeopardized. It was subsequently found that cracking was due to thermally induced stresses, that the crack did not affect the dam safety. Another "early" finite element analysis of dams in which cracking of the foundations was accounted for was published by Zienkiewicz the same year as the previous paper [39]. Following this pioneering study of the early sixties, there has been numerous subsequent finite element analyses of concrete dams. However in most of them, the primary focus was placed on either their dynamic or non-linear.

Despite the numerous research on the fracture mechanics of concrete over the past ten years, and despite the continuous exemplification of dams as a primary type of structure for which fracture mechanics should be applied, there has been surprisingly very little effort to apply this discipline to thoroughly analyze concrete dams.

To the best of the authors knowledge, the first fracture mechanics based analysis of a dam, and the only one in the United States, was performed by Chappell and Ingraffea on the Fontana dam [16]. Following the appearance of cracks on the dam faces, a three dimensional study indicated that the crack mechanism was due to thermal expansion. This analysis generated displacement which were subsequently used as boundary conditions for a two dimensional finite element analysis. This analysis used an early version of the first author's special code to simulate the discrete crack growth inside the dam using a linear elastic fracture mechanics approach. A lower bound prediction for the front location, subsequently proved satisfactory, was made.

Not surprisingly the other reported investigations on the fracture of concrete dams come from two European alpine countries, Switzerland and Austria.

Two research groups from Switzerland leading research institution, Federal Polytechnic of Zurich and Lausanne, independently conducted research on the dynamic behavior of cracked dams. In Zurich, Bachman and Skrikerud [33] [34], used a discrete crack approach with simple remeshing scheme to analyze a dam under dynamic excitation using an explicit approach. However, and surprisingly, despite their adoption of a discrete crack approach, no fracture mechanics based criteria is used for the crack extension, and tensile stresses at the crack tip are simply compared with tensile strength.

In Lausanne, Zimmerman and Chapuis [9] adopted a different approach for this same problem. Instead of a discrete crack, a smeared crack approach was used. While the smeared crack model facilitated the adoption of an existing finite element code, an innovative technique for the extraction of the stress intensity factors from such a model and the subsequent determination of the angle of crack propagation was developed [40], [27]. Initial results were found to be highly encouraging, and currently Mr. Droz is completing his doctoral thesis on this approach [11].

In Austria [21], Linsbauer used a simplified approximation to the dam geometry in order to obtain stress intensity factors (restricted to pure mode I) as a function of water pressure. This approach was subsequently adopted as a methodology to determine critical crack lengths in dams.

Finally, and to the best of the authors knowledge, the most comprehensive reported fracture mechanics based study of concrete dams was undertaken by Yaozhong in the People's Republic of China [38]. In this study the following tasks were undertaken:

1. Fracture toughness testing of concrete using 10x10x50 cm plate specimens.
2. Experimental development of a mixed mode crack growth propagation model for concrete (which predicts the angle of crack growth in term of $\frac{K_{II}}{K_I}$).
3. Finite element analysis of a dam subjected to:
   (a) Thermal gradient.
   (b) Upstream water pressure.
   (c) Uplift pressure along the crack (using both constant and triangular pressure distribution)

and thus constitutes an integrated experimental/numerical investigation.

## 1.2 Material Fracture Characterization

A unique fracture characteristic of concrete, caused primarily by its heterogeneity, is the presence of a significant process zone at the tip of a crack in which numerous micro-cracks form ahead of it [17]. This zone has often been modeled by a cohesive zone in which stresses are function of "fictitious" crack opening displacement as characterized by strain softening model (which uniqueness has been questioned by some leading researchers [23]). Clearly this additional "complication" made analyses more difficult and has somehow precluded the rapid application of fracture mechanics to engineering problems. On the other hand it is widely acknowledged that for "long" cracks (such as those found in dams) the effects of the process zone are negligible and Linear Elastic Fracture Mechanics is applicable. Thus a unique fracture property will characterize the resistance of concrete to fracture, namely its fracture toughness $K_{Ic}$ or equivalently its critical energy release rate $G_C$.

A survey of the literature on fracture toughness (or critical energy) of concrete reveals that all tests have been performed on "small" specimen sizes (that is dimensions were not large enough to ensure the determination of plane strain fracture toughness [8]) and great care should thus be exercised in "extrapolating" those quantities to the analysis of dams.

While standards exist for the fracture toughness determination of metals [2], there is not yet an equivalent one for concrete. Hillerborg has recently submitted to RILEM a proposed standard for fracture testing of concrete [14]. However one concern expressed about this standards is that it does not differentiate between notches and cracks, while the extensive testing undertaken at Kansas state by Swartz clearly indicate a difference among the two [35].

## 1.3 Finite Element Fracture Analyses

The low tensile strength of concrete and the resulting cracking, is a major source of the structural response non-linearity under both tensile (macro) and compressive (micro) loading [23]. As such for a finite element analysis to properly model concrete tensile macro-cracking, attention should be given to:

- Crack discretization.
- Crack propagation criteria.

There are two approaches or "schools of thought" to model a crack in a finite element mesh:

- The Smeared crack model.
- The Discrete crack model.

Since much has been written about the phenomenological, and computational advantages of each model [36], [37], [7], [19], the authors would simply add that it is their conviction that there is a need for both models. Smeared cracks are to be used when numerous cracks occur, and the overall non-linear response is not too dependent upon crack profiles and lengths. Discrete crack models should be used when:

- Precise crack profile is sought.
- Crack opening/sliding displacement are used in the analysis (uplift pressure and shear stiffnesses are functions of the opening and sliding displacement respectively).
- Few cracks occur.
- Mixed mode or curvilinear crack propagation is likely to occur (as the smeared crack fails to properly and efficiently model the mixed mode crack cases [24], [22]).

Among the numerous techniques to model the stress singularity and to extract the stress intensity factors the quarter point isoparametric element of Barsoum [4] and the the displacement correlation method of Shih and de Lorenzi [30] are respectively the easiest to implement and probably the most popular. The first author has recently generalized the displacement correlation technique to homogenous anisotropic solids [26].

Cracking in dam is a perfect example where discrete cracks should be used. Smeared crack models should be adopted for all other cases.

Having a crack appropriately modelled in a finite element analysis, a criteria for crack propagation is to be used. Following Bazant and Cedolin early work [6], it is by now universally accepted that this criteria must be fracture mechanics based and not strength based.

Linear elastic fracture mechanics based model for crack propagation is by now well established for isotropic materials, [31], [15], [12]. A recent extension of the maximum circumferential tensile stress model to anisotropic solids was recently proposed by the first two authors and was used in the analysis [28].

# 2 ANALYSIS

## 2.1 Objectives

This preliminary study on the fracture mechanics of concrete dams attempted to address the following basic questions:

1. What is the crack profile?
2. What are the crack opening displacement?
3. Do we have numerous short cracks, or fewer longer ones?
4. What is the effect of initial cracking on the stress redistribution inside the dam?
5. What is the effect of initial thermally induced cracks on the subsequent dam stability, specially when uplift through the crack is present?

Figure 1: Upper Stillwater Dam

6. What are the separate effects of thermal loading, uplift pressure, upstream pressure, and gravity on the crack length?
7. What is the effect of the RCC's anisotropy on the crack profile and length inside the dam?
8. How does cracking evolve with time?
9. How does a fracture mechanics based approach differ from current methodology?

As a vehicle to conduct this study, the Bureau of Reclamation's Upper Stillwater dam, under current construction in the Uinta Mountains in Utah has been selected, Fig. 1. As this dam has been the object of numerous studies, many of its characteristics were well established.

## 2.2 Roller Compacted Concrete Dams

Historically USBR concrete gravity dams have been built in 5-10 foot lifts in vertical blocks and artificially cooled by circulating cold water through coils placed atop each lift, thus the potential for thermally induced cracking was minimized. However, a recent technology uses roller compacted concrete (RCC) for dam construction. While there are numerous advantages to use this technique, primarily economical, its major disadvantage is that it does not lend itself to proper

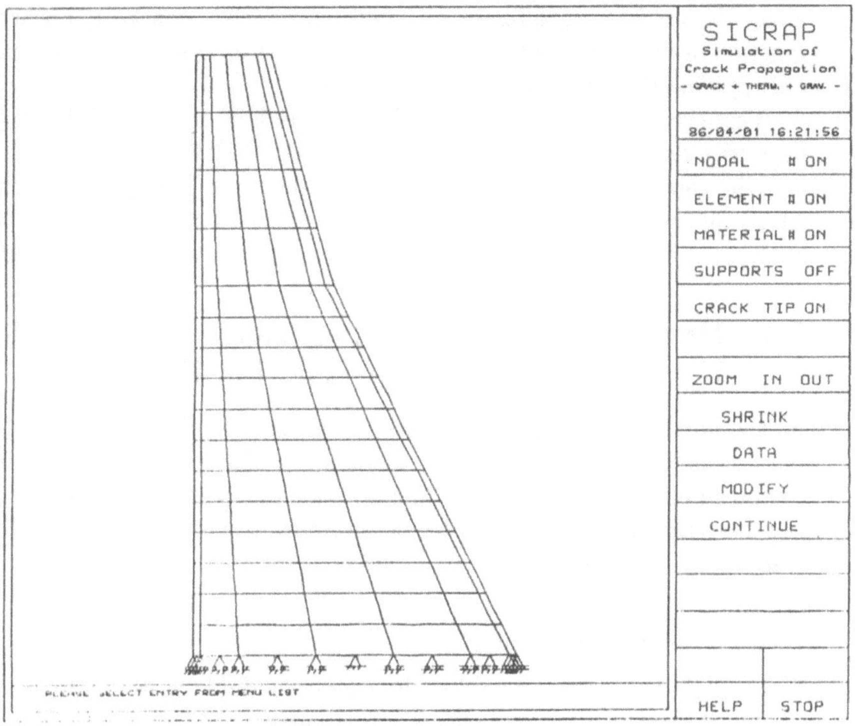


Figure 2: Uncracked Finite Element Discretization of the Dam

"cooling". As such, the major problem associated with RCC is the large residual thermal gradient left inside the dam. A linear elastic finite element analysis of such a structure subjected to gravity and thermal loading shows that most of the dam faces exhibit tensile stresses well above their tensile strength.

Another characteristic of roller compacted concrete is the mechanically induced anisotropy of its elastic properties. Stress-Strain tests conducted on horizontally and vertically cored specimens have shown that $E_h$ is about 3300 ksi while $E_v$ is 1500 ksi, where $E_h$ and $E_v$ are the horizontal and vertical Young's modulii respectively,[10].

## 2.3 Discretization and Initial Analysis

The dam was discretized into 112 quadratic isoparametric elements, Fig. 2, and was loaded in plane strain by its own weight and a thermal load. The thermal load corresponds to an internal temperature distribution computed at an age of 180 days at the the Bureau of Reclamation. The stress free reference temperature was taken as $50°F$.

LEFM analysis was performed by a special purpose finite element code SICRAP [25](Simulation of Crack Propagation) which operates in an interactive graphics environment (using DI-3000) on an Apollo work station, and has the capability to perform automatic mesh modification after each discrete crack extension/nucleation. This program was used to : a) Accommodate fluid pressure along the crack with different distribution models ( constant, linear, partial, function of crack opening displacement ); and b) Interpolate nodal temperature during remeshing for new or displaced nodes.

Three sets of analyses were conducted, the first one was of the dam at 180 days with downstream cracking, the second at 540 days and upstream cracking, and the third one was a parametric investigation on the effect of various load effects on two separate fictitious upstream horizontal cracks. The first and third analyses will be reported here, with further details can be found in [29].

## 2.4 Downstream Crack Propagation Analysis at 180 days Under Gravity and Thermal Loading

As the initial analysis of the uncracked dam revealed higher tensile stresses at the downstream face of the dam, preliminary analysis at 180 days under the combined effects of gravity, and thermal loading assumed cracking to occur on the downstream face only. Isotropic and anisotropic analysis of one and two cracks were performed.

### 2.4.1 Analysis Results

**Isotropic** In this first analysis the elastic modulus,$E$, is 1500 ksi, and only one crack was allowed to nucleate and propagate. The mode I and II stress intensity factors varied from 6.87 and -1.09 to 0.12 and -0.183 (ksi$\sqrt{in}$) as the crack grew. The final crack mouth opening displacement at the dam was 0.335 in.

In the second analysis two cracks were allowed to nucleate and propagate simultaneously. Fig. 3 and Fig. 4 show the upstream and downstream stress distribution with respect to the tensile strength. Again the mode I and II stress intensity factor ranged from 6.44 and -1.17 to 0.50 and -0.03 for the first crack and from 4.61 and -0.19 to 0.69 and -0.02 for the second crack. Final crack mouth opening displacement were found to be 0.14 in and 0.25 in.

**Anisotropic** As indicated earlier, horizontal and vertical coring pointed to different elastic moduli [10]. As this precluded the adoption of existing mixed mode crack propagation criteria, a new one was developed for this project [28].

In this analysis values of 3300 ksi and 1500 ksi were used for $E_h$ and $E_v$ respectively. Poisson's ratio $\nu_{12}$ was assumed to be equal to 0.20 and shear modulus $G_{12}$ was approximated by [5]:

$$G_{12} = \frac{E_h E_v}{E_h + E_v + 2\nu_{12} E_v} \tag{1}$$

yielding a value of 916.7 ksi. Fracture toughness ${K_{Ic}}^h$ in the horizontal direction was *assumed* to be 0.5 ksi $\sqrt{in}$, and the toughness in the vertical direction was

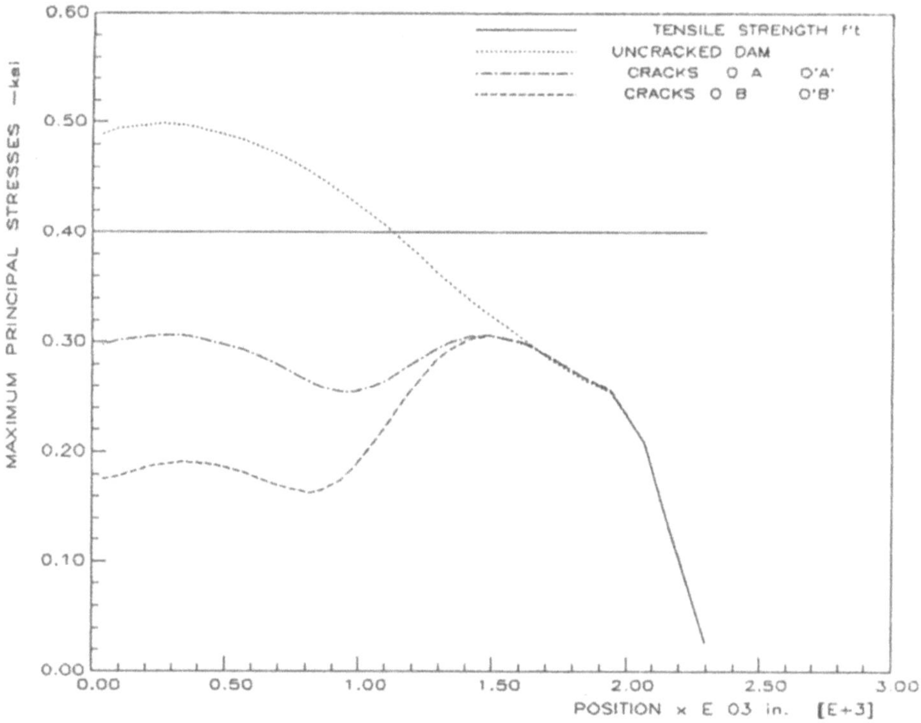


Figure 3: Upstream Stress Distribution in Isotropic Model at 180 Days

again *assumed* to be :

$$K_{\mathrm{Ic}}{}^{v} = K_{\mathrm{Ic}}{}^{h}\frac{E_v}{E_h} \tag{2}$$

which gives a value of 1.1 ksi $\sqrt{in}$.

Two analyses were performed, with one and two simultaneously propagating crack. For the second one the mode I and II stress intensity factor ranged from 10.26 and -2.68 to 0.91 and -2.18 for the first crack and from 7.31 and -2.05 to 1.59 and -2.75 for the second crack. The crack profiles are shown in Fig. 5 and contrasted with those obtained in the isotropic model.

### 2.4.2 Discussion of Results

**Isotropic** From this analysis on the isotropic model with cracking occurring on the downstream face after 180 days it can be concluded that:

1. The cracks tend to align themselves with major axis of compressive stresses.
2. The cracks tend also to propagate in a direction normal to the isothermals lines propagating from lower to higher temperature zones.
3. Substantial stress redistribution takes place after the nucleation and extension of just two cracks. Similar conclusion was found by [3].

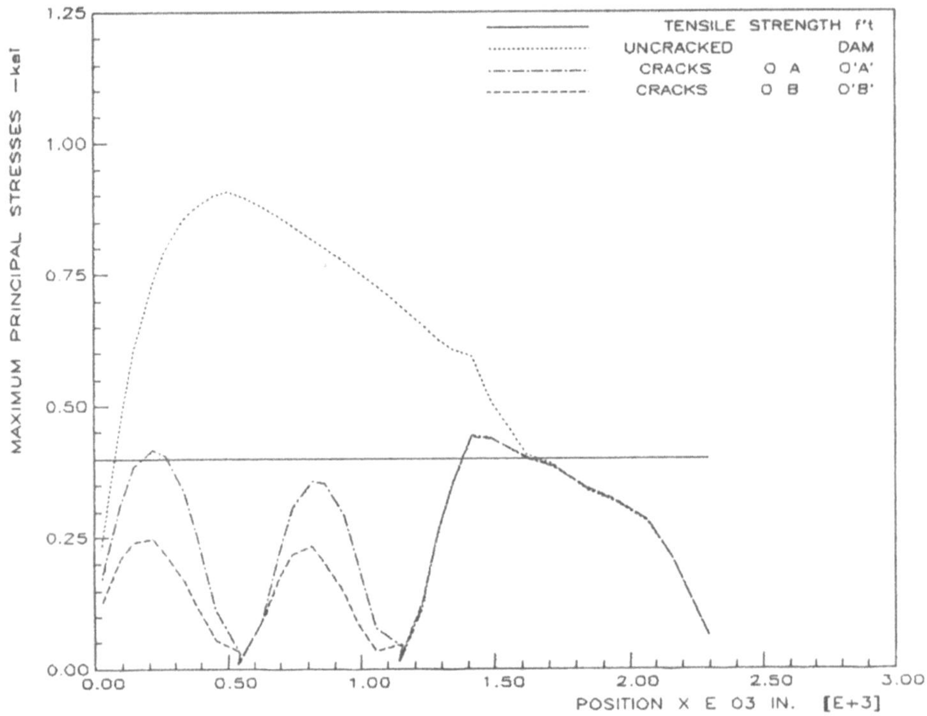


Figure 4: Downstream Stress Distribution in Isotropic Model at 180 Days

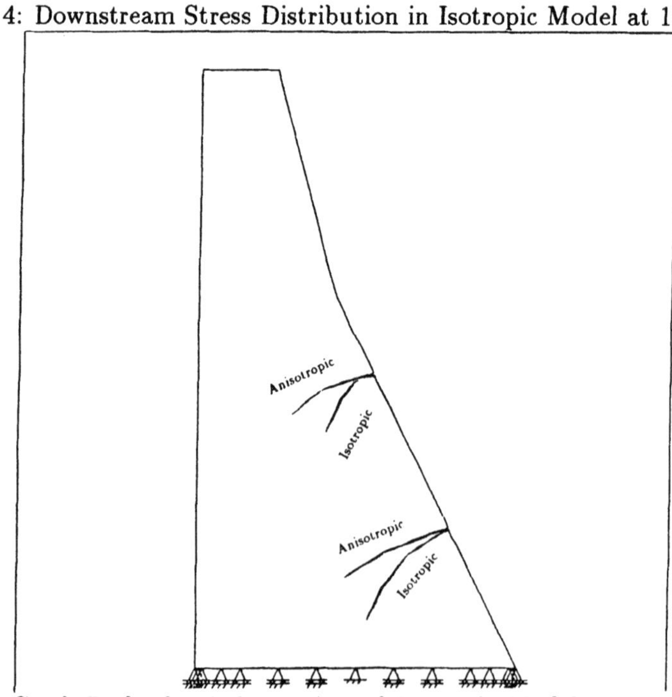


Figure 5: Crack Paths for Anisotropic and Isotropic Model at 180 Days (two cracks)

**Anisotropic** This anisotropic analysis yielded the following conclusions:

1. The stress intensity factors are larger in the anisotropic models than in the isotropic ones. Thus longer cracks are found in this case.
2. The crack profile has been affected by the anisotropy. The crack tends to propagate along a "flatter" angle than in the isotropic case.
3. While the crack tend to align themselves with the large principal compressive stresses caused by gravity loading, this effect is counteracted by the anisotropy which attempts to cause the crack to propagate along the horizontal or stiffer direction.
4. Mode II (sliding) displacement are larger in this analysis than in the isotropic case.
5. Despite the extension of two cracks, and contrarily to the isotropic case, tensile stresses on the on the downstream face are not entirely below the concrete tensile strength $f_t'$.

## 2.5 Effects of Various Loading Components on Dam Cracking at 580 days

The cracked roller compacted cracked concrete dam is simultaneously subjected to various loading components: a) Gravity; b) Reservoir pressure; c) Thermal; d) Hydrostatic uplift within the crack. In order to properly understand the effect of each one of the above loading on the dam's response, and their relative importance with respect to the overall behavior, a parametric study is performed. This study will determine the: a) Crack stress intensity factors $K_{\rm I}$ and $K_{\rm II}$; b) Crack mouth opening displacement (CMOD); c) The horizontal and vertical displacement of the dam's crest (CDD).

The above information will be determined for two crack elevations and four different crack lengths for each crack elevation. Two sets of analysis were performed, the first one for crack at an elevation 35 and the second at elevation 105 from the base.

The hydrostatic uplift was not applied continuously to the entire crack surface, but rather only to a portion of it. This was done to avoid loading singular elements whose lengths are about $LC\%$ of the total crack length.

Finally a negative value of $K_{\rm I}$ means that no cracking occurs.

### 2.5.1 Analysis Results

The finite element discretization for elevation 35 crack is shown in fig. 6. Analysis results are summarized in Table 1 and 2, and a plot of $K_{\rm I}$ caused by each of the loadings and their sum is shown in figs. 7 and 8.

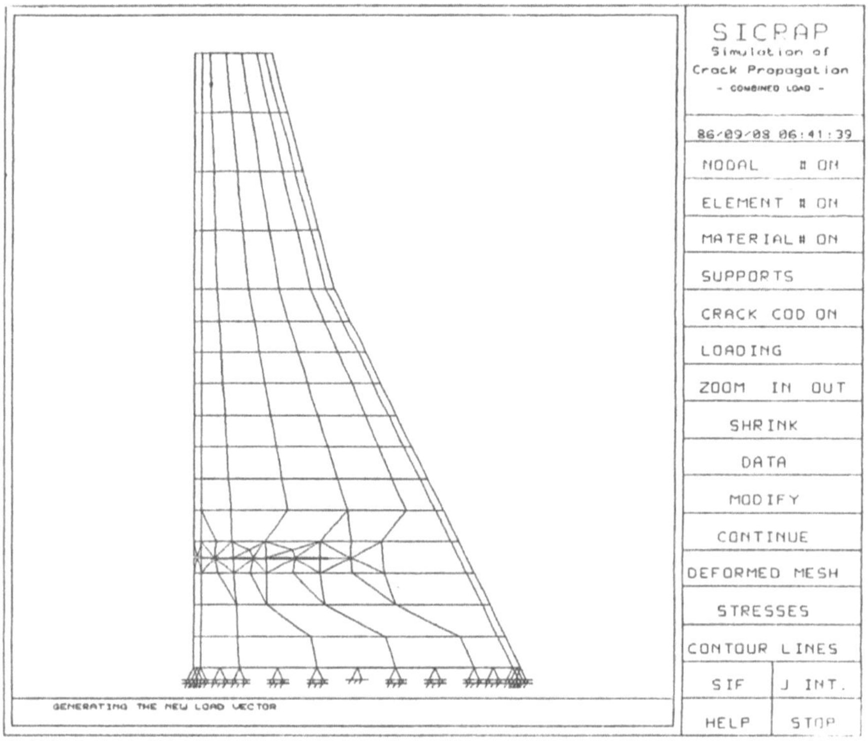


Figure 6: Finite Element discretization for the elevation 35 crack

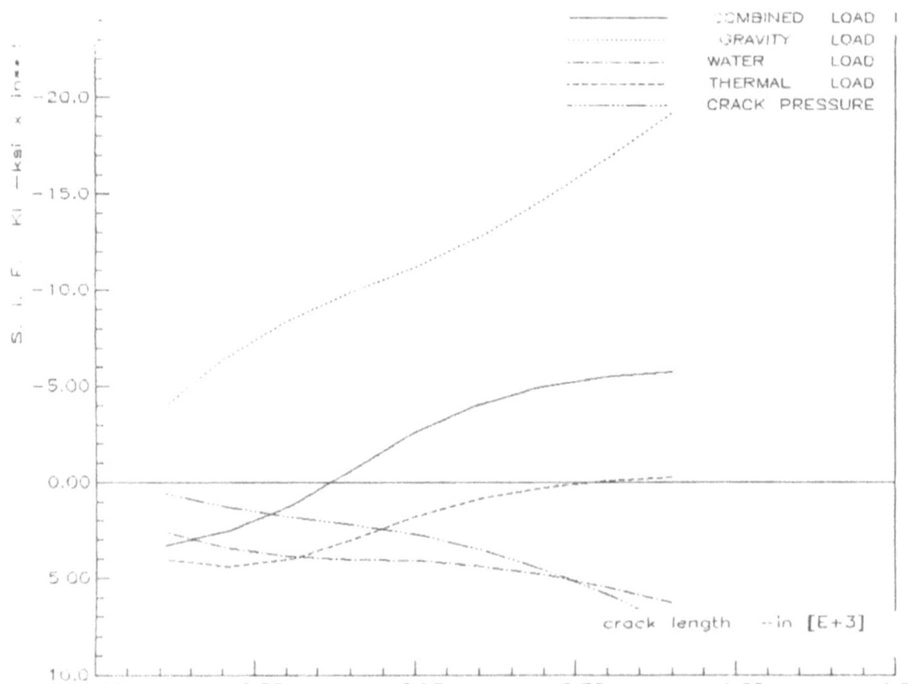


Figure 7: $K_{\mathrm{I}}$ at Elevation 35

Table 1: Results of Parametric Study for Crack Elevation 35

| CRACK LENGTH | GRAVITY LOAD (I) KI ... KII | GRAVITY LOAD (I) CMOD ... CDD (h, v) | WATER LOAD (II) KI ... KII | WATER LOAD (II) CMOD ... CDD (h, v) | THERMAL LOAD (III) KI ... KII | THERMAL LOAD (III) CMOD ... CDD (h, v) | CONST. CRACK PRESSURIZATION KI ... KII | CONST. CRACK PRESSURIZATION CMOD ... CDD (h, v) | TRIANGULAR CRACK PRESSURIZATION KI ... KII | TRIANGULAR CRACK PRESSURIZATION CMOD ... CDD (h, v) | LC % | COMBINED EFFECT I+II + III + CCP KI ... KII |
|---|---|---|---|---|---|---|---|---|---|---|---|---|
| 108.7 | -3.97 | -0.122 | 2.62 | 0.0995 | 4.07 | 0.023 | 0.584 | 0.0800 | 0.425 | 0.0800 | 40 | 3.304 |
| | 0.0295 | -0.254<br>-0.178 | 0.275 | 0.4236<br>0.1039 | 0.116 | 0.516<br>0.0259 | -0.0102 | -<br>- | -0.0078 | 0.0088<br>0.0035 | | 0.4103 |
| 283.2 | -8.07 | -0.314 | 3.83 | 0.200 | 4.18 | 0.214 | 1.68 | 0.0547 | 0.362 | 0.0129 | 25 | 1.62 |
| | 1.46 | -0.686<br>-0.370 | 0.164 | 0.663<br>0.209 | -0.451 | 0.817<br>0.160 | -0.325 | -<br>- | -0.0744 | 0.0299<br>0.0131 | | 0.848 |
| 516.0 | -11.39 | -1.192 | 4.12 | 0.335 | 1.56 | 0.300 | 2.79 | 0.184 | 1.714 | - | 20 | -2.92 |
| | 3.40 | -2.42<br>-1.25 | 0.165 | 1.38<br>0.570 | -0.348 | 1.00<br>0.248 | -0.539 | -<br>- | 0.508 | 1.265<br>0.370 | | 2.68 |
| 900.0 | -19.17 | -5.11 | 6.27 | 1.86 | -0.273 | 0.596 | 7.41 | 1.97 | 4.974 | 1.369 | 10 | -5.76 |
| | 4.87 | -9.30<br>-5.17 | 0.473 | 3.69<br>1.88 | 0.506 | 1.54<br>0.545 | -1.89 | -<br>- | -1.35 | 2.493<br>1.37 | | 3.96 |

Table 2: Results of Parametric Study for Crack Elevation 105

| CRACK LENGTH | GRVITY LOAD (I) KI ... KII | GRVITY LOAD (I) CMOD ... CDD (h, v) | WATER LOAD (II) KI ... KII | WATER LOAD (II) CMOD ... CDD (h, v) | THERMAL LOAD (III) KI ... KII | THERMAL LOAD (III) CMOD ... CDD (h, v) | CONST. CRACK PRESSURIZATION KI ... KII | CONST. CRACK PRESSURIZATION CMOD ... CDD (h, v) | TRIANGL. CRACK PRESSURIZATION KI ... KII | TRIANGL. CRACK PRESSURIZATION CMOD ... CDD (h, v) | COMBINED EFFECT I+ II + III + CCP KI ... KII |
|---|---|---|---|---|---|---|---|---|---|---|---|
| 77.7 | -2.212 | - | 0.657 | 0.164 | 2.22 | 0.104 | 0.287 | 0.00211 | 0.1975 | 0.00156 | 0.954 |
| | -0.054 | -0.221<br>-0.167 | 0.096 | 0.390<br>0.0909 | 0.125 | 0.476<br>0.0121 | -0.00058 | -<br>- | -0.00037 | 0.00300<br>0.00029 | 0.1664 |
| 221.2 | -3.785 | - | 0.886 | 0.182 | 1.46 | 0.156 | 0.714 | 0.0209 | 0.469 | 0.0149 | -0.256 |
| | -0.146 | -0.385<br>-0.252 | 0.317 | 0.4307<br>0.1116 | 0.483 | 0.587<br>0.0691 | -0.00468 | -<br>- | -0.00414 | 0.0310<br>0.0157 | 0.649 |
| 399.5 | -8.71 | - | 1.756 | 0.272 | 0.573 | 0.249 | 2.424 | 0.155 | 1.53 | 0.103 | -3.96 |
| | 0.318 | -1.17<br>-0.695 | 0.506 | 0.599<br>0.206 | 0.660 | 0.766<br>0.168 | -0.139 | -<br>- | -0.0936 | 0.482<br>0.239 | 1.35 |
| 570.2 | -25.10 | - | 4.407 | 0.810 | -0.149 | 0.347 | 8.437 | 1.48 | 5.43 | 1.01 | -12.41 |
| | 3.80 | -5.96<br>-3.701 | 0.282 | 1.473<br>0.750 | 0.530 | 0.933<br>0.268 | -1.292 | -<br>- | -0.889 | 2.01<br>0.79 | 3.32 |

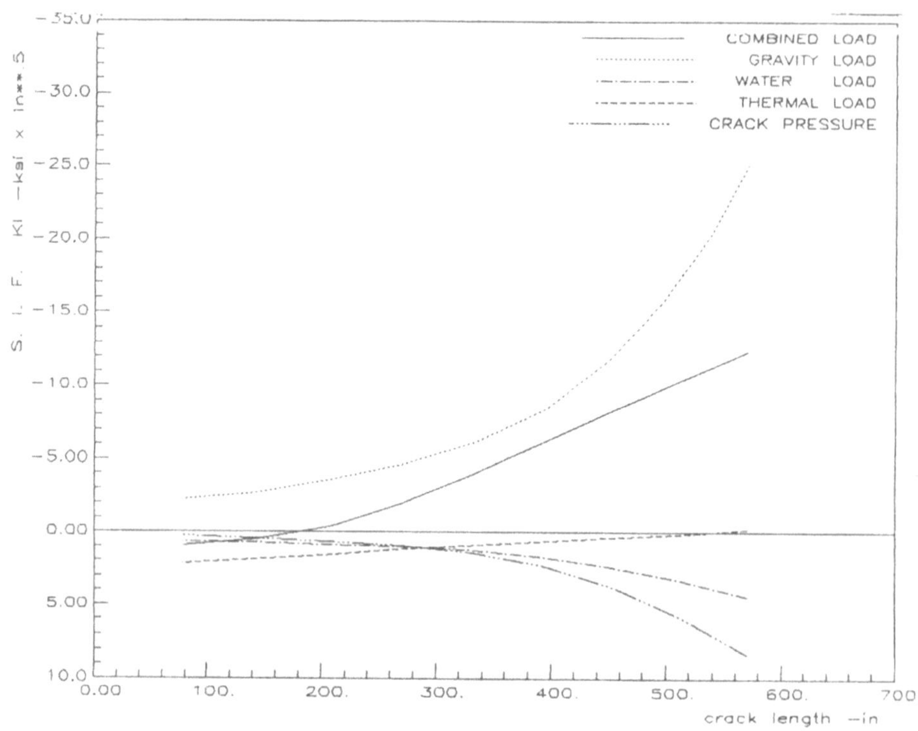


Figure 8: $K_{\rm I}$ at Elevation 105

### 2.5.2 Discussion of Results

On the basis of the preceding analyses, it can be concluded that:

1. $K_{\rm I}^G$ for gravity loading increases much faster than $K_{\rm I}$ caused by the other loading.
2. $K_{\rm I}^T$ decreases rapidly as the crack extends "inside" the dam.
3. $K_{\rm I}^U$ caused by the triangular pressurization in the crack is much smaller than the one caused by a constant one.
4. The numerical value of $K_{\rm I}$ is very sensitive to the presence of a pressure at the very crack tip (i.e. acting over singular elements).
5. When various loading are added together, in different combinations, and at both elevation, the following is found:
   (a) Gravity, reservoir pressure, and uplift pressure in the crack ( $K_{\rm I}^G + K_{\rm I}^R + K_{\rm I}^U$), yield a net negative $K_{\rm I}$ for all crack lengths and at both crack elevations.
   (b) Gravity, thermal, and reservoir pressure, ( $K_{\rm I}^G + K_{\rm I}^R + K_{\rm I}^T$), yield a net value of:
      i. 2.7 for $K_{\rm I}$ at elevation 35 and a crack length of 9 ft.

ii. 0.37 for $K_{\rm I}$ at elevation 105 and a crack length of 6.5 ft.

iii. negative values of $K_{\rm I}$ for subsequent increment of crack lengths for both cases.

(c) Gravity, thermal, and uplift pressure in the crack ( $K_{\rm I}^G + K_{\rm I}^T + K_{\rm I}^U$), yield a net value of:,

i. 0.68 for $K_{\rm I}$ at elevation 35 and a crack length of 9 ft.

ii. 0.30 for $K_{\rm I}$ at elevation 105 and a crack length of 6.5 ft.

iii. negative values of $K_{\rm I}$ for subsequent increment of crack lengths for both cases.

(d) When all loads are considered concurrently, they yield a net positive value of $K_{\rm I}$ through a portion of the dam.

6. When all effects are added together, a critical crack length of 320 in (26.6 ft.) and 180 in. (15 ft) is found for the cracks at elevation 35 and 105 respectively assuming $K_{\rm Ic} = 0.$ .

7. As most of the non-gravity loading is "consumed" in opening the crack, the critical crack length is relatively insensitive to the fracture toughness $K_{\rm Ic}$. This situation is analogous to what is often found in Hydrofracturing where the in-situ stresses play an analogous role to the gravity loading in this analysis.

8. If reliable estimates of the fracture toughness can be obtained a shorter critical crack length would be obtained.

# 3 COMPARISON WITH ALTERNATIVE APPROACHES

In light of the innovative nature of the approach followed in this investigation, it is essential that it be compared with the following alternative ones:

1. The crack profile as determined by the Levy's method discussed by Linsbauer [20].

2. A simplified "strength of material" approach [1].

## 3.1 Linsbauer

Linsbauer [20] has presented:

1. A simplified approach to determine the stress intensity factor for pressurized crack in a dam (for both constant and triangular pressure distribution).

| CRACK LENGTH ft. | $\frac{a}{w}$ | $K_I$ Linsbauer | $K_I$ analysis |
|---|---|---|---|
| 9.1 | 0.082 | 1.3 | 0.58 |
| 23.6 | 0.213 | 2.6 | 1.68 |
| 43.0 | 0.387 | 5.4 | 2.80 |

Table 3: Stress Intensity Variation with crack length at elevation 35.

| CRACK LENGTH ft. | $\frac{a}{w}$ | $K_I$ Linsbauer | $K_I$ analysis |
|---|---|---|---|
| 6.5 | 0.10 | 0.23 | 0.198 |
| 18.4 | 0.29 | 0.65 | 0.469 |
| 33.3 | 0.528 | 2.05 | 1.53 |

Table 4: Stress Intensity Variation with crack length at elevation 105.

2. An approach to determine the critical crack length in a dam when subjected to gravity, upstream water pressure, and internal crack pressurization.

While the dam geometry considered is not identical to the one of our numerical analysis, and this analysis did not extend the pressure all the way to the crack tip (to avoid loading singular elements) a comparison between the two approaches is undertaken.

Using the constant pressure distribution along the crack at elevation 35, the values of the stress intensity factors are found for various ratios of crack lengths to cross section depths " a/w " and shown in table 3.

For the crack at elevation 105, a triangular pressure distribution was used, and the results are summarized in table 4.

For both cracks (at elevation 35 and 105), if the stress intensity factors caused by gravity, water pressure on upstream face, and constant crack pressurization, are added, a net negative value is found for all crack lengths. Also, when Linsbauer's approach to determine the critical crack length in a dam (accounting for gravity, upstream pressure, and constant crack pressurization) is used it is found that no cracking can take place. Thus both approaches yield similar conclusions.

## 3.2 Classical Approach

Using a "classical approach" (based on Strength of Material) as described in [1] a crack of length $a$ at a distance $h$ from the crest is investigated. In this analysis only upstream water pressure, uplift pressure along the crack, and gravity are accounted for (thus neglecting thermal effect). By summation of axial stresses from vertical forces and overturning moments, the stress distribution

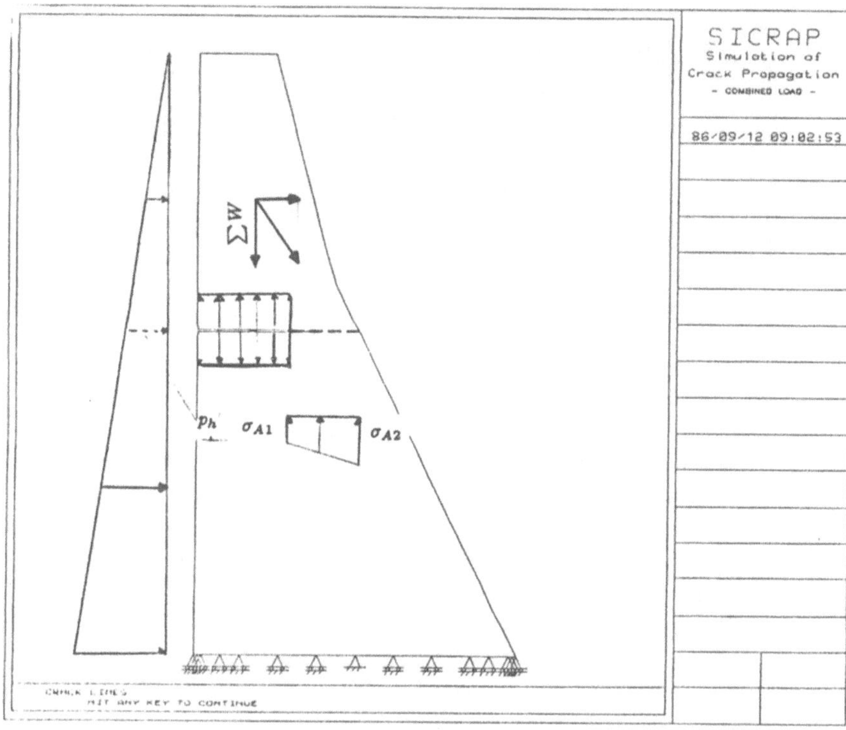


Figure 9: Base Pressure Diagram of Cracked Dam

over the uncracked portion of the dam, at elevation h is determined. This stress distribution varies linearly from $\sigma_{A1}$ to $\sigma_{A2}$ as shown in fig. 9:

$$\sigma_{A1,A2} = \frac{\sum W}{A} \pm \frac{\sum My}{I} \tag{3}$$

From this figure, it is clear that the crack would reach its final equilibrium length when $\sigma_{A1} = p_h - f_t'$ where $p_h$ and $f_t'$ are the hydrostatic pressure and the concrete tensile strength respectively. Assuming $f_t'$ to be zero, the gravity stresses are found to be high enough not to allow any cracking at elevations 35 and 105. The fracture mechanics approach, as summarized in tables 1 and 2, yielded similar conclusion.

In order to qualitatively compare both methods, the following load combination was analyzed:

$$P_{Total} = P_G + \lambda(P_{RP} + P_{UP}) \tag{4}$$

Where: $P_{Total}$, $P_G$, $P_{RP}$, $P_{UP}$ are the total, gravity, reservoir pressure and uplift loads respectively. $\lambda$ is a non-dimensional factor, for which $K_I = K_{Ic}$ in the fracture mechanics approach and $\sigma_{A1} = f_t'$ in the strength approach. The results summarized in table 5 show that the classical approach is much more conservative than the fracture mechanics one, and that results seem to be more sensitive to the material parameters in the strength approach.

| Crack | Classical Approach | | Fracture Mechanics Approach | |
|---|---|---|---|---|
| | $f_t'$=0 psi | $f_t'$=150 psi | $K_{Ic}$=0 psi$\sqrt{in}$ | $K_{Ic}$=500. psi$\sqrt{in}$ |
| 6.5 | 1.35 | 2.61 | 2.34 | 2.87 |
| 17.7 | 1.26 | 1.87 | 2.36 | 2.68 |
| 33.3 | 1.21 | 1.43 | 2.08 | 2.20 |
| 47.5 | 1.21 | 1.26 | 1.95 | 1.99 |

Table 5: Safety Factor $\lambda$ from water pressures and gravity at elevation 35.

# 4 CONCLUSIONS

An initial investigation on the applicability of fracture mechanics to the analysis of cracked concrete gravity dam has been presented. This preliminary study clearly shows that in investigating cracked dam, alternatives exist to the currently "strength of material" approach in which the stress singularity is not accounted for.

It is thus concluded that through our improved understanding of the fracture mechanics of concrete, coupled with a new "generation" of powerful special purpose finite element codes, fracture mechanics should be used to assess the safety of cracked concrete dams.

This study was the first attempt to comprehensively investigate the behavior of cracked concrete dams using fracture mechanics concepts implemented into a special purpose interactive graphics finite element program in which discrete crack propagation is simulated.

As a vehicle for our investigation, a roller compacted concrete (RCC) dam was investigated.

The following loads and material variations were considered:

- Thermal loading at two different ages,
- Uplift pressure (caused by water flowing into the crack),
- Reservoir pressure,
- Gravity loading,
- Concrete isotropy and anisotropy,

In addition the stress intensity factors, direction of crack growth, length of the cracks and stress redistribution were assessed leading the following conclusions:

1. The nucleation and extension of few discrete cracks on the faces of the dam would result in a large decrease of the tensile stresses along the faces.
2. Most of the cracking is caused by large thermal gradient near the faces of the dam, for this particular problem.

3. Crack length is relatively insensitive to fracture toughness.
4. If anisotropy is accounted for, the crack path tends to parallel the stiffer direction and is longer than for the isotropic case.
5. Mode II loading is more prevalent for anisotropic models than for isotropic ones.
6. For the investigated cases, downstream cracking is more critical than upstream in the initial stage.
7. $K_I^G$ for gravity loading increases much faster than $K_I$ caused by the other loading.
8. $K_I^T$ decreases rapidly as the crack extends "inside" the dam.
9. $K_I^U$ caused by the triangular uplift pressurization of the crack is much smaller than the one caused by a constant uplift pressure.
10. The value of $K_I$ is very sensitive to the presence of a pressure at the very crack tip, this is a numerical problem which is likely to be a physical one also.
11. When the effects of gravity, upstream pressure, and uplift are added ( $K_I^G+K_I^R+K_I^U$), a net negative $K_I$ is obtained. However when the thermal effect is accounted for, a net positive $K_I$ is obtained. Thus Thermal loading is the main driving force for the crack.
12. As most of the non-gravity loading is "consumed" in opening the crack, the critical crack length is relatively insensitive to the fracture toughness $K_{Ic}$. This situation is analogous to what is often found in Hydrofracturing where the in-situ stresses play an analogous role to the gravity loading in this analysis.
13. When results of this analysis were compared to two alternative ones, results of same order of magnitude were found.
14. Current analytical methods are shown to be more conservative than the one based on fracture mechanics.

# 5 FUTURE RESEARCH

The results reported here are part of a comprehensive research effort undertaken by the University of Colorado on the Fracture Mechanics of concrete dams. Additional topics under current investigation include:

1. Crack growth along an interface of two lift zones.

2. Extension of the code capabilities to include transient dynamic nonlinear analysis.
3. Experimental evaluation of the fracture toughness of roller compacted concrete using large specimens on a million pound MTS machine.
4. Experimental validation of the mixed mode crack growth model for homogeneous anisotropic solid used in this analysis.

# 6 ACKNOWLEDGMENTS

The financial support of the Bureau of Reclamation, Denver for this research is gratefully acknowledged. Computational study was performed on the Apollo Computer network of the Civil Engineering Computer Graphics Laboratory, University of Colorado, Boulder, acquired through partial support from the National Science Foundation.

# References

[1] *Anonym.*, "Design of Small Dams", U.S.Department of the Interior, Bureau of Reclamation, "Design of Small Dams", 1973.

[2] *Anonym.*, "Standard Test Method for Plane Strain Fracture Toughness of Metallic Materials", ASTM Designation E399-74, part 10, ASTM Annual Standards.

[3] Bedard, C., Kotsovos, M. D., " Fracture Processes of Concrete for NLFEA Methods",*Journal of Structural Engineering*, Vol. 112, NO. 3, March 1986.

[4] Barsoum, R. S., "A Degenerate Solid Element for Linear Fracture Analysis of Plate Bending and General Shells", *Int. J. num. Meth. Engng.*, 10, 551-564, 1976.

[5] Batugin, S. A., and Nirenburg, R. K., "Approximate Relationship Among the Elastic Constants of Anisotropic Rocks and Anisotropy Parameters", *Fiz-Tekhn. Probl. Razrabotki Polezn. Iskopaemykh*, Vol. 7, No. 1, pp. 7-11, 1972. (in Russian).

[6] Bazant, Z. P., and Cedolin, L., "Blunt Crack Band Propagation in Finite Element Analysis", *J. of the Engineering Mechanics Division*, ASCE, V. 105, No. EM2, April, 1978.

[7] Bazant, Z. P., "Mechanics of Distributed Cracking", *Appl. Mech. Rev.*, V. 39, No. 5, May 1986.

[8] Broek, D., *Elementary Engineering Fracture Mechanics*, 4th Edition, Martinus Nijhoff Publishers, 1986.

[9] Chapuis, J., Rebora, B. and Zimmermann, Th. "Numerical Approach of Crack Propagation Analysis in Gravity Dams During Earthquakes", *Quinzieme Congres des Grands Barrages*, Q., R. 26, Lausanne, 1985.

[10] Crow, R. D., *et al.*, "Mix Design Investigation - Roller Compacted Concrete Construction, Upper Stillwater Dam, Utah", *Report REC-ERC-84-15, Bureau of Reclamation*, Denver, Colorado.

[11] Droz, P., "Dynamic Fracture Analysis of Concrete Dams", *Doctoral Thesis*, Swiss Polytechnic Institute, Lausanne, to appear.

[12] Erdogan, F. and Sih, G. C., 1963. "On the Crack Extension in Plates Under Plane Loading and Transverse Shear", *Journal of Basic Engineering*, V. 85 pp. 519-527

[13] Griffith, A. A., "The Phenomena of Rupture and Flow in Solids", *Phil. Trans. Roy. Soc. London*, A221, pp. 163-197, 1921.

[14] Hillerborg, A., "The Theoretical Basis of a Method to Determine the Fracture energy $G_F$ of Concrete", *International Conference On Fracture Mechanics of Concrete*, Lausanne, 1-3 october, 1985.

[15] Hussain, M.A., Pu, S. L., and Underwood, J. H., "Strain Energy Release Rate for a Crack Under Combined Mode I and Mode II", *Fracture Analysis*, ASTM, STP 560, 2-28, 1974.

[16] Ingraffea, A. R., and Chappell J. F., " A Fracture Mechanics Investigation of the cracking of Fontana Dam ", *Report No. 81-7*, Cornell University, 1981.

[17] Jenq, Y. S., and Shah, S. P.,"Nonlinear Fracture Parameters for Cement Based Composites: Theory and Experiments", in "Application of Fracture Mechanics to Cementitious Composites", *Nato/ARW, S. P. Shah Editor*, pp. 171-209, 1984.

[18] Kaplan, M. F., "Crack Propagation and The Fracture of Concrete" , *Journal of the American Concrete Institute* , Proc., Vol. 58, No. 5, Mai 1961, pp. 591-610.

[19] Leibengood, L., Darwin, D., Dodds, R. H., "Parameters Affecting Finite Element Analysis of Concrete Structures", *Journal of Structural Engineering*, ASCE, V. 112, No. 2, Feb. 1986.

[20] Linsbauer, H. N., and Rossmanith, H. P., "Back Face Rotation Correction for Trapezoidal Specimens", *Engineering Fracture Mechanics*, V. 19, No.2, pp. 195-205, 1984.

[21] Linsbauer, H. N. "Fracture Mechanics Models for Characterizing crack Behavior in Concrete Gravity Dams", *Quinzienne Congress des grands barrages*, Lausanne, 1985, pp Q57, R16.

[22] Marchertas, A. H., Kulak, R. F., and Pan, Y. C., "Performance of the Blunt Crack Approach within a General Purpose Code", in *Nonlinear Numerical Analysis of reinforced concrete*, L. E., Schwer Ed., ASME Winter Annual Meeting, pp. 107-123., 1982.

[23] Nemat-Nasser, S., "Mechanics of Brittle Failure in Compression", *Computers and Structures*, V. 20, No. 1-3, pp. 235-237, 1985.

[24] Pfeiffer, P. A., Marchertas, A. A., and Bazant, Z. P., "Blunt Crack Band Propagation in Finite Element Analysis of concrete structures", *Proc., 7th Int. Conf. on Structural Mechanics in Reactor Technology, Paper No. H 5/2, 1983.*

[25] Saouma, V. E., "Automated Finite Element Analysis of Discrete Crack Propagatiomn, User's and Technical Manual", (Available from the author), 1986.

[26] Saouma, V. E., and Sikiotis, E. S., "Stress Intensity Factors in Anisotropic Bodies Using Singular Isoparametric Elements", *Engineering Fracture Mechanics*, V. 25, No. 1, pp. 115-121, 1986.

[27] Saouma, V. E., Droz, P., "Mixed Mode Blunt Crack Instability Using a Path Independent Contour Integral", Paper H3/3, *8th Structural Mechanics in Reactor Technology Conference*, Brussels, 1985.

[28] Saouma, V. E., Ayari, M. L., Leavell, D. A., "Mixed Mode Crack Propagation In Homogeneous Anisotropic Solids", Submitted for publication to *Engeneering Fracture Mechanics*, 1986.

[29] Saouma, V. E., Ayari, M. L., Boggs, H., "Fracture Mechanics of Concrete Gravity Dams", *Report under preparation*, 1987.

[30] Shih, C. F., deLorenzi, H. G., and German, M. D., "Crack Extension Modelling with Singular Quadratic Isoparametric Elements", *Int. J. of Fracture*, V. 12, pp. 647 - 651, 1976.

[31] Sih, G. C., "Strain Energy Factors Applied to Mixed Mode Crack Problems", *Int. J. Fracture*, V. 10, p. 305, 1974.

[32] Sims, F. W., Rhodes, J. A., and Clough, R. A., "Cracking in Norfork Dam", *J. of the American Concrete Institute*, March 1964.

[33] Skrikerud, P., "Modelle und Berechnungsverfahren fur das Rissverhalten von Unarmierten Betonbauten unter Erbebenbeanspruchung", (Fracture Model for Concrete Dam under Seismic Loading), *Institute for Static & Construction, Swiss Polytechnic Institute*, Zurich, June 1983.

[34] Skrikerud, P., "Discrete Crack Modelling for Dynamically Loaded, Unreinforced Concrete Structures", *Earthquake Engineering and Structural Dynamics*, V. 14, pp. 297-315, 1986.

[35] Swartz, S. E., "Fracture Toughness Testing of Concrete at Kansas State: is LEFM Acceptable?", in the *Proceedings of the International Conference on Fracture Mechanics of Concrete*, Whittmann, F.H., Editor, Lausanne, 1985.

[36] Task Committee on Finite Element Analysis of Reinforced Concrete Structures, "Finite Element Analysis Of Reinforced Concrete", *Structural Division Committee on Concrete and Masonry Structures*, 1982.

[37] Van Mier, J. G. M., "Multiaxial Strain Softening of Concrete, Part I: Fracture", *Materiaux et Constructions*, V. 19, N0. 111, 1986.

[38] Yaozhong, Y., and Yanqiu, Z., "Stability Analysis of Vertical Cracks on Upstream Face of Diamond Head Buttressed Dam at Zhexi Hydropwer Station", in *Proceedings of the International Conference on Fracture Mechanics of Concrete*, F.W. Whittman, Ed., Lausanne, oct. 1985.

[39] Zienkiewicz, O. C., and Cheung, Y. K.," Buttress Dams on Complex Rock Foundations", *Water Power* V. 16., p.193, 1964.

[40] Zimmerman, Th., Rebora, B., and Droz, P., "Non linear Analysis Of PCRV's with Special Emphasis on Crack Propagation", Paper H2/4, *8th Structural Mechanics in Reactor Technology Conference*, Brussels, 1985.

SEM/RILEM International Conference
on Fracture of Concrete and Rock
Houston, Texas, June 1987
Editors: S.P. Shah, S.E. Swartz

# COMPUTER SIMULATION OF CRACKING IN A LARGE ARCH DAM
## DOWNSTREAM SIDE CRACKING

A. R. Ingraffea*, H. N. Linsbauer**, H. P. Rossmanith***

## ABSTRACT

The history of downstream side cracking of Kölnbrein dam and theories attempting to explain it are described. The stability, trajectory, and opening of the downstream crack are computed and favorably compared to observations. The significant implication of this study is that classical fracture mechanics theory, implemented through modern computer simulation techniques, can be used to explain cracking events in concrete arch dams. This capability can prove valuable in designing against such events, or in specifying effective repair procedures.

## 1. INTRODUCTION

The Kölnbrein arch dam is one of the largest dams in the world and ranks number 24 in the 1985-list of dams. The dam has a crown cantilever height of 200 m and generates a reservoir volume with active storage of $200 \times 10^6$ $m^3$ between levels 1700 m and 1900 m above the Adriatic sea level. The dam was constructed during the period 1973 - 1979 and is designed for a total capacity of 890 MW with a pump capacity of approximately 400 MW. Dam and reservoir are entirely located in an area where massive granitic gneiss on one side and rather schistose gneiss on the other side are separated by a highly schistose zone at the foot of the valley. A double curved arch dam structure was designed with appropriate foundation treatment and installation of an extensive monitoring system for control of dam and foundation movement, deformation, uplift, drainage and temperature behavior. There are about 400 locations for measurement and data acquisition.

Economic requirements dictated partial filling of the reservoir before completion of the dam. The initial and first two follow-up partial fillings created no obvious problems. Deformation control in the dam and in the foundation and measurement of uplift and drainage flow showed no extraordinary behavior and the overall response was acceptable.

Excessive drainage flow was observed when the filling first exceeded the level of approximately 1960 m in 1978. Drainage flow suddenly increased from an unusual 35 l/s to an unacceptable 200 l/s during necessary uplift pressure release. Immediately emptying of the reservoir unveiled an open crack on the upstream face of the dam structure. This obliquely oriented surface crack with a crack mouth extension of 100 m in length forms the visible part of a fracture system surmised to extend into the dam body.

Core drilling in the dam's foundation zone indicated that the fracture system extended to the interface between the dam and its foundation. Preliminary immediate countermeasures employed partial freezing of the foundation region to prevent further water leakage. The very large water leakage was attributed to a cracked and partially leaking grout curtain. The subsequent repairs concentrated on the upstream foundation region and involved a new grout curtain. This grout curtain now stretches at a distance from the dam heal and all hopes concentrated on the impossibility of any further damage to the curtain resulting from possible dam movements. Nevertheless, refilling of the reservoir in 1982 up to a level of 1890 m (approximately 10 m below top level) induced new cracking accompanied by increasing water leakage of 400 l/s. In addition, cracks located on the downstream face of the dam were discovered. The nature and the exact initiation time for these cracks is not known. The reader may find more details in Refs. [1,2].

* Department of Structural Engineering and Program of Computer Graphics, Cornell University, Ithaca, NY, USA
** Institute for Large Dams and Hydraulic Structures, Univ. of Technology, Vienna, Austria
*** Institute of Mechanics, University of Technology, Vienna, Austria

Evolution of Fracture System

A schematic representation of the fracture zone in the arch dam and the fracture system evolved is shown in Figures 1a - 1c, including a longitudinal section of the arch dam (Fig. 1a), the main cross-section (Fig. 1b) and a blow-up of the cracked toe-region of the dam-foundation interface (Fig. 1c).

The visible traces of the fracture systems suggest the following evolution pattern:

The 1978 fracture system surfaced on the upstream side at a rather uniform position of 18 m above foundation level and stretches over a central span of 100 meters. Core drilling showed that the fracture system was inclined toward the foundation as shown in Figure 1c. The 1983 fracture system (which is not an extension of the 1978 fracture system) approximately parallels the 1978 fracture system at a distance of approximately 2 - 5 meters on the uppser side. Both these fracture systems extend all the way to the dam-foundation interface, Figure 1c.

Cracking on the downstream face of the arch dam resembles a crack system centered at a level 12 meters above foundation level. Although the exact depth of this crack system is not known, experts assume the crack to extend approximately across one-third to one-half of the cross section of the dam.

During the course of intensive damage assessment and investigation, several differing theories about the causes of these cracking systems have been developed and partly presented for discussion at several interim and public meetings [2]. In addition, an initial study of slow stable cracking in arch and gravity dams has been performed [3].

Historically, the upstream face cracks were the first to be detected. The downstream face fracture system was observed at a later phase of investigation. It is believed by consulting analysts that the downstream face crack system was introduced by the grouting pressure in the top section of the arch dam. The theory accepted by the dam owners suggests that the upstream crack system was generated as a result of the downstream fracture system.

The authors here suggest a different rationale for the observed cracking as derived on the basis of advanced fracture mechanics as applied to concrete structures. This paper is concerned with the fracture mechanisms on the downstream face of the dam.

## 2. NUMERICAL SIMULATION

A special Finite Element Fracture Analysis Program (FEFAP-G) developed by Ingraffea and his research group [4,5] at Cornell University has been employed in the present study of dam cracking. FEFAP-G is designed to analyze planar and axisymmetric structures for crack initiation and growth. The program combines linear/nonlinear fracture mechanics theory, the use of interactive computer graphics, and a unique, automatic remeshing capability to allow the user to initiate and propagate up to ten discrete cracks simultaneously. The program is meant to be used for basic research into fracture processes in structures and by researchers very familiar with basic finite elements and fracture mechanics theory and techniques. The capabilities of FEFAP have been utilized in solving a wide range of problems in rock and concrete structures.

## 3. FRACTURE-MECHANICS INVESTIGATION OF DOWNSTREAM CRACK SYSTEM IN THE KÖLNBREIN ARCH DAM

Investigations of fractured concrete structures within the framework of three-dimensional analysis are very involved and in addition time-consuming and expensive. Traditionally, two-dimensional models of endangered cross-sectional areas are analyzed and only exceptionally is the analysis extended into three dimensions. Such is the case when a fracture system is a general curved survace in the bulk of the dam and cannot be adequately represented by two-dimensional approximations. In three-dimensional analyses boundary element procedures exhibit greater potential, Ref. [6].

In the case of the cracking in the Kölnbrein arch dam a two-dimensional model associated with modified boundary conditions due to the spatial action of the stress and strain behavior has been performed. In particular, the ratio of the moduli of elasticity between rock and concrete has been selected as 1.39, Poisson's ratio 0.2, and the density of the concrete was assumed to be 0.024 $MN/m^3$.

Generally, the analysts involved in this dam cracking incident believe that the fracture system on the downstream side of the dam was generated as a result of dead load (load case: self-weight of dam with an empty reservoir) superimposed on a stress field due to joint grouting pressure and inelastic deformations in the dam foundation region.

### 3.1 State of Stress in the Uncracked Dam

The first step in the analysis consists of the determination of the stress and deformation field in the uncracked structure. Because of the localization of the fracture system in the dam toe and foundation region, the upper portion of the dam structure has been deleted and its effect in terms of load and stress on the lower portion of the dam is characterized by the action of a resultant force and corresponding moments. This is shown in Figure 2. The reduced forces and moments given in Table 1 correspond to different filling levels of the reservoir are from the result of a three-dimensional trial load method. Additional loads derived from the effect of grouting pressure in the joints and permanent displacements at the base are given in the final row of Table 1. These were obtained by an approximate calculation of the behavior of a central cantilever displaced upstream by assumed additional loads (grouting pressure).

| LOAD COMPONENT | | $F_X$ MN/m | $F_Y$ MN/m | $M_z$ MN-m/m |
|---|---|---|---|---|
| DEAD LOAD | | 0 | - 70 | - 40 |
| WATER LOAD AT LEVEL: (m) | 1767 | - 2 | - 1 | - 46 |
| | 1800 | 0 | - 5 | -149 |
| | 1902 | 16 | - 1 | - 21 |
| ADDITIONAL LOADS | | - 3 | | - 67 |

Table 1. Reduced load components acting on the dam at the level 1760 m.

Results of this stress analysis are shown in Figures 3 and 4. Figure 3 - where FEFAP's menue card is illustrated on the right margin of the display - exhibits the field of principal tensile stresses in the dam toe due to dead load only. In addition the stress distribution along a horizontal section situated at the most critical level in the down-stream part of the dam shows normal tensile stresses of the order of 0.38 MN/m$^2$ which are appreciably lower than the tensile strength of the concrete. It is remarkable that the maximum tensile stress level is calculated to occur in the downstream face of the dam at an altitude of 12 - 15 m above the dam toe. Please observe that the stress concentration in the dam at level 1760 is not realistic but is the result of the approximation of the distributed cross sectional stresses by a single force.

The maximum tensile stresses on the face of the dam, however, do increase appreciably when the additional loads shown in Table 1 are operative. The resulting maximum stress value of 1.6 MN/m$^2$ reaches the value of the tensile strength of the dam material (Fig. 4). In the following detailed fracture mechanics analysis this location of maximum tensile stress is identified with the locus of crack initiation at a level of 13.5 m above the toe.

### 3.2 Crack Initiation

A small flaw was initiated on the downstream face of the dam at this location. The associated finite element mesh surrounding the crack is shown in Figure 5. A fracture toughness value of $K_{IC}$ = 2.0 MNm$^{-3/2}$ was selected in the fracture analysis. The finite-element-based stress analysis of this initial crack yielded stress intensities $K_I$ = 3.45 MNm$^{-3/2}$ and $K_{II}$ = -0.16 MNm$^{-3/2}$. In FEFAP these values are taken into the $\sigma_{\theta\theta max}$ fracture criterion [7] to find that the criterion is not satisfied: $K_{IC}$ is smaller that the effective mixed-mode stress intensity factor and the crack is predicted to propagate. This is easy to see by the following rough calculations: the small value of $K_{II}$ of the order of 0.16 as compared to the $K_I$-value of 3.45 indicates a nearly mode-1 cracking situation with horizontal crack extension. This analytically computed value of 3.45 is appreciably larger than the material's toughness of $K_{IC}$ = 2 MNm$^{-3/2}$.

Automatic mesh-refinement in the vicinity of the crack tip showed only a 1.3 % change of stress intensities as compared to a coarse mesh (see Figure 6). This means that very accurate values of $K_I$ and $K_{II}$ have been computed.

### 3.3 Crack Propagation

After a detailed crack initiation analysis the crack was allowed to advance at discrete steps with the direction conditioned by the fracture criterion. The loading consisted of dead load and additional loads shown in Table 1. Figure 7 shows the deformed finite element mesh associated with the fourth step of crack advancement. One can easily observe that the crack path bends toward the foundation. The length of the crack at this state was about 10 meters and the crack mouth opening displacement was 2.4 mm. These calculated values correspond very well with the actual data. Figure 8 shows that the data point, ($K_I/K_{IC}$; $K_{II}/K_{IC}$), corresponding to this crack tip location is located close to the limit of stability. Therefore crack acvancement in the simulation program came to an arrest.

### 3.4 Crack Closure Due to Refilling of the Reservoir

The second part of the fracture mechanics investigation was concerned with crack closure due to partial refilling of the reservoir. The crack investigated was that described in the previous section and shown in Figure 7.

The water load acting on the cantilever idealization associated with various partial filling levels is shown in Table 1. No crack closure was calculated up to water level 1767 meters and the value of $K_I$ was 1.14 $MN/m^{3/2}$. Increasing the water level in the numerical procedure would induce crack closure and overlapping of crack faces. Therefore, special contact elements had to be placed along the crack face in order to allow for tangential slip without overlap. Appropriate contact conditions can be simulated by attaching different normal stiffness and shear stiffness to these contact elements. Varying the shear stiffness did not alter the size of the crack opening.

Finally, a crack mouth opening displacement of 0.5 mm was computed at a filling level of 1800 meters, Figure 9. This is in excellent agreement with in-situ field data, where complete crack-mouth closure was achieved at a filling level of 1820 meters.

## 4. CONCLUSIONS

It has been shown that basic principles of mixed-mode, linear elastic fracture mechanics can be used to reproduce cracking observed in an arch dam. The stability, trajectory, and mouth-opening of a crack observed near the toe of the Kölnbrein dam have been accurately simulated. This simulation included the effects of dead load, water load, and additional loads such as grouting pressures, etc. The effect of reservoir re-filling on a pre-existing crack was also modelled.

A drawback of this analysis is its current limitation to two-dimensional idealization. Three-dimensional loading effects, and the interaction of the idealized two-dimensional cantilever with adjacent elements of the dam can only be approximated. There exist adequate fracture mechanics theory and finite element analysis capabilities to attack such a problem in three dimensions. The missing element for truly three-dimensional crack propagation modelling is the ability to introduce and to propagate incrementally a crack of arbitrary shape into a three-dimensional model.

This study is the first step in a complete fracture mechanics analysis of the Kölnbrein dam. Multiple cracking events which occurred on the upstream side of the dam, will be investigated in a subsequent paper.

## 5. LITERATURE

[1] Baustaedter, K.; Widmann, R.: The Behavior of the Kölnbrein Arch Dam. 15$^{th}$ ICOLD, Lausanne, 2, 633 - 651 (1985).

[2] Baustaedter, K.: Die Koelnbreinsperre aus heutiger Sicht. Erstes Christian Veder Kolloquium, TU-Graz, Nov. 1985.

[3] Linsbauer, H. N.; Rossmanith, H. P.: On Slow Stable Cracking of Gravity Dams - a Photoelastic/FEM Analysis, ÖIAZ, Juli 1986.

[4] Ingraffea, A. R.; Saouma, V.: Numerical Modelling of Discrete Crack Propagation in Reinforced and Plain Concrete, in Fracture Mechanics of Concrete - Structural Application and Numerical Calculation, Sih, G. C.; DiTommaso, A.; Eds., Martinus Nijhoff Publishers, Dordrecht, 1985.

[5] Gerstle, W.: Finite and Boundary Element Modelling of Crack Propagation in Two- and Three-Dimensions Using Interactive Computer Graphics. Ph.D. Thesis, Cornell University, 1986.

[6] Perucchio, R.: An Integrated Boundary Element analysis System with Interactive Computer Graphics for Three-Dimensional Linear-Elastic Fracture Analysis. Ph.D. Thesis, Cornell University, 1984.

[7] Erdogan, F.; Sih, G. C.: On the Crack Extension in Plates Under in Plane Loading and Transverse Shear. J. Basic Engrg., ASME, Vol. 85, 519 - 527, 1963.


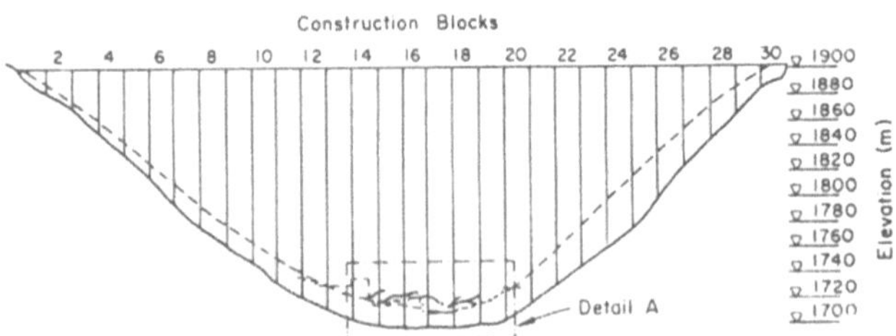


a) Longitudinal section (upstream)

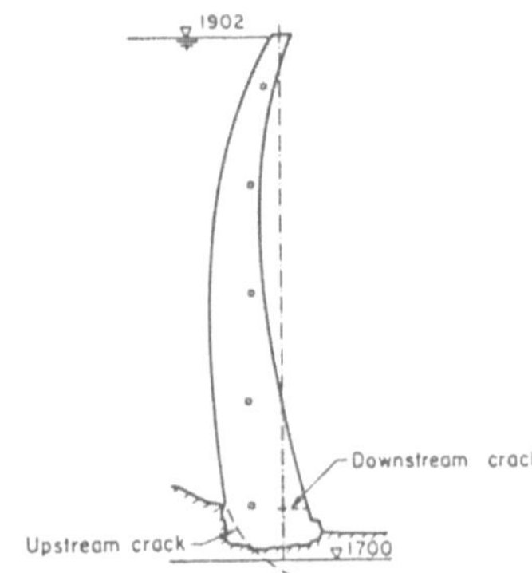


b) Cross-section of dam

Figure 1. Fracture pattern development in the Kölnbrein dam

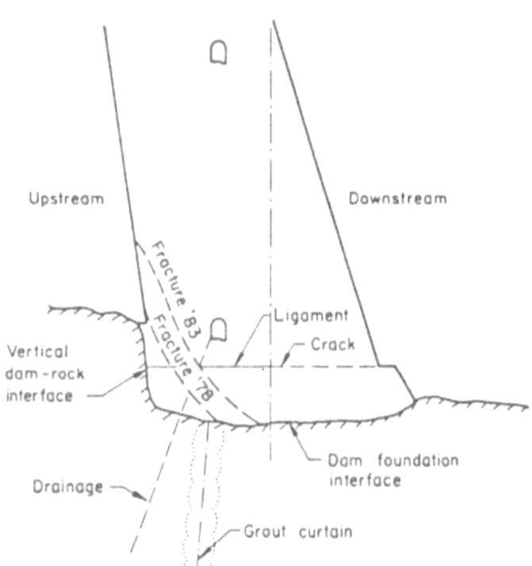


Figure 1. c) Fracture pattern development in the Kölnbrein dam

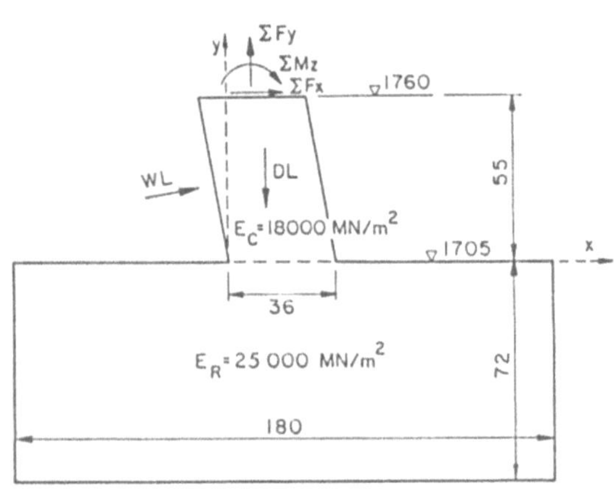


Figure 2. Reduced mechanical load system

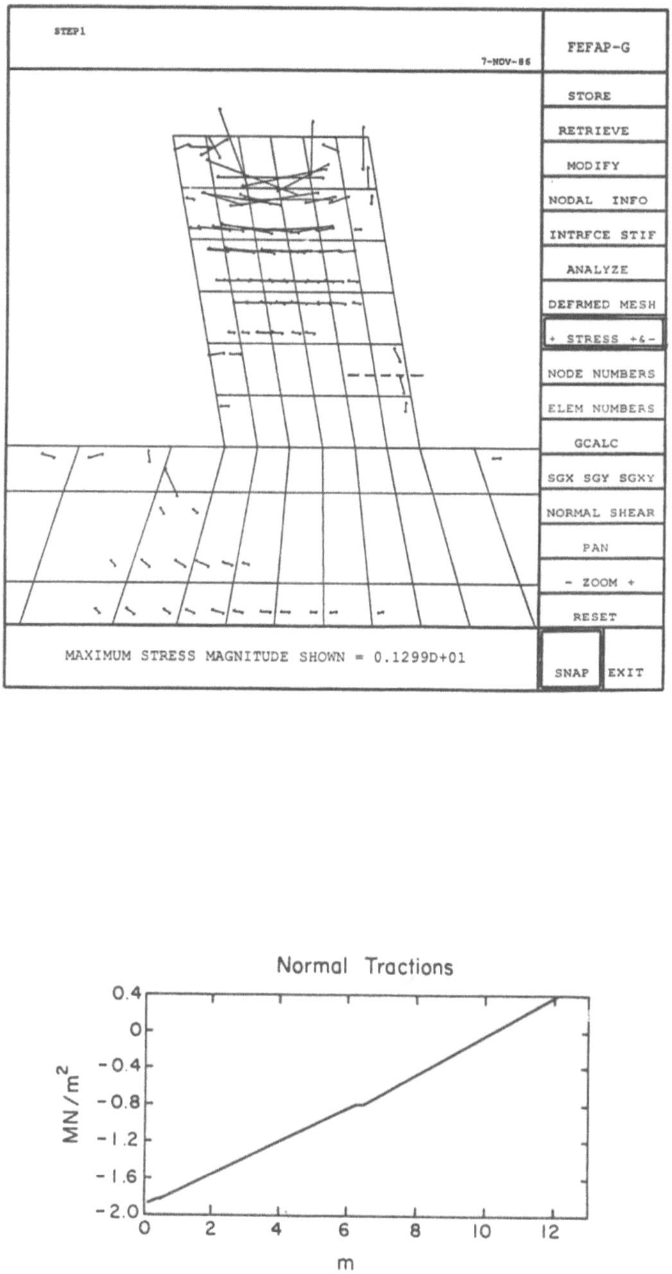


Figure 3. Principal tensile stresses and normal stress distribution in a horizontal section (----) in the downstream region of the dam due to dead load

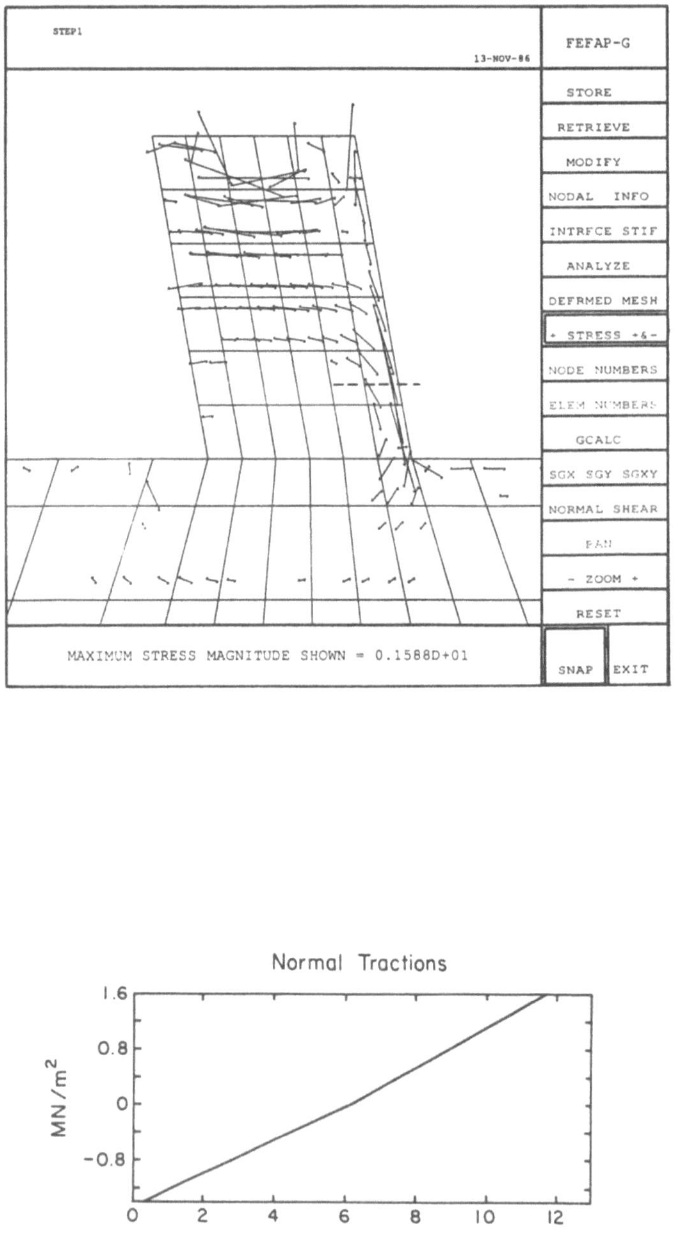


Figure 4. Principal tensile stresses and normal stress distribution in a horizontal section (----) in the downstream region of the dam due to combined loading: dead load and additional load

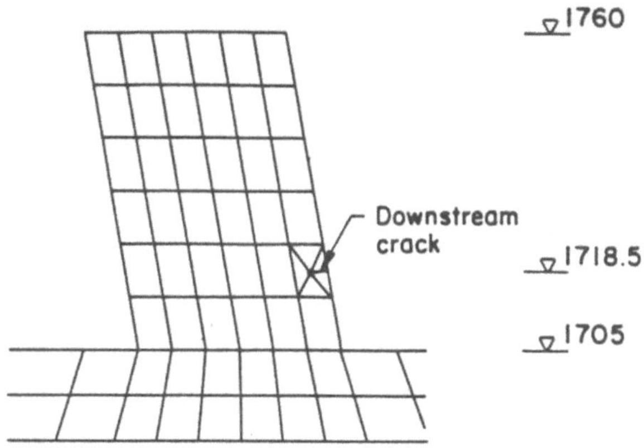


Figure 5. Finite element mesh with initial crack and large crack-tip elements

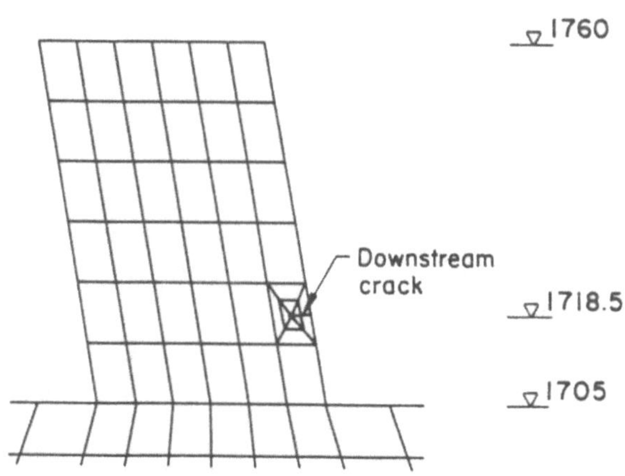


Figure 6. Finite element mesh with initial crack and small crack-tip elements

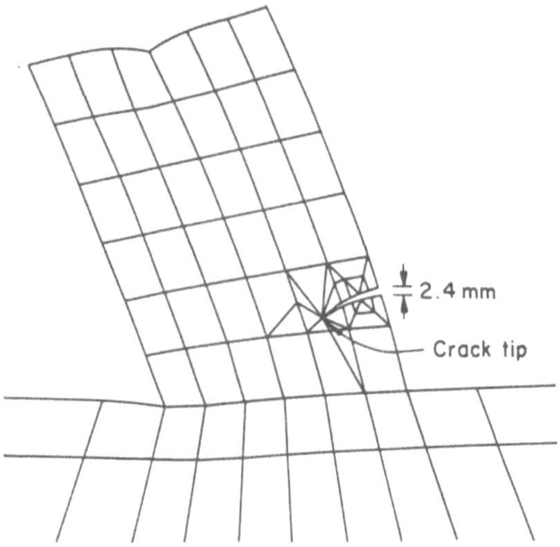


Figure 7. State of deformation of the mesh after the fourth crack advancement

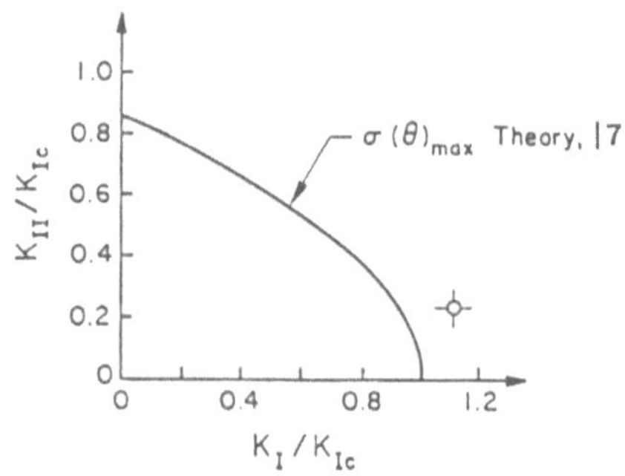


Figure 8. Mixed-mode fracture criterion

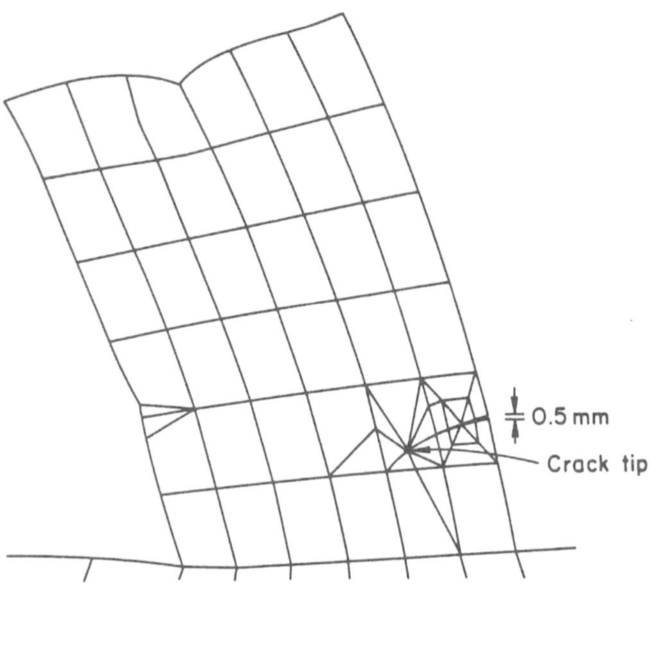


Figure 9. Displacement field associated with a partially closed crack subjected to dead load, additional loading and partial filling at level 1800 m

SEM/RILEM International Conference
on Fracture of Concrete and Rock
Houston, Texas, June 1987
Editors: S.P. Shah, S.E. Swartz



# INFRARED THERMOGRAPHY OF FRACTURE OF CONCRETE AND ROCK

M.P. LUONG

## ABSTRACT

Infrared thermography has been used as a nondestructive and noncontact technique to examine the mechanical response of concrete and rock specimens subjected to given static unconfined compressions and to a superimposed vibratory excitation. The parameter investigated in this paper is the heat generation due to energy dissipation by the material which has been excited beyond its stable reversible limit. This useful technique allows accurate illustration of the onset of unstable crack propagation and/or flaw coalescence when increasing irreversible microcracking is generated by vibratory loading.

## INTRODUCTION

Fatigue and damage behaviour of plain concrete in compressive loading is an important consideration in connection with design or regulations of bridges, offshore concrete structures, earthquake resistant buildings, subjected to high allowable working stresses or in cases where the dead load forms a smaller part of the total load capacity. In such situations, failure of the construction material may occur at stresses below its static strength. Accurate knowledge should therefore be obtained of the mechanical behaviour of concrete subjected to various loadings.

In response to these problems, various fatigue analysis methodologies have been developed in recent years which isolate the factors affecting crack initiation and growth, and enable the prediction of their cumulative effect on the fatigue performance of structural components. They are based on (1) the formulation of analytical models for fatigue crack and growth ; and (2) the acquisition of supporting baseline data and validation of such models by means of a comprehensive testing procedure.

The existing partial damage hypothesis gives unsatisfactory results in evaluating the fatigue behaviour of plain concrete [1]. Volumetric, ultrasonic and acoustic emission measurements have not permitted a complete interpretation of the deterioration of concrete [2]. Research on absorption of energy in fatigue loading of plain concrete has been investigated [3] to aid understanding of the physical process of degradation when concrete is subjected to varying dynamic loads. The absorbed energy is believed to be used in the material in forming microcracks, crushing material, redistributing stress and causing a rise in temperature. Unhappily the difficulties involved with the measurement systems were such that the results obtained by these methods were not completely satisfactory.

In rock mechanics, various geological formations are being investigated as possible sites for underground storage of oil and gas, and underground repositories of radioactive wastes. Much attention has been given to the saline deposits due to their widespread distribution, relative accessibility and ease of excavation. The mechanisms of rock salt deformation and fracture are thus fundamental for processes of strata control, for the support of excavations, and for the construction of salt cavities [4].

Rock salt generally displays failure characteristics intermediate between ductility and brittleness. It is relatively soft and can deform plastically, but it is extremely notch sensitive. In the presence of a flaw, brittle fracture often occurs with the onset of yielding.

The present paper proposes the use of infrared thermography as a nondestructive and noncontact technique to examine the mechanism of the cracking and failure of concrete or rock salt specimens which are subjected to a given unconfined compression and a superimposed vibratory excitation.

The aim of this study is to illustrate the onset of crack initiation by stress concentration in weakness zones, followed by unstable crack propagation and/or flaw coalescence caused by the thermomechanical coupling, when increasing microcracking is generated by vibratory excitation.

---

M.P. Luong is a Research Director (2), Centre National de la Recherche Scientifique UA 317 - Laboratoire de Mécanique des Solides Ecole Polytechnique 91128 Palaiseau Cedex France.

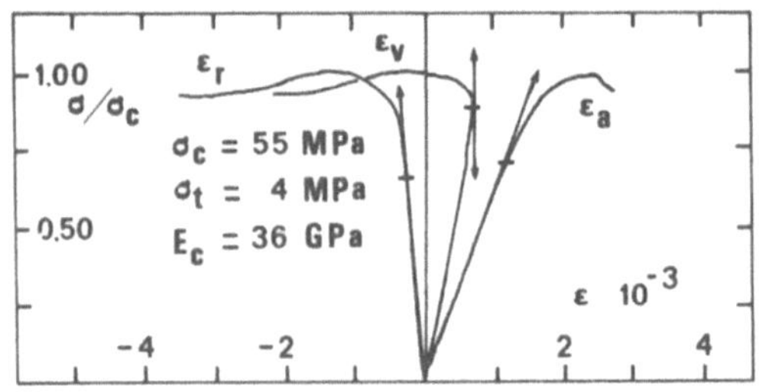


Fig. 1. Axial ($\varepsilon_a$), radial ($\varepsilon_r$) and volumetric ($\varepsilon_v$) strains of a concrete specimen subjected to a uniaxial unconfined compression.

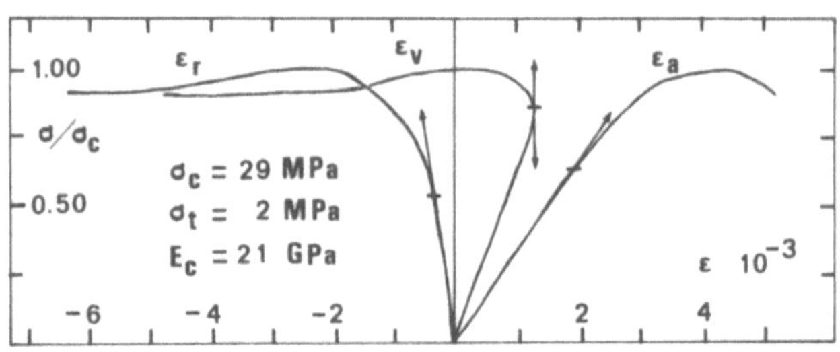


Fig. 2. Uniaxial unconfined compression on a cylindrical specimen of rock salt.

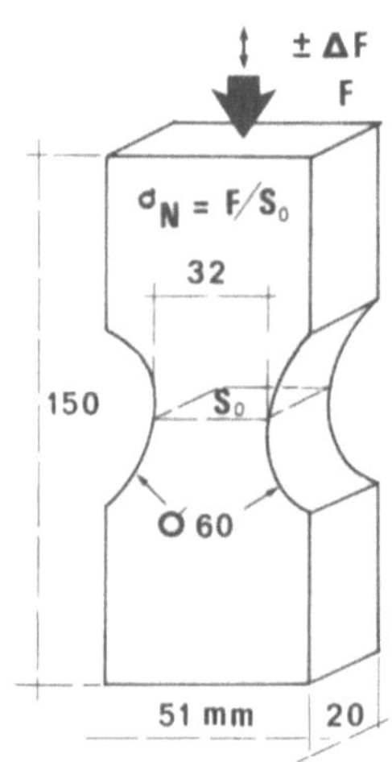


Fig. 3. Geometry and dimensions of the specimen.

In addition, an application of this technique is presented as a nondestructive method for inspection and evaluation of stress concentration on a concrete structure subjected to an earthquake type loading.

## CHARACTERISTICS OF CONCRETE AND ROCK SALT

Concrete is a composite material consisting of coarse aggregates embedded in a continuous matrix of mortar which is a mixture of hydraulic binding materials, additives and admixtures distributed in a suitable homogeneous dosage. This construction material has been extensively used for structural applications because of its excellent processibility, low cost and versatility in manufacture together with desirable mechanical properties for engineering design.

Under applied loading, the concrete as a whole deforms in spite of significant incompatibilities between the aggregates and the matrix which promote further breakdown. At the macroscopic level, breakdown is accompanied by both loss in stiffness and accumulation of irrecoverable deformation. At the structural level, breakdown appears as microcracking and possibly slippage at the aggregate-cement paste interfaces.

Natural rock salt appears to be a discontinuous mass with a medium to coarse grained polycrystal structure. It contains cracks, defects, joints, dispersed clay impurities and eventually bedding planes with varying degrees of cohesion along these discontinuities. It sometimes exhibits variable mineralogy and crystal sizes.

Depending on its crystal structure, bond character, and the temperature, it may be completely brittle, semi-brittle or ductile [5]. Elucidation of the various mechanisms responsible for fracture is related to the plastic resistance associated with grain boundaries and the effects of plastic anisotropy.

When it is loaded, the rock salt deforms also as a whole in spite of significant incompatibilities between the crystal aggregates. Stress concentrations occur and result in localized forces which are sufficient to promote plasticity and crack formation or both. At the macroscopic level, breakdown is accompanied by both loss in stiffness and accumulation of irrecoverable deformation . At the structural level, breakdown appears as microcracking and possibly slippage at the crystal interfaces.

Failure in plain concrete or in rock salt may be viewed as a microstructural process through the activation and the growth of one preexisting flaw or site of weakness, or through the coalescence of a system of interacting small flaws and growing microcracks. The stress level corresponding to the activation of the flaws is related to the flaw size and connected with the encompassing microstructure.Flaw initiating concrete failure may be divided into two classes : the intrinsic flaws develop during the processes of hydration and curing of the cement paste, the extrinsic flaws result from significant incompatibilities between the aggregate and the matrix when the material as a whole deforms under applied loading.

It may be said that fatigue of concrete is associated with the development of internal microcracks, probably both at the cement matrix/aggregate interface and in the matrix itself. For rock salt, it may occur at the crystal interface and in the crystal itself.

The occurrence of microcracking and slippage leads to nonlinearity and softening in the stress-strain response of concrete with a marked dependence on the mean normal stress [6].

The figure 1 presents a monotonic uniaxial compression test on a two years old concrete specimen.

The figure 2 describes the response of a cylindrical specimen of rock salt under uniaxial unconfined compression.

The volume change detected by strain gages in transverse and longitudinal directions of the cylindrical specimen is seen to be highly relevant in characterizing the deformation of concrete under loads. It may be attributed to (1) elastic variations of minerals, (2) or to dilatancy due to growing microcracking.

Under repetitive stresses, fatigue mechanism is a progressive, permanent internal structural changes occurring in the concrete and in the rock salt. These changes result in progressive growth of cracks and complete fracture.

The formation and propagation of microcracks have been detected by means of different measuring methods :

a - The ultrasonic pulse velocity technique involves measurement of the transit time of an ultrasonic pulse through a path of known length in a specimen. The velocity of the ultrasonic pulse in a solid material will depend on the density and elastic properties of the material and it will therefore be affected by the presence of cracks.

b - The acoustic emission method is based on the principle that the formation and propagation of the microcracks are associated with the release of energy. When a crack forms or spreads, part of the original strain energy is dissipated in the form of heat, mechanical vibrations and in the creation of new surfaces. The mechanical vibration component can be detected by acoustic methods and recorded, hence microcracking may be detected by studying sounds emitted from the concrete or the rock salt.

It can be considered that the failure mechanism of concrete and rock salt consists primarily in the formation and propagation of microcracks.

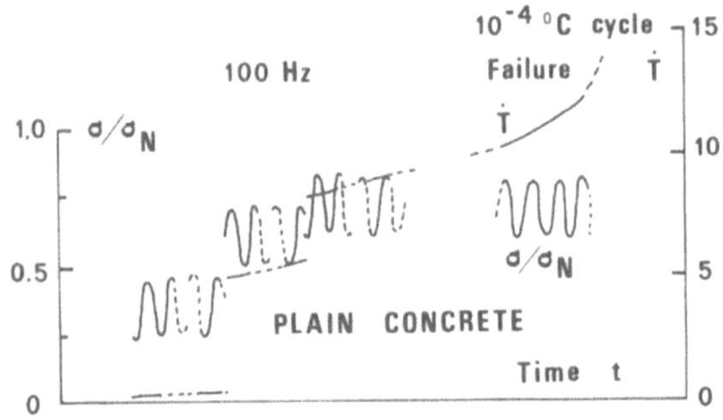


Fig. 4. Vibratory excitation on a concrete specimen subjected to various static uniaxial compression and evolution of the growth rate of heat on the warmest point.

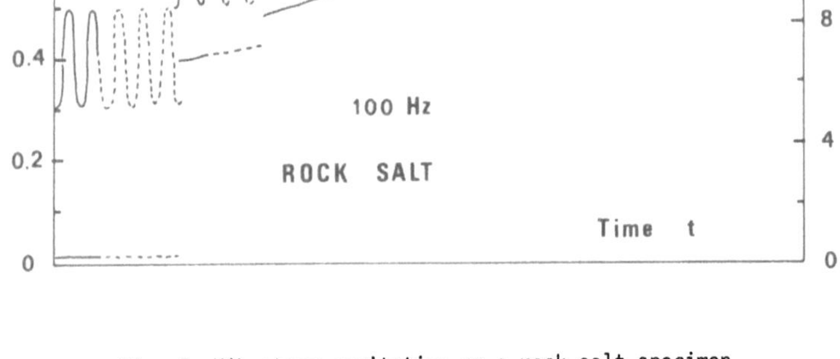


Fig. 5. Vibratory excitation on a rock salt specimen subjected to various static unconfined compression and evolution of the growth rate of heat at the warmest point.

The formations of microcracks are often associated with points of stress concentration in the material. The stress concentrations are located on flaws present in the material, or on existing cracks and notches. Cracks initiate quite early at a site of weakness or stress concentration, then propagate through the plastic zone and into the elastic region.

In other cases, flaws are inherent in the material owing to the process of fabrication of concrete or crystallisation of rock salt. These defects exist prior to the application of any load. There may be some initiation period during which the material at the tip of the flaw undergoes dislocations pile-up, microvoid formation and coalescence, etc..., prior to the onset of progressive growth.

THERMAL DISSIPATION

Infrared thermography has been successfully used as an experimental method for detection of plastic deformation during crack propagation of a steel plate under monotonic loading [7] or as a laboratory technique for investigating damage, fatigue and creep mechanisms occurringin engineering materials [8,9 and 10].

This experimental tool is used to detect the onset of unstable crack propagation and/or flaw coalescence due to the thermomechanical coupling, when increasing irreversible microcracking is induced by vibratory loading.

1 - Thermoplasticity

The heat energy emitted in a loaded element is the main factor associated with the damage process initiating the failure of concrete and rock salt. In the case of concrete or rock salt at room temperature, it can be assumed that the mechanical transformation is small : changes in geometry are neglected. This permits the use of the small strain tensor as the variable of the deformation state [11].

If the material is elastic and perfectly plastic, the density of free energy per unit mass is $W = W(\varepsilon - \varepsilon^p, T)$ and the stress is $\sigma = \rho \frac{\partial W}{\partial \varepsilon}$. The plastic strain is denoted by $\varepsilon^p$ and the absolute temperature by T. The entropy S per unit mass is defined as $S = -\partial W/\partial T$.

The two laws of thermodynamics are written for any volume v :

$$\dot{E} + \dot{C} = \dot{W}_e + \dot{Q} \tag{1}$$

$$\left.\begin{aligned} \dot{Q} &= \int_v \rho\, T\, \dot{S}\, dv - \int_v D\, dv \\ \text{where } &\int_v D\, dv \geqslant 0 \end{aligned}\right\} \tag{2}$$

The letter D denotes the dissipated intrinsic power per unit volume ; Q the heat supply per unit mass ; $E = \int_v (TS) + W)\, \rho\, dv$ is the internal energy per unit mass in v ; $W_e$ the work of external forces and C the kinetic energy.

Eliminating S in (1) and (2), the balance of mechanical energy leads to the equation :

$$\dot{C} + \int_v \sigma\dot{\varepsilon}\, dv = \dot{W}_e \tag{3}$$

and

$$D = \sigma\dot{\varepsilon}^p \tag{4}$$

If q denotes the heat flux, it can be written :

$$\dot{Q} = -\int_{\partial v} q.n\, ds + \int_v \phi dv = \int_v (\phi - \text{div } q)\, dv$$

where n is the exterior normal vector and $\phi$ the eventual specific density of heat sources in v.

The equation (2) becomes locally :

$$\phi - \text{Div } q = \rho\, T\dot{S} - D \tag{5}$$

The Fourier heat conduction law $q = -k \text{ grad } T$ leads to the heat conduction equation :

$$\phi + \text{div } k \text{ grad } T = \rho T\dot{S} - \sigma\dot{\varepsilon}^p \tag{6}$$

Now :

$$\rho T\dot{S} = -\rho T\left(\frac{\partial^2 W}{\partial T\partial\varepsilon}\dot{\varepsilon} + \frac{\partial^2 W}{\partial T\partial\varepsilon^p}\dot{\varepsilon}^p + \frac{\partial^2 W}{\partial T\partial T}\dot{T}\right)$$

Taking into account the form of the free energy $W(\varepsilon - \varepsilon^p, T)$, it follows that :

$$\rho T\dot{S} = -\rho T\frac{\partial^2 W}{\partial T\partial\varepsilon^e}(\dot{\varepsilon} - \dot{\varepsilon}^p) - \rho T\frac{\partial^2 W}{\partial T\partial T}\dot{T}$$

The quantity $c = \rho T\ \partial^2 W/\partial T\partial T$ is the specific heat per unit volume and $l_{ij} = -\rho\ \partial^2 W/\partial T\partial\varepsilon^e_{ij}$ are thermoelastic coefficients.

Finally, the coupled heat conduction equation of thermoelastic plasticity is obtained :

$$c\,\dot{T} = \mathrm{div}\ k\ \mathrm{grad}\ T - T\,\alpha\,\dot{\sigma} + D + \phi \tag{7}$$

where c denotes the specific heat, T the absolute ambient temperature, k the heat conductivity, $\alpha$ the thermal dilatation, $\sigma$ the stress tensor, D the intrinsic dissipation and $\phi$ the heat source per unit volume.

The first term on the right side governs the thermal diffusion which tends to make the temperature uniform in the specimen. The second term represents the thermoelastic effect that may be significant in cases of isentropic loading. The nature of factors k and $\alpha$ is tensorial in case of anisotropic material. The third term is the thermal dissipation generated by viscosity or plasticity. The last term show the existence of heat source in the specimen.

2 - Electromagnetic radiation

Electromagnetic radiation is a form of energy characterized as waves or as particles called photons. Visible light is the most familiar form of electromagnetic energy. Other forms include radio waves, heat radiation, ultraviolet rays, X-rays and gamma rays. All this energy is similar and radiates in accordance with basic wave theory. Electromagnetic radiation is produced by the acceleration of charged particles. More rapid acceleration produces higher energy (shorter wavelength) waves. The electromagnetic spectrum is a categorization by wavelength of electromagnetic energy. The range from 2 to about 100 micrometres is called "thermal infrared".

All matter radiates energy because it contains charged particles being accelerated (changing speed or directions), the higher the temperature the greater the acceleration. The amount of energy radiated depends on the object's temperature and its ability to radiate.

3 - Infrared vibrothermography

Infrared thermography utilizes a photovoltaic detector composed of indium antimonide InSb in a sophisticated electronics system in order to detect radiated energy and to convert it into a detailed real time thermal picture on a video sustem. Response times are shorter than a microsecond. Temperature differences in heat patterns as fine as 0.2°C are discernible instantly and represented by several distinct hues.

This technique is sensitive, nondestructive and noncontact, thus ideally suited for records and observations in real time of heat patterns produced by the heat transformation of energy due to stress concentration and/or plastic strains.No interaction at all with the specimen is required to monitor the thermal gradient.

The quantity of energy $W_r$ emitted by infrared radiation is a function of the temperature and the emissivity of the specimen. The higher the temperature, the more important is the emitted energy. Differencesof radiated energy correspond to differences of temperature, since $W_r = h\ T^4\ \omega$, where h denotes a constant, T the absolute temperature and $\omega$ the emissivity.

Concrete and rock salt presents a low thermomechanical conversion under monotonic loading. Plastic deformation -whereby microcracking and slips occur creating permanent changes globally or locally- is however one of the most efficient heat production mechanisms. Most of the energy which is required to cause such plastic deformations is dissipated as heat. Such heat development is more easily observed when it is produced in a fixed location by reversed applied loads. These considerations define the use of vibrothermography as a nondestructive method for observing the damage process of concrete and rock salt.

4 - Experimental set-up and results

The high-frequency servo-hydraulic test machine Servotest, used at the Laboratoire de Mécanique des Solides, provides a means of vibration and dynamic testing of engineering materials. Control of the machine is provided by a sophisticated closed-loop electronic control system. This utilizes feedback signals from the force and displacement transducers. The programming section comprises a digital function generator and a frequency sweep controller which enables resonant phenomena testing.

The sample is observed in a nondestructive, noncontact manner by means of an infrared thermographic system AGA THERMOVISION 782. The thermal image is shown on the monitor screen.

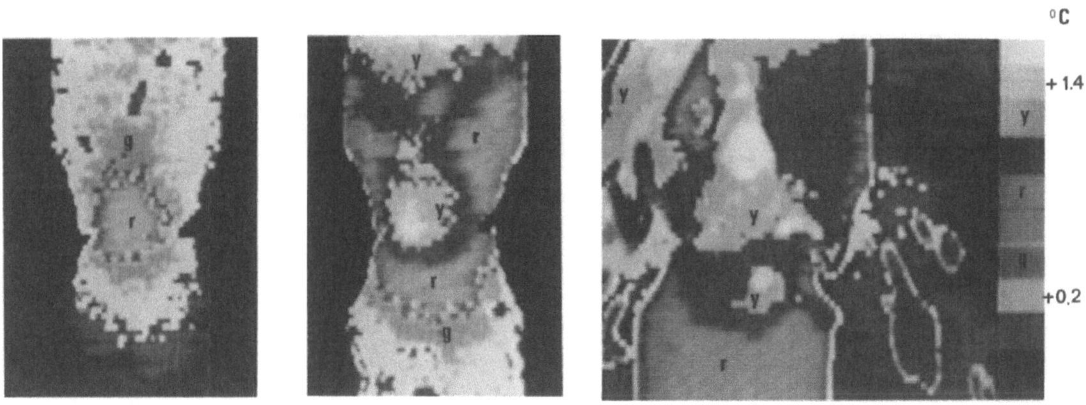


Fig. 6. Thermograms of a plain concrete specimen evidencing the process of unstable failure.

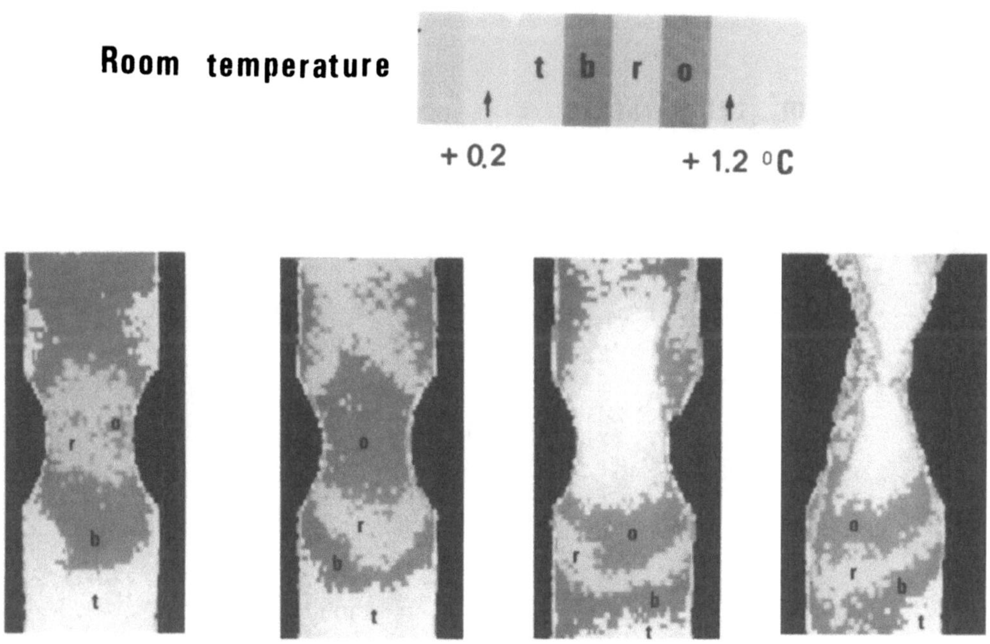


Fig. 7. Heat dissipation describing the process of unstable failure in rock salt specimen.

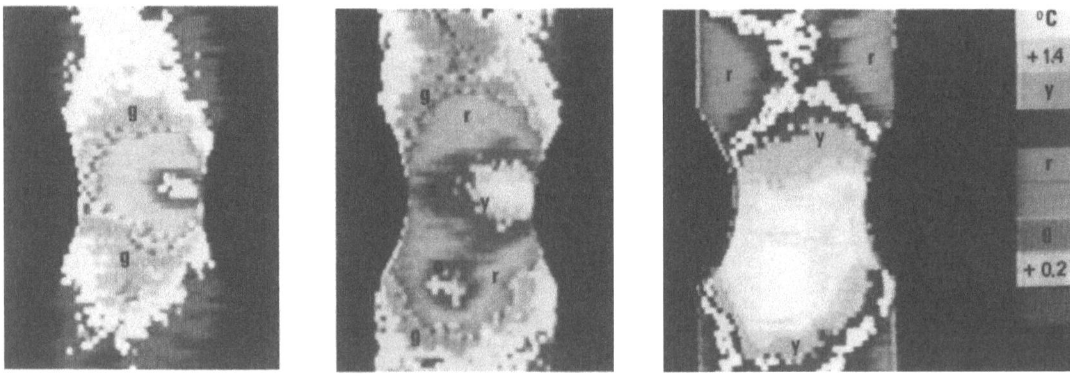


Fig. 8. Flaw coalescence mechanism occurring during concrete failure.

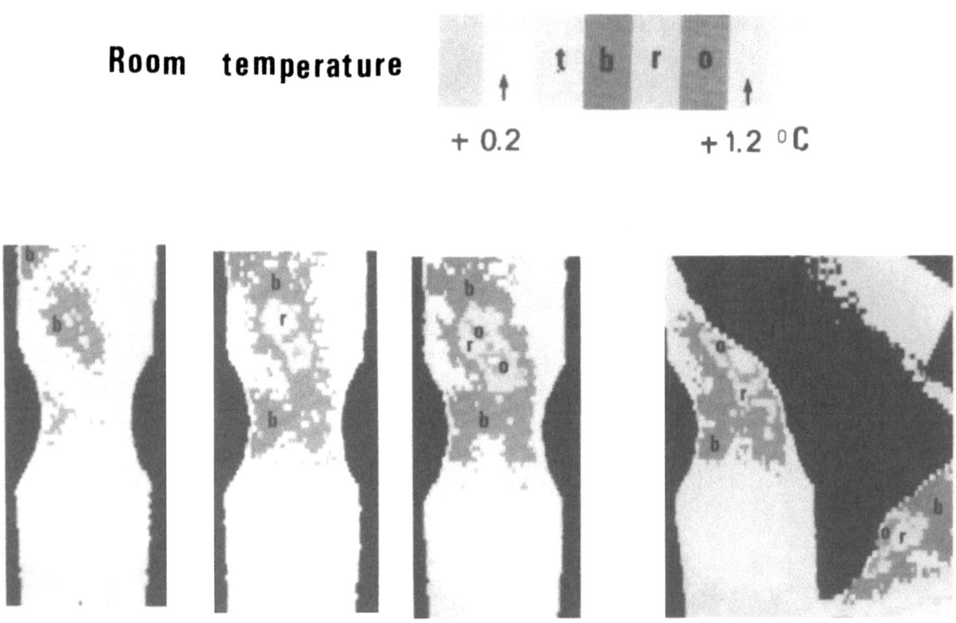


Fig. 9. Flaw coalescence achieving the progressive failure of rock salt specimen.

The parameter investigated in this test is heat generation due to the energy dissipated by the concrete which has been excited beyond the stable reversible domain. A vibratory loading at 100 Hz on the specimen subjected to a given static compression (figure 3) exhibits in a nondestructive manner the irreversible plastic strain concentrations around gaps or cracks generated by stresses exceeding locally the stability limit of the material. The contribution of the plasticity term is revealed by the rapid evolution of heat dissipation once the stable reversible domain has been exceeded.

The figures 4, 5 and 6, 7 show the different stages of heat dissipation describing the process of unstable crack propagation which occurs in specimens of plain concrete and natural rock salt.

The figures 8 and 9 record respectively the flaw coaslescence mechanism in concrete and rock salt.

The growth rate of heat on the warmest point enables the detection and the location of the onset of specimen failure.

Thus the thermomechanical coupling offers a quantitative evaluation for the growth rate of thermal dissipation monitoring the damage evolution of the material. The damaged areas are located and highlighted by heat patterns [12].

## FULL-SCALE EXPERIMENTAL INSPECTION OF A CONCRETE STRUCTURE

Information about the location and significance of structural defects needed as a basis for maintenance decisions, including the extreme case of removal from service, can be obtained through inspection and nondestructive evaluation. The proposed infrared thermographic procedure involves careful examination of areas where defects are most likely to occur. The critical areas can be identified by analyzing the structure and the service histories of similar structures in similar environments.

The application of infrared scanning to inspection of concrete structure relies on the fact that energy is dissipated during the process of failure when internal cracks or flaws develop. It is indeed well known that the fatigue process involves crack initiation and/or nucleation, stable crack propation, and final fracture or unstable crack propagation.

Figure 10 describes an experimental reinforced concrete building frame, intended for earthquake resistance studies. The most severe likely earthquake can be survived if the members are sufficiently ductile to absorb and dissipate seismic energy by inelastic deformations. It is recognized that the connections present adequate ductility in order to undergo large inelastic deformations with little decrease in strength. Under seismic loading, simulated by a rotating mass excitator placed on the top of the building, plastic hinges form progressively at the column bases where heat dissipation can be observed by infrared thermography.

Figure 11, 12 and 13 show the progressive evolution of heat dissipation at a column base before crack line is visible.

## CONCLUDING REMARKS

Owing to the thermomechanical coupling infrared vibrothermography offers the possibility of a nondestructive, noncontact test of concrete and rock salt degradation. It allows a measure of the material damage and permits the detection of the limit of a progressive damaging process under load beyond which the material is destroyed.

This useful and promising technique allows accurate illustration of the onset of unstable crack propagation and/or flaw coalescence when increasing irreversible microcracking is generated by vibratory loading.

## ACKNOWLEDGMENT

The support of the Mission Scientifique et Technique du Ministère de la Recherche et de la Technologie and the interest of the Agence Nationale pour la Gestion des Déchets Nucléaires for this work are gratefully acknowledged.

## REFERENCES

[1] Tepfers, R., Friden, C., Georgsson, L., A study of the applicability of the Palmgren-Miner partial damage hypothesis, Magazine of Concrete Research, vol. 29, n° 100, Sept. 1977, pp. 123-130.

[2] Bergues, J., Terrien, M., Study of concrete's cracking under multiaxial stresses, Advances in Fracture Research, 5th Int. Conf. on Fracture, Cannes, France, 29 March-3 April 1981.

[3] Tepfers, R., Hedberg, B., Szczekocki, G., Absorption of energy in fatigue loading of plain concrete, Matériaux & Construction, vol. 17, n° 97, pp. 59-64, Bordas-Dunod, 1984.

[4] Hardy,R.H.Tr. and Langer, M., The mechanical behaviour of salt, Proc. 1st Conf. Pensylviana Trans. Tech. Pub. 1981.

Fig. 10. Experimental concrete structure under seismic type loading generated by a rotating mass excitat-or on its top.

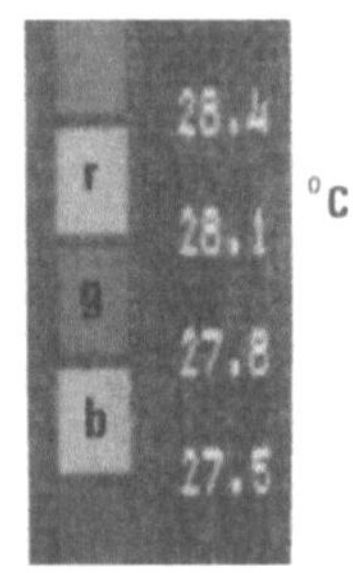


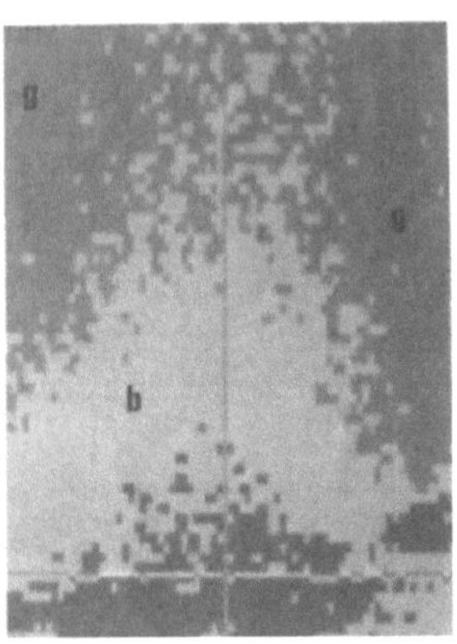


Fig. 11. Thermogram of the base of a column before the start of the excitation

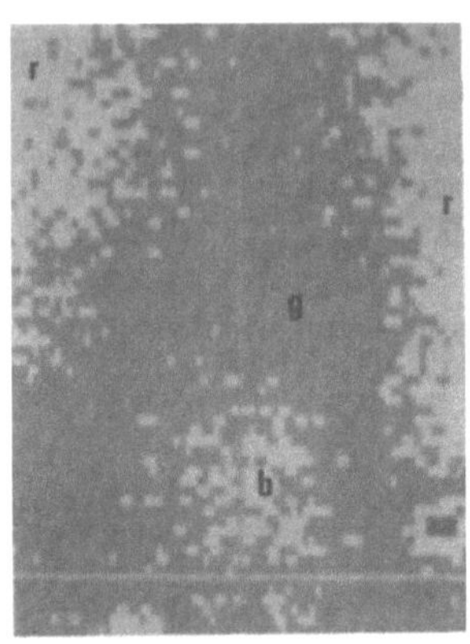


Fig. 12. Thermogram at the same location recorded during loading

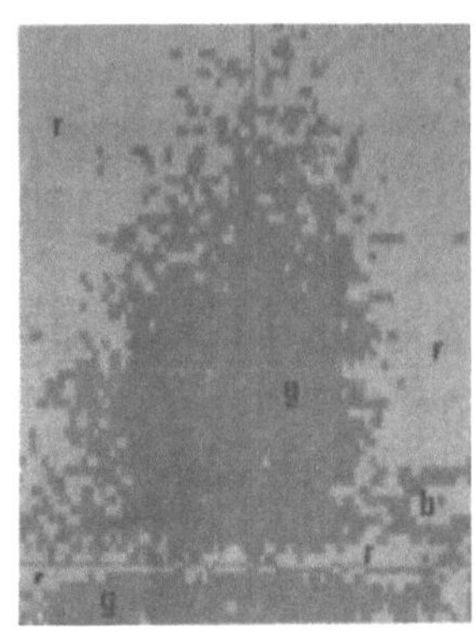


Fig. 13. Thermogram at the same location recorded before cracks are visible.

[5] Stokes, R.J., Fracture of Ceramics, Proc. 4th Symp. on Fundamental Phenomena in the Materials Sciences, Boston Mas. USA. Jan. 31-Feb. 1. 151-175, 1966.

[6] Kotsovos, M.D., Newman, J.B., Generalized stress-strain relationsphips for concrete, Journal Eng. Mech. Div., ASCE, vol. 194, n° EM4, pp. 845-856, 1978.

[7] Bui, H.D., Ehrlacher, A., Nguyen, Q.S., Etude expérimentale de la dissipation dans la propagation de fissure par thermographie infrarouge, Comptes-Rendus Ac. Sci., 293, série II, pp. 1015-1018, 1981.

[8] Reifsnider, K.L., Henneke, E.G. and Stinchcomb, W.W., The mechanics of vibrothermography, Mechanics of Nondestructive Testing ed. by Stinchcomb, W.W., 249-276, 1980.

[9] Luong, M.P., Characteristic threshold and infrared vibrothermography of sand, Geotechnical Testing GTJOD5, vol. 9, n° 2, 80-86, June 1986).

[10] Luong, M.P., Vibrothermographie infrarouge d'un béton endommagé, C.R.A.S. 301 série II, n° 7, Paris 459-464, 1985.

[11] Nguyen, Q.S., Thermodynamique des milieux continus, Cours ENPC DEA, Paris, 1985.

[12] Luong, M.P., Infrared Vibrothermography of plain concrete, magnetic sound 16 mm film, video Umatic and VHS, PAL-SECAM Systems edited by IMAGICIEL, 1984, Ecole Polytechnique, 91128 Palaiseau, France.

SEM/RILEM International Conference
on Fracture of Concrete and Rock
Houston, Texas, June 1987
Editors: S.P. Shah, S.E. Swartz

# STUDIES ON CRACK PROPAGATION RESISTANCE OF ROCKS BASED ON HYDROFRAC DATA OF LARGE SPECIMENS

H. Abé, K. Hayashi, T. Hashida

## ABSTRACT

The crack propagation resistance of rock, which is independent of the size of the crack for large cracks, was determined by analyzing results of the hydraulic fracturing experiments using the 10 m sized granite specimen at a quarry in Fukushima Prefecture, Japan. The method of analysis was the modified BEM, where the crack was assumed to be expanding as a semi-elliptical crack such that the stress intensity factor was approximately constant along its periphery. It was found that the crack propagation resistance of the granite being independent of crack size is 1.93 $MPa \cdot m^{1/2}$.

## INTRODUCTION

For the design of subsurface cracks by means of hydraulic fracturing, the evaluation of the size of the crack should be closely related to rock fracture mechanics. In the actual hydraulic fracturing treatments, the process of the crack growth is difficult to be observed directly. It is therefore necessary to predict the crack propagation resistance of rock from small scale laboratory experiments. However, the crack propagation resistance obtained from small specimens generally depends significantly on the specimen size.

Hydraulic fracturing experiments using a large rock specimen (10.4m × 9.3m × 8.8m, Iidate granite) have been performed at a quarry in Fukushima Prefecture, Japan [1]. The experiments using such large sized specimen may help to bridge the significant gap between the behaviors of small scale laboratory specimens and in the actual subsurface rock mass.

The crack propagation resistance curve was first determined by Hashida [2]. He found that the stress intensity factor $\bar{K}_I$ increased with the crack length $\Delta a$ when $\Delta a$ was small and $\bar{K}_I$ was almost independent of $\Delta a$ for large values of $\Delta a$ ($\Delta a \geq 30$ cm). However, $\bar{K}_I$ values obtained from the experiment were based on the dyed crack geometry, for which $\bar{K}_I$ varied along the periphery of the crack and the fluid pressure was non-uniform on the crack surface.

The fluid pressure is approximately uniform on the crack surface when the crack is expanding with small propagation rate. The stress intensity factor may essentially be constant along the periphery of the crack. Taking these considerations into account, a new method will be proposed to obtain the crack propagation resistance more accurately by using the modified BEM for the analysis of a pair of semi-elliptical cracks around the borehole.

## STRESS INTENSITY FACTOR OF TWO SEMI-ELLIPTICAL SURFACE CRACKS ON A CYLINDRICAL CAVITY

Three-dimensional surface cracks are generally approximated by semi-elliptical surface cracks so that the cracks induced on borehole surface by hydraulic fracturing are approximated by such cracks here.

Consider two semi-elliptical surface cracks on a cylindrical cavity in an infinite elastic body subjected to a constant pressure $-\bar{P}$ acting on the crack surfaces and on the surface of the cavity. The two cracks are lying in a plane symmetrically with respect to the axis of the cylindrical cavity (Fig. 1). A Cartesian coordinate system ($\xi_1$, $\xi_2$, $\xi_3$) shown in Fig. 1 is used. Throughout this section Greek indices take on values 1, 3 and the usual summation convention is employed.

The stress intensity factor of the present problem is identical with that of the problem in which the surface of the cylindrical cavity is stress free and the crack surfaces are subjected to a pressure $-P^*(x_\alpha)$ =

H. Abé is Professor, T. Hashida is Research Associate, Faculty of Engineering, Tohoku University, and K. Hayashi is Associate Professor, Institute of High Speed Mechanics, Tohoku University, Sendai, 980 Japan.

$-\bar{P} - \sigma_{22}^*$, where $\sigma_{22}^*$ is the normal stress on the $\bar{\xi}_2$-plane induced by the constant pressure $-\bar{P}$ acting on the surface of the cylindrical cavity in an infinite elastic body without the cracks. This problem is called the second problem. The second problem can be reduced to solving the following integral equation [3] which is obtained by applying the Betti-Rayleigh reciprocal theorem to the elastic field of the second problem and to that of an infinite elastic body with a cylindrical cavity under a point force acting within the body:

$$- P^*(\bar{x}_\alpha) = \frac{G}{4\pi(1-\nu)}\left[ \int\!\!\!\!\!/_{A_c^+} \bar{F}_\gamma{}^K(\bar{x}_\alpha, \bar{\xi}_\beta)\Delta\bar{u}_{2,\gamma}\, d\bar{\xi}_1 d\bar{\xi}_3 + \int_{A_c^+} \bar{F}_\gamma{}^C(\bar{x}_\alpha, \bar{\xi}_\beta)\Delta\bar{u}_{2,\gamma}\, d\bar{\xi}_1 d\bar{\xi}_3 \right.$$
$$\left. + \int_L \left\{ \bar{F}_1{}^K(\bar{x}_\alpha, \bar{\xi}_\beta) + \bar{F}_1{}^C(\bar{x}_\alpha, \bar{\xi}_\beta)\right\} \Delta\bar{u}_2 d\bar{\xi}_3 \right] \tag{1}$$

where G is shear modulus, $\nu$ Poisson's ratio, $(\ )_{,\gamma} = \partial(\ )/\partial\bar{\xi}_\gamma$ and $\Delta\bar{u}_2$ the crack opening displacement. The slash on the integral sign means that the principal value of the Cauchy integral must be taken. The line L denotes the lines of intersection between the crack surfaces and the surface of the cylindrical cavity and $A_c^+$ denotes the crack surfaces on the plane $\xi_2 = 0^+$. The kernel functions are

$$\bar{F}_\gamma{}^K = \left(1/\sqrt{(\bar{x}_\alpha - \bar{\xi}_\alpha)(\bar{x}_\alpha - \bar{\xi}_\alpha)}\ \right)_{,\gamma} \tag{2}$$

$$\left.\begin{aligned} \bar{F}_1{}^C &= \frac{1}{\pi}\int_0^\infty d\bar{\eta}\ \bar{\eta}\cos\left\{\bar{\eta}(\bar{\xi}_3 - \bar{x}_3)\right\}\sum_{m=0}^\infty e_m\left\{-c_{1m}\frac{\partial}{\partial(\bar{r}\bar{\eta})}K_m(\bar{r}\bar{\eta}) - c_{3m}\frac{m}{\bar{r}\bar{\eta}}K_m(\bar{r}\bar{\eta})\right\} \\ \bar{F}_2{}^C &= \frac{1}{\pi}\int_0^\infty d\bar{\eta}\ \bar{\eta}\sin\left\{\bar{\eta}(\bar{\xi}_3 - \bar{x}_3)\right\}\sum_{m=0}^\infty e_m\left\{c_{1m} + (1-2\nu)c_{2m}\right\}K_m(\bar{r}\bar{\eta}) \end{aligned}\right\} \tag{3}$$

where $r = \sqrt{\bar{\xi}_\alpha\bar{\xi}_\alpha}$, $e_0 = 1/2$, $e_m = 1$ $(m \geq 1)$ and $K_m$ is the modified Bessel function of the second kind. The coefficients $c_{\ell m}$ $(\ell = 1, 2, 3)$ are the solutions of the following linear simultaneous equations:

$$\sum_{k=1}^3 B_{\ell k}c_{km} = D_\ell \qquad (\ell = 1, 2, 3) \tag{4}$$

where

$$B_{\ell 1} = \frac{1}{\bar{\eta}^2}A_{\ell 1}\Big|_{\bar{r}=\bar{c}}\ , \quad B_{\ell 2} = \frac{1}{\bar{\eta}}A_{\ell 2}\Big|_{\bar{r}=\bar{c}}\ , \quad B_{\ell 3} = \frac{1}{\bar{\eta}^2}A_{\ell 3}\Big|_{\bar{r}=\bar{c}} \tag{5}$$

$$\left.\begin{aligned} &A_{11} = \frac{\partial^2}{\partial\bar{r}^2}K_m\ , \quad A_{12} = -2\nu\bar{\eta}K_m + \frac{\partial}{\partial\bar{r}}(\bar{r}\bar{\eta}K_m)\ , \quad A_{13} = \frac{\partial}{\partial\bar{r}}\left(\frac{m}{\bar{r}}K_m\right)\ , \\ &A_{21} = -A_{13}\ , \quad A_{22} = -\frac{m\bar{\eta}}{2}K_m\ , \quad A_{23} = \frac{\bar{\eta}^2}{2}K_m - A_{11}\ , \quad A_{31} = \frac{\partial}{\partial\bar{r}}(-\bar{\eta}K_m)\ , \\ &A_{32} = -\frac{\bar{r}\bar{\eta}^2}{2}K_m + \frac{\partial}{\partial\bar{r}}\left\{-\left(\frac{m}{2} + 2 - 2\nu\right)K_m + \frac{\bar{r}\bar{\eta}}{2}K_{m+1}\right\}\ , \quad A_{33} = -\frac{\bar{\eta}^2}{2}K_{m+1} - \frac{\partial}{\partial\bar{r}}\left(\frac{\bar{\eta}}{2}K_m\right)\ , \\ &A_{41} = \bar{\eta}^2K_m - A_{11}\ , \quad A_{42} = (1-2\nu)\bar{\eta}K_m\ , \quad A_{43} = -A_{13}\ , \\ &A_{51} = -\bar{\eta}^2K_m\ , \quad A_{52} = -(3-2\nu)\bar{\eta}K_m - \frac{\partial}{\partial\bar{r}}(\bar{r}\bar{\eta}K_m)\ , \quad A_{53} = 0\ , \\ &A_{61} = \bar{\eta}^2K_{m+1} + \frac{\partial}{\partial\bar{r}}(\bar{r}K_m)\ , \quad A_{62} = 2(1-\nu)\bar{\eta}K_{m+1} + \frac{\partial}{\partial\bar{r}}\left\{\left(\frac{m}{2} + 2 - 2\nu\right)K_m\right\}\ , \quad A_{63} = \frac{\partial}{\partial\bar{r}}\left(\frac{\bar{\eta}}{2}K_m\right) \end{aligned}\right\} \tag{6}$$

$$\left.\begin{aligned}
&D_1 = d_{1+} + d_{1-}\ , \quad D_2 = d_{2+} - d_{2-}\ , \quad D_3 = d_{3+} + d_{3-} \\
&d_{1+} = (1-2\nu)(K_{m+2}I_{m+2} + K_m I_m) + (K_{m+2} - K_m)\left\{I_m - \frac{\bar{c}\bar{\eta}}{4}(2I_{m+1} + I_{m-1} + I_{m+3})\right\} \\
&d_{2+} = (1-2\nu)K_{m+2}I_{m+2} + \frac{1}{2}(K_{m+2} - K_m)\left\{mI_m + (m+2)I_{m+2}\right\} \\
&d_{3+} = -(1-2\nu)(K_{m+2} + K_m)I_{m+1} + \frac{\bar{c}\bar{\eta}}{2}(K_{m+2} - K_m)(I_m + I_{m+2})
\end{aligned}\right\} \qquad (7)$$

Here $d_{\ell -}$ ($\ell$ = 1, 2, 3) are given by replacing m into -m in $d_{\ell +}$ and $I_m$ is the modified Bessel function of the first kind. The arguments of $I_m$ and $K_m$ are $\bar{c}\bar{\eta}$ and $\bar{\eta}\sqrt{\bar{x}_\alpha \bar{x}_\alpha}$ , respectively.

In order to solve Eq. (1), $\Delta\bar{u}_2$ is set to be

$$\Delta\bar{u}_2 = \frac{\bar{P}(1-\nu)}{G} S(\zeta_1, \zeta_2) g(\zeta_1, \zeta_2) \qquad (8)$$

$$S(\zeta_1, \zeta_2) = \sqrt{1 - \zeta_1^2 - \zeta_2^2}\ , \qquad g(\zeta_1, \zeta_2) = \sum_{K,L=0}^{\infty} C_{KL}\zeta_1^{2K}\zeta_2^{L} \qquad (9)$$

$$\zeta_1 = \frac{\bar{c}}{\bar{a}}\frac{\bar{\xi}_3}{\bar{c}}\ , \qquad \zeta_2 = \begin{cases} \dfrac{\bar{c}}{\bar{b}}\dfrac{\bar{\xi}_1 - \bar{c}}{\bar{c}} & (\bar{\xi}_1 > 0) \\[2ex] \dfrac{\bar{c}}{\bar{b}}\dfrac{\bar{\xi}_1 + \bar{c}}{\bar{c}} & (\bar{\xi}_1 < 0) \end{cases} \qquad (10)$$

The coefficients $C_{KL}$ are determined by a usual collocation method. The stress intensity factor $\bar{K}_I$ along the crack periphery is evaluated by

$$\frac{\bar{K}_I}{P\sqrt{\pi\bar{b}}} = \frac{1}{2}\sum_{K,L=0}^{\infty} C_{KL}\left(\frac{\bar{x}_3}{\bar{a}}\right)^{2K}\left(\frac{\bar{x}_1 - \bar{c}}{\bar{b}}\right)^{L}\left\{1 + \left(\frac{\bar{x}_1 - \bar{c}}{\bar{b}}\right)\left(\frac{\bar{a}^2}{\bar{b}^2} - 1\right)\right\}^{\frac{1}{4}} \Big/ \sqrt{\frac{\bar{a}\bar{b}}{\bar{c}^2}} \qquad (11)$$

Figure 2 is showing the aspect ratio of the crack which gives the stress intensity factor being almost constant along the crack periphery. The maximum difference to the average value is less than 0.8 %. The magnitude of the stress intensity factor at the deepest point of the crack with the aspect ratio given in Fig. 2 is shown in Fig. 3 with respect to the crack depth $\bar{b}/\bar{c}$.

## CRACK EXTENSION DURING HYDRAULIC FRACTURING OPERATION

The test rock was a granite. The rock specimen, 10.4m × 9.3m × 8.8m was quarried as illustrated in Fig. 4. It is simply called 10m sized specimen. The rock has a microcrack fabric typical of granites. Five vertical boreholes, 4.8 cm in diameter (= $2\bar{c}$) were drilled.

Seven hydraulic fracturing experiments were conducted by using straddle packers. In order to analyze the crack extension behavior, three pairs of straddle packers with different pressurization intervals were utilized to extend the crack step by step in such a way that the subsequent injection interval enclosed the crack created at the preceding stage. The pressurization intervals of the packers with small, medium and large intervals (simply called as small, medium and large packers) were 12.5 cm, 73 cm and 203 cm respectively. Dyed water was used as fracturing fluid to delineate the shape of the hydraulically induced crack. During the fracturing treatments the borehole pressure was recorded in terms of time. A typical crack created in the 10m sized specimen is shown in Fig. 5. After the fracturings, the rock specimen was cut along the crack plane and the crack surfaces were examined. Every hydraulically created crack was propagated along a rift plane. As a predictive test, laboratory fracture toughness tests of 3 inch compact tension specimens were performed to measure the crack propagation resistance curve.

The process of the crack growth is shown schematically in Fig. 6. The dyed crack geometries obtained from the experiments using the small and medium packers are Curves ① and ⑤ respectively. Each of them is obtained at the moment when the fluid gets over the straddle packers. Therefore the fluid pressure is non-uniform on the crack surface and the stress intensity factor $\bar{K}_I$ varies along the periphery of the crack. Similarly, the relation between the behaviors of the medium and large packers can also be shown by Fig. 6, if

① and ⑤ correspond to the observed curves for the medium and large packers respectively.

The crack expanding under the uniform pressure is considered. During the crack expansion the stress intensity factor $\bar{K}_I$ should be essentially uniform along the periphery of the crack. Curves ② and ③ in Fig. 6 denote crack shapes where the stress intensity factor is constant along its periphery.

The experimental results for two boreholes, F1 and F3 will be used. Figure 7 shows the relation between borehole fluid pressure $\bar{P}$ and time t for the medium packers at borehole F1. The crack shape is considered to be initially Curve ① and to become Curve ② where the condition of the uniform stress intensity factor along the periphery is just satisfied. The length of the deepest tip of the crack from the borehole is $\bar{b}$ as shown in Fig. 1. It follows from the observed crack shape using the medium packers that $\bar{b}$ = 36.5 cm for the experiment shown in Fig. 7.

The reference time is introduced for time in Fig. 7 when the crack corresponds to Curve ②. The reference time is determined in the following way: For a crack of this size $\bar{K}_I$ is expected to be independent of $\bar{b}$ [2]. First, let a point, for example B, on the $\bar{P}$ - t relation be chosen for the time just stated. The relation of $\bar{K}_I$ versus $\bar{b}$ is obtained. However $\bar{K}_I$ varies with $\bar{b}$ as shown in Fig. 8(a). Several points are chosen and similar computations are performed. As the result of such computations, it is found that the point A should be chosen for the reference time as shown in Fig. 8(c). In this figure Y may correspond to Curve ④ where the fluid begins to get over the straddle packers having medium distance.

In the same way, the point A can be determined based on the $\bar{P}$ - t relations shown in Figs. 9 and 10.

## CRACK PROPAGATION RESISTANCE CURVE

The crack propagation resistance ($\bar{K}_I$ - Δa) curve is obtained from the results of the previous section and the laboratory fracture toughness test using 3 inch compact tension specimens given by Hashida [2], where Δa (≡ $\bar{b}$) is the length of crack growth. In Figure 11, the black circle (●) and triangle (▲) are obtained based on the analyses in the previous section for the medium and large packers, respectively, of borehole F1 and the square (■) are obtained for F3. The solid line in Fig. 11 approximates those results and the laboratory experiments (○). It is found that $\bar{K}_I$ increases with Δa and appears to level off where $\bar{K}_I$ is constant independently of Δa for Δa ≥ 35 cm and this value should be equal to $K_{IC}$ (= 1.93 MPa·m$^{1/2}$). The $K_{IC}$ value thus obtained provides a basic quantity for crack extension in the rock mass by use of linear elastic fracture mechanics.

The dashed line is the crack propagation resistance curve obtained by Hashida [2]. The $\bar{K}_I$ values (Δ), which are close to but below the dashed line, were obtained by him from the calculation based on the observed crack geometry and the corresponding fluid pressure. However, the fluid pressure for the dyed crack geometry was non-uniform on the crack surface and $\bar{K}_I$ varied along the periphery of the crack, as stated in the preceding section. Therefore the dashed line needed some corrections and the solid line just determined is the result after correction.

Hashida proposed a prediction of $K_{IC}$ based on small laboratory rock specimens by the following equation starting from Wnuk's model [4]:

$$\frac{\Delta a}{\Delta} = \ell n \frac{1 - \left(\frac{K_i}{K_{IC}}\right)^2}{1 - \left(\frac{\bar{K}_I}{K_{IC}}\right)^2} \tag{12}$$

where Δ is a constant to characterize the size of the process zone at the crack tip and $K_i$ is the stress intensity factor at the onset of crack growth. The solid line in Fig. 11 can also be expressed approximately by Eq. (12).

All of the cracks found in the 10m sized granite specimen by means of hydraulic fracturing propagated parallel to the rift planes. It is easily expected that the $K_{IC}$ values parallel to and normal to the rift plane are different (Fig. 12). The $K_{IC}$ value normal to the rift plane evaluated from (12) is also shown in the table of Fig. 12, together with $K_i$ obtained from the laboratory specimen having the precracks normal to the rift plane.

## ACKNOWLEDGMENTS

The authors wish to thank Mr. S. Seki for carrying out numerical computations and Mr. T. Shōji for his help in preparing the manuscript. The work presented here was supported by Ministry of Education, Science and Culture under Grant-in-Aid for Special Distinguished Research No. 58065002.

REFERENCES

[1] H. Takahashi, T. Shoji and H. Abé, Recent Progress and Future of "Γ" Project at Tohoku University, Japan, EEC/US Workshop on HDR, Brussels, May 28-30, 1986.

[2] T. Hashida, Fracture Toughness Determination of Rock and Extension Behavior of Hydraulically Created Crack, (in Japanese), Doctoral Thesis, Tohoku University, 1984.

[3] H. Abé, K. Hayashi and S. Takahashi, Stress Intensity Factors for an Embedded Elliptical Crack near a Cylindrical Cavity, (in Japanese), Trans. JSME, A 5-465 (1985), 1350-1358.

[4] M. P. Wnuk, Trans. ASME, J. Appl. Mech., 48 (1981), 500-508.

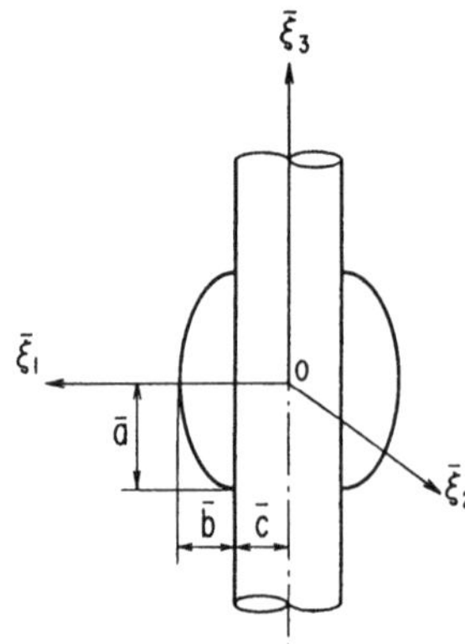


Fig. 1. Semi-elliptical cracks.

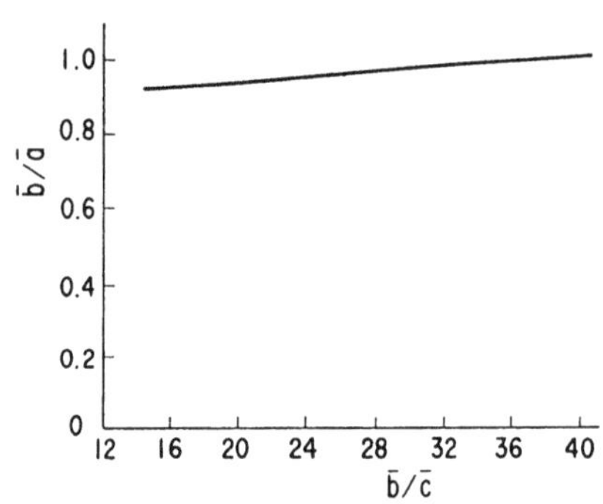


Fig. 2. Aspect ratio giving almost constant stress intensity factor along the crack periphery.

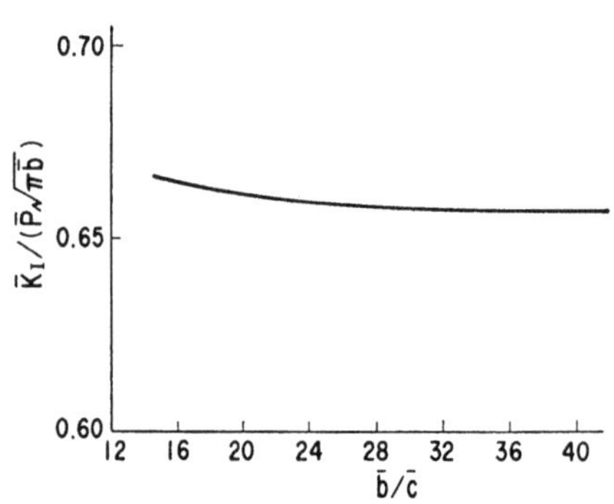


Fig. 3. Stress intensity factor of the crack with the aspect ratio given in Fig. 2.

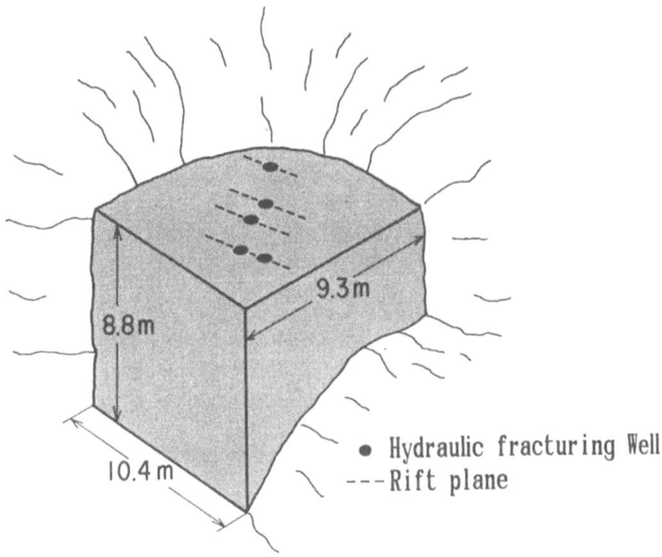


Fig. 4. 10 m sized rock specimen used for hydraulic fracturing experiments.

Fig. 5. Typical crack created in the 10 m sized specimen (borehole diameter = 4.8cm).

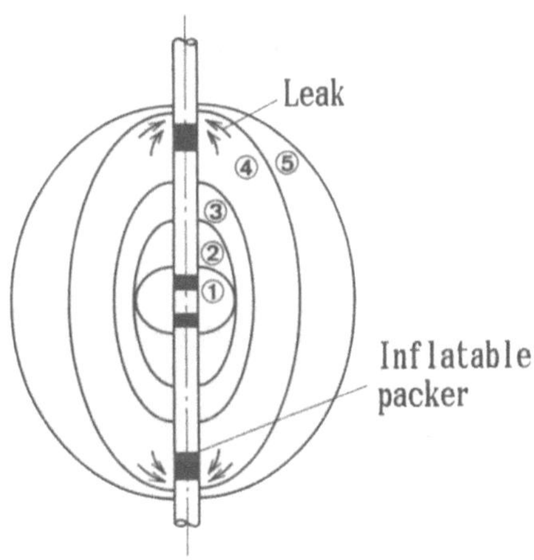


Fig. 6. Process of crack growth.

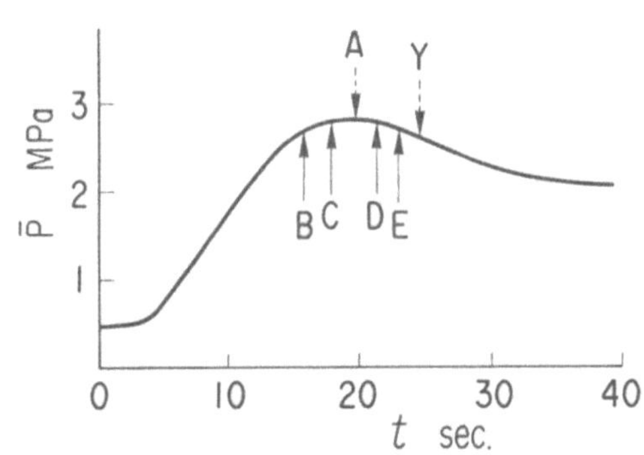


Fig. 7. Borehole pressure $\bar{P}$ versus time t for medium packers (borehole F1).

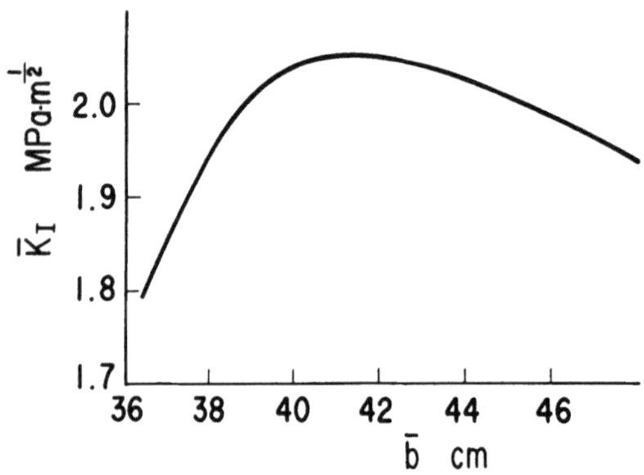


(a) Reference time is chosen for B

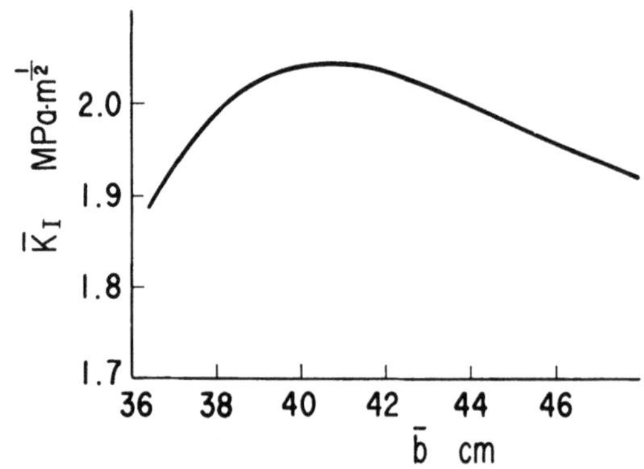


(b) Reference time is chosen for C

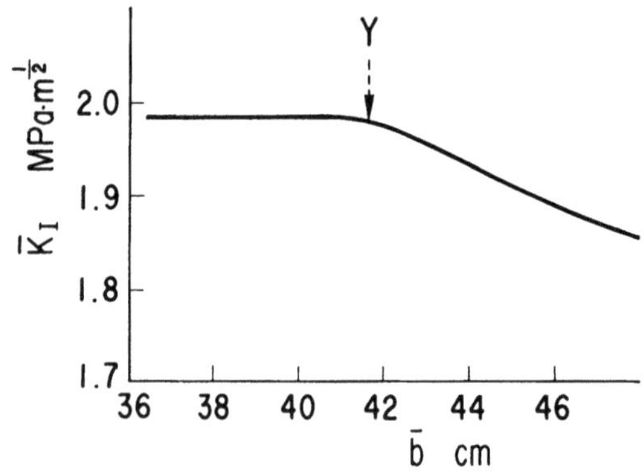


(c) Reference time is chosen for A

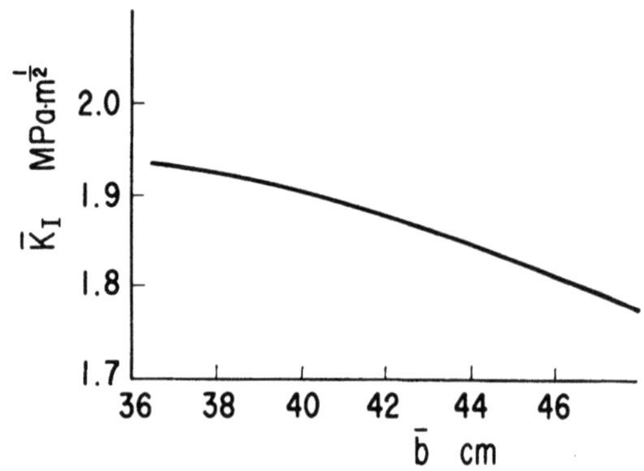


(d) Reference time is chosen for D

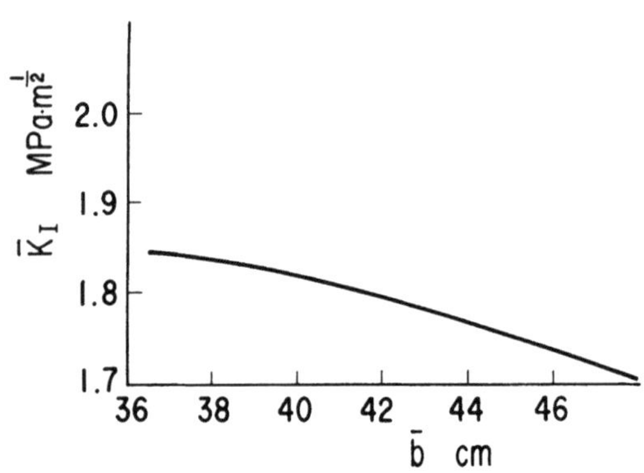


(e) Reference time is chosen for E

Fig. 8. Stress intensity factor $\bar{K}_I$ versus $\bar{b}$ growing from reference time.

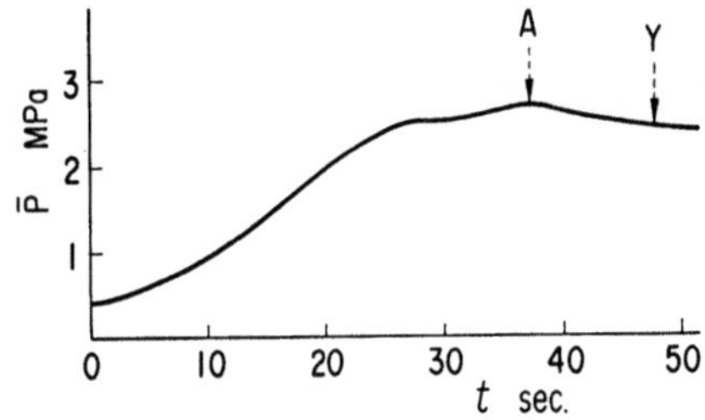


Fig. 9. Borehole pressure $\bar{P}$ versus time t for medium packers (borehole F3).

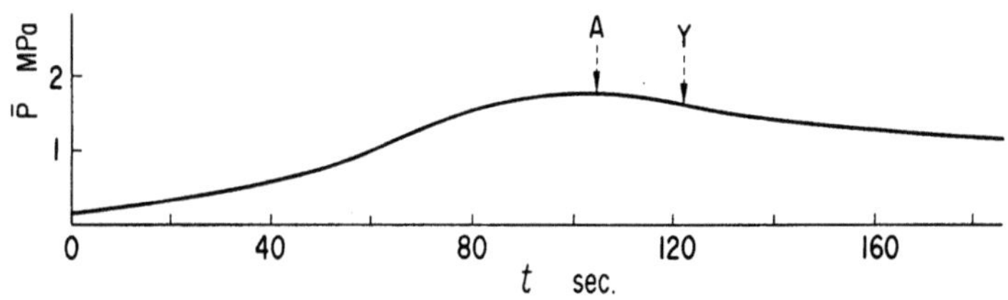


Fig. 10. Borehole pressure $\bar{P}$ versus time t for medium packers (borehole F1).

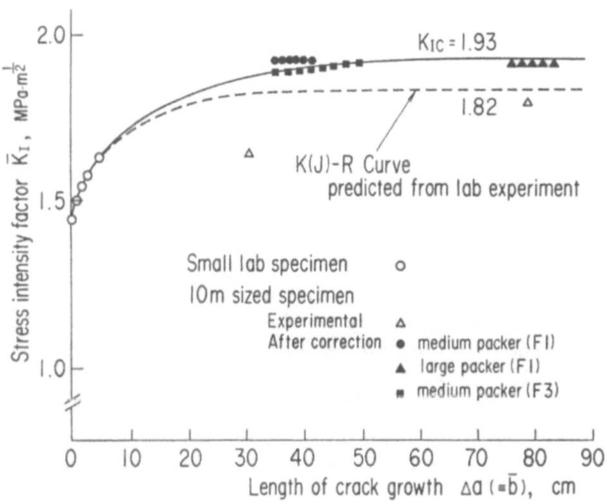


Fig. 11. Crack propagation resistance curve (Iidate granite).

(MPa√m)

| | $K_i$ | $K_{IC}$ |
|---|---|---|
| parallel to rift plane | 1.48 | 1.93 |
| normal to rift plane | 1.87 | 3.42* |

*evaluated from (12)

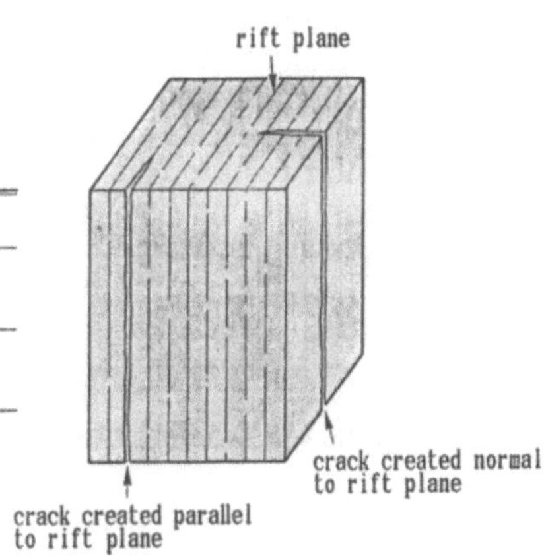


Fig. 12. Fracture toughness parallel to and normal to rift plane.

SEM/RILEM International Conference
on Fracture of Concrete and Rock
Houston, Texas, June 1987
Editors: S.P. Shah, S.E. Swartz

# FRACTURE PROPAGATION IN CONCRETE UNDER COMPLEX STRESS

J.G.M. van Mier [1]

**ABSTRACT:**

The effect of boundary conditions on the softening behaviour of concrete specimens subjected to multiaxial compressive and tensile stress was investigated. When the non-uniformity of deformations in a specimen is taken into account, the shape of the descending branch can be explained with a relatively simple fracture model. Post-peak ductility depends on the degree of heterogeneity of the specimen, which enforces macrocrack growth along a tortuous path. The results indicate that unification of tensile and compressive fracture in a single simple fracture model is possible.

## INTRODUCTION

Present developments in the field of fracture modelling of concrete and the underlying experimental work is mainly concerned with tensile fracture studies of notched specimens (e.g. [1], [2], [3] and [4]). Pure mode I is considered as the most common fracture mode, and modelling the behaviour of specimens subjected to pure mode I loading seems very well possible by using for example the fictitious crack model [3], or the crack band model [5]. The underlying physical model in both approaches is that a relatively large microcrack zone (process zone) preceeds the actual macrocrack tip. In this microcrack zone, a substantial amount of stress can be transfered, and the influence of the process zone on the overall specimen behaviour will be more or less pronounced depending on the actual size of the specimen. The stress distribution over the process zone length is derived directly from a direct tension test, which is assumed to yield basic material properties (see Fig.1). Localization of cracking is taken into account (in x-direction): different measuring lengths yield different post-peak response. The macrocrack or crack band is always assumed to open uniformly over the specimen's cross-section.

Currently, the experimental and theoretical efforts are extended for mixed mode (combined mode I and II) loadings at several laboratories around the world. In these types of experiments, the crack path is no longer straight (e.g. [6]), and the boundary conditions under which the specimen is loaded may become extremely complicated (see for example the mixed mode equipment which is currently being developed at the Stevin Laboratory of Delft University). From standard compression tests on 'unnotched' specimens, it is quite well known that the experimental environment may have a more or less pronounced influence on the observed specimen response. Especially when multiaxial loadings are applied, the boundary conditions under which the experiment is carried out may have a significant influence on the measured 'material response'.

---

1) Stevin Laboratory, Delft University of Technology, Stevinweg 1, 2628 CN Delft, The Netherlands.

In this paper some results of fracture tests using 'unnotched' specimens (cubes d= 100 mm), subjected to uni- and multiaxial tensile and compressive stress are discussed. The results were obtained with a servo-controlled 'true triaxial' device developed at Eindhoven University of Technology [7], [8].

Similarities between tensile and compressive fracture are discussed, and the influence of the experimental boundary conditions on the load-displacement behaviour of a plate subjected to direct tension is shown in a numerical simulation. The results suggest that the physical explanation for softening may be different than previously assumed in the widely accepted process zone models. Furthermore, unification of tensile and compressive behaviour in a single theory seems possible.

## EXPERIMENTAL OBSERVATIONS

Only some limited experimental results are reported, for detailed information about the experimental technique, the reader is refered to [7] and [8]. Similarities with observations from experiments carried out by others are also discussed.

*(1) Uniaxial compression*

When concrete prisms with varying height (h = 50, 100 and 200 mm) and constant cross-sectional area (A = 100 x 100 $mm^2$) were loaded in uniaxial compression between brushes, non-homogeneous deformations were measured as soon as the overall stress-strain curve described a descending branch. Strains measured at the specimens surface showed unloading, whereas the average strain, which was derived from the relative displacement between the opposite brushes, increased continuously. Some results of uniaxial loaded cubes are shown in Fig.2.

Moreover, by comparing the post-peak (nominal) stress - displacement response of uniaxially loaded prisms of different slenderness, as mentioned above, it was concluded that fracture in uniaxial compression is a localized phenomenon. A comparison between overall stress-strain curves and post-peak stress-displacement relations is shown in Fig.3.

The uniaxial compression tests were carried out with the lower platen effectively fixed, while the upper platen could rotate freely. The shear type of failure that was observed (see Fig.3.a) seemed to be induced by these boundary conditions. Furthermore, the effect of a rotating end platen on the propagation of the shear fracture plane may be accelerated by testing a specimen with an intrinsic weakness on one side. This may occur, for example, when a specimen is loaded perpendicular to the direction of casting (see Fig.2). The stength and deformational characteristics will be different at the upper side of the specimen (when casted), from those at the lower side.

Experiments by Kotsovos [9] revealed that the slope of the shear plane is not an intrinsic material property, but rather a direct result of the 'boundary shear' between the loading platen and the specimen. The inclination of the shear plane decreased when the friction between loading platen and specimen decreased (Fig.4).

*(2) Multiaxial compression*

Stable displacement controlled tests on cubes of size d = 100 mm showed that fracture in multiaxial compression is a localized phenomenon, as long as the confining pressure is such that the brittle-ductile transition is not exceeded. The localized shear band developed gradually through the specimens cross-section while the overall stress-displacement diagram described a descending branch (see Fig.5). Consequently, non-homogeneous stress and strain distributions are present in the specimen. Similar results were obtained by Hallbauer et.al. [10] for quartzite specimens.

When loading was applied to a specimen (along the $x_3$ -axis, after it was loaded to a certain point in the descending branch in a different direction (e.g. along the $x_1$ -axis), behaviour as depicted in Fig.6 was observed. In such 'rotation experiments' the descending branch of the second loading in the $\sigma_3$-direction seemed merely an extension of the softening branch measured in the first loading when the lateral confinement was symmetric ($\sigma_c$ = *const.* = $-1$ N/ mm$^2$ in this particular test, Fig.6.a). When the confinement was not symmetric, however, the same fracture load was measured and the inclination of the descending branch was the same in both the first and second loading (Fig.6.b).

*(3) Uniaxial tension*

In uniaxial tension, also a stable descending branch was measured, as shown in Fig.7. In some of the experiments, the descending branch was smooth (e.g. test 9B2-6 in Fig.7), whereas in other experiments a pronounced plateau was measured (see test 9B1-6). Additional deformation measurements showed that in the latter case the deformation field was non-homogeneous (Fig.7.b. for test 9B1-6). Similar behaviour was observed in experiments carried out on double notched specimens by Cornelissen et.al. [11]. In the latter experiments the boundary conditions were similar as in the cube test: the specimen was loaded between non-rotatable end platens, with the only difference being the special guiding system in the experiments of the Delft group. Experiments by Gopalaratnam & Shah [2] and Labuz et.al. [12] showed that differences between the response of notched and unnotched plates is not very large. The notched specimens showed a somewhat larger curvature before peak, yet the peak stress and the shape of the descending branch were almost identical (cf. Fig.5 in [12]).

Interesting results were obtained by Labuz et.al. [12] in a series of uniaxial tension tests on double notched granite plates (see Fig.8). In these experiments the optical crack length was monitored, and as can be seen from Fig.8, the 'macrocrack' gradually traversed the specimens cross-section during the steep part of the descending branch. The boundary conditions in these latter experiments were such that the loading platens could rotate freely in the plane of the specimen.

The above experimental results all indicate that fracture in multiaxial compression (below the brittle ductile transition) as well as in tension is a localized phenomenon. The only difference between compressive and tensile fracture seems to be the different orientation of the macroscopic fracture plane, whereas of course frictional aspects will be more important in the compressive case (i.e. shear transfer in the inclined crack plane). Yet, it should be realized that the experimental context has a significant influence on the observed post-peak response. The results of Kotsovos are very illustrative in this context (Fig.4). However, also allowable rotations of the loading platen are very important, [13]. The latter aspect will be investigated in the following sections for the case of direct tension.

## FRACTURE LOCALIZATION IN A PLATE SUBJECTED TO UNIAXIAL TENSION

Consider a plate with length $H$, width $W$ and thickness $D$. The plate is double notched at half height and is made of a heterogeneous material: aggregates with a maximum size $d_a$ (it is assumed that $d_a$ << D) are embedded in a matrix with a significant lower stiffness.

A slice of unit thickness ($D$=1) is taken from the double notched plate. The slice will intersect with numerous aggregate particles as indicated in Fig.9.a. The aggregates are assumed to be circular inclusions, having the same diameter over the slice thickness. When external loading is applied, microcracks will develop in regions close to the notches. However, these microcracks will immediately be arrested in the steep stress-gradient near the notches, but also due to the presence of tough aggregate particles. The results of Labuz et.al.[12], and also those of Gopalaratnam & Shah [2], as was mentioned before, indicate increased microcracking in notched specimens when compared to unnotched plates.

At a sufficient high stress level, a macrocrack will start propagating from one of the notches (assume at point A in Fig.9). The first growth of the crack occurs almost undisturbed until a first aggregate particle is met. Due to this initial crack extension, the effective load carrying area is reduced and a decrease of macroscopic stress will be observed (Fig.9.b). The crack is arrested by the stiff aggregate, and when the bond between aggregate and matrix is still intact, the external load probably has to be increased again in order to extend the cracktip to point B (it is assumed that the crack propagates along the interface, rather than through the aggregate). After point B is reached, a more or less spontaneous crackgrowth is possible until another stiffer region is met (aggregate C). Following this pattern, the crack will eventually reach the other side of the specimen. Occasionally crack growth will be accelerated due to the presence of microcracks along the macrocrack path, although recent research by Kachanov [14], indicates that also shielding of the macrocrack may occur due to the presence of microcracks near the macrocrack tip. As a result of this shielding, delayed macrocrack growth will occur.

During crack propagation, the geometry of the specimen is considerably changed: the deformation field is non-uniform, and a 'displacement gradient' will develop over the intact area. The load-displacement curve of the slice is not smooth in the descending branch (Fig.9.b) as a direct result of the arresting effect of the larger aggregates. By averaging the effects of all slices a smooth descending branch will be found as normally is observed in experiments, e.g. [1], [2], [12] and Fig.7.

The rate of crackgrowth, or the displacement $u$ needed to obtain a continuous crack obviously depends on the number and size of the aggregates. When many large aggregates are present, delayed crackgrowth occurs and the crackpath will become increasingly more tortuous. A few large aggregates in a homogeneous matrix will probably lead to a load-displacement relationship with a 'bump' in the softening branch. No increase of external load is needed to pass around an aggregate as was discussed before for the case of a thin slice. The crackfront is arrested only partially by a large aggregate particle, yet is allowed to propagate further in the surrounding matrix material. Only when the inclusion is very large compared to the total cross-sectional area of the specimen, a pronounced plateau may be observed in the overall load-displacement curve.

Experimentally, it has been shown that the aggregate size has a considerable influence on the slope of the descending branch of the load-displacement curve [4], [15]. An increase of aggregate size, while maintaining a constant aggregate proportion by weight, results in a decreasing slope of the softening branch [4]. Peterson reasoned that the increased tortuosity of the crackpath leads to a higher frictional resistance in a crack. In contrast, the new physical model suggests that the increased post-peak ductility with increasing aggregate size may be the result of crack arrest mechanisms in the heterogeneous material structure.

The post-peak response of a direct tension test may be described with a simple linear elastic fracture mechanics model for a single notched specimen (see also in [8]). The reason for this is that due to the heterogeneity of the material crack extension will start from the 'weakest' notch. During macrocrack propagation, the critical stress-intensity factor remains constant, which assumption also was made in the two parameter fracture model of Yenq & Shah [16]. The actual value of the critical stress-intensity factor depends on the pre-peak microcrack process. The relation between the nominal stress and the relative crack-length can be calculated using

$$\frac{\sigma_1}{\sigma_p} = \alpha \,/\, \left(Y\sqrt{\frac{a_1}{W}}\right) \tag{1}$$

where $\sigma_p$ is the peak stress, $\sigma_1$ is the nominal stress for a specimen with a crack length $a_1/W$, and Y is a geometrical factor depending on the specimen configuration and boundary conditions. The factor $\alpha$ depends on the initial notch size $a_0/W$ which triggers macrocrack growth. Experimentally stable crack growth can now only be observed when special measures are taken (e.g. the application of a closed loop servo-control). In figure 10.a, the decrease of nominal stress as a function of the relative crack length is shown for specimens with different initial notch size. The actual size of the initial notch depends on the specimen geometry as well as on the local material structure.

The slope of the curve becomes slightly less steep when the initial notch size is increased. The shape of the diagram seems to correspond with the physically observed tensile softening diagrams for concrete and rock, which may indicate that the proposed mechanism indeed occurs. Yet, the stress - crack length diagram has to be translated to a stress - crackwidth diagram in order to compare with experiments. While not much information is available about the non-uniformity of the displacements in a direct tension test (only in [1] some information is found, yet for the case with non-rotatable end platens only), only a relation between the average displacement over the 'crack-band' and the crack length $a_1/W$ may be derived. Some examples of such relations are shown in Fig.10 for cementitious materials (Fig.b.) and for two types of granite (Fig.c). When the initial notch size is assumed to be $a_0/W = 0.05$, most $a/W - u$ relations appear to be straight lines. More information regarding the derivation of these relations is given in [8].

Of course, this latter fit is an enormous simplification of reality. If the mechanism which was described in the previous section (see Fig.9) is valid, combined mode I and II cracking occurs at the submacroscopic level, and a mathematical description will become extremely difficult for such a 'tortuous' crack geometry.

Yet the simple formulation is in agreement with experimental observations. Fine grained materials generally display a steep descending branch associated with a rather brittle behaviour. The $a/W - u$ relation for such materials runs almost parallel to the $a/W$ – axis, indicating a lack of crack arresters. In Fig.10, the experiments with high $D/d_a$ values display a rather brittle behaviour (cement paste and charcoal granite). The only exception is the concrete tested by Gopalaratnam & Shah [2], but they used crushed limestone aggregates, which are not very effective crack arresters. On the other hand, when the degree of heterogeneity of the specimen increases (by decreasing the $D/d_a$ value), a more ductile behaviour is observed.

Experiments on double notched plates also were carried out by Reinhardt [1], and by Cornelissen et.al. [11]. Yet as mentioned before, the boundary conditions in their experiments (uniform boundary displacement, special attention was given to avoid rotation of the loading platens) were different from those in the experiments by Gopalaratnam & Shah [2] and by Labuz et.al.[12]. Consequently the experession for the SIF will change, and eq.(1) will not apply. This implies that for each combination of geometry and boundary conditions a different formulation will be needed. But this is of course to be expected while softening is a 'structural property' rather than a 'material property'.

## NUMERICAL SIMULATION OF A DIRECT TENSION TEST

More insight in the structural behaviour of a tensile specimen can be obtained from a numerical simulation of an experiment. In Fig.7, the non-uniformity of deformations in a tensile specimen beyond peak was shown, and the overall stress-displacement curve showed a remarkable plateau in the descending branch. Such behaviour also was observed in the tensile experiments on double notched prismatic specimens by Reinhardt [1] and Cornelissen et.al.[11]. In contrast, the results of Labuz et.al.[12] and Gopalaratnam & Shah [2] did not show such a characteristic shape of the softening branch. The only difference between the two sets of experiments seemed to be the boundary conditions to which the specimen was subjected. More specifically, the (allowable) rotations of the end-platen seemed to be the cause for the observed response.

An attempt was made to investigate this behaviour numerically. To this end, the DIANA finite element package [17] was used with the standard smeared crack model developed by Rots et.al.[18]. Basically this crack-model consists of a 2x2 crackmatrix of which only the diagonal terms are non-zero. The first diagonal term describes the relationship between the crack-stress and the crack-strain normal to the crack plane. The crack is smeared over a band with width $h$, in a way similar to the model proposed by Bažant & Oh [5]. The second diagonal term describes the shear stiffness of the crack via a constant shear-retention factor $\beta$.

In the standard crack-model, various shapes of the softening branch can be adopted (see Rots 1985b), but in the current analyses, only the linear softening variant has been applied. In this case the ultimate strain in the tensile softening diagram is determined following,

$$\varepsilon_u = 2\frac{G_f}{f_t . h} \tag{2}$$

For details of the crack model, the reader is refered to Rots et.al. [18]. They showed that localization of cracking can be simulated with the smeared crack model, which aspect makes is usefull for the current analyses as well.

In the numerical simulations, the geometry of the plates as tested by Labuz et.al.[12] has been adopted. These plates of size 280 x 75 mm$^2$, and thickness 13 mm, were double notched at half height. The element discretization is shown in Fig.11. Eight noded isoparametric elements, numerically integrated by means of a nine point Gaussian scheme, were used. The mesh was refined in the region with the notches. The transition from the coarse to the fine mesh was done by using numerically integrated (three point Gaussian scheme) six noded triangular isoparametric elements. The boundary conditions were such that the far edges of the plate displaced parallel to eachother.

The material properties were largely varied. Basically a tensile strength of $f_t$ = 3.0 N/mm$^2$ was used. Experience of Rots.et.al. shows that the crack localized in a band extending over two rows of Gauss-points, and consequently $h$ = 5/3mm in the current analyses. The shear-retention factor was constant and equal to 0.20. In the analyses, the value of $G_f$ was varied substantially, ranging from very low values (3.75 - 10 N/m) to normally accepted values for concrete (50 - 100 N/m).

By decreasing the tensile strength of four elements (effective length 10 mm) near the left notch, non-symmetrical cracking of the plate, as observed in direct tension tests, is simulated. It is not a necessary condition that the perturbation should be situated near one of the notches. This seemed however the most likely location for crack initiation.

In Fig. 12 to 14, some of the results are gathered. In Fig.12, the influence of the fracture energy parameter $G_f$ on the tensile (nominal) stress - displacement diagram is shown. The tensile strength of the four elements near the left notch was equal to 4.35 N/mm$^2$. The influence of the fracture energy on the computational result is very clear from this figure. For low values of $G_f$ (viz. $<$ 25 N/m), the numerical analysis was rather unstable in the softening region. The analyses converged slowly (a modified Newton iterative scheme was used), and the norm energy was not reached in some of the load steps. For the cases with $G_f$ $<$ 25 N/m, the softening branches were almost parallel to one another, and only the peak stress was increased by increasing $G_f$. The softening branches in these three analyses were almost parallel to one another, and only by further increasing $G_f$, the slope of the softening branch was affected. Macrocrack growth occured identical in the analyses with low $G_f$ (see Fig.14), and the descending branch seemed to be merely the result of decreasing the effective load carrying area.

The macrocrack gradually propagated through the specimens cross-section, and non-uniform deformations were calculated (Fig.13). In fact, both Fig.13 and 14 indicate what may have happened in the experiments with non-rotatable end platens. The crack starts at one of the notches and in regions above and below this crack unloading takes place. Some bending will appear in the specimen, and at a certain moment, the crack will jump to the other side of the specimen, leading to 'local waggling' (Fig.14). In fact this is also the impression from the deformation distributions published by Reinhardt (e.g. see Fig.16 in [1]). In the analyses, the characteristic plateau was not found. However, it should be mentioned that the loading (i.e. the displacement increment) was applied to the plate at the upper edge (see Fig.11). In fact, the conditions in the experiments are slightly different, and usually the average displacement measured with two LVDTs near the notches is used for 'test-control'. Probably when the analysis is controlled similar, the convergence will improve [19].

The fracture mechanism in the analyses with high $G_f$ was completely different, but was in agreement with the assumptions of the process zone model (as mentioned in the introduction). Due to the large size of the 'process zone', the complete cross-section of the specimen was cracked (with relatively small 'crack strains'), whereafter the crack band opened uniformly. The numerical solution was very stable in these latter analyses.

The numerical simulations indicate that the widely accepted process zone model does not seem to comply with physical observations. In fact, pre-peak energy dissipation due to 'microcracking' seems to be modelled by applying a 'stress-closing profile' derived from the softening branch measured in a direct tension test. Also the occurence of non-uniform deformation distributions cannot be simulated when using the $G_f$ concept.

## CONCLUDING REMARKS

The notion of non-uniform stress and strain distributions in a specimen loaded in (multiaxial) compression are quite well known for some time (e.g. [7]). In this contribution it is shown that also in tension non-uniformity of deformations occurs. The appearance of non-uniform deformations seems to be a direct concequence of the hetrogeneous nature of the material. Depending on the actual boundary conditions under which an experiment is carried out, typical post-peak response may be observed (e.g. a characteristic plateau in the softening branch when a sufficently long specimen is loaded between non-rotatable end-platens). If indeed the material is responsible for these non-uniform deformations in the softening region, it seems more logical to perform experiments on single notched specimens for studying the propagation of a macrocrack through heterogeneous material structures. In such experiments, the arresting effects of large aggregates might be studied as well.

Based on the results presented in this paper, the following conclusions can be drawn:

1.) Localized fracture modes are observed for all multiaxial loading combinations (including both tensile and compressive components) below the brittle-ductile transition. Differences in the loading combinations are manifested in different properties and orientations of the macroscopic fracture planes.

2.) The macroscopic fracture plane develops non-uniformly through the specimens cross-section, while the overall stress-displacement curve describes a descending branch. This is contradictory to assumpations made in the widely accepted process zone models.

3.) The notion of a fracture process zone for describing softening behaviour as observed in displacement controlled uniaxial and multiaxial tensile and compressive experiments seems not correct. The shape of the softening branch can be explained from the decreasing effective 'load carrying area' during macrocrack propagation.

4.) The physical explanation for delayed crack growth in heterogeneous materials is that at the submacroscopic level combined mode I and mode II cracking occurs. Increased ductility (in the softening region) in specimens containing large aggregates (with higher fracture toughness than the surrounding matrix material), must be explained from the arresting effects of the aggregates, which enforce macroscopic crcak growth along a tortuous path.

**ACKNOWLEDGEMENTS**

The numerical simulations were carried out at the Institute TNO for Building Materials and Structures, Dept. of Concrete Structures, and was supported by a joint grant of the Ministry of Public Works and TNO, which is gratefully acknowledged.
The numerical analyses were carried out on the GOULD S.E.L. 9000 computer system operated by the Software Engineering Department.

**REFERENCES**

[1] Reinhardt, H.W. (1984), "Fracture Mechanics of an Elastic Softening Material like Concrete", *HERON*, Vol. 29, no.2.

[2] Gopalaratnam, V.S. and Shah, S.P. (1985), "Softening Response of Plain Concrete in Direct Tension", *ACI Journal*, Vol. 82, no.3, pp.310-323.

[3] Hillerborg, A., Modéer, M. and Peterson, P.E. (1976), "Analysis of Crack Formation and Crack Growth in Concrete by means of Fracture Mechanics and Finite Elements", *Cement & Concrete Research*, Vol. 6, no.6, pp.773-782.

[4] Peterson, P.E. (1981), "Crack Growth and Development of Fracture Zones in Plain Concrete and Similar Materials", *Report TVBM-1006*, Division of Building Materials, University of Lund, Sweden.

[5] Bažant, Z.P. and Oh, B.H. (1983), "Crack Band Theory for Fracture of Concrete", *Materials and Structures*, RILEM, Vol.16, no.94, pp.155-177.

[6] Area, M. and Ingraffea, A.R. (1982), "Mixed mode Crack Propagation in Mortar and Concrete", *Report no.81-13*, Dept. of Struct. Eng., Cornell University, Ithaca, N.Y.

[7] Van Mier, J.G.M. (1986a), "Multiaxial Strain-Softening of Concrete, part I: Fracture" and "part II: Load-histories", *Materials and Structures*, RILEM, Vol.19, no.111, pp.179-200.

[8] Van Mier, J.G.M. (1986), "Fracture of Concrete under Complex Stress", *HERON*, Vol.31, no.3.

[9] Kotsovos, M.D. (1983), "Effect of Testing Techniques on the Post-Ultimate Behaviour of Concrete in Compression", *Materials and Structures*, RILEM, Vol.16, no.91, pp.3-12.

[10] Hallbauer, D.K., Wagner, H. and Cook, N.G.W. (1973), "Some Observations Concerning the Microscopic and Mechanical Behaviour of Quartzite Specimens in Stiff, Triaxial Compression Tests", *Int.J.Rock Mech.Min.Sci.&Geomech.Abstr.*, Vol. 10, pp.713-726.

[11] Cornelissen, H.A.W., Hordijk, D.A. and Reinhardt, H.W. (1986), " Experimental Determination of Crack Softening Characteristics of Normalweight and Lightweight Concrete", in " *Fracture Mechanics and Structural Aspects of Concrete*" *HERON*, Vol. 31, no.2.

[12] Labuz, J.F., Shah, S.P. and Dowding, C.H. (1985), "Experimental Analysis of Crack Propagation in Granite", *Int.J.Rock Mech.Min.Sci.& Geomech.Abstr.*, Vol. 22, no.2, pp.85-98.

[13] Paterson, M.S. (1978), " *Experimental Rock Deformation - The Brittle Field*" Springer Verlag.

[14] Kachanov, M. (1986), "Elastic Solids with many Cracks and Crack-Damage Interactions", lecture, presented at Delft University of Technology, Delft, The Netherlands, June 25, 1986.

[15] Moavenzadeh, F. and Kuguel, R. (1969), "Fracture of Concrete", *Journal of Materials*, Vol. 4, no.3, pp.497-519.

[16] Jenq, Y.S. and Shah, S.P. (1985), "Two Parameter Fracture Model for Concrete", *J. Eng. Mech.*, ASCE, Vol. 111, no.10, pp.1227-1240.

[17] De Borst, R., Kusters, G.M.A., Nauta, P. and De Witte, F.C. (1983), " DIANA - A Comprehensive but Flexible Finite Element System", *Finite Element Systems Handbook*, 3rd edition, Springer Verlag, Berlin, Germany.

[18] Rots, J.G., Nauta, P., Kusters, G.M.A. and Blaauwendraad, J. (1985), "Smeared Crack Approach and Fracture Localization in Concrete", *HERON*, Vol. 30, no.1.

[19] De Borst, R. (1986), "Non-Linear Analysis of Frictional Materials", *PhD dissertation*, Delft University of Technology, Delft, The Netherlands.

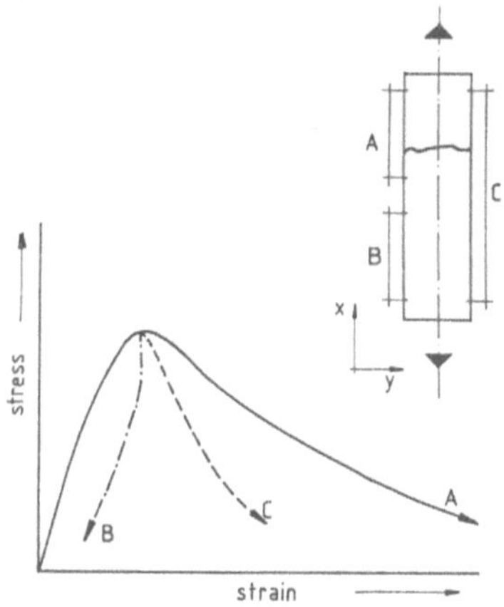


1. Direct tension test.

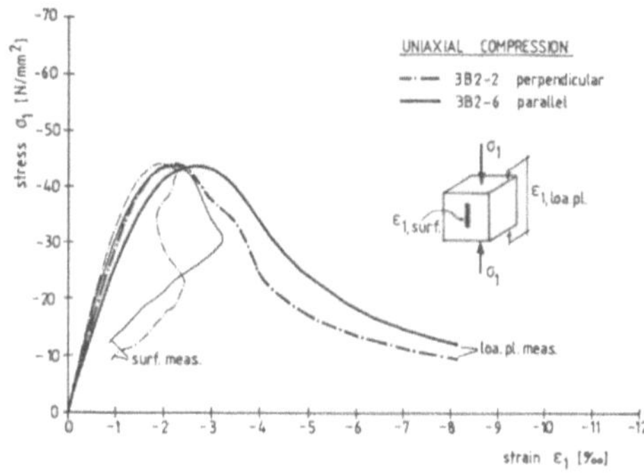


Stress-strain response in uniaxial compression. The surface measurements are the average value of six 60 mm strain gauges mounted at two opposite surfaces of a 100 mm cube parallel to the direction of loading. The loading platen "strain" is derived from the relative travel of the loading platens. The phrases parallel and perpendicular refer to the direction of loading with regard to the direction of casting.

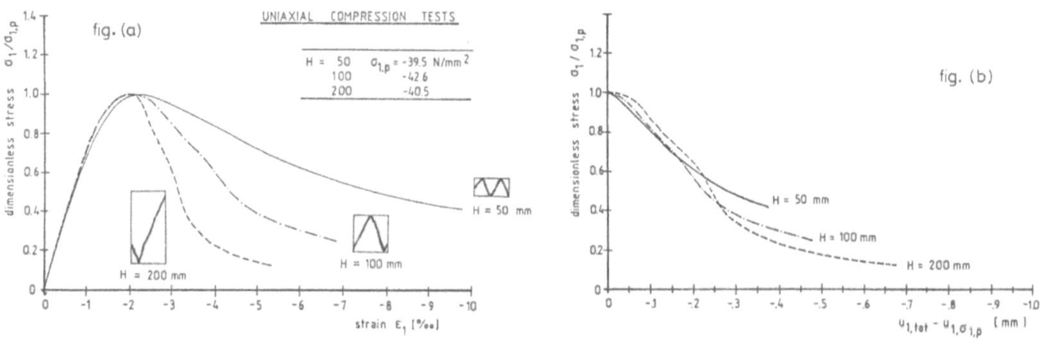


3.a. Stress-strain curves for prisms with different slenderness ratio H/D (H/D = 0.5, 1.0 and 2.0, $A = D^2 =$ 100 x 100 mm²) loaded in uniaxial compression between brushes.
b. Post peak stress-displacement curves for the three tests of figure 3.a.

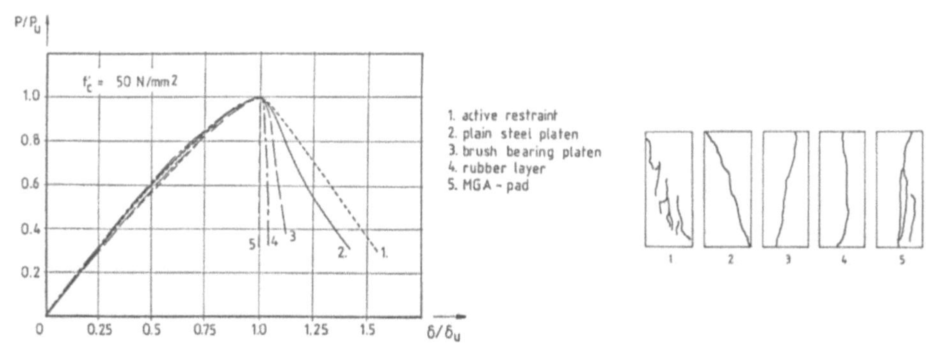


4. Influence of the boundary conditions on the uniaxial compressive stress-displacement behaviour and on the inclination of the macroscopic fracture plane.

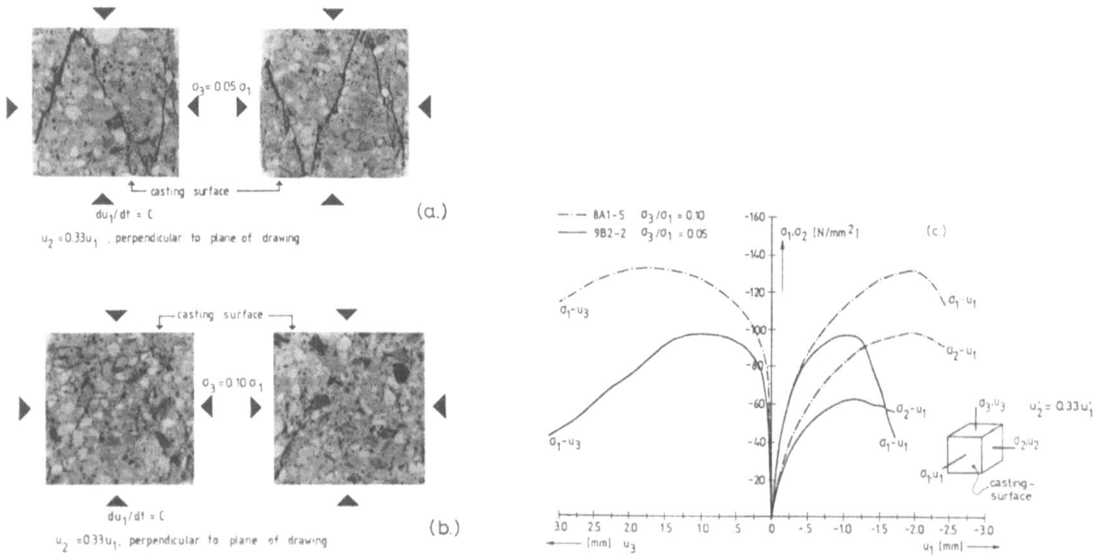


5. Fracture modes and stress-displacement behaviour for two constant displacement ratio experiments. Fig.(a).: Fracture pattern for specimen 9B2-2, $u_1/u_2 =$ -1.0/-0.33, $\sigma_1/\sigma_3 =$ -1.0/-0.05. Fig.(b).: Fracture pattern for specimen 8A1-5, $u_1/u_2 =$ -1.0/-0.33, $\sigma_1/\sigma_3 =$ -1.0/-0.10. Fig.(c).: Stress-displacement curves.

6. Two examples of 'rotation experiments'. In both experiments, the specimen was loaded a second time after it was loaded to some extent in the descending branch in a direction perpendicular to the second loading. In fig.(a), a result is shown of a specimen loaded under symmetric confinement: first loading, $\sigma_2 = \sigma_3 = -1\text{N/mm}^2$, and $du_1 / dt$ = const., in the second loading the 1 and 3 directions were exchanged (see inset).
Fig.(b) shows a result of a specimen loaded under non-symmetric confinement. The compressive componenent in the intermediate direction $\sigma_2$ was taken relative to the major compressive stress (i.e. $\sigma_2 = 0.50^* \sigma_i$, with i=1 in the 1st loading, and i=3 in the 2nd loading).

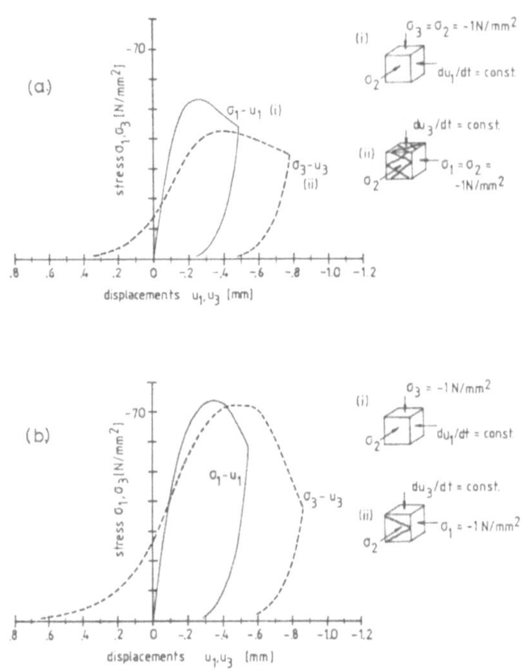

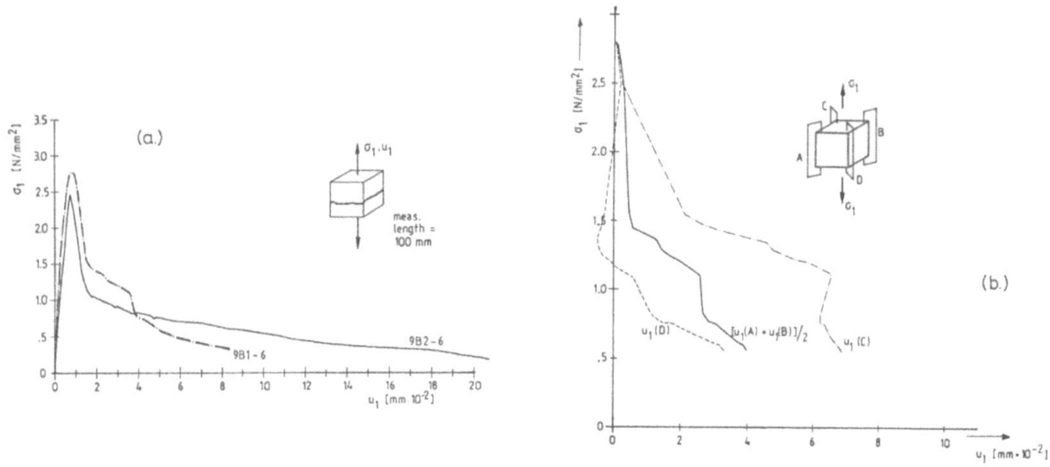


7.a. Two examples of uniaxial tensile stress-displacement curves. The cubical specimens (D = 100 mm) were loaded between brushes.

b. Non-uniform deformations observed in one of the tensile tests of fig.a. (9B1-6). The positions of the LVDTs are indicated. The average displacement measured with transducers A and B was used for test control. The curve appearing in Fig.a for 9B1-6 is the average signal of A and B.

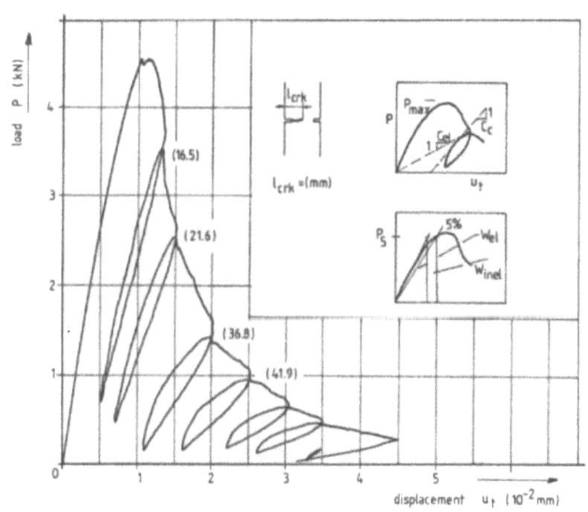


8. Load-displacement diagram obtained in a direct tension test on a double notched charcoal granite plate (after Labuz et.al.(1985)).

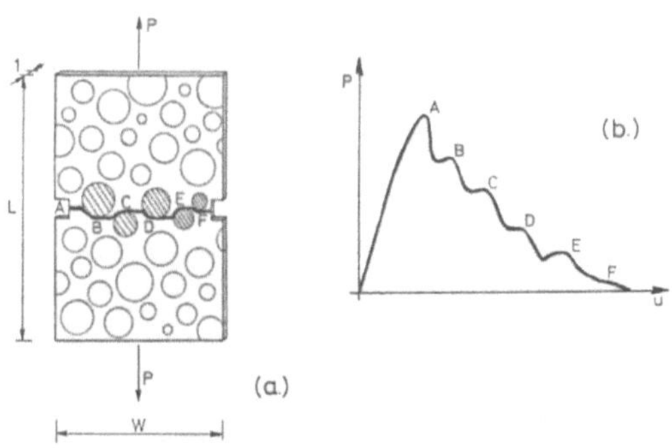


9. Stable macrocrack propagation in a thin slice containing rigid disks: crack-path (a), and load-displacement curve (b).

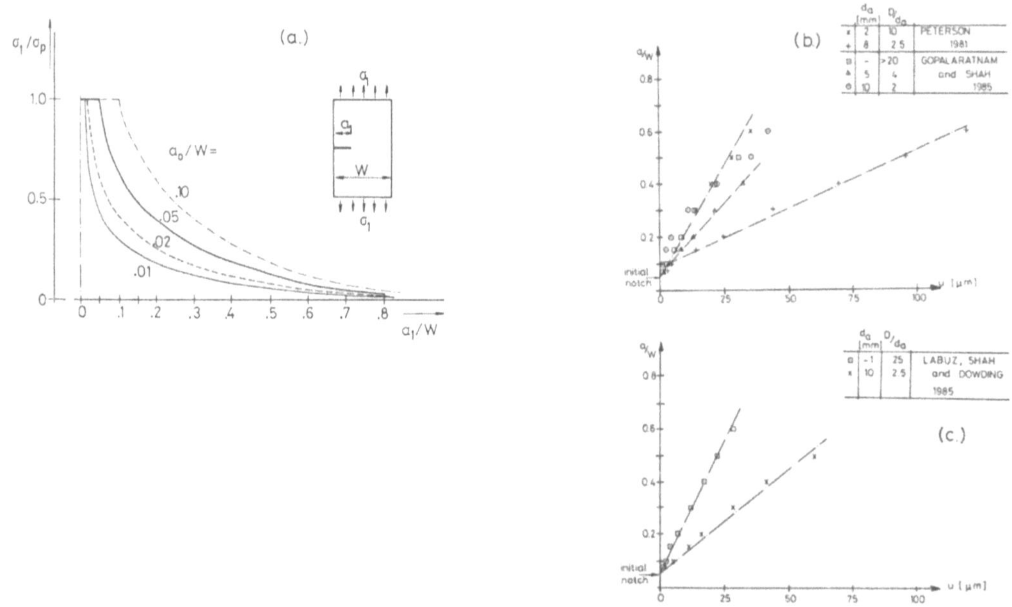


10. (a). Decrease of load carrying capacity with increasing crack length in a single notched specimen subjected to direct tension. Four different cases are shown for different size of the 'initial notch'.
(b). Crack width - crack length relationships for direct tension for concrete, mortar and cement paste. The initial notch size was assumed to be $a_0/W = 0.05$.
(c). Crack width - crack length relationships for direct tension for charcoal and rockville granite. Initial notch size $a_0/W = 0.05$.

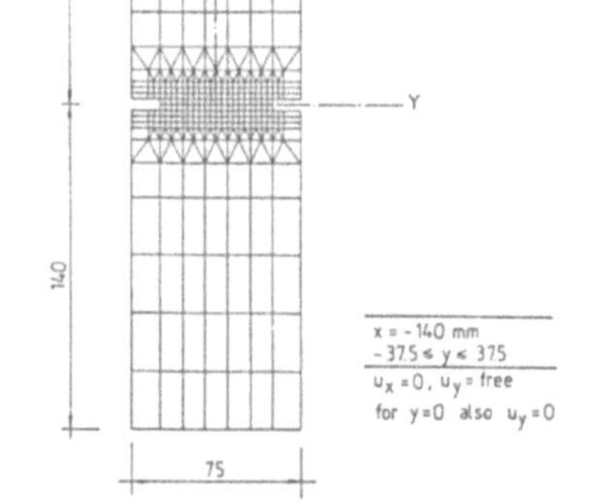


11. Element discretization and boundary conditions.

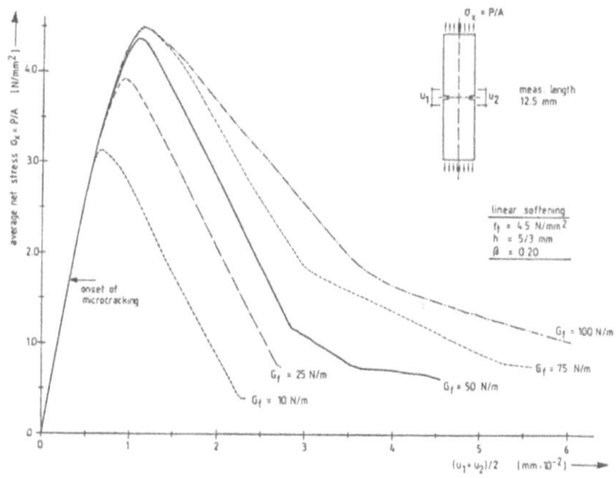


12. Influence of the fracture energy parameter $G_f$ on the load-displacement diagram in direct tension. The displacement is the average of the displacements $u_{x,1}$ and $u_{x,2}$ which were calculated from the relative displacements of two points at mutual distance of 12.5 mm at each side of the plate near the notches as indicated.

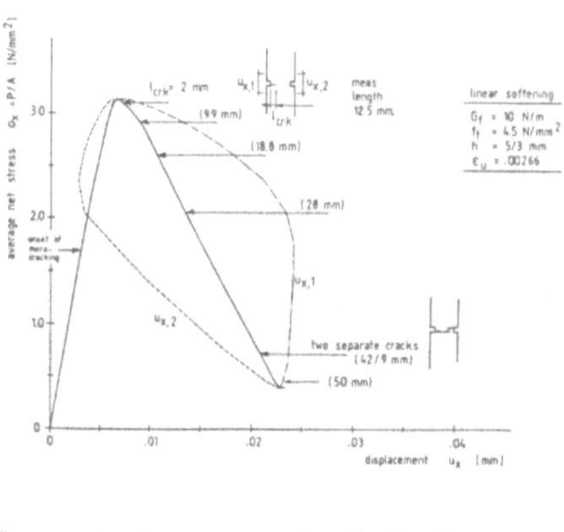


13. Computed load-displacement diagrams for a direct tension test ($G_f$ = 10 N/ mm$^2$, $f_t$ = 4.5 N/ mm$^2$, and $\beta$ = 0.20).

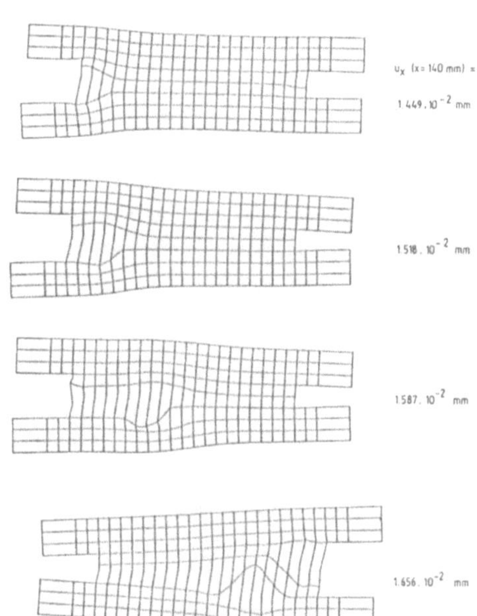


14. Incremental deformations, case $G_f$ = 3.75 N/ m, $f_t$ = 3.0 N/ mm$^2$, and $\beta$ = 0.20. Local 'waggling' occurs due to non-symmetric cracking of the plate.

SEM/RILEM International Conference
on Fracture of Concrete and Rock
Houston, Texas, June 1987
Editors: S.P. Shah, S.E. Swartz

# DELAYED CRACK INSTABILITY OF CONCRETE

Zhang Chunyuan[1]

ABSTRACT

A theory of delayed crack instability of viscoelastic bodies is outlined and is applied to concrete in this paper. Several rheologic models are recommended for the description of concrete. An approach for determining critical load or crack size or critical time for delayed crack instability is proposed.

## INTRODUCTION

The equilibrium of a crack in concrete or rock may become unstable not only instantaneously at t=0, but also at a certain time $t_c$ after the application of the load. Delayed unstable crack propagation may take place in a concrete dam on account of the time-dependent behavior of the material even if $K_I < K_{IC}$ at the initial time. Such time-dependent problems can no longer be treated by elatic or elastic-plastic fracture mechanics. Recently, a theory of delayed crack instability of viscoelastic bodies has been developed [1-4]. This theory is a direct generalization for the Griffith-Irvin approach of linear elastic cracked bodies. It plays the same important role as the linear elastic theory does for elastic bodies. This theory is applied to concrete in this paper.

From the point of view of mechanics, concrete is a heterogeneous, anisotropic, viscoelastic aging material. Its mechanical properties change with time, temperature, humidity and the loading history. The fracture feature of concrete is also very complicated. Because of the large quantities of preexisting defects, flaws and airholes, the application of fracture mechanics to concrete becomes doubtable. However, if we are dealing with a crack, for which the characteristic size is far larger than the characteristic size of the defects and aggregates, then we can still view it as a homogeneous and isotropic one, and fracture mechanics is still applicable. Moreover, it is assumed that the age of concrete is so large that the mechanical properties of concrete does not vary with time anymore. Furthermore, for our purpose, it is more interesting to know what level of load could be applied in order to avoid the instantaneous or delayed catastrophic failure than to know the velocity of the stable crack growth. In fact, the stable quasi-static subcritical crack growth of concrete is not too appreciable, disregarding the failure zone ahead of the crack-tip does not change the critical value as we will see later. Thus, concrete may be viewed as a linear viscoelastic material. As we know, for an idealized linear viscoelastic cracked body, there is no slow stable subcritical crack growth under constant load for those cracks with dK/da > 0. The critical state (the initiation of unstable crack propagation) occurs either instantaneously at t = 0, or at a certain time $t_c$ after the application of the load. It should be noted that until $t = t_c$, the initial size of the crack length remains unchanged, but the displacement on the crack surface increases gradually on account of the time-dependent behavior of the material. As a result, the energy release rate $G_I(t)$ also increases with time. As $G_I(t)$ reaches $G_{Ic}$, the equilibrium of the crack becomes unstable and the crack will begin to

[1]Zhang Chunyuan is a professor in the Mechanical Engineering Department at the Research Institute for Rheology at Xiangtan University, Hunan, People's Republic of China.

propagate with an unbounded velocity in the absence of inertial constraints. If the creep compilance of a material possesses an upper limit, then there exists a lower limit load/crack size below which a crack will not propagate. We shall refer to this load/crack size as the critical load/crack size of delayed crack instability.

Since when $t < t_c$ , the crack problem does not involve a time-dependent boundary region, the viscoelastic correspondence principle is then applicable only for $t < t_c$ . By using this principle the crack-tip fields for concrete can be obtained. According to the equation of the thermodynamic power balance, the main results of LEFM have been generalized to the idealized viscoelastic cracked bodies, especially to concrete.

## CONSTITUTIVE EQUATIONS FOR CONCRETE

The constitutive equations for concrete may be written in the form of the creep integral law

$$e_{ij}(x,t) = J_1 * ds_{ij} , \qquad (1)$$

$$\varepsilon_{kk}(x,t) = J_2 * d\sigma_{kk} ,$$

or in the form of the relaxation integral law

$$s_{ij}(x,t) = G_1 * de_{ij} , \qquad (2)$$

$$\sigma_{kk}(x,t) = G_2 * d\varepsilon_{kk} ,$$

which reduces to the following form for uniaxial stress state

$$\varepsilon(x,t) = \sigma * dD, \qquad \text{or} \qquad \sigma(x,t) = \varepsilon * dE. \qquad (3)$$

where * denotes the Stieltjes convolution defined by

$$[f*dg](x,t) = f,t)g(x,0 + \int_0^t f(x,t - \tau)[\partial g(x,\tau)/\partial\tau]d\tau ,$$

x and t denote the position vector and time respectively, $e_{ij}$, $s_{ij}$ denote the deviatoric components of the strain tensor $\varepsilon_{ij}$ and the stress tensor $\sigma_{ij}$ , which are defined through

$$e_{ij} = \varepsilon_{ij} -(1/3)\delta_{ij}\varepsilon_{kk} , \qquad (4)$$

$$s_{ij} = \sigma_{ij} -(1/3)\delta_{ij}\sigma_{kk} ,$$

$\delta_{ij}$ is Kronecker's delta, $J_1$, $J_2$, D (or $G_1$, $G_2$, E) are the creep (or relaxation) functions in shear, in isotropic compression and in uniaxial tension respectively. $\phi_1(t) = \partial J_1(t)/\partial t$, $\varphi_2(t) = \partial J_2(t)/\partial t$, $\phi(t) = \partial D/\partial t$ are the corresponding creep kernels. $\nu$, $\mu$, k, E stand for Poisson's ratio, the instantaneous shear, bulk and Young's moduli respectively, while $1/(2\mu) = J_1(0)$, , $1/(3k) = J_2(0)$, and $1/E = D(0)$. In order to describe concrete, several rheologic models have been propsed [5,6,7]. For our purpose, the following models are recommendable:

1. Generalized Kelvin Solids

The shear and uniaxial creep functions make the form

$$J_1(t) = J_1(0)[1 + \sum_{i=1}^{m} uW_i(1-e^{-\beta_i t})] , \tag{5}$$

$$D(t) = D(0)[1 + \sum_{i=1}^{n} EC_i(1 - e^{-\lambda_i t})] .$$

when n = 1, which becomes the standard linear body:

$$D\ (t) = D\ (0)[1 + EC\ (1 - e^{-\lambda t})] . \tag{6}$$

2. Solids for which Creep Kernel has a singularity at t = 0

The creep kernel has the form

$$\phi(t) = \lambda e^{-\beta t}/t^{1-\alpha} , \quad 0 < \alpha < 1, \ \lambda > 0, \ \beta > 0 . \tag{7}$$

3. Rabotnov Solids

Rabotnov fractional exponential kernel has the form

$$\phi(t) = \lambda t^{-\alpha} \sum_{n=0}^{\infty} [(-\beta)^n t^{n(1-\alpha)}]/\{\Gamma[(n + 1)(1 - \alpha)]\} = \lambda E_\alpha(-\beta, t) , \tag{8}$$

where $\lambda > 0$, $\beta > 0$ , $0 < \alpha < 1$ , which can be determined from the creep testing, and $\Gamma(\ )$ is the gamma function. When $\alpha = 0$ , $\phi(t) = \lambda e^{-\beta t}$, which becomes the ordinary exponential function. The corresponding uniaxial creep function for Equation (8) is

$$D(t) = D(0)[1 + E \int_0^t \phi(t)dt] = D(0)[1 + E\lambda \int_0^t E_\alpha(-\beta, t)dt]$$

The function $\phi(t)$ and $D(t)$ for Rabotnov solids can be calculated by means of the tables of the appendix in [8].

4. Abel Bodies

The creep function of the models mentioned above are bounded at infinity. Otherwise, Abel kernel is useful:

$$\phi(t) = \lambda[t^{-\alpha}/\Gamma(1 - \alpha)] = \lambda I_\alpha , \quad \lambda > 0, \quad 0 < \alpha\ 1.$$

Figure 1 shows a comparison between the experimental points pertaining to the creep of 150# concrete for a conncrete dam and the creep function described by two different rheologic models, with $D(0)=1/E=4.69 \times 10^{-5}$ 1/MPa. $D(\infty) = 11.18 \times 10^{-5}$ 1/MPa . Curve 1 is the creep function according to Equation (9), with $\alpha = 0.65$, $\beta = 0.0658$ $1/h^{0.35}$ , $\lambda = 0.426 \times 10^{-5}$ $1/(\mathrm{MPa}\ h^{0.35})$ . Curve 2 is the creep function according to Equation (6), with $C = 6.487 \times 10^{-5}$ 1/MPa,$\lambda = 1.875 \times 10^{-4}$ 1/h . We see that if we want to describe the whole creep process, then Rabotnov solid is much better than the standrard linear body. Equation 9 can fit the experimental points fairly well especially at the stage immediately after the applicaion of the load.

## CRACK BORDER STRESS, DISPLACEMENT FIELDS

According to the classical viscoelastic correspondence principle [9], the necessary substitution of material constants are [1]

$$2\mu^* = 1/[pJ_1^*(p)] = 2\mu/(1 + 2\mu W^*_1) , \qquad \underset{\sim}{E}^* = 1/[pD^*(p)] = E/(1 + E\phi^*) . \tag{11}$$

Here $\phi^*(p)$ denotes the Laplace transform of $\phi(t)$. Since the crack border stress, displacement fields of elastic bodies are well known, the corresponding viscoelastic solution can be obtained by an inversion [3].

If the boundary condition takes the form of assigning traction and if the loading processes are such that the stress intensity factors can be written in the form

$$K_i(t) = K_i f_{i\sigma}(t) \qquad (i = I, II, III) \tag{12}$$

then the stress fields in the vicinity of the crack-tip at $\theta = 0$ are

$$\sigma_\theta(t) = [K_I/(2\pi r)^{1/2}]f_{I\sigma}(t) , \qquad \tau_{r\theta}(t) = [K_{II}/2\pi r)^{1/2}]f_{II\sigma}(t). \tag{13}$$

$$\tau_{\theta z}(t) = [K_{III}/2\pi r)^{1/2}]f_{III\sigma}(t)$$

The displacement fields in the vicinity of the crack-tip at $\theta = 0$ are

$$v(t) = v(0)f_{Iu}(t) , \qquad u(t) = u(0)f_{IIu}, \qquad w(t) = w(0)f_{IIIu}(t) \tag{14}$$

where

$$f_{iu}(t) = f_{i\sigma}(t)*L^{-1}(\underset{\sim}{E}'/\underset{\sim}{E}^{*\prime}) = [f_{i\sigma}(t)*dC'(t)]/C'(0) , \qquad (i = I, II) \tag{15}$$

$$f_{IIIu}(t) = f_{III\sigma}(t)*L^{-1}\underset{\sim}{u/u^*}) = [f_{III\sigma}(t)*dJ_1(t)]/J_1(0) ,$$

and $v(0)$, $u(0)$, $w(0)$ have same values as the corresponding elastic solutions. $L^{-1}$ denotes the inverse Laplace transformation. The usual convolution is defined by

$$f_1(t)*f_2(t) = {}_0\int^t f_1(\tau)f_2(t - \tau)d\tau$$

$C'(t)$ denotes the creep function for plane stress $D(t)$ or that for plane strain $C(t)$, which are defined through

$$pC'^*(p) = 1/\underset{\sim}{E}^* = pD^*(p) , \qquad \text{for plane stress,}$$

$$= (1 - \underset{\sim}{\nu}^{*2})/\underset{\sim}{E}^* = pJ_1^*(p)[2 - J_1^*(p)/D^*(p)] , \qquad \text{for plane strain,} \tag{16}$$

so that

$$C'(t) = L^{-1}D^*(p) = D(t), \qquad \text{for plane stress,}$$

$$= L^{-1}\{J_1^*(p)[2 - J_1^*(p)/D^*(p)]\} = C(t), \qquad \text{for plane strain,} \tag{17}$$

when $f_{i\sigma}(t) = U(t)$, Equation (15) reduces to

$$f_{iu}(t) = U(t)*[dC'(t)/C'(0)] = C'(t)/C'(0) , \qquad (i = I, II) \tag{18}$$

$$f_{IIIu}(t) = U(t) *[dJ_1(t)/J_1(0)] = J_1(t)/J_1(0).$$

If the boundary condition takes the form of assigning displacement and if the displacement components at the vicinity of the crack-tip can be written in the form

$$v(t) = vf_{Iu}(t), \qquad u(t) = uf_{IIu}(t), \qquad w(t) = wf_{IIIu}(t), \tag{19}$$

then the stress fields in the vicinity of the crack-tip can be determined by the same formulae as the corresponding elastic solution except that the stress intensity factors must be replaced by

$$K_i(t) = K_i(0)f_{i\alpha}(t) \qquad (i = I, II, III) \tag{20}$$

where

$$\begin{aligned} f_{i\sigma}(t) &= f_{iu}(t)*L^{-1}(\underset{\sim}{E^*/E'}) = [f_{iu}(t)*dR't)]/R'(0) , \qquad (i = I, II) \\ f_{III\sigma}(t) &= f_{IIIu}(t)*L^{-1}(\underset{\sim}{u^*/u}) = [f_{IIIu}(t)*dG_1(t)]/G_1(0) \end{aligned} \tag{21}$$

and $K_i(0)$ has the same value as the corresponding elastic solution. $R'(t)$ denotes the relaxation modulus for plan stress $E(t)$ or the for plane strain $R(t)$, which are defined through

$$\begin{aligned} pR^{*\prime}(p) &= \underset{\sim}{E^{*\prime}} = \underset{\sim}{E^*} = pE^*(p) , && \text{for plane stress} \\ &= \underset{\sim}{E^*}/(1 - \underset{\sim}{v}^{*2} = pG_I^*(p)/[2 - E^*(p)G_I^*(p)] , && \text{for plane strain} \end{aligned} \tag{22}$$

so that

$$\begin{aligned} R'(t) &= L^{-1}E^*(p) = E(t), && \text{for plane stress,} \\ &= L^{-1}\{G_I^*(p)/[2 - E^*(p)/G_I^*(p)]\} = R(t) , && \text{for plane strain,} \end{aligned} \tag{23}$$

when $f_{iu}(t) = U(t)$ Equation (21) reduces to

$$\begin{aligned} f_{i\sigma}(t) &= U(t)*dR'(t)/R'(0) = R'(t)/R'(0) \qquad (i = I, II) \\ f_{III\sigma}(t) &= U(t)*dG_1(t)/G_1(0) = G_1(t)G_1(0) , \end{aligned} \tag{24}$$

CONDITION FOR THE CRITICAL EQUILIBRIUM OF A CRACKED BODY

The thermodynamic power balance law for quasistatic cracked body can be written in the form [10]

$$\dot{W}-\dot{\psi}-\dot{D}- {}_R\int \rho\ \eta\ \theta\ dv \ + {}_{s_f(t)}\int \underset{\sim}{q}\ \ \underset{\sim}{n}da = \dot{\Gamma} , \tag{25}$$

where $W = {}_S\int \sigma_{ij}u_in_jds$ denotes the time rate of work of the applied forces, $\dot{\psi} = {}_R\int \rho\ \dot{\Psi}\ dv$, $\dot{D} = {}_R\int \Lambda\ dv$ and $\Gamma = \frac{D}{Dt}\,{}_{s_f(t)}\int \gamma^*ds$ the time rate of change of free energy, dissipated energy and fracture energy respectively, and $\underset{\sim}{n}$ the exterior unit normal, $\rho$ the mass density, $\eta$ the entropy per unit mass, $\psi$ the free energy per unit mass, $\Lambda$ the rate of internal dissipated energy per unit volume, q the heat flux vector, $s_f(t)$ the newly fractured surfaces, R the material volume excluding those material points lying on $s_f(t)$, $\theta$ the local absolute temperature and $\gamma^*$ the thermodynamic surface energy associated with the fractured surfaces introduced by Eftis et. al. [11]. If the heat flow through $s_f(t)$ can be neglected for the idealized isothermal case, then we can rewrite Equation (25) in the form

$$\dot{W}-\dot{U}-\dot{D} = \dot{\Gamma} , \tag{26}$$

where $\dot{U} = \dot{\psi}$ is the rate of strain energy. Following the approach of Erdogan [12], we can easily show that the rate of change of the crack closure energy E is equal to the time rate of change of the energy of the system

$$\dot{E} = \dot{V}-\dot{U}-\dot{D} , \tag{27}$$

it follows that

$$\dot{E} = \dot{\Gamma} . \tag{28}$$

We see that Equation (26) and Equation (28) are equivalent to each other. As the dissipation term does not appear in Equation (28) directly, it is more convenient for us if the crack-tip fields are already known. From Equation (27), (28), we have

$$\dot{E} = (\partial E/\partial l)(\partial l/\partial t) + (\partial/\partial t)(W-U-D) , \tag{29}$$

where dl/dt is the time rate of the parameter of the crack length. The second term on the right-hand side of Equation (29) is equal to zero, on the basis of the law of conservation of energy for a nonmoving crack, Equations (28) and (29) become

$$\dot{E} = (dE/dl)(dl/dt) = 2\gamma^* dl/dt . \tag{30}$$

Restricting our discussion for crack initiation and assuming that the properties of the material does not vary with time, we can take the fracture toughness $G_{Ic} = 2\gamma^*$ as a constant. For mode I cracks, Equation (30) reduces to

$$G_I(t) = dE/da = G_{Ic} , \tag{31}$$

where a is the crack length.

Using the crack-tip fields derived in the preceeding section (Equations (13), (14) or (19), (20)) and following the approach of Irwin for calculating the energy release rates (i.e., the crack closure energy rates) $G_i(t)$ at any fixed time t for a given virtual crack extension, we arrived at a result in the form

$$G_i(t) = G_i(0)f_{ig}(t), \qquad f_{ig}(t) = f_{i\sigma}(t)f_{iu}(t), \qquad (i = I, II, III) \tag{32}$$

which is different from that of Nuismer [13]. Here $f_{i\sigma}(t)$, $f_{iu}(t)$ can be determined by Equations (12), (15) or (19), (21) according to the form of the boundary condition as mentioned in the preceeding section. Substitution of $G_I(t)$ from Equation (32) into Equation (31) leads to the following fracture criterion

$$G_I(t) = G_I(0)f_{Ig}(t) = G_{Ic}, \tag{33}$$

where $G_I(0) = K_I^2/E'$ is the instantaneous energy release rate. Equation (33) denotes the condition for critical equilibrium of a viscoelastic cracked body.

## CONDITION FOR CRACK INSTABILITY

Crack propagation can be viewed as an irreversible thermodynamic process [11, 14]. Let $\delta s_f$ be the infinitesimal perturbation of the crack surface, and $\delta U$, $\delta u_i$, $\delta D$ are infinitesimal perturbations of the corresponding quantity, then for isothermal cases we have [18]

$$\int_s \sigma_{ij}\delta u_i n_j ds \geqslant \delta U + \delta D + \gamma^*\delta s_f , \tag{34}$$

where the equality sign corresponding to the equilibrium state, and the inequality sign results from the irreversibility of infinitesimal perturbations. The left side is larger than the right side by the value of kinetic energy and dissipations of other type. Let

$$\begin{aligned} P &= U - \int_s \sigma_{ij}\, u_i n_j ds + D , \qquad && (\sigma_{ij}n_j = \text{const}) \\ P &= U + D, && (u_i = \text{const}) \end{aligned} \tag{35}$$

From Equation (34) we see that only processes accompanied by a decrease in function $\Pi$, $\Pi = P + \gamma^* s_f$ can occur spontaneously in the system:

$$\delta\Pi \geqslant 0 . \tag{36}$$

This corresponds to the case of crack instability, the function $\Pi$ must be a maximum,

$$d\Pi = 0 , \qquad d^2\Pi < 0 . \tag{37}$$

For mode I cracks from Equations (31), (35) we see that Equation (37) requiers that

$$G_I(t) = -\partial P/\partial A = G_{Ic}', \tag{38}$$

$$\partial G_I(t)/\partial a = \partial^2 P/\partial a^2 > 0. \tag{39}$$

Equation (38) is the condition for local critical equilibrium of the viscoelastic cracked body, derived earlier. Equation (39) is the condition for local instability of the equilibrium. For brittle or quasi-brittle fracture, restricting our discussion for crack initiation, we may assume that $\partial G_{Ic}/\partial a = 0$. For those cracks with $\partial G_I/\partial a > 0$ in idealized viscoelastic bodies, as soon as the critical state of equilibrium is achieved, the equilibrium of the crack becomes unstable and the crack growth mode becomes dynamic. The point of crack initiation coincides with the point of crack instability.

## DELAYED CRACK INSTABILITY

From Equation (33) we see that for those cracks with $\partial G_I/\partial a > 0$, there are several possible cases:

If $G_I(0) \geqslant G_{Ic}$ , then crack instability occurs instantaneously at $t = 0$. The equality sign corresponds to the critical state of instantaneous crack instability.

If $G_I(0) < G_{Ic}$ and $G_I(t_c) = G_I f_{Ig}(t_c) = G_{Ic}$, (40)

then at a certain time $t_c$ after the application of the load, delayed crack instability will occur.

The critical state for delayed crack instability is determined by the condition

$$G_I(\infty) = G_I f_{Ig}(\infty) = G_{Ic}. \tag{41}$$

If $G_I(\infty) < G_{Ic}$, the crack will not propagate forever. If the rack size/load is assigned, then from (41) the critical load/crack size for delayed rack instability can be determined. If the actual crack size/load is less than this critical value, then the state of delayed crack instability can never be reached. If the crack size and the load are all assigned, then from Equation (40), the time $t_c$ for the occurance of delayed crack instability can be determined. For Maxwell bodies, standard linear bodies and Burgers bodies, the time factors can be found in the tables of [1, 2].

In particular, if the load is assigned and $f_{I\sigma}(t = U(t)$, where $U(t)$ is the unit Heaviside step function, according to the theory of final value and Equations (32), (18), (17), (11), we have

$$\begin{aligned} f_{Ig}(\infty) &= \lim_{t\to\infty} D(t)/D(0) = \lim_{p\to 0} pD^*(p)/D(0) = \lim_{p\to 0} (1 + E\phi^*) \qquad \text{for plane stress,} \\ &= \lim_{t\to\infty} C(t)/C(0) = \lim_{p\to 0} pC^*(p)/C(0) \\ &= \lim_{p\to 0} pJ_1^*(p)[2-J_1^*(p)/\overset{*}{D}(p)[2\mu/(1-\nu)] \\ &= \lim_{p\to 0} (1 + 2\mu\,\phi_1^*)\{1 + [(1+\nu)/(1-\nu)][1-(1 + 2\,\mu\phi_1^*)/(1 + E\phi^*)]\} \qquad \text{for plane strain.} \end{aligned} \tag{42}$$

For such material as concrete, Poisson's ratio $\nu$ is approximately constant. In this case

$$C(t) = (1-\nu^2)D(t), \qquad R(t) = E(t)/(1-\nu^2), \qquad \text{for plane strain} \tag{43}$$

so that the time factor $f_{Ig}(t)$ for plane strain is the same as that of plane stress.

Substitution of $\phi(t)$ from Equations (5), (7), (8) into Equation (42) leads to the following results:

For generalized Kelvin solids

$$f_{Ig}(\infty) = 1 + E\sum_{i=1}^{n} C\ , \tag{44}$$

For solids with creep kernal as Equation (7)

$$f_{Ig}(\infty) = 1 + E\lambda\Gamma(\alpha)/\beta^{\alpha}\ , \tag{45}$$

For Rabotnov solids

$$f_{Ig}(\infty) = 1 + E\lambda/\beta \tag{46}$$

For Abel solids, $f_{Ig}(\infty) \to \infty$ . $G_I(t)$ will increase with time unboundedly because of the existence of viscous flow. In this case, conditional critical load/crack size for delayed crack instability, defined by a given life time, may be useful.

As an example, let us calculate the energy release rate $G_I(t)$ for a double-beam-model crack (Figure 2), assuming that the material is a standard linear body with $\nu$ = const, and $P_i = P_i U(t)$ $(i = 1, 2)$ . From Equations (32), (18), (17), and (6) we have

$$G_I(t) = G_I(0)f_{Ig}(t) = G_I(0)f_{Iu}(t) = G_I(0)[1 + EC(1 - e^{-\lambda t})],$$

where $G_I(0)$ is the same as the energy release rate for the corresponding elasstic solution, which can be determined by [15]

$$G_I(0) = M_A^2/(EJ), \qquad (a/h > 10)$$

$$G_I(0) = [M_A^2/(EJ)][1 + Q_A/(mM_A)]^2 , \qquad (a/h > 1.5,\ 1//h > 1.5)$$

where $m = 1.565/h$, EJ is the flexural rigidity of the beam of unit thickness and $M_A$, $Q_A$ are the bending moment and shear force of a single beam at the cross-section A respectively.

$$W_A = P_1 a + P_2 a/2 , \qquad Q_A = P_1 + P_2.$$

For the case of constant load, $f_{I\sigma}(t) = U(t)$ and constant Poisson's ratio, $\nu$ = const, from Equations (32), (33), (18), (43) we have

$$G_I(t)/G_I(0) = C'(t)/C'(0) = D(t)/D(0) = G_{Ic}/G_I(0) \tag{47}$$

It can be seen that, as long as we obtain the experimental creep curve $D(t)/D(0)$, we can determine $f_{Ig}(\infty)$, so that the critical load/crack size for delayed crack instability graphically by no means of any analytical constitutive equations. On the other hand, if the load and crack size are all assigned, we can also determine the time $t_c$ for the occurance of delayed crack instability graphically as shown in Figure 3.

If the displacement is assigned, and $f_{Iu}(t) = U(t)$, according to the theory of final value and Equations (32), (24), (23), (11), we have

$$\begin{aligned} f_{Ig}(\infty) &= \lim_{t\to\infty} E(t)/E(0) = \lim_{p\to 0} pE^*(p)/E = \lim_{p\to 0} 1/(1 + E\Phi^*) , \qquad \text{for plane stress,} \\ &= \lim_{t\to\infty} R(t)/R(0) = \lim_{p\to 0} pR^*(p)/R(0) \\ &= \lim_{p\to 0} \{pG_1^*(p)/[2-E^*(p)/G_1^*(p)]\}[1 - \nu)/(2u)] \\ &= \lim (1 - \nu)(1 + E\phi^*)/\{(1 + 2\mu\phi_1^*)[(1 - \nu) + E(2\phi^* - \phi_1^*)] , \qquad \text{for plane strain.} \end{aligned} \tag{48}$$

When $\nu$ = const, substitution of $\phi(t)$ from Equations (5), (7), (8) into (48) leads to the following results:

For generalized Kelvin solids

$$f_{Ig}(\infty) = 1/1 + E\sum_{i=1}^{n} C) \tag{49}$$

For solids with creep kernel as Equation (7)

$$f_{Ig}(\infty) = 1/[1 + E\lambda\Gamma(\alpha)/\beta^{\alpha}] \tag{50}$$

For Rabotnov solids

$$f_{Ig}(\infty) = 1/[1 + E\lambda/\beta] \tag{51}$$

For Abel solids, $f_{Ig}(\infty) = 0$. We see that under the condition of constant displacement, energy release rate decreases with time. Accordingly, if instantaneous crack instability does not occur, then delayed crack instability will not occur either. Thus, restricting the displacements on the crack surface by some means is one of the effective measures for the avoidance of delayed crack instability.

## COMPARISON WITH EXPERIMENTS

We now give a comparison of the present theory with experimental results, which were performed on specimens of Solithane 113 (50/50), a polyurethane rubber, by Knauss [16]. In Figure 4, Log $(\sigma/\sigma_c)^2$ is plotted against Log $t_c$, where $t_c$ is the failure time and $\sigma_c$ is determined by

$$\sigma_c^2\pi a/E_\infty = G_{Ic}, \tag{52}$$

where $E_\infty$ is the long time Young's modulus. $\sigma_c$ is in fact the same as our critical stress for delayed crack instability, since Equation (52) can be written in the form

$$\sigma_c^2\pi a/E_\infty = (\sigma_c^2\pi a/E)(D(\infty)/D(0)) = G_1(0)f_{Ig}(\infty) = G_{Ic}. \tag{53}$$

From the definition of critical stress for delayed crack instability, when $\sigma/\sigma_c < 1$, or Log $(\sigma/\sigma_c)^2 < 0$, delayed crack instability will not occur. According to the present theory delayed crack instability will take place only when Log $(\sigma/\sigma_c)^2 > 0$ . From Figure 4 we see that all the experimental points are above the critical line Log $(\sigma/\sigma_c)^2 = 0$, the abscissa. The asymptotic line of the experimental curve coincides with the critical line of the present theory. It can be seen that even though Solithane 113 (50/50) has noticeable subcritical crack extension, so hat we cannot accurately determine the failure time by the present theory of idealized linear viscoelastic fracture mechanics, the critical stress/crack size for delayed crack instability determined by this theory is still applicable. The existence of the small failure zone ahead of the crack-tip, which was ignored in the present model, does not affect the critical value for delayed crack instability.

## MIXED MODE CRACK PROBLEMS

We now adopt the equivalent maximum energy release rate criterion, which has taken account of the biaxial effect and different amounts of energy consumption for different fracture mechanisms, as the fracture criterion for mixed mode cracks. By using the same approach, we can generalize the results of reference [17] to concrete.

We adopt Rabtonov solid as the rheologic model for concrete. If $K_i(t) = K_iU(t)$ (k = I, II), then the average energy release rate at any fixed time t for a given virtual extension $r_o$ along the direction $\theta$ for a I - II mixed mode crack is determined by

$$G_{xd}(\theta, r_o, t) = G(\theta, r_o, 0)[1 + E\lambda \int_0^t E_\alpha(-\beta, t)dt] \tag{54}$$

where $G_{xd}(\phi, r_0, 0)$ has the same expression as the elastic result [17]. According to the equivalent maximum energy release rate criterion, fracture will take place in the direction $\theta_0$, where $G_{xd}(\phi_0, r_0, t)$ is a maximum. It can be seen that for viscoelastic bodies, $\theta_0$ is the same as that for elastic bodies. Substitution of $\theta_0$ into Equation (54), we have

$$G_{xd_{max}} = G(\theta_0, r_0)[1 + E\lambda \int_0^t E_\alpha(-\beta,t)dt] = G_{Ic} . \tag{55}$$

From (55) we can determine the critical value for delayed crack instability.

REFERENCES

[1] Zhang Chunyuan, Science exploration. 1 (1), 29-42 (1981).
[2] Zhang Chunyuan, Science exploration, 1 (2), 35-50 (1981).
[3] Zhang Chunyuan, Science exploration, 2 (3), 93-104 (1982).
[4] Zhang Chunyuan, Proc. of ICF Int. Symp on Fracture Mech., (Beijing) 1038-1044 (1983).
[5] M. Reiner, Lectures on theoretical Rheology, North-Holland, Amsterdam (1960).
[6] T. C. Hansen, Creep and Stress Relaxation of Concrete, Stockholm (1962).
[7] N. Kh. Arutyunyan, Some Problems of Creep Theory, Gostekhizdat, Moscow-Leningard (1952).
[8] Yu N. Rabotnov, Elements of Hereditary Solid Mechanics, Mir Publishers, Moscow (1980).
[9] E. H. Lee, Quart, Appl. Match., 13, 183-190 (1955).
[10] Zhang Chunyuan, On the basic balance laws of fracture mechanics, Proc. of ICFFM Int. Conf. on Fracture and Fracture Mech., (Shanghai) 837-839 (1987).
[11] J. Eftis, H. Liebowitz, Engng. Fracture Mech., 8 (3), 459-485 (1980).
[12] F. Erdogan, In "Fracture," Vol. II, H. Liebowitz, Ed., 524-526, (1968).
[13] R. J. Nuismer, J. Appl. Mech., 41, 631-634 (1974).
[14] Zhang Chunyuan, Science exploration, Journ. of Xiangtan University 2, 44-59, 1987.
[15] Zhang Chunyuan, Acta Mechanica Solida Sinica, 1, 117-122 (1980).
[16] K. G. Knauss, Int. Journ. of Fracture Mech., 6 (1), 7-20, (1982).
[17] Zhang Chunyuan, Engng. Fracture Mech., (Abs.) 16 (1), 94-139 (1979).
[18] Zhang Chunyuan, Journ. of Xiangtan University 4, (1987).

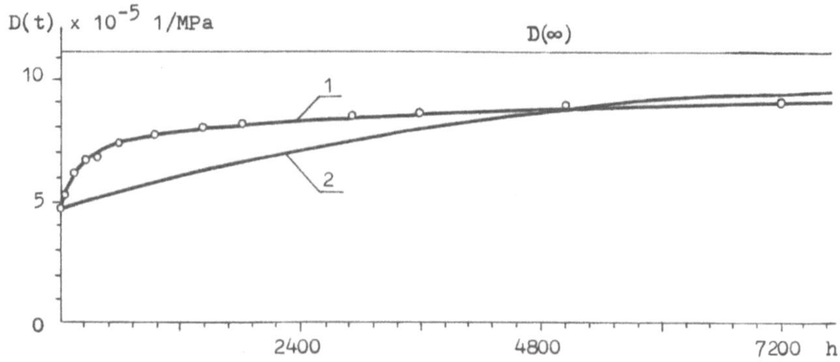


Fig. 1 Comparison of experimental points and Eq.(9) and Eq.(6)

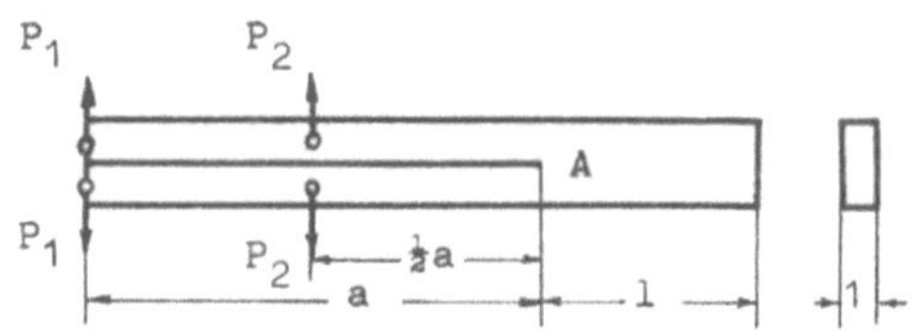


Fig. 2 Double-beam-model crack

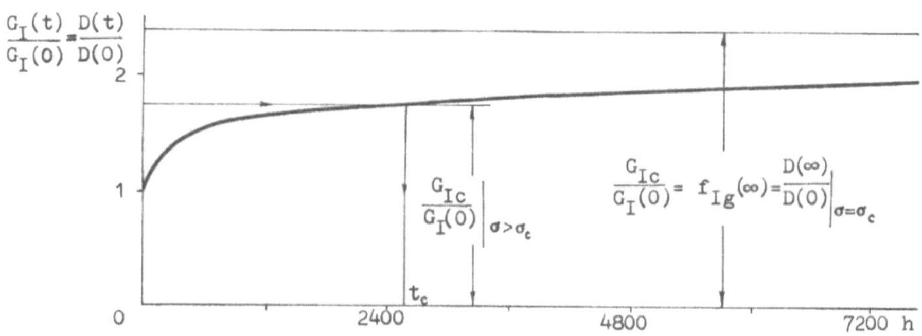


Fig. 3 Determination of $f_{Ig}(\infty)$ and $t_c$

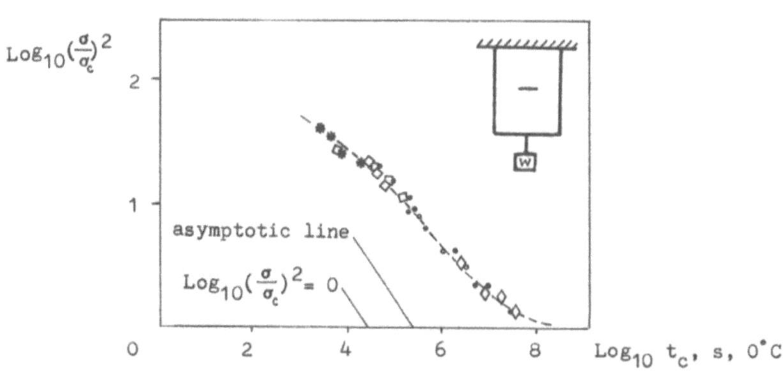


Fig. 4 Comparison with experiments [16]

SEM/RILEM International Conference
on Fracture of Concrete and Rock
Houston, Texas, June 1987
Editors: S.P. Shah, S.E. Swartz



# THEORETICAL ANALYSES OF TEST METHODS

M. Hassanzadeh, A. Hillerborg

## ABSTRACT

By means of a model based on tension softening it is possible to make a theoretical analysis of a test and of evaluation methods used for the determination of fracture mechanics properties of concrete and rock. In this way it is rather easy and inexpensive to check whether a test may be expected to give valid results, e. g. if the size of the specimen is large enough. A negative result of such an analysis, i. e. if it shows that the results cannot be expected to be valid, can be regarded as reliable and should lead to a change in the the test method. A positive result may, on the other hand, be misleading if the assumptions used during the analysis differ too much from the real material properties of the tested material. A number of examples are given regarding the possibility to use pure LEFM, or LEFM with corrections, for the evaluation of tests.

## INTRODUCTION

The fracture properties of a material can be described in many different ways, e. g. by means of pure LEFM, LEFM with certain corrections, R-curves etc. A number of test methods are used or proposed for the determination of the corresponding material properties. Often the evaluation of the test is based on some assumption, which involves an approximation. Thus e. g. tests where the evaluation is based on LEFM formulas are size dependent. It is as a rule assumed that this size dependency can be disregarded if the chosen size of the specimen is larger than a given limit.

In many cases it may be difficult to decide whether a test method, based on certain assumptions, can be expected to produce acceptable results. This is mostly studied by means of tests on different specimens, e. g. specimens of different sizes, in order to check that they give consistent results. This method is time-consuming and expensive, and due to scatter it is often difficult to interprete the results.

As an alternative to tests it is possible to perform a theoretical analysis of the relevance of a test method. One advantage of the theoretical analysis is that it is free from scatter. Another advantage is that a wide area of variations in properties and sizes can easily be covered.

## THE ANALYTICAL MODEL

The analytical model used by the authors is based on the assumption that the material is homogenous and that it is strain-softening. All non-elastic deformations are assumed to be due to the opening of microcracks. The practical implication of this is that non-elastic deformations in one direction have no major influence on the deformations or stresses in the perpendicular directions. Thus plane stress conditions are assumed. This may be regarded as a realistic assumption for tensile failure of concrete and rock.

These assumptions result in a material behaviour where the strain softening in tension is associated with strain localizations to thin bands. A model based on such a material behaviour is known as the fictitious crack model [1, 2, 3, 4]. This model will be used for the examples demonstrated below.

The basic idea of the fictitious crack model is that the deformation properties of a material is divided into one stress-strain curve and one stress-deformation curve, Fig. 1. The stress-strain curve is valid for all the material, and it contains loading as well as unloading branches, but no descending branch. The stress-deformation curve i valid for the zone of strain localization and thus it takes care of the descending branch and the strain softening behaviour. It shows the relation between the stress $\sigma$ and the additional deformation w within the strain localization band. It also contains unloading branches.

---

Research Assistant and Professor, respectively, Div. of Building Materials, Lund Inst. of Tech., Box 118, S-221 00 Lund, Sweden.

Numerical calculations are much simplified if the stress-strain relation is assumed to be linear, Fig. 2. Thus, this assumption is used in all the examples shown below. Variations in the shape of the stress-deformation curve are not difficult to cope with in the numerical calculations. However, only two simplified curves will be used here, one straight line (SL) or a bilinear relation (C) according to Fig. 2. The bilinear relation approximates the shape which has been found to be valid for concrete.

In Fig. 2 $f_t$ is the tensile strength, E the modulus of elasticity and $G_F$ is the fracture energy, i. e. the area below the stress-deformation curve. By means of these material properties, a property called the characteristic length $\ell_{ch}$ can be calculated according to the following formula

$$\ell_{ch} = EG_F/f_t^2$$

By means of the characteristic length it is possible to present analytical results in a dimensionless form by relating all sizes to the characteristic length.

The value of the characteristic length is normally between 100 and 1000 mm for concrete, with an average value of about 400 mm [4,5]. The value for rock is so far not so well known, but for some usual rocks it seems to be of the order 10 to 100 mm, i. e. about one tenth of the value for concrete. This figure is however very uncertain.

THREE-POINT BEND TESTS WITH PURE LEFM EVALUATION

The three-point bend test on a notched beam was first analysed by Petersson [4]. He mainly analysed the evaluation of such a test by means of pure LEFM formulas. His results for a beam with a notch depth equal to 0.4 times the beam depth are shown as full lines in Fig. 3. In this diagram $G_c$ and $K_{Ic}$ are the values which can be expected as a result of a test evaluation according to LEFM with the crack depth equal to the original crack (or notch) depth. The two different assumptions according to Fig. 2 regarding the stress-deformation relation are used, as well as a third assumption with a rectangular shape of this diagram, corresponding to the well known yield assumption of Dugdale. The assumption of Dugdale may be assumed to be approximately valid for steel.

On the assumption that the ascending part of the stress-strain curve is linear it can be demonstrated that $G_F$ is equal to the correct value of $G_C$, determined on a large specimen. Thus the correct value of $K_{IC}^2$ for plane stress conditions is equal to $EG_F$, and $\ell_{ch}$ equal to $(K_{IC}/f_t)^2$. The depth d of the specimen should be at least $5(K_{IC}/f_t)^2$, according to the normal requirement regarding the size of a three-point bend test for $K_{IC}$-determination for steel. It can be seen from Fig. 3 that this requirement is in good agreement with the analytical results for a material following the assumption of Dugdale, as a specimen with $d/\ell_{ch}=5$ in this case will give a value of $K_{IC}$ which is only a few percent too low.

If we wish to perform the same type of test with concrete or rock, represented by the curve marked C in Fig. 2, we have to use a value of $d/\ell_{ch} > 10$ in order to obtain a reasonable accuracy. As $\ell_{ch}$ is about 400 mm for ordinary concrete and probably about 40 mm for rock, this would require beam depths greater than 4 m for concrete and 0.4 m for rock, and still the accuracy is not at all as good as is normally required for steel. Such sizes are not suitable to use in practice. The accuracy of the result will be quite unacceptable, when making use of more moderate sizes, such as d = 400 mm for concrete or 40 mm for rock. The practical conclusion of this is that this type of test and test evaluation is not suitable for concrete or rock.

THREE-POINT BEND TEST, COMPLIANCE-LEFM EVALUATION

One reason why the LEFM evaluation gives too low values for small beams is that a process zone exists in front of the crack. The whole process zone takes part in the energy dissipation, and not only the crack tip, as it is assumed in LEFM. Therefore, some point within the process zone is more representative than the original crack tip. One possibility to calculate such a representative point and a corresponding crack length is to measure the compliance of the specimen at maximum load, and to calculate the crack length which corresponds to this compliance. This calculated crack length is used in the LEFM evaluation instead of the original crack length.

The values which can be expected from such an evaluation can also be theoretically analysed, and the result is shown in dashed line in Fig. 2 for a material with the stress-deformation curve of type C. It can be seen that such an evaluation can be expected to give more accurate results than an evaluation based on the original crack length. However, the values of $G_c$ and $K_{IC}$ are still too low according to this method, so it cannot be recommended unless the demand on accuracy is low.

CHEVRON NOTCHED BEAM WITH PLASTICITY CORRECTION

A proposal is under discussion for the determination of $K_{IC}$ for rock by means of a cylindrical test specimen with a chevron notch, Fig. 4. As the specimen often can be expected to be too small for a simple evauation by means of LEFM, it has been proposed that a "plasticity correction" is made, which is intended to correct for the size of the process zone. In order to determine this plasticity correction the specimen is unloaded to a certain extent and reloaded at least twice after the maximum load has been passed. From the slopes of the unloading-reloading curves in the load-deflection diagram conclusions are drawn regarding the plasticity correction.

This type of test has also been theoretically analysed. In order to be able to perform such an analysis it is necessary to assume the shape of the unloading curve in the stress-deformation diagram. From tests on concrete the general shape of the unloading curve is known, and it can be expected that it will be about the same for rock. Based on this knowledge assumptions have been made regarding the unloading curves.

For this analysis the $\sigma$-w relation for increasing w-values has been assumed to be a straight line. Two different assumptions have been used regarding unloading, see Fig. 5. Assumption I is that w is unchanged during unloading, i. e. a vertical unloading line. Assumption II is that the unloading curve is a straight line following the equation

$$\frac{d\sigma}{dw} = 9f_t \left[\frac{1}{w_1} - \frac{1}{1.2w_c}\right]$$

Analyses have been performed for two cases, viz. $d/\ell_{ch}$=0.5 and $d/\ell_{ch}$=0.2 respectively.The resulting load-deformation curves are shown in Fig. 6. The application of the proposed evaluation method gave values of $K_{IC}/\sqrt{EG_F}$ according to Table 1.

Table 1. Evaluated values of $K_{IC}/\sqrt{EG_F}$

| $d/\ell_{ch}$ | Unloading curve I | Unloading curve II | Compliance evaluation |
|---|---|---|---|
| 0.5 | 1.04 | 2.87 | 0.89 |
| 2.0 | 1.11 | 1.16 | 0.97 |

The correct value of $K_{IC}/\sqrt{EG_F}$ should be equal to 1. From the table, it is evident that the values from the proposed evaluation method are not always so good. It can also be seen that they largely depend on the assumption regarding the unloading curve, particularly for the small specimen. However, there seems to be no real connection between these unloading curves and the values of $K_{IC}$. Therefore the theoretical basis for the evaluation method can be questioned.

The possibility of evaluating $K_{IC}$ by means of the compliance method has also been analysed. The compliance has then been evaluated from the slope of the line from the origin to the maximum point in the load-deflection curve, se Fig. 6. The value of $K_{IC}$ has then been calculated from this compliance and the maximum load. It seems like this simple evaluation method gives better results than the method based on a plasticity correction. It is also independent of the unloading behaviour.

R-CURVES

One way of handling the influence of the size of the process zone is the use of R-curves, i. e. curves which show how the values of $K_{IC}$ and $G_C$ vary with the growth of the crack. It is then assumed that an R-curve is a material property, so that the same R-curve is valid independent of the size of the specimen and the length of the original crack.

The use of R-curves is based on the behaviour of metals at crack growth. Metals show substantial lateral deformations and formation of shear lips as a crack grows. These deformations are small near the original crack tip, but increase as the crack front advances. The lateral deformations involve energy dissipation, which thus increases with increasing crack growth. Threre is a good reason to believe that an R-curve can be accepted as a formal material property for a metal specimen with a constant width (length of the crack front).

It has been suggested that R-curves may be a useful tool also for concrete in order to overcome the problem of size dependency, which for concrete is due to the length of the fracture process zone. The validity of such an approach can also be theoretically analysed. The result of such analyses are shown in Figs. 7-8.

The growth of the crack, i. e. the position of the crack front, has to be measured in the determination of R-curves by means of tests. This position is normally determined either by visual inspection or by compliance measurements. Both these methods can be applied also in the theoretical analysis. A visually observable crack then corresponds to a certain deformation within the process zone.

Figs. 7-8 are based on the assumption that the crack length is determined from compliance measurements, although the use of visual inspection leads to the same general results.

Fig. 7 shows theoretical R-curves based on tests with beams of a constant depth, but with varying notch depths. It shows that the R-curves then turn out to be approximately independent of the notch size. Test made in this way may lead to the conclusion that R-curves can be accepted as material properties, which are not size dependent.

Fig. 8 shows the corresponding theoretical R-curves based on tests with beams with a constant ratio between notch depth and beam depth, but with varying beam depths. Fig. 8 clearly shows that the R-curves are size dependent and that they therefore cannot be regarded as material properties for materials lika concrete and rock.

INFLUENCE OF SIDE-GROOVES

In some types of tests, e. g. the double cantilever beam, side grooves are sometimes used in order to direct the crack growth. Provided that the crack follows the groove, the behaviour is practically the same as if the tensile strength $f_t$ and the fracture energy $G_F$ were reduced in the proportion to the ratio between the width at the groove and the width of the ungrooved specimen. This means that the formal characteristic length $\ell_{ch} = EG_F/f_t^2$, referred to the groove, becomes proportional to the ratio between the specimen width and the width at the groove. The performance of a specimen depends on the ratio $d/\ell_{ch}$, where d is the size of the specimen. A consequence of this is that the performance of a grooved specimen corresponds to that of a smaller specimen without a groove. This has to be taken into account if tests are evaluated by means of LEFM. If e. g. 1 m is supposed to be a sufficient size for LEFM evaluation of a test without a groove, the corresponding size must be 2 m if the ratio between the specimen width and the width at the groove is 2.

THE DOUBLE TORSION TEST

The double torsion test has been theoretically analysed [6]. This analysis shows that ordinary LEFM formulas for evaluation are only acceptable if the specimen size is very large, in the order of 3 m times 0.6 m for ordinary concrete. Thus the double torsion test is not suitable for practical application to concrete if the aim is to determine reliable values of $K_{IC}$ or $G_C$. For pure cement paste acceptable results may be reached with a specimen length of about 0.2.

FURTHER EXAMPLES

A number of different test specimens have been theoretically analysed by Gustafsson [7]. In his report, a more comprehensive analysis and discussion can be found.

DISCUSSION

The theoretical analyses make it possible to draw valuable conclusions regarding the validity of material properties evaluated from a certain test. It can give a warning that a test can be expected to give invalid results. It can also show how changes in material properties influence the validity of the test results.

It can be argued that the theoretical analysis does not tell the truth of the behaviour of the real matereial, at the assumptions are simplified and do not properly represent a real material. However, a valid test method should be valid for all materials which have reasonable and possible properties.

Even if the assumed curves according to Fig. 2 do not exactly represent real concrete or rock, it represents a hypothetical material with properties which are similar to those materials. Thus a test method which is valid for concrete or rock must also be valid for this hypothetical material. The conclusions which have been drawn for the hypothetical material can be regarded as valid also for concrete and rock in so far as they indicate that a test cannot be expected to give valid results.

On the other hand it is not possible to draw the conclusion that a test gives valid results because a theoretical analysis seems to indicate that this is the case. It is possible that the assumed curves, e. g. according to Fig. 2, are not in a sufficiently good accordance with the properties of the real material. An examples of this is that the stress-strain relation of a real material is not quite linear all the way up to the tensile strength. The greater this nonlinearity is, the less valid is the result of a theoretical analysis where linearity is assumed.

It is of course also possible to perform a theoretical analysis of the influence of the nonlinearity on the test results. Such an analysis is numerically more complicated.

It is essential to notice that a theoretical analysis can easily show if a test and evaluation method cannot be expected to give valid results, but that it is much more difficult to show that it can be expected to give valid results.

CONCLUSIONS

All test and evaluation methods which are used for the determination of the material properties used in fracture mechanics of concrete and rock can easily be theoretically analysed for a hypothetical material with properties which approximate those of the real material in question.

If such an analysis shows that the test cannot be expected to give valid results, the test method should not be used, at least not without a thourogh discussion and motivation.

If the analysis indicates that the test may be expected to give valid results, it cannot be taken for granted that this conclusion is correct. The assumptions upon which the analysis is based may be too approximate with regard to the real material. Thus e. g. an assumption regarding a linear stress-strain relation leads to unvalid conclusions if the real material shows a pronounced nonlinearity.

No test method ought to be used before it has been theoretically analysed. Such an analysis is rather inexpensive and it can save much cost for invalid test results.

REFERENCES

1. Hillerborg, A., Modéer, M., Petersson, P. E.: Analysis of crack formation and crack growth in concrete by means of fracture mechanics and finite elements. Cem. and Concr. Res., Vol. 6, pp 773-782, 1976.

2. Hillerborg, A: A model for fracture analysis. Report TVBM-3005, Div. of Building Materials, Lund Inst. of Tech., 1978.

3. Modéer, M.: A fracture mechanics approach to failure analyses of concrete materials. Report TVBM-1001, Div. of Building Materials, Lund Inst. of Tech., 1979.

4. Petersson, P. E.: Crack growth and development of fracture zones in plain concrete and similar materials. Report TVBM-1006, Div. of Building Materials, Lund Inst. of Tech., 1981.

5. Hillerborg, A.: Results of three comparative test series for determining the fracture energy $G_F$ of concrete. Materials and Structures, Vol. 18, No 107, pp 407-413, 1985.

6. Hillerborg, A.: Theoretical analysis of the double torsion test. Cem. and Concr. Res., Vol, 13, pp 69-80, 1983.

7. Gustafsson, P J: Fracture mechanics studies of non-yieldning materials like concrete. Report TVBM-1007, Div. of Building Materials, Lund Inst. of Tech., 1985.

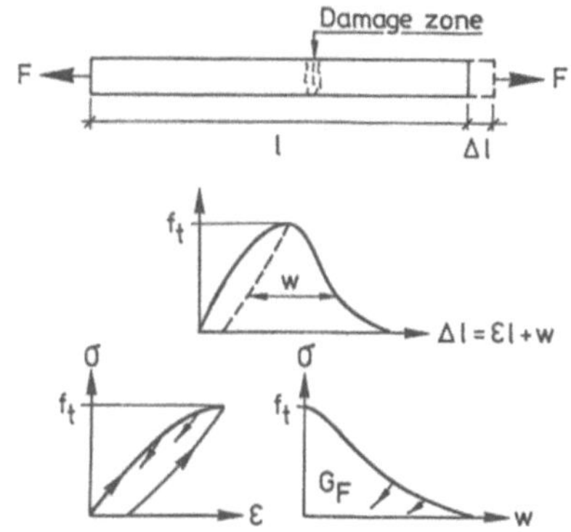


Fig. 1 The division of the tensile behaviour into one stress-strain relation (σ-ε) and one stress-deformation relation (σ-w), where w denotes the additional deformation due to the strain localization in the damage zone (fracture process zone).

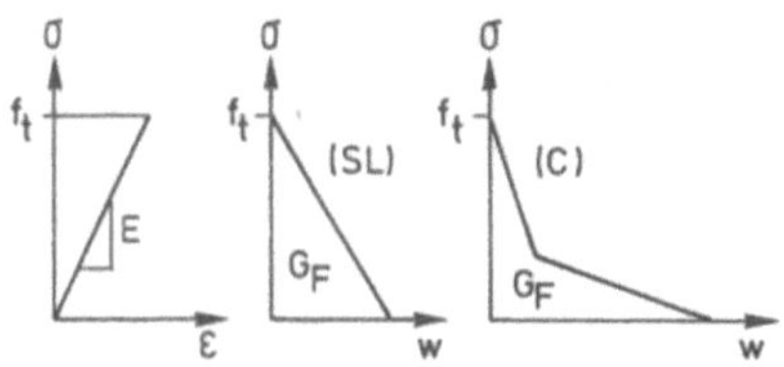


Fig. 2 Simplified assumptions regarding the stress-strain and stress-deformation relations, used in the numerical analyses.

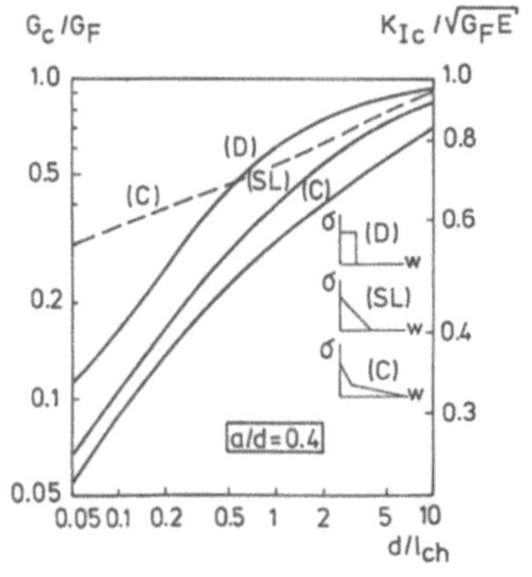


Fig. 3 Theoretical influence of $d/\ell_{ch}$ on evaluated values of $G_c$ or $K_{Ic}$ from three-point bend tests on notched beams. Full lines show results of LEFM evaluation assuming that the crack length is equal to the original crack or notch length. The different curves result from different assumptions regarding the shape of the σ-w-curve. The dashed line shows results of LEFM evaluation where the crack length has been calculated from compliance measurements. Only the σ-w-curve type C is used in this case.

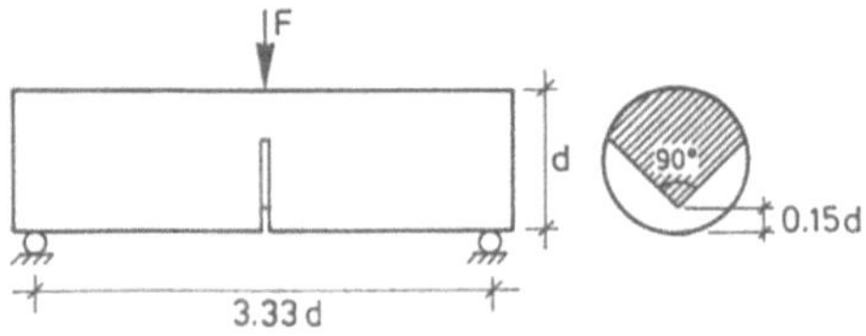


Fig. 4 Chevron notched cylindrical test specimen.

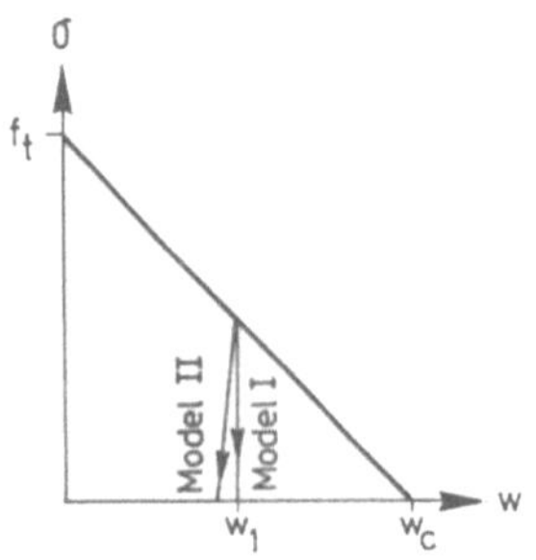


Fig. 5 Assumptions regarding the σ-w-curve used for the analysis of the specimen in Fig. 4, including unloading curves.

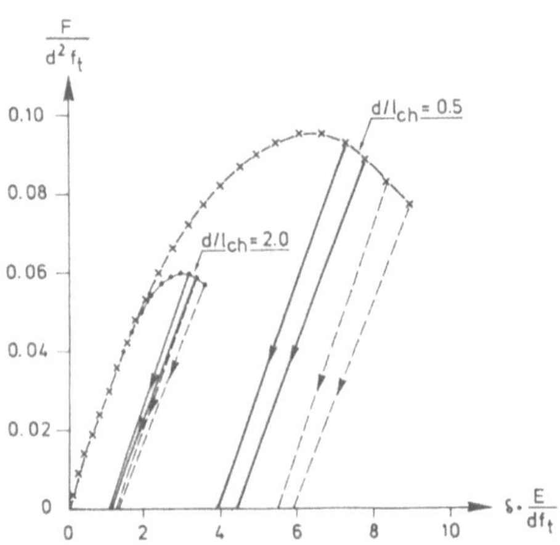


Fig. 6 Theoretical load-deflection curves for the specimen of Fig. 4 for $d/\ell_{ch}$=0.5 and 2.0, and different assumptions regarding unloading curves according to Fig. 5. Dashed unloading lines correspond to model I and full unloading lines to model II.

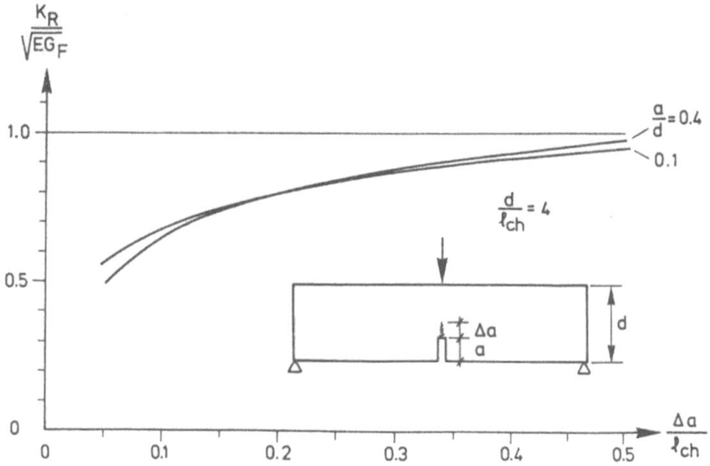


Fig. 7 Theoretical $K_R$-curves for beams with a constant depth d but with a varying initial notch depth a.

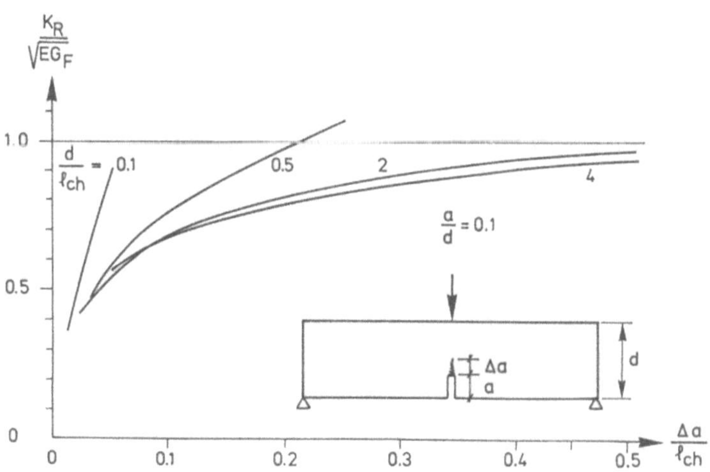


Fig. 8 Theoretical $K_R$-curves for beams with a constant ratio between notch depth a and beam depth d, but with a varying depth.

SEM/RILEM International Conference
on Fracture of Concrete and Rock
Houston, Texas, June 1987
Editors: S.P. Shah, S.E. Swartz

# CONCERNING THE MEASUREMENT OF THE FRACTURE ENERGY OF A MICRO-CONCRETE ACCORDING TO THE CRACK GROWTH IN A THREE POINTS BENDING TEST ON NOTCHED BEAMS

A. BASCOUL, F. KHARCHI, J.C. MASO

## ABSTRACT

We used an impregnation method to determine the shape of the cracked area in three points bending tests on notched beams with different sizes and different notch lengths. By loading and unloading with controlled displacement of the applied load we measured, through integration of the load displacement curve of each specimen, the work necessary to create some cracked area ; then the fracture energy was derived from the work-crack relationship. A steady propagation rate was observed that was not noticeably modified by the initial notch length. Taking into account the influence of the strain gradients on the cracking limite state threshold, we proposed an analysis of the microcracking evolution.

## INTRODUCTION

Through linear fracture mechanics, the resistance to crack extension of a material is determined by either a critical stress intensity factor or a critical energy release rate. These terms are considered intrinsic physical characteristics of the material. According to a linear elastic model in mode I, the stress is infinite at the crack tip. Concerning real materials, it cannot be so. A material damaging occurs in the neighbourhood of the crack front such as a microcracking for concrete.

The volume and the intensity of the microcracked zone which is generally called process zone are connected to the shape of the strain field applied to the material.

We already showed that the cracking limite state threshold by stretching in concrete for both the stress and strain is an asymptotic increasing function of the strain gradient for non-uniform loading |1|,|2|. This would be due to the stable development of the microcracking in the highly strained zone. This result allowed us to think that the conditions of propagation were not the same according to the strain field in which the crack propagates.

If there is an influence of the strain gradients on the cracking limite of concrete, the propagation mechanisms must certainly vary with the specimen geometry according to the size, the initial notch length and the crack growth too.

Many authors have dealt with the influence of these parameters. They often have different points of view. According to Hillerborg |3| the size effect in bending would be negligible for concrete. In the other hand Bazant and Pleiffer |4| proposed a law to express the size effect. Concerning the notch effect Shah and Mc.Garry |5| concluded that there was a slight notch sensitivity. But according to Ziegeldorf, Muller and Hilsdorf |6|, this effect cannot be neglected and they proposed a law connecting critical stress intensity factor and notch sensitivity. At present concerning concrete, it seems that the resistance to crack extension could not be characterized by a constant R which would be a material characteristic. Sok and Baron |7| propose that a R-curve (according to the crack growth) is characteristic of the material. In that spirit, Wecharama and Shah |8| give a mathematical model to express the R-curve.

We used three points bending tests on notched beams in order to study the propagation conditions according to various strain fields. We considered specimens with different depths and different notch lengths. But we tried to define a specific fracture energy according to the crack growth for every kind of specimens. The crack growth following is a difficult problem. It is not usually treated because the specimens are regarded as plane models. In fact, it is difficult to define the shape of the crack front on finite dimension specimens because any of plane strain or plane stress assumption cannot

A. Bascoul, Maître de Conférences, F. Kharchi, Research Worker, J.C. Maso, Professor, Director of the Civil Engineering Laboratory of the National Institute of Applied Sciences and Paul Sabatier University, Toulouse, France.

be adopted. There is a three dimensional loading. For a given load, the crack area value cannot be deduced from a direct observation of the specimen surface. In order to measure this area we used a method of impregnation under vacuum. For every tested specimen, we measured the energy necessary to create the crack area. Then it was possible to derive the specific fracture energy according to the crack growth and to propose an analysis of the microcracking evolution.

## EXPERIMENTS

### a - Material and Specimens

Mortar specimens have been used. The 55 R cement proportion was 555 g per litre of moulded mortar. The aggregates were made with a crushed marble with 3,15 mm maximum grain size. The water cement ratio was 0,53.

We used several kinds of specimens, the dimensions of which lxhxb were : 160x40x25 mm and 380x80x50 mm. Concerning the first series of specimens, there was only one notch length a = h/4 and such specimens are designated as P 1/4. Concerning the second series, three notch lengths have been considered, a=h/2, a=h/4, a=h/8. These specimens are respectively designated as M 1/2, M 1/4, M 1/8. The notch was made by means of a PVC plate, 3,5 mm thick, the tip of which was cut according to a 30° angle. As soon as they were cast, the specimens were kept until the test at 28 days in an atmosphere with constant hygrometry (H = 100 %) and constant temperature (T = 20°C).

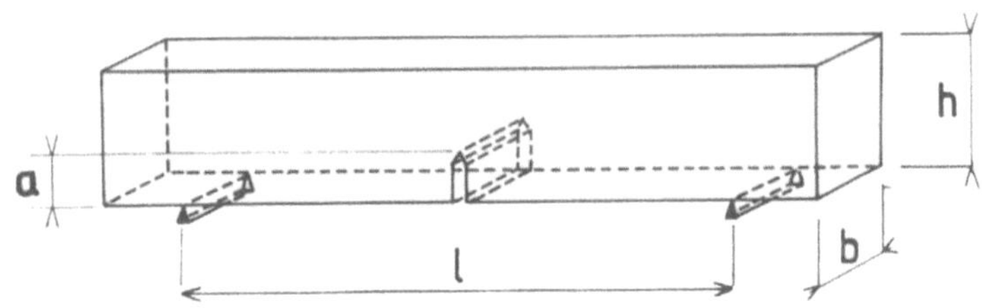


Fig. 1 - Three points bending notched beam

### b - Test proceeding

A 50 KN hydraulic press was used. It can be controlled as well in load as in displacement. A PDP 11 computer governs the press and allows the recording of the measurements and their exploitation. Here the tests were carried out by controlling the displacement of the applied load F. The imposed displacement rate was 10 μm/min. The moving bearing A is connected to the crosspiece B fig. (2). A displacement transducer measures the displacement of the crosspiece in comparison with a fixed point independent of the specimen.

Fig. 2 - Testing apparatus

**c - Cracked area measurement**

The crack propagation following presents a fundamental problem with respect to this kind of test because the specimen thickness (b) is large. Concerning the loading field, it is impossible to retain a plane stress or a plane strain hypothesis. A direct observation on the specimen side does not allow to define the actual cracked area. This problem has been held up by some authors |9|,|10|,|11|.

Then we decided to measure directly the cracked area with respect to several crack growth states on a great number of identical specimens. For a test, a specimen was loaded up to a certain displacement level. Next, it was unloaded at the same displacement rate and impregnated with a coloured resin. After resin hardening, the specimen was broken and the cracked area which corresponds to the surface of the product penetration was found by difference. This area is designated as $A_m$. $A_m$ was measured between the initial notch tip and the outline of the coloured area. So, every specimen gave only one cracked area value.

Table I illustrates the followed method. The various load-displacement curves are presented in the first column. The second column shows the median cross section of the specimens after impregnation, hence the cracked area. In the third column, we give the ratio between the total area A = b(h-a) and the cracked area $A_m$.

The specimen is impregnated under vacuum to allow the product to reach the crack tip.

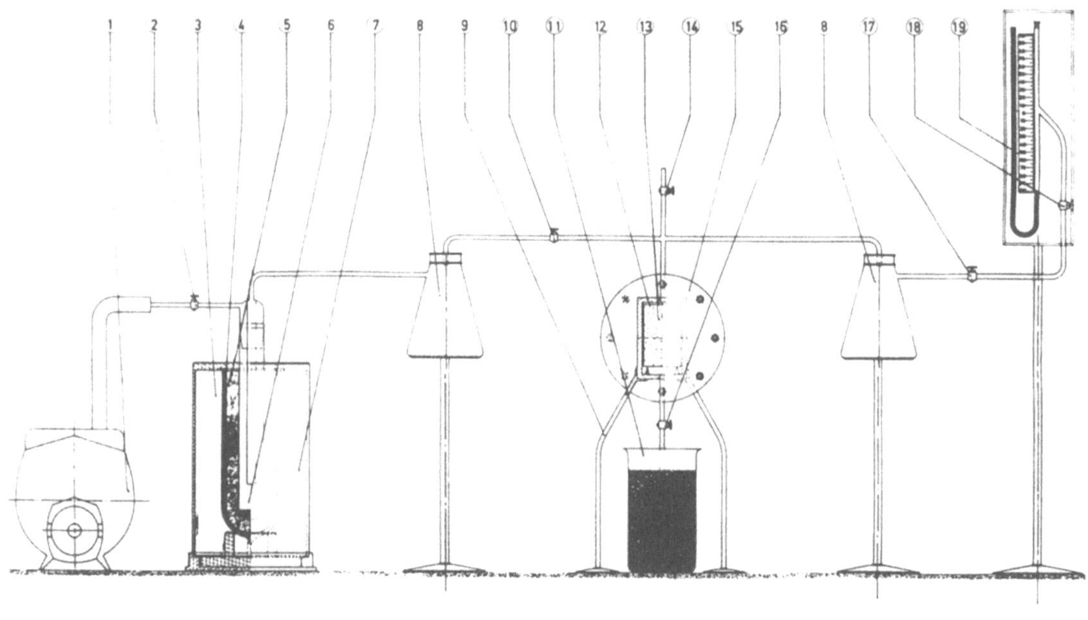


Fig. 3 - Impregnation device

After the mechanical test, the specimen 13 is put into the vessel 12 . The colouring product is introduced through 16 . The vacuum is created by 10 . In order to avoid some infiltration into the pump 1 , a condenser trap 6 is placed between the impregnation vessel and the vacuum pump.

Impregnation is performed in two stages. First, a vacuum is created before introducing the product, next 16 is opened while maintaining depression. This allows to eliminate the colouring product vapour in the trap 6 and all the free space of the vessel to be occupied by the liquid.

The impregnation liquid is made of a polyester resin (stratyl 116) diluted with styrene. A hardening accelerator and a catalyst are added. The proportion in volume are :

| load-displacement curve | cracked area | cracked area-cross section ratio $A_m/A$ % | mean crack length cm | |
|---|---|---|---|---|
| | | | along the section axis | on the specimen side |
| 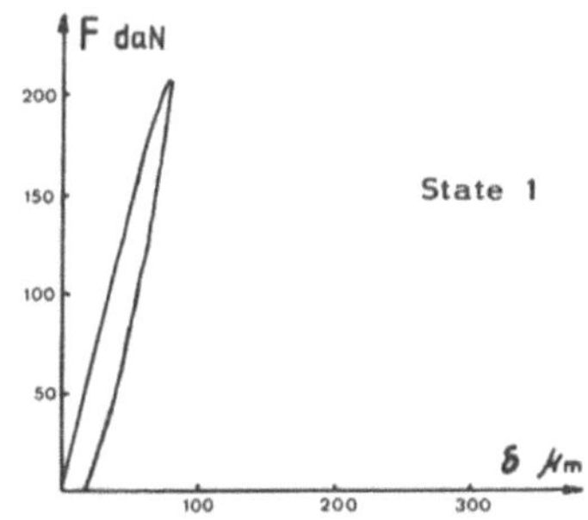 State 1 | 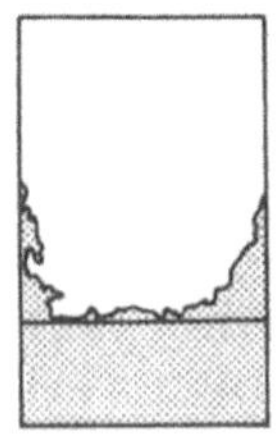 | 9,2 | 0,25 | 2,65 |
| 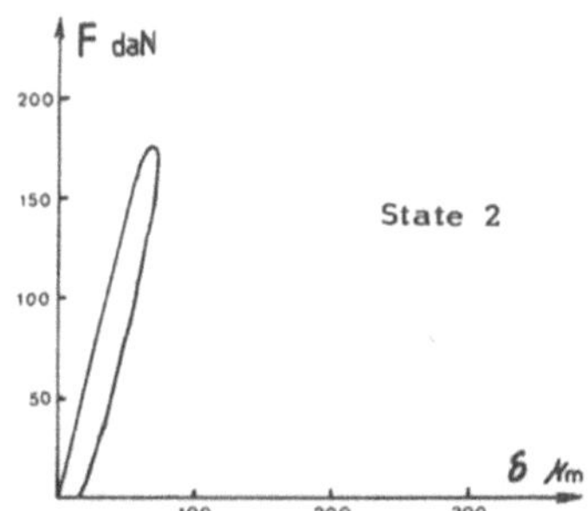 State 2 | 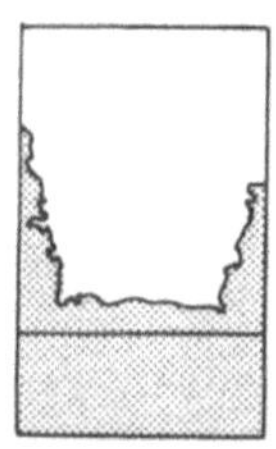 | 18,7 | 0,65 | 3,5 |
| 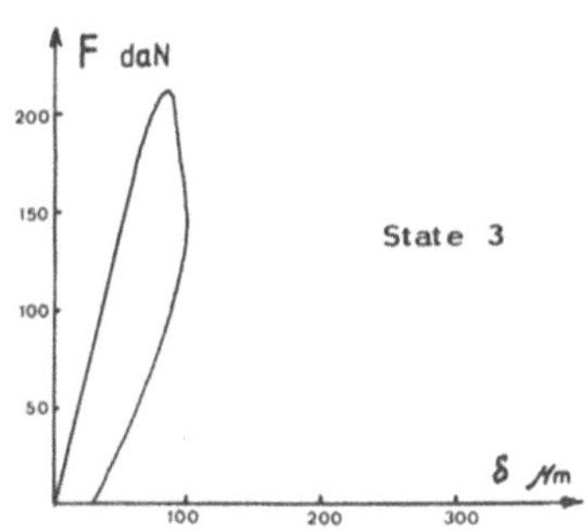 State 3 | 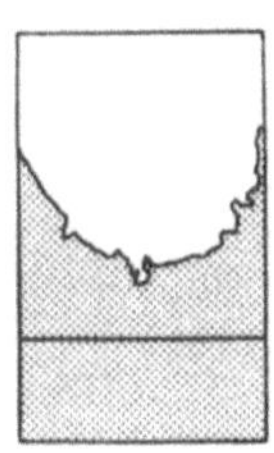 | 34,2 | 1,65 | 4 |

<table>
<tr><td>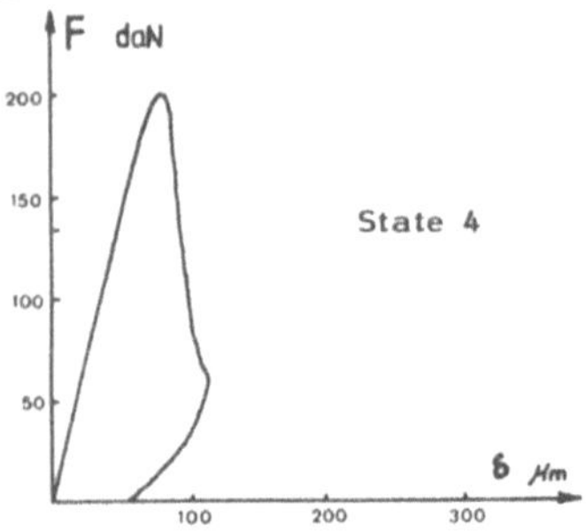</td><td>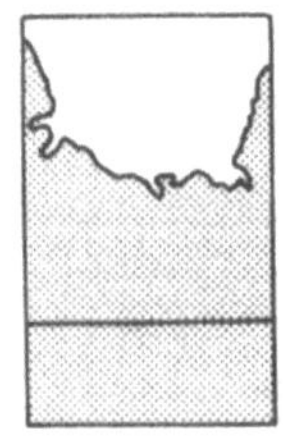</td><td>53,6</td><td>2,85</td><td>5,3</td></tr>
<tr><td>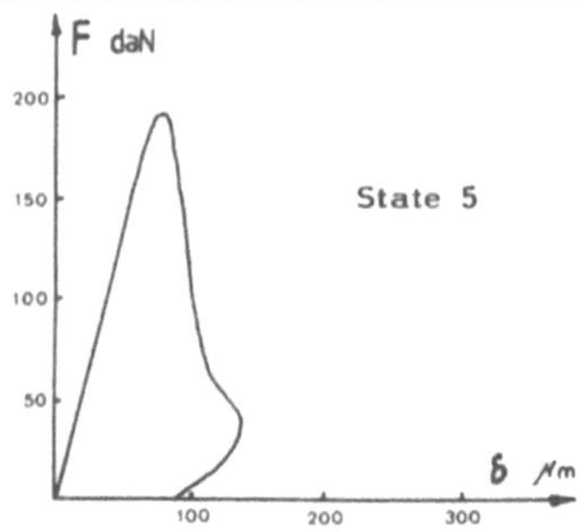</td><td>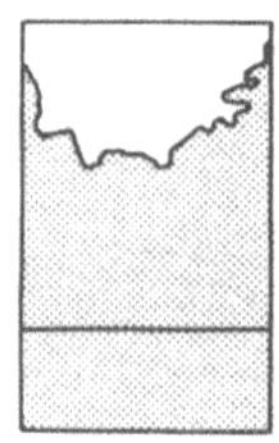</td><td>70,0</td><td>3,5</td><td>5,45</td></tr>
<tr><td>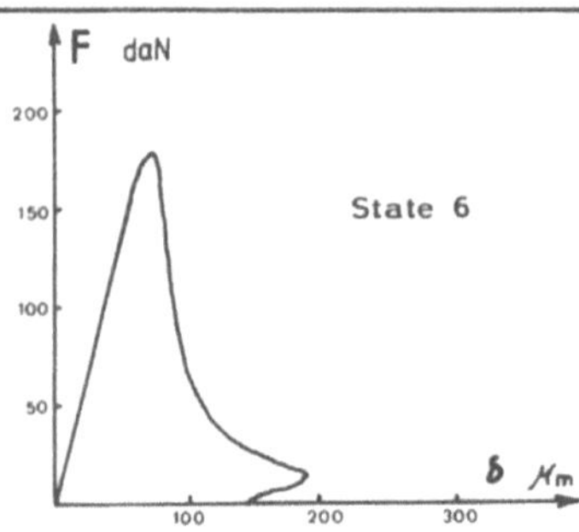</td><td>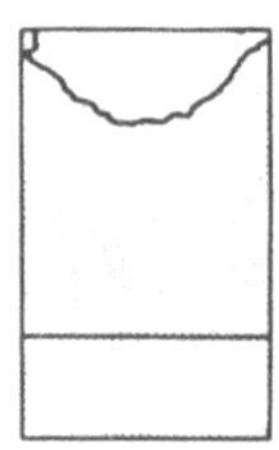</td><td>78,5</td><td>4,3</td><td>6</td></tr>
<tr><td>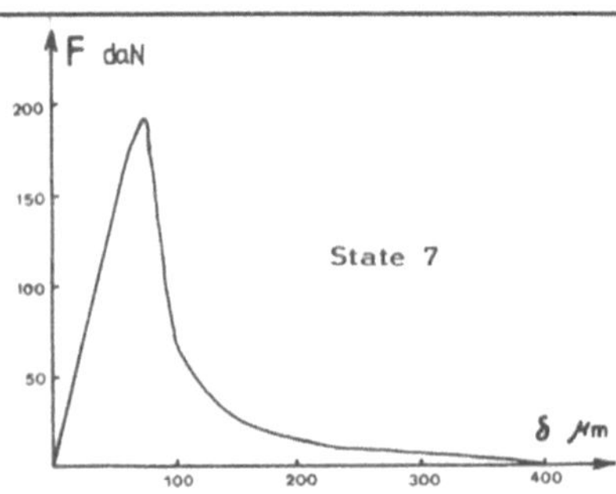</td><td>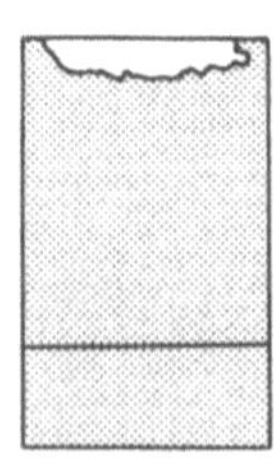</td><td>91,5</td><td>5,3</td><td>6</td></tr>
</table>

Table I - Crack front shape evolution with crack growth (M1/4 specimen)

| | |
|---|---|
| resin | 49,68 % |
| diluter | 49,68 % |
| catalyst | 0,6 % |
| accelerator | 0,05 % |

The proportions give a product the viscosity of which is close to $10^{-2} P_o$. The hardening begins five hours after the mixture preparation.

**RESULTS**

The integration of the load displacement curve gives the total energy dissipated during a test. It is designated as W. We present, in the figures 4,5,6,7, the experiment points which show the W evolution according to $A_m$ for the various kind of specimens.

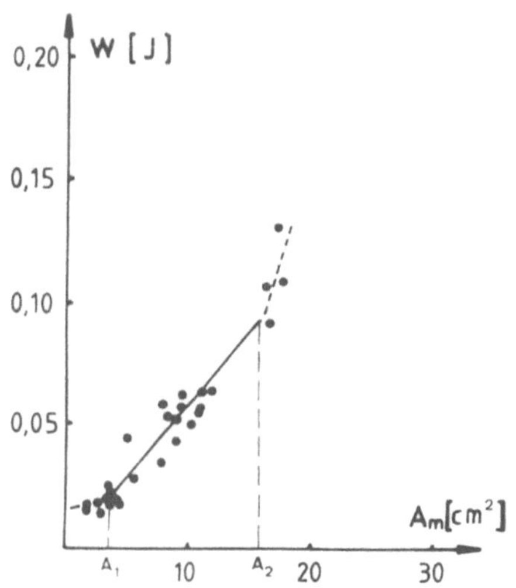


Fig. 4 - M1/2 specimen

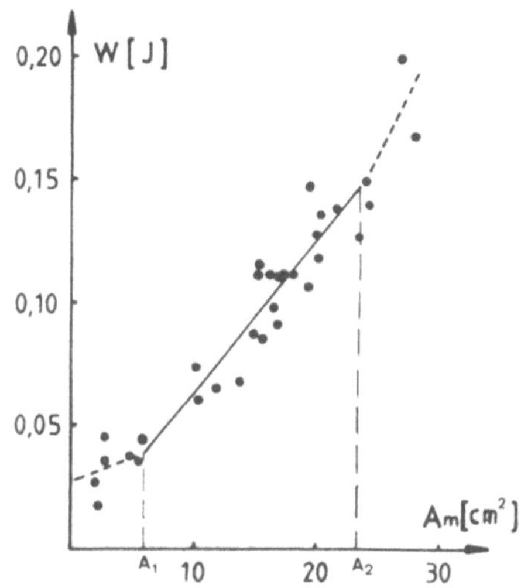


Fig. 5 - M1/4 specimen

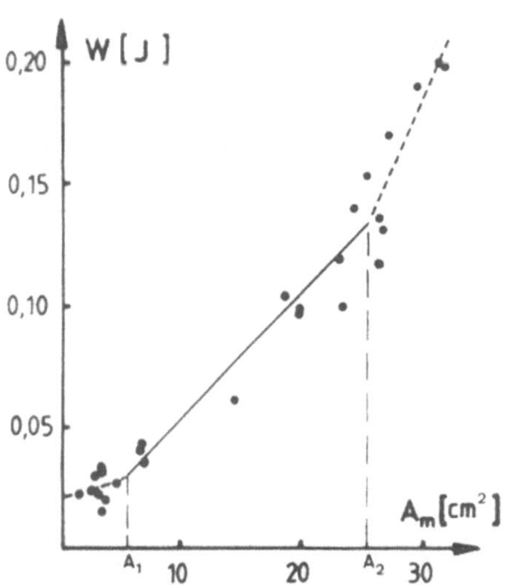


Fig. 6 - M1/8 specimen

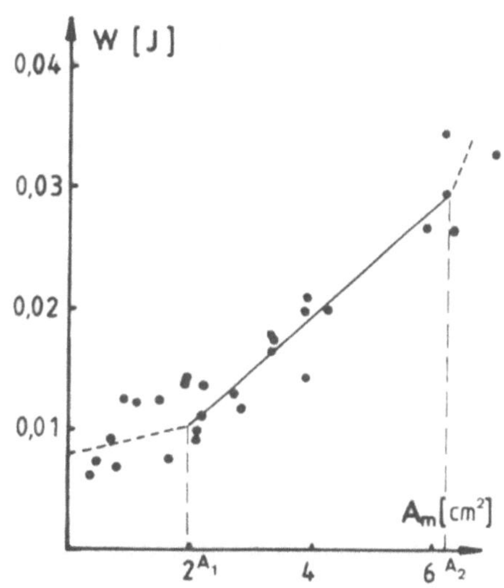


Fig. 7 - P1/4 specimen

Fig. 4,5,6,7 - Variation of the dessipated energy according to the measured cracked area

Three domains can be distinguished to describe the W variation law according to $A_m$. In the first domain corresponding to the low $A_m$ values (as the states 1 and 2 in table I), there is a slight increase of W with $A_m$ and the W values are scattered. The last domain corresponds to the high $A_m$ values (as the states 6 and 7 in table I) with a more important W variation rate. There is initial microcracking to initiate the crack propagation from the artificial notch and an energy $W_o$ is dissipated. Then, concerning the first points, the actual crack grows through a highly damaged zone and the increase in energy necessary to make the actual crack move forward is small. Concerning the last points the increase in energy has to be more important because it is used at the same time to make the crack move and to create consolidations in the highly compressed zone |12|.

If all the points corresponding to the three kinds of tests M1/2, M1/4, M1/8 were drawn in a same graph, there would be observed a slightly scattered zone where the points are around a straight line. If we assume that far away from the zone disturbed by the crack initiation and the compression under the bearing, the propagation mechanisms are identical for the three kinds of specimens, then it is logical to find a common band for the tests with a low scattering.

Taking into account this observation, we chose to represent the W variation in the central part of the graphs, bounded by the points $A_1$ and $A_2$, by a straight line segment. Therefore, the energy W is written as : $W = W'_o + G_r A_m$

The slope of this line $\partial W / \partial A_m$ represents the energy rate necessary to an increase in $A_m$ equal to the unit. Then, the term $G_r$ is an apparent fracture energy if $A_m$ does not represent all the created discontinuity areas.

We give in table II the $W'_o$ and $G_r$ values for all the kinds of specimens. r is the correlation coefficient.

| | P 1/4 | M 1/2 | M 1/4 | M 1/8 |
|---|---|---|---|---|
| $W'_o$ [J] | $17.10^{-4}$ | $4,2.10^{-4}$ | $12,2.10^{-4}$ | $10,3.10^{-4}$ |
| $G_r$ [J/m$^2$] | 44 | 59 | 62 | 53 |
| r | 0,92 | 0,96 | 0,94 | 0,94 |

Table II - Terms of linear law

Table II shows that the $G_r$ values for M 1/2, M1/4, M1/8 specimens are of the same order and that the P1/4 specimen gives a different value.

These results show that there is a steady mechanism of propagation for the specimens which have the same size. Concerning the M specimens, $G_r = 58 \pm 5$ J/m$^2$. But we find again that there is size effect since the value provided by the tests on P1/4 specimens is 44 J/m$^2$.

In addition we have carried out two DCB tests on large size plates (fig. 8). Displacements are imposed at the points A and B.

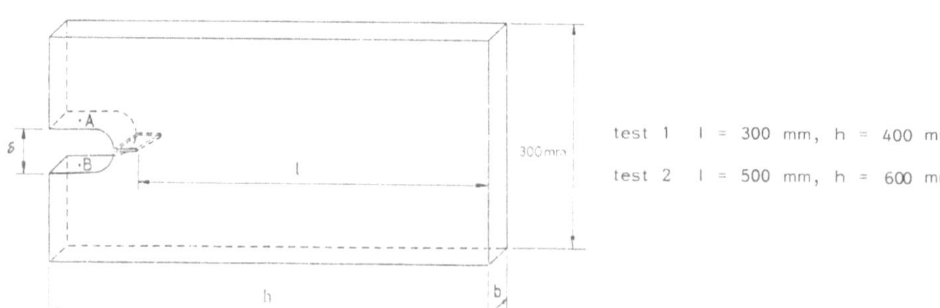


Fig. 8 - DCB test specimen

The specific fracture energy is calculated through the expression :

$$G_r = \frac{W_{tot}}{b.l}$$

where $W_{tot}$ is the area under the load displacement curve and b is the plate thickness. The results are given in table III. The difference with the three points bending tests is very significant. We find again the size effect according to the parameter h.

$$G_{rP} < G_{rM} < G_{rDCB}$$

| | h = 400 mm | h = 600 mm |
|---|---|---|
| $G_{rDCB}$ [J/m²] | 80 | 81 |

Table III - DCB test results

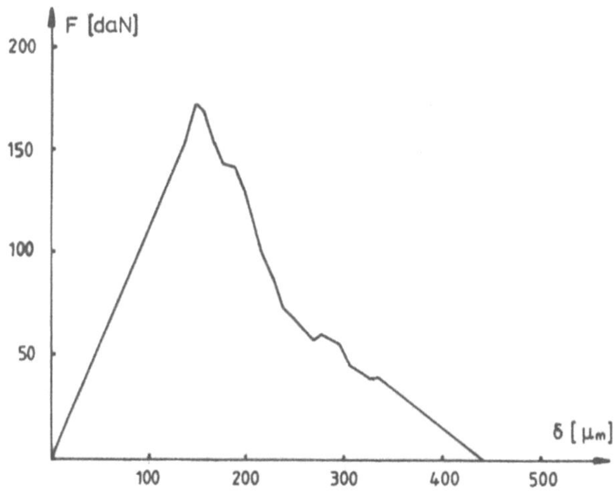


Fig. 9 - DCB load-displacement curve (h = 400 mm)

It appears interesting to compare the results of table II with the ones obtained through an overall calculation where the specific fracture energy (here called $G_{rg}$) would be derived from the expression :

$$G_{rg} = \frac{W}{A_m}$$

The results are presented in figure 10 according to the M1/8 specimens. Concerning the other tests (P1/4, M1/4, M1/2), the curves are similar. Figure (10) shows that there are high $G_{rg}$ values at the propagation beginning. Next, $G_{rg}$ decreases down to a steady value between 0,25 A and 0,75 A. It increases for the last $A_m$ values.

The $G_{rg}$ steady values are given in table IV according to the various specimens.

| | P1/4 | M1/2 | M1/4 | M1/8 |
|---|---|---|---|---|
| $G_{rg}$ [J/m²] | 46 | 58 | 63 | 52 |

Table IV - $G_{rg}$ steady values

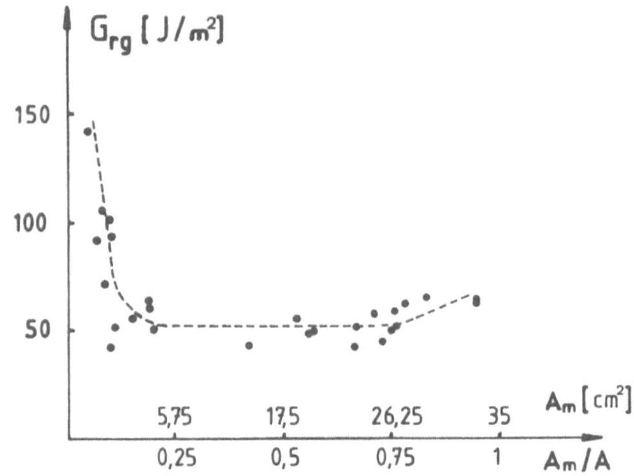


Fig. 10 - Fracture energy derived from an overall calculation (M1/8 specimen)

Such a variation is obvious, because of the previously proposed analysis of crack propagation conditions. Since an energy $W_o$ is dissipated before propagation, the curve has to be asymptotic for the low $A_m$ values. The flat level corresponds to a steady mechanism which is slightly influenced by the initial energy $W_o$. $G_{rg}$ value is close to the $G_r$ one. As $A_m$ tends towards A, an additional energy is consumed by the consolidations in the highly compressed zone.

## MICROCRACKING EVOLUTION WHILE STEADY PROPAGATION RATE

An overall cracked area corresponds to each growth state. If there is a microcracked zone above the crack, which is usually called process zone, the overall area may be written :

$$S = A_m + A_{mf}$$

$A_{mf}$ is the microcracked area in the process zone. Let $dw_s$ be the work necessary to an increase dS in S. If we make it a matter of principle that there is a parameter G characteristic of the material, then the energy dissipated while creating dS would be :

$$dW_s = GdS$$

where $dS = dA_m + dA_{mf}$

From the experimental results, we proposed a modelisation of the law W while steady propagation rate. Then we calculated an apparent fracture energy $G_r$ that we assignated to the only measured crack $A_m$ :

$$dW = G_r dA_m$$

$$\text{if} \quad dW = dW_s$$

$$\text{then} \quad dA_{mf} = \left(\frac{G_r}{G} - 1\right) dA_m$$

$$\text{and} \quad W = G_r A_m + GA'_{mfo}$$

where $A'_{mfo}$ would be the created area above an ideal initial crack such as :

$$W'_o = GA'_{mfo}$$

According to our results, $W'_o$ has a very low value. For information only, we give the $A'_{mfo}$ value corresponding to $G = 60\ J/m^2$ (table V).

| | P1/4 | M1/2 | M1/4 | M1/8 |
|---|---|---|---|---|
| $A'_{mfo}(cm^2)$ | 0,28 | 0,07 | 0,2 | 0,17 |

Table V - Cracked areas above ideal initial cracks

The difference in $G_r$ value for various specimens is connected to the way the microcracking develops with crack extension. Three cases are possible :

$G_r < G$ The microcracking propagates decreasing in absolute value,

$G_r = G$ The microcracking propagates in a steady way,

$G_r > G$ The microcracking propagates increasing in absolute value.

Some authors |13|,|7| proposed that an estimate G value might be obtained with large sized specimens in which the volume of the process zone is small in comparison to the overall volume of the specimen. Thus, $G_{rDCB}$ would be the closest to G value. Concerning the three points bending tests, we are in the first case.

We showed that, under non uniform loading, concrete undergoes limite tensile strains higher than the limit strain $\mathcal{E}_{to}$ in uniform traction and that there is an asymptotic increase in the limit strain with the strain gradient |1|. This fact was connected to a stable microcracking in relation with the strain gradient. Microcracking is all the more developed since the gradient has a high value.

Tacking into account these observations, here we assume that there are microcracks in the tight volume which undergoes longitudinal strain higher than $\mathcal{E}_{to}$.

Through a finite element computation in linear mode and using the crack front shapes that were found by impregnation, we calculated the strains in the median cross section. Here, we give an example of equal strain curve map that was obtained. Figure 12.

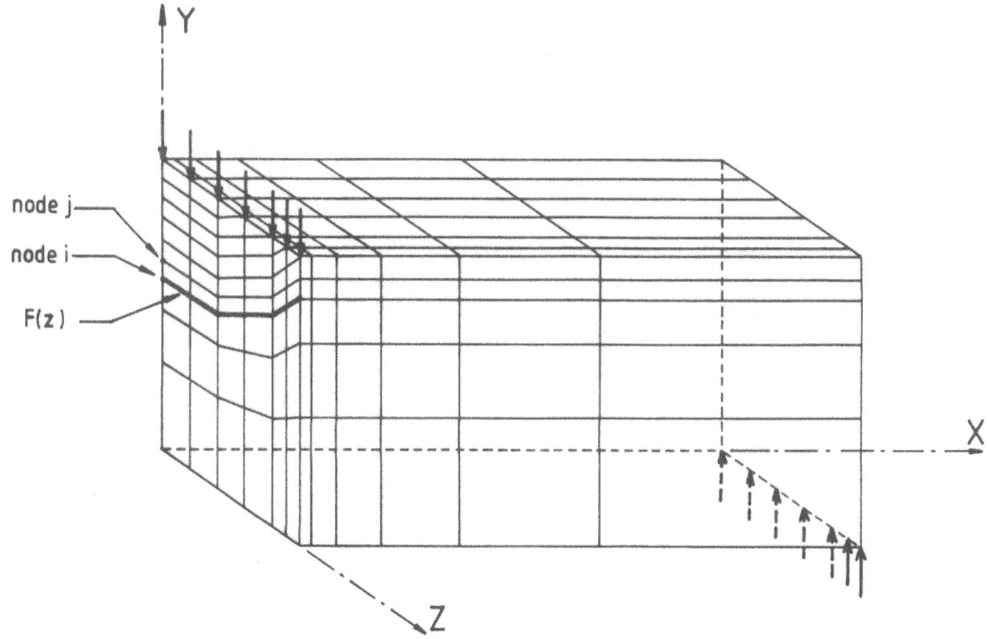


Fig. 11 - Finite element model

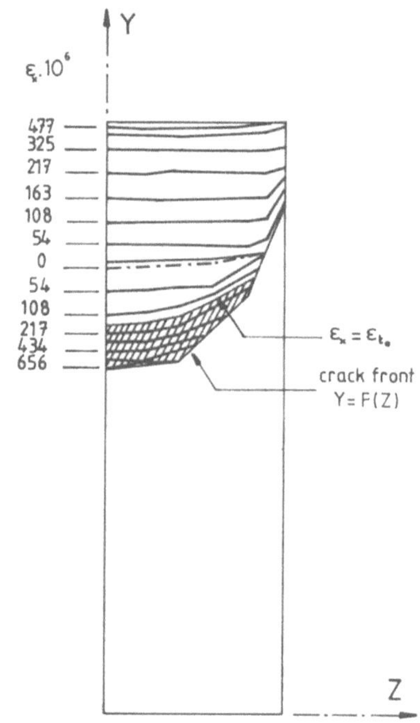


Fig. 12 - equal strain curves according to $A_m/A = 0{,}53$ (M1/4 specimen)

$\mathcal{E}_{to}$ = limit strain in uniform traction

Let $A_{\varepsilon t}$ be the area of the median cross section that is delimited by the crack front and the equal strain curve $\varepsilon_{to}$. $A_{\varepsilon t}$ is showed by the hachured zone in figure 12. Here we assume that $dA_{mf}$ is connected to $A_{\varepsilon t}$ by the relation :

$$dA_{mf} = d(DA_{\varepsilon t})$$

where D is the microcrack density in the process zone.

In view of |1|,|2|, D must be an increasing function of the strain gradient component $\partial \varepsilon x/\partial y$ at the crack front.

According to a theoretical elastic linear calculation, the component of the strain gradient $g = (\partial \varepsilon x/\partial y)_{x=o,y=f(z)}$ has an infinite value. Physically, it cannot be so in a real material.

Refering to the finite element model, we give the g evolution in function of $A_m$, at least in the qualitative plane. g is the strain variation rate between the nodes i and j of the figure 11.

The curves of figure 13 show that $\frac{dg}{dA_m} > o$.

Since $\frac{dD}{dg} = \frac{dD}{dA_m} \times \frac{dAm}{dg} > o$ , (from |1|,|2|)

Then $\frac{dD}{dA_m} > o$

Thus the microcrack density increases as $A_m$ advances.

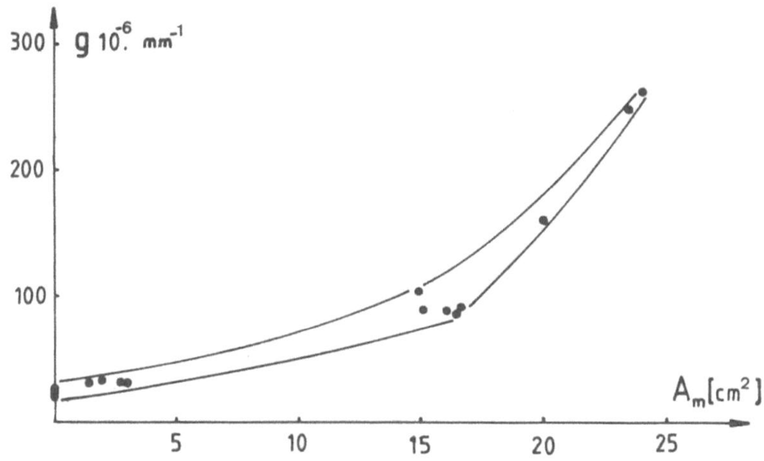


Fig. 13 - Variation of the strain gradient according to the measured cracked area (M1/4 specimen)

**CONCLUSION**

It has been shown in a previous work that a cracking criterion, in stress or in strain, cannot be defined for concrete under tensile conditions without taking into account the strain gradients. From this point, we thought that it was logical to study the crack propagation mechanisms under variable strain field conditions.

In this work, we have dealt with the determination of an apparent fracture energy $G_r$ under three points bending tests by varying the depth of the specimens and the specimen depth-notch length ratio. In every case, we have tried to define a specific fracture energy according to the crack growth, and thus the strain field variation.

In order to satisfy this condition, it was necessary to know the actual shape of the crack front and consequently the cracked area so as the supplied work in accordance.

The tests were carried out with controlled displacement and we used an impregnation proceeding on a great number of specimens. The specimens were loaded up to some levels of displacement $\delta$ of the load F and next unloaded. The work was calculated by integration of the (F - $\delta$) curve.

Every unloaded specimen was impregnated under vacuum with a coloured resin. After resin hardening, the specimens were broken and it was then possible to measure the cracked area by difference.

The results show that, in every case, there is a steady value of the apparent specific fracture energy $G_r$. The depth of the initial notch does not modify noticeably the $G_r$ value. But it can be observed that $G_r$ is higher for the great depth specimen. DCB tests on large size plates give the greatest value.

From these results and a strain study by finite elements, we proposed an analysis of the microcracking evolution according to the crack growth. We assumed that the microcracked zone is contained in the tensile domain in which the strains are greater than the pure traction limit strain $\varepsilon_{to}$. This analysis shows that the microcracking decreases in absolute value as the crack grows but, in the other hand, the microcracking density increases. Such a result has to be confirmed through direct visualizations.

|1| Bascoul A. : "Influence des gradients de déformation sur les états limites d'écoulement ou de fissuration des matériaux". Thèse docteur d'Etat, Université Paul Sabatier, Toulouse, 1985.

|2| Bascoul A., Maso J.C. : "On the effects of strain gradients on the limit state of cracking of concrete and the stable development of microcracking". International Conference on Fracture Mechanics of Concrete. Lausanne, 1-3 october 1985.

|3| Hillerborg A. : "Influence of beam size on concrete fracture energy determined according to a Draft Rilem recommandation". IUTVDG/(TVBM - 3021)/1-15/(1984).

|4| Bazant Z.P., Pleiffer P.A. : "Shear fracture tests of concrete". Matériaux et Constructions, 1986, n° 110, p. 111-121.

|5| Shah P., Mc.Garry F.J. : "Griffith fracture criterion and concrete". Journal of Engineering Mechanics Division, December 1971, p. 1663-1677.

|6| Ziegeldorf S., Muller H.S., Hilsdorf H.K. : "A model law for the notch sensitivity of brittle materials". Cement and Concrete Research. vol. 10, 1980, p. 589-599.

|7| Sok C., Baron J. : "Mécanique de la rupture appliquée au béton hydraulique". Cement and Concrete Research. vol. 9, 1979, p. 641-649.

|8| Wecharama M., Shah S.P. : "Experimental methods to determine fracture parameters for concrete". Seminar "Fracture mechanics". Collège International des Sciences de la Construction. I.T..B.T.P. France, 1982/06/08-11.

|9| Entov V.M., Yagust V.I. : "Experimental investigation of flaws governing quasi-static development of macrocracks in concrete". Mechanics of Solids. vol. 10, n° 4, 1975, p. 87-95.

|10| Mazars J. : "Prevision de la rupture des structures en béton par la mécanique de la rupture". Thèse de docteur de 3ème cycle. Université Pierre et Marie Curie, Paris VI, 1976.

|11| Swartz S.E., Go C.G. : "Validity of compliance calibration of cracked concrete beams in bending" Experimental Mechanics, June 1984, p. 129-134.

|12| Pons G., Maso J.C. : "Comportement mécanique du mortier soumis à des chargements harmoniques à fréquence faible : influence de l'amplitude et des bornes de la sollicitation cyclique". Matériaux et Constructions, RILEM, Novembre-Décembre 1983.

|13| François D. : "Mécanique de la rupture et lois de comportement". Cahiers du Groupe Français de Rhéologie, Tome V, n° 2 juin 1979.

SEM/RILEM International Conference
on Fracture of Concrete and Rock
Houston, Texas, June 1987
Editors: S.P. Shah, S.E. Swartz

# DEVELOPMENT OF STRENGTH AND DEFORMABILITY OF VERY YOUNG CONCRETE

W. Brameshuber, H.K. Hilsdorf

ABSTRACT

A description of experiments concerning the development of strength and deformability including fracture mechanics parameters of very young concrete is given. It is shown that the strength of concrete increases during the observed period from 4 hours to 28 days after casting whereas the deformability - e.g. represented by the deflection of notched beams at maximum load - decreases until a minimum is reached and then increases again. The characteristic length $l_{ch}$ may be influenced by hydration temperature but not the notch sensitivity; $l_{ch}$ reaches a maximum around 8 hours and then decreases continuously up to an age of 28 days.

## 1. INTRODUCTION

Concrete passes through different stages from the time of mixing until a more or less stable state is reached after several months. In the fresh state the properties of concrete are classified primarily by its consistency. For the hardened concrete its properties are mainly described by strength and elasticity parameters as well as by parameters to express time dependent strains. In standard testing no information on the development of strength and fracture mechanics parameters at a very early stage is obtained.

Nevertheless, a substantial number of studies described in the literature deal with the strength and deformation characteristics of young concrete e.g. [1, 2, 3, 4]. A summary of references is given in [5]. Only a few authors reported on the mechanical properties of young concrete subjected to tensile stresses e.g. [6]. The main results of these experiments show, that the deformability of very young concrete reaches a minimum some hours after casting when the strength still is very low. This is valid for compression as well as for tension.

During the last few years the behavior of very young concrete has been studied by testing specimens subjected to nearly 100 percent constraint in order to simulate the situation in large concrete members [7, 8, 9]. Soon after casting compressive stresses arise in the specimens due to a temperature increase as a consequence of heat of hydration. These stresses are reduced by the high relaxation capacity of the very young concrete. Subsequently, the temperature decreases and tensile stresses are developed in the concrete. Depending on mix proportions and type of cement [8], the tensile strength of concrete may be exceeded and cracking will occur. Information about the relaxation behaviour of very young concrete is given in [9], but a criterion for cracking of young concrete is still missing.

It has been shown that fracture energy is under certain conditions a valid parameter to describe the fracture behaviour of concrete [10, 11, 12, 13]. In this paper some preliminary results of experiments concerning the development of strength, deformability and fracture mechanics parameters of very young concrete are presented.

## 2. DESCRIPTION OF EXPERIMENTS

### 2.1 TEST PROGRAM

#### 2.1.1 Materials and mix design

Concrete beams and cylinders have been cast using a portland cement PZ 35 F, Rhine River sand and

Dipl.-Ing. W. Brameshuber and Prof. Dr.-Ing. H.K. Hilsdorf,
Institut für Massivbau und Baustofftechnologie,
University of Karlsruhe, Am Fasanengarten, 7500 Karlsruhe, West-Germany

gravel with a maximum aggregate size of 16 mm. The grading curve was chosen according to DIN 1045. The concrete had a cement content of 315 $kg/m^3$ and a w/c-ratio of 0.54.

### 2.1.2 Specimens

To evaluate the mechanical properties of very young concrete in compression cylinders Ø 75 mm, l = 150 mm have been cast. The uniaxial tensile strength of concrete cylinders with the same dimensions was determined only for concretes with an age of 3 days up to 28 days.

The mechanical properties of concrete in tension were evaluated by testing beams in bending. The span was 500 mm, the depth and the width of the beams 100 mm. Notched and unnotched specimens with a notch/depth ratio of 0.5 have been tested.

## 2.2 MANUFACTURING OF THE SPECIMENS

Cylinders and beams have been cast in steel molds. The concrete was mixed in a continuous pan mixer and compacted on a table vibrator.

The notches were cast by attaching a metal plate to the top of the molds. At the tip of the notch the plate had a width of 1 mm. The tension side during testing was the top side as cast. To avoid moisture loss of the beams during handling and testing, plastic sheets were fixed in the molds before casting. These plastic sheets were never removed until testing of the specimens was completed.

After compaction the specimens were covered with wet burlap and plastic sheets and stored at a temperature of 23°C. The notch plates were removed as soon as possible - in this case 3 hours after mixing - to avoid cracking caused be chemical shrinkage. At the same age steel plates were fixed with gypsum plaster to the cast surface of the concrete cylinders while the cylinders were not yet demolded.

The specimens to be tested at an age older than one day were demolded 24 hours after mixing. After wetting the surfaces, the specimens were sealed with plastic sheets and stored at a temperature of 23°C.

For uniaxial tension tests steel plates were glued to the cylinders one day before testing. The plastic sheets were only removed from the top and the bottom side of the specimens in order to assure a dry surface for good bonding between the steel plates and the concrete.

## 2.3 EXPERIMENTAL PROCEDURES

### 2.3.1 Compression tests

With exception of cylinders to be tested at an age older than one day, the specimens were demolded immediately before testing and covered with a thin plastic sheet. A special equipment to measure the deformation during the test with inductive strain gages was fixed to the concrete cylinders. The specimens were subjected to a constant rate of straining so that the maximum load was reached in about 90 sec. The concrete age at testing was 4, 5, 6, 7, 8, 12, 24 hours and 3, 7 and 28 days, respectively. For each concrete age 3 cylinders have been tested.

### 2.3.2 Uniaxial tension tests

The cylinders were kept in the plastic sheets until the tension test had been completed. The same strains gages used in the compression tests were also used for the tensile tests. The strain rate was controlled such that the maximum load was reached in about 90 sec. For younger concretes the tensile strength was determined in tensile splitting tests. However when testing very young concrete (4 to 6 hours) no reliable results could be obtained because the specimens failed in compression in the region adjacent to the line load.

### 2.3.3 Bending tests

Standard equipment is not suitable to test very young concrete in bending. Especially notched beams will break during handling or under their dead weight. In addition, the ratio of dead weight to maximum load is very high for young concretes so that uncertainties will occur when extrapolating according [14] to calculate fracture energy. Therefore, a special testing equipment was developed to measure the complete load-deflection-curve of concrete beams in bending which is shown in Fig. 2.1. The notch in the beam is on its top side and the applied load acts against the dead weight. It has been shown in previous studies on older concrete that for such a set-up fracture energy is nearly independent of beam depth [13].

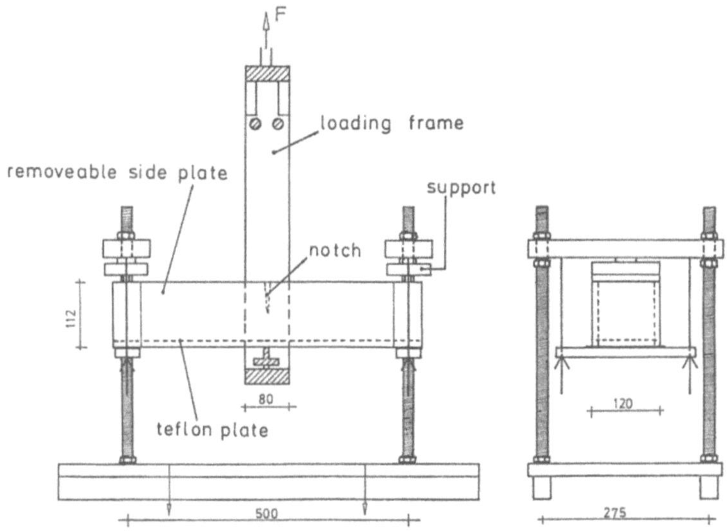


Fig. 2.1: Testing equipment to determine tensile properties of very young concrete

A constant rate of deflection was chosen such that the maximum load was reached in about 90 sec. Only for concrete beams with an age of 4 and 5 hours, the maximum load was reached in 60 and 70 sec, respectively.

Demolding of notched concrete beams for an age at testing of 4 to 8 hours is quite difficult because of the risk of cracking the specimen during handling. To avoid demolding very young concrete beams were tested in a special mold.

This mold is constructed such that the side plates can be removed immediately before testing. The bottom of the mold is made of teflon with a thickness of 1 mm. Teflon was chosen to avoid bonding between the concrete and the bottom of the mold. The properties of the teflon plate were determined to calculate the fraction of the total load which is carried by the teflon plate (see also chapter 2.4).

A special equipment directly mounted to the specimen was used to measure the deflection of the beams with inductive strain gages. It was designed such that possible deformations at the supports could be eliminated.

The bending experiments carried out on notched beams were always stable. Unnotched beams older than 7 hours subjected to a constant rate of deflection would break in an unstable manner. To record the complete load-deflection curve compliance tests have been carried out on unnotched specimens. The deflection is increased until the maximum load is reached. Then the specimen is unloaded and reloaded again. Normally up to 20 cycles are necessary to obtain a complete load-deflection-curve. In [13] it is shown, that no significant difference exists in $G_F$ calculated from stable bending tests with a constant rate of deflection and those calculated from compliance tests.

## 2.4 EVALUATION OF DATA

Fracture energy was calculated according to [12, 13]. In cases where the special mold had to be used a correction of the observed load-deflection-curve was necessary. This is shown in Fig. 2.2. Before calculating the flexural strength and fracture energy the portion of the teflon plate must be substracted. The slope of the teflon curve can be estimated from calibration tests and from Young's modulus of the teflon.

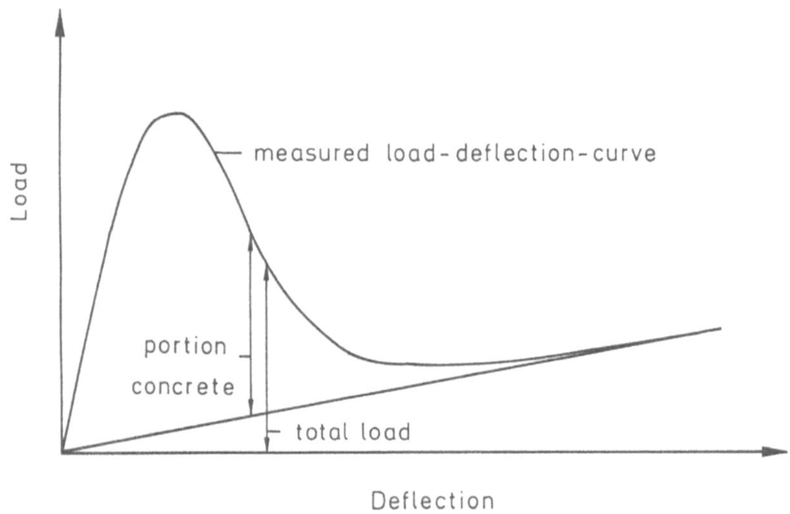


Fig. 2.2: Correction of load-deflection-curve when testing beams in the mold

For compliance tests, fracture energy was determined from the envelope of the load-deflection-cycles. To get information on the brittleness of concrete at varying ages the characteristic length and the notch sensitivity have been evaluated. Notch sensitivity k is defined as the ratio of the net flexural strength $f_{net}$ of a notched beam to the flexural strength $f_f$ of an unnotched beam:

$$k = \frac{f_{net}}{f_f} \leq 1 \tag{1}$$

According to [10, 11] the fracture resistance of a brittle material can be described by the characteristic length

$$l_{ch} = \frac{G_F \cdot E}{f_t^2} \tag{2a}$$

Since it was not possible to determine experimentally the uniaxial tensile strength $f_t$ of very young concrete, the flexural strength was used to calculate a characteristic length $l_{ch}^*$:

$$l_{ch}^* = \frac{G_F \cdot E}{f_f^2} \tag{2b}$$

It is well known that the flexural strength is affected by the ratio $d/l_{ch}$. Nevertheless, the tendency of the development of characteristic length with age should become apparent using $l_{ch}^*$.

## 3. EXPERIMENTAL RESULTS

### 3.1 SETTING OF CONCRETE AND TEMPERATURE DEVELOPMENT

Fig. 3.1 shows the variation of temperature due to heat of hydration measured in the center of a beam. The beam was stored in a steel mold at an ambient temperature of 23°C. The maximum value of 30°C was reached 8 1/2 hours after casting. The dashed vertical lines show the initial and final set of the concrete. They were measured on concrete samples using an equipment similar to the Vicat-apparatus but on a larger scale.

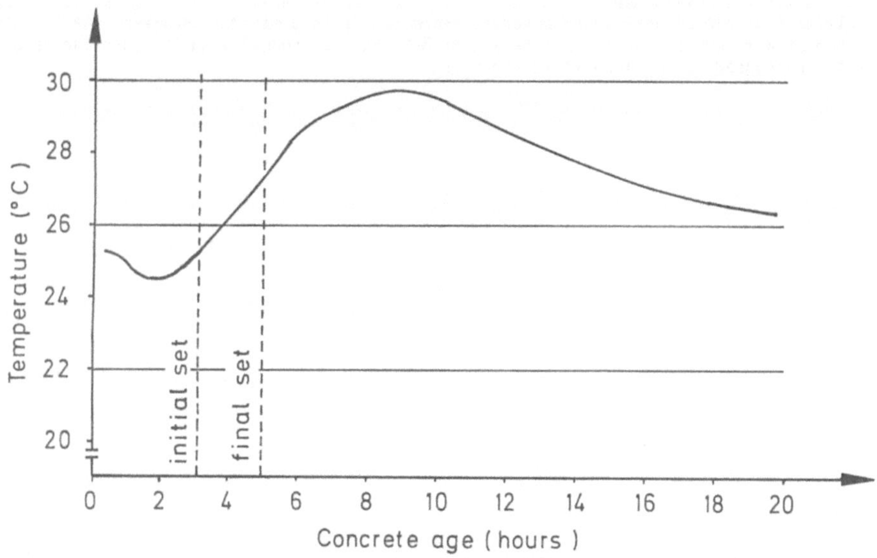


Fig. 3.1: Hydration temperature and setting of concrete

From Fig. 3.1 it can be seen that the concrete specimens tested at an age of 4 hours had surpassed the initial set. The concrete specimens tested after 5 hours where close to final set. However, we should keep in mind that the definitions of inital and final set are to some extent arbitrary.

The difference between maximum temperature in the concrete and ambient temperature is not very large. Nevertheless it may influence the development of the characteristic length as will be shown in chapter 3.3.2.

## 3.2 MECHANICAL PROPERTIES

### 3.2.1 Compression

Fig. 3.2 shows the development of compressive strength $f_c'$, Young's modulus $E_c$ i.e. the tangent to the stress-strain diagram in the origin and the ratio of $f_c'/E_c$ as a measure of the deformability of the concrete under compression [1]. Strength and Young's modulus develop at different rates. The rate of increase of $E_c$ up to a concrete age of about 12 hours is higher than that of compressive strength. At an age of 6 hours $f_c'/E_c$ passes through a minimum.

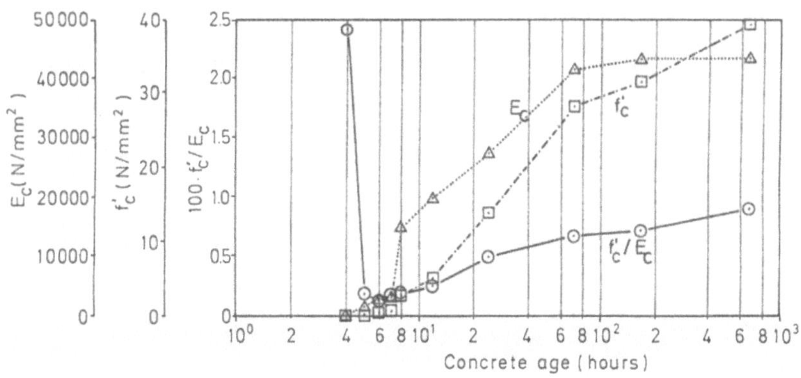


Fig. 3.2: Development of compressive strength $f_c'$, Young's modulus $E_c$ and ratio of $f_c'/E$

Fig. 3.3 gives the development of the ultimate strain $\varepsilon_u$ at maximum load. Similar to the ratio $f_c'/E_c$ the ultimate strain of very young concrete decreases up to a minimum. However, for $\varepsilon_u$ this minimum is reached at a somewhat higher concrete age of 8 hours. Subsequently an increase can be observed which is less pronounced for $\varepsilon_u$ than it is for $f_c'/E_c$.

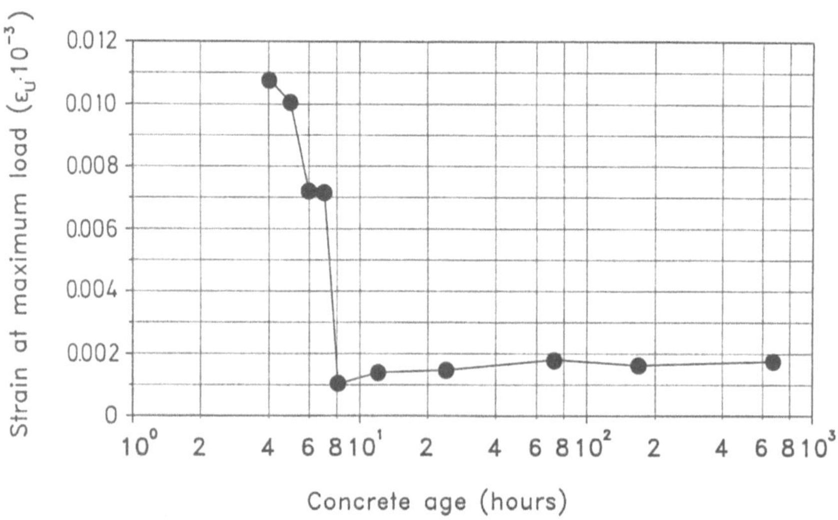


Fig. 3.3: Development of strain $\varepsilon_u$ at maximum compression load

In Figs. 3.4 and 3.5 compressive stress-strain diagrams of concrete cylinders at an age of 6 hours and 24 hours, respectively, are shown. The dashed lines represent the scatter of the experiments. Whereas, the $\sigma$-$\varepsilon$-curve of the older concrete has the usual appearance, a different curve is observed for the very young concrete. The first part of the $\sigma$-$\varepsilon$-diagram is characterized by a nearly linear increase of stress with strain up to about one third of the compressive strength. This limit of linearity decreases with decreasing age. Similar curves have been reported in [1].

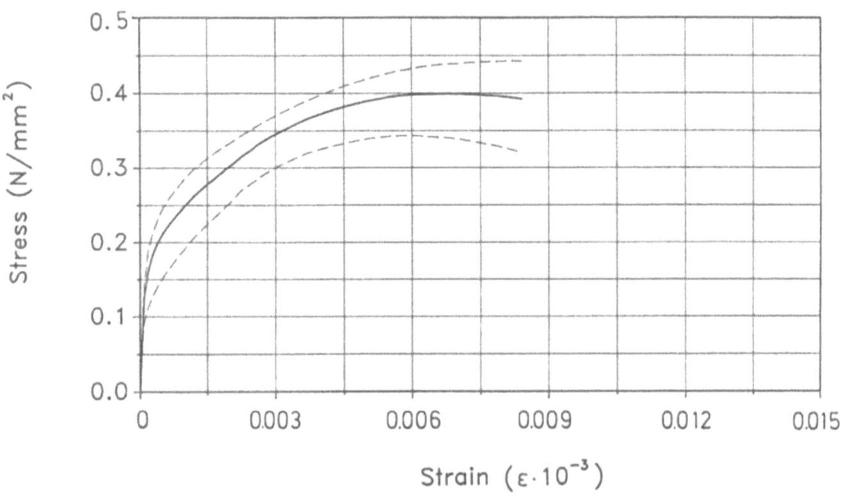


Fig. 3.4: Stress-strain-diagram, concrete age 6 hours, compression

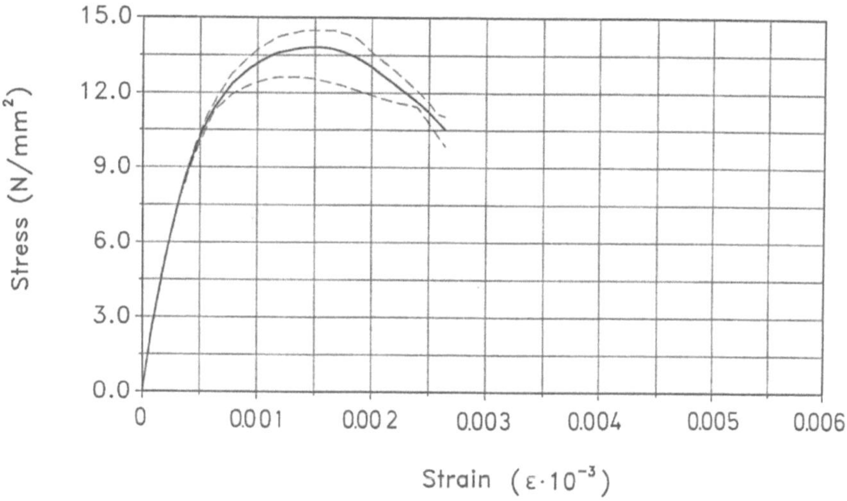


Fig. 3.5: Stress-strain-diagram, concrete age 1 day, compression

3.2.2 Bending and uniaxial tension

In Fig. 3.6 a typical load-deflection-curve of an unnotched beam tested at an age of 6 hours is shown. In this case a stable crack growth has been obtained.

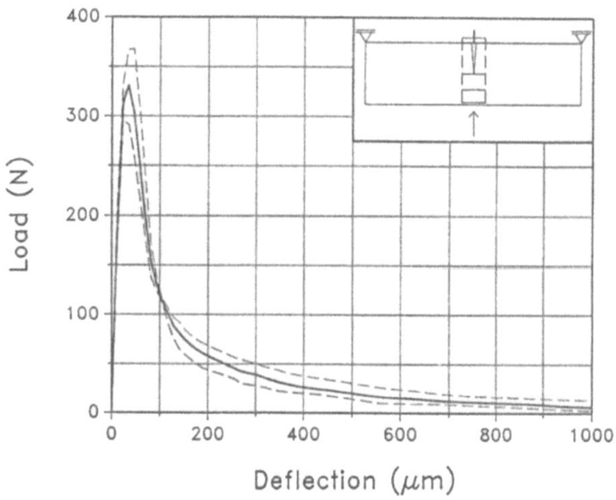


Fig. 3.6: Load-deflection diagram, concrete age 6 hours, unnotched beam.

Fig. 3.7 shows the development of flexural strength $f_f$ with concrete age. The development of the ratio $f_c/E_c$ is given in Fig. 3.8. This ratio reaches a minimum at a concrete age of 7 hours. For higher ages a continuous increase is observed. After 28 days it is similar to that after 4 hours. This behaviour has also been observed in [1].

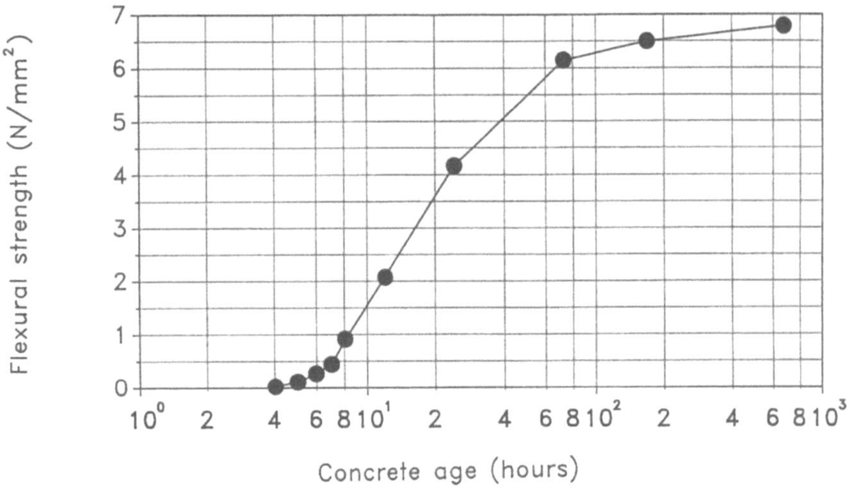


Fig. 3.7: Development of flexural strength

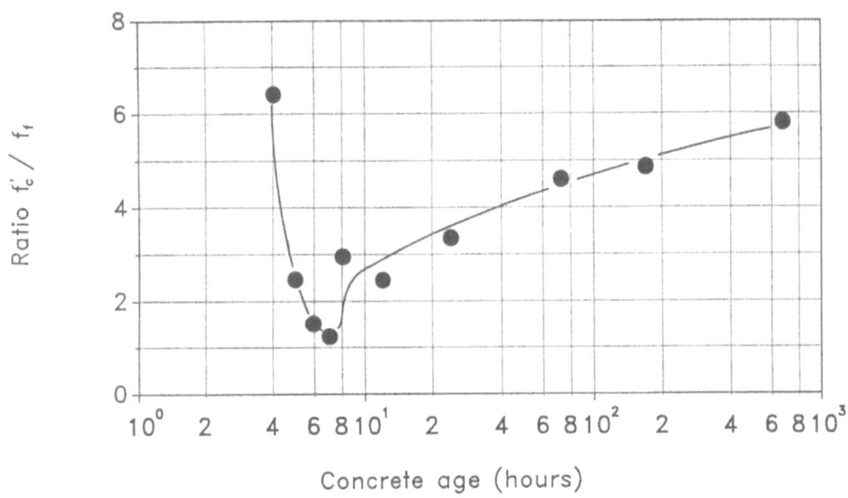


Fig. 3.8: Effect of concrete age on $f_c'/f_f$

The ratio of flexural to uniaxial tensile strength $f_f/f_t$ for concretes tested at an age of 3, 7 and 28 days was unaffected by concrete age. A nearly constant value of 2.0 was observed.

In Fig. 3.9 the development of deflection at maximum load determined from tests of unnotched and notched beams, respectively, is shown. Again a minimum can be observed which is reached after 5 to 6 hours i.e. earlier than the minimum of ultimate strain in compression. The deflection at maximum load in the range of 12 hours to 28 days appears to be unaffected by concrete age.

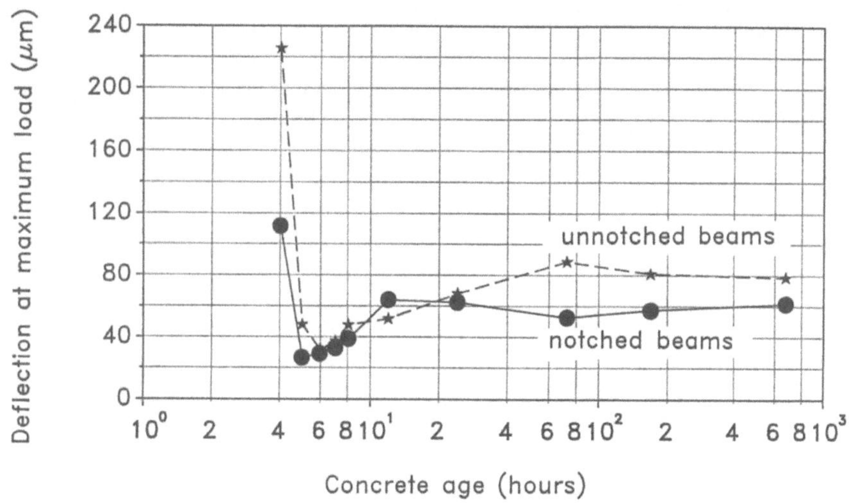


Fig. 3.9: Development of deflection at maximum load

## 3.3 FRACTURE MECHANICS PARAMETERS

### 3.3.1 Fracture energy

Fig. 3.10 shows the development of fracture energy of notched beams (solid line) and of unnotched beams (dashed line). It can be seen that fracture energy $G_F$ of unnotched beams is always larger than that of notched beams. This can be explained by a larger amount of microcracks developing at the tension side of an unnotched beam until a fracture process zone develops. Additional nonlinear deformations outside the fracture plain will increase the value of fracture energy. Similar behavior has been observed previously [12]. The ratio of $G_F$(notched)/$G_F$(unnotched) varies between 0.12 and 0.79. No systematic influence of the age of concrete is apparent. The development of $G_F$ and of flexural strength $f_f$ with age are similar.

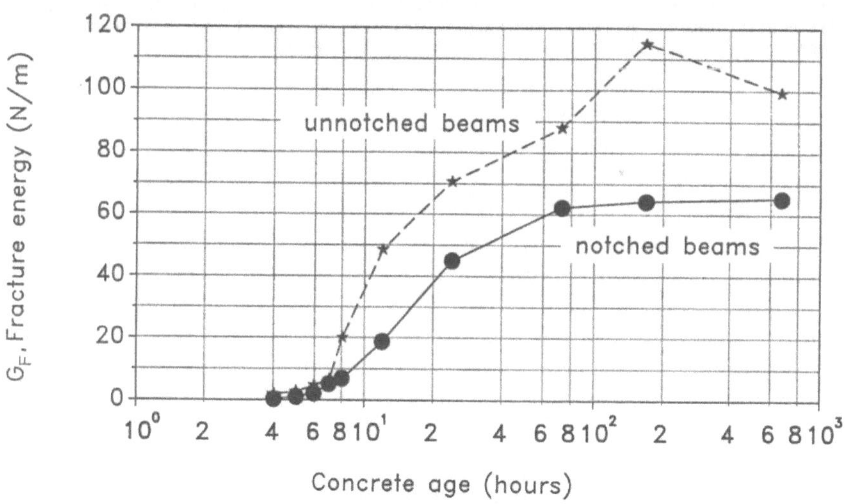


Fig. 3.10: Development of fracture energy

### 3.3.2 Notch sensitivity

Fig. 3.11 shows the influence of concrete age on notch sensitivity. This curve exhibits some important characteristics. Soon after initial set - in this case 4 hours - the concrete is highly notch sensitive. After a period of no notch sensitivity from 5 to 8 hours k decreases continuously and approaches a value of about 0.65. When comparing Figs. 3.1 and 3.11 there is no apparent relation between notch sensitivity and temperature.

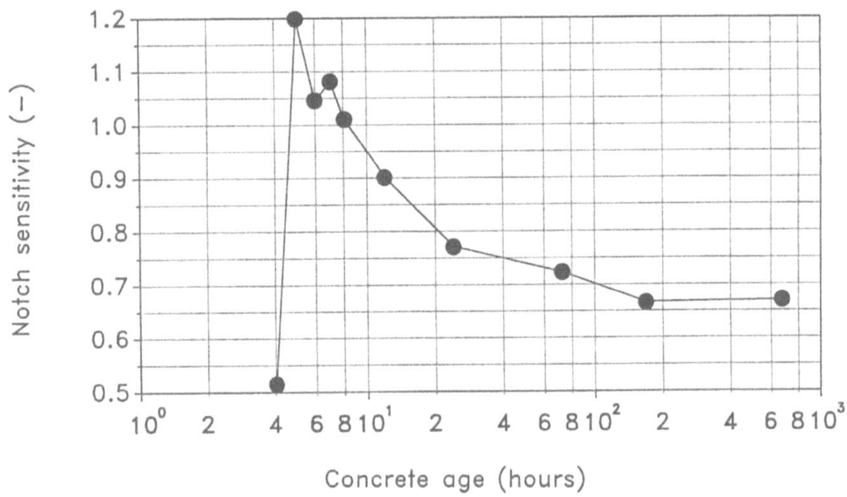


Fig. 3.11: Development of notch sensitivity k

### 3.3.3 Characteristic length

Figs. 3.12 and 3.13 show the development of the characteristic length calculated from $G_F$ for the notched and the unnotched beams, respectively. Except for the values at ages of 6 and 7 hours, the characteristic length develops similar to the notch sensitivity. No qualitative difference between $l_{ch}$(notched) and $l_{ch}$(unnotched) is obtained. The minimum of characteristic length at concrete ages of 6 and 7 hours may be related to the temperature development in the concrete specimens. At an age at which $l_{ch}$ is low the temperature in the concrete is substantially increased. It is likely that fracture energy decreases with increasing temperature resulting in a reduction of $l_{ch}$. However, temperature also affects rate of hydration and thus strength development so that no conclusions regarding temperature effects can be drawn without some systematic studies. At least the experiments give an idea of the sensitivity of fracture energy when temperature effects are likely to play an important role.

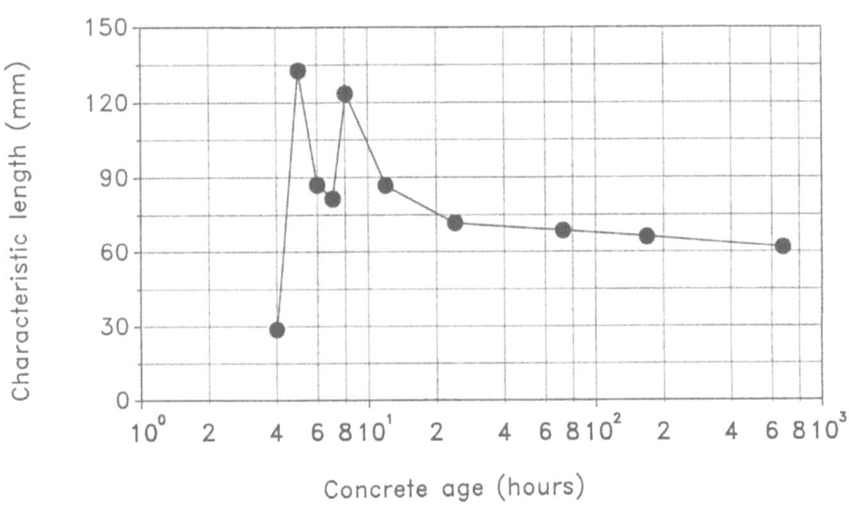


Fig. 3.12: Dependency of characteristic length on the age of concrete, notched beams

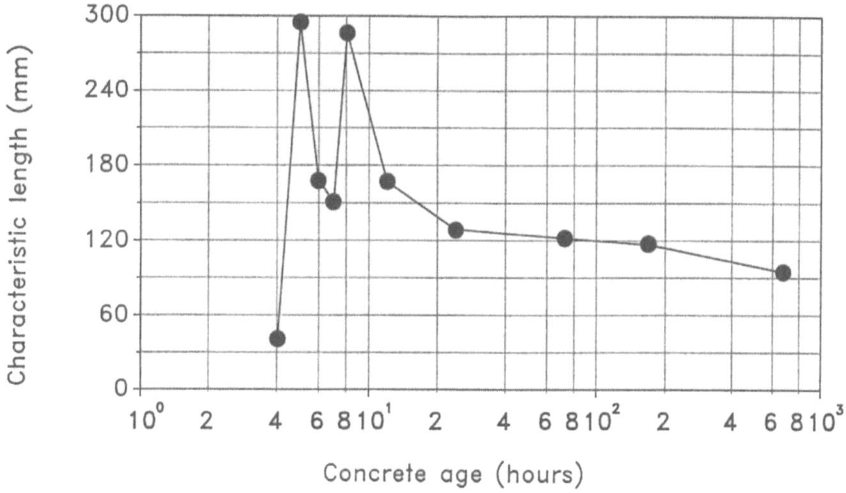


Fig. 3.13: Characteristic length as a function of age of concrete, unnotched beams

### 3.3.4 Relation between notch sensitivity and characteristic length

According to [15] notch sensitivity of brittle materials can be described as follows:

$$k = \frac{f_{net}}{f_f} = \frac{K_{Ic}}{f_f} \cdot \frac{1}{\sqrt{a_0}} \cdot \frac{1}{F^*(a_0/d)} \leq 1 \qquad (3)$$

where $K_{Ic}$ = fracture toughness
$a_0$ = initial notch length
$F^*(a_0/d)$ = function of geometry

For three-point bending:

$$F^*(\frac{a_0}{d}) = (1 - \frac{a_0}{d})^2 \cdot (1.93 - 3.07\,(\frac{a_0}{d}) + 14{,}53\,(\frac{a_0}{d})^2 - 25.11\,(\frac{a_0}{d})^3 + 25.80\,(\frac{a_0}{d})^4 \qquad (4)$$

Substituting $K_{Ic} = \sqrt{G_{Ic} \cdot E}$ (5)

where $G_{Ic}$ = critical strain energy release rate
$E$ = Young's modulus

and introducing $l_{ch} = G_F \cdot E/f_t^2$ in eq. 3

We obtain

$$\frac{\sqrt{l_{ch}}}{k} = \frac{n}{\sqrt{m}} \cdot \sqrt{d} \cdot F^{**}(\frac{a_0}{d}) \qquad (6)$$

where $n = f_f/f_t$
$m = G_{Ic}/G_F$

$$F^{**}(\frac{a_0}{d}) = \sqrt{\frac{a_0}{d}} \cdot F^*(\frac{a_0}{d}) \qquad (7)$$

Thus, a relation exists between notch sensitivity k and characteristic length $l_{ch}$. This has also been shown in [10, 11]. The ratios n and m are not constant but vary with $d/l_{ch}$. Theoretical relations for n, m = $f(l_{ch})$ are also given in [11]. Using the definition for $l_{ch}$ expressed by eq. 2b, the ratio of $d/l^*_{ch}$ varies between 0,75 and 1.6 with the exception of the value obtained for a concrete age of 4 hours. According to [11] this results in variations of $1.20 \leq n \leq 1.32$, of $0.25 \leq m \leq 0.34$ and of $2.06 \leq n/\sqrt{m} \leq 2.64$. Thus, the influence of age on $n/\sqrt{m}$ is not too pronounced.

In Fig. 3.14 the ratio $\sqrt{l^*_{ch}}/k$ is given as a function of concrete age. For a range of 8 hours to 28 days this ratio is nearly constant. Between 4 and 7 hours $\sqrt{l^*_{ch}}/k$ decreases. Further studies should clarify if this can be explained by the afore mentioned temperature effects.

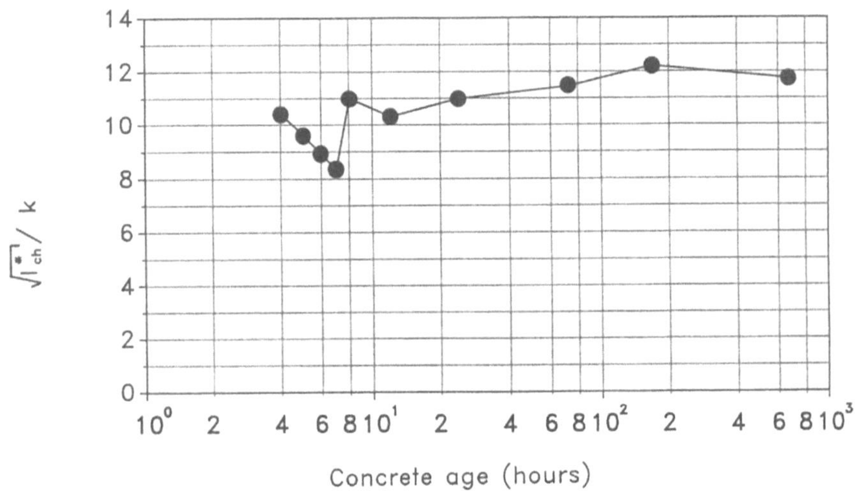


Fig. 3.14: Development of the ratio $\sqrt{l^*_{ch}}/k$

For ideally brittle materials $G_{Ic} = G_F$ and $f_f = f_t$. Then the ratio of $n/\sqrt{m}$ is eqal to unity and $\sqrt{l_{ch}}/k$ depends only on the geometry of the specimen, $a_o$ and $a_o/d$, respectively. Then the characteristic length $l_{ch}$ can be determined from e.q. 6 by testing notched and unnotched specimens and calculating the notch sensitivity k.

The advantage of using eq. 6 to evaluate $l_{ch}$ would be that no stable tensile or bending test had to be carried out. Presently, various analytical approaches are studied to estimate $l_{ch}$ from notch sensitivity if $n/\sqrt{m}$ is not a constant.

## 4. SUMMARY

4.1 A testing equipment is described for the determination of mechanical and fracture parameters of very young concretes. Testing in the mold with the tension side up proved to be advantageous with regard to handling the specimens and determining of fracture energy.

4.2 Compressive and flexural strength develop at different rates. The ratio $f'_c/f_f$ depends on the age of concrete and reaches a minimum about two hours after final set.

4.3 The deformability in compression and bending reach a minimum at an age at which low values of strength are observed.

4.4 Fracture energy increases monotonically with increasing concrete age. The characteristic length $l_{ch}$ does not develop continuously and passes through a minimum prior to final set.

4.5 Immediately after final set the very young concrete is not notch sensitive. With increasing age an increase of notch sensitivity is observed.

4.6 Since the concrete temperature increases due to heat of hydration during the critical periods where $l_{ch}$, notch sensitivity and $f_c'/f_f$ exhibit a minimum the effect of temperature on fracture parameters has to be investigated in further studies.

4.7 It is shown that the ratio $\sqrt{l_{ch}}/k$ appears to be nearly age independent with the exception of a drop during the period of increased concrete temperature.

LITERATURE

[1] H.J. Wierig, E. Gollasch
Untersuchungen über das Verformungsverhalten von jungem Beton,
Mitteilungen aus dem Institut für Baustoffkunde und Materialprüfung, Universität Hannover, Heft 47, 1982

[2] H. Weigler, S. Karl
Junger Beton - Beanspruchung - Festigkeit - Verformung
Betonwerk- und Fertigteiltechnik, Heft 6, 1974

[3] Y. Kasai
Method of estimation for compressive strength of concrete at early ages
Colloque International sur le Bêton Jeune, Vol. 1, 1982

[4] S. Kral, F. Becker
Zur Entwicklung mechanischer Betoneigenschaften im Frühstadium der Erhärtung, Beton 9/1976

[5] Properties of set of concrete at early ages - state-of-the-art-report -
Commission 42-CEA, Materiaux et Concstruction, Vol. 14

[6] Y. Kasai, K. Yokoyama, I. Matsui
Tensile properties of early-age concrete
Proceedings of the International Conference on Mechanical Behaviour of Materials, Vol. 4, 1971

[7] R. Springenschmid, G. Adam
Mechanical properties of set concrete at early ages - Test methods
Materiaux et Construction, Vol. 13, 1980

[8] R. Springenschmid, R. Breitenbücher
Über die Ursache und das Vermeiden von Rissen im Beton von Tunnelauskleidungen
Felsbau 3, Nr. 4, 1985

[9] M. Emborg
Temperatur stresses in massive concrete structures
Division of Structural Engineering, Lulea University, Sweden, 1985

[10] A. Hillerborg
Analysis of one single crack
Report to RILEM-Committee TC 50-FMC, January 1981

[11] P.-E. Petersson
Crack growth and development of fracture zones in plain concrete and similar materials
Report TVBM-1006, Lund, Sweden 1981

[12] H.K. Hilsdorf, W. Brameshuber
Size effects in the experimental determination of fracture mechanics parameters
Application of Fracture Mechanics to Cementitious Composites, NATO-ARW-September 1984

[13] H.K. Hilsdorf, W. Brameshuber
Probengrößenabhängigkeit bruchmechanischer Parameter
Baustoffe '85, Karlhans Wesche gewidmet, Bauverlag 1985

[14] RILEM draft recommendation
Determination of the fracture energy of mortar and concrete by means of three point bend tests on notched beams
Materiaux et construction, Vol. 18, No. 106, 1985

[15] S. Ziegeldorf, H.S. Müller, H.K. Hilsdorf
A model law for the notch sensitivity of brittle materials
Cement and Concrete Research, Vol. 10, No. 5, 1980

SEM/RILEM International Conference
on Fracture of Concrete and Rock
Houston, Texas, June 1987
Editors: S.P. Shah, S.E. Swartz

# FRACTURE PROPERTIES OF EPOXY POLYMER CONCRETE

by

C. Vipulanandan[1] and N. Dharmarajan[2]

## ABSTRACT

Epoxy polymer concrete beams with varying polymer content and notch depth were tested at two temperatures. The results were analysed to examine the applicability of various fracture parameters such as critical stress intensity factor, J-integral and fracture surface energy to characterize the fracture behavior of epoxy polymer and polymer concrete. The notch sensitivity of polymer and polymer concrete are also investigated.

## INTRODUCTION

Polymers are being increasingly used in civil engineering applications as adhesives, modifiers and matrix materials in concrete (6,8,14,15). There are three categories of concrete which contain polymers: polymer-impregnated concrete (PIC), polymer concrete (PC) and polymer-portland cement concrete (PPCC). PIC is a hydrated portland cement concrete that has been impregnated with a monomer which subsequently polymerizes in-situ; PC is a composite material formed by combining a mineral aggregate such as sand and gravel with a polymerizing monomer; and PPCC is produced by adding either a monomer or polymer to a fresh concrete mixture either with water or cement which subsequently polymerizes during curing of the material in place. From these definitions, the concrete-polymer composites differ by polymer to portland cement substitution and the way the polymer is introduced into the concrete. Of these concrete-polymer composites PC has the most selective control of the mechanical properties to suite the needs of the applications. The versatility in formulation and processing has led PC to many applications such as flooring, cast articles of various kinds, patching and overlays for highways and bridge decks (7,8,14). PC is a high strength, rapid-setting material and its current applications warrant mainly fine aggregate fillers with grain size no more than 5 mm (6). Both thermosetting polymers (cross linked polymers) such as epoxies and polyesters and thermoplastic polymers such as acrylics (methamethyl acrylates) are commonly used as binders in PC. Generally the high strength PC exhibits rather brittle behavior and the properties are highly sensitive to temperature (20). Emergence of this new family of construction materials has dictated the need for better and more efficient characterization of materials and standardization of testing procedures. As structural and repair material, polymer and its composites which are invariably brittle, must be able to withstand high stresses under extreme service conditions. Hence a knowledge of their fracture properties is vitally important in aiding their efficient utilization.

One potential approach to better understand the fracture behavior of brittle and ductile materials is given by fracture mechanics. The fracture mechanics parameter $K_{IC}$ (fracture toughness) is a critical value of the stress intensity factor, which can be regarded as a one-parameter description of the crack tip region for a single crack and considered as a material property for linear elastic, homogeneous and isotropic materials. Since some of these assumptions are incorrect for polymers, cement paste, mortar, cement concrete and rock several different approaches have been used in recent years to characterize the crack behavior of these materials. Of these different approaches fracture surface energy, critical value of J-integral, critical crack opening displacement, compliance technique for determining slow crack growth and R-curve analysis are being frequently used for characterizing polymers and geomaterials (3,9,10,11,22,23). The J-integral, a path-independent line integral, has been used as a failure-criterion for materials with elastic-plastic behavior (23). For both linear and nonlinear materials J is equivalent to the energy necessary to increase a unit width crack surface by an infinitesimal amount. For linear elastic materials $J_{IC}$ is equal to the strain energy release rate ($G_{IC}$). Some assumptions and limitations have to be respected when the J-integral is used as a failure criterion for cementitious materials. $J_{IC}$ applies only to on set of crack propagation with unloading not permitted (23). Furthermore, LEFM and J-integral concepts are no more applicable when local cracking occurs in the material before rapid crack propagation begins. In cementitious materials, such cracking is a well known fact. One approach in studying the fracture behavior of particulate composite is to study the response of the composite, binder, aggregate and their interface separately under similar environmental conditions to better understand the contribution of the constituents to the overall behavior of the composite.

1 Assistant Professor, Department of Civil Engineering, University of Houston, Houston, Texas 77004.
2 Research Associate, Department of Civil Engineering, University of Houston, Houston, Texas 77004.

Many attempts have been made in recent times to apply LEFM and elastic-plastic fracture mechanics concepts to epoxy and polyester polymers, cement paste, polymer impregnated cement paste, mortar, cement concrete, PIC and rocks (1,2,3,17,23). Mindess, Lawrence and Kesler (12) have used $J_{IC}$ as well as linear fracture parameters $K_{IC}$ and $G_{IC}$ while studying the fracture properties of cement paste, concrete and fiber reinforced concrete systems. They concluded that $J_{IC}$ appears to be a more promising fracture parameter for all the materials considered and that $K_{IC}$ was only good for cement paste. Velazco, Visalvanich and Shah (19) have observed that the R-curve analysis was the most promising fracture criterion for fiber reinforced systems. Fracture studies on concrete-polymer composite materials have been primarily limited to PPCC and PIC (1,2,5). Cook and Crookham (5) observed that $K_{IC}$ and stiffness decreased in PPCC, irrespective of polymer type. Polymer impregnation increased the fracture toughness and failure strain of the concrete. Polymer impregnated concrete was found to be notch sensitive with $K_{IC}$ having a limiting value between notch-to-depth ratio of 0.35 and 0.42 and beyond which $K_{IC}$ generally decreased. Auskern and Horn (2) while studying the fracture surface energy ($\gamma$) of polymer-impregnated hardened cement paste observed that the energy to fracture increased by a factor of 8 due to impregnation with methyl methacrylate polymer materials. The increase in fracture energy was attributed to the polymer phase. Alezka and Beaumont (1) have also observed that the work to fracture was enhanced in the concrete systems when impregnated with an acrylic polymer. In an earlier study Tazawa and Kobayashi (18) noted that the improvements in PIC properties could be reasonably described using the Griffith theory. Several investigators have reported widely varying fracture parameters for geomaterials and only the results from 4-point bending tests (unless otherwise stated) are summarized in Table 1. In general the cement paste has the least fracture toughness compared to mortar and cement concrete with the cement concrete having the highest fracture toughness. The cement paste is highly notch sensitive compared to mortar and cement concrete. Studies on fracture surfaces of concretes made of normal weight aggregates and having a unconfined strength of 30 $N/mm^2$, indicate that the crack normally propogates only in the matrix or at the interfaces. For light weight aggregate concretes and high strength concretes the cracks also propagate through aggregate particles (23). Unfortunately there are only a few investigations on the fracture mechanics parameters of aggregates which are linked to crack propagation of some concretes. However, there are several investigations on the fracture of rocks and typical values for a granite rock are shown in Table 1. Fracture toughness of granite rock is larger than cement paste, mortar or concrete. There are only a few experiments concerning fracture at the cement paste - aggregate interface and $K_{IC}$ of 0.21 and 0.16 $MNm^{-3/2}$ for cement-quartzide and cement-limestone interface have been reported from asymmetric CT-specimens (23). Since cracking at the cement paste-aggregate interface will not satisfy homogeneity and isotropy the applicability of fracture mechanics parameters have not been well established. The validity of fracture mechanics parameters to cement concretes and its constituents are still under investigation.

Epoxy resins are very versatile thermosetting polymers and can be hardened by many different curing agents. An important variable that affects the fracture behavior of thermosetting polymer such as epoxy is the chemical structure, and hence the degree of cross-linking, which is in part controlled by such factors as method of curing and type and amount of curing agent or hardener used. Study into crack propagation in epoxy resins showed that crack propatation tended to take place by means of either continuous stable propagation or discontinuous unstable stick-slip mechanism which resulted in load displacement curves of saw-tooth appearance (22). Further studies have shown that these crack propagation mechanisms are temperature and loading rate dependent. Using epoxy epikote 828 cured with triethylene tetramine and double torsion specimen Yamini and Young (21) determined the plane strain fracture toughness at conditions of both crack initiation ($K_{Ici}$, corresponding to the maximum of the saw-tooth curve) and crack arrest ($K_{Ica}$, corresponding to the minimum of the saw-tooth curve). The results indicate the $K_{Ica}$ decreased with increase in temperature and $K_{Ici}$ increased with increase in temperature. Researchers have also measured $G_{IC}$ for different epoxy systems and the reported values ranged from 86 $Jm^{-2}$ to 1000 $Jm^{-2}$ (22). For a particular resin/hardener system there is usually one particular hardener content which gives the maximum value of $G_{IC}$ and it was found that there was different relationships between Young's modulus and hardener content than between $G_{IC}$ and hardener content (22). This means that although one particular hardener content will give a maximum value of $G_{IC}$ there may not be a corresponding maximum value of $K_{IC}$ at this composition.

It is well known that glassy polymers will readily undergo deformation by crazing. This is a process whereby small crack-like entities which are normal to the maximum principal tensile stress are formed during deformation. However, crazes differ from cracks in that they are internally bridged by fibrils. Although there is little direct evidence for crazing in epoxy resin, Lilley and Holloway have reported seeing small crazes in the vicinity of crack tips in some epoxy resins and Van den Boogaart has observed crazes at the tips of moving cracks in under cured epoxy resins (22). Recent work on strained thin films of epoxy resins examined in the transmission electron microscope, has shown that crazes are formed and they are relatively short (approximately 1 μm) and the size of the crazes would explain why they are not normally observed at the crack tip by optical techniques (22).

The application of LEFM to cementitious materials was justified because crack branching was treated analogeous to the plastic zone at the crack tip and assuming that the crack branching is confined to a small region. It was also shown that on metallic and ceramic materials application of LEFM leads to reasonable results even if not all the requirements of LEFM had been fulfilled (23). The validity of LEFM parameters and/or J-integral as criteria for fracture of polymer concrete can only be established if it can be shown that tests on specimens with significantly different dimensions (crack length, specimen sizes and notch width) yield identical results. Fracture behavior of thermosetting polymers such as epoxy and polyester are affected by type and amount of hardener (degree of cross linking), temperature, and mode and rate of loading. With thermosetting polymers LEFM can normally be applied well below their glass transition temperatures (22). But, with increase in temperature the polymer behavior is nonlinear and it may be more appropriate to use J-integral to characterize fracture (10). Hence, influence of temperature and resin content on the fracture behavior of epoxy polymer concrete is investigated and the applicability of LEFM parameters and J-intergal to characterize the fracture behavior of polymer concrete are verified. Due to its economical fabrication and relatively simple testing procedures SEN beams loaded in 4 point bending were used for this study. The notch sensitivity of epoxy polymer and polymer concrete are also investigated.

Table 1. Summary of Fracture Parameters from Bending Tests (11)

| Material | $\gamma_{IC}$ ($J/m^2$) | $K_{IC}$ ($MNm^{-3/2}$) | $G_{IC}$ ($J/m^2$) | $J_{IC}$ (N/m) |
|---|---|---|---|---|
| Cement | *1.3 - 9.1 | 0.3 - 0.66 | 1 - 15 | 11 - 15 |
| Mortar | *4.2 - 18 | 0.2 - 0.87 | 8 - 40 | 15 - 35 |
| Cement Concrete | *3.2 - 3.5 | 0.3 - 1.7 | 10 - 34 | 14 - 57 |
| PIC | 2,000 - 20,000 | 0.2 - 0.4 | - | - |
| Granite Rock (17)** | - | 2.6 | - | 114 |

* 3-point bending
**Compact Tension

TESTING PROGRAM

Epoxy resin is a widely used polymeric binder in PC formulations (6,8,15). The epoxy resin has a viscosity between 110 and 150 poise at room temperature and a specific gravity of 1.1. The versamid 140 has a viscosity of 150 poise at 20°C. The composition of the polymer concrete systems used in this study are summarized in Table 2. Ottawa 20-30 sand with exclusively spherical particles are composed of pure quartz and has a specific gravity of 2.65. Both polymer and polymer concrete beams specimens were tested in 4-point bending (third-point loading). In order to investigate the effect of resin content on the fracture parameters the resin content was varied within practical limits and the complete mix proportions are given in Table 2. The aggregate and the polymer were hand mixed in small disposable containers. Each polymer concrete specimen was cast in teflon lined aluminum mold of dimensions 300 mm x 50 mm x 50 mm and compacted in three layers before compressing it to a maximum pressure of 100 psi. The PC specimens were 38 mm wide, 50 mm deep and 300 mm long and were loaded so that the length of the central constant moment zone was 84 mm. The epoxy polymer was cast in aluminum trays of dimensions 250 mm x 250 mm x 25 mm with the inside lined to facilitate demolding. The cured epoxy polymer was cut to 25 mm wide, 25 mm deep and 200 mm long beam specimens and loaded so that the length of the central constant moment was 58 mm. The polymer and PC specimens were first cured at room temperature for a day and then at 60°C for another day before testing. At least two specimens were tested at each condition of notch depth and temperature and the specimens were notched using a diamond saw. The notch depths were varied from 12.5 to 37.5 mm for PC and 6 to 19 mm for polymer specimens.

Table 2. Composition of Polymer Concrete Systems

| | |
|---|---|
| Polymeric Matrix | (By Weight) |
| Epoxy: | |
| Resin: Epon 828 (Shell Chemical Co., Houston, TX) | 10-18% |
| Initiator: Versamid 140* | 35% |
| Aggregate | |
| Sand: Ottawa 20-30 (Coefficient of uniformity = 1.08) | 82-90% |

* By weight of resin.

Flexural testing was done on both polymer and polymer concrete at room temperature and at 80°C to investigate the effect of temperature at a constant strain rate of $10^{-4}$ strain per minute. The upper limit of 80°C was selected based on the glass transition temperature of epoxy and its nonlinear behavior. For the elevated temperature test the specimens were placed in the temperature chamber for at least a day before testing. All dimensions were measured and the PC specimens were weighted before testing. These measurements were then used in calculating the density of polymer concrete and the volume fraction of voids and polymer in the specimen.

## BEHAVIOR OF POLYMER

It is imperative to characterize the fracture properties of the epoxy polymer under the curing and loading conditions used in this investigation. The load deflection relationship for unnotched and notched epoxy polymer at 22°C and at 80°C are shown in Figure 1a and 1b. The specimens failed in about 20 minutes and the fractured surfaces were not plane and hence appropriate corrections will have to be made in calculating the fracture surface area. With the increase in temperature, the linear behavior of both unnotched and notched specimens become nonlinear. At room temperature the tensile failure is brittle but with increase temperature it becomes ductile. The flexural modulus changes from 2.90 x $10^3$ MPa at 22°C to 0.89 x $10^3$ MPa at 80°C. This is a reduction of about 70% in modulus compared to the modulus at 22°C. For the testing conditions adopted in this study no saw-tooth appearance was observed in the load displacement curves. Since the epoxy polymer is homogeneous and exhibits linear and brittle behavior it appears that LEFM will be applicable to this material at room temperature.

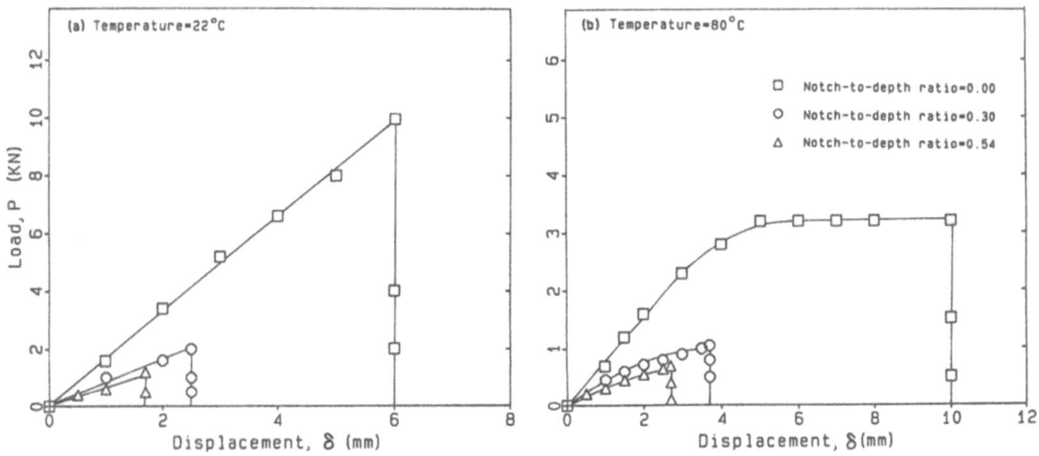


Figure 1. Load-Displacement Curves for Epoxy Polymer at Different Notch-to-Depth Ratios and Temperatures.

According to ASTM E 399 the critical load for evaluating a valid $K_I$ value is determined by the secant line with a slope of 0.95 times the initial tangent slope. Instead of this the maximum load is utilized in determining $K_I$ and hence this would lead to higher $K_{IC}$ values. Similar approach has been adopted for cementitious materials (23). Since the determination of the true crack length at the onset of rapid crack propagation will require detailed instrumentation of specimen with appropriate data acquisition system, notch depth is used as the initial crack length in evaluating the fracture parameters. Assuming a beam of cross-section b x w with an initial crack length a, the stress intensity factors were calculated by using the equation developed for 4 point-bending by Brown and Srawley (4)

$$K_I = \frac{6M\sqrt{a}}{bw^2}[1.99 - 2.47\left(\frac{a}{w}\right) + 12.97\left(\frac{a}{w}\right)^2 - 23.17\left(\frac{a}{w}\right)^3 + 24.80\left(\frac{a}{w}\right)^4] \quad (1)$$

where M is the applied pure bending moment. Figure 2a shows the variation in $K_I$ with crack length at two different temperatures for epoxy polymer. Note that the data points for each temperature could be approximated by a horizontal straight line. $K_I$ remains constant with increase in crack length and hence may be considered as a material property. Increase in temperature results in decrease in fracture toughness and when the temperature changes from 22°C to 80°C the fracture toughness is reduced by 30%. It can be seen from Figure 2a that ASTM recommendation (plane-strain fracture toughness of metallic materials) for a ($>2.5(K_{IC}/\sigma_y)^2$) is valid for epoxy, where $\sigma_y$ is the flexural strength of epoxy. The $K_{IC}$ for the epoxy used in this study is much higher than the values reported for cement paste and is comparable to the fracture toughness of granite rock (Table 1).

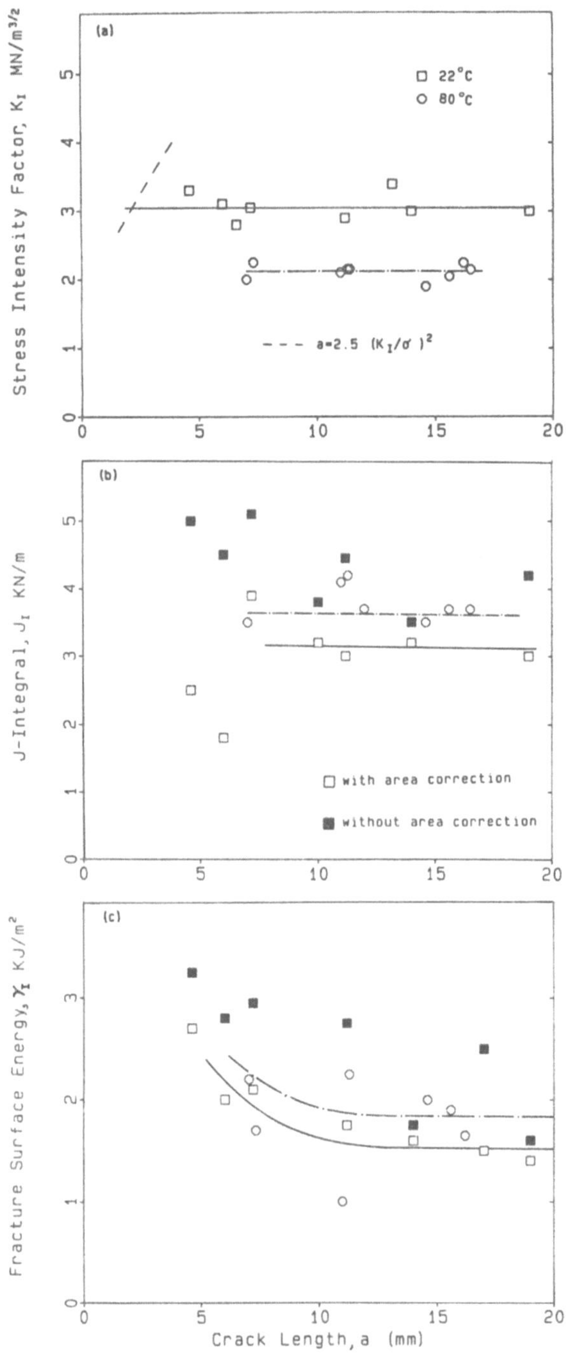


Figure 2. Variation of Epoxy Polymer Fracture Parameters with Crack Length and Temperature.

The J-integral is defined as the change in potential energy when the crack is extended by a small amount and various methods have been proposed to evaluate J experimentally. Rice, Paris and Merkle (16) have developed a simplified method to evaluate J using load vs. load-point-displacement curve of notched and unnotched specimens. In adopting this procedure the following relationship is used

$$J_I = \frac{2}{b(w-a)} (A_T - A_u) \tag{2}$$

where $A_T$ is the strain energy stored in the notched specimen at maximum load and $A_u$ is the strain energy in the unnotched specimen at the corresponding load level. Since the fracture surface was not a plane, both nominal and corrected value for the cracked length (w-a) were used in the calculation and compared in Figure 2b for 22°C. There is a larger scatter in the $J_I$ values calculated from the nominal cracked length (w-a). Data obtained from the corrected cracked length will be used in determining $J_{IC}$. As shown in Figure 2b the data could be approximated by horizontal straight lines. The $J_I$ value for epoxy polymer increases with increase in temperature but decreases with increase in crack length as shown in Figure 2b. The $J_{IC}$ value for epoxy is about 200 times higher than cement paste and about 25 times greater than the granite rock. When linear elastic behavior predominates $J_{IC}$ is directly related to the $K_{IC}$ and $\gamma_{IC}$ through following equations.

$$K_{IC} = \sqrt{\frac{J_{IC} E}{(1-\nu^2)}} \tag{3a}$$

$$\gamma_{IC} = J_{IC}/2 \tag{3b}$$

Hence $K_{IC}$ could be estimated from $J_{IC}$ and is compared to the $K_{IC}$ measured in Table 3. Since the Poisson's ratio ($\nu$) for epoxy is not available, the results are compared at two values of Poisson's ratios. There appears to be greater disagreement between calculated and measured $K_{IC}$ at higher temperatures.

Assuming that all the energy expended is changed into creating new surface area, the fracture surface energy is then obtained by dividing the total work to fracture U (area under the load-deflection curve) by the new fracture surface area 2A.

$$\gamma_I = \frac{U}{2A} \tag{4}$$

Since the failure surface for polymer was not planar, as shown in Figure 2c actual area and planar areas were used in the calculations. Critical value of $\gamma_I$ is the limiting value of $\gamma_I$ with crack length. The critical fracture surface energy $\gamma_{IC}$ increases with increase in temperature. When the temperature changes from 22°C to 80°C $\gamma_{IC}$ increases by 19%. Comparing $\gamma_{IC}$ to $J_{IC}$ it appears that Equation (3b) holds good for epoxy polymer (Table 3). Like other fracture parameters $\gamma_{IC}$ for epoxy is much higher than cement paste.

Table 3. Temperature Effects on Fracture Parameters

| Polymer Content (W%) | Temperature (°c) | $\gamma_{IC}$ J/m² | $K_{IC}$ MNm$^{-3/2}$ | $J_{IC}$ (N/m) | $K_{IC}$ (Eq. 3a) MNm$^{-3/2}$ $\nu$=0.1 | $\nu$=0.3 |
|---|---|---|---|---|---|---|
| | 22 | 250 | 1.5 | 420 | 1.54 | 1.61 |
| 14% | 80 | 350 | 1.2 | 560 | 1.44 | 1.50 |
| | 22 | 1600 | 3.0 | 3100 | 3.01 | 3.14 |
| 100% | 80 | 1900 | 2.1 | 3600 | 1.80 | 1.88 |

BEHAVIOR OF POLYMER CONCRETE

The variation of flexural strength and modulus of epoxy polymer concrete with polymer content (by weight) is shown in Figure 3. The method of preparation adopted in the study resulted in void content of 0.17, 0.11, and 0.05 for 10%, 14% and 18% polymer concrete systems. The modulus increases with increase in polymer content; but on reaching a maximum it decreases with further increase in polymer content. The flexural strength increases at a decreasing rate with increase in polymer content. Figure 4a compares the effect of notch depth on the load-displacement relationship for 14% polymer concrete system. The response becomes nonlinear for all systems close to peak load. Inspection of the failure surface shows both cohesive and adhesive failure of epoxy. Figure 4b compares the effect of temperature on the load-deflection behavior of 14% polymer concrete at a notch to depth rate of 0.65. As shown in Figure 5a, $K_I$ remains constant with increase crack length and also satisfies the ASTM recommendation for a. $K_{IC}$ increases with increase in

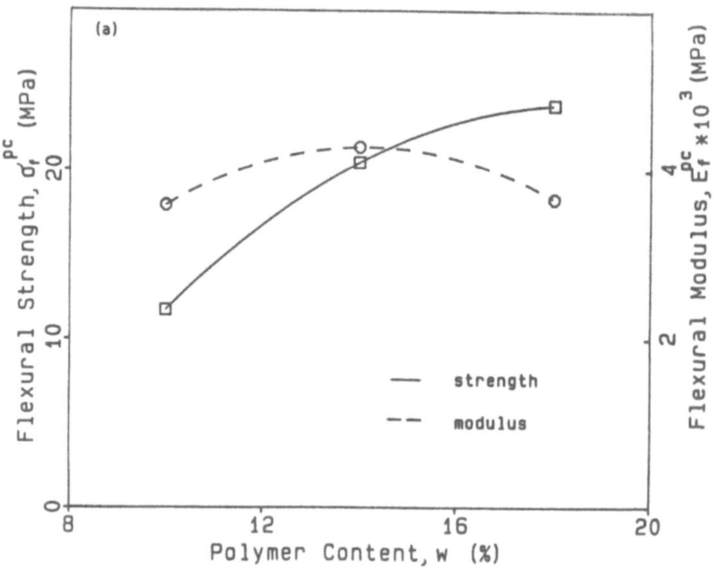


Figure 3. Variation in Epoxy Polymer Concrete Strength and Modulus with Polymer Content.

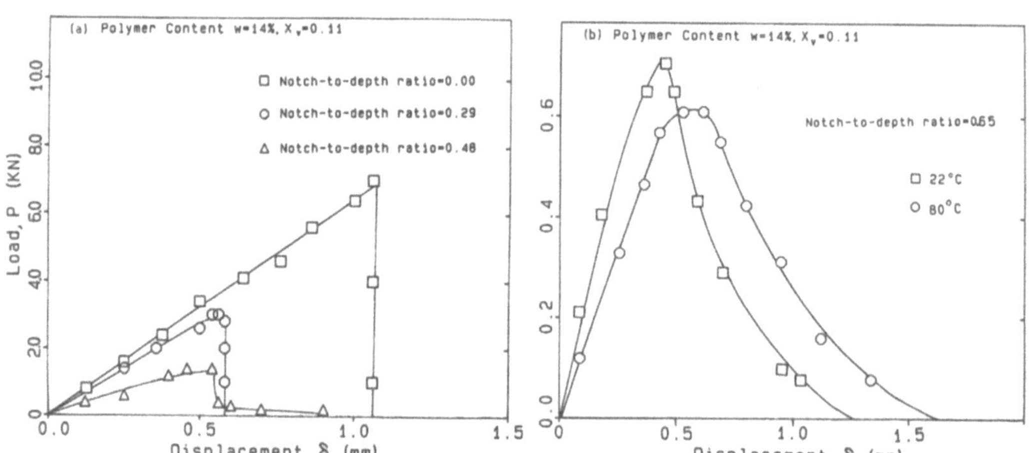


Figure 4. Load-Displacement Curves for Epoxy Polymer Concrete at Different Notch-to-Depth Ratios and Temperature.

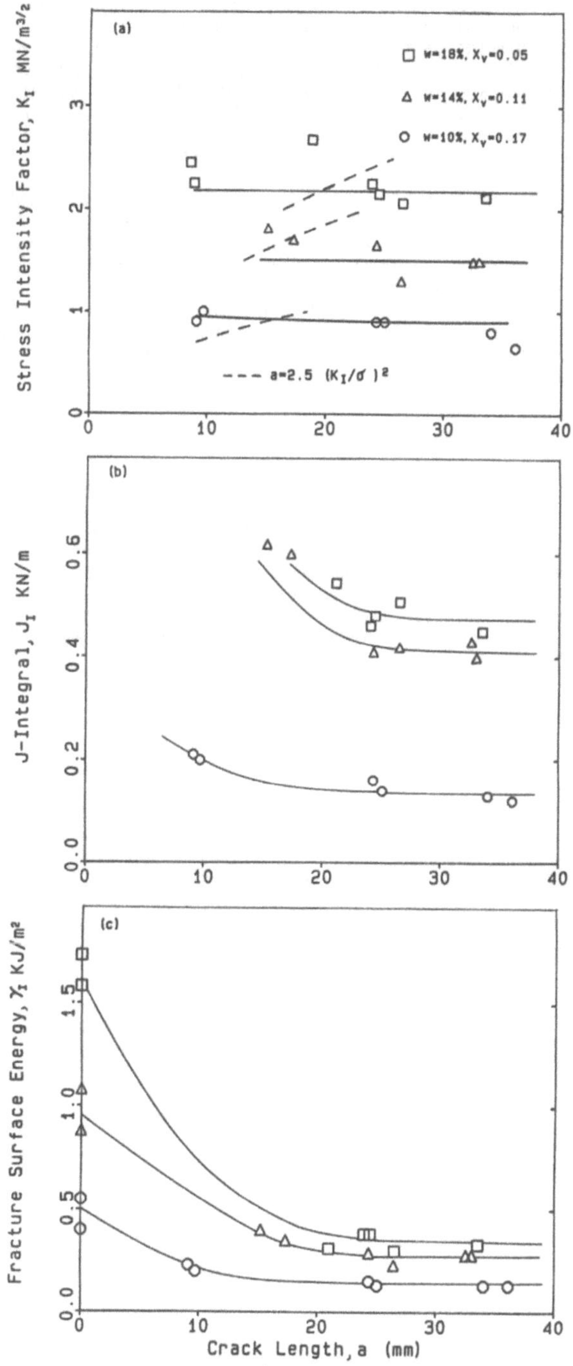


Figure 5. Variation of Epoxy Polymer Concrete Fracture Parameters with Crack Length and Polymer Content.

polymer content and is 2.1, 1.5 and 0.9 $MNm^{-3/2}$ for 18%, 14% and 10% polymer content respectively. The change in $K_{IC}$ is proportional to change in polymer content and void content. While $K_I$ depends only on the maximum load, $J_I$ depends not only on the maximum load but also on the shape of the load-deflection curve. $J_I$ decreases with increase in notch depth and approaches a constant value beyond a crack length of about 19 mm (notch to depth ratio of 0.5). $J_{IC}$ increases with increase in polymer content and changes from 140 N/m at 10% polymer content to 400 N/m at 14% and 450 N/m at 18% polymer content (Figure 5b). $\gamma_I$ decrease with increase in crack length but approaches almost a constant value near notch to depth ratio of 0.5. Like $K_{IC}$ and $J_{IC}$, $\gamma_{IC}$ also increases with increase in polymer content. At room temperature $\gamma_{IC}$ has values of 100, 250 and 300 $Jm^{-2}$ at polymer contents of 10%, 14% and 18% respectively (Figure 5c).

The values summarized in Table 3 for 80°C temperature were obtained from similar series of tests. Estimated $K_{IC}$ (Equation 3a) is compared to the measured $K_{IC}$ in Table 3. With increase in temperature the disagreement between estimated and measured $K_{IC}$ becomes greater, indicating the non-applicability of LEFM at higher temperatures. Comparing $\gamma_{IC}$ to $J_{IC}$ indicates that Equation 3b is not applicable to epoxy polymer concrete, even at room temperature. Similar observations have been made on cement paste, motar and cement concrete when $\gamma_{IC}$ is determined from $G_{IC}$ (13). The fracture parameter ratios of PC to polymer at corresponding temperature increase with increase in temperature, which implies that the cohesive fracture properties of the epoxy polymer is not the only property that is governing the behavior of PC; and hence adhesive fracture properties may have to be considered for better characterization of fracture properties of polymer concrete.

NOTCH SENSITIVITY OF POLYMER AND POLYMER CONCRETE

A material may be considered to be notch sensitive if the presence of a notch causes a change in the strength of the material (calculated on the reduced cross section, but neglecting the stress concentration effect of the notch). For metals the introduction of a notch may lead to notch strengthening or notch weakening. But for cementitious materials notch sensitivity refers only to the possible reduction in strength due to presence of a notch. Figure 6 shows the variation in notch sensitivity (ratio of notched strength to unnotched strength) with relative notch depth. The epoxy polymer is much more notch sensitive than the polymer concrete. Similar trends have been observed between cement paste and cement concrete. Notch sensitivity of polymer concrete is reduced with increase in polymer content.

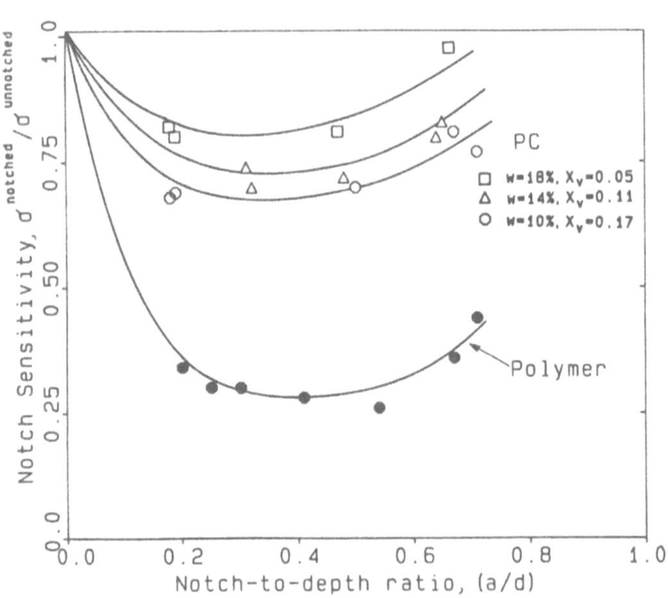


Figure 6. Effect of Relative Notch Depth on Notch Sensitivity of Epoxy Polymer and Polymer Concrete.

## CONCLUSION

Based on the experimental study the following conclusions can be advanced:

1. Stress intensity factor ($K_I$) and J-intergal for epoxy polymer are insensitive to the inital crack length. While $K_I$ for epoxy polymer concrete is independent of the crack length, $J_I$ and $\gamma_I$ vary with crack length and reach limiting values around notch to depth ratio of 0.5.

2. $K_{IC}$, $J_{IC}$ and $\gamma_{IC}$ of polymer concrete increase with increase in polymer content (reduced void content). While $K_{IC}$ reduces with increase in temperature, $J_{IC}$ and $\gamma_{IC}$ of epoxy polymer concrete increase with increase in temperature for the range of temperature considered in this study. $J_{IC}$ and $K_{IC}$ for epoxy polymer concrete do not correlate well at 80°C. The fracture parameter ratios of PC to polymer increase with increase in temperature.

3. Epoxy polymer is much more notch sensitive than polymer concrete. Notch sensitivity of polymer concrete is reduced with increase in polymer content.

## ACKNOWLEDGMENT

This study was supported by the Texas Advanced Technology Research Program grant to the University of Houston.

## REFERENCES

1. Alezka, J.C. and Beaumont, P.W.R., "The Work of Fracture of Concrete and Polymer Impregnated Concrete Composites," Proceedings, 1st International Congress on Polymer in Concrete, England, pp. 269-275, 1975.

2. Auskern, A. and Horn, W., "Fracture Energy and Strength of Polymer Impregnated Cement," Cement and Concrete Research, Vol. 4, pp. 785-795, 1974.

3. Brown, J.H. and Pomeroy, C.D., "Fracture Toughness of Cement Paste and Mortars," Cement and Concrete Research, Vol. 3, pp. 475-480, 1973.

4. Brown, W.F. Jr. and Srawley, J.E., "Plane Strain Crack Toughness Testing of High Strength Metallic Materials," ASTM, STP 410, 1966.

5. Cook, J.D. and Crookham, G.D., "Fracture Toughness Measurements of Polymer Concretes," Magazine of Concrete Research, Vol. 30, No. 105, pp. 205-214, 1978.

6. Czarnecki, L., "The Status of Polymer Concrete," Concrete International, Vol. 17, No. 7, pp. 47-53, 1985.

7. Fairweather, V., "Filling the Cracks," Civil Engineering, ASCE, Vol. 56, No. 10, pp. 38-41, 1986.

8. "Guide to the Use of Polymers in Concrete," ACI Committee 548, ACI Journal, Vol. 83, No. 5, pp. 798-829, 1986.

9. Halvorsen, G.T., "J-Integral Study of Steel Fiber Reinforced Concrete," International Journal of Cement Composites, Vol. 2, pp. 13-22, 1980.

10. Hashemi, S. and Williams, J.G., "Fracture Characterization of Tough Polymers Using the J Method," Polymer Engineering and Science, Vol. 26, No. 11, pp. 760-767, 1986.

11. Mindess, S., "Fracture Toughness Testing of Cement and Concrete," Fracture Mechanics of Concrete: Material Characterization and testing, Martinus Nijhoff Publishers, pp. 67-110, 1984.

12. Mindess, S., Lawrence, F.V., and Kesler, C.E., "The J-Integral as a Fracture Criterion for Fiber Reinforced Concrete," Cement and Concrete Research, Vol. 7, pp. 731-742, 1977.

13. Moavenzadeh, F. and Kuguel, R., "Fracture of Concrete," Journal of Materials, JMLSA, Vol. 4, No. 3, 1969.

14. "Polymers in Concrete-State of the Art Report," Report of ACI Committee 548, ACI, 1977.

15. Potter, W.G., Epoxide Resins, Springer-Verlag New York Inc., London Iliffe Books, pp. 248, 1970.

16. Rice, J.R., Paris, P.C., and Merkle, J.C., ASTM, STP 536, pp. 231-245, 1973.

17. Schmidt, R.A. and Lutz, T.J., "$K_{IC}$ and $J_{IC}$ of Westerly Granite Effects of Thickness and In-Plane Dimensions," Fracture Mechanics Applied to Brittle Materials, ASTM, STP 678, pp. 166-182, 1979.

18. Tazawa, E. and Kobayashi, S., "Properties and Applications of Polymer Impregnated Cementitious Material," Polymers in Concrete, SP-40, ACI, p. 57, 1973.

19. Velazco, G., Visalvanich, K. and Shah, S.P., "Fracture Behavior and Analysis of Fiber Reinforced Concrete Beams," Cement and Concrete Research , Vol. 10, p. 41-51, 1980.

20. Vipulanandan, C. and Dharmarajan, N., "Flexural Behavior of Polyester Polymer Concrete," Cement and Concrete Research, Vol. 17, No. 2, pp. 219-230, 1987.

21. Yamini, S. and Young, R.J., "Stability of Crack Propagation in Epoxy Resins," Polymer, 18, pp. 1075-1080, 1977.

22. Young, R.J., "Fracture of Thermosetting Resins," Development in Polymer Fracture, edited by E.H. Andrews, Applied Science Publishers, London, pp. 183-222, 1979.

23. Ziegeldorf, S., "Fracture Mechanics Parameters of Hardened Cement Paste, Aggregates and Interfaces, Fracture Mechanics of Concrete, Edited by F.H. Wittmann, Elsevier Science Publishers, pp. 371-409, 1983.

SEM/RILEM International Conference
on Fracture of Concrete and Rock
Houston, Texas, June 1987
Editors: S.P. Shah, S.E. Swartz

# Fracture Characteristics of Concrete under Static Loading

Byung Hwan Oh

## ABSTRACT

A method to determine the fracture energy of concrete is investigated. Several series of concrete beams have been tested. The present experimental study indicates that the fracture energy decreases as the initial notch-to-beam depth ratio increases. Some problems to be observed to employ the three-point bend method are discussed. A simple and accurate formula to predict the fracture energy of concrete is proposed.

## INTRODUCTION

The growing attention is paid to explore the complex fracture behavior of concrete. It is increasingly recognized that the linear elastic fracture mechanics developed for metals cannot be directly applied to concrete without any appropriate modification. This is due to the fact that large size of nonlinear zone exists in front of the crack tip in concrete. This nonlinear zone in concrete, called the fracture process zone, is caused by microcracking ahead of the crack tip. The fracture process zone in concrete is known to be very large compared with those of metals or other engineering materials [2-7, 11, 27, 28]. This causes difficulties to measure the fracture toughness of concrete. Another difficulty lies in measuring the crack length increments during loading process since the crack tip is blurred by a microcracking zone in concrete.

One possible remedy is to use the complete load-deflection diagram to determine the fracture energy of concrete. The fracture energy can be calculated by measuring the area under the complete load-deflection curve. This method was used recently by Petersson[21]. There are, however, some problems to be resolved to employ this method. First, the energy consumption taking place outside the fracture zone must be negligible. This requirement may roughly be satisfied by selecting an appropriate value of the initial notch-to-beam depth ratio. If the initial crack is too small compared with the beam depth, the concrete material outside the fracture zone would be strongly stressed. This will result in a certain additional amount of energy consumption in the test beam, in addition to the energy comsumed in the fracture zone. In this case, the calculated fracture energy may obviously be higher than the actual one. The second problem to be considered is the stability of fracture. This means that the fracture must be stable so that the energy consumption due to dynamic effects may be eliminated. This in turn means the smooth and complete load-deflection curve during the fracture test.

The purpose of this paper is, therefore, to investigate the method of three-point bend test to determine the fracture energy of concrete and then to propose a simple formula to predict the fracture energy of a given concrete. The prediction equation for the fracture energy is important since the fracture energy is a required value to perform the fracture analysis of concrete Structures.

## FRACTURE TESTS

Several series of concrete beams have been tested in the present study to measure the maximum failure loads of the beams and to determine the fracture energy of concrete.

Materials and Test Specimens. - The mix proportions and compressive strengths for each series of concrete are summarized in Table 1. The maximum aggregate size was 25mm. The ordinary Portland cement was used. Note here that the mix proportion for Series-5 concrete was the same as Series-1 concrete and it was used for cyclic loading test.

---

Byung Hwan Oh is Assistant Professor, Department of Civil Engineering, Seoul National University, Kwanak-ku, Seoul, Korea.

The dimension of the test beams was 100 x 100 x 400mm. The specimens were cast in steel moulds and stored in 100% humidity during the first two days after casting. The test beams were then cured in the curing water-tub until one day before the tests. The concrete beams for each series had four different initial notch depts, i.e., $a_0/H$ = 0.0, 0.2, 0.4, 0.6 in which $a_0$ = initial notch depth and H = beam depth (Fig. 1). The reason for the different initial notch depths was to investigate the effect of initial notch depth ratio on the fracture energy of concrete. The initial notch was made in advance by inserting a thin steel sheet in the mould before placing the concrete.

Test Method. - The concrete beams were tested in three-point loading condition. The schematic diagram for the present three-point bend test is shown in Fig. 1. The test specimens were loaded in the Instron testing machine. The cross head speed (or loading velocity) was 0.05mm/min and was maintained constant throughout the tests. The load-deflection curves were generated by the plotter attached to the machine. The loading-unloading-reloading scheme was empolyed for some of the specimens in order to see the change of compliances for the unloading and reloading.

Test Results. - The results of present experimental study are shown in Figs. 2-6. Each figure is classified according to the initial notch-to-beam depth ratios. The almost complete load-deflection curves were possible due to the stable loading arrangements.

## DETERMINATION OF FRACTURE ENERGY OF CONCRETE

The fracture energy of concrete, $G_F$, can be determined as the area under the complete load-deflection diagram. The area under the curve represents the amount of energy consumed when the crack propagates through the concrete beam. The area under the curve in Figs. 2-6 was measured with a planimeter.

Now, the contribution from the self-weight of the beam must be considered to calculate the fracture energy. The fracture energy per one unit of area may then be calculated by the following equation.

$$G_F = \frac{\int_0^{\delta_0} P(\delta)d\delta + mg\,\delta_0/2}{B(H - a_0)} \tag{1}$$

in which $\delta_0$ = maximum deflection [see Fig. 8(c)], m = mass of the beam, g = gravitational acceleration, B = beam width, H = beam depth, and $a_0$ = initial crack length.

The fracture energy for each series of concrete has been calculated according to Eq. 1 and the results are shown in Fig. 7. As can be seen in Fig. 7, the fracture energy is decreased as the initial notch-to-beam depth ratio increases. Fig. 7 also indicates that the reduction in fracture energy becomes smaller as the relative notch depth becomes larger.

The reason having the larger fracture energy for the case of beams with a small initial crack is probably due to the fact that if the relative initial crack is small, the material outside the fracture zone would be greatly stressed and this would cause an additional energy consumption in the beam.

The beams with relatively large initial cracks would, however, yield somewhat smaller fracture energy than the actual one since the fracture process zone in front of the crack tip would not be well-developed in this case due to the short ligament size [see Ref. 6]. It is known from R-curve that the fracture energy increases somewhat as the crack propagates [6, 8, 23]. Therefore, higher value of initial notch-to-beam depth ratio may not be suitable for determining the fracture energy of concrete.

It is, therefore, reasonable to assume from Fig. 7 that the value 0.5 of initial notch-to-beam depth ratio is adequate to determine the fracture energy of concrete. This value is reasonable in view of Fig. 7.

Table 2 shows the average values of fracture energy obtained from the present experimental study for various series of concrete. It was found in this study that the contribution of the self-weight of a beam to the fracture energy was very small and it was only about 2-6 percent of the total $G_F$ value.

## DERIVATION OF A FORMULA FOR FRACTURE ENERGY OF CONCRETE

The fracture energy, $G_F$, must be known in advance as a material property in order to perform the fracture analysis of concrete structures. This requires a certain prediction formula to determine the fracture energy of concrete.

The progressive microcracking in the fracture process zone may be described by a stress-strain relation that exhibits strain-softening. The strain-softening uniaxial stress-strain relation may be well-idealized as a bilinear stress-strain diagram [see Fig. 8(a)][6]. The fracture energy may then by expressed as

$$G_F = w_c A = \frac{1}{2} w_c \varepsilon_0 f'_t \tag{2}$$

in which $w_c$ = width of microcrack band = $c_0 d_a$[3, 6], $d_a$ = maximum aggregate size, A = area under the stress-strain curve, $\varepsilon_0$ = strain at zero stress, $f'_t$ = tensile strength of concrete, and $E_c$ = its elastic modulus.

By assuming $\varepsilon_0 = c_1 \varepsilon_p$ and $w_c = c_0 d_a$[3, 6], and noting that $\varepsilon_p \simeq f'_t/E_c$ [Fig. 8(a)], Eg. (2) may be rewritten in the following form.

$$G_F = \frac{1}{2} c_0 c_1 f'^2_t d_a/E_c = c_2 f'^2_t d_a/E_c \quad (3)$$

Eq. (3) indicates that the fracture energy increases very rapidly as the tensile strength increases. The present experimental study with the recent results of Gopalaratnam and Shah[14] and Petersson[21], however, indicates that the fracture energy increases slowly with the increase of the tensile strength of concrete. This means that one should take the relation $\varepsilon_p = c_3/E_c$, which is reasonable and coincident with experimental data [14, 22][see Fig. 8(b)]. By using this relation, Eqs. (2) and (3) may be rewritten as

$$G_F = \frac{1}{2} c_0 c_1 c_3 f'_t d_a/E_c = c_4 f'_t d_a/E_c \quad (4)$$

in which $c_0$, $c_1$, $c_2$, $c_3$ and $c_4$ represent the empirical constants to be determined from experimental data.

The recently-published good experimental data [12, 14, 21] for the fracture energy of concrete have been also used to determine the fracture energy equation. Fig. 9 shows the regression analysis of test data [12, 14, 21] and shows the dependence of fracture energy on the tensile strength. Fig. 9 clearly shows that the fracture energy varies almostlinearly with the tensile strength. Namely, the linear form [Eq. (4)] is much superior to the quadratic form [Eq. (3)] (see Fig. 9). The best prediction equation for the fracture energy of concrete was found from the test data as follows.

$$G_F = 56.24 f'_t d_a/E_c \quad (5)$$

in which $G_F$ is in N/mm, $f'_t$ and $E_c$ in N/mm² (or MPa), and $d_a$ in mm. When the value of $E_c$ is not given, the usual relation $E_c = 4{,}733 \ f'_c$ may be used [1] in which $f'_c$ is the compressive strength of concrete in N/mm² (or MPa). For the tensile strength of concrete, the follwing relation proposed recently by Raphael[22] on the basis of the comprehensive experimental data may be used.

$$f'_t = 0.324 f'^{2/3}_c \quad (6)$$

or

$$f'_t = 0.74 f'_r \quad (7)$$

in which $f'_t$ and $f'_c$ are in N/mm² (or MPa), $f'_r$ = modulus of rupture = $0.438 f'^{2/3}_c$ (N/mm²). Here 1 MPa = 1 MN/m² = 1 N/mm² = 145 psi. The fracture energy equation without $E_c$ was also tried, but it did not give better results. Eq (5) may now be used to calculate the fracture energy of any given concrete. The fracture analysis of concrete structures requires a correct value of fracture energy which may be obtained from Eq. (5).

## CONCLUSIONS

1. The fracture energy of concrete can be determined from a stable three-point bend test. The fracture energy is then calculated from the area under the load-deflection curve. This method is relatively simple and accurate.

2. The present experimental study indicates that the fracture energy is decreased as the initial notch-to-beam depth ratio increases. The appropriate ratio of initial notch-to-beam depth to determine the fracture energy of concrete is found to be 0.5.

3. It is found that the influence of the self-weight of a beam to the fracture energy is very small and it amounts to only several percent of the fracture energy.

4. A simple and accurate formula for the fracture energy of concrete is proposed. The fracture energy is a required value to perform the fracture analysis of concrete structures. The fracture energy of concrete depends on the aggregate size and tensile strength and is about 80-130 N/m in value for normal aggregate concrete.

## ACKNOWLEDGMENT

The financial support from the Ministry of Education of Korea to this research is gratefully acknowledged.

REFERENCES


1. ACI 318-83, "Building Code Requirements for Reinforced Concrete," American Concrete Institute, 1983.

2. Bazant, Z.P., "Crack Band Model for Fracture of Geomaterials," Proceedings, 4th International Conference on Numerical Methods in Geomechanics, Edmonton, Alberta, Canada, Vol. 3, Z. Eisenstein, ed., June, 1982, pp. 1137-1152.

3. Bazant, Z.P., "Size Effect in Blunt Fracture : Concrete, Rock, Metal," Journal of Engineering Mechanics, Vol. 110, No. 4, April, 1984, pp. 518-535.

4. Bazant, Z.P., and Cedolin, L., "Approximate Linear Analysis of Concrete Fracture by R-curves," Journal of Structural Engineering, Vol. 110, No. 6, June, 1984, pp. 1336-1355.

5. Bazant, Z.P., and Cedolin, L., "Blunt Crack Band Propagation in Finite Element Analysis," Journal of the Engineering Mechanics Division, ASCE, Vol. 105, No. EM2, Apr., 1979, pp. 297-313.

6. Bazant, Z.P., and Oh, B.H., "Crack Band Theory for Fracture of Concrete," Materials and Structures (RILEM, Paris), Vol. 16, 1983, pp. 155-177.

7. Bazant Z,P., and Oh, B.H., "Rock Fracture via Strain-Softening Finite Elements," Journal of Engineering Mechanics, ASCE, Vol. 110, No. 7, 1984, pp. 1015-1035.

8. Broek,D., Elementary Engineering Fracture Mechanics, Noordhoff International Publishing, Leyden, Netherlands, 1974.

9. Brown J.H., "Measuring the Fracture Toughness of Cement Paste and Mortar," Magazine of Concrete Research Vol. 24, No. 81, Dec., 1972.

10. Carpinteri, A., "Static and Energetic Fracture Parameters for Rocks and Concretes," Report, Istituto di Scienza delle Costruzioni-Ingegneria, University of Bologna, Italy, 1980.

11. Cho, K.Z., Kobayashi, A.S., Hawkins, N.M, Barker, D.B., and Jeang, F.L., "Fracture Process Zone of Concrete Cracks," Journal of Engineering Mechanics, ASCE, Vol. 110, No. 8, 1984, pp. 1174-1184.

12. Entov, V.M., and Yagust, V.I., "Experimental Investigation of Laws Goverining Quasi-Static Development of Microcracks in Concrete," Mechanics od Solids (Translation from Russian), Vol. 10, No. 4, 1975, pp. 87-95.

13. Gjørv, O.E., Sørenson, S.I., and Arnesen, A., "Notch Sensitivity and Fracture Toughness of Concrete," Cement and Concrete Research, Vol. 7, 1977, pp. 333-344.

14. Gopalaratnam, V.S., and Shah, S.P., "Softening Response of Plain Concrete in Direct Tension," Journal of the American Concrete Institute, May-June, 1985, pp. 310-323.

15. Huang, C.M.J., "Finite Element and Experimental Studies of Stress Intensity Factors for Concrete Beams," thesis presented to the Kansas State University in partial fulfillment of the requirements for the degree of Doctor of Philosophy, 1981.

16. Hutchinson, J.W., and Paris, P.C., "Stability Analysis of J-Controlled Crack Growth; Elastic-Plastic Fracture," ASTM-STP668, American Society for Testing and Materials, 1979, pp. 37-64.

17. Kaplan, M.F., "Crack Propagation and the Fracture of Concrete," American Concrete Institute Journal, Vol. 58, No. 11, Nov., 1961.

18. Knott, J.F., Fundamentals of Fracture Mechanics," Butterworth, London, England, 1973.

19. Nadeau, J.S., Mindess, S., and Hay, J.M., "Slow Crack Growth in Cement Paste," Journal of the American Ceramic Society, Vol. 57, No. 2, 1974, pp. 51-54.

20. Parker, A.P., "The Mechanics of Fracture and Fatigue," E.& F.N. Spon, Ltd. London, 1981.

21. Pertersson, P.E., "Fracture Energy of Concrete: Method of Determination," Cement and Concrete Research, Vol. 10, 1980, pp. 78-89, and "Fracture Energy of Concrete: Practical Performance and Experimental Results," Cement and Concrete Research, Vol. 10, 1980, pp. 91-101.

22. Raphael, J.M., "Tensile Strength of Concrete," Journal of the American Concrete Institute, March-April, 1984, pp. 158-165.

23. Shah, S.P., and McGarry, F.J., "Griffith Fracture Criterion and Concrete," Journal of the Engineering Mechanics Division, ASCE, Vol. 97, No. EM6, Proc. Paper 8597, Dec., 1971, pp. 1664-1676.

24. Sok, C., and Baron, J., "Mecanique de la Rupture Appliquee au Beton Hydraulique," Cement and Concrete Research, Vol. 9, 1979, pp. 641-648.

25. Visalvanich, K., and Naaman, A.E., "Fracture Model for Fiber Reinforced Concrete," Journal of American Concrete Institute, Vol. 80, Mar.-Apr., 1983, pp. 128-138.

26. Walsh, P.E., "Fracture of Plain Concrete," The Indian Concrete Journal, Vol. 46, No. 11, 1979, pp. 469, 470, and 476.

27. Wecharatana, M., and Shah, S.P., "Slow Crack Grwoth in Cement Composites," Journal of Structural Engineering Division, ASCE, Vol. 108, No. ST6, June, 1982, pp. 1400-1413.

28. Wecharatana, M., and Shah, S.P., "Predictions of Nonlinear Fracture Process Zone in Concrete," Journal of Engineering Mechanics, ASCE, Vol. 109, No. 5, 1983, pp. 1231-1246.

29. Wittmann, F.H., "Fracture Mechanics of Concrete," Elsevier, 1983.

Table 1. Mix Proportions and Compressive Strengths

| | Number of Specimens | Mix Proportions (C : S : G : W ) | Compressive Strength (N/mm²) | Monotonic(M) or Cyclic(C) |
|---|---|---|---|---|
| Series 1 | 8 | 1:1.88:2.86:0.41 | 25.8 N/mm² | M |
| Series 2 | 4 | 1:1.88:2.90:0.41 | 27.5 | M |
| Series 3 | 4 | 1:2.44:3.44:0.49 | 21.5 | C |
| Series 4 | 8 | 1:3.07:4.09:0.58 | 20.3 | M, C |
| Series 5 | 8 | 1:1.88:2.86:0.41 | 26.2 | C |

Note : C = Cement, S = Sand, G = Gravel(Coarse Aggregate), W = Water,

1 N/mm² = 145psi.

Table 2. Fracture Energy for each Series of Concrete

| | Average Fracture Energy |
|---|---|
| Series 1 | 127.0 N/m |
| Series 2 | 129.6 N/m |
| Series 3 | 124.0 N/m |
| Series 4 | 112.1 N/m |
| Series 5 | 127.8 N/m |

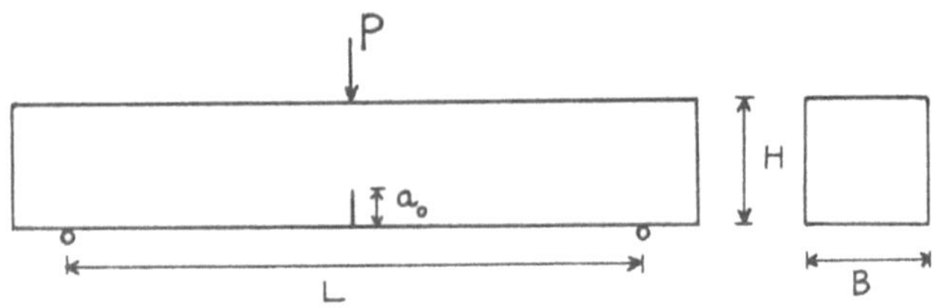


Fig. 1 Schematic Diagram for the Present Three-Point Bend Test

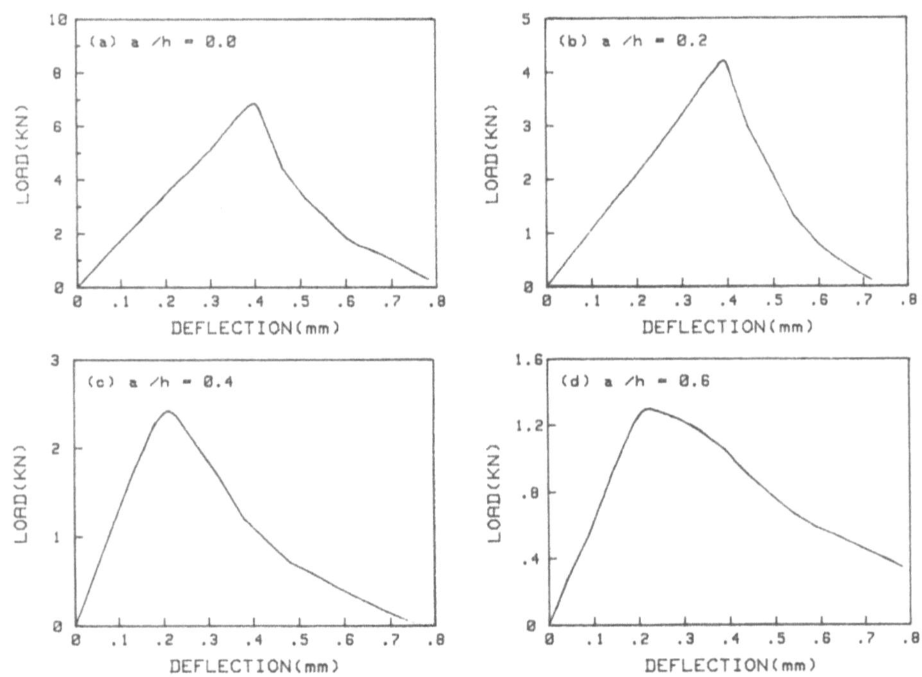


Fig. 2 Load-Deflection Curves for Series 1 Concrete

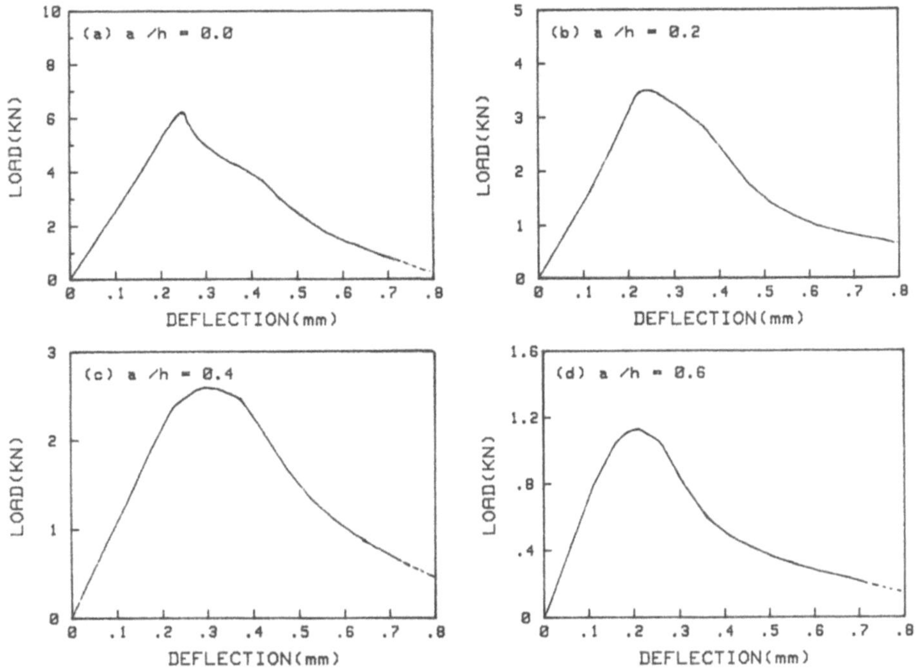


Fig. 3 Load-Deflection Curves for Series 2 Concrete

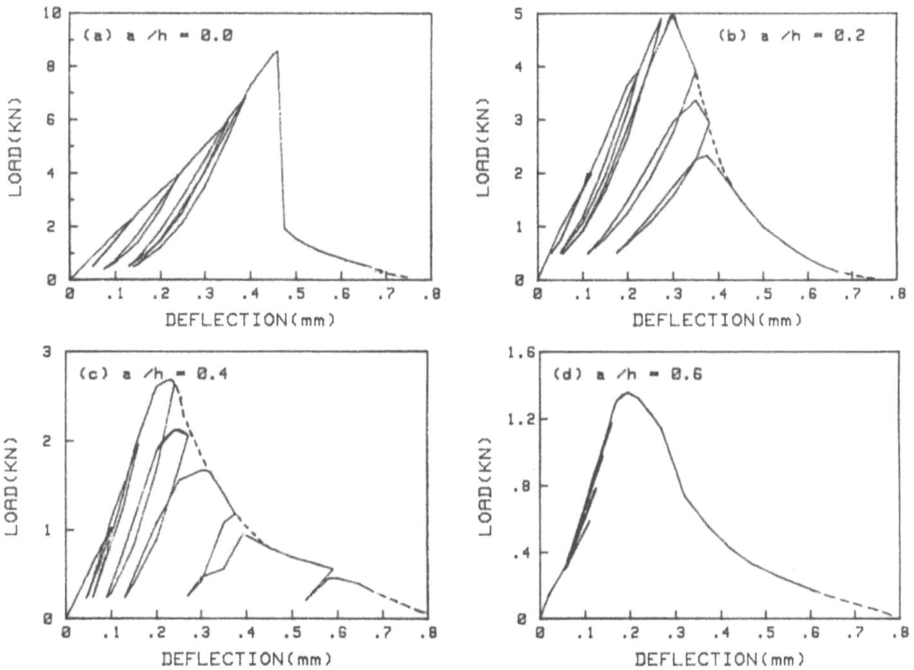


Fig. 4 Load-Deflection Curves for Series 3 Concrete

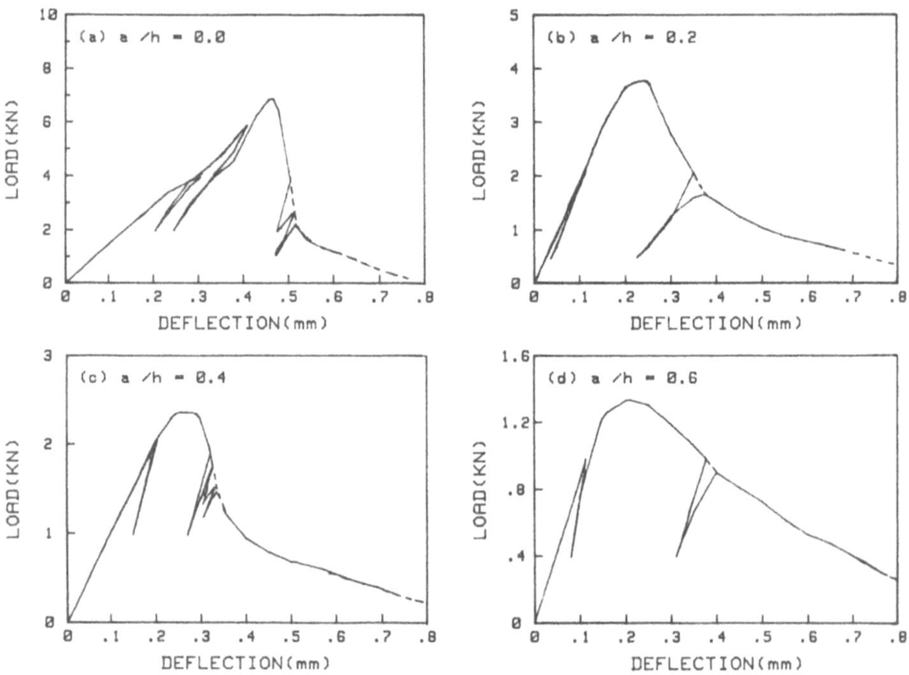


Fig. 5 Load-Deflection Curves for Series 4 Concrete

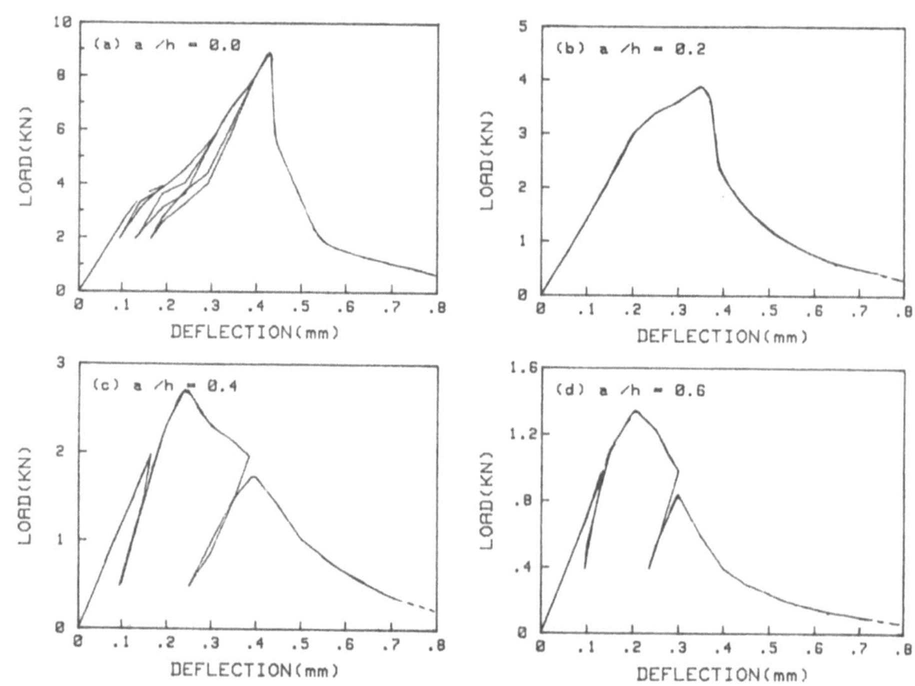


Fig. 6 Load-Deflection Curves for Series 5 Concrete

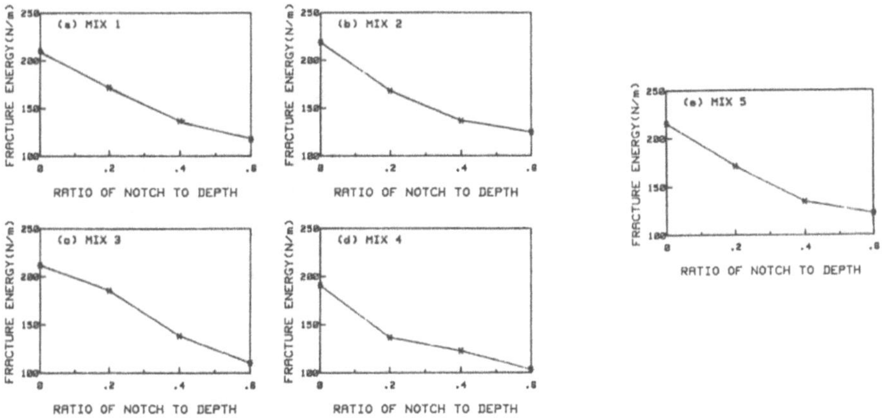


Fig. 7 Variation of Fracture Energy According to the Initial Notch-to-Beam Depth Ratio

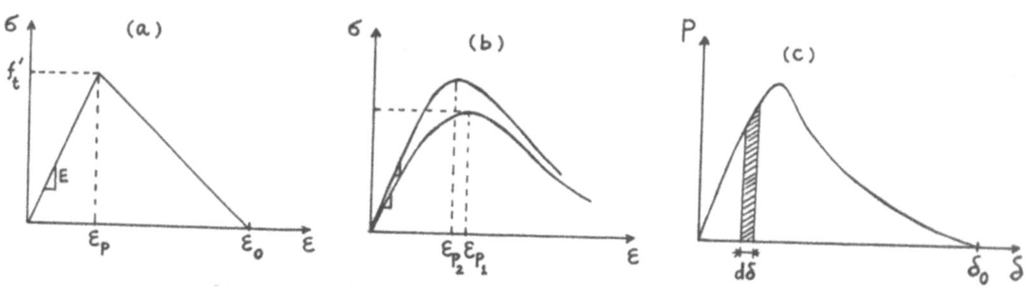


Fig. 8 (a) Stress-Strain Diagram for Fracture Process Zone;
(b) Stress-Strain Curves for Different Strengths;
(c) Load-Deflection Curve

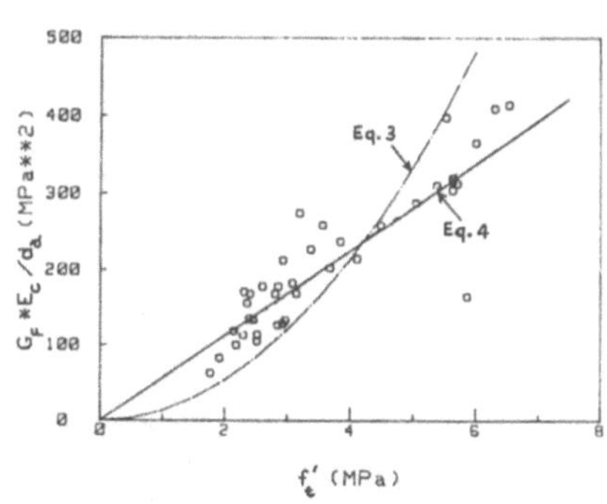


Fig. 9 Fracture Energy As a Function of Tensile Strength

# Author Index

O



P



Q



R



S

## Subject Index

MIX
Papier aus verantwortungsvollen Quellen
Paper from responsible sources
FSC® C105338

If you have any concerns about our products,
you can contact us on
**ProductSafety@springernature.com**

In case Publisher is established outside the EU,
the EU authorized representative is:
**Springer Nature Customer Service Center GmbH**
**Europaplatz 3, 69115 Heidelberg, Germany**

Printed by Libri Plureos GmbH
in Hamburg, Germany